PASSENGER SAFETY AND CONVENIENCE SYSTEMS

Other SAE books in this series:

Sensors and Transducers
by Ronald K. Jurgen
(Order No. PT-68)

Object Detection, Collision Warning and Avoidance Systems
by Ronald K. Jurgen
(Order No. PT-70)

Electronic Instrument Panel Displays
by Ronald K. Jurgen
(Order No. PT-71)

Navigation and Intelligent Transportation Systems
by Ronald K. Jurgen
(Order No. PT-72)

Electronic Engine Control Technologies
by Ronald K. Jurgen
(Order No. PT-73)

Actuators
by Ronald K. Jurgen
(Order No. PT-74)

Automotive Microcontrollers
by Ronald K. Jurgen
(Order No. PT-75)

Electronic Braking, Traction, and Stability Control
by Ronald K. Jurgen
(Order No. PT-76)

Electronic Steering and Suspension Systems
by Ronald K. Jurgen
(Order No. PT-77)

Multiplexing and Networking
by Ronald K. Jurgen
(Order No. PT-78)

Electronic Transmission Controls
by Ronald K. Jurgen
(Order No. PT-79)

On- and Off-Board Diagnostics
by Ronald K. Jurgen
(Order No. PT-81)

PASSENGER SAFETY AND CONVENIENCE SYSTEMS

Automotive Electronics Series

PT-83

Edited by

Ronald K. Jurgen

Published by
Society of Automotive Engineers, Inc.
400 Commonwealth Drive
Warrendale, PA 15096-0001
U.S.A.
Phone (724) 776-4841
Fax: (724) 776-5760
www.sae.org

ISBN 0-7680-0683-X
Library of Congress Catalog Card Number:

SAE Order No. PT-83

Introduction

Electronics: The Key Enabler for Safety and Convenience

The British Parliament's "Red Flag Act" of 1861 was passed in the name of safety. It limited auto speeds to 6.5 km/h and required a flagman to precede each vehicle at a distance of about 50 meters.[1] In the nearly 140 years since then, passenger safety measures have become much more sophisticated, as have systems for passenger comfort and convenience. Initial progress in these areas was made without the benefit of electronic devices and systems, but the greatly accelerated progress since the 1970s resulted mainly from new and innovative applications of electronics. It can be said realistically that electronics has been the key enabler for both passenger safety and passenger convenience systems.

This book presents papers on a variety of electronically controlled safety and convenience systems, including smart air bags, roof air bags, seat belt pretensioners, night vision, crash sensors, the X-by-wire concept, biometric systems, keyless entry systems, automatic climate control with neural networks or fuzzy logic, and central tire inflation systems. Purposely not included are papers on electronic braking, traction, and stability controls or object detection, collision warning, and avoidance systems, because two other books in this automotive electronics series[2,3] are devoted to these topics.

The papers in this book have been divided into two categories: safety systems and convenience systems, but not all systems are easily categorized as strictly one or the other. Some systems could, arguably, be in both categories. A case in point is the Cadillac DeVille thermal imaging night vision system (Paper 2000-01-0323). It has been placed in this book with other papers on safety systems, but conceivably it could have been classified just as logically as a passenger convenience system. Keyless entry systems also present a similar categorizing dilemma. Here they have been placed under convenience systems, but they are often called security systems by their designers. The categorizing used here represents our best efforts.

The final paper in this book, on future developments in electronically controlled body and safety systems, considers these systems in the overall context of vehicle manufacturers' needs to create brand identity. It then considers trends in networking, software, safety systems, architecture ownership, and vehicle simulation.

References

1. Special issue on The Automobile, *IEEE Spectrum*, November 1977, p. 81.
2. PT-76, *Electronic Braking, Traction, and Stability Controls*, Ronald K. Jurgen, Ed., Society of Automotive Engineers, Warrendale, Pa., 1999.
3. PT-70, *Object Detection, Collision Warning and Avoidance Systems*, Ronald K. Jurgen, Ed., Society of Automotive Engineers, Warrendale, Pa., 1998.

* * * * * * * * * * * *

This book and the entire automotive electronics series is dedicated to my friend Larry Givens, a former editor of the SAE monthly publication, *Automotive Engineering.*

Ronald K. Jurgen, Editor

Table of Contents

Introduction

Safety Systems

Convenience Systems

Future Outlook

SAFETY SYSTEMS

2000-01-0346

An Integrated Approach to Automotive Safety Systems

Stephen N. Rohr, Richard C. Lind, Robert J. Myers,
William A. Bauson, Walter K. Kosiak and Huan Yen
Delphi Automotive Systems

ABSTRACT

The industry strategy for automotive safety systems has been evolving over the last 20 years. Initially, individual passive devices and features such as seatbelts, airbags, knee bolsters, crush zones, etc. were developed for saving lives and minimizing injuries when an accident occurs. Later, preventive measures such as improving visibility, headlights, windshield wipers, tire traction, etc. were deployed to reduce the probability of getting into an accident. Now we are at the stage of actively avoiding accidents as well as providing maximum protection to the vehicle occupants and even pedestrians. Systems that are on the threshold of being deployed or under intense development include collision detection / warning / intervention systems, lane departure warning, drowsy driver detection, and advanced safety interiors.

In this paper, we will discuss the concept of *the safety state diagram*, a unified view of the automotive safety system, and the technologies that are required to implement this vision. Advanced ideas such as pre-crash sensing, anticipatory crash sensing, X-by-wire systems, advanced safety interiors, integrated vehicle electrical/electronics systems, data networks, and mobile multimedia (telematics) will also be addressed.

INTRODUCTION

The expanded use of electronics, microcontrollers, sensors, actuators, high-speed data busses, X-by-wire technologies, etc. in the automotive industry will have a major impact on the architecture of future safety systems. Many traditional safety technologies are beginning to merge. Looking at the vehicle as a personal safety system, one can describe five vehicle driving scenarios: normal driving state, warning state, crash avoidable state, crash unavoidable state, and post event state. The first three states focus on accident avoidance while the last three states focus on damage mitigation (with an overlap of the third state). Using the state diagram, it is apparent that automotive safety concerns should be addressed with an integrated system approach.

THE INTEGRATED SAFETY SYSTEM STATE DIAGRAM

As depicted in Figure 1, an integrated automotive safety system may be thought of as a series of interdependent safety states.

Figure 1. Integrated Safety System State Diagram

THE AVOIDANCE ZONE – The three states that comprise the avoidance zone include: normal driving state, warning state, and collision avoidable state. It is important to note that all the safety actions occurring in these states reduce the probability of a collision. This is also known in the industry as *active safety*.

Normal Driving State – Under the normal driving state, a driver enjoys many of the comfort and convenience features afforded by modern automotive electronics. For instance, a millimeter wave or laser-based adaptive cruise control (ACC) system maintains either a constant vehicle cruising speed or a constant headway between vehicles. In addition to the conventional AM/FM radio broadcasts, the on-board telematics system provides cell phone and wireless data capability, as well as GPS, map, and navigational aids. If desired, real-time traffic information can also be accessed through either the Internet or a preferred Call Center.

These features not only provide the needed information in a timely manner, but also offer added safety and security protection for vehicle occupants. For example, the navigation system with its turn-by-turn instructions lets the driver concentrate on the task of negotiating the traffic because there is no need to take his eyes off the road to look for his destination. Additionally, the telematics system allows the vehicle occupants to stay in touch with the dealer or repair shop should anything go wrong with the car.

The next generation of ACC systems will be able to handle low-speed stop-and-go traffic situations as well. This is especially significant since a large percentage of accidents take place under these driving conditions. Some vehicles are already equipped with the night vision system that provides enhanced vision at nighttime for the driver. The system is implemented in such a way that the infrared (IR) image of the scene is projected via the head up display (HUD) unit and superimposed on the driver's natural field of view. With this implementation, the driver enjoys enhanced vision without having to take his eyes off the road.

Some driver monitoring systems have made their way into commercial fleet vehicles because of the desire to ensure public safety and to minimize costly accidents. Most likely, these systems will soon appear in passenger cars as well. Using a combination of biological sensors, eye tracking devices, and vehicle steering information, it is possible to infer the degree of driver alertness. Appropriate countermeasures then can be employed to stimulate the driver or warn the driver that it is time to pull off the road and rest.

A roadway condition sensor that can reliably indicate to the driver whether the roadway ahead is wet, dry, icy, rough, etc. is a valuable safety tool especially during the wintertime. With this sensor information, the vehicle could issue an advisory to the driver to adjust his speed accordingly. Ultimately, the vehicle may automatically adjust its parameters to help the driver maintain control of the vehicle. In short, the vehicle development allows drivers and passengers under the normal driving state to be well protected by a large array of safety, comfort and convenience features. The interior of the vehicle will not only provide the maximum level of protection possible but will adapt according to each occupant's preference. The telematics functions will be working in concert with the safety features of the vehicle through a well designed Human Machine Interface (HMI).

Warning State – Sensing systems are the key in the warning state. The Integrated Safety System (ISS) must maintain full awareness of the driving situation in order to detect potential crash situations. Sensing needs range from external object detection (other vehicles, trees, signs, etc.) to internal vehicle states (tire pressure, vehicle stability, etc.). Sensor and data-fusion algorithms will combine information from various sensors to form a model of the current situation. Ultimately, the ISS needs to be able to sense objects around the entire (360 degrees) vehicle.

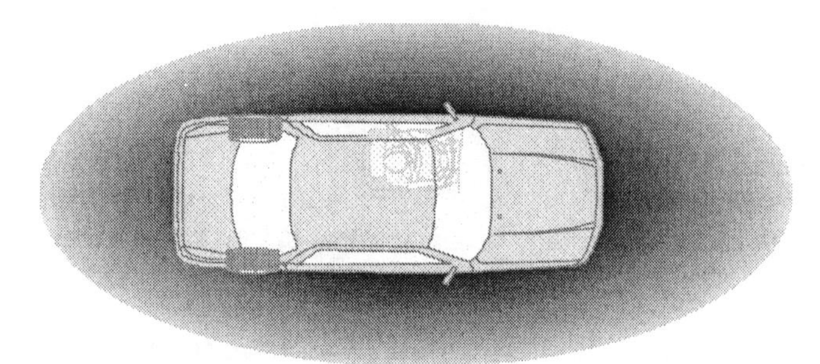

Figure 2. ISS 360 degree sensing

Once the situation is understood by ISS, the appropriate warning can be delivered to the driver. At this point, the driver must take action to avoid the collision.

Practical vehicle applications involved in the warning state include:

- Low tire pressure warning
- Impending rollover warning
- Lane / Roadway departure warning
- Parking assistance/warning
- Back-up assistance/warning
- Blind-spot warning
- Rear-end collision warning
- Lane change warning

One particularly difficult problem arises when sensing objects at longer ranges: the issue of path prediction. If the ISS is unable to accurately project the host vehicle's path, it will be impossible to determine which objects in the field of view represent a threat to the host vehicle. If, in the scenario shown below, the ISS vehicle is unaware that the road is curving, then there are several objects that would appear to be directly in the path of the host vehicle. The ISS would incorrectly warn the driver of all the supposed threats. For the ISS to be effective, it is important to provide warning or control signals at the appropriate time and to minimize false alarms and nuisance alerts.

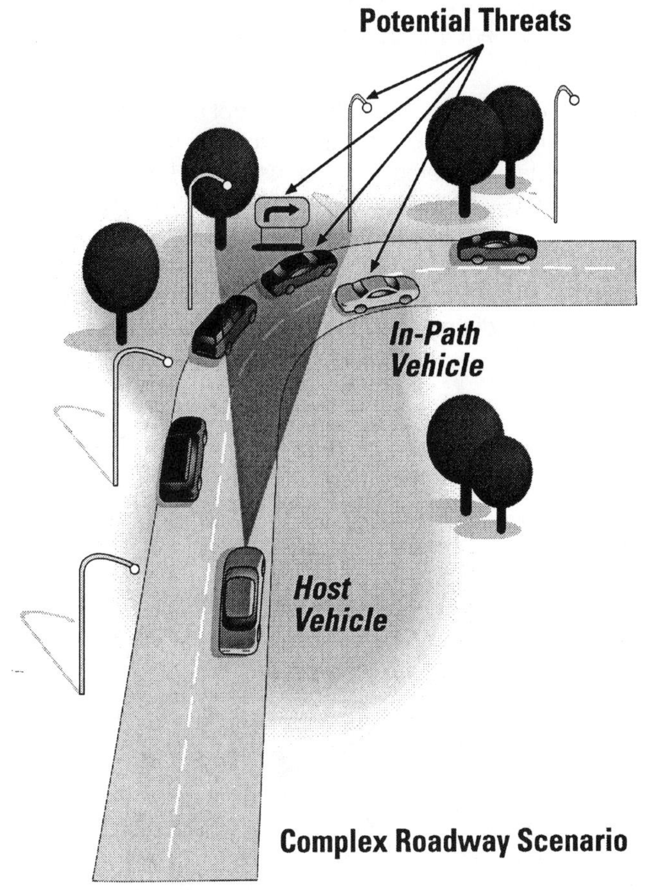

Figure 3. Complex Roadway Scenario

Assuming the driver follows the road, the only object actually in the path of the host vehicle is the vehicle around the curve in the same lane. Depending on the vehicle spacing and relative speeds, this vehicle may or may not represent a threat. Correct detection of road geometry would enable the ISS to correctly ignore all objects except the vehicle in the same lane as the host vehicle.

The solution to the path estimation problem is highly dependent on roadway geometry, inter-vehicle kinematics, driver reaction times, and braking behaviors. Yaw-rate sensors can be used to determine roadway curvature, but only when the host vehicle is already in the curve. For scenarios such as curve entry or curve exit, other sensors such as solid-state cameras will be needed. GPS, digital road maps, and roadside transponders can also be used to increase the accuracy and robustness of the ISS path estimation algorithms.

Several other issues surface in the warning state:

- Threat Assessment: The ISS must determine whether a particular situation poses a threat and merits warning the driver. Driver preferences must be taken into account as well. One driver may not want to be warned until the threat is severe while another driver may want to be notified at the slightest hint of trouble.
- False alarms and nuisance alerts: It is important to eliminate or minimize false alarms to win the driver's confidence in the system.
- Human Factors: In a critical situation, the ISS must notify the driver in a way that is quickly recognized and that encourages the driver to take the proper action. It is vital that warnings not distract or confuse the driver during an impending collision.

Collision Avoidable State – The collision avoidable state is our last opportunity to avoid an accident and return to the normal driving state. Everything that is done in this state is with the intention to completely avoid the accident. Reaction time as well as vehicle stability and control are extremely important in this state. The ability to stop or steer clear of harm is the most important. Examples of features included in this state are:

- Automatic stopping
- Automatic lane change
- Lane keeping
- Chassis and suspension control
- Vehicle rollover prevention
- X-by-wire (steer, brake, throttle)

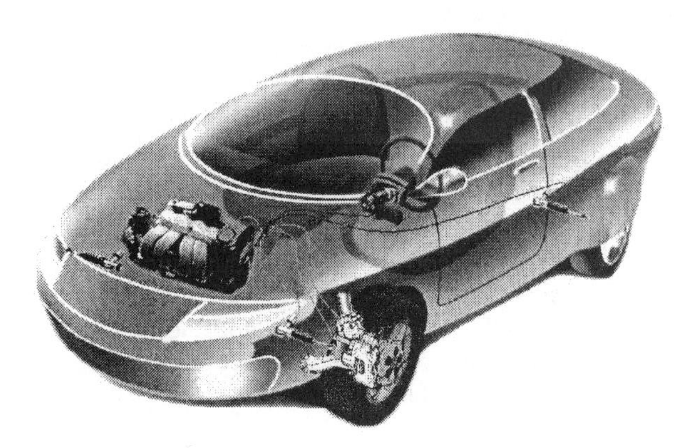

Figure 4. Future "by wire" vehicle control

The implementation of the collision avoidable state is primarily dependent upon two things: a suite of sensors and sensor fusion algorithms that provide information about the state of the vehicle and its surroundings as defined in the warning state; and a suite of X-by-wire products (steering, braking, throttle, and suspension) that de-couple the actuation from the mechanical input provided by the driver as shown in the illustration above. This de-coupling is a key enabler as it allows the vehicle to be commanded to perform various maneuvers without direct driver input.

Consider the current anti-lock brake systems that are prevalent on passenger vehicles and light trucks today. In these systems, the braking function is augmented by a computer controlled brake release and then re-applied to mitigate the effect of wheel slippage. It is important to note that the initiation of the brake function today still requires the driver's input.

Collision avoidance features will evolve into three modes:

- Driver initiated
- Vehicle initiated
- A blend of both

Consider again a simple braking maneuver in a vehicle equipped with the necessary sensors and a brake-by-wire system. Under any driving conditions, the vehicle will know its speed, closing speed of approaching objects, road surface conditions, driver intended path, and vehicle attitude (pitch and yaw). Once the driver requests braking, the system will provide the appropriate level of braking effort to effect a normal stop.

What will happen if the ISS vehicle has detected a slowed or stopped object in its path and the driver has ignored all of the warnings or is unable to command the brake function by stepping on the brake pedal? Because of the brake-by-wire product the vehicle will have the ability to initiate braking without input from the driver.

The above examples can all be accomplished with only a brake-by-wire system in the ISS vehicle. The options to enhance vehicle performance and stability are greatly increased when by-wire steering, throttle, and suspension are added. If the ISS vehicle detects that it is closing upon an object too quickly, then a simple vehicle commanded throttle reduction might be a viable response.

Or, perhaps the driver has initiated a very quick turn that will take the vehicle into an unstable condition. This could be from lack of driving experience, or over reaction to a driving situation. With an effective ISS system, the steering angle may be reduced, throttle reduced, independent differential braking applied and the suspension stiffened all simultaneously without input from the driver to keep the vehicle stable.

THE MITIGATION ZONE – The three states that comprise the mitigation zone include: collision avoidable state, collision unavoidable state, and the post event state. All safety actions that occur in these states focus on reducing the effects of a collision.

Collision Avoidable State – Note that this state appears in both the avoidance and mitigation zones. It's obvious that the best way to protect an occupant, pedestrian, or property is to avoid an accident. If the accident cannot be avoided, then the goal is to reduce its effects. Many of the systems used in collision avoidance come into play for damage mitigation. With automatic braking, for example, we can slow down the vehicle as much as possible to minimize injury and damage.

Using the same sensors and fusion algorithms described in the warning state, the ISS vehicle could prevent the driver from directing the vehicle to cross the roadway centerline and into the path of an oncoming vehicle. Or, it could inhibit the driver from a collision with a utility pole or bridge abutment that can possibly be at the right side of the vehicle.

In essence, when all the vehicle control authority of braking, steering, throttle, and suspension has been used, the ISS vehicle will attempt commands for the "softest possible landing."

Collision Unavoidable State – Since this is the point of no return, everything should be done immediately before as well as immediately after the crash to reduce the effects of the accident.

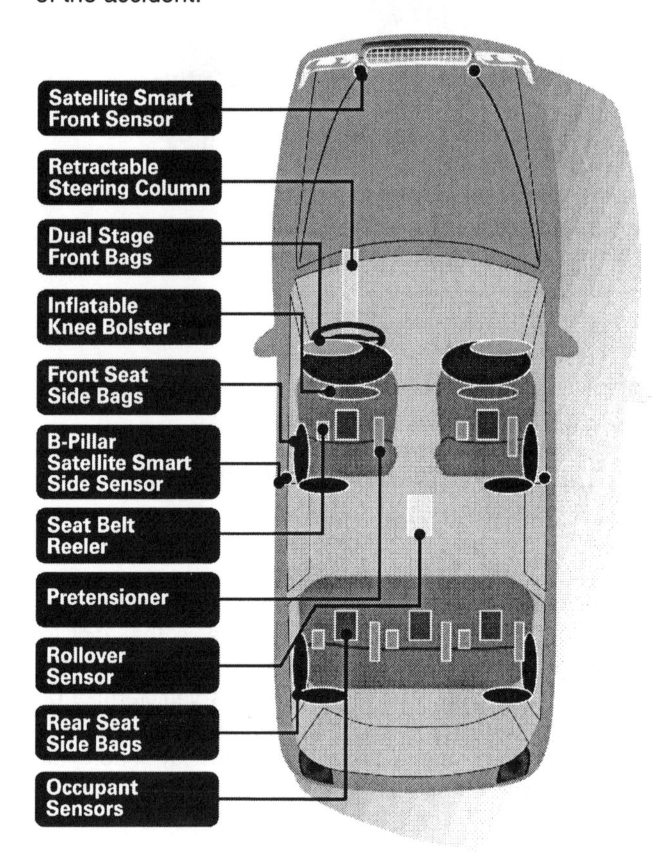

Figure 5. Near-term, high-content occupant restraint system.

The collision unavoidable state has traditionally encompassed the realm of occupant protection. These features include everything from crashworthy vehicle structures and interior padding to seatbelts and airbags. Interest in "smart" or advanced airbag systems has intensified recently in order to provide improved occupant protection under a variety of real-world accidents, as well as to minimize the potential adverse effects caused by airbag deployments. "Smart" restraint systems are intended to be more adaptable to various real-world factors such as crash type, crash severity, seat belt usage, and occupant type and position.

As occupant protection countermeasures increase in number and sophistication, electronic sensing requirements continue to grow. The finer the sensing resolution of both vehicle dynamics and occupant kinematics, the "smarter" the complete occupant protection system becomes.

Although the above technologies provide significant benefits in the area of occupant protection, the advent of collision avoidance technologies has now made it possible to incorporate new functionality into the vehicle.

Anticipatory or pre-crash sensing is a key enabler not only to post-impact countermeasures (such as variable stage airbags and seatbelt tensioners), but also for resettable, pre-impact countermeasures (such as adaptable interior and exterior structures and pedestrian protection countermeasures).

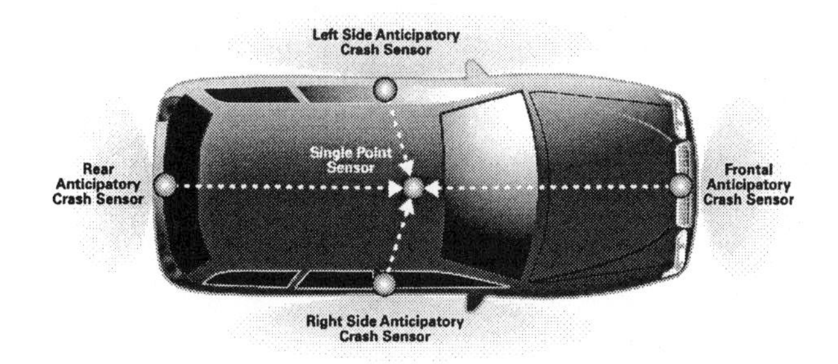

Figure 6. Joint discrimination using anticipatory crash sensors

Post Event State – After an accident has occurred, the ISS vehicle will automatically assess the severity of the event based on a number of sensory indications:

- Did the airbags deploy?
- Did the vehicle roll over?
- What is the rest position of the vehicle?
- How many occupants are there?
- What are the vital signs of the occupants?
- Is there a fuel leak?
- Is there a fire?

In the case of a severe accident (i.e., airbags were deployed), the vehicle's telematics system will automatically dial 911 (or equivalent) and summon help. If the occupants are still able to communicate with the dispatcher, the extent of injuries can be obtained and passed on to the paramedics. If the occupants suffered severe injuries and could not communicate with the dispatcher, then on-board biosensors can be used to assess the situation. If the vehicle is equipped with a video camera in the passenger compartment, video images can be used to aid in situation assessment.

In addition, the vehicle will have enough intelligence to detect and extinguish fires, release seatbelts, and unlock car doors (allowing easy egress as well as greater ability to be reached by rescuers).

To prevent fires caused by a ruptured fuel line, the fuel pump can be shut off automatically, the engine can be turned off remotely, and unnecessary electrical power can be disconnected. If the event takes place at night, the vehicle can also provide illumination and road hazard flashing warning lights.

In the post event state, there are many possibilities for enhancing the survivability of the victims. Many of these features are enabled by the telematics system. In order for these features to be available, the telematics system must survive the accident. Therefore, there is an implied level of robustness that must be built into the system.

A CASE STUDY – COLLISION AVOIDANCE AND OCCUPANT PROTECTION

An interesting phenomenon occurred as we looked at the future of pieces to the integrated safety systems puzzle. No matter how we approached the problem, it was evident that feature/function sets, as well as technology building blocks, tended to merge as time passed. In the following example, we'll discuss how collision avoidance and occupant protection tend to *blend together.*

FOREWARN® COLLISION AVOIDANCE – Collision avoidance systems depend on short- and long-range sensors to characterize the location and motion of objects around the vehicle. A typical system determines object attributes with a suite of sensors, develops a model of the scene around the vehicle, and issues a set of vehicle control commands depending on the desired system function.

In its simplest form, the function would be to issue a warning to the driver so that the driver can take the appropriate avoidance action. In a full collision avoidance system, the desired function involves automatic lateral or longitudinal control of the vehicle.

To develop this capability, the industry has been following a path similar to the one shown below. Several enabling technologies are needed to achieve the vision of a true collision avoidance system: long- and short-range object detection sensors, x-by-wire systems, appropriate human-machine interfaces, etc.

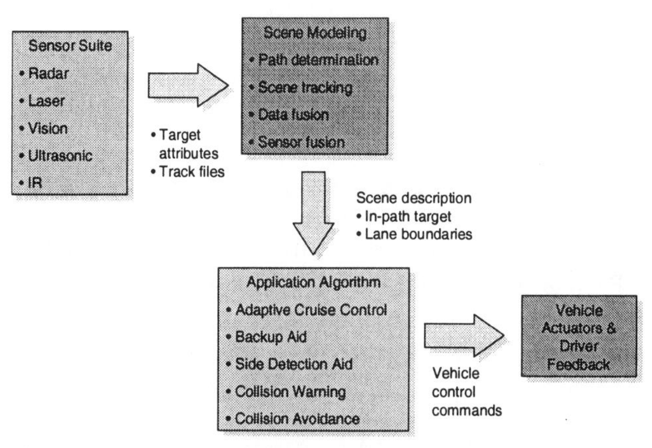

Figure 7. Collision Avoidance System Mechanization

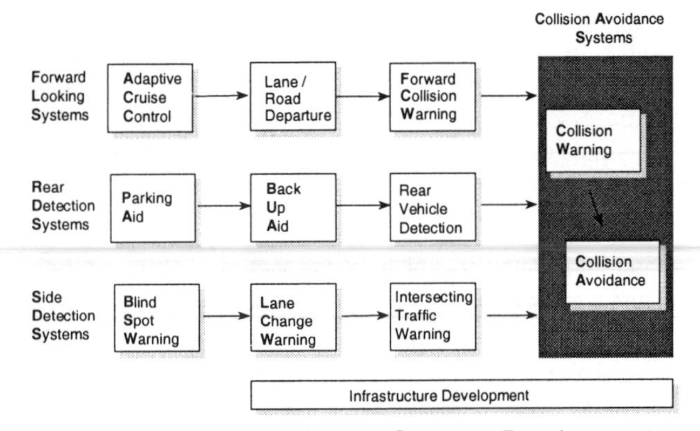

Figure 8. Collision Avoidance Systems Development

SHORT-RANGE PROXIMITY SENSING NEEDS – Collision avoidance systems are being developed to do as the name implies – avoid accidents. As we've shown, short-range proximity sensing is one of the key enablers. We know, however, that there is a long list of needs for short-range sensors as well. Some of these potential applications include:

- Power doors and liftgates
- Express close window and sunroof systems
- Occupant position sensing
- Security system applications
- Pre-crash sensing systems
- Active pedestrian protection systems
- Parking / Back-up aid
- Close cut-in detection
- Anti-trap trunks

The main point to understand is that features that traditionally have been looked at separately by product-focused teams basically need the same fundamental technical solutions.

ANTICIPATORY (PRE-CRASH) SENSING – By integrating long-range cruise control and collision warning sensors with short-range sensors needed for urban automatic cruise control (stop-and-go driving), pre-crash sensing systems will have the ability to perform object detection and tracking up until the actual time of impact. This ability now provides the capability to calculate angle-of-impact and region-of-impact, both critical parameters to understand crash type and to deploy the appropriate occupant protection countermeasures.

The logical next step is to integrate the vision systems utilized for lane tracking/lane departure into the system to provide object classification. This feature will provide the remaining information needed to truly predict crash severity.

CONCLUSION

No matter how you approach it, the future of automotive safety systems is certainly an integrated, vehicle systems-level approach.

Safety technology roadmaps are beginning to look alike. Collision avoidance sensors and occupant recognition sensors employ basically the same technologies. The same can be said about vehicle dynamic control sensors and vehicle crash sensors, as well as distributed safety architectures and distributed mobile multimedia architectures.

Safety features and functions are blending. The best way to protect an occupant is to avoid the accident. Subsystem information can and should be shared (vehicle dynamic state estimation information, occupant information, airbag status, scene information, etc.). Subsystem blending can enhance vehicle (and integrated safety) performance (e.g. a collision threat can mute a radio and cell phone, etc.). Proximity sensors can be used for a multitude of applications including:

- Security
- Collision avoidance
- Occupant position and recognition
- Pre-crash sensing
- Pedestrian protection systems

As a result, a systems approach to integrated safety is driving our future developments.

ACKNOWLEDGMENTS

The authors would like to acknowledge the Delphi Automotive Systems Advanced Vehicle Systems, Collision Avoidance, AED, Chassis, and Restraints Electronics teams for their assistance and hard work in making the contents of this paper a reality.

REFERENCES

1. W.G. Najm, "Comparison of Alternative Crash Avoidance Sensor Technologies," SPIE Vol. 2344 Intelligent Vehicle Highway Systems, 1994.
2. W.K. Kosiak and G. Nilson, "Benefits and Issues of Anticipatory Crash Sensing for Enhancing Occupant Protection," 3rd International Symposium of Sophisticated Car Occupant Safety Systems, 1996.
3. W.K. Kosiak and S.N. Rohr, "Future Trends in Restraint Systems Electronics," SAE Brazil, 1999.
4. Y. Shimizu, T. Kawai, J. Yuzuriha, "Improvement in driver-vehicle system performance by varying steering gain with vehicle speed and steering angle: VGS (Variable Gear-ratio Steering system)," SAE 1999-01-0395, 1999.

5. H. Kuzuya, H. Nakashima, K. Satoh, "Development of Robust Motor Servo Control for Rear Steering Actuator Based on Two-Degree-of-Freedom Control System," SAE 1999-01-0402, 1999.
6. S.M. El-Demerdash, A.M. Selim, "Vehicle Body Attitude Control Using an Electronically Controlled Active Suspension," SAE 1999-01-0724, 1999.
7. G.R. Widmann, "Delco Electronics' Activities in Collision Avoidance – The Path to Deployment of a 'Smart Vehicle'," SAE Toptech, 1997.

2000-01-0323

Cadillac DeVille Thermal Imaging Night Vision System

Nancy S. Martinelli and Scott A. Boulanger
General Motors Corp.

ABSTRACT

The purpose of the Cadillac DeVille Night Vision System is to provide drivers with visual information beyond with the range of their headlamps. It can also help drivers see beyond the glare of oncoming vehicle's headlamps. With increased visual range the driver may have more time to react to potentially dangerous situations.

The system consists of a thermal imaging camera, a head-up display, and image controls. The camera senses temperature differences of objects in the road scene ahead and creates a thermal image of the scene. The head-up display projects this image onto the windshield creating a virtual image that appears at the front edge of the vehicle's hood just below the driver's line of sight.

This paper will describe the system requirements and parameters of the 2000 Cadillac DeVille Night Vision system.

INTRODUCTION

Thermal imaging night vision systems have been used by the military for over 30 years. The early systems were cryogenically cooled. Their cost, size and maintenance requirements made them impractical for the automobile. General Motors began investigating an application for the automotive industry in the mid-80's when the first uncooled systems were available. Recent advances in electronics and infrared (IR) transmissive materials have made it affordable and practical to develop a thermal imaging night vision system for the automobile.

The main purpose of the Cadillac DeVille system is to provide drivers with visual information beyond their headlamps and to help the driver "see" beyond the glare from headlamps of oncoming vehicles. The system is not intended to be used as a "drive-by" system.

The Cadillac system consists of a thermal imaging camera, a head-up display and image controls. The intent of this paper is to describe these system components and the development of critical system parameters.

SYSTEM COMPONENTS

CAMERA – The camera in the Cadillac system (see Figure 1) detects energy in the 8 to 14 micron wavelength band or long wavelength infrared (LWIR) band. This band is ideal for the automotive application since objects of interest have peak radiation in the LWIR band e.g. humans have peak radiance at 9.3 microns. Since objects in the environment naturally emit this thermal energy, the system does not require a light source or ambient light - it can work in total darkness.

Figure 1. Thermal Imaging Camera

The camera's refractive optics focus the thermal energy onto an uncooled focal plane array (UFPA) detector. The array is 320 pixels wide by 240 pixels tall. Each pixel is a temperature-dependent capacitor that changes capacitance depending upon the amount of energy it is receiving. The detector material is barium-strontium-titanate (BST), a ferroelectric material that is sensitive to LWIR. The material is most sensitive to temperature differences at its phase transition, 35C, which is near room temperature. A thermal electric cooler maintains the detector temperature at its phase transition. A chopper disc rotates at 30 cycles per second in front of the detector to modulate the scene's energy. The camera electronics perform all the video processing and transmit the image to the head-up display.

Ideally, the camera should be located as close to the driver's line of sight as possible. The size of the camera, 110 mm x 125 mm x 118 mm, and the requirement for an unobstructed field of view made it difficult to package. Since glass is not IR transmissive, the camera could not easily be packaged inside the passenger compartment. The only practical location that met the criteria and provided an aesthetically pleasing appearance was the center of the radiator grille.

HEAD-UP DISPLAY (HUD) – An important factor in the Cadillac system design was the manner in which the night vision image was to be presented to the driver. It was necessary for the image to be easily viewed and interpreted without requiring drivers to remove their eyes from the road or obstructing their direct line of sight. This was a perfect application for a head-up display.

The HUD, which located in the instrument panel in front of the driver (see Figure 2), was designed to project the image onto the lower portion of the windshield creating a virtual image at the front of the vehicle just above the hood line (see Figure 3). In this position, the image would be in the driver's lower peripheral view. Drivers need only glance down slightly to view the image while still keeping the road scene in their upper peripheral view. This allows drivers to use the system the way they would the rearview mirror. Also, having the virtual image projected out in front of the vehicle minimizes the refocusing required to go from the road to the image. This reduces eye fatigue and is an added benefit for drivers with bifocal corrective lenses.

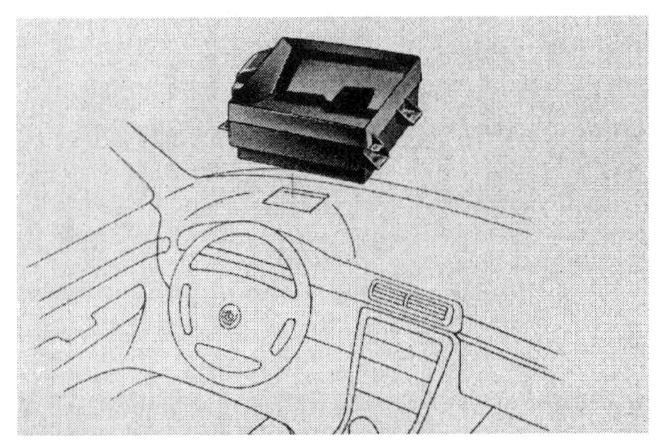

Figure 2. HUD location in vehicle.

The system powers up when the vehicle's ignition switch is in the run position, the front headlamps or front fog lamps are active and the vehicles' ambient light sensor indicates darkness. The requirement that the headlamps or fog lamps must be on to operate the system prevents driving at night with only the night vision system.

The HUD image is generated by an active matrix liquid crystal display (AMLCD) which produces a monochromatic image with 16 shades of gray. The AMLCD image is focused onto an aspheric mirror, which reflects the image onto the windshield. The aspheric mirror is specifically designed to magnify the image and correct for windshield surface curvature. The AMLCD and mirror are precisely positioned relative to the windshield to focus the virtual image approximately 2.5 meters in front of the driver at the front edge of the vehicle's hood.

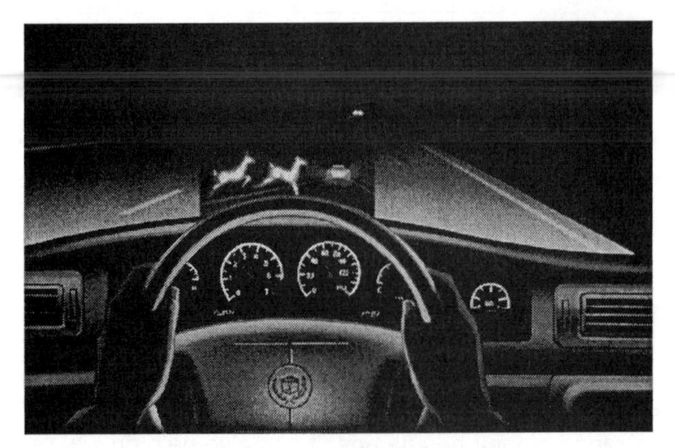

Figure 3. View of scene with thermal image.

CONTROLS – Controls are available to the driver for adjusting the image. These controls include a dimmer switch for turning the image on or off and adjusting the intensity. A rocker switch is also available for adjusting the vertical position of the image to accommodate variations in driver-seating heights.

SYSTEM PARAMETERS

In developing the night vision system for the Cadillac DeVille, there were several parameters that were critical in making this system easy to use. These parameters include magnification, field of view, alignment, polarity and detection range. The following paragraphs describe these parameters and the rationale for decisions made on the DeVille program.

MAGNIFICATION – To help the driver correlate objects in the image to objects in the road scene, it was critical to make objects in the image appear the same size as they would through normal vision. This 1-to-1 image to object ratio was achieved by controlling the magnification and projection distance of the HUD image. With the 1-to-1 image to object ratio the driver has the proper perspective or sense of distance.

FIELD OF VIEW (FOV) – There is a tradeoff between FOV and depth of field. The wider the system's FOV, the shorter the depth of field. Likewise, the narrower the system's FOV, the longer the depth of field. A key system objective was to provide a view of adjacent lanes of traffic before the far edge of the headlamp range and to provide as much resolution as possible beyond the headlamps. It was not necessary to provide a thermal image of the area illuminated by the headlamps. As a result, a narrow FOV was chosen for this system. A system with a wider FOV would not only have provided an inadequate depth of field, but would also have required a much larger HUD to maintain the 1-to-1 image to object ratio. A larger HUD would have been very difficult to package due to the limited space in the instrument panel.

The system's horizontal FOV of 11.25× provides coverage of the adjacent lanes at 68 meters (see Figure 4). The vertical FOV is 4× which provides enough road scene information without obstructing the driver's view of the road.

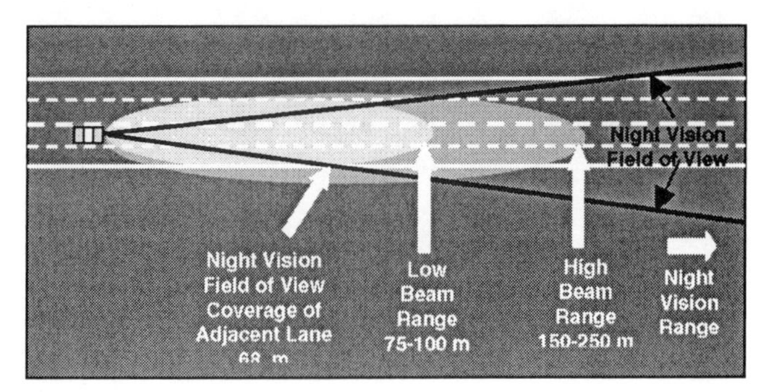

Figure 4. System Field of View

FOCAL LENGTH AND DEPTH OF FIELD – The camera optics are focused at 125 meters to optimize the resolution of objects beyond the range of the headlamps. The depth of field is 25 meters to infinity.

ALIGNMENT – The position of the HUD's virtual image and the alignment of the camera were key elements in helping the driver correlate objects in the image to objects in the road ahead of the driver. It was determined that objects in the HUD image should be aligned horizontally with the objects in the road scene to give the driver a sense for the object's location. In order to achieve this alignment, the HUD image was centered horizontally with the driver's line of sight, and the camera was aimed 0.7× to the left to compensate for being packaged on the vehicle's centerline as opposed to the driver's centerline.

The vertical alignment of the camera was chosen such that the horizon appeared in the lower one-third of the HUD image. This provides a view of the road in the image even when the vehicle is driven up and down hills.

DETECTION RANGE – Using the night vision system under clear conditions, the driver can detect a 0.5 meter wide by 1.8 meter tall person at a distance of 300 meters and a running vehicle at 500 meters. Using only the vehicle's headlamps, the driver would not be able to see the object until it entered the headlamp range assuming low ambient light conditions. The typical range for low beam headlamps is 75-100 meters and for high beam headlamps is 150-250 meters.

POLARITY – Since the image is created from thermal information, the 16 shades of gray produced by the camera are based on temperature. The system can be configured to produce a "white hot" image where hotter objects appear whiter or a "black hot" image where hotter objects appear darker. A "white hot" image configuration was chosen for this system because objects of interest, such as people, animals, and other vehicles, which are typically warmer than their surroundings, become more noticeable as white objects on a darker background in the night vision display.

CONCLUSION

The parameters on the 2000 Cadillac DeVille have been optimized within the vehicle constraints to provide the driver with the ability to "see" beyond their headlamps and beyond the glare of oncoming vehicle headlamps.

CONTACT

Nancy S. Martinelli
100 Renaissance Center
M/C 482-A08-B25
PO Box 100
Detroit, Michigan 48265-1000
Nancy.Martinelli@gm.com

Scott A. Boulanger
4100 South Saginaw
M/C 485-303-117
Flint, Michigan 48557
Scott.Boulanger@gm.co

2000-01-0431

Adaptive Frontlighting Systems for Optimum Illumination of Curved Roads, Highway Lanes and Other Driving Situations

Henning Hogrefe
Automotive Lighting

ABSTRACT

Modern adaptive lighting techniques offer the chance to improve the vision of car drivers in front of their car in situations which are not well served by conventional static headlamps. Especially curved roads, intersections and the far zone on straight roads and highways at higher speed need to be illuminated better by the automotive headlamps. This can be achieved by well-designed and tested adaptive light modules. The application of headlamps with adaptive light pattern requires special attention to a new set of questions: Which light pattern should be used to illuminate curves and other areas in the field of vision, how do they fit to the main beam and how should these functions be activated? Technical alternatives to realize the light modules will be discussed and performances will be presented.

INTRODUCTION

The nocturnal traffic scene ahead on which a driver must concentrate his view depends on 4 major factors:

- geometrical / topological layout of the street and it's surroundings
- other traffic participants
- ambient illumination of the scene
- vision, influenced by weather conditions

All these factors together represent a wide variety of possible scene classes to say nothing of the complexity of real scenes. Automotive lighting should provide good illumination for these scenes. Although there has been considerable technical progress in the field of automotive lighting throughout the last decade applying better light sources, new headlamp techniques and achieving better manufacturing quality a headlamp with static beam pattern is only able to deliver a compromise illumination. Adaptive frontlighting will be the next innovation step towards optimum illumination and driver information for many driving situations. With this concept a new dimension will be opened: Instead of a static illumination beam for the whole diversity of occurring situations, in future, advanced systems will have situation-adaptive sensor-controlled light pattern. Lighting experts distinguish different adaptive lighting functions: adverse weather lighting, curve lighting, country road lighting, motorway lighting, city lighting and overhead lighting. In all these lighting domains there is a more or less pronounced need for an improved vision by means of specific adapted light pattern which are supplementary or different compared to the conventional static light pattern. This paper focuses on those functions which primarily handle certain geometrical and topological layouts of the street, in particular curve and motorway lighting. In order to find the appropriate light pattern for these functions it is neccessary to get some knowledge about the location of characteristic scene points in the driver's field of vision. This loacation depends on the type of street and it's course in front of the car. It is obvious that there must be some kind of relationship between different street parameters, like street width, curve radius and street class (eg. interstate highway, rural road etc.) but there is little information about the numerical distribution of these parameters in public driveways. Some hints are given in references 1 and 2 about the design guidelines and the statistical distribution of street parameters in Germany. Figure 1 illustrates the street geometries which are most relevant for the driver's vision on the street level (bird's eye view). Beginning with perpendicular intersections we further have rural roads with 5-6m width and typical curve radii between 50m and 150m and motorways with width of 7m and curve radii of 200m and more. The conventional static light pattern cannot satisfactory cover all these scene geometries! This is demonstrated in figure 2, where the gray shaded area represents a good static light pattern. In this figure the hatched areas indicate those zones in the vision field which need to be better illuminated. It can be seen that there is a considerable potential for lighting improvement by application of the adaptive headlamp concept.Therefore, the conventional beam pattern must be modified or supplemented by additional light according to the actual driving situation. Especially, the side illumination for curves and crossroads can be significantly improved. For curve and crossroads scenes appearing in front of the car a much broader field of vision is needed which either is not covered or is covered only with low intensity

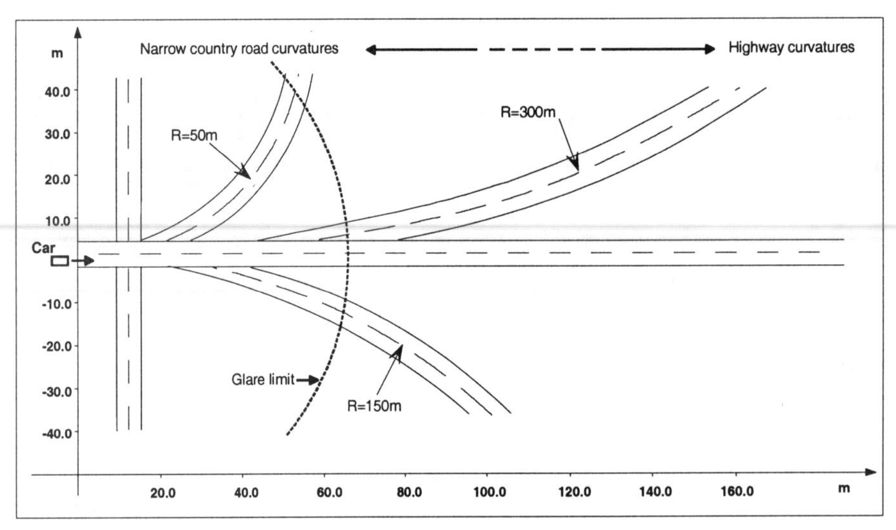

Figure 1. Street geometries in the vision zone of a car driver

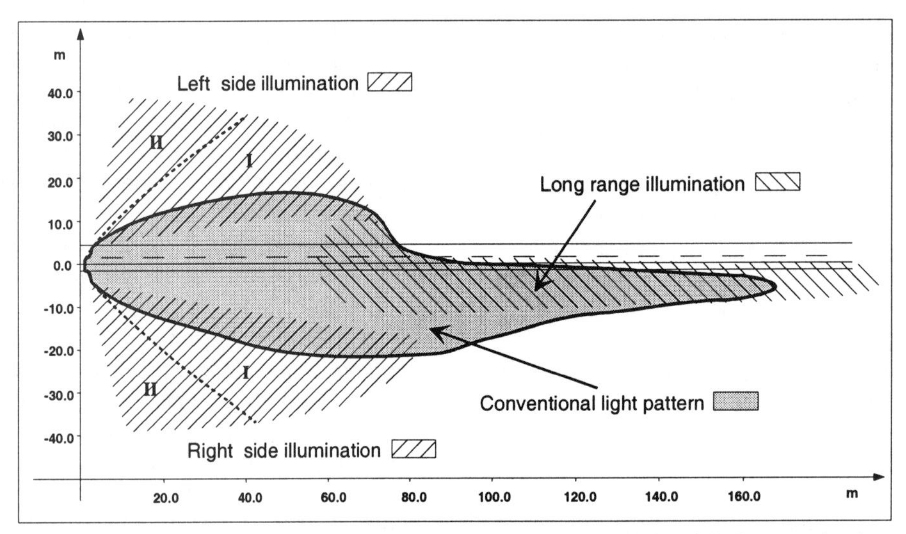

Figure 2. Vision zones in front of a car for right hand traffic (bird's eye view):

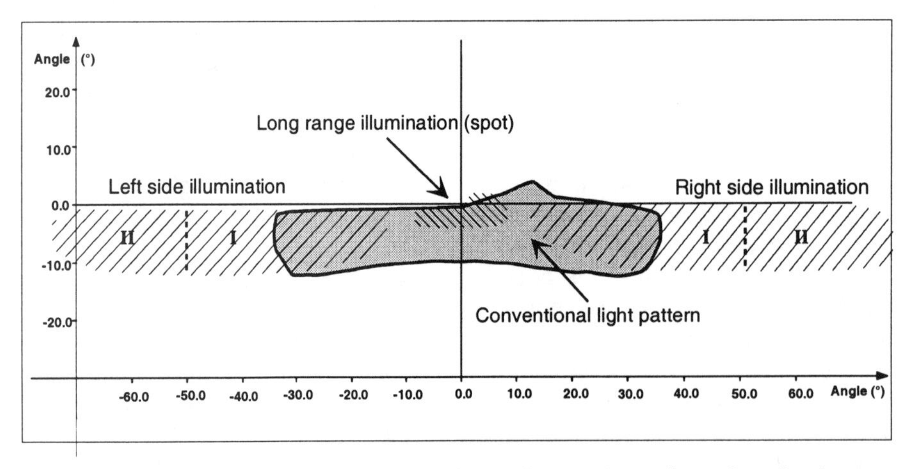

Figure 3. Vision zones (corresponding to figure shown in wall projection

by the conventional static light pattern. The side illumination zone has been subdivided in this figure into two different parts where zone I accounts for normal road curves and zone II for crossroads and sharp bends. Another zone of interest for adaptive lighting is the long range vision zone. An adaptively increased long range illumination (spot illumination) is important when driving at higher speed on straight roads or roads with little curvature (motorway lighting, country road lighting). A large part of this zone is already illuminated by the basic light pattern but in these situations the amount of light is insufficient. Hence, the visibility at long distance should be improved temporary in certain driving circumstances by enhanced intensities in this zone. When the high speed situation or the curve is over the headlamp must return into it's basic state in order to direct the attention of the driver to his main vision tasks. For the lighting engineer it is important to transform the graphical representation of the vision zones of figure 2 from the bird's eye view to the wall projection view. The wall projection is directly related to illumination optics of the system and it is used to analyze a headlamp in the laboratory (see fig. 3). When the light pattern of the conventional light pattern - used as "basic component" for the adaptive headlamp - is quantitatively known, then a more detailed elaboration of the wall projection will be used to determine the contributions of the other adaptive components in order yield a homogeneous and effective total adaptive light pattern.

CURVED ROAD ILLUMINATION

It is a common idea, that adaptive lighting primarily consists of a basic lighting component (Ref. 3), a sort of conventional low beam pattern. This can either be static or variable. Now, a curve lighting beam must enhance this basic light pattern and illuminate a horizontal angular width of ±60 – ±80 degree from the center as shown in figure 3. The curve lighting zone may even reach straight ahead close to the center of the basic beam (± 5 –20 degree) depending on the width of the light pattern of the basic headlamp (gray area) and on the technical concept chosen in a particular case. Table 1 gives an overview of the major curve lighting realization concepts. The first concept simply applies a static basic beam and a static curved road beam from a separate static reflector. Next, a separate horizontally rotatable reflector can be used for a continuously variable curve illumination.

In the third concept the basic beam is horizontally rotable to fulfil the curve lighting function. Finally, the variabilty may be incorporated together with the basic light function into one single "vario" module. Of course, combinations and variations of these major realization strategies are possible. There have been layout proposals for all of these basic concepts (Ref. 3 - 6). Within the AFS working group (Advanced Frontlighting Systems, Ref. 7), which is set up by various European companies to make a

Table 1. Major curve lighting realization alternatives

	System	Mechanics	Light Pattern
1	Separate curve reflector	Static	
2	Separate curve reflector	Movable	
3	Basic reflector also used for curve light	Movable	
4	Basic reflector also used for curve light	Internal optics variable	

proposal for the homologation of adaptive headlamps, there have been test drives with curve lighting components of type 2 and 3 featuring movable reflectors controlled by evaluation of the turn wheel angle signal. On a curvy country road these systems yield a good impression. They have the ability to direct the hot spot of the curve light pattern dynamically on the driver's lane generating a comfortable vision of the scene ahead. This ability is bought at the price of higher mechanical complexity as compared to solution 1. The same lighting properties can in principal also be expected by curve components of type 4. This type comes close to the ideal of the adaptive headlamp which is able to deliver all functions by variations of one component. But so far in practice the optical efficiency and the mechanical complexity of type 4 systems seem to be even more delicate. For adaptive illumination of sharp bends or crossroads (area II in figures 1 and 2) when the whole scenery from left to right shall be homogeneously illuminated curve lighting type 1 is advantageous. The advantages in each category can be summarized as follows:

- **Optical efficiency**: advantage with concept 3 (the others can also have good light pattern depending on available space)
- **Adaptive function**: advantage with concept 2-4
- **Expenditure/Complexity**: advantage with concept 1
- **Space needed**: advantage with concept 3 or 1 (depending on motor drive needed for 3)

For a complete comparison of the different systems the electronic control must also be considered. Common to all advanced adaptive headlamp systems is the need for sensor signals and an electronic ballast.

Here, we report about tests with three experimental curve lighting systems applying the above-mentioned basic curve lighting types No. 1 and 3:

1. In an adaptive headlamp with multiple static components using the superposition of individual static light pattern (see Ref. 4) we incorporated one static reflector on each side with a broad curve light pattern . The performance of this static system (type 1) is shown in the upper part of figure 4.
2. The same headlamp was equipped with two static curve lighting reflectors (occupying approximately the same space) each covering a certain angular light pattern width, splitting the width which is covered by version 1) into two separately switchable beams. The performance is shown in the middle part of figure 4.
3. In the next experiment, we combined static and movable systems (types 1 and 3): One static reflector for the extreme angles (for narrow curves) and a rotatable basic component for the medium angular range (for moderate curves) were used.

In order to realize adaptive headlamps with a rather simple, cost-effective curve lighting function a static curve lighting reflector is an interesting alternative. Tests drives showed that both systems of type 1 added a substantial lighting benefit to the basic light in curve driving situations. With the first layout one could only illuminate the whole left side in total. This was not optimal for medium curve situations (zone I in figures 3 and 4) because the viewing direction of the driver may be focused on a zone which is not important in a particular moment. Hence, version 2 is superior because it is able to illuminate specifically only the zone which is corresponding to the actual range of curve radii. Anyway, we can consider static curve lighting components as a valuable component for adaptive headlamps. Optimum lighting performance and visual comfort is obtained with the third setup. With this combination of type 1 and type 3 narrow curves and intersections are well served by the static reflector whereas medium and slight curves are very well illuminated by continuous small horizontal adjustments of the basic beam. This combination is better than a pure type 3 system: the larger angular rotation needed to illuminate even narrow curves and intersections only with the basic beam would generate a temporary illumination reduction or gap in central direction. Therefore, a combination of static and movable components, as outlined here, can be recommended for the curve lighting function of an adaptive headlamp which allows the expenditure of state-of-the-art technique.

LONG DISTANCE ILLUMINATION

An optimum illumination of the long distance zone becomes very important when driving with elevated speed on highways and straight country roads. One has to distinguish different circumstances:

- Street class: Separate lane for oncoming traffic available ?
- Traffic density

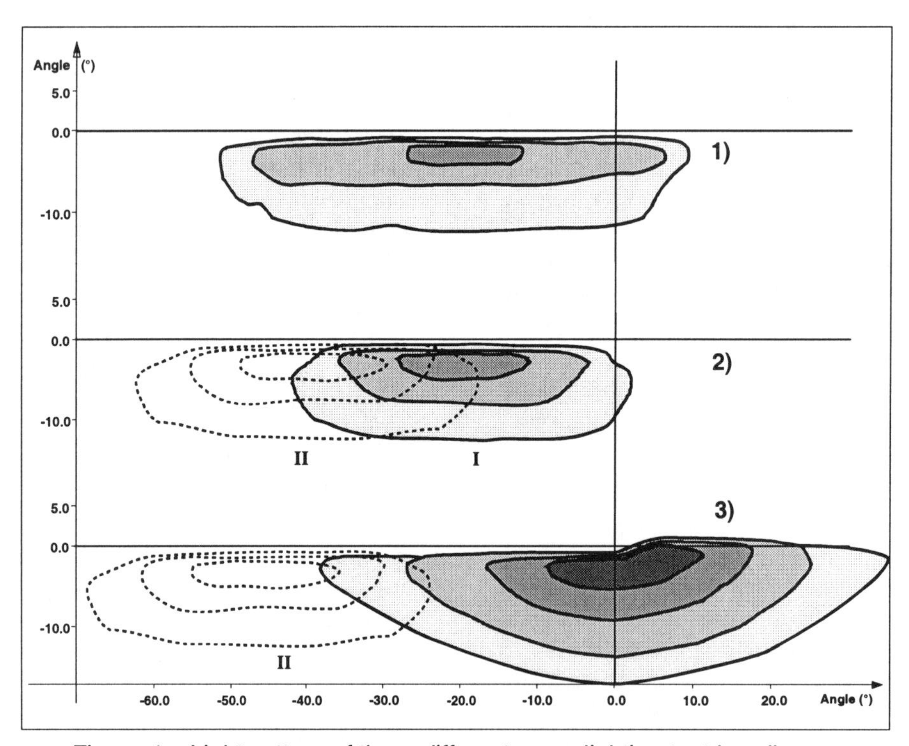

Figure 4. Light pattern of three different curve lighting test headlamps

Depending on these circumstances additional enhancements of the far zone illuminance can be applied. If there is a high traffic density with oncoming traffic on a two-lane-road then an additional long distance illumination is not useful even when driving at high speed. If the traffic is not as dense then a higher intensity in the far zone right below the standard low beam cutoff makes sense. Going one step further we could also envision to change the right hand 15-degree ECE cutoff to form a z-beam cutoff which improves the vision around the right hand street border. If there is no oncoming traffic or the lane for this traffic well separated from the own lane (e.g. on highways) then another alternative can be envisioned: the cutoff of the basic beam may be rised slightly vertically to allow for a longer visual range. This method is under discussion in the AFS working group (Ref. 7) to be introduced in a future ECE adaptive lighting regulation.

Table 2. Adaptive long distance illumination alternatives

	System	Light Pattern
1	Basic light pattern	
2	Enhanced spot illumination (same cutoff)	
3	Enhanced spot illumination (z-cutoff)	
4	Raised cutoff of basic light pattern	

In the wall projection the long distance zone extends over a rather small angular area (Fig. 3). Hence, a spot lighting component should not contribute any intensity into the foreground zone near to the car (below 3 degree down). Furthermore, to obtain a significant supplementary illumination effect with a spot at long distance, a module with high illuminances is necessary. These two goals – restricted angular area and high illuminance output – are very demanding technical requirements for a headlamp component of limited size. If they are successfully realized then the vision on straight roads, motorways, wet roads etc. can be supported by this function. The hot spot intensity of such a concentrated beam is proportional to the size of the exit pupil of the spot module. But the available space in a headlamp including the basic low beam component and supplementary adaptive components is very tight. To take this into account we have realized spot lighting systems using ellipsoid projection systems (PES, Ref. 8) which have small lateral dimensions. With this kind of optics the spot illumination intensity is proportional to the lens pupil area. Using an H7 bulb we could add up to 70 lux to the hot spot maximum of the basic light pattern (Fig.5). This spot has a z-shape cutoff in order to deliver additional light even above the cutoff of the basic light pattern. First driving tests showed a very good performance if the layout of the basic light pattern and spot lighting pattern match together well.

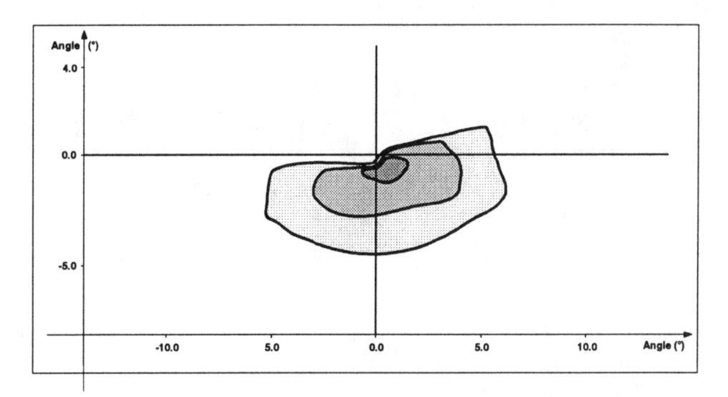

Figure 5. Spot light pattern for enhanced long distance illumination

Tests including a raised cutoff were also performed through vertical adjustment of the basic component. This must be done in a very careful moderate way in order to prevent glare. Hence, the margin for upward tilt is less than 0.5 degree which is enough to obtain a significant increase of long range vision. This can be seen in practical tests but also already verified in advance by appropriate simulation methods. In order to design and evaluate the different adaptive modules modern simulation tools are mandatory. We used our optical design software for the optical layout and light pattern calculation of the modules and for evaluation of the visual impression our visualization tool was used (Ref.9). The simulation images of the basic-, spot- and motorway function are shown in figures 6-8.

Figure 6. Visualization of the basic light pattern component (with standard angular adjustment, zoom onto long distance zone)

Figure 7. Visualization of the basic light pattern component together with the spot lighting function (zoom onto long distance zone)

Figure 8. Visualization of the basic light pattern component with raised cutoff (zoom onto long distance zone)

CONCLUSION

Adaptive headlamps for better curve illumination and enhanced long range illumination have been investigated. Application of such systems is not yet allowed in Europe but new regulations are in preparation. Different types of curve lighting modules are available for the wide variety of practical applications. Static curve reflectors allow good light performance with a simple mechanical setup. Full adaptive lighting performance can be reached with combinations of movable basic beam for slight curves and static curve lighting beam for narrow curves. The long range illumination can be comfortably enhanced without glaring oncoming traffic by separate spot modules or by means of a raised basic beam cutoff in specific driving situations.

REFERENCES

1. Forschungsgesellschaft für Straßen- und Verkehrswesen, Arbeitsgruppe Straßenentwurf: Richtlinie für die Anlage von Straßen RAS-L, 1995

2. W.Leutzbach, J. Zoellmer: Zusammenhang zwischen der Verkehrssicherheit und den Elementen des Straßenbaus, Forschung Straßenbau und Straßenverkehrstechnik Herausgeber BMV, 1989, Heft 545, p. 1-56

   J.Damasky: Geometry of the Road Area and Effects on Motor Vehicle Lighting, Progress in Automobile Lighting (PAL), Symposium, Darmstadt, 26.-27. Sept. 1995

3. H.Hogrefe: Future Lighting Concepts–Headlamps with Adaptive Light Pattern, Progress in Automobile Lighting (PAL '97), Darmstadt,1997, p. 254

4. H.Hogrefe, R.Neumann: Adaptive Light Pattern–A New Way to Improve Light Quality, SAE Technical Paper Series 970644, 1997

5. S.Kobayashi, M.Hayakawa: Beam Controllable Headlighting System, SAE Technical Paper Series 910829, 1991

6. E.Groh: Intelligent Lighting, SAE Technical Paper Series 970645, 1997

7. AFS – Advanced Frontlighting Systems, Eureka Project 1403

8. G.Lindae: Improvements of Low-Beam Pattern by Use of Polyellipsoid Headlamps (PES), SAE 850228, 1985

9. H.Hogrefe: Modern Simulation Tools for Headlamp Lighting, Progress in Automobile Lighting (PAL '97), Darmstadt, 1997, p. 260

980558

Smart Airbag Systems

Helmut E. Mueller and Burghard Linn

ABSTRACT

Paper introduces cost effective design alternatives for "out of position" occupant safety, including:

- Extractable passenger side lids which move out of the way before airbag deploys.
- Multi-stage tether controlled airbag deployment dynamically senses "out of position" resistance and controls deployment of further stages.
- Airbag back pressure activated steering column forward and down movement.
- Two stage thorax bag with tether control to prevent catapulting "out of position" occupants.
- Two stage headrest that expands forward and sideways to protect occupant heads in rear end crashes. Second stage activates only if no "out of position" is detected.

INTRODUCTION

Over the years thousands of lives have been saved by airbags, however in some instances airbags have caused fatalities. These fatalities occur mostly when occupants are close (less then 100mm) to the airbag and lid when it deploys. Publicity of these fatalities, especially in the US, has fueled the demand by customers for "smart airbags" which can detect "out of position" situations and control deployment accordingly. Sophisticated ultrasonic and infrared systems combined with seat load distribution sensing are under development. They require computer control and they will be expensive. A significant engineering challenge exists to make these systems fast enough to deal with the dynamics of the crash.

This paper introduces a number of inexpensive solutions to protect "out of position" occupants, addressing airbag lids, multi-stage frontal and thorax bags as well as headrest airbags for rear end crash and front crash rebound. Many of these solutions don't require sophisticated technology and can be brought into production rapidly.

1. EXTRACTABLE AIRBAG MODULE LIDS

The deployment pressure of the first generator stage will move an inner housing ("pressure housing"), that surrounds the generator on one side, forward and away from the instrument panel. The pressure housing is connected on all 4 corners to the outer edges of the two lids (fig. 1). As the pressure housing moves, it pulls the lids into the module. The inner edges of the lids are guided in slots inside the airbag module housing. The lids should be as small as possible since they need to fit inside the airbag module. The instrument panel can be laser cut on the inside to create clean separation lines for the lids when deploying. When the lids are opening, the airbag unfolds and deploys through its first soft stage.

2. TETHER CONTROLLED AIRBAG DEPLOYMENT

To prevent "out of position" airbag injuries, the bag must deploy in stages. The first stage inflates at lower pressure "soft". At this stage a tether connected to a spring loaded electrical switch is extended with the deploying bag. If there is resistance, the second or subsequent stages are not deployed. The time delay between the soft/hard deployment stage is around 5 msec with some tuning possible. (Fig. 2)

Electrically triggered two stage generators (fig. 3) already exist and can be adapted to this application. As an alternative, fig. 4 shows a low cost mechanical trigger directly activated by the tether. With a dual mechanical ignition, the reliability is higher than with an electrical ignition.

3. TWO STAGE BAG FOR DRIVER'S SIDE MOVES STEERING COLUMN

The same tether controlled airbag system principle can be used on the driver's side as well. Two stage generators have already been developed. However, the performance of this safety system can be improved significantly by moving the steering column/wheel forward and down (fig. 5). The first "soft" stage will be ignited by a pressure pulse from a cartridge which is located on the outside of the steering column. This will eliminate the expensive "clock spring" used in current systems. The pressure of the first stage generator will fill the airbag with low pressure and at the same time will move a piston down in the steering column. This will activate a bowden cable through a coupling sleeve to unlock the steering wheel adjustment mechanism. The back pressure of the bag then moves the steering column forward and down, of course against a damper to avoid the need for a larger bag. Moving the steering column forward and down can be traded off for less frontal structure while maintaining the same injury indexes.

4. TETHER CONTROLLED THORAX BAG

Thorax bags can be dangerous if an unbelted child is standing in close proximity. A two stage thorax bag, controlled by a tether, will only fully deploy after the first stage has moved the bag in place. (Fig. 6)

5. TETHER CONTROLLED HEADREST AIRBAG

To prevent head and neck injuries from rear end crashes and front crash rebound the head has to be supported in the proper location ideally with an airbag surrounding it. In the past airbags were not considered for this location because of close proximity to the head and "out of position" concerns. With the advent of a multi-stage airbag these concerns can be eliminated. Fig. 7 shows the concept for such an airbag. A dual stage generator is placed inside the headrest (fig. 8). The first stage is mechanically triggered via a bowden cable that is connected to a spring loaded strap in the seat back (fig. 9). When the first stage deploys and the three tethers don't sense resistance, the second stage fills the bag to a higher pressure. For improved reliability a dual mechanical ignition is used in both stages. To minimize the unfolding power of the airbag, it should be made of sealed cloth, folded and afterwards foamed from the inside. Noise absorption is important in this application and can be achieved through perforated tubes and steel wool.

An even better release of the first headrest deployment stage can be achieved by a bowden cable connected to the movement of the seat rest attachment (fig. 10). This design will move the seat rest rearward and up resulting in the forward movement of the upper seat back in a rear end crash. This movement will prevent the occupant to slide up on the seat back. Four links with different length guide the movement accordingly. A built in load limiter prevents activation under normal operating conditions.

6. TEST RESULTS

A number of tests have been performed on a tether controlled dual stage passenger airbag with positive results. Up to date information will be provided at the February '98 SAE conference.

SUMMARY

- Passenger side airbag lids pose a danger to "out of position" occupants. Extracting these lids into the instrument panel before the airbag deploys can eliminate this problem.
- Tether controlled multi-stage airbags promise to be a simple and cost effective solution to reduce airbag-inflicted injuries.
- A steering column that moves away from the occupant in a crash can further enhance safety for the driver.
- Headrest bags and forward tilting seat backs can be a valuable addition to improve rear end crash protection.

ABOUT THE AUTHORS

Helmut E. Müller was manager of interior design and development including passive safety at Opel for 12 years and at VW for 6 years. He is now working as an independent consultant on safety innovations.
Burghard Linn has been an engineer and manager with Opel, GM, and EDS. He now works in cooperation with H. Müller on safety innovations..

For further information, please contact:

Helmut E. Müller, Buchenweg 42
38550 Isenbüttel, Germany
Tel: 011.49.5374.5684, Fax: 011.49.5374.5185

Burghard Linn, 5248 Milroy
Brighton, Michigan 48116-9727
Tel: 810.227.1223, Fax: 810.227.5179

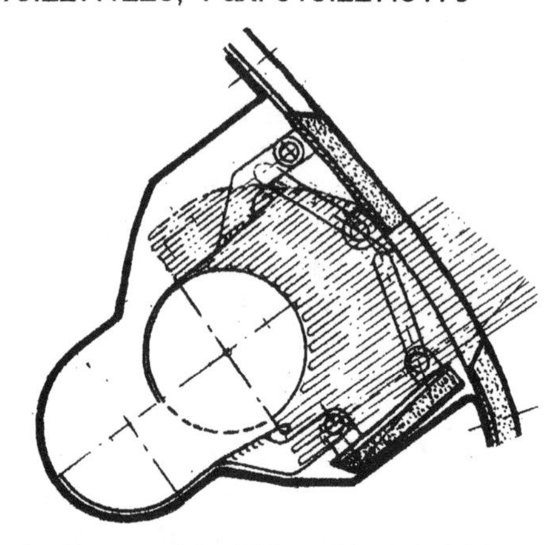

Figure 1. Extractable Airbag Module Lids

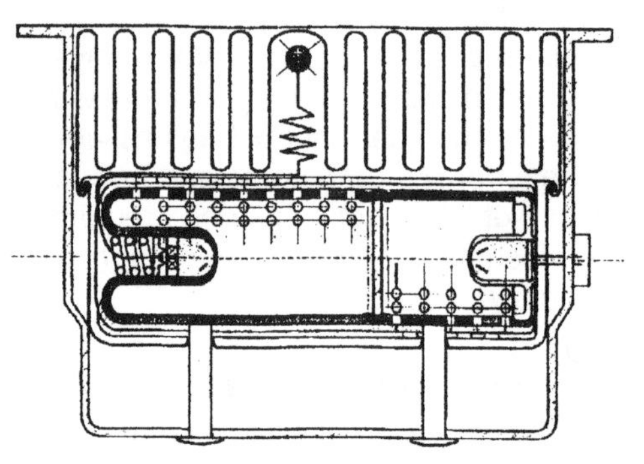

Figure 2. Tether Controlled Airbag With Dual Mechanical Ignition And Pressure Housing To Open The Lids

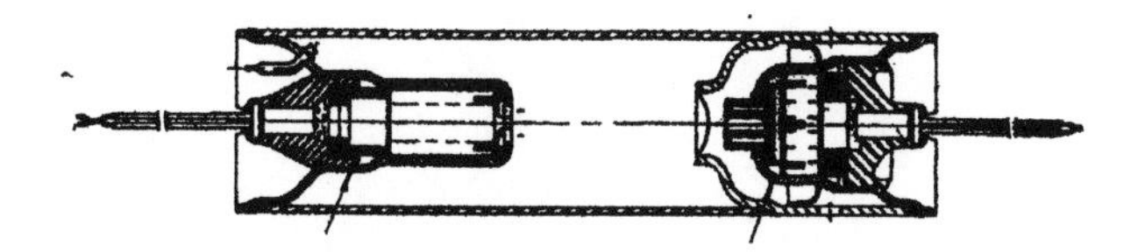

Figure 3. Electrically Triggered Two Stage Generator

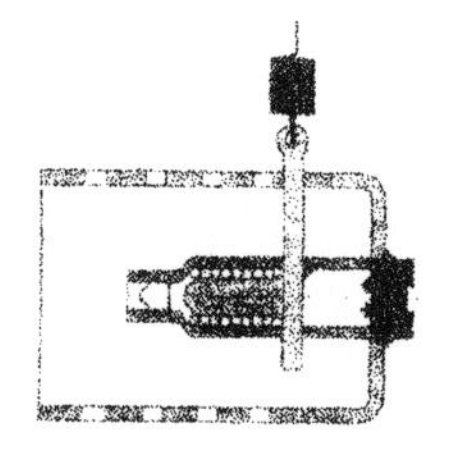

Figure 4. Low Cost Dual Mechanical Trigger

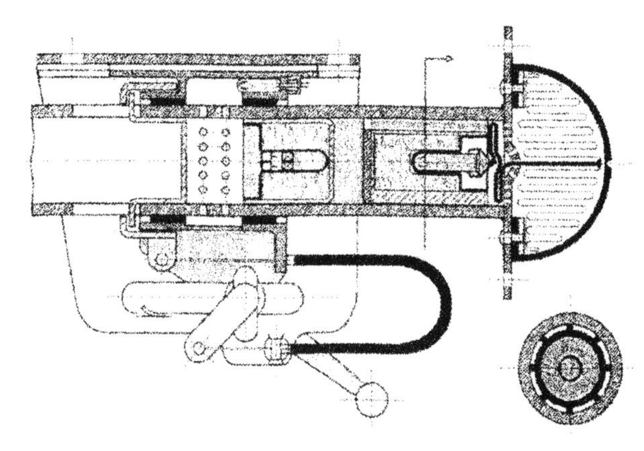

Figure 5. Two Stage Driver Side Bag That Moves Steering Column Forward And Down

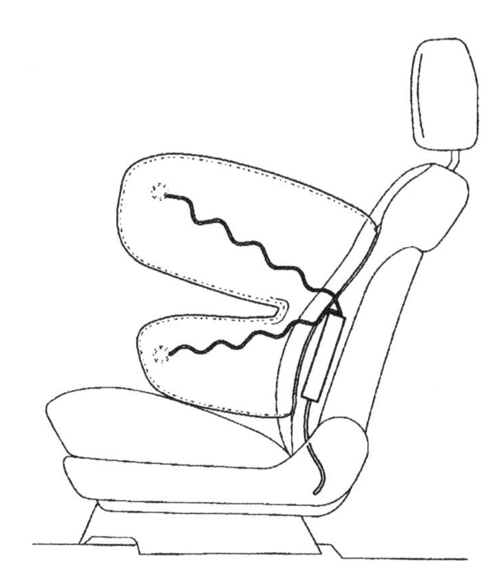

Figure 6. Tether Controlled Thorax Bag

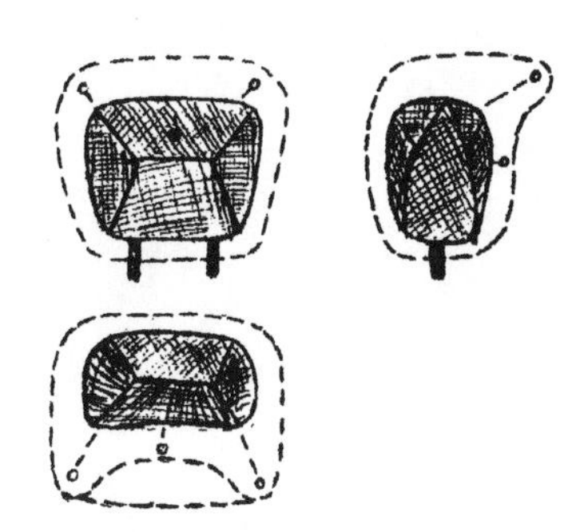

Figure 7. Headrest Airbag Concept

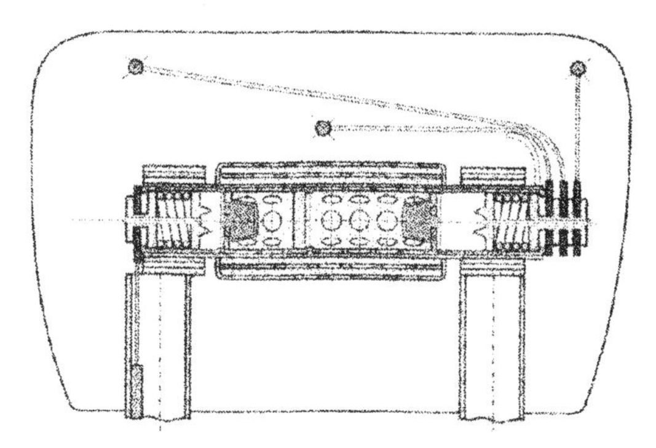

Figure 8. Headrest Dual Stage Generator

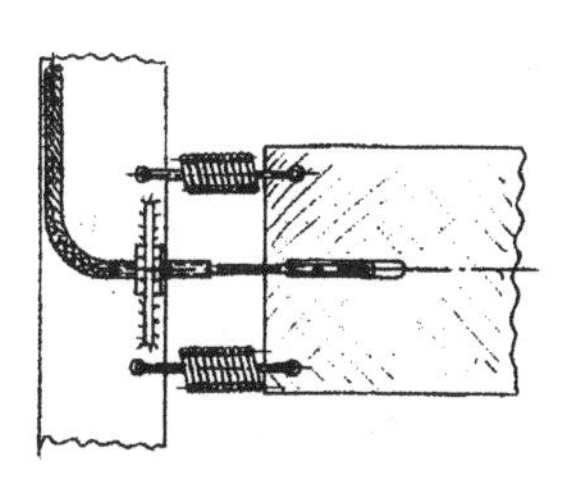

Figure 9. First Stage Headrest Airbag Trigger Mechanism In Seat Back

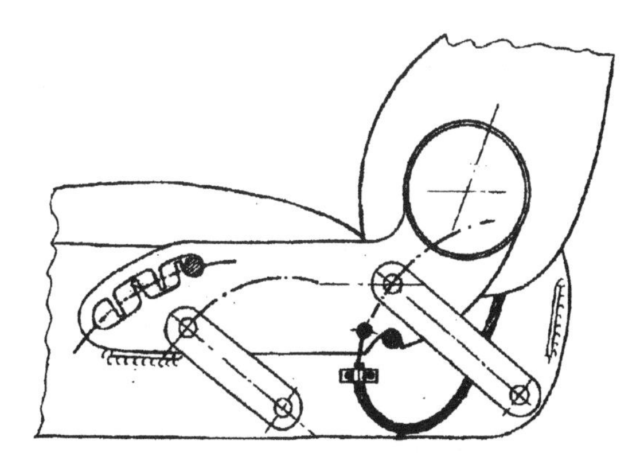

Figure 10. Seat Rest Release Mechanism To Tilt Seat Back Forward And Up And To Trigger First Stage Headrest Airbag

1999-10-0043

The Development of An Impact Simulator and The Study of a Side Airbag Algorithm

K. A. Sung
Dept. of Mechanical Engineering,
Kangwon State University,
Republic of Korea

ABSTRACT

Improved passive safety measurement and analyses are required to minimize injuries to passengers in vehicles when collisions arise. Impact tests using the developed impact simulator have been carried out to validate proposals for the specifications of a new electronic single point-sensing module (ESPS). This paper describes the structure and control of the simulator that includes a servomechanism, a signal acquisition, sensor interface circuits, and a crash recognition algorithm for a side airbag system. The aim of the tests using this equipment is to make the output signals from sensors equal to those on real vehicle collision tests. Using the simulator, the output signals have been used to define some thresholds governing the firing or non-firing conditions for the inflators. The results of tests undertaken provide data applicable to the development of an algorithm for a side airbag system.

1. INTRODUCTION

The realization of design targets and the requirements relate to more passive safety of transportation have become a specialty in vehicle design applicable to a growing weight of passenger car bodies. The variety of the possible collisions, that have to be considered in design are extreme as well as the steps to be considered in the character of the load-deflection curves of the structures involved [1].

In the front of the passenger compartment in side collision cases, an energy absorbing structure is arranged in the direction of the expected introduction of load. In order to protect serious injuries, a side airbag system is mounted on the side bucket of a front seat. The development for an airbag system of a brand new car needs a lot of money and a long time, because the price of a proto-car could be 15 or 20 times more than that of a mass-production car in general. Hyundai Motor Company, therefore, has developed the side impact simulator to simulate a crash instead of real crash tests and to reduce development costs and finds.

2. HARDWARE SYSTEM DESCRIPTION

A model that accurately predicts impact acceleration, impact force and intrusion displacement is required for accurate impact simulation and airbag controller design. The de-acceleration, the contact force and the crush energy that will be denoted by A_c, F_c, and E_c, respectively depend on the impact. When a striking vehicle first makes contact with a struck vehicle, the impulsive impact forces act for a short time. A modeling assumption could be that for the duration of impulse the striking vehicle and struck vehicle are free of each other but have a common applied impact force F_c. For times of contact without impulse, however, the striking vehicle and struck vehicle could be modeled as the same lumped mass [2].

A simple energy analysis following is used to model the impact [3, 9]. This approach was chosen because it predicts the overall motion changes from a global viewpoint and local details are unimportant for control of sensed acceleration and force. The impact collision is equivalent to one half oscillation of a lumped parameter spring-mass-damper system shown in Fig.1, Fig.2.

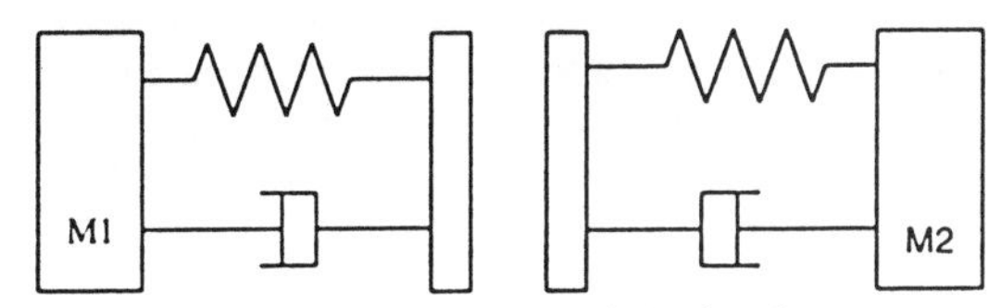

Fig.1 collision energy model: A simple system

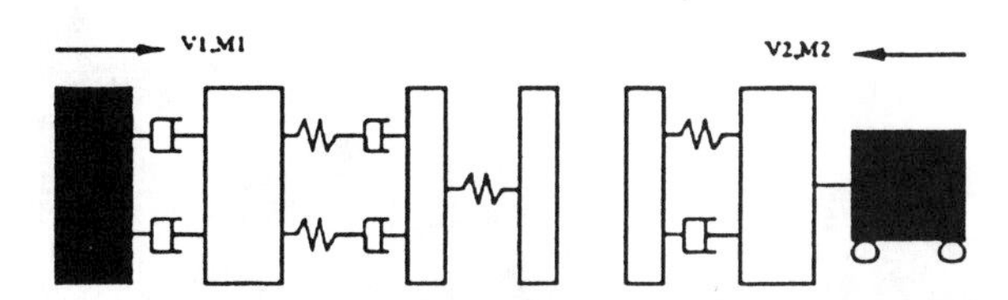

Fig.2 A higher order lumped mass model of a striking vehicle and a struck vehicle

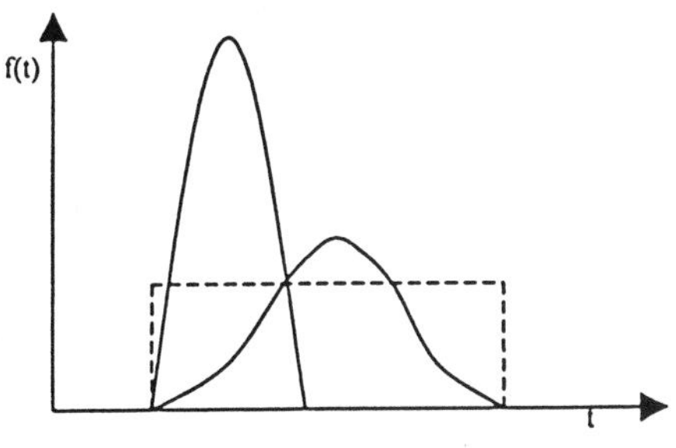

Fig.3 Impact model

The impulse, defined in terms of the time integral of the force, is the change in momentum,

$$\int_{t1}^{t2} F\,dt = \int_{v1}^{v2} m\,dv$$

Further information regarding the force involved during impact defines the collisions. A general form of the force-time plot is shown in Fig.3. The force is zero before impact. It rises to a maximum value then decreases when the two bodies leave each other and collision is over. If the two bodies are relatively rigid then the maximum force is large and the duration of collision is short. On the other hand if the bodies are more yielding then the maximum force is less and the duration of collision is larger. The measure of energy loss is given by the coefficient of restitution e,

$$e = \frac{Kinetic\ coenergy\ released\ during\ restitution}{Elastic\ energy\ stored\ during\ collision}$$

The impact momentum is then

$$\int_0^{\tau} F\,dt = (e+1)\frac{M_1 M_2}{M_1 + M_2}(V_{1i} - V_{2i})$$

$$e = \frac{V_{2f} - V_{1f}}{V_{1i} - V_{2i}}$$

where subscripts *i* and *f* represent the states before and after collision and the variable *V* designates velocity. *M* and τ denote the mass and the duration of impact impulse, respectively.

The side impact simulator consists of a servomotor, a transmission, striking and struck vehicles, accelerometer and force sensors, and overall controller IBM PC included the airbag main algorithm [4]. The overall side impact simulator is shown in Fig.4 and is represented schematically in Fig.5.

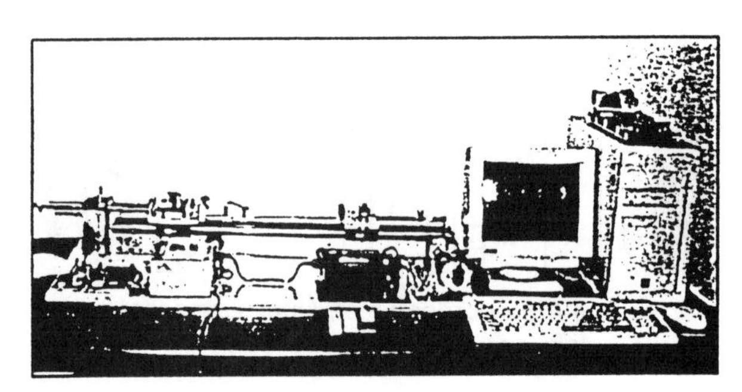

Fig.4 Overall side impact simulator

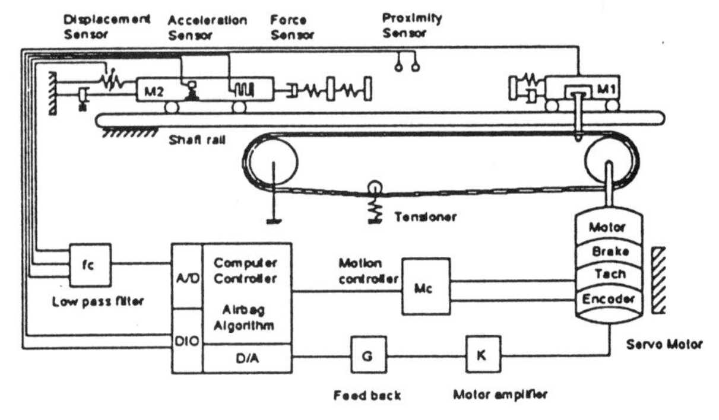

Fig.5 System schematic representation

The motor is a high torque and high speed DC servomotor, MicroMo model number GNM4150 with brake, tachometer, and encoder feedback. The servomotor amplifier is a pulse width modulated (PWM) amplifier with an input voltage limited to ±10 volts. The pulse width modulated current has a 33 kHz frequency. This servomotor has no gearheads because the servomotor has 94 watts power and 270 oz-in torque. The simulator could not allow gearhead friction and motor damping. An AT-MIO-16E-10 model by National Instruments is used as a data acquisition board (DAQ). Its maximum sampling rate is 100kHz with a resolution of 12 bit. The output signals of the sensors such as a strain gauge, an accelerometer, a linear potentiometer, and a safety contact sensor connect to the A/D input ports of the DAQ board. A 200Hz low pass filter was used to eliminate much of the sensor noise. The filter also has an amplifier necessary for isolating the computer circuits from sensor circuits. The inputs and outputs of the sensors like proximity, limit switch, and solenoid connect to the DIO ports of the DAQ board. The servomotor controller, Technology 80 model number TE5650A, is applied to control the striking vehicle and position. Its specifications have a trapezoidal motion profile, a 25MHz DSP controller chipset, and a 1.75MHz incremental quadrature encoder for the position feedback interface. The striking vehicle consists of a solenoid, a simulated bumper as a moving deformable barrier (MDB), a supplemental mass, and a Thomson linear dual shaft carrier. The struck vehicle consists of a dual shock absorber, ACE model number SC190HD-1, a Bimba T024 air cylinder, a linear potentiometer, Maurey Instruments model number P1613, a strain gauge on a deformable steel sheet, an accelerometer sensor, and a mechanical safety contact sensor. An Analog Devices LM741C chip processes a bridge voltage and amplifies a strain gauge to a usable force signal. An Analog Device ADXL50 and its peripheral circuit convert an impact vibration into a structure de-acceleration signal. The chips are mounted on the sensor to reduce noise. The three stages crush zone is modeled to make a structure de-acceleration signal that simulates an equivalent output of a real crash test. The four proximity sensors, Honeywell model number 922ABIY-A4N-L, are used for calculating the accurate impact speed of an approaching striking vehicle. The striking vehicle simulated as a bullet vehicle is shown in Fig.6 and the struck vehicle simulated as a target vehicle is shown in Fig.7.

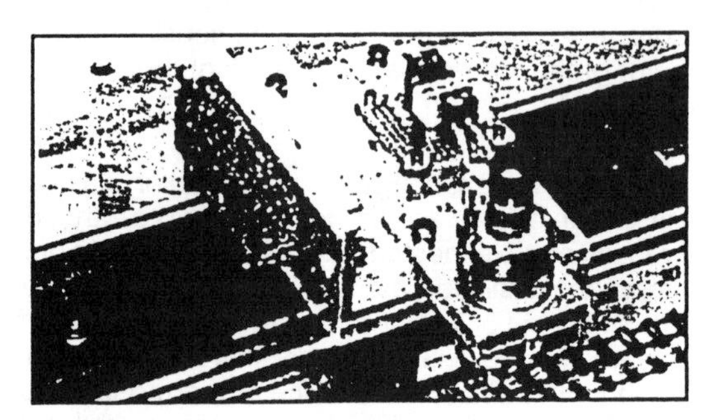

Fig.6 The striking vehicle

Fig.7 The struck vehicle

3. SYSTEM CONTROL AND SIGNAL PROCESSING

In order that the simulator may get the lowest deviation error between the results of real crash tests and the ones presented by the simulator crash tests during collision, the system must be controlled very accurately. The results to be compared are de-accelerations, crush displacements, and forces. The measures of performance are the impact speed of the striking vehicle and the peak and slope of the de-acceleration and force with time. Thus, an impact approach speed and a damping coefficient must be optimized. In what follows, we evaluate the system performance using a control system involving proportional, integral, derivative and velocity feedback actions. Note that in this paper, the modeling and experimental results are of prime importance. The tuning of controller gains was mostly experimental. The servomotor is notorious for operating with some degree of following error. Following error is the condition where the actual motor position lags behind the desired position. In order to attain smooth servo motion with following error, it is usually necessary to use a digital filter that conditions the servo signals with certain gains. The servomotor controller, TE5650A, is applied to the gain algorithms – Proportional Integral with Velocity Feed foreward (PIVFF) to control the impact speed of the striking vehicle. The applied PIVFF filtering uses filter parameters *Kp, Ki, Il, Kd,* and *Kv* for more stubborn position error problems. Velocity feed foreward filtering by adding an additional gain structure to the algorithm. PIVFF is really second-order PID filtering. PIVFF enables smoother filter by adding the element of velocity errors as shown in Fig.8 below. Just as with PID, there are many ways to combine the different gain elements in the PIVFF filter. Fig.9 shows the PIVFF filter block diagram [5].

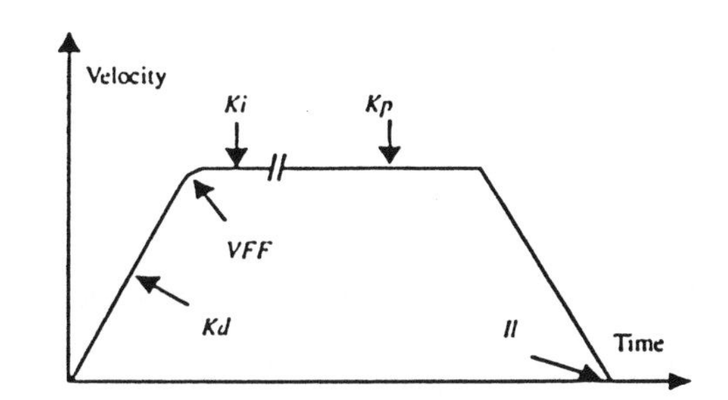

Fig.8 Trapezoidal profile with velocity feed foreward

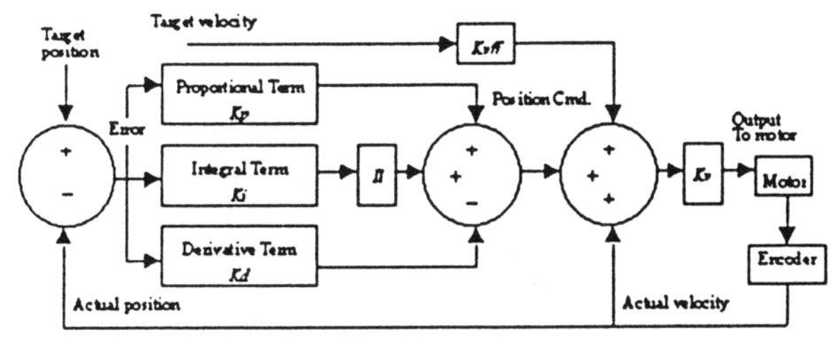

Fig.9 PIVFF filter block diagram

The equations for the PIVFF are shown below.

$$PositionCommand_n = PC_n = (Kp * E_n) + \left[Ki * \left(\frac{\Sigma En}{256}\right)\right] + [Kp * (E_n - E_{n-1})]$$

$$Position\ Error_n = E_n = TP_n - AP_n$$

Where: subscript n is the sample time
PC_n is the position command at n
TP_n is the target position at n
AP_n is the actual position at n
ΣE_n is the integration sum at n

The velocity error is calculated as follows

$$VelocityError_n = VE_n = 32\left(Kvff * \frac{TV_n}{16384} - AV_n\right) + \frac{PC_n}{256}$$

$$Output_n = VE_n * Kv$$

Where: VE_n is the velocity error at n
TV_n is the target velocity at n
AV_n is the actual velocity at n

The ADXL50 accelerometer sensor mounted on the struck vehicle has a 3rd-order low pass filter. The cutoff frequency is set at 210Hz. That analog circuit is shown in Fig.10 [6].

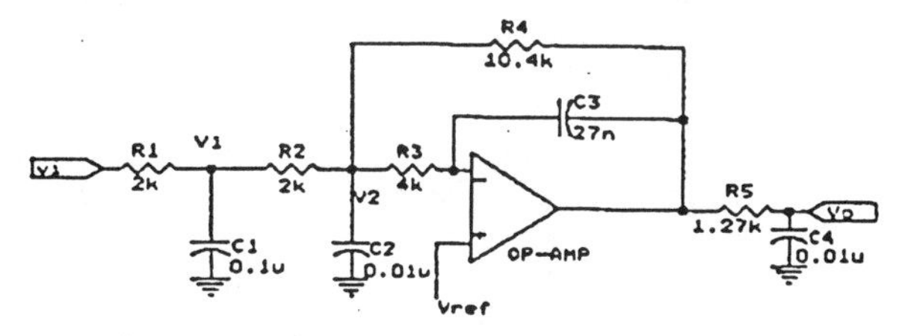

Fig.10 3rd-order low pass filter circuit

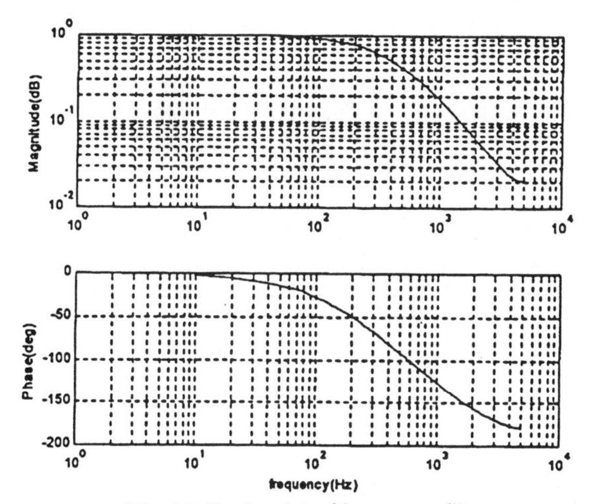

Fig.11 Bode plot of low pass filter

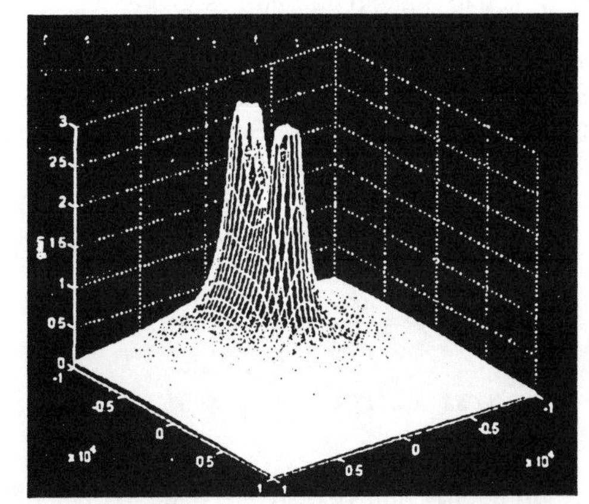

Fig.12 Characteristic diagram of S-plane amplitude

The corresponding Bode plot is shown in Fig.11 and the characteristic diagram of the filter S-plane amplitude is shown in Fig.12. The transfer function from V_i to V_o is like that below.

$$\frac{V_o(s)}{V_i(s)} = \frac{-2.315*10^{12}}{s^3 + 95615s^2 + (6.852*10^8)s + 8.915*10^{11}}$$

Pole = -1687, -6085, -86848

4. THE ALGORITHM FOR A SIDE AIRBAG & MODEL VALIDATION

Automotive side impact is one of the major safety issues. In North America and Europe, all passenger vehicles have to satisfy their own national vehicle safety standard. The crashworthiness and the occupant protection, therefore, are the most important factors at the early design stages. To protect occupant injury, a side airbag system will tend to assemble into a vehicle gradually. The ideal crash recognition algorithm must need to fire a gas inflator within a proper firing time to be deployed for the crash condition. Therefore it should be a necessary condition that an exact crash type and an impact speed is decided [7]. In this simulator, the algorithm for a side airbag calculates the several physical values to use as the output signal of the sensors mounted on the struck vehicle. The adapted sensors in this algorithm are an accelerometer, a strain gauge, a potentiometer, and a safety contact sensor. The block diagram of a single point-sensing system is shown in Fig.13, and the flow chart of an airbag algorithm is shown in Fig.14.

In general, a side airbag must deploy within 12 millisecond from initial impact crumpling for high speedy crash cases and 15 millisecond for low speed crash cases when a passenger vehicle is involved in side impact accidents. A side airbag must not be deployed for rough roads, misuse, and light impact. A gas inflator for the side airbag system needs about 7 millisecond to make full development of the air cushion. The electronic single point-sensing module (ESPS), therefore, must decide proper firing time within 5 millisecond for high speedy crash cases and 7 millisecond for low speedy ones. This is very short time in which to make a decision as to the degree of severity or what kind of crash is occurring during crashing [8]. In this simulator, the DAQ board is used to achieve conversion of the digital data from the analog signals of sensors every 0.1 millisecond. Some physical values are calculated along the times set out like below [4, 5].

$$Vibration\ \ energy = \int_0^t acceleration\ dt$$

$$Velocity = \frac{d}{dt}\ displacement$$

$$Stress\ \ energy = \int_0^t stress\ dt$$

Vibration energy and velocity are used to decide what kind of crash it was. Stress force, besides, is used to decide what time to fire the gas inflators. The initial thresholds of some physical value are very important to control the airbag firing condition and time. These initial threshold parameters are

Acceleration G_th = 50
Vibration energy E_th = 1500
Stress energy S_th = 3000
Velocity V_th = 150
Pre-displacement P_th = 10000

According to these values [4], the performance of an electronic single point-sensing module (ESPS) can be evaluated. Also these values depend on the occupant injury criterion. The impact performance is now explored by simulations and verified by experiments. Fig.15 shows test scenes captured by high-speed digital camera. Fig.16 shows good agreement between the accelerometer signal of the simulator and the one of a high speed crash test. The impact test condition of the simulator is on 21 KPH. The actual vehicle crash test condition is on 20 KPH and the accelerometer sensor was located on B-Pillar upper position. Fig.17 shows the airbag algorithm results of the simulator for the high speed crash case. This leads to conclusion that this simulator can simulate on an equal status to the actual vehicle crash tes

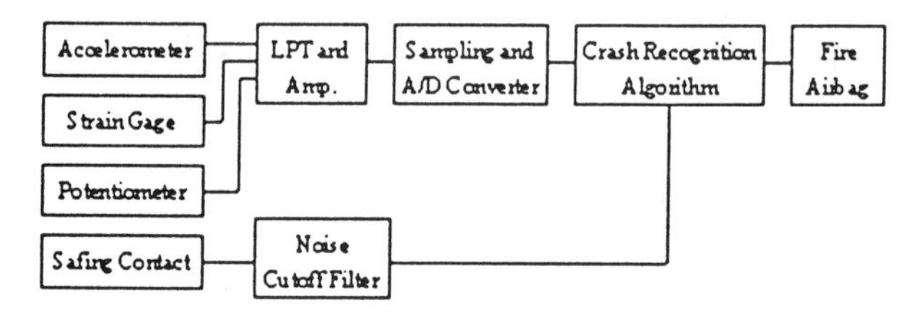

Fig.13 Block diagram of a single point-sensing system

Fig.18 shows a good agreement between the accelerometer signal of the simulator and the one of an actual crash test for a low speed crash test. The impact test condition of the simulator is on 13 KPH. The actual vehicle crash test condition is on 15 KPH and the accelerometer sensor was located on B-Pillar upper position. Fig.19 shows the airbag algorithm results of the simulator for the low speed crash case.

5. CONCLUSION

The purpose of this study was to make a contribution to the development of a side impact simulator and crash recognition algorithm for a side airbag system. The impact simulator was developed in order to simulate i.e. to be equal to crash data when a vehicle had a collision accident. The experimental tests were conducted on the simulator system that consists of a servomotor, a transmission, striking and struck vehicles, accelerometer and force sensors, and a controller. The developed model was shown to accurately simulate the acceleration for both high and low speed actual vehicle crash tests. The impact simulator was so effective for tests that the development cost and time taken for a side airbag system was reduced to around 30%. The simulator could generate arbitrary curves and progress the reliability of main algorithm using an electronic single point-sensing module (ESPS).

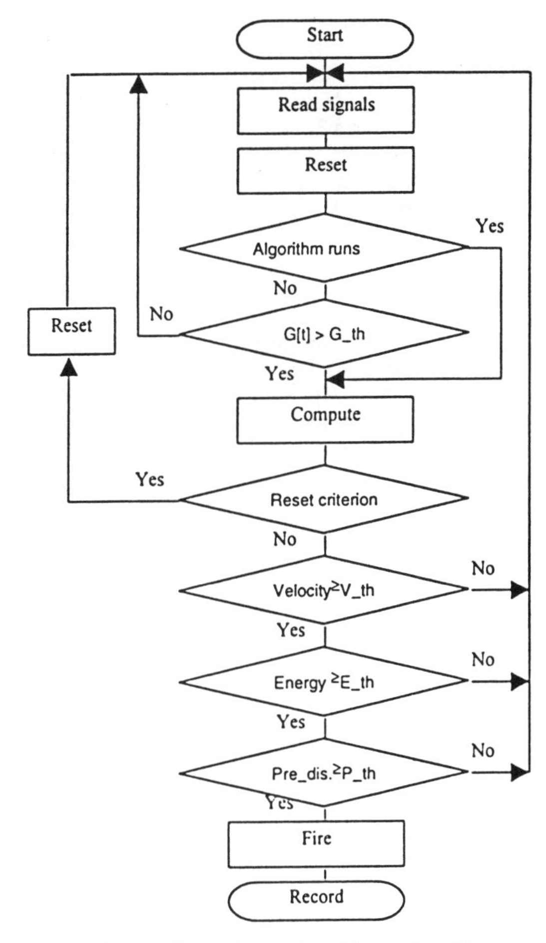

Fig.14 Flow chart of a airbag algorithm

(Step 1: Before Impact)

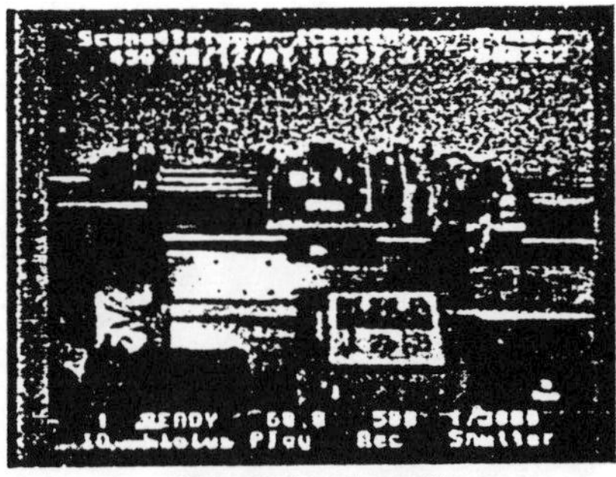

(Step 2: Just Impact)

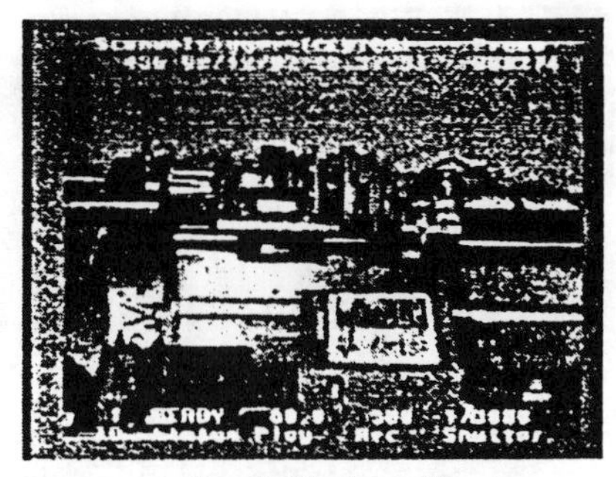

(Step3: After Impact)

Fig.15 Photographs captured by high-speed digital camera (Step1, Step2, and Step3)

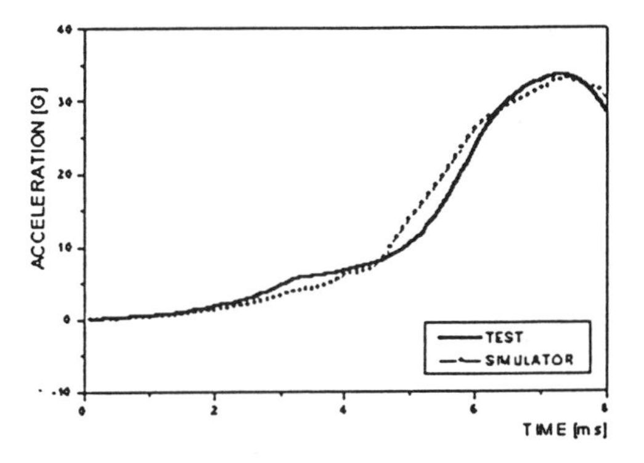

Fig.16 The comparison of a simulator impact to an actual vehicle crash for a high speed crash

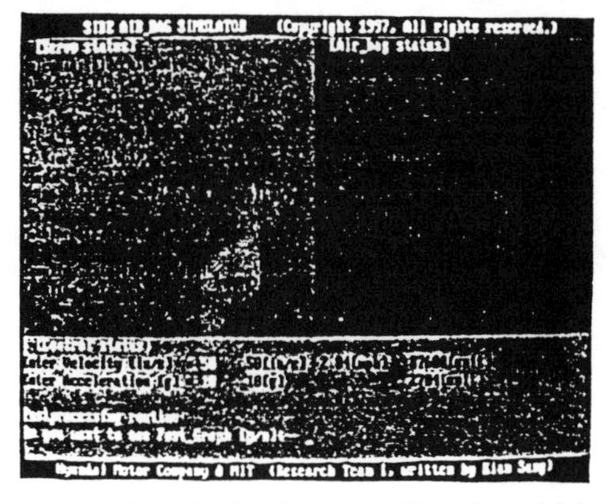

Fig.17 Result of airbag algorithm for a high speed crash

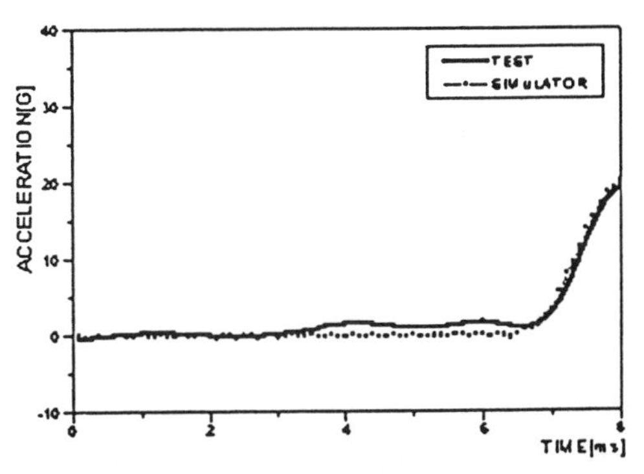

Fig.18 The comparison of a simulator impact to an actual vehicle crash for a low speedy crash

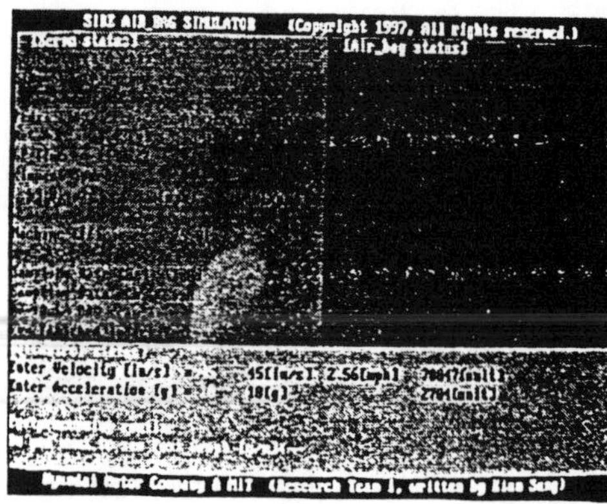

Fig.19 Result of airbag algorithm for a low speedy crash

REFERENCES

1. Wallentowitz, H., "Predicting the crashworthiness of vehicle structures made by lightweight design materials and innovative joining methods," *ASME Crashworthiness and Occupant Protection in Transportation Systems*, 1995, pp.331-354.
2. Youcef-Toumi, K., and Gutz, D., "Impact and Force Control: Modeling and Experiments," *Journal of Dynamic System, Measurement, and Control*, Vol. 116,Mar. 1994, pp.89-98
3. Goldsmith, W., *Impact: The Theory and Physical Behavior of Colliding Solids*, Edward Arnold Publishers, London, 1959
4. Sung, K. A., "Report of A Side Impact Simulator and Side Airbag Algorithm," *Joint Project Report of Hyundai Motor Co. And Massachusetts Institute of Technology*, 1997.
5. Raskin, C., P. E., "Designing with Motion Handbook 5th," *Technology 80, Inc.*, Minneapolis, 1997.
6. Coughlin, R., "Operational Amplifiers and Linear Integrated Circuits 3rd, *Prentice-Hall*, London, 1987.
7. Riling, J. J., "Sensing and Diagnostic Module for Airbags," *SAE 952682*, 1995.
8. Kelley, J. P., "Sensing Considerations and Tradeoffs for Single Point Sensing," *SAE 932916*, 1993.
9. Wang, Y., and Mason, M., "Modeling Impact Dynamics for Robotic Operations," *Proc. of the IEEE International Conference on Robotics and Automation*, Apr. 1987.

1994-20-0039

Side Impact Airbag System Technology

David S. Breed
Automotive Technologies International, Inc.

Abstract

The side impact airbag systems now being implemented won't save many lives but nevertheless are an important step forward in safety technology. For such systems, inertially damped mechanical or electro-mechanical sensors or electronic sensors using a movable mass, are preferred over crush or pressure sensors.

Future side impact systems will use anticipatory sensors and show promise of saving a significant percentage of those injured or killed in side impacts. For these sensors, ultrasonic or radar technologies are usable since only a few meters from the vehicle need be sensed. Sophisticated and inexpensive pattern recognition technology is required.

Frontal impacts are the number one killer of vehicle occupants in automobile accidents with about 16,000 fatalities each year. Side impacts are the second cause of automobile related deaths with about 8,000 fatalities each year. The number of fatalities in frontal impacts is now decreasing due to the introduction of airbags and mandatory seatbelt use laws. It is natural now that a considerable effort be applied to saving lives in side impacts.

Several automobile manufacturers are now considering the use of side impact airbags to attempt to reduce the number of people killed or injured in side impacts. The side impact problem is considerably more difficult to solve in this way than the frontal impact problem due to the lack of space between the occupant and the side door and to the significant intrusion which typically accompanies a side impact.

Some understanding of the severity of the side impact problem can be obtained by a comparison with frontal impacts. In the Federal Motor Vehicle Safety Standard (FMVSS) 208 49 kph crash test which applies to frontal impacts, the driver, if unrestrained, will impact the steering wheel at about 30 kph. With an airbag and a typical energy absorbing steering column, there is about 40 to 50 cm of combined deflection of the airbag and steering column to absorb this 30 kph difference in relative velocity between the driver and vehicle interior. Also there is usually little intrusion into the passenger compartment to reduce this available space.

In the FMVSS 214 standard crash for side impacts, the occupant, whether restrained or not, is impacted by the intruding vehicle door also at about 30 kph. In this case there is only about 10 to 15 cm of space available for an airbag to absorb the relative velocity between the occupant and the vehicle interior. In addition, the human body is more vulnerable to side impacts than frontal impacts and there is usually significant intrusion into the passenger compartment. A more detailed discussion of side impacts can be found in a paper by Breed et al, "Sensing Side Impacts", Society of Automotive Engineers No. 940651.

Sensing Side Impacts

An occupant of a vehicle is injured in proportion to the velocity of the object which strikes him. In the case of a side impact that object is the vehicle inner door panel. A crash sensor for side impacts must initiate inflation of the airbag when the occupant is likely to be impacted at some velocity above the chosen threshold. The crash sensor therefore must be able to predict the velocity at which the inner door panel will strike the occupant. A second principal of sensing side impacts is that the sensor must trigger as fast as possible since there is little time, compared with frontal impacts, to sense and inflate the airbag. If the sensor is close to the outer door panel then it will be struck by the intruding vehicle, in the case of a vehicle to vehicle impact, at the speed of the impacting vehicle. As the target vehicle crushes, the speed of the bullet vehicle decreases and finally the occupant is struck by the inner door panel at a velocity which is substantially below the initial speed of the bullet vehicle. A crash sensor, therefore, must trigger at a higher velocity when struck directly than the airbag deployment threshold velocity. On the other hand, when the sensor is not initially struck, such as the case of an impact with a tree at the A-pillar, the pulse is stretched and the sensor experiences nearly the same velocity change as the inner door panel. Thus, a sensor must be less sensitive to short duration pulses than to longer duration pulses.

As discussed in detail in the above referenced SAE paper, a crush sensor does not respond to velocity change and therefore cannot have the proper response. The conventional frontal impact electro-mechanical sensors, such as the ball-in-tube and spring mass sensors like the rolemite, also do not have the proper response since they are not less sensitive to short duration pulses. Of the electro-mechanical sensors, only the inertially fluid damped sensors have the proper response.

Electronic sensors can be used. An accelerometer based electronic sensor combined with a proper algorithm can be made to have the proper response. A pressure measuring sensor, on the other hand, has also been proposed but it does not have the proper response and the response will vary over the life of the vehicle.

For most if not all vehicles, the sensor must be mounted in the door or otherwise be impacted before the door has crushed more than a few centimeters. This precludes the type of single point sensors used for frontal impacts. In addition generally there is not sufficient signal in the passenger compartment for even a safing sensor and, therefore, if a safing sensor is used it also must be close to the outer door panel.

Future Side Impact Airbag Systems

Ideally an airbag for side impact protection would displace the occupant away from the intruding vehicle door in an accident and create the required space for a sufficiently large airbag. Sensors now being used for side impact airbags, however, begin sensing the crash at the beginning of the impact at which time there is insufficient time remaining to move the occupant before he is impacted by the intruding door. Even if the airbag were inflated instantaneously it is not possible to move the occupant to create the desired space without causing serious injury. The problem is that the sensor which starts sensing the crash when the impact has begun, is already too late.

There has been discussion over the years in the safety community about the use of anticipatory sensors so that the side impact accident could be sensed before it occurs. Heretofore this has not been practical due to the inability to predict the severity of the accident prior to the impact. A heavy truck, for example, or a tree can be a much more severe accident at low velocity than a light vehicle or motorcycle at high velocity. Until now it has not been possible to differentiate between these different accidents with a high degree of certainty.

Once a sufficiently large airbag is deployed in a side impact and the driver displaced away from the door and the steering wheel, he will no longer be able to control the vehicle which could in itself cause a serious accident. It is critically important, therefore, that such an airbag not be deployed unless there is great certainty that the driver would otherwise be seriously injured or killed by the side impact. Anticipatory sensors have heretofore not been used because of their inability to predict the severity of the accident. Now new pattern recognition technology solves this problem and therefore makes anticipatory sensing practical. This permits side impact airbag systems which can save a significant percentage of the people who would otherwise be killed as well as significantly reducing the number and severity of injuries.

These technologies are capable of pattern recognition with a speed, accuracy and efficiency heretofore not possible. It is now possible, for example, to recognize that the front of a truck or another car is about to impact the side of a vehicle when it is one to three meters or more away. This totally changes the side impact strategy since there is now time to inflate a large airbag and push the occupant out of the way of the soon to be intruding vehicle. Naturally not all side impacts are of sufficient severity to warrant this action and therefore there will usually be a dual inflation system as described in more detail below.

Although the main application for anticipatory sensors is in side impacts, frontal impact anticipatory sensors can also be used to identify the impacting object before the crash occurs. Prior to going to a full frontal impact anticipatory sensor system, pattern recognition can be used to detect many frontal impacts using data in addition to the output of the normal crash sensing accelerometer. Simple radar or acoustic imaging, for example, can be added to current accelerometer based systems to give substantially more information about the crash and the impacting object than possible from the acceleration signal alone.

Some Examples

Figure 1 shows an overhead view of a target vehicle about to be impacted in the side by an approaching bullet vehicle where the target vehicle is equipped with an anticipatory sensor system with a transmitter transmitting waves toward the bullet vehicle. A perspective view of the target vehicle is shown in Figure 1a and illustrates the transmitter connected to an electronic module. The electronic module contains circuitry to drive transmitter as well as circuitry to process the returned signals from the receivers. This latter circuitry contains a microprocessor or ASIC which performs the pattern recognition determination based on the received signals.

An ultrasonic transmitter operating at a frequency of approximately 40 khz can be used. This technology has the advantage that it doesn't require FCC approval of the frequency and doesn't interfere with other devices such as police radar. Ultrasonics will be attenuated by heavy rain and some noise may result from reflections off rain drops but initial studies indicate that this is not a serious problem even in heavy rain for detecting objects up to a few meters from the vehicle.

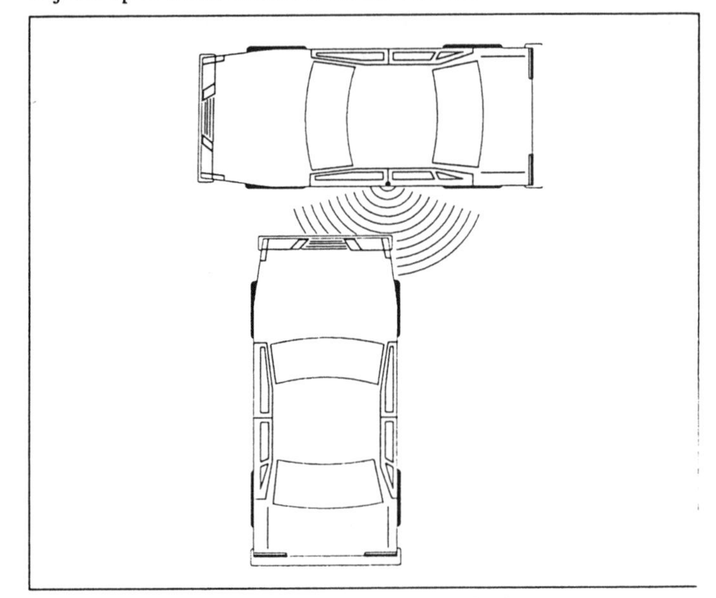

Fig. 1a: Overhead view of the target vehicle

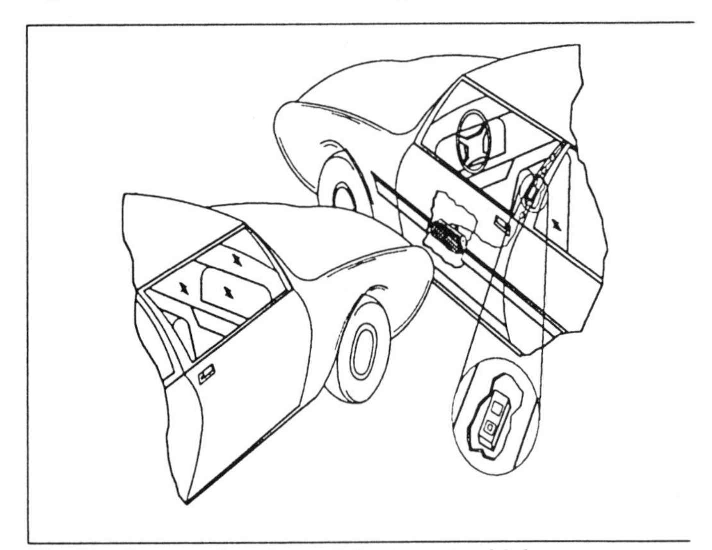

Fig. 1b: Perspective view of the target vehicle

ATI has been developing an acoustic based system called VIMS for Vehicle Interior identification and Monitoring System. This system is capable of identifying the presence of a vehicle occupant and of a rear facing child seat. If a rear facing child seat is placed on the front passenger seat and the passenger side airbag deploys, the child can be seriously injured or killed. It is therefore necessary to determine that such a seat is present and distinguish this from a forward facing child seat where airbag deployment is desired. The VIMS system uses pattern recognition techniques and has been proven capable of this task. This developed technology can now be applied to the exterior of the vehicle to identify that the object about to impact the vehicle is a tree, a small truck, a motorcycle, a car or a large truck. Each of these impacting objects requires a different response from the airbag system.

Pattern recognition" is any system which processes a signal that is generated by an object, or is modified by interacting with an object, in order to determine which one of a set of classes that the object belongs to. Such a system might determine only that the object is or is not a member of one specified class, or it might attempt to assign the object to one of a larger set of specified classes, or find that it is not a member of any of the classes in the set. The signals processed are generally electrical signals coming from transducers which are sensitive to either acoustic or electromagnetic radiation and if electromagnetic they can be either visible light, infrared, ultraviolet or radar.

In addition to the acoustic system described above, a laser optical system using an infrared laser beam can be used to momentarily illuminate the bullet object. In some cases a charge coupled device (a type of TV camera) is used to receive the reflected light. The laser can either be used in a scanning mode, or, through the use of a lens, a cone of light can be created which covers a large portion of the object. In each case a pattern recognition system is used to identify and classify the illuminated object and its constituent parts. This system provides the most information about the object and at a rapid data rate. Its main drawback is cost which is considerably above that of ultrasonic or radar systems and the attenuation which results in bad weather conditions such as heavy rain, fog or snow storms. As the cost of lasers comes down in the future, this system will become more competitive. The attenuation problem is not as severe as might be expected since the primary distance of concern for anticipatory sensors as described here is usually less than three meters.

Radar systems have similar properties to the laser system discussed above with the advantage that there is less attenuation in bad weather. The wave length of a particular radar system can limit the ability of the pattern recognition system to detect object features smaller than a certain size. This can have an effect in the ability of the system to identify different objects and particularly to differentiate between different truck and automobile models.

The ultrasonic system is the least expensive and potentially provides less information than the laser or radar systems due to the delays resulting from the speed of sound and due to the wave length which is considerably longer than the laser systems. The wave length limits the detail which can be seen by the system. In spite of these limitations, as demonstrated in the VIMS application, ultrasonics can provide sufficient timely information to permit the position and velocity of an approaching object to be accurately known and, when used with an appropriate pattern recognition system, it is capable of positively determining the class of the approaching object.

The application of anticipatory sensors to frontal impact protection systems is shown in Figure 2 which is an overhead view of a target vehicle about to be impacted in the front by an approaching bullet vehicle. In a similar manner as in Figure 1, a transmitter transmits waves toward the bullet vehicle. These waves are reflected off of the bullet vehicle and are then received by the receiving transducers.

Figure 3a illustrates the front of an automobile and shows some of the distinctive features of the vehicle which cause a distinct pattern of reflected waves which will differ from that of a truck, for example, as shown in Figure 3b. In some pattern recognition

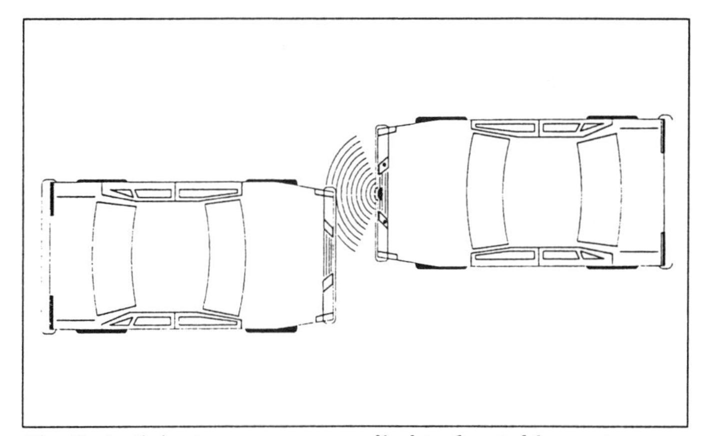

Fig. 2: Anticipatory sensors applied to frontal impact protection

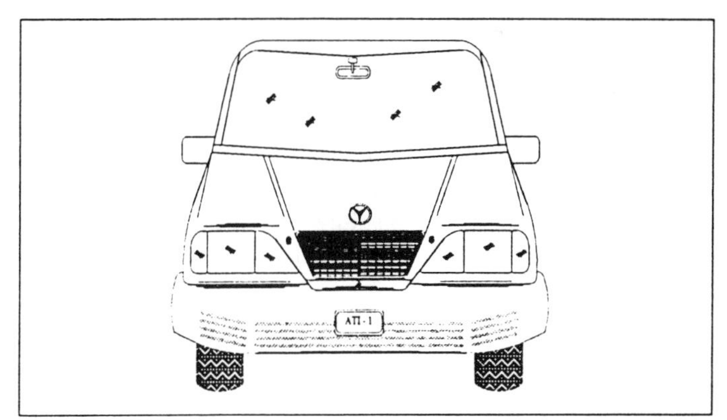

Fig. 3a: Distinctive frontal passenger car surfaces

Fig. 3b: Distinctive frontal truck surfaces

technologies, the researcher must determine the distinctive features of each object to be recognized and form rules which permit the system to recognize one object from another of a different class. In an alternative technology the identification system is trained to recognize different classes of objects. In this case a training session is conducted where the system is presented with a variety of objects and told to which class each object belongs. The system then learns from the training session and, providing a sufficient number and diversity of training examples are available, it is able to categorize other objects which have some differences from those making up the training set of objects. The system is quite robust in that it can still recognize objects as belonging to a particular class even when there are significant differences between the object to be recognized and the objects on which the system was trained.

Once the pattern recognition system has been sufficiently trained, it is possible to analyze the system and determine the "rules" which evolved. These rules can then sometimes be simplified or generalized and programmed as a fuzzy logic algorithm.

The anticipatory sensor system must also be able to determine the distance, approach velocity and trajectory of the impacting object in addition to the class of objects to which it belongs. This is easily done with acoustic systems since the time required for the acoustic waves to travel to the object and back determines its distance based on the speed of sound. With radar and laser systems, the waves usually need to be modulated and the phase change of the modulation determined in order to determine the distance to the object.

Figure 3a is a frontal view of the front of a car showing the headlights, radiator grille, bumper, fenders, windshield, roof and hood and other objects which reflect a particular pattern of waves whether acoustic or electromagnetic. Similarly, Figure 3b is a frontal view of the front of a truck showing the headlights, radiator grill, bumper, fenders, windshield, roof and hood illustrating a significantly different pattern. Pattern recognition techniques can be used to positively classify trucks as a different class of objects from automobiles and further to classify different types of trucks giving the ability to predict accident severity based on truck type and therefore likely mass, as well as velocity.

The anticipatory sensor system described here is mainly used when the pending accident will cause death or serious injury to the occupant. Since the driver will no longer be able to steer or apply the brakes to the vehicle after deployment of an airbag which is sufficiently large to protect him in serious accidents, it is important that this large airbag not be deployed in less serious accidents where the driver's injuries are not severe. Nevertheless, it is still desirable in many cases to provide some airbag protection to the driver. This can be accomplished as shown in Figure 4 which is a view of a dual inflator airbag system. Although a single inflator having a variable inflation rate capability can be used, Figure 4 illustrates the system using two discrete inflators. One Inflator and associated airbag are controlled by the anticipatory sensor and the other inflator and associated airbag could also be initiated by the same system. In a less severe accident, the smaller inflator can be also initiated by the anticipatory sensor without initiating the larger inflator or, alternately, the smaller inflator could be initiated by another in door crash sensor system such as described above.

When the large airbag is inflated from the driver's door it will attempt to displace the occupant away from the vehicle door. If the seatbelt attachment points do not also move, the occupant will be prevented from moving by the seatbelt and some method is required to introduce slack into the seatbelt or otherwise permit him to move.

One such system, shown in Figure 5 has an occupant being restrained by a seatbelt having two anchorage points and on the right side of the driver. The one closest to the occupant is released prior to the crash allowing the occupant to be laterally displaced to the left.

In another system, the vehicle seat is so designed that in a side impact it can be displaced or rotated so that both the seat and occupant are moved away from the door. In this case, if the seatbelt is attached to the seat, there is no need in induce slack into the belt. Figure 6 illustrates an occupant being restrained by a seatbelt integral with seat so that when the seat moves during a crash with the occupant, the seatbelt and associated attachments also move with the seat allowing the occupant to be laterally displaced during the crash.

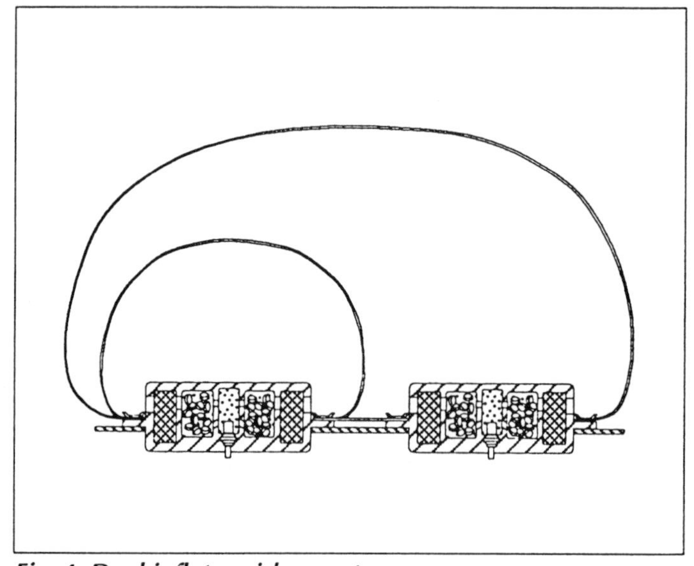

Fig. 4: Dual inflator airbag system

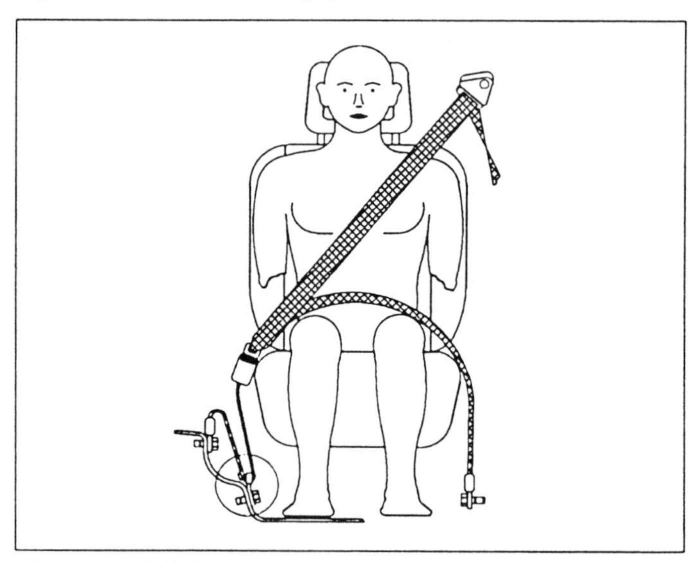

Fig. 5: Seatbelt having two anchorge points (right side of the driver)

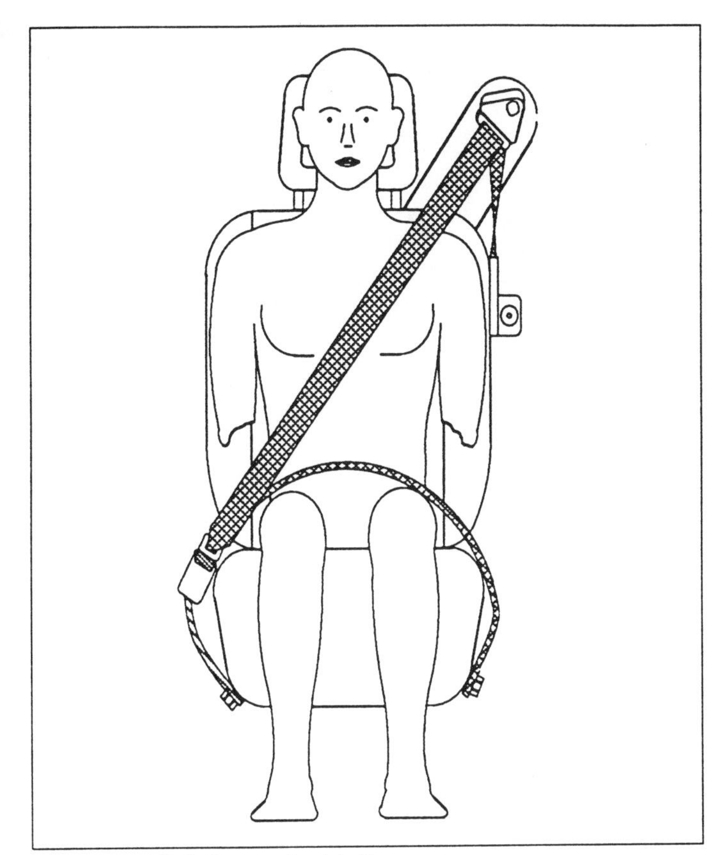

Fig. 6: Seatbelt integral with the seat

Once an anticipatory sensor system is in place, it becomes possible to consider deployment of an airbag external to the vehicle. This possibility has appeared in the automobile safety literature in the past but it has not been practical until the impacting object can be identified and an assessment of the probable severity of the accident made. For prior art systems, it has not been possible to differentiate between a construction barrier or a cardboard box, for example, neither of which would result in a serious accident and a concrete pillar, tree or wall which would. With the development of the pattern recognition systems, this problem has been solved and the use of an external airbag now becomes feasible.

Such a system adapted for side impact protection is shown in Figure 7a which is a view of a target vehicle about to be impacted in the side by a bullet vehicle. An airbag module is shown mounted to the side door of the vehicle prior to inflation. A portion of the side door of the vehicle has been cutaway to permit viewing of the airbag module. The vehicle contains a strong support beam which provides a reaction surface along with the vehicle door for the airbag. Upon initiation by the anticipatory sensor, a deployment door is opened in the external door panel by any of a number of methods such as pyrotechnically, permitting the airbag to emerge from the vehicle door as shown in Figure 7b. Through a system such as illustrated in Figures 7a and 7b, the accident can be substantially cushioned prior to engagement between the vehicle and the impacting object. By this technique, an even greater protection can be afforded the occupant especially if an internal airbag is also used. This has the further advantage that the occupant may not have to be displaced from behind the steering wheel and thus the risk to causing an accident is greatly reduced.

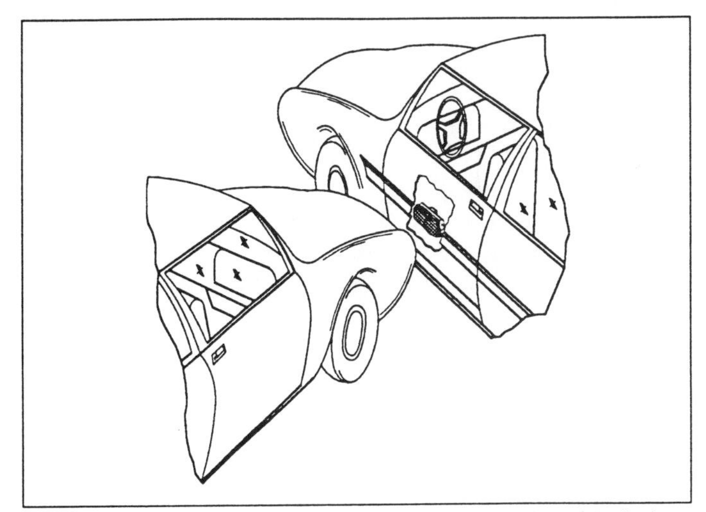

Fig. 7a: Airbag module, mounted to the door prior to inflation

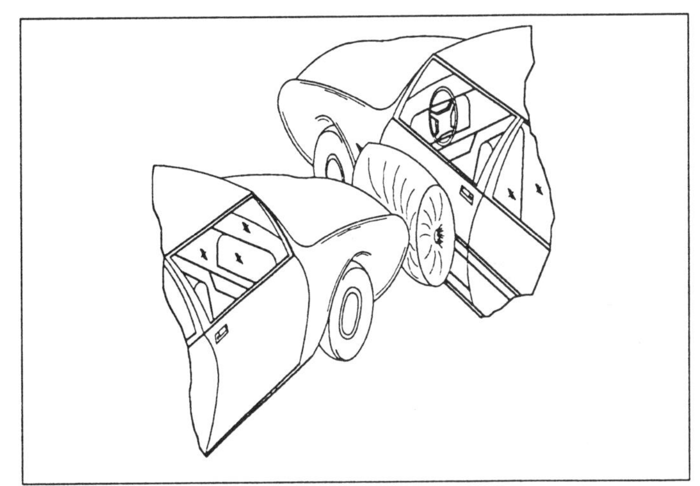

Fig. 7b: A deployment door is opened showing the airbag to be deployed

Fig. 7: Potential future side impact protection

Summary

The side impact airbag systems now being implemented won't save many lives but are nevertheless an important step forward in safety technology. For such systems, electromagnetic or electronic sensors using a movable mass and inertially damped, in the case of electromagnetic sensors, are preferred over crush or pressure sensors.

Future side impact systems will use anticipatory sensors and show promise of saving a significant percentage of those injured or killed in side impacts.

Reference

1. Breed, David S., *"Sensing Side Impacts,"* SAE Technical Paper 940651.

Biography

Dr. David S. Breed received his Ph.D. Degree in Mechanical Engineering from Columbia University.

Currently Dr. Breed is President of Automotive Technologies International Inc., prior to this he was Vice President, Director of Research, then President of Breed Corporation.

99SC27

The Influence of Occupant and Vehicle Characteristics on Risk of Pediatric Air Bag Injury

KB Arbogast, DR Durbin, BF Resh, FK Winston
The Children's Hospital of Pennsylvania and The University of Pennsylvania

ABSTRACT

A case-comparison study was conducted of children between one and twelve years of age exposed to passenger air bag (PAB) deployment. Cases were children fatally injured by PAB exposure and were investigated by the Special Crash Investigation Program of NHTSA. For comparison, children exposed to PABs, but suffering minor injury were identified through the Partners for Child Passenger Safety (PCPS) Study, a system utilizing insurance claims data for crashes involving children.

The crash severity as measured by Delta V was not significantly different between the two groups. Restraint status in conjunction with pre-impact braking highly influenced injury outcome indicating the importance of pre-crash positioning as a risk factor in child exposure to PAB deployment. Other related variables such as child size and age reinforced the importance of restraint. No vehicle characteristics or interior vehicle space measurements were significantly different between the two groups. Current vehicle designs cannot be differentiated with respect to their potential for producing serious child injuries due to PAB deployment. Researchers must continue to examine the entire spectrum of occupant injury severity in order to fully understand injury risk and ensure that future design of vehicles considers the safety of all occupants, including children.

INTRODUCTION

Much attention has been devoted to the interaction between children and passenger air bags (PAB). Highly publicized cases of child fatalities have focused the efforts of vehicle manufacturers and regulators in redesigning the vehicle air bag system to be safe for child occupants.[1]* In September 1998, the National Highway Traffic Safety Administration (NHTSA) issued a Notice for Proposed Rulemaking for Advanced Airbags accelerating this redesign process.[2]

In response to the crisis of child deaths due to air bag deployment, the role of the specific vehicle parameters, for example air bag mounting location, has been brought into question. Some research groups have suggested that certain vehicles may be predisposed to causing life-threatening injuries to children when the air bag deploys. [3] As vehicle manufacturers are investing much time and effort redesigning their air bag systems, it is important to examine vehicle characteristics such as air bag location and space available to the right front occupant for a range of injury severities. Information relating the vehicle to injury risk would be crucial to this redesign process.

Until recently, however, a comparison group of children exposed to air bags and experiencing minor injuries did not exist. Most current injury surveillance systems have not identified children experiencing a range of injury severities when exposed to passenger air bags. Geographically based and trauma center based systems have focused their efforts on seriously or fatally injured children.[4, 5] A study by the University of Michigan Transportation Research Institute described a limited number of children suffering only minor injuries from air bag exposure and suggested that children in lap-shoulder belts may be protected.[6] This study highlights the need for a more comprehensive child-specific surveillance system to systematically describe the range of injury severities of children exposed to PABs.

A direct comparison of cases of injured and uninjured children exposed to PABs would provide a more complete understanding of the performance of air bags with child occupants and assist in the design process.

*Numbers in brackets designate references at the end of the paper.

The Partners for Child Passenger Safety (PCPS) Study, based on State Farm Insurance claims data, is a newly developed crash surveillance system focused exclusively on children. Using this system, several children with minor injuries when exposed to PABs have been identified.

The objective of this study was to 1) use the PCPS surveillance system to identify children exposed to passenger air bags and experiencing no or minor injuries, 2) compare these cases with children fatally injured by air bag deployment, and 3) determine the influence of child, crash, and vehicle characteristics on the child's injury severity.

METHODS

A case-comparison study was conducted of children between one and twelve years of age exposed to air bag deployment during motor vehicle crashes. Cases were defined as children fatally injured by air bag exposure and were identified through the NHTSA Special Crash Investigation (SCI) Program. Once a case is identified by the SCI Program through national voluntary reporting of unusual crash injuries, crash investigation teams are dispatched to the scene to measure and document the crash environment, damage to the vehicles involved, and occupant contact points according to a standardized protocol.

For comparison, children exposed to air bags, but suffering only minor injuries were identified through the PCPS Surveillance System. PCPS identified crashes involving child occupants less than sixteen years of age using State Farm Insurance claims data from fifteen states and Washington, D.C. On a daily basis, data on qualifying claims were electronically transferred to The Children's Hospital of Philadelphia (CHOP).[1] For this study, claims data were reviewed at CHOP daily to identify potential cases of children exposed to the deployment of a passenger airbag. The claims file included crash information such as the vehicle identification number (VIN) and a four level code estimating the vehicle damage (none, minor, moderate, and severe). The file also included occupant information such as age and a four level code identifying the treatment status of each occupant (no treatment, seen in a doctor's office or emergency department, admitted to the hospital, or died). In addition, text fields were used by claims representatives to provide a brief description of the damage to the vehicle and the injuries sustained by the occupants.

In the PCPS study, cases of children exposed to PAB deployment were selected by either of two mechanisms: 1) text was present in the injury description that indicated air bag induced injuries to a child occupant less than twelve years of age, or 2) a child less than twelve years of age was in a moderate to severe frontal crash in a vehicle with a VIN that indicated the presence of a passenger side airbag. Cases that met either of these criteria were sent to a crash investigation firm and screened to confirm case status. Once confirmed, a full-scale on-site crash investigation was conducted in a similar manner to the SCI studies using modified child-specific data collection forms. In each group, delta v was calculated using the SMASH program based on the crush profile obtained in the crash investigation. When the non-case vehicle's crush profile was unknown, the MISSING VEHICLE routine of the SMASH program was used. [7]

An exemplar vehicle for each case in both subgroups was located and used to collect detailed interior dimensions. A total of five measurements were made inside the vehicles using a tape measure with standard protocol. (Figure 1) All measurements were taken with the seat back in the most forward locked position. These measurements included seat cushion length (AB), seat cushion height (BF), height of the midpoint of the instrument panel (CE). In addition, the horizontal distance between the seat back and rear most edge of the air bag module door was measured with the vehicle seat in the rear most track position and the foremost track position and averaged (seat to airbag distance). The width of the occupant compartment was measured from the B-pillar to the midpoint of the center console (HI). The location of the air bag module (mid mount versus top mount) was recorded. In addition, each vehicle's weight and wheelbase were obtained from the Insurance Institute for Highway Safety's Vehicle Features Database. [8]

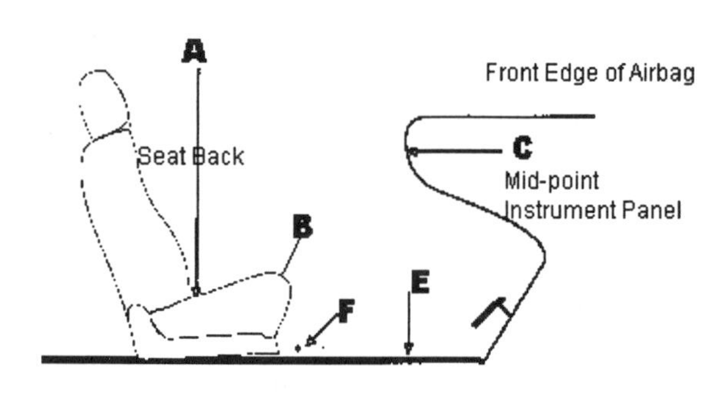

Figure 1a:Interior measurements – side view. Measurements taken include seat cushion length (AB where B is the forward most and highest point of the seat cushion), seat cushion height (BF), height of the midpoint of the instrument panel (CE), the horizontal distance between the seat back and rear most edge of the air bag module door (seat to airbag distance).

[1] This research was approved by the Institutional Review Board of The Children's Hospital of Philadelphia and the University of Pennsylvania.

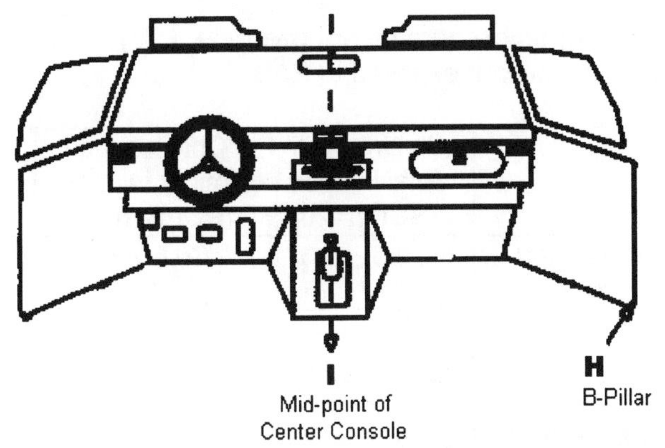

Figure 1b: Interior measurement – top view. Measurements taken include the width of the occupant compartment from the B-pillar to the midpoint of the center console (HI).

Height and weight of the children were included in each of the crash investigation reports. Specific body segment lengths were collected as part of the PCPS crash investigations, however, body segment lengths were not collected as part of the SCI program. For this reason, the height and weight measurements were converted to population-based percentiles using standard growth charts.[9] Using this percentile, sitting height and thigh length of the children were obtained from published body segment data.[10] (Figure 2)

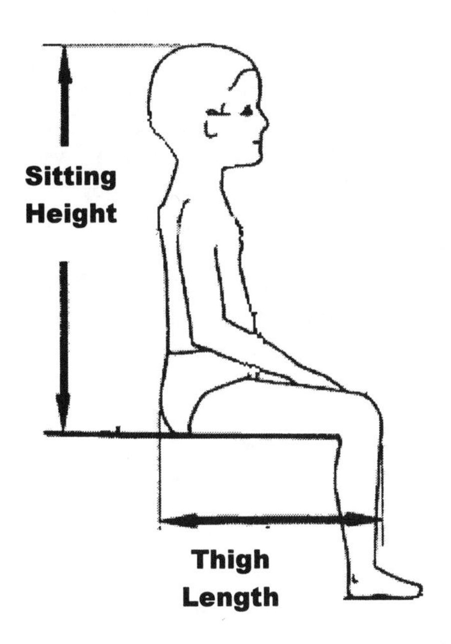

Figure 2: Body segment lengths used in the ratio calculations.[10]

The following ratios were calculated to compare the child body segment lengths to vehicle measurements:

$$relative\ thigh\ length = \frac{thigh\ length}{AB} \quad (1)$$

$$relative\ head\ height\ vertical = \frac{sitting\ height + BF}{CE} \quad (2)$$

$$relative\ head\ height\ horizontal = \frac{sitting\ height}{seat\ to\ airbag} \quad (3)$$

The volume available to the right front occupant was also calculated.

$$occupant\ space\ volume = HI \times CE \times (seat\ to\ airbag) \quad (4)$$

Descriptive statistics on both case samples were calculated, including frequencies for categorical variables, and mean and standard deviation for continuous variables. The Mann Whitney U test was used to compare vehicle measurements (seat to airbag distance, relative head height, relative thigh length, occupant space volume, vehicle weight, and wheelbase), crash severity, and child characteristics (child age, child height and child weight) between the two groups. A Chi squared test was used to compare restraint status, the occurrence of pre-impact braking, and air bag module location between the two groups.

The actual position of the right front seat on the vehicle seat track during the crash (foremost, between middle and foremost, middle, between middle and rear most, and rear most) was extracted from the case reports. The seat to airbag distance measurement was then adjusted for actual seat position and the statistical test repeated. An adjusted occupant space volume was calculated based on this measurement and tested for difference between groups.

RESULTS

As of February 1999, 55 children over the age of one were identified via the NHTSA SCI program as having died as a result of an interaction with a deploying PAB. Twenty of these cases were publicly available and were the subject of the current study. Fatal injuries were primarily to the head and cervical spine, including cerebral hemorrhage, cerebral edema, and contusion of the spinal cord with an associated fracture, disruption, dislocation, or transection.

Between March 1998 and February 1999, 38 confirmed cases of children under twelve years of age exposed to PAB deployment were identified through the PCPS system and sent to the crash investigation firm for investigation. Eleven vehicles were repaired before case initiation and in two cases the owners refused to participate. These cases were subsequently excluded.

Fourteen investigations have been completed and are the comparison cases of the current study.[2] Most injuries to these children included facial contusions and abrasions (AIS=1). One child sustained a fractured wrist and jaw (AIS=2), however, there were no spinal cord injuries. The single head injury was not attributed to the air bag; the child impacted the window rail during a side impact and suffered a concussion. The range of ISS was one to six with a median of one. Table 1 presents child-specific and crash-specific characteristics of the two groups. Table 2 presents vehicle characteristics of the two groups. Summary data tables for each group are contained in the Appendix.

Table 1 - Child-specific and crash-specific parameters for the two groups of children exposed to PABs

	Children with fatal injuries (SCI, n=20)	Children with minor injuries (PCPS, n=14)	p-value
Age: mean (years)	5	8	0.006
Height: mean (cm)	115	132	0.075
Weight: mean (kg)	23	30	0.03
Delta V: mean (kph)	18.5	17.7	0.45
Presence of pre-impact braking	95% (19/20)	43% (6/14)	0.003
Restraint: none	85%	14%	<0.001
lap only	10%	0 %	
lap shoulder	5%	86%	

Table 2 - Vehicle parameters (average ± sd) for the two groups of children exposed to PABs

	Children with fatal injuries (SCI, n=20)	Children with minor injuries (PCPS, n=14)	p-value
Seat to airbag distance (cm)	70.5 ± 3.5	70.6 ± 5.8	0.49
Volume (m^3)	2.7 ± 0.4	2.8 ± 0.5	0.71
Relative head height - vertical	1.7 ± 0.1	1.7 ± 0.2	0.43
Relative head height - horizontal	0.91 ± 0.09	0.95 ± 0.11	0.34
Relative thigh length	0.7 ± 0.1	0.8 ± 0.2	0.29
Vehicle weight (kg)	1347.5 ± 232.1	1477.1 ± 215.7	0.12
Vehicle wheelbase (cm)	267.8 ± 17.9	272.4 ± 16.3	0.56
Seat to airbag adjusted (cm)	70.4 ± 10.6	71.6 ± 12.9	0.44
Volume adjusted (m^3)	2.7 ± 0.6	2.8 ± 0.7	0.75
Airbag module location	mid 35% top 65%	mid 28% top 72%	0.61

The crash severity as measured by Delta V was not significantly different between the two groups (p=0.45). The children exposed to PABs with minor injuries were significantly older (p=0.006), heavier (p=0.03) and somewhat taller (p=0.075) than the children who were fatally injured by air bags. In the child fatality cases, the children were more likely to be unrestrained (p<0.001) and the drivers more often applied the brakes prior to impact (p=0.003) allowing the occupants to move out of position.

No vehicle characteristics showed a statistically significant difference between the two groups. A trend was evident in vehicle weight; although not statistically significant, PCPS cases were heavier vehicles as compared to the SCI cases. Of note, no difference was present in the location of the air bag module or the seat to airbag distance between the two groups.

DISCUSSION

As the safety community strives to create a child friendly air bag system, it is crucial that they incorporate the knowledge obtained from real- world crashes in which children experience a range of injury severities when exposed to a passenger air bag deployment. Looking at a complete picture will allow a full understanding of the interactions between a child and an air bag. With the creation of the Partners for Child Passenger Safety surveillance system, such analyses are now possible.

[2] The remaining 11 cases were still in active investigation at the time of manuscript preparation.

The objective of this study was to determine the influence of specific child, crash, and vehicle characteristics on injury severity in crashes involving children and air bags by comparing a group of fatally injured children with those experiencing minor injuries.

The most striking difference between the two groups was the pattern of restraint usage. Use of both segments of the lap shoulder belt was much more common in the group of children with minor injuries than those with fatal injuries. Use of both portions of the vehicle safety belt restricts the forward movement of the occupant reducing the chance that the child would enter the path of the deploying air bag. Although using only the lap belt restricts movement of the pelvis, the upper torso and more importantly the vulnerable head and neck are permitted to move forward unrestricted. Contact with the air bag during the deployment phase can result in high forces being transmitted to the head and neck.

Pre-impact braking, occurring more often in the crashes that resulted in fatal injuries, contributes to the forward movement of the occupant. Upon application of the brakes, an unrestrained occupant moves forward unimpeded toward the instrument panel and the air bag module. Upon impact, the occupant may be dangerously close and in some case directly in contact with the air bag as it deploys. Lap shoulder belt use restricts this forward movement when the brakes are applied. The importance of pre-crash positioning on injury risk in air bag deployment has been replicated in experimentally controlled lab tests. Static deployments with three- and six-year-old dummies show significant reductions in head and neck injury risk with distances of as little as three inches between the dummy and the air bag module at time of deployment.[11] In the fatally injured group, the combination of unrestrained children with pre-impact braking is a crucial determinant. Pre-impact braking alone, regardless of restraint use, is not necessarily a risk factor. Lap-shoulder belted children are kept out of the air bag deployment zone even with pre-impact braking as indicated by the presence of pre-impact braking with many of the restrained children who had favorable outcomes.

The data indicate that child weight, and to a lesser extent, child height, also factor into the child's risk. In particular, the children in the minorly injured group were significantly heavier and somewhat taller than those in the fatally injured group. This result supports previous work that identified height and weight, rather than age alone, as important factors in correct seat belt fit and thus, improved occupant protection [12]. Child size and restraint use are strongly interrelated – because almost all children with fatal outcomes were unrestrained and almost all children with favorable outcomes were restrained, one cannot distinguish the independent effects of child size from restraint use on risk of injury. Older, heavier children may be more likely to keep the shoulder belt properly positioned in front of their chest and over their shoulder. The average height (132 cm, 52 inches) and weight (30 kg, 69 pounds) of the children in the PCPS group correspond to at least a fifth percentile eleven-year-old (132 cm, 27 kg).[9] These data suggest that properly restrained children who are older and heavier may be at less risk of fatal injury from exposure to PABs. In addition, this result supports the current NHTSA recommendation that a child twelve years or older may sit in the right front seat in a PAB equipped vehicles.

Although actual body sizes showed some variation between the two groups, specific body segment lengths as measured by the relative head height and relative thigh length were not different between the two groups. This dichotomous finding may be attributed to the fact that the body segment lengths of the children were based on percentiles determined from their weight and height, not on the segment lengths of the actual child. Specific segment lengths were collected as part of the PCPS crash investigations, however, body segment lengths were not collected as part of the SCI program. In addition, it is important to note that there is a greater difference in actual child height than relative child height between the two groups. Since the vehicles in the minor injury group are slightly larger, examining a relative height may dilute the effect of height in general.

The data demonstrated that injury outcome was not dependent on the vehicle characteristics under investigation. Even though each group contained a wide variety of vehicle types, vehicle size and right front occupant space were remarkably similar among vehicles. The absolute differences in virtually all measurements between the two groups were very small. Such small differences most likely have little relevance to the vehicle design process and a dependence on injury severity would not influence vehicle manufacturers to alter interior or exterior dimensions. Only a limited number of cases were available for evaluation and these cases may not be representative of all vehicles and air bag exposures. A range of vehicle types was represented in each group, however, and the marked homogeneity of vehicle measurements overwhelmingly point to the importance of restraint status rather than vehicle characteristics as a determinant of injury severity.

Although not statistically significant, a trend was visible in the vehicle's weight. The children suffering minor injuries were in vehicles of greater weight than those with life-threatening injuries. This result suggests that heavier cars may be safer for a child passenger during a crash. Further studies would have to be conducted to support this theory.

The values adjusted for actual seat position may be limited in that the case reports did not provide an exact measurement of the child's seat location on the seat track during the crash. Rather, they classified the position into five categories. Also, this classification was missing in some cases. These deficits may mask actual differences between groups in the adjusted values of the seat back to the airbag.

Past studies have suggested that top mount air bags may be safer when interacting with children.[3] The data from this study does not support this assertion. The distribution of air bag location was similar in both groups of crashes with well over 50% of the vehicles having top mount air bags. Due to the limited data set, this result does not prove that all orientations are equal in terms of injury potential but suggests that serious injuries are not specific to one type of module mounting orientation. A slight overrepresentation of mid-mount air bags was evident in the fatally injured group, however, this trend was not statistically significant. Additional cases must be added to each dataset to further explore relationship between injury severity and air bag mounting orientation.

The current vehicle crashworthiness testing procedures may, in part, explain the results of this study. Among other measurements, an unrestrained average-sized adult male dummy exposed to a PAB in a frontal crash must experience neck and head injury parameters below established injury tolerance criteria. The current regulation does not include testing with child occupants. All vehicles, regardless of size, must meet these performance standards. As a result, the front passenger compartment, the location of the air bag, and other design criteria must be optimized in order to prevent serious injuries to the adult occupant as a result of a frontal crash. Consequently, a small car is not necessarily less safe than a larger car simply because of size. Other parameters such as dashboard stiffness or air bag inflator characteristics may be varied to maximize occupant protection. For example, vehicles with stiffer crash pulses may need air bags with higher inflation pressure and rate to meet current standards.

This process has resulted in tremendous progress in the safety of vehicles for adult passengers as demonstrated by the limited number of right front seat adult occupants (six since 1993) who have died due to air bag deployment and the reduction of overall adult occupant fatalities. Because vehicles are not optimized for child passengers, a parallel benefit has not been seen for children. This supports our results that current vehicle designs cannot be differentiated with respect to their potential for producing serious child injuries due to air bag deployment. Current vehicle, restraint, and air bag design and testing efforts that focus on reducing child injury risk may produce vehicles that can differentiated by their potential for protection of child occupants. Even with the efforts to emphasize rear seating for children, many children remain front seat passengers. For this reason, it is imperative that occupants of all sizes be considered in vehicle design.

CONCLUSIONS

Overwhelmingly, this study indicates that, in the current fleet of vehicles, restraint status plays a much more crucial role in injury severity than the vehicle characteristics themselves. Nearly 90% of the fatally injured children were not restrained appropriately – either by not using a restraint at all or by misusing the seat belt system. In contrast, nearly 90% of the children with no or minor injuries were using the lap shoulder belt. None of the vehicle characteristics showed such dichotomous results. Other related variables such as child size and the presence of pre-impact braking reinforced the importance of restraint.

The availability of data provided by PCPS allows for the comparison of children with a spectrum of injuries from air bag interactions. In order to gain a complete picture of the complex interaction between the air bag system, the child, and injury risk, a range of injury severities needs to be examined. These comparisons should continue, as more cases become available. Future studies should examine the role of specific air bag characteristics such as inflation rate and inflation pressure between the two groups. The future design of vehicles must consider the safety of all occupants, including children.

ACKNOWLEDGMENTS

This study was performed as part of the Partners for Child Passenger Safety Study at The Children's Hospital of Philadelphia funded by State Farm Insurance Companies. The authors acknowledge the assistance of the entire Partners for Child Passenger Safety research team especially the contribution of the crash investigation teams of Dynamic Science Inc. under the direction of Fran Bents.

REFERENCES

1. anonymous, From the Centers for Disease Control and Prevention. Update: fatal air bag-related injuries to children--United States, 1993-1996 [published erratum appears in JAMA 1997 Feb 5;277(5):372]. JAMA, 1997. **277**(1): p. 11-2.

2. National Highway Traffic Safety Administration, Federal Motor Vehicle Safety Standards: Occupant Crash Protection; Proposed Rule. Federal Register, 1998. **63**(181): p. 49957-50021.

3. Kido, M., Occupant Crash Protection: Air Bags, 1996, Center for Auto Safety: Washington DC.

4. Kleinberger, M., N. Yoganandan, and S. Kumaresan. Biomechanical considerations for child occupant protection. in 42nd Annual Proceedings of the Association for the Advancement of Automotive Medicine. 1998. Charlottesville, VA: Association for the Advancement of Automotive Medicine.

5. Gotschall, C., et al. Nonfatal air bag deployments involving child passengers. in Second Child Occupant Protection Symposium. 1997. Orlando, Florida: Society of Automotive Engineers.

6. Huelke, D. Children as front seat passengers exposed to air bag deployments. in Second Child Occupant Protection Symposium. 1997. Orlando, Florida: Society of Automotive Engineers.

7. US Department of Transportation and National Highway Traffic Safety Administration, National Accident Sampling System 1997 Crashworthiness Data System Data Collection, Coding, and Editing Manual. 1997, Washington DC: US Department of Transportation,.

8. Insurance Institute for Highway Safety, Vehicle Features Database. 1999: Arlington, VA.

9. Hamill, P., et al., Physical growth: National Center for Health Statistics percentiles. American Journal of Clinical Nutrition, 1979. **32**: p. 307-629.

10. Snyder, R.G., et al., Anthropometry of Infants, Children, and Youths to Age 18 for Product Safety Design, . 1977.

11. Rains, G., et al. Assessment of advanced air bag technology and aggressive air bag designs through performance testing (98-S5-O-06). in 16th International Conference on the Enhanced Safety of Vehicles. 1998. Windsor, Ontario: National Highway Traffic Safety Administration.

12. Klinich, K. and H. Pritz, Study of older child restraint/booster seat fit and NASS injury analysis. 1994, National Highway Traffic Safety Administration, Vehicle Research and Test Center: East Liberty, Ohio.

APPENDIX

SCI cases – fatal injuries

Vehicle	Vehicle weight (kg)	Air bag module orientation	Pre-impact braking	Delta V (kph)	Child age (years)	Child weight (kg)	Child height (cm)	Restraint usage	Injuries
93 Volvo 850	1439	mid	Yes	11	6	23	112	None	Brain
94 Dodge Caravan	1623	top	Yes	13	4	16	104	None	Skull fx
94 Ford Mustang	1366	mid	Yes	18	4	25	112	None	Brain, neck
95 Plymouth Grand Voyager	1602	top	Yes	26	9	30	140	2 point	Brain, neck
93 Lexus LS400	1754	top	Yes	Low	7	35	135	None	Atlanto-occipital dislocation, cervical spine transection
94 Chevrolet Camaro	1519	top	Yes	16	5	25	104	None	Cervical spine
95 Dodge Caravan	1602	top	Yes	18	7	25	130	2 point	Closed head injury
95 Ford Contour	1259	top	Yes	21	5	20	107	None	Cervical spine, atlanto-occipital dislocation, brain
93 Dodge Intrepid	1466	top	Yes	24	5	20	117	None	Brain, neck
95 Toyota Camry	1331	mid	Yes	16	6	Unk	Unk	None	Closed head injury
95 Dodge Caravan	1491	top	Yes	24	9	41	137	None	Fx/disclocation of spinal cord at atlanto-occipital joint
95 Chevrolet Lumina	1514	top	Yes	19	5	19	Unk	None	Fractured neck
95 Dodge Caravan	1602	top	Yes	27	4	23	122	None	Basilar skull fx
96 Mitsubishi Galant	1252	mid	Yes	19	5	18	117	3 point	Closed head injury, depressed skull fx
95 Ford Mustang	1399	mid	Yes	16	6	23	117	None	Neck fx
96 Plymouth Neon	1115	top	Yes	Minor	6	30	119	None	Cervical fx
95 VW Jetta	1203	mid	Yes	11	1	18	107	Unsecured FFCSS	Decapitation, arm fx
95 Geo Metro	882	top	Yes	Low	4	18	114	None	Neck fx
95 Geo Metro	882	top	Yes	16	2	11	91	On lap	Decapitation
95 Saturn SL2	1056	mid	No	21	2	10	81	On lap	Transection of cervical spinal cord

Key: FFCSS – forward facing child safety seat; fx – fractured; on lap – on lap of other passenger

PCPS cases – minor injuries

Vehicle	Vehicle weight (kg)	Air bag module orientation	Pre-impact braking	Delta V (kph)	Child age (years)	Child weight (kg)	Child height (cm)	Restraint usage	Injuries
93 Dodge Intrepid	1466	top	No	11	9	34	124	3 point	Facial burns
96 Kia Sephia	1125	top	Yes	21	8	41	147	3 point	Contusions/abrasions upper abdomen
94 Pontiac Firebird	1467	top	Yes	20	4	18	102	3 point	Facial abrasions around eyes/nose
97 Nissan Altima	1298	top	Yes	17	6	27	122	3 point	Eyelid abrasion, fat lip
94 Ford Mustang	1366	mid	Yes	12	5	16	104	3 point	Fx clavicle, chest contusions, facial contusions
95 Ford Windstar	1695	mid	No	12	11	34	157	3 point	Facial contusions, thermal burns
97 Dodge Caravan	1632	top	Yes	15	9	32	170	3 point	Facial abrasions
96 Pontiac Grand Am	1310	top	No	34	8	37	119	3 point	Closed head injury – side door
98 Ford Ranger	1595	mid	No	12	8	37	178	None	Abrasion, contusions Fractured wrist
96 Mercury Villager	1730	top	Unknown	20	7	36	122	3 point	Abrasion – head
96 Nissan Sentra	1062	top	Yes	20	7	19	117	3 point	Lacerations- hand, face Abrasion – face
97 Plymouth Breeze	1368	top	No	12	11	52	160	3 point	None
96 Ford Taurus	1512	top	No	12	9	29	130	None	Wrist fx, teeth fx, jaw fx, cut above lip
98 Chevy Blazer	1675	mid	No	30	3	13	91	3 point	Lacerations; abrasions on legs

Key: FFCSS – forward facing child safety seat; fx – fractured; on lap – on lap of other passenger

940906

The Use of Signal Processing Techniques in an Occupant Detection System

Edward J. Gillis and Tony Gioutsos
Automotive Systems Lab.

***Abstract*- An occupant detection system must employ advanced signal processing techniques if it is to work throughout a full range of environmental variations, and make use of all potentially relevant information. This review begins with a discussion of the kinds of signals that one expects from an occupant position sensor described in [1]. Potential noise sources are then examined. The kinds of information that can be extracted from those signals are described. Finally, an example of the application of signal processing techniques to a complex problem is presented.**

1.0 INTRODUCTION

In [1], an occupant position sensing system (OPSS) capable of monitoring the position of a person or object at a better than 1 Khz rate was described. That paper included a discussion of the reasons for developing such a system, the specific technology used, and the possibilities for enhancing its capabilities. The system described in [1] uses an active (modulated), infrared source. The signals reflected from the target enable us to determine the distance to the target, and gain other relevant information.

The need for such a system to incorporate advanced signal processing techniques arises for two kinds of reasons: (1) the wide range of environmental conditions over which it must operate; (2) the variety of information that one needs to extract from it.

In his classic paper Shannon[2] showed how the use of signal processing techniques enables one to extract information from "noisy" channels. In potential applications of an occupant position sensor, there are many possible sources of noise, e.g., stray optical signals (sunlight), stray motions (hand waving, paper blowing by), electrical noise, and optical interference from dirt, dust, smoke, etc. There is good reason to believe that these difficulties can be overcome by employing appropriate processing techniques.

In addition to measuring the distance to the reflecting surface, one would like the occupant detection system to be able to distinguish people from objects, child seats from grocery bags, estimate the size of the occupant, determine whether they are wearing a bulky coat, and, eventually, to assess the severity of a crash. By applying appropriate signal processing techniques to patterns of motion and other available information (such as variations in reflectivity), these goals can be attained.

Section 2 includes a brief review of the potential applications of such a system. Section 3 gives a description of the kinds of signals that one expects to see from the sensor. Section 4 is a survey of potential difficulties due to environmental factors. Section 5 contains a discussion of the specific types of information that can be extracted from the signals. Section 6 presents an example of how advanced signal processing techniques enable one to extract relevant information from a complex signal.

2.0 POSSIBLE APPLICATIONS OF AN OCCUPANT DETECTION SYSTEM

The interest in occupant detection systems has stemmed mainly from a desire to avoid deploying an air bag when a person is close enough to the bag to be seriously injured by it. There is a secondary interest in avoiding unnecessary deployments, as in cases when no one is present (on the passenger side), or when a driver is already adequately restrained by a seat belt in a marginal crash. Further applications for an occupant position sensing system were discussed in [1]. These included:
(1) optimizing the deployment decision in all cases;
(2) tailoring the inflation profile of the air bag to provide the greatest protection for the occupant (this assumes

the presence of a bag with a variable rate inflator); (3) using the position, velocity, and acceleration of the occupant relative to internal structures to make the primary deployment decision (thus eliminating the need for accelerometers).

3.0 SIGNALS FROM AN OCCUPANT POSITION SENSOR

The system described in [1] can employ either a single beam or multiple scanning beams. The reflected signal can be used to determine the distance to the surface and the reflectivity of that surface. It can determine both spatial and temporal variations in reflectivity.

In order to best apply signal processing techniques, it is important to understand the physical context of the signal. We can choose the best noise filtering techniques if we understand the physical sources of the noise. Similarly, one can more easily decode the "message" in the signal by first developing an understanding of the physical processes that give rise to the characteristic features of a given "message".

In the occupant detection case, the primary signal is the measurement of distance between the structures on which the sensor is mounted and the occupant or object in front of it. Obviously, it can be affected by a displacement of either the sensor or the occupant. To interpret the signal correctly, one must be able to characterize the typical motions of both the mounting structures and the sensed targets. There can be important information in the motion of either.

In tracking changes in relative position over time, we want to decompose the signal into sensor and occupant components. Our ability to do this depends on the fact that the sensor and target motions fall into essentially different frequency ranges. The vibrational or crash motions of the mounting structures have a substantially higher frequency content than the relatively slow movements of an occupant.

As a simple illustration of the efforts to characterize various types of motion, several figures are included. Figures 1a - 1b are simple, conceptual sketches of some motion patterns. These reflect an attempt to anticipate what these patterns would look like. For example, an object moving in and out in front of the sensor should produce a quasi-sinusoidal pattern. A hand moving laterally across the beam should produce more abrupt changes in detected position. In this case, the sensor first sees a surface at a relatively large distance. As the hand passes in front of it, the reflecting surface is measured to be much closer. Then, when the hand passes out of the beam, the more distant surface is again seen. A pen flicked across the beam generates a pattern similar in shape, but on a shorter time scale. Figures 2a - 2c show the actual measured traces of these motions.

These patterns were easy to anticipate and interpret because the target motions were simple, and the sensor was rigidly fixed in the laboratory. Obviously, the analysis of real, in-vehicle patterns will be much more complex. Before we discuss these (in section 5), we must first deal with the problem of reducing noise in the signal.

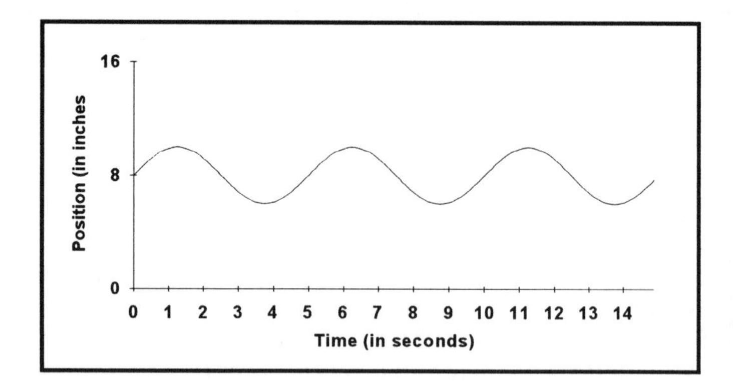

Figure 1a - Object moving in and out

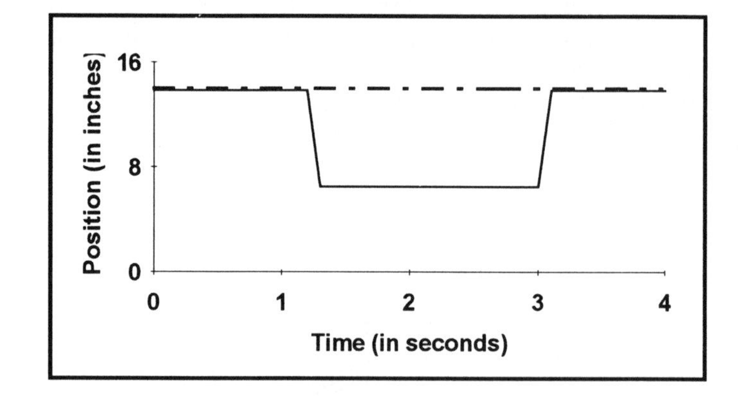

Figure 1b - Lateral hand motion

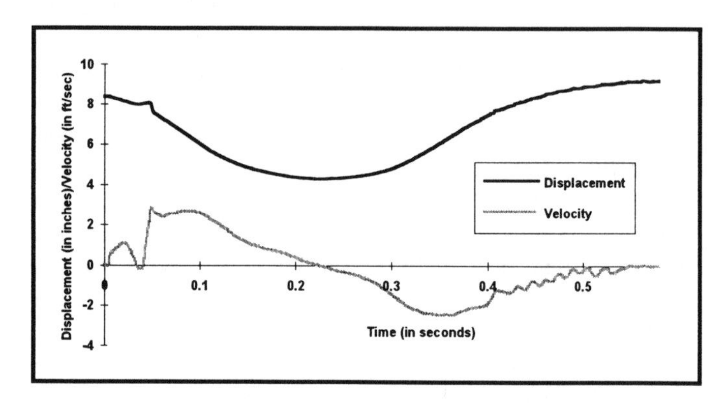

Figure 2a - Object moving in and out relative to the sensor

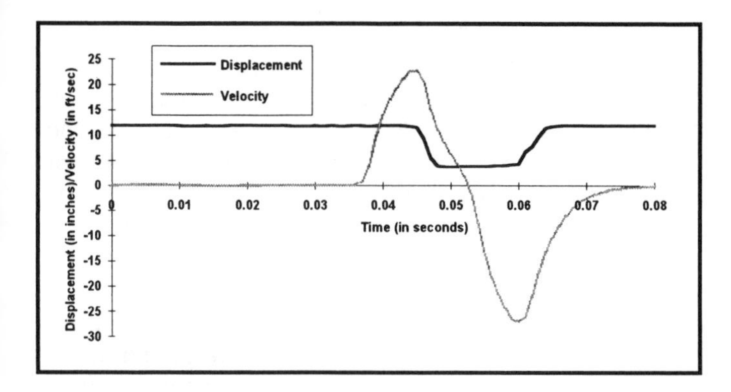

Figure 2b - Object moving in front of background at 12 inches.

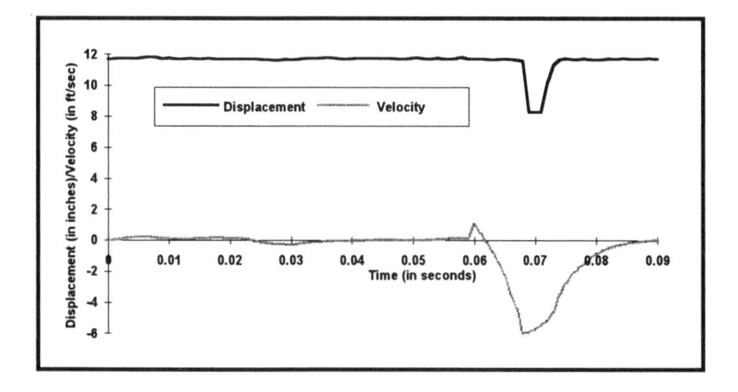

Figure 2c - Pen flicked in front of background object at 12 inches

4.0 ENVIRONMENTAL NOISE AND CONSTRAINTS ON THE SIGNAL

The most obvious sources of noise are from reflected light outside of the incident beam, electrical circuit noise, outside electromagnetic interference, optical interference from dirt, dust, film, etc. Stray or "irrelevant" motions are also a concern, but it is somewhat arbitrary whether these are classified as noise or signal.

The reduction of noise is essential for a number of reasons. Noise can obscure the signal, particularly in situations in which relatively little reflected energy is received. It can also lead to the misinterpretation of the signal. It is especially troublesome when one is trying to take meaningful derivatives of the signal.

The generation of the traces shown in the figures above involved some noise suppression techniques. Naturally, the incident beam had to be separated from ambient light. It was also necessary to suppress electrical noise (both from the circuit and the environment) using a combination of hardware and software techniques. These traces were generated, using these noise-reduction methods, in real time.

In the early stages of research, other, more complex noise suppression techniques were applied in a post-processing mode. These techniques helped to establish the feasibility of our distance-measuring method over a wide range of distances and materials. For example, it was possible to show that we could measure the distance to a very soft, black velvet (highly absorbent) material with good accuracy up to 30 inches away. One should be able to duplicate these measurements in real time by adopting a more complex scheme.

The possibility of taking derivatives (at least first derivatives) is illustrated by the figures in the previous section. In fact, careful examination of those figures will show that the start of the motion appears in the velocity trace slightly before it appears in the position trace. This is a consequence of the filtering technique that was used. As those familiar with signal processing know, the application of filters involves some time lag or "group delay". This means that the velocity at a given time is actually computed a few time slices after the computation of position at that time. The important thing is that this lag can be kept quite small on the relevant (crash diagnostic) time scale.

One of the most frequently raised concerns is that of interference by dirt, dust, film, or frost. There are some simple steps that one can take in packaging and mounting the sensor to reduce these problems, but it is still necessary that one be able to interpret signals that are scattered or partially blocked. By combining a physical understanding of how light is scattered in these situations with the application of signal processing techniques, it is possible to deal with these concerns.

5.0 INFORMATION FROM AN OCCUPANT POSITION SENSOR: INTERPRETING THE SIGNAL

In Section 2, we listed some of the functions of an occupant position sensing system. These require one to be able to determine what type of object is in front of the sensor (seat back, grocery bag, child seat, or person), and to characterize the situation (normal driving, rest, rough road, minor collision, severe crash). In this section, we will briefly sketch some of the approaches that could be taken to address these issues.

How does one distinguish a person from an inanimate object? Obviously, one way is through motion. Even a sleeping passenger will be breathing. Because the characteristic frequencies of human motion are different from those of the vibrating structure, it should be a straightforward task to classify relative motions as those of a person or an object.

In figures 3a-3c, we present some conceptual sketches, comparing possible states of motion of humans and objects. Note that it is possible that auxiliary inputs to the system indicating the position of the seat and seat back, will be available. These could make the interpretive problem somewhat easier.

One question that is often asked is how one can infer the position of the person's body from the distance to their outermost clothing. After all, in winter people might be wearing bulky sweaters or jackets that can be several inches in front of their chest. It could happen that the clothing would be judged to be too close to fire the air bag, while the person's body is actually far enough away from the point of deployment to avoid serious injury from the exploding bag.

To deal with such a classification problem, one should ask what characteristic physical aspects of this situation reveal themselves to the sensors. In this case one can note that heavy outer clothing worn by a seated person is almost always rumpled to some extent. The breathing and other motions of the person, (possibly combined with the scanning action of the sensor) will reveal variations in distance and maybe also reflectivity that are characteristic of heavy, rumpled clothing. By recognizing these variations, it will be possible to make a reasonably accurate estimate of the distance from the outer clothing to the person's chest.

We have already touched on several of the techniques that might allow us to use the occupant position sensor as an accelerometer. These depend first on the recognition that the accelerometer is mounted on interior structures that are undergoing essentially the same accelerations as the mounting structures for a single point electronic accelerometer. It is also critical that the characteristic frequencies of these motions are much different from those of human initiated motions. On occasion, the question has been raised as to whether an occupant position sensing system could be fooled into thinking that a crash was occurring by a rapid hand movement toward the sensor or across the beam. Although a human movement might (in rare circumstances) be able to duplicate the magnitude of the velocity or acceleration, it simply is not possible for a human to replicate the jerkiness of a crash. Of course, for the forseeable future the occupant position sensor will be used in conjunction with an electronic accelerometer. The comparison of the directly measured acceleration with the acceleration trace from the position sensor will provide an additional signal processing tool. Of course in order to obtain the velocity and acceleration from a position signal we must take derivatives. As already stated in the previous section, to get a meaningful derivative it is necessary to apply an appropriate noise filtering technique.

A matter of serious concern to a number of people is the ability to identify a rear-facing infant seat. Many vehicles have no rear seat, and in many cases parents place the infants where they can see them while driving. This results in the back of the infant seat being only a few inches from the point of deployment. Fortunately, the seriousness of this concern is not matched by the difficulty of the diagnostic task. In most situations, the closeness of the object, the smoothness of the surface, the (probable) constancy of the reflectivity, and the characteristic motion of an infant in a seat will make the job of recognizing these situations amenable to analysis.

Of course, it is easy to concoct slightly more complicated scenarios that make the identification problem more difficult. Suppose, for example, in the infant seat situation, that the parent throws a jacket or towel over the back of the baby seat. We would then be looking at a rough surface, of possibly varying reflectivity, perhaps with an additional oscillatory motion superimposed. It is important to understand that such additional complications do not make the problem insoluble. Complex signals can be interpreted by a systematic approach that decomposes them into their component parts. In the next section we describe a complex crash analysis problem. This problem yielded to an approach that combines an understanding of the underlying physics with the application of appropriate signal processing techniques.

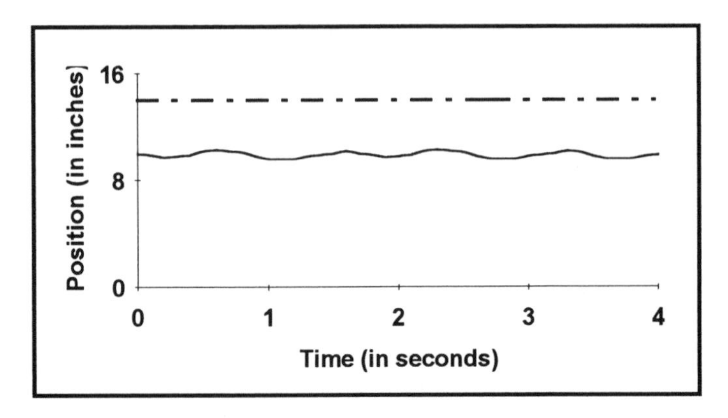

Figure 3a - Person Breathing

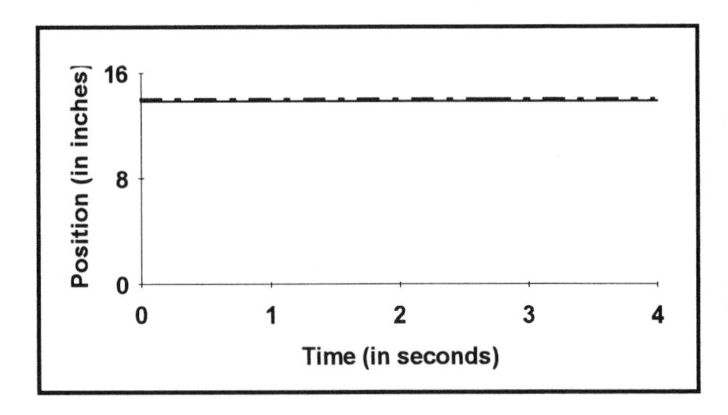

Figure 3b - Stationary object in front of sensor

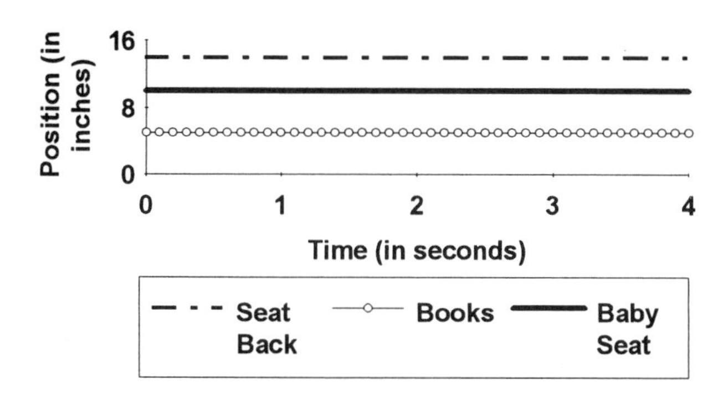

Figure 3c - Miscellaneous objects

6.0 A SIGNAL PROCESSING EXAMPLE: EXTRACTING INFORMATION FROM A COMPLEX SIGNAL

A reader who considers the wide range of situations which the OPSS is intended to characterize is apt to be skeptical that the application of signal processing techniques will make such a system workable. The example of this section is intended to demonstrate the feasibility of the approach described. [This example was used in [1] to make a related point.]

The example offered is of a roughly similar level of complexity to the occupant detection problem. The problem is to distinguish a crash requiring deployment of an air bag (a 17 mph pole crash waveform shown in Figure 4a) from a roughly similar looking crash waveform in which the bag should not be inflated (an 8 mph frontal crash shown in Figure 4b). The signal traces are from an electronic accelerometer mounted in the passenger compartment.

Figure 5 shows the velocity change comparison of the two crashes, obtained by integrating the acceleration signal. The upper, broken trace is from the 8 mph crash (off condition). Clearly the velocity change in the off-condition crash is greater than that in the on-condition crash until the time to fire.

Virtually everyone involved in developing crash algorithms uses velocity change as a significant measure of crash severity. It is one of the measures used in the algorithm referred to here. Of course, the problem is to *predict* the total velocity change that will result from the event. Velocity change up to the current time is not, by itself, an adequate predictor. To deal with this difficulty, several additional measures of crash severity were developed (using the principles outlined in [3]). In combination, these measures made possible the recognition of the 17 mph pole crash as a fire condition (before the required time-to-fire - 55 milliseconds), while also indicating that the 8 mph crash was a no-fire situation.

The crash measures used have two key characteristics: (1) they are closely related to physical aspects of the crash, and (2) they employ advanced signal processing techniques. The general approach described here has made it possible to distinguish fire and no-fire situations with a very high degree of confidence. This approach involves the use of noise reduction techniques to separate relevant information from irrelevant noise, and the development of crash measures relating to *all* the relevant physical aspects of the crash (instead of just one). It combines these measures to get an overall picture of the crash. By bringing to bear a similar type of physical understanding, and applying the appropriate signal processing techniques it should be possible to solve the occupant detection problem.

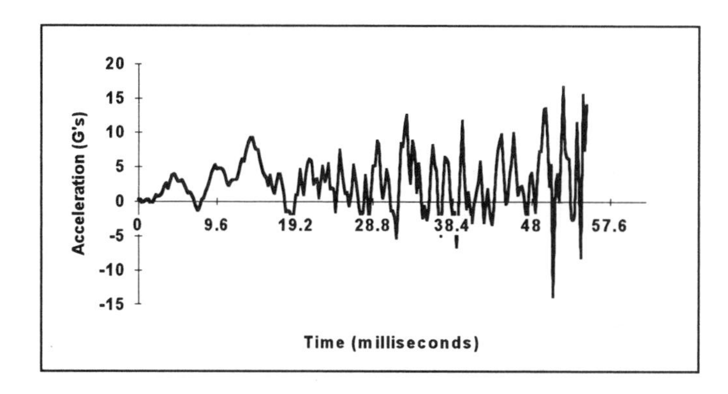

Figure 4a - 17 MPH Pole (Required TTF = 55 ms)

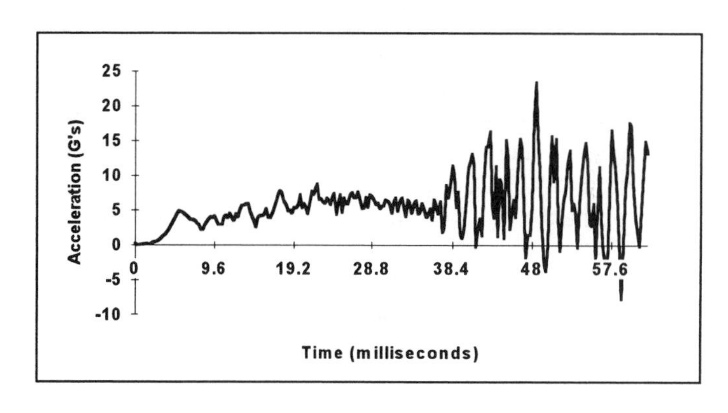

Figure 4b - 8 MPH Frontal Barrier (Req TTF = OFF)

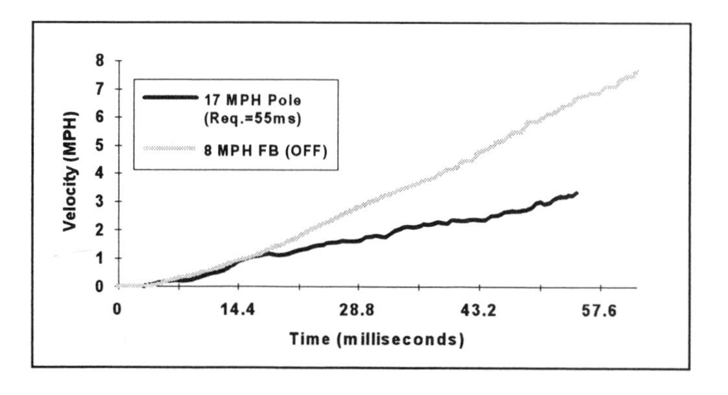

Figure 5 - Velocity Comparison

7.0 SUMMARY

We have described an approach to the problem of interpreting the signal from an occupant position sensor. This approach involves the application of appropriate signal processing techniques based on an understanding of the underlying physics of the sensor-environment interaction. We believe that such an approach is required in order to both suppress the noise in the signal and to extract all the information required to confidently and accurately interpret the signal. We have discussed the basic physics of the signal, described some noise suppression techniques, displayed their efficacy, and given examples of how different kinds of information can be extracted from the signal. Finally, we presented an example of how a similar approach was used to solve an interpretive problem of comparable complexity. Although a great deal of work remains to be done, particularly in gathering data, we believe that the occupant detection problem will ultimately yield to this approach.

REFERENCES

[1] Gillis, Ed and Tony Gioutsos, "Occupant Position Sensing Systems: Functional Requirements and Technical Means", SAE Worldwide Passenger Car Conference and Exposition, Paper # 932914, 1993

[2] Shannon, Claude, "A Mathematical Theory of Communication", Bell System Technical Journal, Vol. 27 (July & Oct, 1948), pp 379-423 and 623-656. Reprinted in D. Slepian, editor, Key Papers in the Development of Information Theory, IEEE Press, NY,1974.

[3] Gioutsos, Tony, "A Predictive Based Algorithm for Actuation of an Airbag", SAE International Congress and Exposition, Paper # 920479, 1992

980648

Airbag Technology: What it is and How it Came to Be

Donald E. Struble
Collision Safety Engineering

ABSTRACT

Since air bags emerged as an occupant protection concept in the early '70s, their development into a widely-available product has been lengthy, arduous, and the subject of an intense national debate. That debate is well documented and will not be repeated here. Rather, operating principles and design considerations are discussed, using systems and components from the developmental history of airbags as examples.

Design alternatives, crash test requirements, and performance limits are discussed. Sources of restraint system forces, and their connection with occupant size and position, are identified. Various types of inflators, and some of the considerations involved in "smart" systems, are presented. Sensor designs, and issues that influence the architecture of the sensor system, are discussed.

INTRODUCTION

When a vehicle crashes, it is acted upon by collision forces that tend to change its velocity. In a direction opposite to these forces, everything in the vehicle that can move does so. This includes the occupants, with their various articulated segments. A restraint system has the purpose of intercepting these motions and managing or eliminating the "second collisions" of an occupant's parts with potential contact surfaces, so that injuries can be mitigated to the extent possible.

An occupant restraint is a system, and works in harmony with the vehicle structure. The restraint system may include an air bag, which itself is a collection of components designed to work with each other and in cooperation with other parts of the vehicle.

RESTRAINT SYSTEMS IN FRONTAL CRASHES

In a crash, potential contact surfaces in the vehicle experience changes in their velocities, and to some extent, their directions of travel. Figure 1 shows the velocity-time history, in an actual crash, for the instrument panel. In contrast, a free particle would keep moving as before the crash. This behavior would be represented in Figure 1 as a horizontal line at 35 mph, in comparison to the descending lines associated with a belted occupant. As time proceeds, the velocity differential between the occupant and the vehicle builds up. One might think of this as an accrual of a "velocity debt," which has to be paid back in order for the vehicle and its occupants to achieve a final common velocity (perhaps zero). In Figure 1, this debt would be the vertical distance between lines representing a potential contact surface, such as the instrument panel, and those representing the occupant. The debt is due and payable when contacts are made between the occupant and one or more contact surfaces. Figure 1 shows that debt management starts earlier for a belted occupant, and that the debt itself is much reduced.

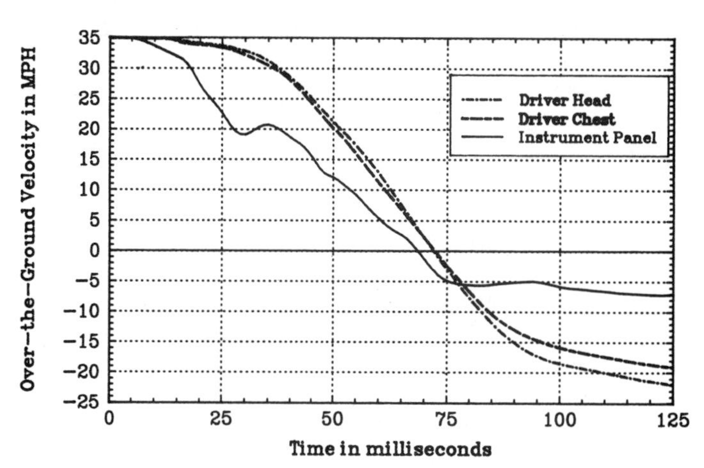

Figure 1. Velocity Histories in 35 mph Barrier Crash.

In this particular vehicle, there was 10.71 inches of space between the sternum and the steering wheel, and 23.18 inches between the head and the windshield, at the beginning of the crash. Figure 2 shows that a free particle would use up these distances in 54 and 75.6 milliseconds, respectively. If an unrestrained occupant behaves similarly, we would expect debt repayment to occur in earnest at about 76 milliseconds, as the head reaches the windshield, and the debt amounts to about the final velocity change, or delta-V (ΔV), of the vehicle.

Of course, these numbers will vary, depending on the crash, the initial position of the occupant, etc. Figures 1 and 2 show estimated occupant velocities and displacements, based on accelerometer data. These indicate that debt management began at about 25 milliseconds for a belted occupant. For an unbelted occupant, the restraint process would have to begin by 50 milliseconds at the latest.

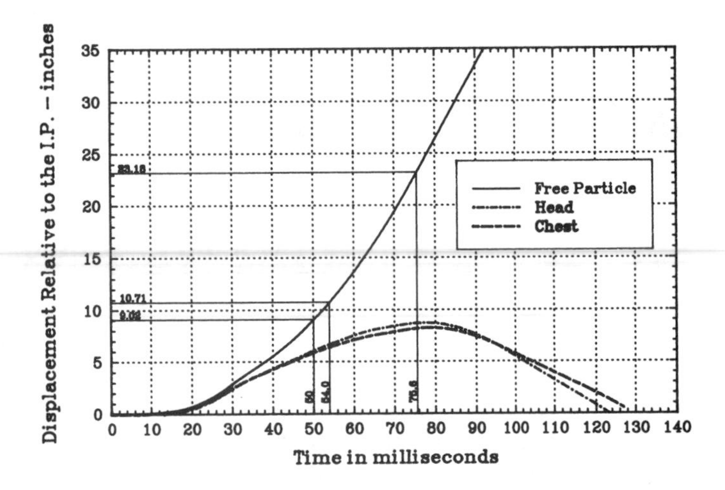

Figure 2. Relative Displacements in a 35 mph Barrier Crash.

The purpose of the restraint system is to intervene early, before the velocity debt becomes unmanageable, and to initiate a repayment program in the form of occupant accelerations - or decelerations (negative accelerations), as the case may be. An early start reduces the debt that has to be paid back, and at the same time, a good restraint system will tend to lengthen the payback period (to perhaps 150 milliseconds or more). If the crash is not excessively severe, the restraint system can keep the debt payments down to a level the human body can tolerate. The action of the restraint system in meting out accelerations to the occupant while the vehicle itself is still accelerating is known as "ride down."

The earliest restraints tended to be lap belts, which were thought to promote occupant retention within the vehicle, but which obviously could not provide much restraint for the upper body during severe frontal crashes. In the U.S., the usage of such systems was disappointingly low, and the prospects for shoulder belt usage were even more discouraging [DOT 78]. Thus, in the late '60s, air bags were seen as forcible intervention (debt management) for the vast majority occupants who would otherwise be unrestrained. This forcible intervention became known as "passive protection," and airbag requirements were promulgated by the NHTSA with unrestrained occupants in mind. The U.S. thus embarked on a technological odyssey in which the occupant protection standard, FMVSS 208, did not require dynamic (crash) testing of the most effective restraint system then and now available - seat belts. Meanwhile, the airbag designer was saddled with the task of protecting those who failed to use their available seat belts, and who would presently be in violation of the law in 49 states.

DEVELOPMENT OF THE AIRBAG CONCEPT

In this historical context, it is easy to see how the concept of air bags came into being: the forcible intervention would come in the form of an inflatable bag which would leap into the gap between the occupant and the vehicle interior. Once there, the bag's internal pressure would be uniformly applied to the occupant over a large area, not only meting out accelerations in controlled doses, but minimizing the stress on the occupant by distributing restraint the forces over as much area as possible. The result would be a more gradual pay down of the velocity debt, ideally with a minimum interest charge (in other words, rebound velocity). Unfortunately, the question of how to accomplish all this would not be a trivial one.

Of course, bags could not be pre-inflated because of the possibility of actually causing an accident. Since inflation had to begin after the crash started, the first step was to determine whether a bag could be inflated quickly enough. Figure 3 shows the time budget used by General Motors in their airbag vehicles of the mid-70s. These were very large sedans, so the time budget for today's vehicles would tend to be shorter, depending on such variables as the vehicle size, object struck, structural engagement, etc. In actual fact, the air bag must be inflated in a considerably shorter time, typically 25 to 30 milliseconds, because at the beginning of the event, the crash detection system requires a certain amount of time to do its work.

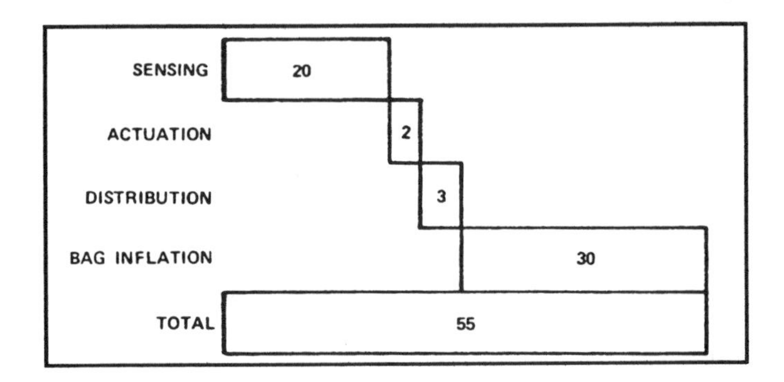

Figure 3. Typical Time Budget for Large Cars -GM

In addition to filling the air bag quickly, the inflation gas itself should not present hazards to the occupants or any one else involved with the vehicle from production to salvage. High-pressure air, stored in a tank and released with a quick-action valve, probably seemed an obvious choice to early airbag developers.

DRIVER-SIDE AIRBAG SYSTEMS

Design of a driver-side airbag system is heavily influenced by the available volume in the steering wheel hub. This volume is entirely insufficient to house a high-pressure tank, stowed air bag, and other hardware. Consequently, the only way to implement a stored-gas system on the driver side was to locate the tank elsewhere, and pipe the gas up the steering column. A prototype system, in which a single bottle inflated both driver and passenger systems, was developed by Volkswagen for its Type I Beetle, but without satisfactory results [Seiffert 72]. At Ford, after a long development effort to achieve adequate inflation time, a cooperative program was started with the automotive supplier Eaton, Yale, and Towne to develop a system for the right front passenger only [Frey 70]. The justification for this approach was that the driver was already protected by the combination of compressible steering column, padded hub, and yielding rim; it was the right front passenger who had the greatest need of an air bag. (On the other hand, there is almost always a driver present, whereas this seat is occupied only about 40 percent of the time, reducing the available benefit there.) In any case, a fleet of 831 1972 Mercury Montereys was fitted with passenger-side air bags for field testing, of which 126 were delivered to the U.S. Government.

SOLID PYROTECHNIC INFLATORS - In the meantime, General Motors and Chrysler were trying to find a gas source that could be fitted in the steering wheel hub. Chrysler's efforts utilized smokeless powder, but did not result in hardware being integrated into vehicles. GM, on the other hand, developed a driver system using the combustion of sodium azide to produce inflation gas. Sodium azide is a highly toxic and unstable compound and thus requires special care in production and disposal, but it has chemical reaction characteristics suitable for inflating air bags, and its primary combustion product is harmless nitrogen gas. Sodium azide has thus been the main ingredient of gas generants for driver's systems from the earliest times until the present, although it seems destined to be replaced by more environment-friendly materials.

Gas generant comes in solid form and is like a solid rocket propellant, in that the chemical reaction rate at any instant depends on the exposed surface area and the temperatures created by the reaction itself. These are controlled by the size, shape, and sodium azide content of the "grain" -- a rocket propellant term. Typically, the grain is about 60 per cent sodium azide, it is packaged in pellet form, and there is 75 to 100 grams of it in a driver's system.

The sodium azide pellets are part of a unit known as the inflator. The inflator housing is hermetically sealed to keep the pellets isolated and protected from vandalism until they are needed. Holes around the circumference of the housing allow the nitrogen gas to escape and provide axisymmetric bag filling, without creating any net thrust (for safety during handling). The inflator also contains a labyrinth of screens and perhaps baffles, to filter out particulates and to cool the hot nitrogen gas. Finally, the inflator contains a squib with electrical connections, so that the ignition of the pellets may be started. Figure 4 shows an inflator produced by Morton Thiokol.

AIRBAG MODULE - The inflator is part of a larger assembly called the airbag module, or simply module, which is the unit that actually gets installed in the vehicle. In addition to the inflator, the module contains the stowed air bag, which is securely fastened to the side of the module away from the occupant, so as to contain the pressure of the nitrogen gas when inflated. The exterior surfaces of the module include the bag cover, which is typically plastic. The cover has molded-in lines where the material is weaker, allowing it to be split by the pressure of the inflating bag. The cover opens up, typically like petals of a flower, allowing the air bag to unfurl. A typical airbag module, installed in an airbag steering wheel, appears in Figure 5.

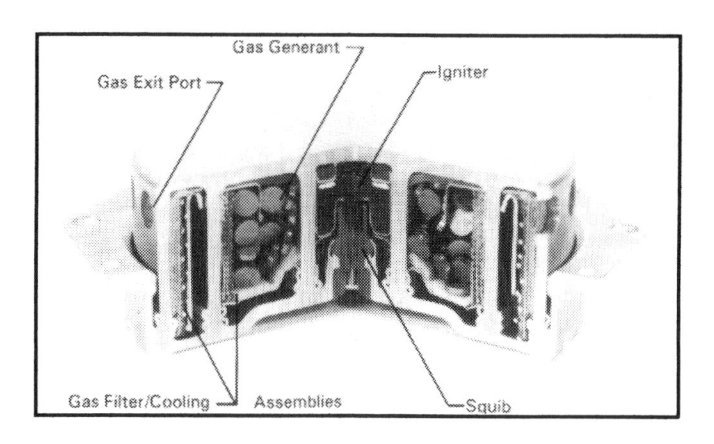

Figure 4. Typical Driver System Gas Generator - Morton Thiokol

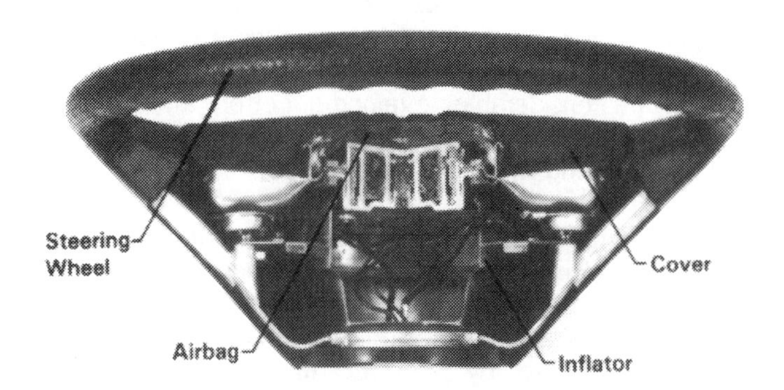

Figure 5. Driver Airbag Inflator and Module Assembly.

Clearly, the airbag module occupies precious real estate in the steering wheel hub, and significantly increases its mass. The volume must be kept to a minimum so as to avoid blocking the driver's view of the instrument cluster, or being in the way during emergency steering maneuvers. The added mass adds to the rotational inertia of the steering wheel; since this could adversely affect steering system returnability, the module mass must also be minimized. These factors have fostered the development of lighter and smaller inflators, and the use of thinner airbag material (so that the stowed air bag can be smaller and lighter).

The airbag cover must be durable, since it is often contacted by the driver. Since it is one of the most prominent features of the driver's station, it is important to the visual appeal of the vehicle, so the bag cover must also be attractive. It must protect the air bag from moisture, spilled liquids, etc. Most importantly, it must not impede the unfurling of the air bag, nor pose a hazard to vehicle occupants during deployment.

ELECTRICAL CONNECTIONS - Where the steering wheel meets the column, there is relative rotational motion between the airbag module electrical leads and the stationary wiring harness in the column. In the first-generation GM system, electrical contact throughout the steering wheel motion was maintained via a special slip ring assembly. Redundant electrical contacts were used to handle threats to reliability posed by electrical noise due to friction. In the 1985 Ford Tempo/Topaz, and in many later designs, a spiral wire (like the main spring in a clock), having enough travel to accommodate steering wheel turns from lock to lock, was used. A down side of the clock spring design is that repair personnel must be sure it is properly "rewound" when a steering wheel is replaced.

STEERING WHEEL - The steering wheel is not only the "launch pad" for the air bag; it (and to a lesser extent, the windshield) also serves as the reaction surface. This terminology means that restraint forces developed in the air bag itself have to go somewhere, and the primary load path is through the steering wheel and into the column. Consequently, the design of an airbag wheel goes beyond merely accommodating the volume, mass, and electrical connections of the module. The strength and stiffness of the spokes and rim must be sufficient to provide a stable reaction surface, but as yielding as possible in non-deployment accidents involving unbelted occupants. The General Motors first-generation airbag wheel was the result of a considerable development effort, and seems to have served as the point of departure for subsequent airbag steering wheel designs.

STEERING COLUMN - In general, steering columns are designed to limit rearward displacement (relative to the compartment) in frontal crashes, as regulated by FMVSS 204, and they are also designed to limit the forces due to occupant contact, by being able to stroke forward in the event of such a contact. In this sense, the steering column is part of the restraint system, even in a non-airbag vehicle.

In an airbag vehicle, however, contact is deliberately made with the driver. If the driver is unbelted (as was assumed to be the case in the early '70s), virtually all of the upper body restraint forces pass through the column. A conventional column might stroke forward under these conditions, but at such a low force that the stroking element reaches the end of its available travel and abruptly "bottoms out" against some mechanical limit. An airbag column, on the other hand, is designed to move forward in a controlled fashion, which means that its inertia, and static friction in the column, must be dealt with carefully. Typically, this is done by starting the static stroke vs. force characteristic at a reduced level (perhaps by "pre-stroking" the energy-absorption unit), and then increasing the static force as stroke progresses. This generally means a redesign of the stroking element, and a strengthening of the cowl structure and bracketry through which the column passes the loads on to the rest of the vehicle. GM's structural modifications are shown in Figure 6.

The legacy of research done at GM in the mid- to late-'60s on energy-absorbing steering columns is clearly seen in the design philosophy of their first-generation airbag systems. The steering column was the primary energy-absorbing element; the air bag's function was to leap into the gap between the driver and the steering wheel, couple the driver to the mechanical energy-absorbing elements in the column, and provide a more uniform application of restraint loads on the occupant's upper body.

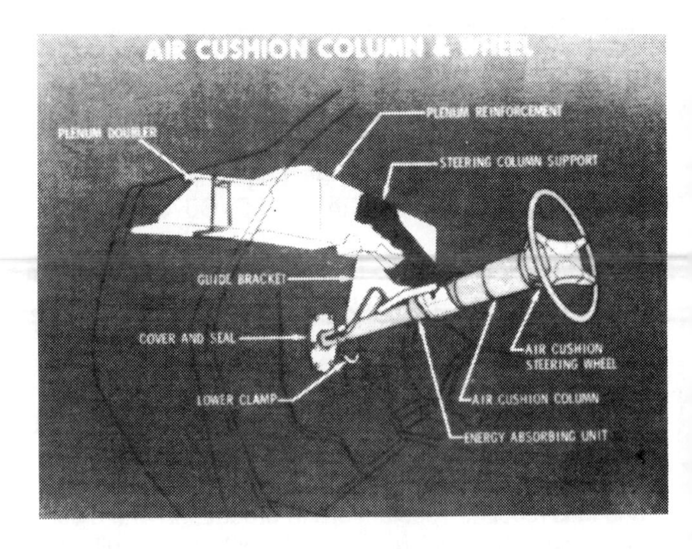

Figure 6. Structural Modifications Required for Driver System - GM

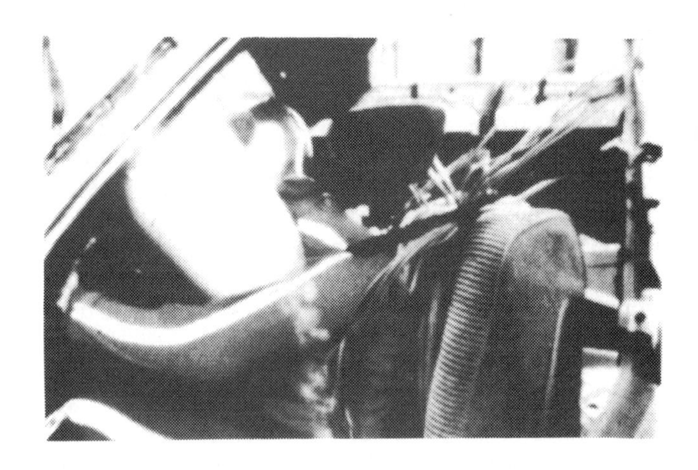

Figure 7. Belted Occupant Kinematics - Mercedes-Benz System

The column is aided and abetted in performing this duty if the torso is perpendicular to it. If a driver has a lap belt, the lap belt will limit the forward stroke of the pelvis and cause the upper body to pivot forward, thus helping to achieve the desired alignment. See Figure 7. For unbelted drivers, however, this does not occur; the whole body tends to translate forward with little articulation until interior contacts are made (typically, with the knees first). In this situation, we may well have a mostly erect torso moving straight forward toward an inclined steering wheel. As seen in Figure 8, a GM illustration of its first-generation airbag system, this geometry can cause the air bag to develop a wedge shape in side view. More importantly, it can cause a significant upward force component to be applied to the end of the column. See Figure 9.

These (off-axis) up-loads considerably increase the force levels required to stroke the column. The situation is rather like trying to close a chest of drawers by pushing on one side of the drawer. It may "jam," and resist closing

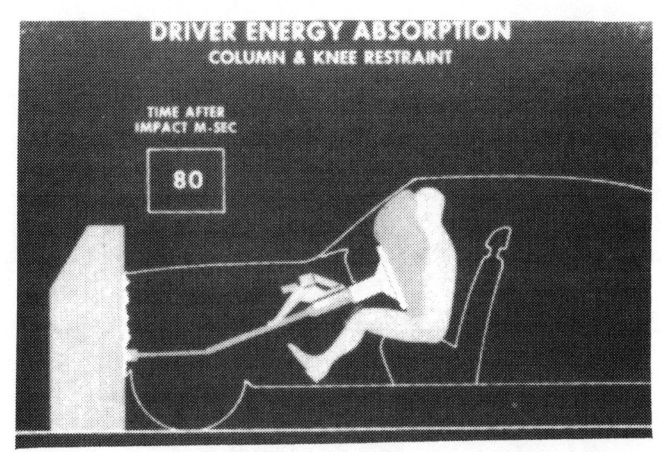

Figure 8. Unbelted Driver Kinematics - GM System

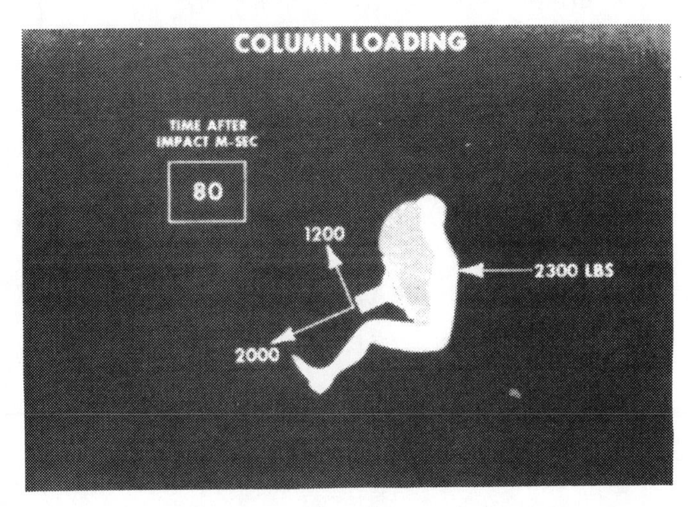

Figure 9. Column Loading - First-Generation GM System

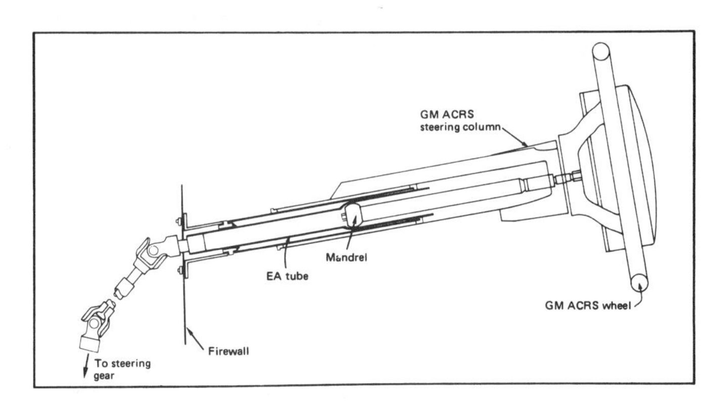

Figure 10. Minicars RSV Wheel and Column

altogether. To overcome such difficulties, one tends to design the stroking element for greater lateral stiffness, and less sensitivity to the minor misalignments that result from off-axis loads. An example of such a design is the steering column found in the Minicars Research Safety Vehicle (RSV), shown in Figure 10 [Struble 79]. Despite the lack of front-seat belt restraints, this system achieved dummy injury measures well below the FMVSS 208 criteria set for 30 mph, at delta-Vs in excess of 50 mph. At these speeds, one needed to use all the interior occupant stroking space that one could get. Reducing the articulation of the body segments (recall the lack of a lap belt) tended to keep the head away from the windshield and header, and was helpful in increasing the available stroking distance.

Another approach to dealing with column up-loads is to reduce the angle between the torso and the steering wheel. For belted occupants, this is achieved by the lap belt, as described above. For unbelted occupants, particularly in the extreme crash conditions addressed in the RSV program, the need to reduce the angle resulted in a steering column rather more horizontal than most. Of course, such architecture causes the steering shaft to penetrate the dash panel at a water line much higher than the steering gear, so an intermediate shaft with double U-joints was required to make the connection. It is unknown how most drivers would have liked the more vertical steering wheel, or the closer positioning to the sternum.

Air bag - A third approach to up-load difficulties is to reduce the dependence on column stroking, which brings us to the design of the air bag itself. The first-generation GM air bag was about 22 inches in diameter when deployed, and extended about 10 inches rearward of the wheel rim [Campbell 72]. It was pressurized to approximately 3 psi, absent occupant loading, and was about 2.8 cubic feet in volume. This was larger than most current designs, such as the Mercedes-Benz bag shown in Figure 11, which tend to run about two cubic feet, or 60 liters, in volume.

Figure 11. Inflated Mercedes-Benz Air Bag

The GM bag material itself was neoprene-coated nylon. The neoprene coating served to reduce the porosity of the bag, thus extending the duration of bag inflation. It was discovered, however, that the bag tended to act as a pneumatic spring, storing energy as the occupant stroked forward into it, but then returning the energy to the occupant later [Klove 72]. This resulted in undesirable amounts of occupant rebound into the seat back and head restraint. To counteract this tendency, the neoprene coating was subsequently removed; forcing gas through the fabric pores increased the energy dissipated (as opposed to energy being stored and subsequently returned).

Subsequent bag designs have included actual holes, or vents, which took on a larger share of the task of absorbing occupant energy. The vents are generally placed on the back side of the bag, away from the occupant, and are so located as to avoid blockage (by the steering wheel spokes, instrument cluster brow, windshield, etc.). Figure 12 is typical. The total vent area is determined during the development process so as to provide the best compromise among the disparate demands being made on the restraint system.

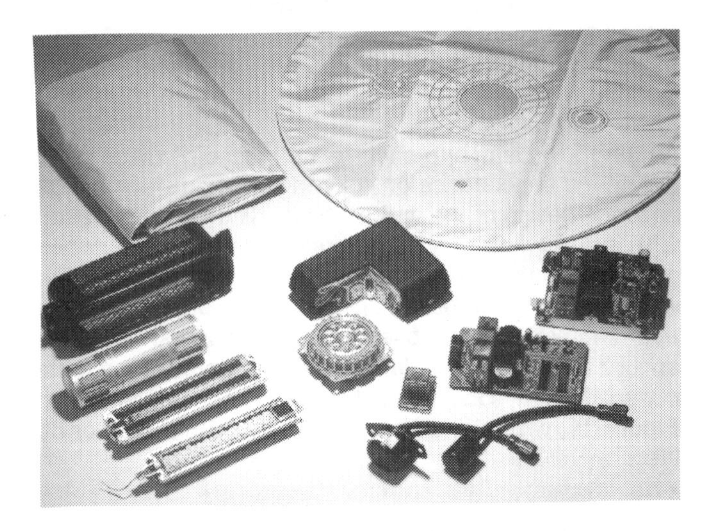

Figure 12. Typical Components, Including Vented Air Bag - Takata

As production volumes have increased, the bag has become perhaps the most labor-intensive component to manufacture and inspect. Typically, it has at least two major sections, which have to be sewn together in the traditional manner, by a worker at a sewing machine. In addition, reinforcements have to be added at the vent holes, the edge or edges where it is attached to the module, and at the tethers (if any). More recent developments indicate that bags may be woven, by machine, in a single piece.

Tethers are straps that connect the front and back surfaces of the bag. They have become more widespread in recent years, and their purpose is to limit the travel of the bag front during deployment. For small occupants seated close to the steering wheel, tethers can reduce the occupant accelerations generated when the bag front contacts the sternum. Tethers also increase the bag's aspect ratio (diameter divided by depth), improving the ability of the bag to protect occupants who load one side of the bag more than the other (due to angularity of impact or being out of position, for example).

The air bag used in the Minicars RSV had two chambers, as indicated in Figure 13 [Struble 79]. The inner chamber was connected directly to the inflator and filled first. Its relatively small size (28 l, or 1.0 cubic feet) provided a quick fill time, so it could jump into the gap between the steering wheel and the chest as soon as possible. It was aimed directly at the sternum, and this geometry was instrumental in establishing the column angle. The chest bag was vented to the outer bag (48 l, or 1.7 cubic feet), which filled more slowly and provided restraint to the head. The timing of the chest and head bag inflation could be tuned somewhat by the size of the vent in the chest bag. Once the gas had been re-used in the head bag, it was vented to the atmosphere.

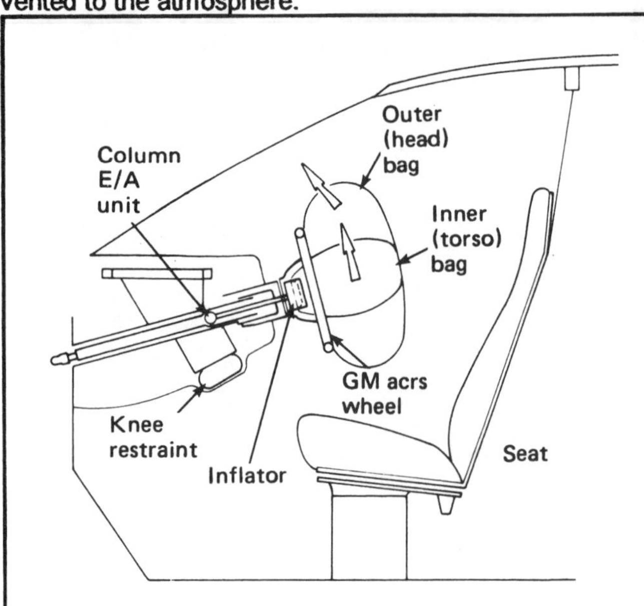

Figure 13. Minicars RSV Driver Airbag System

While the air bag inflates faster than the blink of an eye, a look at high-speed films will reveal that it is hardly instantaneous. Indeed, there is an inflation sequence, and a review of a bag pressure time history, as seen in Figure 14, will reveal some distinct phases. First, there is a relatively high pressure spike of relatively short duration, associated with inflation gas being pumped into an extremely confined space behind the folded air bag. The bag is pressed against the bag cover with sufficient force to split the cover seams, and the cover opens. In very short order, there is a volume increase behind the air bag - a high percentage increase because the volume was so low to start with. Consequently, the bag pressure drops, typically to zero. This is known as the punch-out phase,

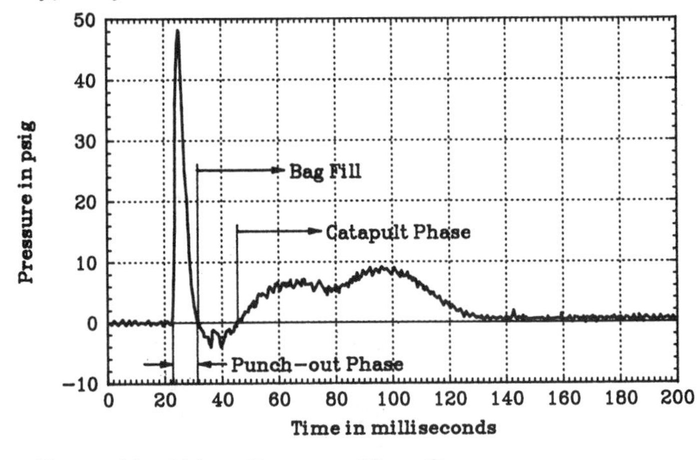

Figure 14. Airbag Pressure-Time Curve

because the bag is punching out through the airbag cover [Lau 93]. In the test from which Figure 14 was derived, the occupant was not close enough to interfere with bag deployment, and the punch-out phase lasted about nine msec. If the bag has to move the occupant out of the way to achieve fill, both the peak pressure and the duration of the punch-out phase can increase.

The end of the punch-out phase marks the beginning of the bag fill phase. Now the airbag material has some velocity, and hence some momentum, as the cover doors swing open. Thus the fabric keeps moving, and as a result a negative gage pressure is created in the bag. During this time, the bag is seen in high speed films to be unfolding rapidly, with numerous sharp creases in the fabric. Bag unfurling motions are highly complex and three-dimensional, and the pattern depends on the folding process. The lateral portions of the bag may even appear to be sucked in. Of course, the inflator is continuing to produce gas throughout this time.

Finally, the bag material reaches the geometric limits of its travel, and the pressure climbs back up through zero. This is the earliest point at which bag pressure is available as an occupant restraint mechanism (just over 45 msec in Figure 14). In very short order, the creases come out of the material, and the bag assumes its inflated shape. The actual time budget used in the design for the system of Figure 14 is unknown, but it appears that 50 msec would have been an attainable goal.

AIRBAG PERFORMANCE - Airbag design is driven by a number of considerations, among which are the usual villains of cost, weight, and size. Occupant protection performance is dictated by the requirements of FMVSS 208, which has specified that dummy injury measures for the head, chest, and femurs be within certain limits for 30 mph frontal barrier crashes at any horizontal angle up to plus or minus 30 degrees from vehicle center line. FMVSS 208 has required that tests be run with the dummy occupants, representing 50th percentile males, unbelted. Generally, the angled barrier crashes involve longer stopping distances and softer crash pulses (vehicle acceleration versus time) and in that sense are less severe, but they do pose potential difficulties associated with the occupant moving into the bag at an angle. Another crash condition is the 35 mph frontal barrier test used in NHTSA's New Car Assessment Program (NCAP), in which the dummy occupants are belted. This test is not required by the safety standards, but it has nevertheless become a de facto design requirement. These dichotomous test conditions mean that a single bullet (i.e., one system design) has to be fired at two targets. It is thus not surprising to find, in the earlier airbag designs NCAP-tested at 35 mph, that the FMVSS 208 injury criteria were exceeded. In fact, while some vehicles equipped with belts only have been meeting all the 208 criteria since the beginning of the 35-mph NCAP tests, it was not until 1988 that an airbag-equipped vehicle did so.

LOWER BODY RESTRAINT - The lack of a lap belt in the 30 mph test means that lower body restraint has to be provided by other means, typically by resisting the forward movement of the knees. The hardware involved is variously known as a knee bolster, knee restraint, or knee blocker. In any case, the design concept involves the knees engaging a deformable structure which limits knee movement to some extent, while maintaining femur loads within acceptable limits. These femur loads are transmitted to the pelvis, providing pelvic restraint. Attention has to be paid to the possibility of knee contact with the steering column, particularly during angular impacts.

Needless to say, the action of the knee restraint, and its effect on occupant kinematics, depends on the initial spacing between the knees and the restraint. Generally, the designer would like the spacing to be small, but care must be taken not to interfere with the operation of the foot pedals -- particularly the brake pedal. On the other hand, the need to reach the pedals tends to cause the driver to position the seat so that the knees are placed at a fairly uniform distance from the knee bolster, regardless of occupant size.

In a 35 mph NCAP test, pelvic restraint provided by the knee bolster, when combined with a lap belt, could be excessive. One option for dealing with this is to sew in one or more loops in the lap belt, which can pull out at a force level sufficient to provide occupant retention in non-frontal accidents, but allow enough pelvic motion to avoid unacceptable occupant kinematics. Webbing material with varying stretch characteristics can also be chosen. It's a bit of a balancing act.

UPPER BODY RESTRAINT - Similarly, the non-use of the torso belt in the FMVSS 208 test results in all the upper-body restraint being provided, in that test, by the air bag. The safety standard therefore effectively establishes the force and stroke requirements for the air bag, which may not be optimal for smaller occupants seated closer to the bag (see below). Even for a normally-seated 50th percentile male, the addition of a shoulder belt in the 35 mph NCAP test may mean an excessive amount of restraint force in that test, and indeed this could be the cause of the chest accelerations exceeding 60 Gs in NCAP tests of the earlier airbag cars. One approach to dealing with this situation is to adjust the timing of restraint forces from the air bag and the belt so that the peaks do not coincide.

An insight into this timing is provided in Figure 15, which shows the timing of chest accelerations and torso belt loads. We see that the belt loads reach their peak at about 50 msec, which closely corresponds to time budgeted for airbag inflation that was mentioned earlier. After 50 msec, we see the torso belt loads falling off while the chest accelerations continue more or less level until about 80 msec. The likely explanation is that the timing of the belts and airbag inflation has been adjusted so that the air bag picks up where the torso belt leaves off, in terms of providing upper torso restraint.

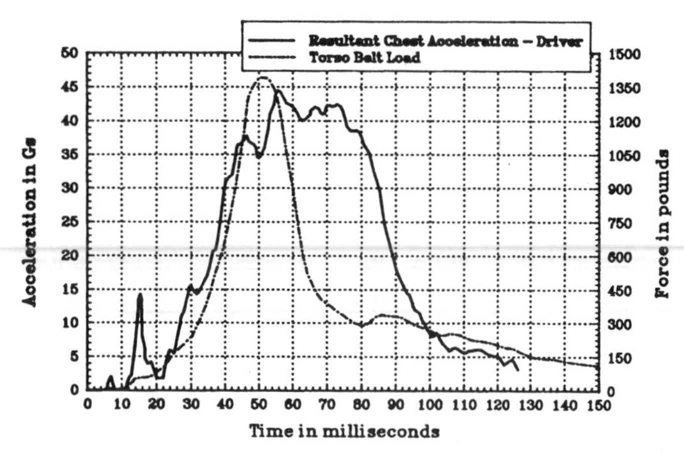

Figure 15. Belt Load and Chest Acceleration in 35 mph Barrier Test - 1994 Volvo 850

VARIABLES AFFECTING PERFORMANCE LIMITS - There are other variables not addressed by government standards or tests. Factors considered by the manufacturers probably vary, but some that come to mind are occupant size (5th percentile female through 95th percentile male), distance between the airbag cover and the sternum, and object struck (e.g., pole, offset barrier, etc.). Generally, protection of larger occupants is stroke-limited, which is to say that as crash severity is increased, some specified injury criteria limit is reached when the occupant stroke becomes excessive, resulting in bottoming out of the air bag or steering column, or contact with the interior. Protection of smaller occupants tends to be acceleration-limited. This is because restraint systems have to generate enough force to protect the many larger occupants in the population. In the same crash, these same force levels applied to smaller occupants will result in larger occupant accelerations. Smaller occupants may thus reach acceleration limit values at lower crash severities than larger occupants will.

When the object struck is not a barrier, the crash pulse may be softer, but compartment intrusion may also be greater than in a barrier crash at the same speed, due to the concentration of crash forces on only a part of the structure. This could be reflected in displacement or rotation of the steering column, which could affect the "aim" of the airbag restraint forces, and the resulting occupant kinematics. It could also affect the sensing time, possibly producing a later deployment. This is a function of the sensor system design, to be discussed later.

If the driver is sitting closer to the air bag than in the design condition, contact between the deploying bag front and the occupant can occur earlier, and at a higher contact velocity. This could increase the accelerations experienced by the occupant, particularly if he or she is small.

DESIGN PARAMETERS - To deal with all these requirements, some of them in conflict with one another, the designer and developer of airbag systems has a number of parameters to work with. The crash pulse is an important one, albeit one that the airbag engineer may have little control of. Another one is the inflator characteristic, which is generally expressed in terms of the time histories of pressure when the inflator is discharged into a fixed, closed volume (the so-called "tank test"). Typical tank test results are shown in Figure 16. Other variables include the column stroke characteristic (force vs. distance), steering wheel location, airbag vent area, bag volume, bag diameter, tether length, seat belt anchorage points, webbing stretch characteristics, and seat cushion stiffness.

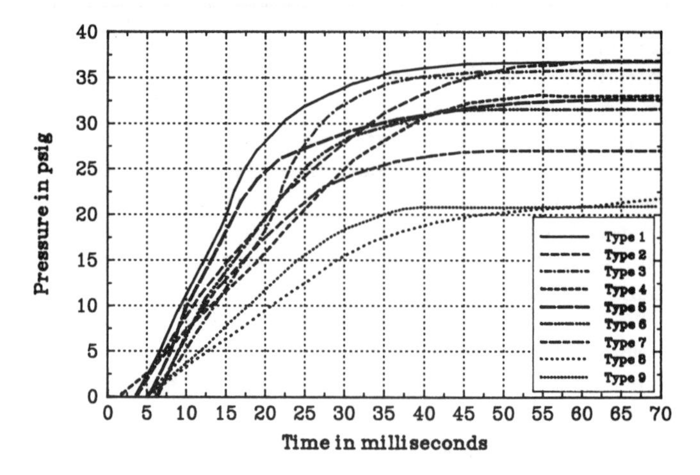

Figure 16. Tank Test Curves for Morton-Thiokol Inflators - 60 l, 22°C

PASSENGER-SIDE AIRBAG SYSTEMS

Perhaps the feature that most consistently distinguishes passenger-side systems from their driver-side counterparts is the location of the launch pad: driver systems are mounted on the steering wheel, and passenger systems are mounted on, and part of, the instrument panel. This distinction causes a significant difference in the longitudinal distance between the airbag mount and the occupant's sternum. Moreover, the passenger side lacks a stroking element, like the steering column, that could be used for absorbing energy.

One approach to transferring driver's side airbag technology to the passenger's side would be to blur these distinctions. In other words, move the launch pad aft and mount it on a stroking element. This concept led directly a design using a "bag-bolster" positioned rearward about even with the steering wheel. A bag-bolster system was sled-tested at Calspan [Romeo 75] at 75 kph (47 mph), but concerns remained regarding public acceptance of the appearance and ease of ingress and egress of such a design. Subsequent concepts for a passenger-side air bag have avoided significant changes in the instrument panel location.

In the Minicars RSV program, the objectives for the right front passenger system included the ability to undergo a crash at 50 mph delta-V, and still provide occupant protection within the 30 mph FMVSS 208 criteria. Originally, it was thought that a stroking dash (albeit conventionally-positioned) would be required to meet such ambitious goals. However, it turned out that the performance goals could be achieved, with room to spare, using venting alone. All other passenger-side airbag systems, as far as is known, have similarly relied upon venting for energy absorption.

BAG GEOMETRY - In the development process, it may be tempting to extrapolate from a driver side system by starting with a deeper (in the longitudinal direction) version of a driver bag, mounted on the instrument panel. This approach would not be valid because the aspect ratio of such a design would not be nearly high enough; the bag would buckle or be pushed aside by the occupant's motion. The bag needs to be wider (laterally) and taller (vertically) for two reasons: (1) to avoid instability (buckling) when loaded in compression, and (2) to handle the greater lateral deviation in an occupant's path during an angular collision, due to the occupant being farther from the bag when the crash event starts. This is true even when there is not a designated center seated position in the front seat. If there is a center position, and airbag protection is provided for that occupant, the bag must be wider still.

These considerations lead to a considerably larger air bag. The first-generation GM bag, designed to protect both center and right front occupants, had a volume of about 14 cubic feet. Most present bags are much smaller. The inflated shape is typically a lateral cylinder with vertical ends, but the cross-section is not necessarily circular. See Figure 17.

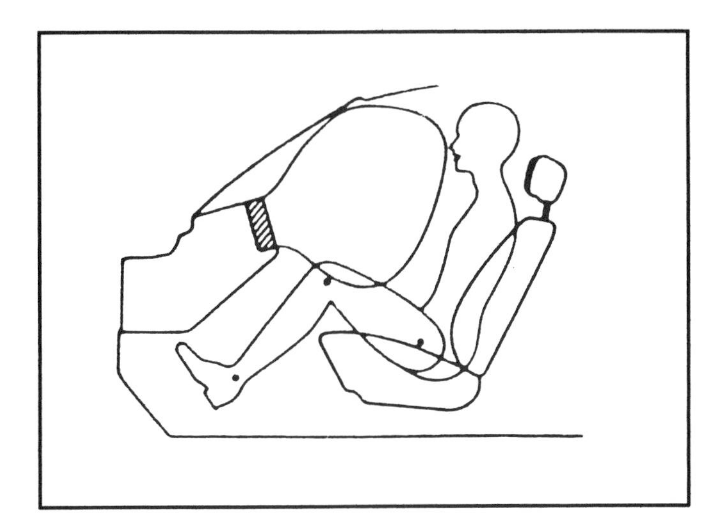

Figure 17. Mid-Mount Airbag Configuration

The Minicars RSV passenger-side air bag was dual-chambered, for reasons that can be understood by comparing it to its counterpart on the driver's side. Refer to Figure 18. In this design, the inflator emptied directly into the lower, or torso bag, where earlier application of restraint forces was needed. This bag had a volume of about 2.75 cubic feet (78 l). The torso bag was vented to the upper, or head bag, to pick up the head somewhat later. The volume of the head bag was about 3.0 cubic feet (85 l). Gas from the head bag was then vented to the atmosphere.

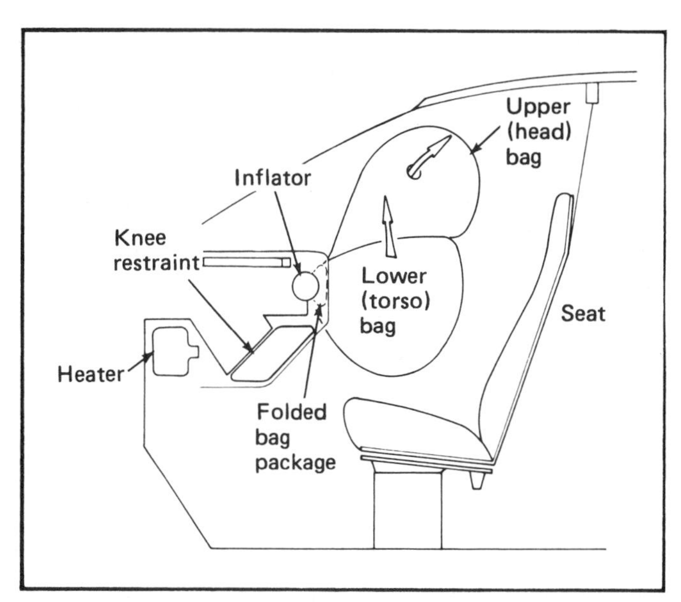

Figure 18. Minicars RSV Dual-Chambered Passenger Airbag System

The first-generation GM system also had a dual-chambered bag, but for entirely different reasons. One chamber was intended specifically for knee restraint. Since knee restraint loads are more concentrated than, say, the loads applied to the chest or the head, the knee bag operated at a much higher pressure. Due to its smaller size, it filled first. The other chamber, being larger, filled relatively more slowly and provided restraint to the torso and head. See Figure 19.

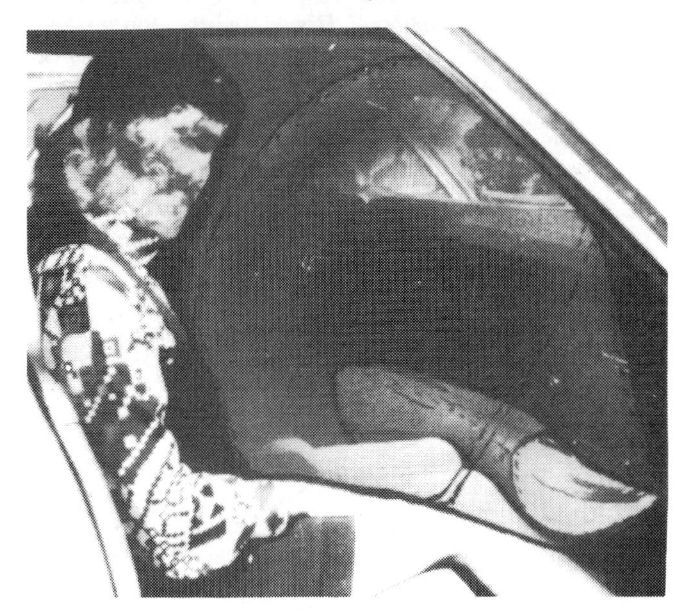

Figure 19. First-Generation GM Low-Mount Airbag with Knee Bag

TYPES OF AIRBAG MOUNTS - Compared to driver systems, passenger air bags have many more design options regarding their integration with the vehicle. The location and orientation of the module gives rise to some terminology regarding the type of mount. At the time it was designed, the Minicars RSV air bag was called a high-mount system, but as indicated in Figure 17, it would be called a mid-mount configuration in today's parlance. The airbag module is located on the aft face of the instrument panel, about where a glove box might traditionally be located. In this position, the air bag cannot, and is not intended to, provide lower-body restraint. The Minicars system was typical of many in that it was designed to be tested without the use of lap belts. Thus the lower-body restraint had to be provided by other means, such as a knee bolster. The module and the knee bolster make it very difficult to provide a traditional glove box in the instrument panel. Obviously, there is also a significant impact on the location and routing of heating, ventilation, air conditioning, and other components in the instrument panel. The bolster itself is similar in concept to those used on the driver's side, except that the designer does not have to deal with potentially hard contact surfaces presented by the steering column.

In another contrast to the driver's side, there is no requirement for the passenger to reach the pedals. For bench-seat vehicles, the passenger seat position may be determined by the driver. With bucket seats, the position may be a function of who is seated in the right rear seat, and how big they are. Variability in the size of the right front passenger provides (literally) another dimension. In any event, one can expect considerably more variation in the knee-to-knee-bolster distance at the beginning of the crash. A small occupant, seated far from the bag, could tend to submarine under the bag.

General Motors, in the design of their mid-1970s system, used an inflatable knee restraint to provide added tolerance to such variations. (Recall the very large size of the vehicles involved.) To effect the design, the airbag module was located lower on the instrument panel in what has come to be known as a low-mount configuration, illustrated in Figure 20.

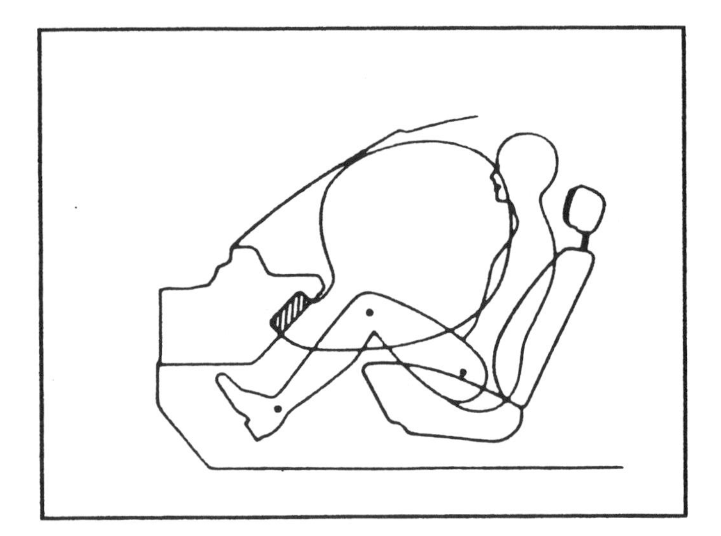

Figure 20. Low-Mount Airbag Configuration

A third design is known as a top-mount (or dash-top) configuration. In this system, the deploying bag is not directed aft at the chest or down at the knees, but rather upward toward the windshield. The airbag cover is typically on the top of the instrument panel, as seen in

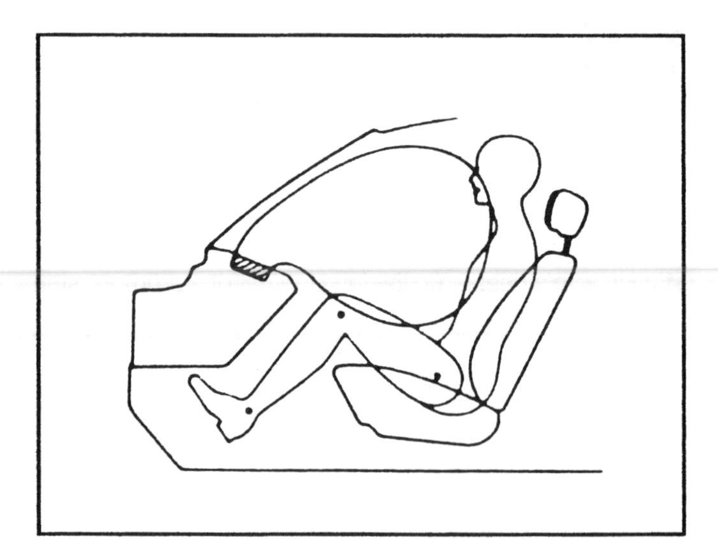

Figure 21. Top-Mount or Top-Dash Configuration

Figure 21. Top-mount designs were proposed in the early '70s, but were not used in either the GM or Minicars systems. More recently, this configuration has become widely used, for reasons to be discussed below.

HOW RESTRAINT FORCES ARE DEVELOPED - When the occupant presses against an air bag, the pressure in the bag is transmitted directly across the layer of fabric, onto the occupant's body. Restraint forces are thus generated, but it would be a mistake to attribute all such forces to this mechanism (gas pressure).

Consider, for example, that with a 50th percentile male dummy and the seat in its middle position, there may be 560 mm (22 in) between the chest and the instrument panel. If the air bag is to fill this gap in 25 msec (say), the deploying air bag surface must move rearward at an average speed of at least 50 mph. Of course, the instantaneous speed varies, so one would expect the peak to be much higher. Indeed, film analysis of the deployment of various driver-side systems showed peak speeds ranging from about 100 mph to over 200 mph [Kossar 92]. Speeds at the time of facial contact were lower, of course, and would depend on where the seat is positioned.

When the air bag contacts the occupant, there is a momentum transfer between the two that depends on the portion of the air bag's mass brought to rest against the occupant, its velocity at contact, and the portion of the occupant's mass involved in the contact. This phenomenon is called bag slap, and can generate restraint forces when the bag pressure is low or even negative. See Figure 22. If multiple layers of fabric are involved (as when the bag is still partly folded, for example) and the brunt of the impact is taken by just the sternum or the head, the increased effective mass of the bag and the reduced effective mass of the occupant will combine to produce higher occupant accelerations. Obviously, if the occupant is initially positioned closer to the air bag, contact occurs sooner, earlier in the bag unfolding sequence, and possibly at a higher contact velocity.

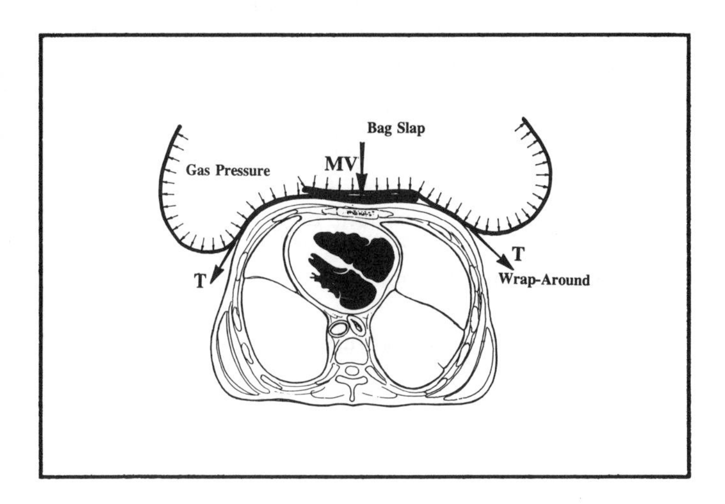

Figure 22. Sources of Restraint Forces

Clearly, bag slap is a greater concern for occupants having smaller mass, seated closer to the bag, and so positioned that the deployment forces are directed higher on the body.

Once momentum has been transferred to the occupant, the bag slap phase of the restraint process gives way to the catapult phase, wherein the occupant and aft bag material move together. As the occupant strokes into the bag, the pressure builds and generates restraint forces, as mentioned before. At the same time, membrane tension is created in the bag fabric, which is partially wrapped around the occupant. Because of occupant penetration, the bag tension has a rearward component at the locations where the fabric and the occupant cease to be in contact. These rearward components, or wrap-around forces, contribute significantly to restraint action during the catapult phase. This is particularly true on the passenger side, because the bag tends to be wider and deeper.

EFFECTS OF OCCUPANT SIZE AND POSITION - Since passengers don't operate foot pedals, they can choose to sit closer to the instrument panel (or farther away, of course) than they might on the driver side. In fact, the position of a bench seat may be controlled by the driver; with bucket seats, the position might depend more on the presence and size of a rear occupant than anything else. At the same time, the front passenger doesn't have to be of driving age. One finds, therefore, a much wider range of occupant sizes, weights, ages, and seat positions on the passenger side.

If the occupant is short in stature, the bag may be directed more at the head and neck, and less at the torso. In this way, the relatively small effective mass of a child, for example, could be reduced still further. If the membrane tension is sufficiently high, coupled with a small effective mass, the bag fabric may actually snap taut, propelling the head rearward.

If restraint performance is optimized for a normally-seated 50th percentile male with the seat in mid-position, one must expect some degradation of performance when any, or some combination of, these conditions is varied. Clearly, a system acting as if an unbelted 160-pound occupant is present is going to be much too powerful to be optimal for a child weighing a fifth as much. While the Minicars RSV system may have presented the most potential for occupant protection at the highest severities, at the same time it may also have posed a higher risk to small and out-of-position occupants. Short of tailoring the restraint system for the conditions (see below), the best one can do is make the greatest accommodation for the occupants most at risk (e.g., children, small adults, the elderly), while keeping performance for the "nominal" occupant and seat position within the limits imposed by Government testing.

Within a given airbag configuration, one can vary such parameters as the inflator charge, the angle and position at which the inflator is mounted, the fabric weight and vent area, and the folding pattern. Obviously, lighter fabric weight tends to reduce bag slap and stowage volume. Typically, the folding pattern has been found to be very important in addressing the needs of small or out-of-position occupants.

Another approach tried in the '70s was the so-called aspirated system [Katter 75]. This design concept stemmed, at least in part, from early concerns regarding overpressure in the compartment due to bag deployment. (Consider, for example, that the U.S. ESV family sedan designs of the early '70s had air bags in both front and rear seats [Alexander 74].) It was thought that overpressure might be alleviated if a supersonic ejector concept could be adapted to pump compartment air into the air bag. Such devices worked through viscous mixing, in a diffuser, of the air streams from the primary source (the inflator) and the secondary source (the compartment). The secondary flow continued until the bag pressure became high enough to stall the diffuser, at which time a check valve in the secondary air stream must close. Since the occurrence of stall depended on what the bag contacted, the system would naturally adjust the degree and rapidity of fill if the occupant were closer to the bag. Another advantage was the reduced inflation requirements for the inflator.

As it turned out, bag folding technique had a much larger influence than aspiration on the results for out-of-position children [Romeo 78]. As of this writing, there has been no known further development of aspirated systems.

TYPES OF INFLATORS - The volume of a passenger-side air bag is much larger than that of the driver-side bag, so one might expect a passenger-side inflator to need proportionately more gas generant. Of course, the packaging constraints are altogether different, meaning that there is little incentive to make a passenger-side inflator look much its driver-side counterpart. Rather, it is typically a circular cylinder, mounted so that its long axis is lateral to the vehicle. As with a driver-side inflator, it typically has a hermetically-sealed metal housing, for the same reasons. Generally, nitrogen gas passes through a complex of screens and perhaps baffles, and exits through a series of holes or slots. These openings are evenly distributed from one end of the cylinder to the other, so as to provide an even fill across the width of the air bag.

Of course, other inflator configurations are possible. In the late '70s, when the inflator business was at its nadir, Calspan Corporation developed a design using two ganged driver inflators to fill a passenger bag [Romeo 78]. The primary motivation was the lack of a production passenger inflator, but utilizing two driver units with a common manifold has certain other advantages, discussed below.

Finally, we return to the type of inflator thought of first -- stored gas. Packaging constraints permit the storage bottle to be sufficiently close to the air bag to allow rapid inflation, but the same constraints cause the pressure in the bottle to be on the order of 3000 pounds per square inch. (The smaller the bottle, the higher the pressure required to store an amount of gas sufficient to inflate the bag.) Such pressures raise concerns about leakage (and thus reliability), when one considers that the system must remain absolutely leak-proof for perhaps 20 years. Nevertheless, the mid-70s passenger side systems by General Motors utilized stored-gas inflators, and successful deployments have occurred in cars of relatively advanced age.

The passenger-side module, shown in Figure 23, employed a membrane that was pierced to start the inflation process, and a manifold to distribute the inflation gas to the air bag. A characteristic of stored-gas systems is that the pressure in the tank is at its highest at time zero; as gas escapes from the tank (at the speed of sound), the remaining volume expands adiabatically (without heat loss), and as it does, the temperature drops. At the same time, the tank pressure drops.

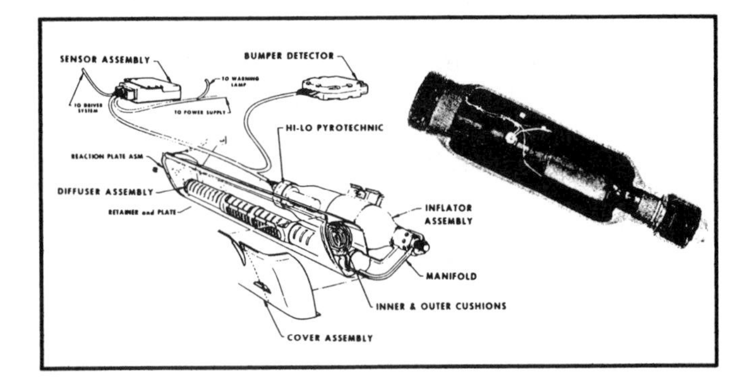

Figure 23. First-Generation GM Passenger Air Bag, with Inflator

The higher the storage pressure and the smaller the tank volume, the more steeply the pressure declines. At the same time, the peak noise level increases [Jones 71]. Questions have been raised regarding hearing loss, but the concerns seem to have subsided with the advent of pyrotechnic inflators on the passenger side. This is because pyrotechnic systems provide a more even gas flow, which reduces to some extent the concern about inflation noise. Because the gas generant grain can be modified to some extent without changing the inflator housing, the gas flow is more readily tailored or adjusted than with a stored-gas system.

At the same time, the combustion temperature in a pyrotechnic inflator is much higher than ambient, such that ambient temperature variations do not have much effect on gas generation. Stored-gas systems are thought to be more sensitive in this regard.

Since the inflation gas expands as it passes through the opening and into the air bag, it cools (and in fact gets very cold). If the gas pressure within the bag were insufficient, it could be increased by heating the gas. This thought gives rise to the concept of an "augmented" or "hybrid" inflator, in which heat is applied to the gas on its way to the air bag.

The GM passenger-side systems of the mid-1970 utilized this concept. A small charge of pyrotechnic materials could be ignited, not for the purpose of materially increasing the amount of inflation gas, but rather to add energy to the inflation gas by raising its temperature, and causing it to expand more. The increased temperature reduced the sensitivity to variations in ambient temperature [Seiffert 72].

DUAL-LEVEL INFLATORS - The engineers at Calspan were not the only ones to design a passenger system using two driver inflators. A 1979 paper described a Mercedes-Benz design in which "incremental deployment" of one or both inflators could advantageous in low-speed impacts, or for out-of-position occupants or children [Reidelbach 79]. The wording of the paper suggests that such a feature was not implemented, nor was it at Calspan. With GM's design, however, the choice could be made between a "low-level" deployment in which just the stored gas was released, or a "high-level" deployment which also involved the ignition of the augmented inflator charge.

Of course, to make a choice there must be some logic employed, and there must be sensors to provide the inputs. With the GM system, the choice involved the nature of the crash pulse. GM's design employed two impulse detectors on the bumper near the frame attachments, plus sensors on the dash panel. Generally, the trigger level was lower for the bumper units than for the dash panel sensors. Lower-speed crashes, underrides, center pole impacts, etc. might cause the lower-level units to trigger, but not necessitate a maximum-level inflation (as would be achieved by igniting the augmented pyrotechnic charge). Therefore, the logic was this: to release the stored gas if the lower-level sensors triggered, but to ignite the augmented charge only if the higher-level sensor triggered as well. It is noteworthy that the logic involved only the nature of the crash as experienced by the vehicle; no decisions were made on the basis of conditions inside the compartment.

FROM DUAL-LEVEL INFLATORS TO "SMART" SYSTEMS - The decision whether to have a high- or a low-level deployment could be made on the basis of factors other than the crash characteristics seen by the vehicle. For example, Is the seat occupied? If the seat is occupied, is the belt being worn? In this electronic age, we could envision a so-called "smart" system, in which the deployment logic could be based on detecting the presence of a child seat (particularly a rear-facing one), discriminating between humans and various objects in the seat, and/or detecting the size of the occupant, his or her proximity to the instrument panel, etc. Again, it is noteworthy that all considerations involve the occupant and conditions inside the passenger compartment, in contrast to the logic utilized in the first-generation General Motors air cushion system of the mid-70s.

The term "smart" has more recently acquired an official definition, thanks to the NHTSA. In a rule issued in late 1996 [NHTSA 96], an airbag system is considered "smart" if:

- It does not deploy if the mass on the seat is 30 kg or less.
- It does not deploy if a rear facing child seat or out-of-position occupant is present.
- It does deploy if a properly belted child is present and there is no risk of injury.

Obviously, the ability to adjust the air bag's deployment to these and other factors is highly dependent on sensor technology, which is still in a state of intensive development at this writing.

Among the competing sensor technologies are the following:

- Sensing occupant weight by detecting the seat cushion deflection. Such a device could be fooled by heavy objects, would not be sensitive to occupant position, and would require extensive cushion redesign to incorporate the sensor.
- Infrared sensors to measure body heat. Here, the challenge is to distinguish the body heat "signal" (about 37°C) from the high "noise" levels due to variables such as compartment temperature (which can vary from -20°C to +70°C), heated seats, and heat-absorbing clothing.
- Detecting changes in capacitance due to an occupant. Such technology could detect occupant position, and sensors could be located in a variety of places, but at the same time they could be fooled by conductive materials, including moisture.
- Ultrasound could be used, as during pregnancy, and sensors could be located in a variety of places, but would be sensitive to temperature and humidity. Large objects could block sound waves.
- A semiconductor-based "seeing eye" could discriminate between the visual appearance of an occupied seat and an empty one, and thus could detect occupant presence and position. However, high resolution plus pattern recognition equals high cost.

All of these systems are "passive" in the sense that occupant does not have to (and indeed, cannot be expected to) wear a reflector or transponder to "talk back" to the sensor or sensors. An "active" system, by way of contrast, includes such a device, which makes the sensor's task easier and improves the quality of communication between sensor and object. This concept would be practical for specific hardware designed to fit on an automobile seat, such as a child seat. A rear-facing child seat could have a reflector or transponder which could cause the air bag to be depowered or deactivated if the child seat is placed in the front seat, despite warnings to the contrary. Alternatively, a reflector or transponder on a forward-facing child seat could provide quality information on the child seat's presence and distance from the airbag module. The problem, of course, is the installed base of millions of child seats not so equipped.

CRASH SENSORS

At the instant of contact, the vehicle will not yet "know" that it is in a crash. The best that can be done is to have sensors and associated electronics continually on "sentry duty," keeping track of velocities and accelerations, and looking for the telltale signs that a crash, and not just a hard bump, has begun. The sensor system will also have to distinguish the direction of the crash, and whether the crash severity warrants a deployment. This is a tall order and takes some time, called sensing time -- perhaps on the order of 20 to 25 milliseconds. What is left of the first 50 milliseconds can be devoted to filling the air bag.

DEPLOYMENT THRESHOLD - Clearly, there are circumstances in which airbag deployment is undesirable. In non-frontal impacts (e.g., side, rear, and rollovers), the injury hazards may not be amenable to reduction by an air bag. In any case, airbag deployment necessitates replacement, which may add considerably to the repair bill. In property-damage-only accidents, airbag replacement cost could represent a significant part of the total. If airbag deployment were perceived as unnecessary, and if the replacement cost were high, negative reaction could hinder the public acceptance of air bags.

Similarly, the injury potential may be low in low speed non-deployment accidents, particularly if the belts are being worn; at the same time, the energy of airbag deployment could actually increase the probability of injury. Therefore, for frontal impacts a deployment threshold is created.

Below the threshold the air bag should never deploy, and above the threshold it always should. Of course, every crash has a different signature, so for design purposes the threshold is described in terms of barrier impacts. At one time, the Government was proposing a 15 mph barrier impact as the minimum to initiate airbag deployment [NHSB 70], probably with an eye to the 15 mph interior impact requirements of FMVSS 201 [FMVSS 201]. In any event, a requirement for deployment threshold was never promulgated, but similar considerations for protecting unrestrained occupants has tended to cause thresholds to be set at about 12 mph in frontal barrier impacts.

Of course, variability in crashes means that the threshold has to have some thickness. In other words, there has to be a "gray area" within which deployment may or may not occur. The specification for this threshold can, and probably does, vary with vehicle design, but generally the specification does reflect a gray area. The low end of the gray area, below which the bag should never deploy, may be set to 8 mph in a barrier crash, and the upper end, above which it should always deploy, may be 14 mph, again in a barrier crash. The sensor system is designed to respond only to the frontal component of an accident, so deployment occurs only if the frontal component exceeds the threshold.

The setting of the deployment threshold can depend on many factors, including the usage of seat belts. Some systems have the ability to adjust the threshold up or down automatically, based on whether the belts are being worn, since for seat belt users the potential benefits of air bags are reduced at low speeds. Such decision-making capability is one aspect of a "smart" restraint design.

RELIABILITY CONSIDERATIONS - Deployment threshold is related to another topic - reliability. Air bags are unique among automotive systems. They are different from brakes, for example, which can be disassembled for inspection or maintenance, and which can give can give cues regarding their condition whenever they are used. On the other hand, air bags may remain unused for long periods -- perhaps 20 years -- but they must remain fully ready to perform when needed, and not deploy when not needed. Therefore, an airbag system must have a built-in readiness tester that can immediately inform the operator or the repair technician when there is a problem, and a diagnostic system that will indicate where the problem is. Even so, the consequences of an error (especially a failure to deploy when needed) can be so serious that reliability targets have been set at levels comparable to man-rated space missions [Jones 70]. These levels are achieved by quality assurance inspection and testing at a 100 percent level, as opposed to a statistical sampling process. In other words, every step in the manufacture and assembly of every airbag system entails inspection, testing if appropriate, and documentation, and every vehicle is subjected to a complete diagnostic test procedure before it is offered for sale. Thereafter, every system is electronically tested every time the engine is started.

ELECTRICAL CONNECTIONS - If the airbag system wires were manipulated by service personnel, there would be increased odds of a wire being cut, incompletely re-connected, or not re-connected at all. Therefore, the airbag wiring is contained in a separate and independent harness, wrapped with a material of a distinctive color (typically yellow), and routed inconspicuously. Harness routing is also chosen so as to minimize the likelihood of being pinched or crushed during the crash.

The connections to this harness are crucial to system reliability. Typically, each connector employs dual contacts, gold plated to prevent any compromise of electrical continuity due to corrosion. As shown in Figure 24, part of a bar code may be printed on each of the mating connector housings, so a bar-code reader can verify and record that connector is correctly assembled [Kobayashi 87].

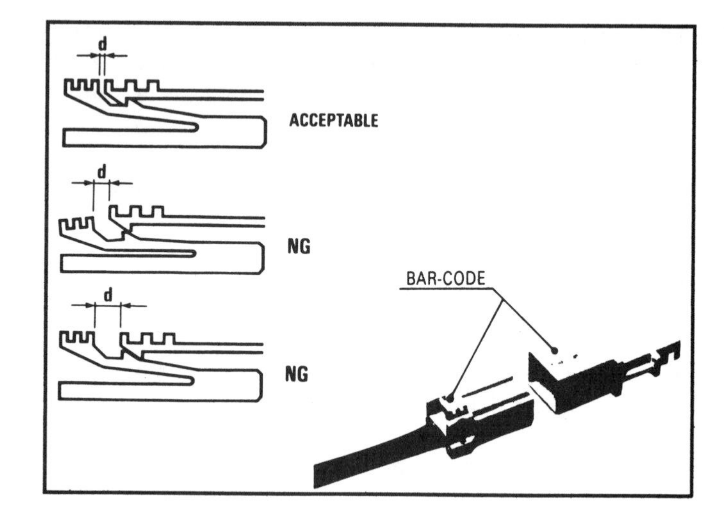

Figure 24. Machine-Checkable Airbag Harness Connector - Honda

THE NEED FOR FIELD TESTING - Because of the extremely stringent reliability requirements, and the relative scarcity of accidents warranting an airbag deployment, it is virtually impossible to design a test protocol, prior to vehicle introduction, which would give adequate insights regarding reliability issues [Jones 71]. Thus we find manufacturers being very cautious, with the early airbag designs being introduced in limited quantities, in fleet settings. A notable example is GM's 1000-car "green fleet" in 1973 [Smith 73].

Another example is the NHTSA-sponsored airbag retrofit program for 539 police cars in 1983-85 [Romeo 84].The intent of this program was to design, test, and evaluate a driver airbag retrofit system, and using such production hardware as existed at the time, to manufacture and install retrofit kits into Highway Patrol vehicles in various states. Sensors and diagnostic systems were supplied by the Technar Corporation, the gas generator came from Bayern Chemie, and the steering wheel and airbag module were made by Takata Corporation. Important lessons were learned, but plans to offer a retrofit kit on a wider scale were thwarted by the inability to obtain adequate products liability insurance [DeLorenzo 86].

Other important field test fleets in the U.S. were the 5300 Ford Tempos for the U.S. Government in 1985 [Maugh 85], further described below, and Chrysler's introductory fleet in 1988 [Edwards 91]. Deployment accidents in all these fleets were investigated in detail [Mertz 88], and valuable insights were obtained from this field experience.

HOW RELIABILITY REQUIREMENTS AFFECT SENSOR SYSTEM ARCHITECTURE - Basically, reliability entails the avoidance of two kinds of errors: false negatives and false positives. In other words, a sensor should avoid a failure to trigger when it is supposed to, and avoid triggering when it is not supposed to. The probability of a false negative can be reduced by placing more sentries on duty, and empowering any one of them to sound the alarm. In engineering terms, one would have multiple sensors connected in parallel. However, the probability of a false positive (i.e., an unwanted deployment) increases with the number of sensors in parallel, since the probability of a false positive for the system as a whole is a combination of the individual false positive probabilities.

On the other hand, the system-wide probability of a false positive can be reduced by placing two or more sensors in series. In this system, any one sentry can squelch the alarm. Consequently, the probability of a false negative increases with the number of sensors in series.

If several sensors are employed, many combinations of series and parallel connections are possible. In the General Motors system of the 1970s, the driver and low-level passenger circuits were connected through sensors on both the positive and the negative (ground) side; i.e., in series. On each side, there was a low-level G switch and a bumper switch in parallel. The same parallel combination was used on the negative side of the high-level passenger circuit, but the positive side was connected through a separate high-level G switch [Louckes 73]. See Figure 25. The G switches, actuation circuits, backup energy source, diagnostics, and crash monitoring and recording devices were included in a single component mounted in the passenger compartment.

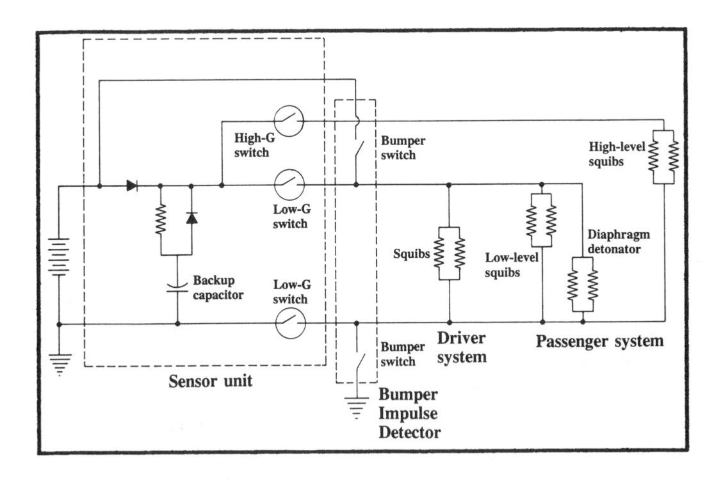

Figure 25. Sensor Circuit, First-Generation GM System

The result of this logic was that a low-level crash would trigger a driver airbag inflation and a low-level passenger system inflation. A high-level passenger system inflation would occur only if the low-level circuit were completed on the negative side, and the high-level sensor closed (on the positive side).

The switch used in GM's Bumper Impulse Detector was a prototype of sensor designs to come, in that it consisted of an inertial mass and a spring; for the switch to close, the acceleration would have to be strong enough to overcome the resistance of the spring and would have to last long enough for the mass to travel the requisite distance to reach an electrical contact. See Figure 26. Consequently, this device was a mechanical integrator of accelerations, in which switch closure depended, for the most part, on the velocity change. It was calibrated to close in an 11 mph barrier crash, and its purpose was to achieve early detection of the crash. (Recall that the G switches, in contrast, were well aft, in the passenger compartment.) The Bumper Impulse Detector employed dual contacts and springs in series.

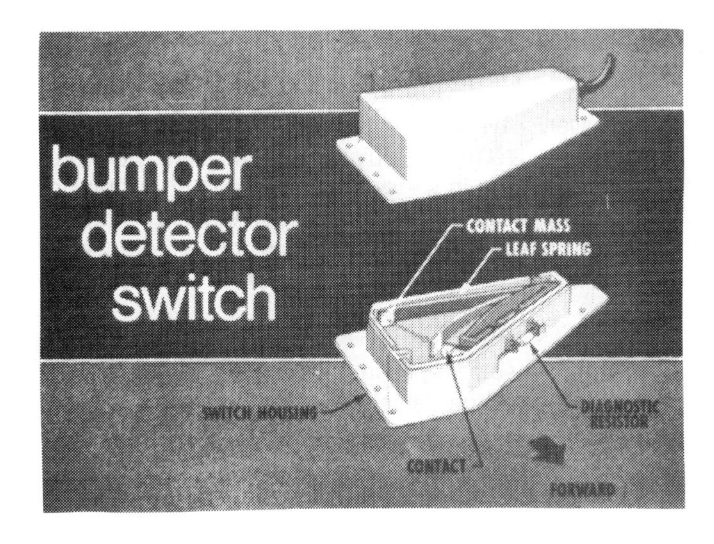

Figure 26. Bumper Impulse Detector, First-Generation GM System

The G switches consisted of a mass on a wire, which formed a pendulum, as shown in Figure 27. The mass was held in its aft most position by a magnet; if the vehicle acceleration were strong enough, the mass could break free of the magnet and move forward in a wedge-shaped groove which extended to 37 degrees either side of straight ahead. The low-level G switch was calibrated (by means of the magnet strength) to close at an acceleration level corresponding to an 11 mph barrier crash. Obviously, the acceleration level would vary with the vehicle and the location of the switch, but in the GM design the level was 18 Gs. The high-level G switch was reportedly calibrated to close at 30 Gs [Louckes 73].

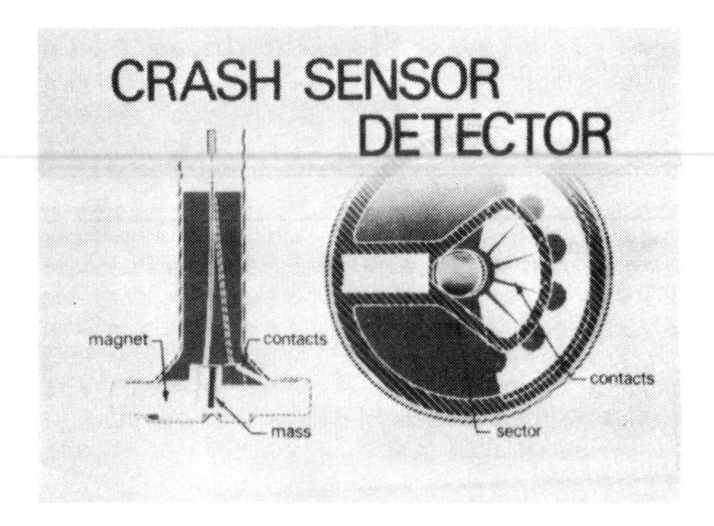

Figure 27. G-Switch Sensor, First-Generation GM

The bumper presented a very harsh environment for any electrical device, including the Bumper Impulse Detector. Designers of subsequent systems have refrained from placing sensors in such a location.

The 1985 Ford Tempo/Topaz system (driver side only) employed five sensors, which were functionally divided into two groups. One group, the secondary or "safing" sensors, were intended to minimize the probability of false positives (inadvertent deployments), and were therefore, as a group, wired in series with the group of primary or "discriminating" sensors. To avoid raising false negatives (failure to deploy), the safing sensors were set to trigger at lower crash severities than the discriminating sensors were.

The safing sensors were wired in parallel with each other and were located on vehicle center line at two locations: the top of the radiator support and on the dash panel. The discriminating sensors were connected in series with the safing sensors, which could thus switch "on" frequently

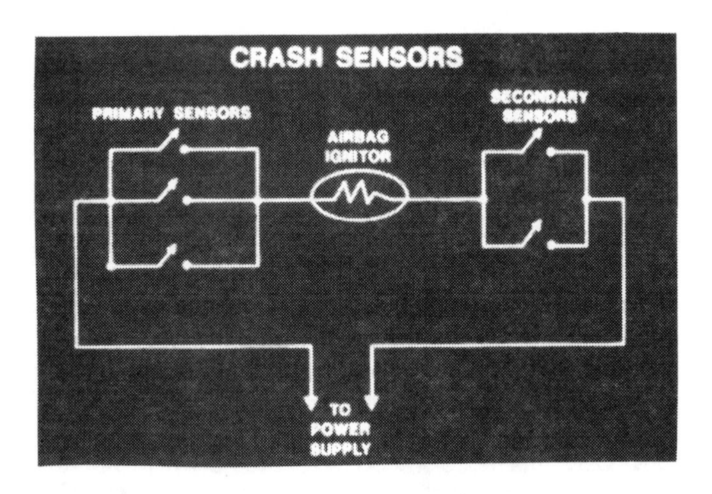

Figure 28. 1985 Ford Tempo/Topaz Sensor Circuit

without necessarily resulting in a deployment. See Figure 28. The discriminating sensors, wired in parallel with each other, were located on the left and right front fender aprons, and at the top center of the radiator support, as indicated conceptually in Figure 29.

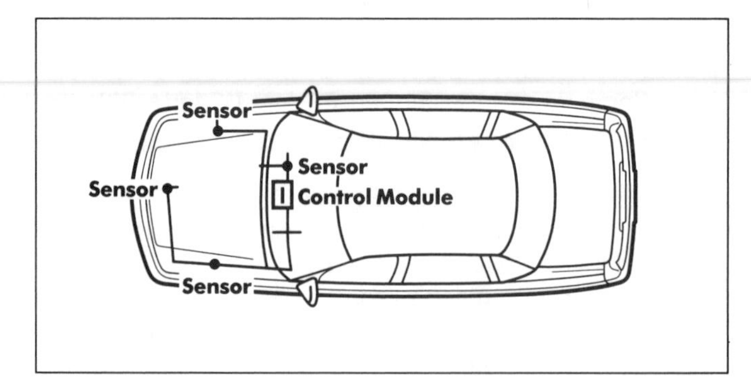

Figure 29. Schematic of Distributed-Sensor System

The sensors were manufactured by the Breed Corporation, and employed a ball that, at a specified deceleration, would pull away from a bias magnet and move through a tube toward a set of electrical contacts, which would have to be bridged to activate the inflator. A small clearance gap between the ball and the tube damped the response to higher-frequency accelerations. This concept is shown in Figure 30. It was simple and reliable, but sensitivity to cross-axis accelerations and temperature variations made its performance more difficult to predict in low speed impacts.

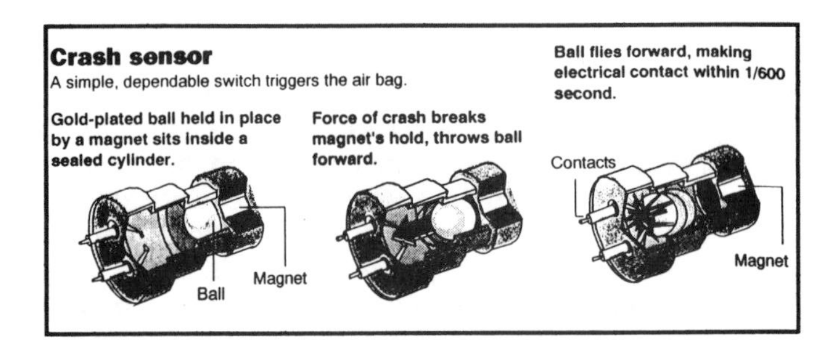

Figure 30. Ball-in-Tube Sensor Function

SENSOR CALIBRATION - Beyond meeting the specifications for deployment threshold, there is a second requirement that influences how sensors are calibrated: sensing time, or time-to-fire. Recall that the sensing time budget is typically 25 or 30 milliseconds. In this time frame, the discriminating sensors must make their decision. Figure 31 shows the time history of ΔV accumulation for four hypothetical situations involving different speeds and objects struck. The upper curve describes an obvious deployment accident, and it has distinguished itself from the others by 15 msec or so. The remaining three curves include one other deployment event and two for which deployment is probably not warranted; yet in the first 30 msec of the crash it is difficult to tell which is which.

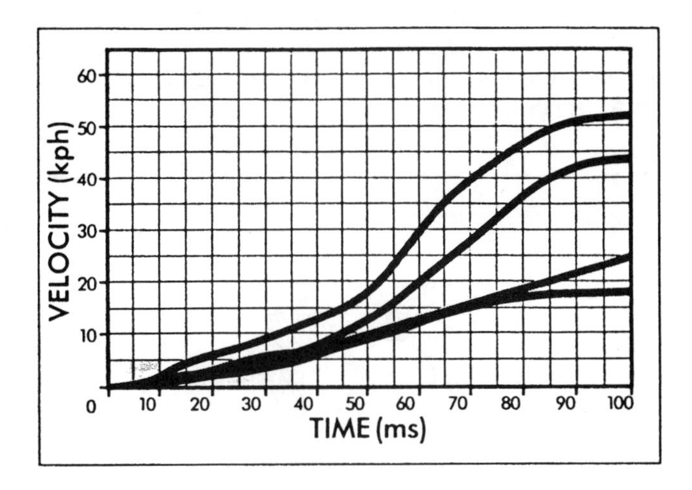

Figure 31. Typical Deployment and Non-Deployment Events - NEC

Recognizing that the crash pulse, and hence the ΔV accumulation, depend on where the sensor is placed, one can attempt to sort out the curves of Figure 31 in a shorter time by placing a the discriminating sensors at a variety of locations on the vehicle. Note, however, that the ΔV required to trigger the sensors will be well below the ΔV ultimately experienced in the crash.

We can illustrate the considerations involved by assuming that a sensor is a ΔV switch; i.e., that its trigger decision is made solely on the basis of the ΔV seen at the sensor location. (As indicated previously, this is not a bad assumption for traditional mechanical sensors.) Figure 32 shows the ΔV accumulation in a typical 35 mph barrier crash. If one needed a 25 msec sensing time for such conditions, and if one had a sensor at a rear location (such as near the B-pillar), one would need about a 10 mph trigger level. If that level proved to be too low (for deployment threshold considerations, say), then one could choose to move the sensor to a forward location; there, it could achieve a 25 msec sensing time with a trigger level of just over 16 mph. Alternatively, the same trigger level, in the forward location, would reduce the sensing time to just over 19 msec.

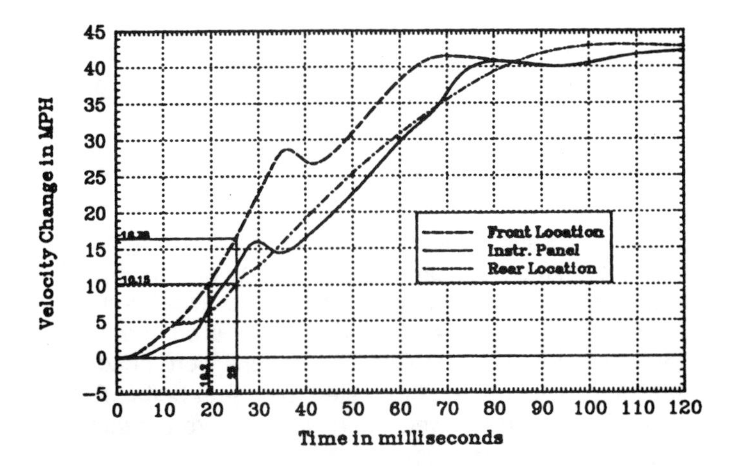

Figure 32. Typical Structural Responses for 35 mph Barrier Crash

The results of these considerations are seen in the design of the Ford Tempo/Topaz system. Calibration variables for the sensors included the distance between the ball's rest position and the contacts, its clearance between the guide tube inner walls, and the strength of the bias magnet. These variables were tuned by subjecting them to haversine acceleration pulses of various amplitudes and durations, worked out by Breed to meet Ford's requirements regarding deployment threshold and sensing time. For the discriminating sensors, the "must deploy" threshold in 14 mph barrier crashes was tested with "soft," "medium," and "hard" haversine pulses at 10 mph, and the "no deploy" threshold in 8 mph barrier crashes was tested with "soft," "medium," and "hard" pulses at 6.5 mph. Safing sensor calibration pulses were at 3.5 mph for the forward sensor and 1.1 mph for the aft one. At Ford, each and every sensor was tested twice: once during the manufacturing process and again after final sensor assembly.

This distinction between safing and discriminating sensors, and the use of haversine pulses to test their calibration, has become typical of sensor system design to the present.

TYPES OF SENSOR DESIGNS - Airbag sensor concepts since the '70s have shown their origins in the GM Bumper Impulse Detector: an inertial mass being required to overcome a bias force (from a magnet, for example) and travel a certain distance before electrical contact is made. One implementation of this concept was the ball-in-tube device described previously. Another was the widely used "rolamite" sensor developed by the Technar Corporation (now a part of TRW). See Figure 33. A roller was wrapped with a spring band, which tended to keep the roller pushed against a stop. In a crash, the roller moved forward without slipping (or damping), unrolling the spring band in so doing, and covering the distance to the electrical contacts (again, gold-plated) if the crash was severe enough. The rolling action reduced friction to a minimum; the surface on which the rolling occurred was slightly arched. The calibration level was determined by the spring band and the mass of the roller. For a particular application, each sensor would be calibrated via a set screw, adjusting the position of the roller at rest.

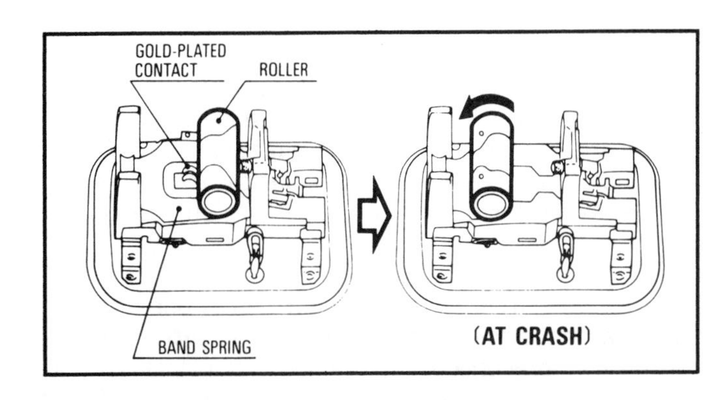

Figure 33. Rolamite Sensor Action - Honda

Another electro-mechanical sensor is the gas-damped diaphragm type, shown in Figure 34. A diaphragm, to which the inertial mass is mounted, provides both the bias force and segregation of the gas volumes. Gas is permitted to pass from one chamber to another through one or more orifices, creating a damped response.

As noted previously, such systems are electrical switches that respond to velocity change, by perform a mechanical integration of acceleration. Another approach would be to measure the strain in the bias spring, by making that spring a piezoresistive or piezoelectric element in a solid state electrical circuit. This results in a very small device in which the only motion is beam bending. The acceleration can be integrated digitally, or processed in other ways. For example, acceleration and velocity could be used together in some mathematical formula or algorithm [NEC 95].

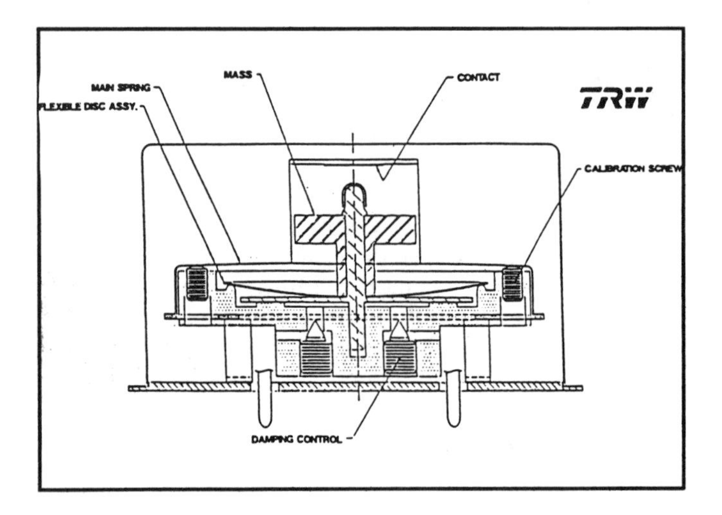

Figure 34. Gas-Damped Sensor - TRW

Electronic sensor systems could be tuned to a particular vehicle by adjusting the parameters or code stored in firmware. This approach opens the door to a great deal of design flexibility, and does so with reduced weight, cost, and complexity (and thus increased reliability). At this writing, such systems are common, and there is every reason to expect them to predominate in the future.

FROM MULTI-POINT TO SINGLE-POINT SYSTEMS - Both the early GM and Ford systems involved multiple sensors at multiple locations, and we have seen how such a strategy results in different calibrations for different sensors. It also results in a considerable wiring harness, which, as we have seen before, entails special care to maintain reliability.

It is desirable, from both cost and reliability perspectives, to decrease the amount of wiring in the air bag system. A Breed Corporation concept carried this objective to its logical conclusion: a purely mechanical system, located at the inflator, that both sensed the crash and initiated the bag inflation [Breed 85]. In this system, a spring-loaded firing pin was held in place by a lever, which itself was held in place by a bias spring. See Figure 35. The lever could be moved by a sensing mass, and if the motion were sufficient, the firing pin would be released, and propelled into a stab primer.

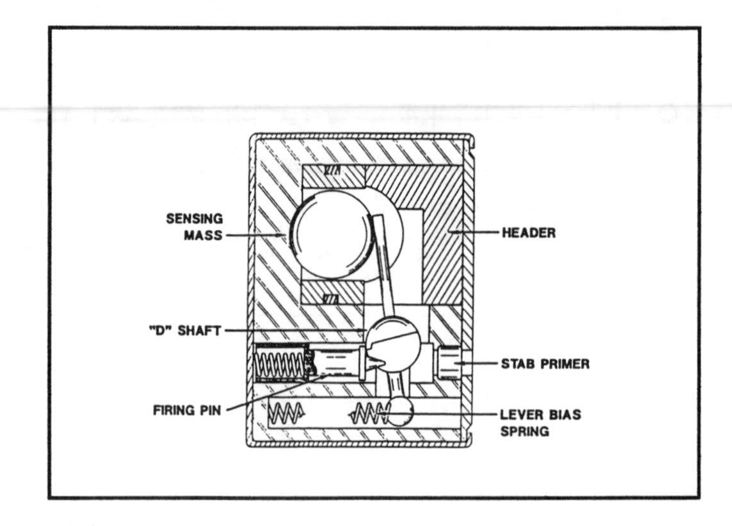

Figure 35. All Mechanical Sensor - Breed

Having no sensors any closer to the crush zone than the steering wheel hub caused some concern regarding sensing time. However, the reduction of axial play in the column produced sensing times on the order of 30 milliseconds in 30 mph frontal barrier tests. The system was intended to be retro-fitted into Chevrolet Impala police vehicles.

The general trend, however, has been to stay with electrical sensing elements, so that the increasing computational power, cost-effectiveness, space efficiency, and reduced weight of digital electronics can be used to fullest advantage. This approach allows the integration of sensors, diagnostics, backup power supply, and the logic elements into a single sealed unit. These are known as single-point systems, although separate safing and discriminating sensors are generally employed. Since the electronic environment is much less hostile in the occupant compartment than in the engine compartment, the air bag electronic module is often placed on the tunnel, near or at the dash panel. Because no information comes directly from the crush zone, increased emphasis is placed on signal processing and digital logic.

CONCLUSION

Among all safety system concepts that have ever found their way into production automobiles, the air bag arguably represents the most dramatic departure ever taken from traditional automotive technology. It started in a time of technological optimism when Americans were heading to the Moon, yet refusing to fasten their seat belts. At that time, the air bag seemed like a way to solve a behavioral problem by technological means.

Since then, many nations have shown the way in achieving widespread belt use, and American belt use, while still far behind, has risen to levels not imagined in the '70s or early '80s. Belt systems have thus garnered much of the safety benefit that was assumed to be available to air bags. At the same time, the operating airbag concept has changed from being the only restraint widely used, to being strictly supplementary to the belts. We find that air bags are playing a more limited role, and are contending for a smaller portion of the available safety benefit.

Notwithstanding these developments, air bags were greeted enthusiastically by the public (with perhaps some help from the various passive belt designs). They became almost a litmus test of automotive safety, The development of these devices, and their availability in large numbers, represent astounding technological achievements, and are tributes to the many thousands of individuals, beyond the few cited in this paper, who made it all happen. Air bags are now busily saving lives and reducing injuries, though not necessarily in the numbers originally envisioned.

However, there have also been continued warnings of technological problems and potential adverse side effects, and some of these have come to pass as vast numbers of airbag-equipped vehicles have taken to the road. We are now engaged in redoubled efforts to address these, and the near future will contain many new developments. The work is not yet finished.

CONTACT

Donald E. Struble, Ph.D.
Collision Safety Engineering, Inc.
2320 West Peoria Avenue, Suite B-145
Phoenix, AZ 85023 Phone: (602) 395-1011

REFERENCES

[Alexander 74] *An Evaluation of the U.S. Family Sedan Experimental Safety Vehicle (ESV) Project*, GH Alexander, RD Vergara, JT Herridge, W Millicovsky, and MR Neale, Final Report, Contract DOT-HS-322-3-621-1, October 1974.

[Breed 85] "The Breed All-Mechanical Airbag Module," A Breed, SAE Paper 856014, *Proceedings, Tenth International Technical Conference on Experimental Safety Vehicles*, 1985.

[Campbell 72] "Air Cushion Restraint Systems Development and Vehicle Application," DD Campbell, SAE Paper 720407, May 1972.

[DeLorenzo 86] "Supplier cuts off air bag retrofitter - Lack of adequate liability insurance cited," M DeLorenzo, Automotive News, 10 October 1986.

[DOT 78] "New DOT Study Finds Only 14% of Drivers Use Auto Safety Belts," Press Release, U.S. Department of Transportation, 15 December 1978.

[Edwards 91] "A Preliminary Field Analysis of Chrysler Airbag Effectiveness," WR Edwards, *Proceedings, Thirteenth International Technical Conference on Experimental Safety Vehicles*, November 1991.

[FMVSS 201] National Highway Traffic Safety Administration, FMVSS 201, *Interior Impact Protection.*

[Frey 70] "History of Air Bag Development," SM Frey, *Proceedings, International Conference on Passive Restraints*, May 1970.

[Jones 70] "Inflatable Passive Air Restraint System Crash Sensors," TO Jones, *Proceedings, International Conference on Passive Restraints*, May 1970.

[Jones 71] "Crash Sensor Development," TO Jones and OT McCarter, SAE Paper 710016, January 1971.

[Katter 75] *Development of Improved Inflation Techniques*, LB Katter, Final Report, Contract DOT-HS-344-3-690, September 1975.

[Klove 72] "Special Problems and Considerations in the Development of Air Cushion Restraint Systems," EH Klove, Jr. and RN Oglesby, SAE Paper 720411, May 1972.

[Kobayashi 87] "Reliability Considerations in the Design of an Air Bag System," S Kobayashi, K Honda, and K Shitanoki, *Proceedings, Eleventh International Technical Conference on Experimental Safety Vehicles*, May 1987.

[Kossar 92] "Air Bag Deployment Characteristics," LK Sullivan and JM Kossar, Final Report No. DOT HS 807 869, February 1992.

[Lau 93] "Mechanism of Injury from Air Bag Deployment Loads," IV Lau, JD Horsch, DC Viano, and DV Andrzejak, *Accident Analysis and Prevention*, Pergamon Press, Vol. 25, No. 1, February 1993.

[Loukes 73] "General Motors Driver Air Cushion Restraint System," TN Louckes, RJ Slifka, TC Powell, and SG Dunford, SAE Paper 730605, May 1973.

[Maugh 85] "Supplemental Driver Airbag System - Ford Motor Company Tempo and Topaz Vehicles," RE Maugh, SAE Paper 856015, July 1985.

[Mertz 88] "Restraint Performance of the 1973-76 GM Air Cushion Restraint System," HJ Mertz, SAE Paper 880400, February 1988.

[NEC 95] Advertising material from NEC Technologies, Inc., 1995.

[NHTSA 96] *Final Rule on Labels*, National Highway Traffic Safety Administration, 27 November 1996.

[NHSB 70] National Highway Safety Bureau: Proposed Amendment to Motor Vehicle Safety Standard 208, *Occupant Crash Protection*, 3 November 1970.

[Reidelbach 79] "Advanced Restraint System Concepts," W Reidelbach and H Scholz, SAE Paper 790321, February 1979.

[Romeo 75] "Front Passenger Passive Restraint for Small Car, High Speed, Frontal Impacts," DJ Romeo, SAE Paper 751170, November 1975.

[Romeo 78] *Front Passenger Aspirator Air Bag System for Small Cars*, DJ Romeo, Final Technical Report, Phase II, Contract DOT-HS-5-01254, March 1978.

[Romeo 84] "Driver Air Bag Police Fleet Demonstration Program - A 15-Month Progress Report," DJ Romeo and JB Morris, SAE Paper 841216, October 1984.

[Seiffert 72] "Development Problems with Inflatable Restraints in Small Passenger Vehicles," UW Seiffert and GH Borenius, SAE Paper 720409, May 1972.

[Smith 73] "The 1,000 Car Air Cushion Field Trial Program," GR Smith and MR Bennett, *Proceedings, Automotive Safety Engineering Seminar*, June 1973.

[Struble 79] "Status Report of Minicars' Research Safety Vehicle," DE Struble, *Proceedings, Seventh International Technical Conference on Experimental Safety Vehicles*, June 1979.

950338

Investigation of Improving Energy Absorption Performance and Reducing Weight of Passenger Air Bag Modules Using Computer Aided Analysis

Edward Wilson
Mitsubishi Motors Corp.

ABSTRACT

A computer analysis method using software for nonlinear dynamic finite element calculations has been developed to perform impact simulation on a passenger air bag module. The effects of the following parameters on energy absorption performance are investigated:

1) Air Bag Module Door Bead and Thickness
2) Reaction Can Material and Plate Thickness

From the results of this parameter study, design criteria are determined for achieving a low weight design with optimum energy absorbing characteristics.

INTRODUCTION

In the 1960's and early 1970's many instrument panels were modified by adding padding and reducing structural rigidity to increase the passive safety afforded to an unrestrained front passenger. According to a study done by the National Highway Traffic Safety Administration, these improvements reduced fatality risk and serious injury risk by nearly 25 percent for unrestrained right front passengers of cars in frontal crashes. In the car of the 90's, passenger side air bags further reduce the risk to occupants involved in a severe frontal crash. Nevertheless, it is still necessary to provide the passenger with protection in the case where the collision is not severe enough for the air bag to deploy.

A passenger side air bag, while increasing the passive safety capability provided by an automobile involved in a major collision, presents a formidable task for the engineer concerned with developing an energy absorbing instrument panel. The passenger air bag is located in what is called an air bag module. This module houses the air bag and actually forms part of the instrument panel surface. It has to be strong enough to withstand the forces of the deploying air bag and at the same time, be able to absorb impact energy from an occupant hitting the instrument panel in the case where the air bag does not deploy.

When designing an air bag module, many development tests are performed to ensure appropriate safety levels. To meet these performance criteria, several parameters of basic air bag module design are investigated by actual tests, but due to time limitations, an easier method is needed. For this reason, a computer analysis method is developed where design parameter changes can be easily analyzed without having to actually conduct the tests. Next, a parameter study is performed to investigate improving energy absorption performance and reducing weight of a representative passenger air bag module.

THE AIR BAG MODULE

The air bag module used as the base for this analysis consists of: a steel cage called the reaction can which controls the trajectory of the air bag upon deployment; an inflator used to inflate the air bag; the folded air bag itself; and an aluminum door coated with urethane foam which sits flush with the instrument panel surface, allowing an opening for the deploying air bag. The reaction can consists of pressed steel parts that are either welded or bolted together. The inflator is held in place with a steel frame and bolts that connect at the rear of the reaction can. The air bag itself is folded and bolted to the reaction can with pressed steel plates. The door is screwed to the upper part of the reaction can and the lower edge is held in place with two plastic tabs which tear at inflation to enable the air bag to deploy. See Figure 1.

Part Name	Number In Figure
aluminum door	1
air bag	2
inflator	3
plastic tear	4
reaction can	
upper retainer component	5
lower retainer component	6
middle retainer component	7
retainer component top A	8
retainer component top B	9
air bag plate (2) top, bottom	10
air bag plate (2) side	11
reaction can support brackets	
A (to crossbeam)	12
B (to instrument panel core)	13
C (to instrument panel core)	14

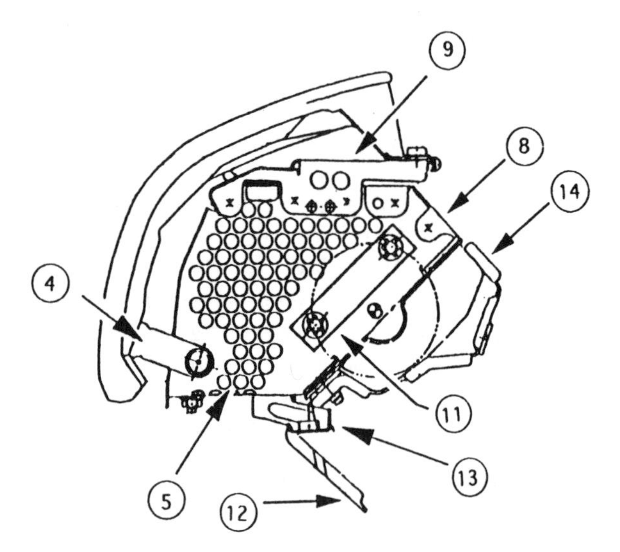

Side View

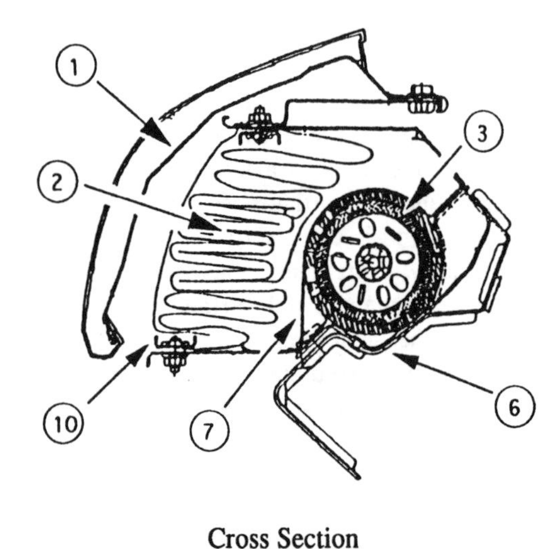

Cross Section

Figure 1. Air Bag Module

ANALYSIS MODEL

The analytical model for this analysis was developed so that it would behave like the actual physical system upon impact. This was done by modeling and defining the following criteria as accurately as possible: physical properties, material properties, boundary conditions and loads. The analysis was done with nonlinear dynamic finite element calculation software.

PHYSICAL PROPERTIES - The finite element modeling was done using 4 node shell elements to provide accuracy upon deformation. Also, 3 node shell elements were used where necessary to maintain geometrical accuracy. The total number of elements is about 10,000. The complete model is shown in Figures 2 and 3.

Air Bag Module Model - Each part of the air bag module as shown in Figure 1 was modeled separately and to the respective dimensions for each part. Bolts, nuts and spotwelds were not modeled physically, but were represented by nodes.

This model does not include the following: the inflator, the actual air bag, the two plastic tears, the urethane coating on the door and the surrounding instrument panel. The inflator is left out of the model because it was not considered to be in the deformation region due to its location at the rear of the reaction can.

The air bag, the plastic tears and the urethane coating on the door are not included because their effect on energy absorption is rather small and because of the difficulties encountered in modeling these components. In order to specifically investigate the influence of the reaction can on the energy absorbing properties of the air bag module, the surrounding instrument panel is also not included in this model.

Head Form Model - The head form model is a model of the head form that is used in development testing. It does not deform upon impact, so it is defined as a rigid body.

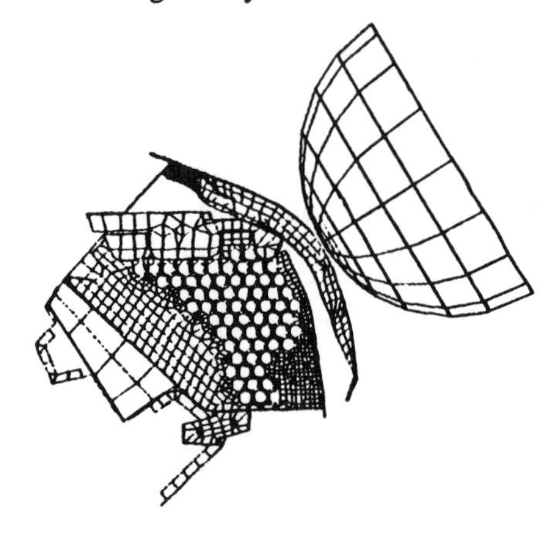

Figure 2. Analysis Model with Head Form

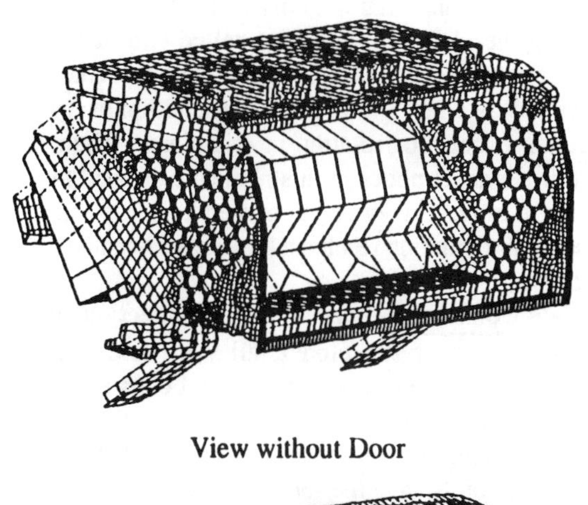

View without Door

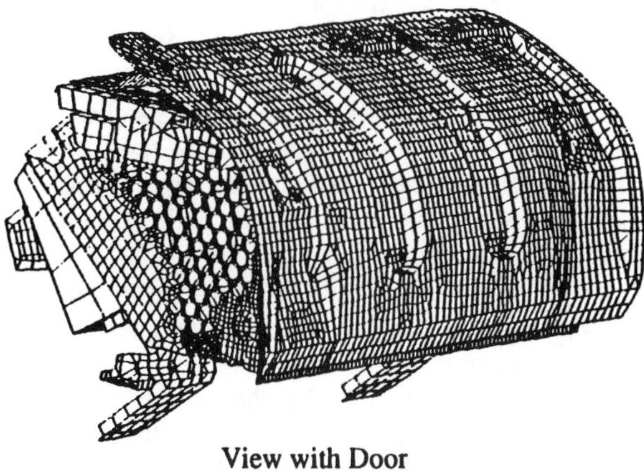

View with Door

Figure 3. Air Bag Module Analysis Model

Reaction Can Hole Modeling - The reaction can has punched holes for the purpose of weight reduction and energy absorption. These proved a challenge to model correctly with elements small enough to give realistic deformation, but big enough to keep the calculation time below an acceptable level. In reality, correctly modeling the holes involves modeling the area around the holes. Therefore it is important to remember that the weakest point of this area will be the closest space between each hole. This has been represented by having the elements join at this point in the model.

Calculations using a trial model were performed to determine the best configuration of elements to use to model these holes. Several different configurations of elements were tried, including those shown in Figure 4.

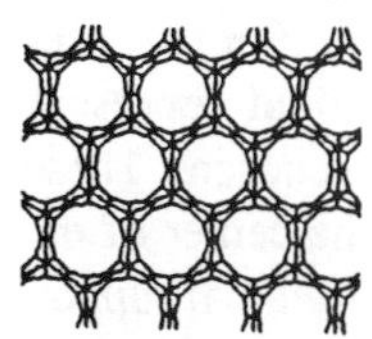

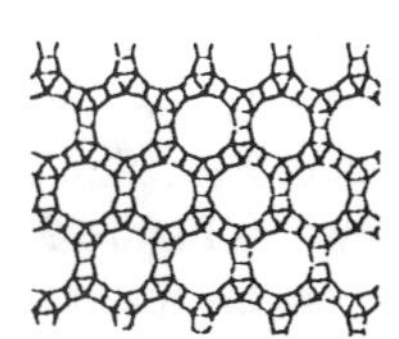

HOLE MODEL A HOLE MODEL B

Figure 4. Hole Models

The trial model consisted of a model of the reaction can without brackets, rear inflator support structures or air bag retainer brackets. The rear of the model was restrained to prevent any movement. This was impacted with the head form for each hole model. The rigid head form was constrained to move in the impact direction only, at an initial impact speed of 12 m.p.h. (5.4 m/s). This impact direction is similar to that used for development testing purposes. The head form weight is 6.8 kg. The trial model for Hole Model B is shown in Figure 5.

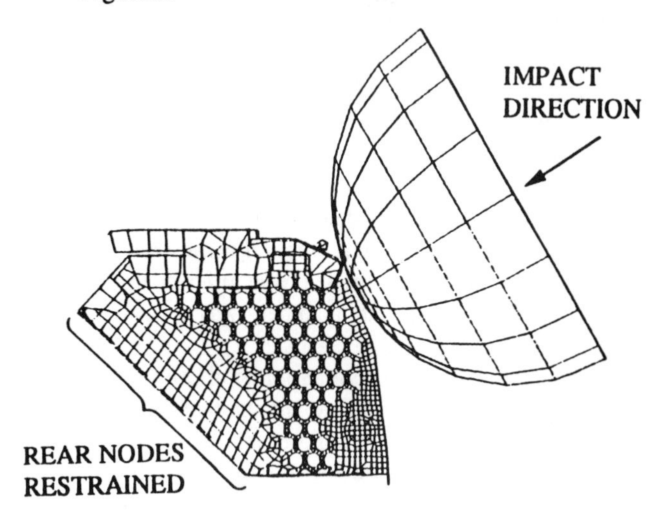

Figure 5. Trial Model for Hole Model Comparison Testing

The results for the trial models for each hole design are shown in Figure 6. The calculation times for each model for impact up to about 15msec are also shown.

Calculation Time

Hole Model A: 15 hrs.
Hole Model B: 5 hrs.

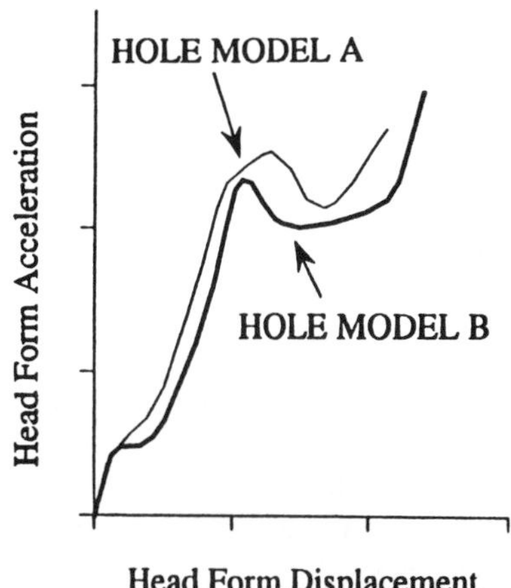

Figure 6. Trial Model Analysis Results

As can be seen from these curves, Hole Model A with two elements between each hole supports greater force, but the difference is small. A parameter study is conducted for this investigation, so it is necessary to have a model with as short a calculation time as possible. In this case, the trial model for Hole Model B can be solved in one-third of the time it takes to solve Hole Model A. For the full analysis model, Hole Model A would result in unrealistic calculation times to be able to conduct a parameter study, so Hole Model B was determined to be the best for this study.

MATERIAL PROPERTIES - The material properties for the aluminum and steel parts of the air bag module were defined using data taken from actual testing or from recognized data sources. These properties include: modulus of elasticity, yield stress, plastic tangent modulus, plastic stress, Poisson's ratio and mass density. The Cowper-Symonds strain rate law was applied using strain rate parameter values determined from in-house test data. Also, the actual plate thickness was defined in the input data file.

In regard to the head form, a material property is used which does not calculate the internal forces in the head form which are unnecessary for this analysis. This results in a savings in calculation time.

BOUNDARY CONDITIONS - The boundary conditions were defined in the data input file.

Reaction Can Bracket Restraints - The reaction can brackets were restrained at the contact location between the bracket and supporting test jig.

These boundary conditions are a bit complex due to rotation around the connection bolt as shown in Figure 7 for the reaction can support bracket B.

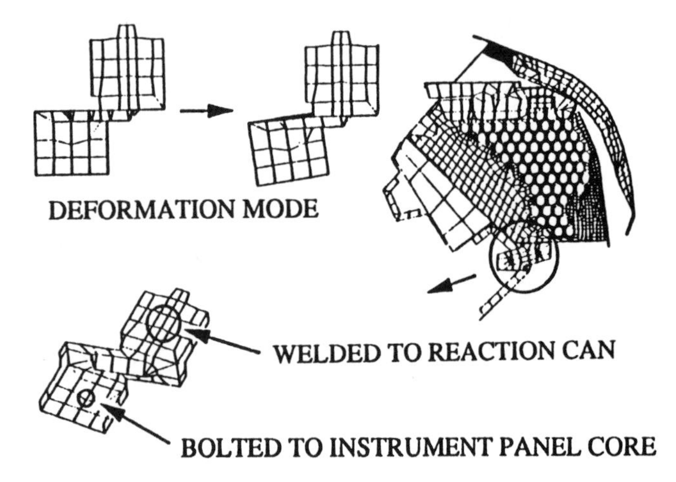

Figure 7. Close-up of Reaction Can Support Bracket and Deformation Mode (Analysis Results)

The rotation above occurs in the actual test, so it was deemed necessary to reproduce this in the calculation. This rotationability is defined by first defining the plane in which the bolted end of the bracket contacts the test jig. Next, the bolt center is defined by a node which is constrained in every direction except for rotation in this plane. Then, the other bracket nodes in this plane are defined to be free to move anywhere but perpendicular to this plane. This definition results in the desired boundary condition.

Plate Connections - As for the connection of different parts of the model, all weld and bolt connections were represented by nodal constraints or rigid bodies which are defined in the calculation input data file. These constraints ensure that the connected points have the same motion.

Contact Definitions - The contact between the different parts is accomplished by means of slide contact definitions. These contacts are also used between the air bag module and the head form. A self-contact sliding interface algorithm is used. This allows for all of the contacts to be defined with only one contact card.

Head Form Motion - The head form was constrained to move in the impact direction, normal to the aluminum door.

LOADS - A mass of 6.8 kg, the same as in the test, is applied to the head form. Next, the head form is given an initial velocity of 12 m.p.h. (5.4 m/s).

By recording the displacement vs. time history of the head form, the force vs. displacement curve for the corresponding analysis can be computed.

TEST CONDITIONS FOR MODEL VERIFICATION

The test conditions used to verify the model for this analysis are the same conditions used when doing compliance testing for FMVSS 201: "Occupant protection in interior impact." The purpose of this standard, as defined by the National Highway Traffic Safety Administration, Department of Transportation, is "to afford impact protection for occupants," and applies to certain interior components, including portions of instrument panels of all passenger cars, MPVs and Trucks. Figure 8 depicts these tests.

For development and compliance tests, many impact points are chosen for testing. For purposes of this study, two points that would represent different deformation scenarios are chosen. These are shown in Figure 9. A2 is in the center of the door where the reaction can offers minimal support for impact. A1 is on the door at the point where the reaction can is directly behind the door, representing a structurally stiff point.

Impact speed: 12 m.p.h. (5.4 m/s)
Impactor: 6.5" (0.165 m) diameter head form
Impact angle: normal to a plane tangent to the surface at the point of contact

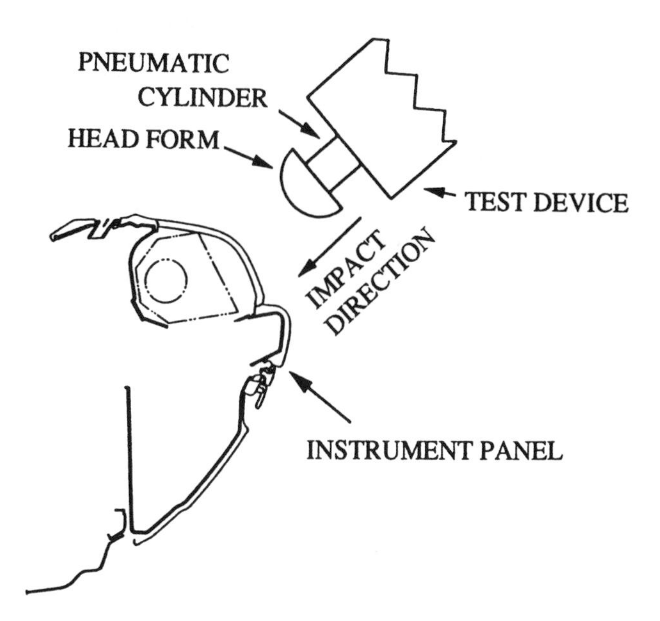

Figure 8. Schematic of Test Procedure

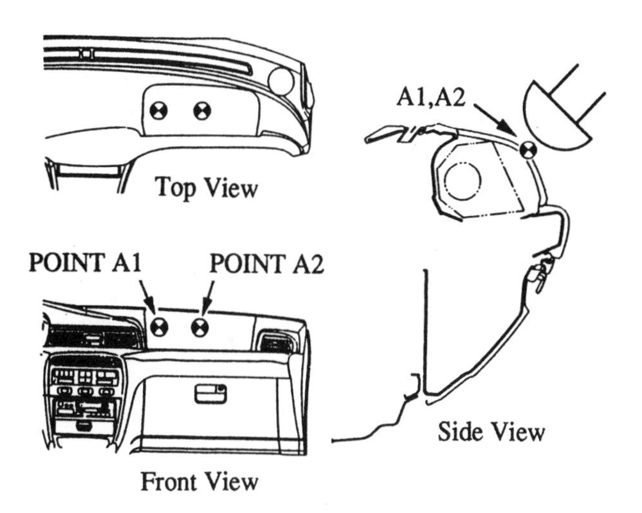

Figure 9. Impact Points for Model Verification

TEST VS. ANALYSIS RESULTS

Graphs of the analysis and test results are shown in Figures 10 and 11. These graphs show the acceleration vs. displacement curves of the head form for the corresponding impact points. Although a small disparity can be observed, these results are very good for this calculation model. Also, the deformation mode of the analysis is very similar to that of the test as can be seen in Figures 12 and 13. Therefore, this model is judged to be adequate for conducting a parameter study.

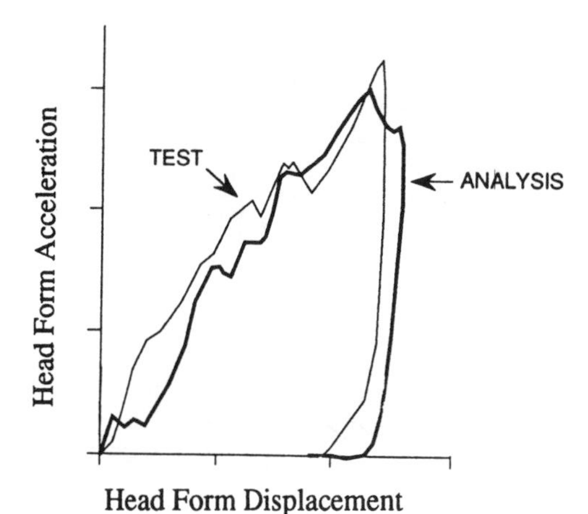

Figure 10. Acceleration vs. Displacement Curve for Impact Point A1

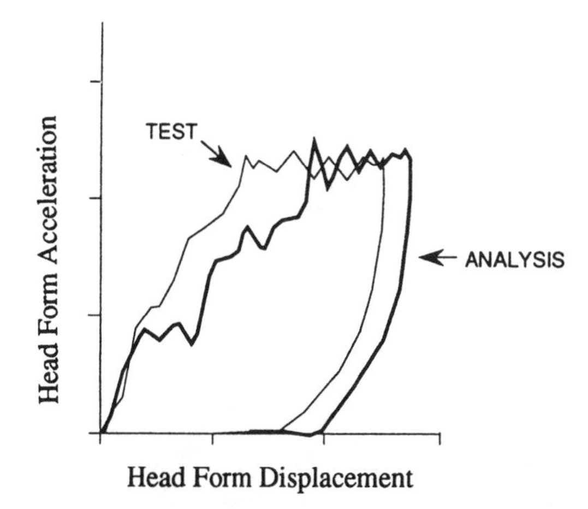

Figure 11. Acceleration vs. Displacement Curve for Impact Point A2

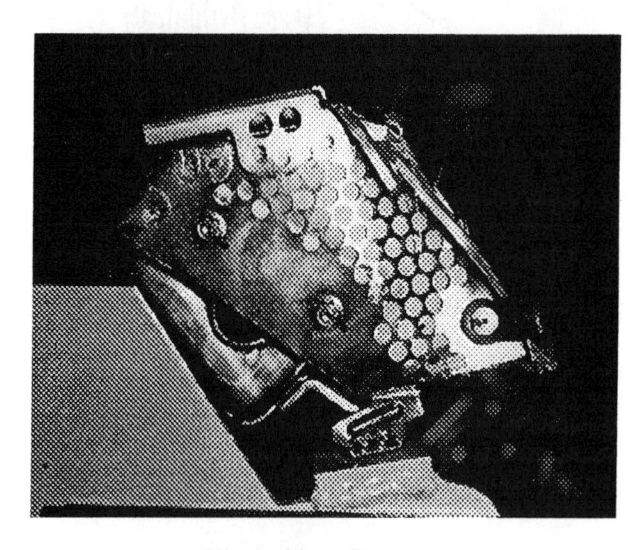

View without Door

Figure 12. Post Impact Deformation of the Air Bag Module for Impact Point A1 (Test)

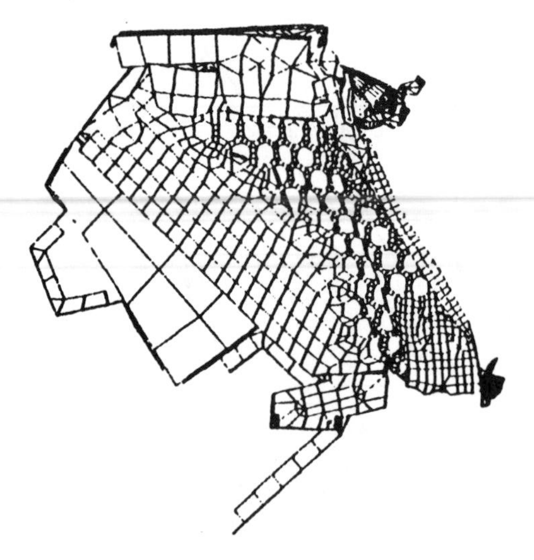

View without Door

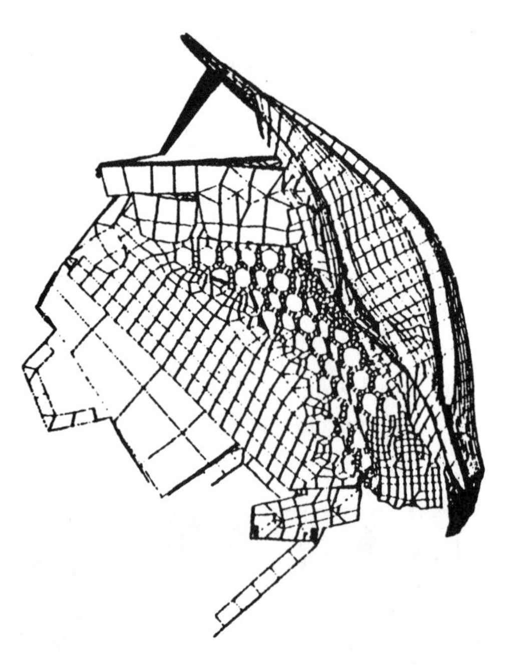

View with Door

Figure 13. Post Impact Deformation of the Air Bag Module for Impact Point A1 (Analysis)

TEST RESULTS OF THE FULL INSTRUMENT PANEL VS. THE AIR BAG MODULE ONLY

Development testing is usually done on a complete passenger side instrument panel. For this analysis, only the effect of changes made to the air bag module in the instrument panel is analyzed. Therefore, in order to see how these changes would relate to a test on the full instrument panel, it is important to compare the difference between impact tests done on a complete instrument panel to those done on only the air bag module.

For the impact test on the full instrument panel, the instrument panel is placed similarly to how it would be in an actual vehicle. The air bag module tests on the other hand are done with the air bag module in a fixed jig, without the air bag or the inflator. Also, the surrounding instrument panel core is not included in this test configuration, so any data obtained from testing or analysis of just the air bag module has to be interpreted for the case where the full instrument panel is tested.

From this comparison, the desired impact characteristics of the air bag module are determined. These are then used to analyze the results of the parameter study.

The results of these tests for the corresponding impact points are shown in Figures 14 and 15.

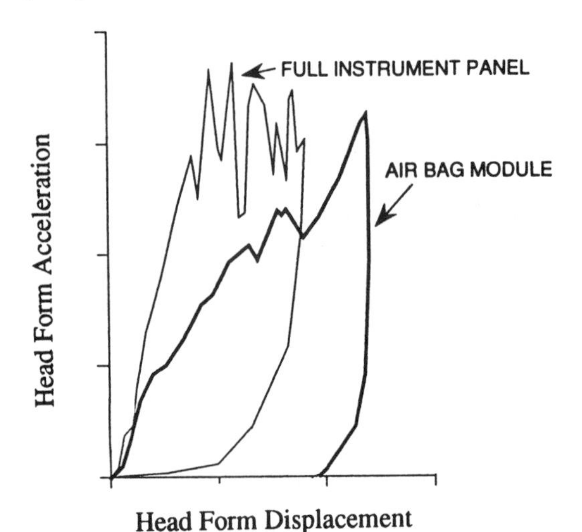

Figure 14. Full Instrument Panel vs. Air Bag Module only Test Results for Impact Point A1

From this graph, it can be seen that there is a significant difference between these two tests. The impact into the full instrument panel is more severe. For impact point A1, the greater acceleration value is attributed to the fact that the aluminum door of the air bag module is also supported by the instrument panel core which it partially overlaps.

This accounts for the greater displacement of the head form upon impacting the air bag module only. For this impact point, sufficient energy has to be absorbed with minimal displacement and with as low an acceleration level as possible.

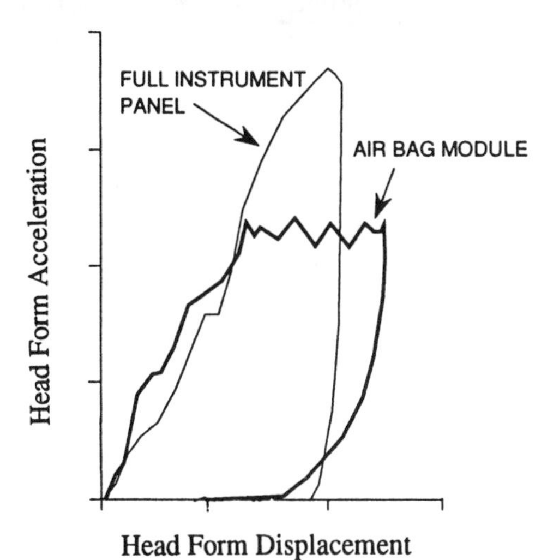

Figure 15. Full Instrument Panel vs. Air Bag Module only Test Results for Impact Point A2

For impact point A2, in the center of the air bag module, the instrument panel core adds some support at the sides and the lower edge of the door, and there is also an air bag present. For impact at the center of the door, the door most be strong enough to absorb impact energy so that contact with the air bag is limited, but have sufficient energy absorption characteristics to prevent the acceleration level from being too high.

PARAMETER ANALYSIS

Now that the model has been verified and the desired energy absorbing characteristics have been determined, a parameter study is conducted. This study shows the ease with which this model can be used to analyze various impact conditions or design changes. The following parameters are investigated for possible ways to improve the energy absorption performance. Refer to Table 1.

Parameters:
1) Air Bag Module Door Bead and Thickness
2) Reaction Can Material and Plate Thickness

For purposes of this study, the following evaluation criteria are used to investigate the effect of a certain design change. These are used to determine the energy absorption performance of the air bag module. As for weight reduction, if less material or a lighter weight material can be used to get the same or better results, then this change could result in substantial weight savings.

1) Amount of Energy Absorption - Using the force vs. displacement curve of the head form, this is defined as the total energy absorbed up to a certain displacement value that varies with the respective parameters. This is outlined in Figure 16.

2) Maximum G - Examined to ensure that the maximum G level will not exceed that necessary to meet regulation (<80G for 3msec). For this purpose, given that the G level will probably be higher when the air bag module is tested with the full instrument panel, a much lower maximum level is necessary.

Table 1. Parameter Study Items

Item	Parameter		Contents
Air Bag Module Door	Number of Beads	0,3,4,5	
	Thickness	±50%	
Reaction Can	Material	Steel Aluminum	
	Plate Thickness	±50%, - 25%	

3) Maximum Displacement - For a design to be beneficial, it must absorb the most energy with the least displacement while not exceeding a specific G level. For this analysis, since the head form should not bottom out upon impact, a maximum displacement value is necessary.

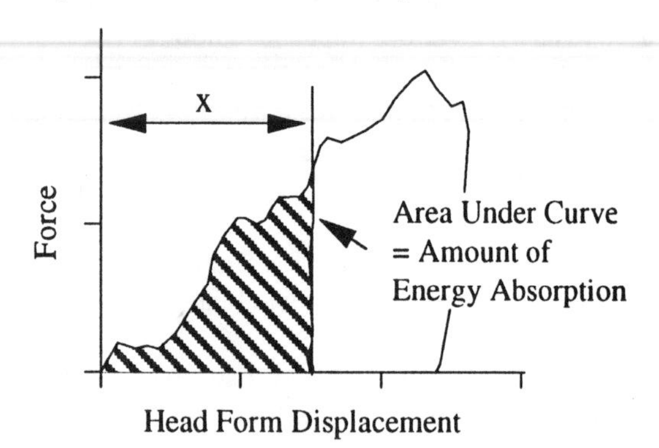

Figure 16. Amount of Energy Absorption Example

PARAMETER ANALYSIS RESULTS

1) AIR BAG MODULE DOOR BEAD AND THICKNESS - First, the effect of the number of beads in the air bag module door, while keeping the depth and width of these beads as constant as possible, is investigated. Next, the effect of the door thickness is determined. These items are analyzed with respect to the amount of energy absorbed for the displacement region where the effect of the door is the greatest for this air bag module, as shown in Figure 16. This analysis is based on results from impact point A2 in the center of the door, where changes to the door will have the greatest effect.

The air bag module doors are shown in Table 1. The analysis model door has 4 beads. For comparison, doors with no beads, 3 beads and 5 beads were modeled.

Door Bead Influence - The relationship between the number of door beads and amount of energy absorption is shown in Figure 17.

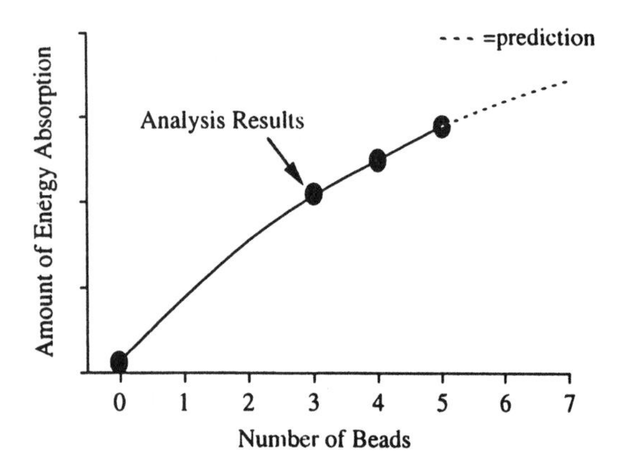

Figure 17. Door Bead Influence

From this graph, it can be seen that the relationship between the number of door beads and the amount of energy absorption is almost linear. By increasing the number of beads, the energy absorption performance can be improved.

Door Thickness Influence - The relationship between door thickness and amount of energy absorption is shown in Figure 18.

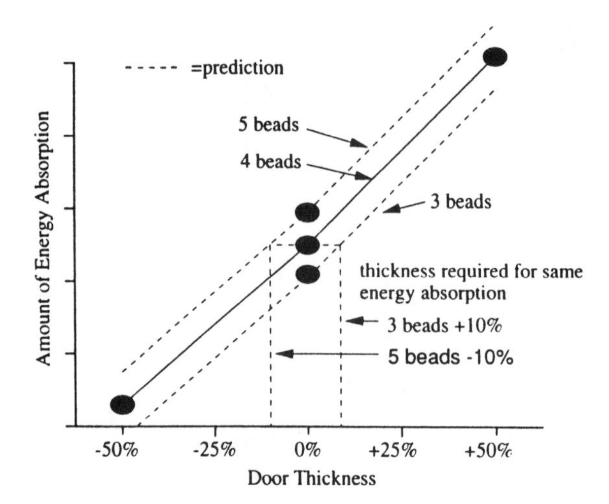

Figure 18. Door Thickness Influence

From these results, it can be seen that the relationship between door thickness and amount of energy absorption is almost linear for the door with 4 beads. Considering the difference in energy absorption between the 3, 4 and 5 bead doors, the energy absorption in regard to door thickness can be predicted for the 3 and 5 bead doors, as shown by the slanted dotted line. From this prediction, the door thickness required for the same amount of energy absorption can be determined. This is also shown in Figure 18. By increasing the number of beads, the thickness of the door can be reduced without any loss in energy absorption performance. For the case with 5 beads, this results in a weight reduction of 10% for the air bag module door.

2) REACTION CAN MATERIAL AND PLATE THICKNESS - The effect of using aluminum instead of steel for construction of the reaction can is investigated. Also, the effect of changing the reaction can plate thickness, without changing the air bag door or bracket properties, is determined. This analysis is based on results from impact point A1 where the reaction can is directly behind the door. In regard to the amount of energy absorption, this is the total energy absorbed for each case when the head form displacement value is equal to that of the case with the least displacement.

For the steel reaction can, the plate thickness is increased by 50%, and decreased by 25% and 50%.

The calculations using aluminum material definitions are done for the unchanged thickness, +50% and -50%.

Relationship with Amount of Energy Absorption - These results are shown in Figure 19.

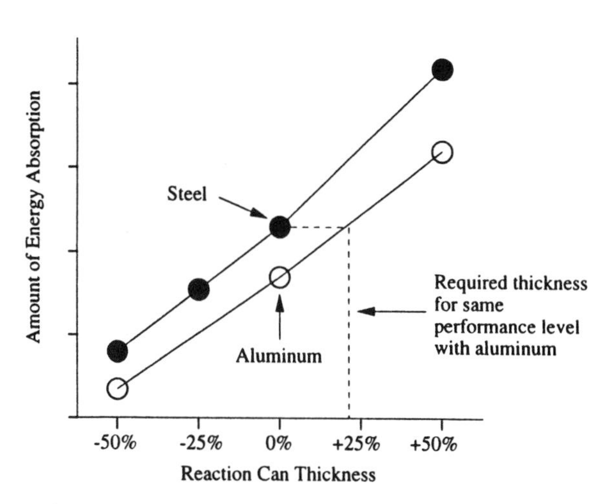

Figure 19. Influence on Energy Absorption

From this graph, it can be seen that as the plate thickness of either the aluminum or steel reaction can is increased, the amount of energy absorption also increases.

An analysis example showing the effect of plate thickness on force/head form displacement curves is shown in Figure 20.

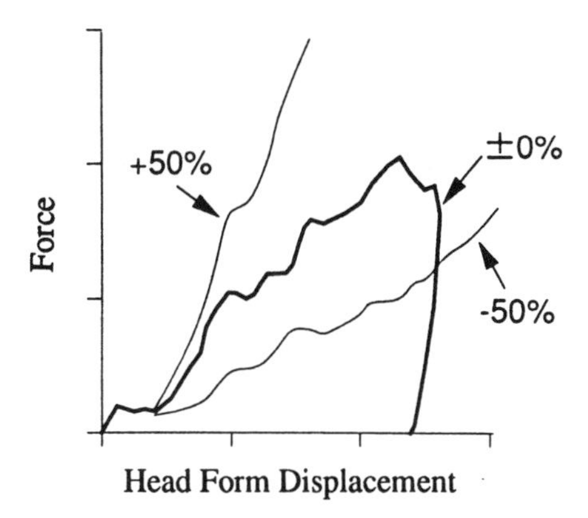

Figure 20. Analysis Example for steel

From this analysis example, it can be seen that the relationship between reaction can plate thickness and force is almost linearly proportional. Therefore, as shown in Figure 20, if aluminum is used for construction of the reaction can and the plate thickness is increased to approximately 20% that of steel, then it is possible to maintain the same performance level. In this instance there would be a weight savings of approximately 60% in the reaction can.

Relationship with Maximum G and Head Form Displacement - These results are shown in Figure 21.

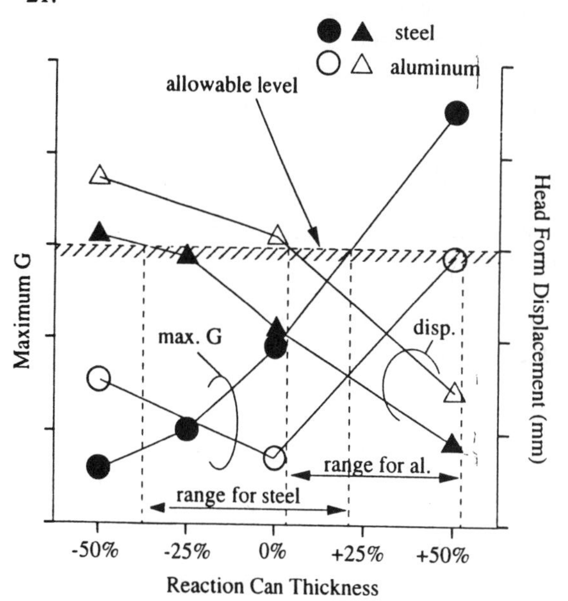

Figure 21. Influence on Maximum G and Head Form Displacement

Ignoring the case where the head form bottoms out (aluminum, -50%), the maximum G increases when the reaction can plate thickness is increased for both the aluminum and steel reaction cans. Furthermore, the displacement decreases in response to this increase in plate thickness.

The allowable level for the maximum G and displacement is shown by the horizontal line in this graph. From this, the acceptable range for the plate thickness of the reaction can can be determined for both the steel and aluminum reaction cans. For steel, this results in an acceptable range from about 60% to about 110% of the current thickness. For aluminum, this results in an acceptable range from about 100% to about 160% of the current thickness. For these minimum values, a substantial weight reduction could be made.

CONCLUSION

When conducting air bag module development tests, it is time consuming to investigate many methods that could ensure a lightweight air bag module with high energy absorption performance. In this study, an analytical method has been established which allows the necessary parameters to be analyzed to achieve this goal. Furthermore, the analysis model for this study was verified by actual tests.

The quantitative effect of the following parameters on the energy absorption performance and the weight of the air bag module has been determined.

1) Air Bag Module Door Bead and Thickness

2) Reaction Can Material and Plate Thickness

From the results of this parameter study, the following can be concluded for the analysis model in this investigation:

1. The amount of energy absorbed by the air bag module door has an approximately linear relationship to the number of beads in the door and also the door thickness. Therefore, increasing the number of beads or the door thickness is an effective way to increase the energy absorption performance of the air bag module.

2. When aluminum is substituted for steel in construction of the reaction can, and the plate thickness is increased by 20%, the same energy absorption performance can be maintained. For this case, it is possible to reduce the weight of the reaction can by approximately 60%.

3. The deformation force and amount of deformation of the reaction can are approximately linearly related to the plate thickness. With the crash stroke kept under a reasonable level, the thickness of the steel reaction can can be reduced by 38%. For aluminum, the appropriate value could be 105% of the current steel thickness. Either method would result in a proportional weight reduction.

REFERENCES

1. "An Evaluation of Occupant Protection in Frontal Interior Impact for Unrestrained Front Seat Occupants of Cars and Light Trucks," NHTSA Publication DOT HS 807 203.

2. Gerault, P., Huber, M.. and Rosser, M., "Structural Thermoplastic Composites for Interior Energy Management to Meet FMVSS 201 and 208," SAE Paper No. 910045, 1991.

3. Pendergast, P.L., "Response of Instrument Panel Retainer Materials to Impact Loadings," SAE Paper No. 860258, 1986.

4. Sounik, D.F., McCullough, D.W., Clemons, J.L. and Liddle, J.L., "Dynamic Impact Testing of Polyurethane Energy Absorbing (EA) Foams," SAE Paper No. 940879, 1994.

1999-01-0757

Side Airbag Sensor in Silicon Micromachining

D. Ullmann, G. Bischopink, M. Schöfthaler, R. Schellin,
B. Maihöfer, J. Seibold and J. Marek
R. Bosch GmbH

ABSTRACT

For side airbag systems it is necessary to measure the acceleration within a time of less than 3 ms in order to inflate the side airbag in time. A new generation of side airbag sensors that uses a linear accelerometer is presented. The evaluation circuit includes amplification, temperature coefficient compensation, two wire unidirectional current interface, and a zero-offset compensation. The sensing element for the measurement of acceleration is a surface micromachined accelerometer. In order to minimise the production costs the surface micromachined sensor element and the corresponding evaluation ASIC are packaged into a standard PLCC28 housing. For the entire function only few external components are necessary. During the power-on cycle an internal selftest is carried out and the result is transmitted to the airbag control unit. Most important results of the characterisation are presented.

INTRODUCTION

The market for automotive side airbag sensors is rapidly growing. The driving force for this fast development is the requirement for low cost and high production volume devices with high reliability. A technology that enables such a low cost accelerometer is surface micromachining together with a capacitive sensing technique. The surface micromachining technique has been transfered to volume production in the recent years [1-2].

This paper describes the complete electrical function as well as the advantages of the sensor. There is also a brief overview about the necessary technology for producing the surface micromachined accelerometer. Additionally, some measurement results of important parameters will be presented.

The Peripheral Acceleration Sensor (PAS) is realised in PC-board technology. The main part is a PLCC28 housing that contains the micromachined sensor element and the evaluation IC that is also used for the calibration of the sensitivity. The PC-board is mounted inside a customer specific plastic housing for direct assembly into the car. In order to obtain a waterproofed outer housing we use a lid with a sealing ring (Figure 1). This results in a high insensitivity against environmental conditions like water and dust.

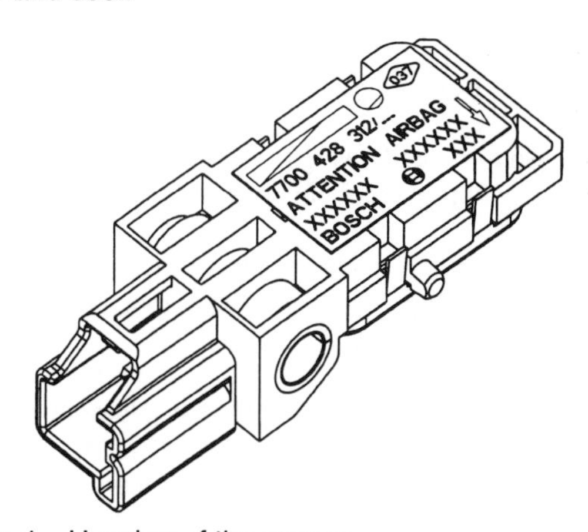

Figure 1. Housing of the sensor

The sensor measures the real acceleration value. The amplified and filtered analog signal (three pole Bessel filter at 400 Hz) is converted into a digital signal. This value is transmitted via a two wire current interface to the airbag control unit. The sensor supply voltage is 6 V and the current consumption is less than 40 mA. The full scale acceleration range is normally ±100 g and can be adapted to ±50 g by using an external resistor. The accelerometer is tested by a real deflection of the sensing element with electrostatic forces.

The interface from the airbag control unit to the PAS can be realised by using the Peripheral Integrated Circuit (PIC). The PIC can provide the power supply for the control unit and PAS as well as the communication interface for two PAS.

SENSOR COMPONENTS

ACCELEROMETER – The sensor element is constructed using surface micromachining technology and consists of a comb structure as shown in Figure 2. Two fixed outer electrodes and an electrode connected to the seismic mass form the capacitors C_1 and C_2. The seismic mass is suspended by springs, and as a result of a linear acceleration the gap between the electrodes is changed. This movement yields a change of the two capacitors C_1 and C_2 and is evaluated by using an open-loop measuring principle [3-5].

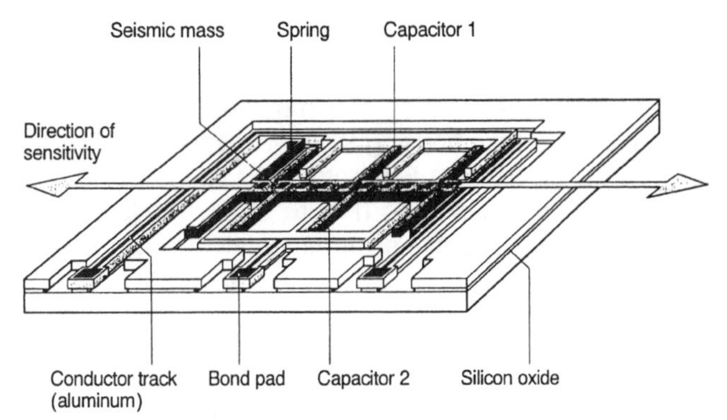

Figure 2. Schematic drawing of the surface micromachined accelerometer

The gap between the electrodes which form the capacitors is approximately 2 - 3 μm. The capacitance of the comb structure is about 1 pF.

Sensing Element Fabrication – The key element of surface micromachining is the deposition of poly silicon on top of a sacrificial layer. At elevated temperature an 11 μm thick polycrystalline silicon layer is deposited [4, 6].

The deposited conductive and isolating layers that are patterned by using several lithographic mask steps. By removing the sacrificial oxide, free-standing cantilevers are obtained. Figure 3 shows the process flow chart schematically.

A special trench technique allows the formation of almost vertical sidewalls with large aspect ratio. Thus, the trench process ensures excellent vertical dimensions of the capacitor plates. Figure 4 shows a detailed view of the completed sensor device.

The flexible sensing structure is protected by using a silicon cap. The permanent connection between the cap and sensor is obtained by means of a waferbonding technique based on a seal glass.

Finally the wafer is diced and the individual acceleration sensors are assembled into the PLLCC28 housing.

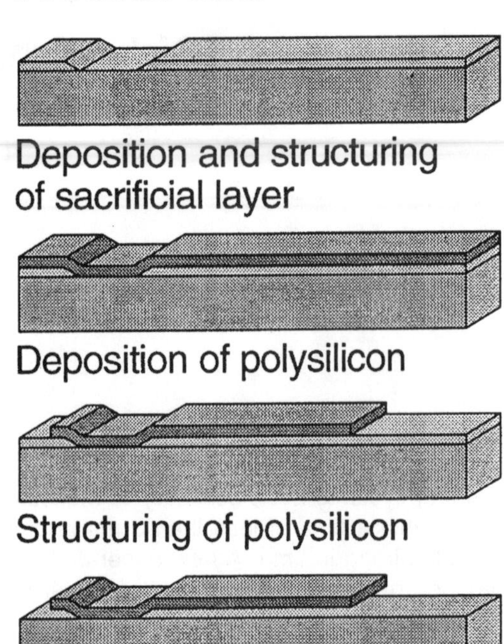

Figure 3. Technical principle of the surface micromachining process

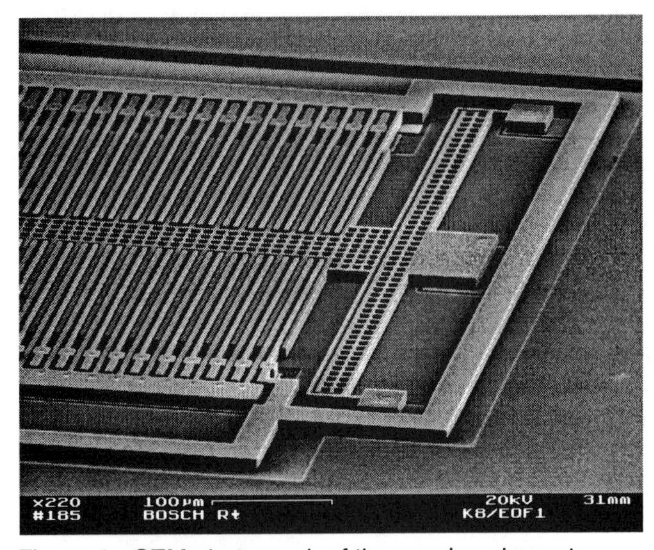

Figure 4. SEM photograph of the sensing element

INTERFACE CIRCUIT – The interface circuit is fabricated in a mixed analog and digital process. The ASIC includes a part for evaluation of the acceleration signal, a voltage regulator for the internal power supply, an on-chip-oscillator and the state machine for the whole control and timing of the startup and initialisation. In this circuit are also implemented the function for the selftest, error handling, and data transmission. Figure 5 shows the principle structure of this component.

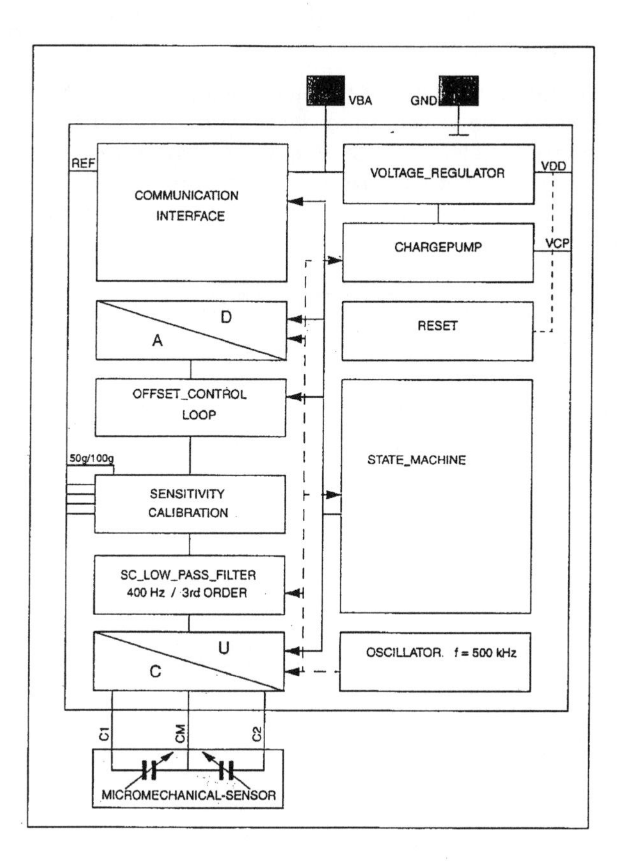

Figure 5. Block diagram of the whole evaluation circuit

The primary function of this circuit is to evaluate the change of capacitors C_1, C_2. This change is converted into an analog voltage by a C/U-converter. The output of the converter is filtered by a SC-low-pass-filter with a corner frequency of 400 Hz (three pole Bessel).

For the intended application, the sensitivity adjustment has to ensure a accuracy of ± 4%. Thus a 4 bit sensitivity calibration is performed. An additional bit selects the ± 50 g version.

The amplified and filtered signal is then digitized by using an 8 bit A/D-converter. The actual acceleration value is transformed into a special data word and is transmitted to the airbag control unit via a two wire current interface in an 11-bit frame with a timing of 250ms. Each message consists of 2 start bits, 8 data bits (the real acceleration signal) and 1 parity bit. For the transmission a Manchester Code is used. A logic "0" or "1" is represented by a rising or falling slope in the middle of each bit. Thus the average current consumption is minimised.

The logic levels are realised by a modulation of the current on the power supply lines. A "low"-level (I off) is represented by current of approximately 10mA, a "high"-level by approximately 30mA (see Figure 6).

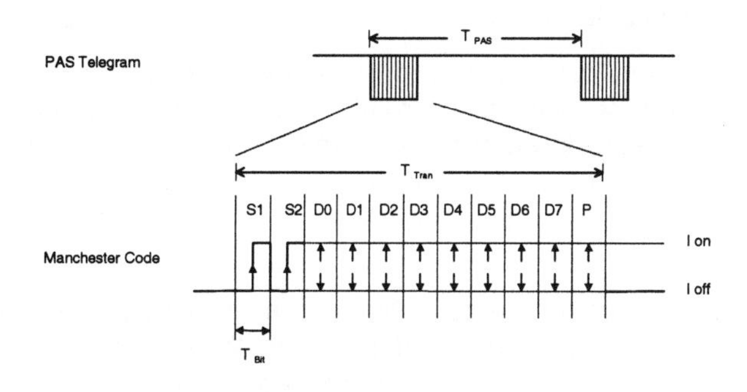

Figure 6. Data transmission structure

The acceleration value is transmitted in a range of ±120 LSB. The remaining bits are used for transmission of status code (± 50 or ± 100 g version) and error codes.

The chip also provides compensation for sensor offset and a selftest during the start-up phase. The selftest feature allows a real deflection of the micromechanical structure by using electrostatic forces. The response of this test mimics the behaviour during a real acceleration event.

PACKAGING

The packaging technology is driven by low cost requirements, high reliability and high volume manufacturing.

The PAS mainly consists of 3 parts. Figure 7 shows a schematic drawing with the cap, the PC-board, and the plastic outer housing.

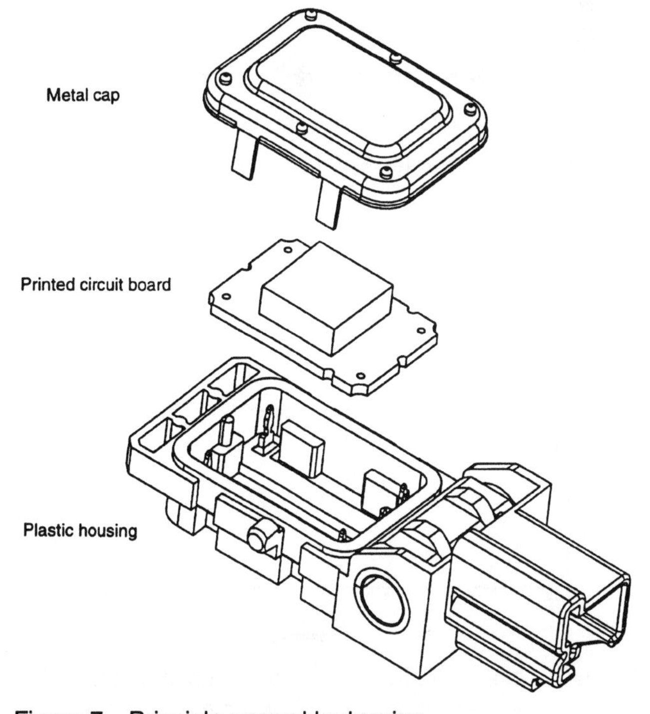

Figure 7. Principle assembly drawing

The PC-board with the PLCC28-package and some external components is directly assembled into the customised outer housing.

The metal cap is provided with a sealing ring to obtain a waterproof housing. In order to obtain a good and easy connection between the cap and the housing, a simple bending process is used. The cap is pressed onto the housing and the four fingers are bent around the plastic housing.

The concept of the outer housing takes into account the possible mounting position. Normally the sensor is located at the crossmember or B/C-pillar.

The connector has just two terminals to realise the connection between the airbag central unit and the airbag sensor.

CHARACTERISATION RESULTS

The function of the sensor has been verified on a shaker. Figure 8 shows the corresponding sensitivity of the PAS as a function of the acceleration within a range from –40 g to +40 g.

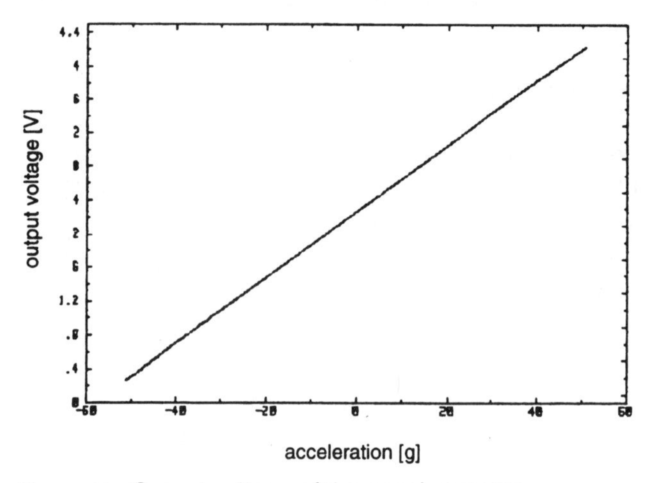

Figure 8. Output voltage of the accelerometer

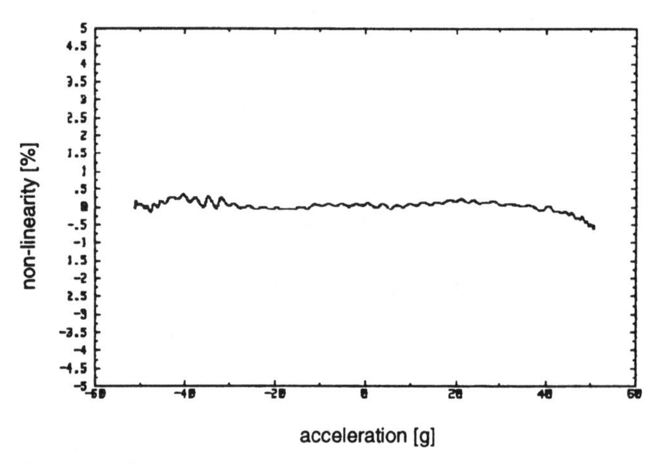

Figure 9. Non-linearity (relative deviation of output voltage in percent of full scale)

Figure 9 shows the measured non-linearity of the whole sensor, indicating that the drift is very small and reaches a value of less than ±1%.

A feature important for the application in the automotive section is the temperature drift. Figures 10 and 11 show the results of the measurements. As a conclusion there is no additional compensation necessary to reduce the temperature dependence of the output which is less than ±1% in this case.

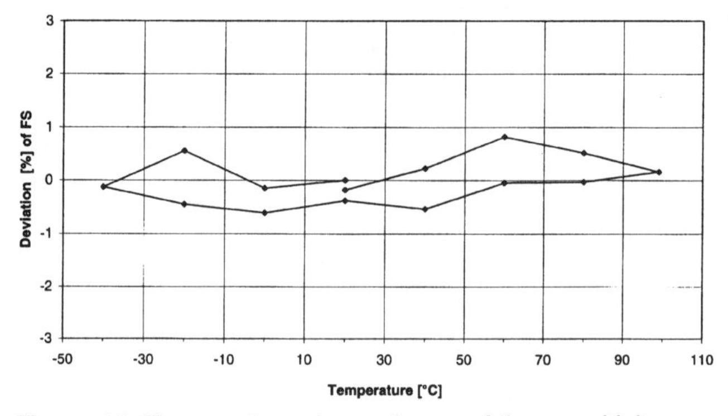

Figure 10. Temperature dependence of the sensitivity, reference at 20°C

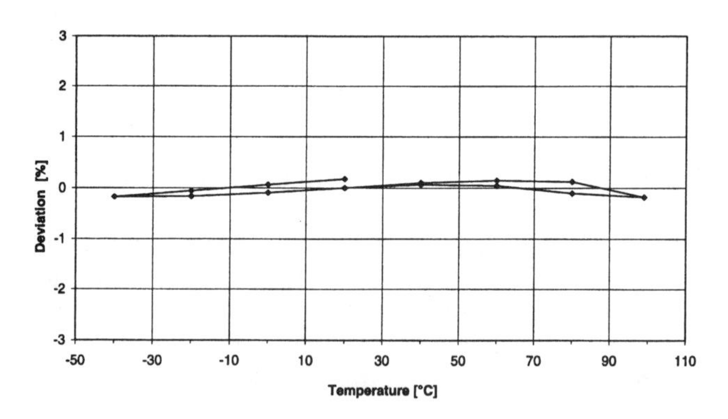

Figure 11. Temperature dependence of the zero-g Level without offset regulation, reference at 20°C

Additional characterisations not shown here indicate a small cross sensitivity and a low dependence of ageing effects.

CONCLUSION

A side airbag sensor in surface micromachining with full mechanical selftest capability and high accuracy is presented. The accelerometer complies with all requirements demanded by the automotive sector. An low cost packaging concept makes the sensor suitable to different mounting locations. The robust design of the sensor allows a simple production process flow and leads to a cost optimised product. This sensor is designed for high-volume production and will start series production by the end of 1999.

ACKNOWLEDGMENTS

Many people were involved in the realisation and tests of the sensor. The authors would like to thank their colleagues at Robert Bosch GmbH, particularly F. Haag, K. Weiblen, and G. Hopf, for their help in developing and characterising the accelerometer.

REFERENCES

1. L. Ristic, D. Koury, E. Joseph, F. Shemansky, and M. Kniffin: "A two –chip accelerometer system for automotive application", Micro System Technology ´94, Berlin, Germany, October 19-21, 1994.
2. J. Marek: "Silicon Microsystems for Automotive Applications", ESSDERC `97.
3. M. Lutz, W. Golderer, J. Gerstenmeier, J. Marek, B. Maihöfer, S. Mahler, H. Münzel, U. Bischof: "A Precision Yaw Rate Sensor in Silicon Micromachining", SAE Technical Paper Series, 980267.
4. M. Offenberg, H. Münzel, D. Schubert, O. Schatz, F. Lärmer, E. Müller, B. Maihöfer, J. Marek,: "Acceleration Sensor in Surface Micromachining for Airbag Applications with High Signal/Noise Ratio", SAE Technical Paper Series, 960758.
5. K. H.-L. Chau, S. R. Lewis, Y. Zhao, R. T. Howe, S. F. Bart, R. G. Marcheselli: "An integrated force-balanced capacitive Accelerometer for Low-G Applications", Transducers 95, 149 – C4.
6. M. Offenberg, F. Lämer, B. Elsner, H. Münzel, W. Riethmüller: "Novel process for a monolithic integrated accelerometer", Transducers ´95 Eurosensors IX, 148 –C4.

940561

Sensing Side Impacts

David S. Breed, W. Thomas Sanders, and Vittorio Castelli
Automotive Technologies International, Inc.

Abstract

Now that airbags are the accepted solution for protecting occupants in frontal impacts, and now that safety sells cars, it is natural to look closely at the second largest killer of automobile occupants, side impacts. This paper develops the theory of sensing side impacts based on the assumption that airbags will soon be used for side impact protection. The trade-offs between the various sensor technologies are discussed including electronic and mechanical sensors. For mechanical sensors, fluid damped, undamped and crush sensing switches are compared. Finally, the requirements for a successful predictive sensor will be presented.

APPROXIMATELY ONE QUARTER of all injury-producing accidents are side impacts in which the direction of the force is within 45 degrees of the lateral axis of the vehicle. Airbags are alleviating injuries from frontal impacts and it is important now to focus on the next largest cause of fatalities, side impacts.

Summary

To satisfy the various requirements for a side impact sensor, it is concluded below that a sensor having an inertial mass is required; if electro-mechanical, a damped sensor is desired; and, an inertially damped sensor is most adaptable to properly sensing side crashes. Most cars need only two discriminating sensors per side, one adjacent each occupant. For these cases, an all mechanical airbag system is possible where the sensor is integral with the rest of the airbag system. A crush sensitive safing sensor in series with, and located proximate to the velocity change sensor, could be used to minimize the chance of inadvertent airbag deployments providing it covers a sufficient portion of the vehicle side that its involvement in a crash can be assured.

Comparison With Frontal Impacts

In frontal impacts, the crush zone of the vehicle changes velocity early in the crash and a sensor located within 30 cm of the front of the vehicle, or an electronic sensor mounted elsewhere, can sense the crash and initiate the inflation of the airbag long before the occupant has begun to move relative to the passenger compartment. Also, for most cases, there is little intrusion in the passenger compartment, and the entire space between the occupant and the instrument panel or the steering wheel, as well as the stroke of the steering column, is available to cushion the occupant. In contrast, in side impacts there is almost always significant intrusion into the vehicle and the motion of the occupant relative to the vehicle interior begins immediately after impact. In addition, there is far less space for a restraint system and thus the injury-reducing potential of an airbag, even if it were deployed in time, is substantially less for side impacts than for frontal impacts.

In the standard FMVSS 208 48 kph frontal barrier crash test mandated by the Government for frontal impacts, an unrestrained occupant would typically impact the steering wheel at about 30 kph. When an airbag is deployed, this impact velocity of the driver with the vehicle is absorbed through the collapse of the airbag plus the stroking of the energy absorbing steering column giving a total energy absorbing displacement of about 46 cm. In the standard FMVSS 214 side impact crash test, the occupant is impacted by the intruding door also typically at about 30 kph however there is no more than a few cm of space available in the door and inner door panel to absorb energy and much of this is consumed by the intruding vehicle.

Ultimately, the necessary space must be created either by making the vehicle door much thicker or by an airbag which pushes the occupant away from the door before the start of the crash. Such an airbag requires an anticipatory sensor system as discussed below.

The Marginal Sensor Response

To design a crash sensor for side impacts, the side-structure of a vehicle must be understood. In an SAE paper "Trends in Sensing Side Impacts" (1), the side intrusion problems and general vehicle response characteristics are analyzed. As discussed in this paper, once the marginal occupant protection velocity has been chosen (usually 16 to 20 kph), the desired response curve of a side impact crash sensor can be determined by the impact conditions, such as vehicle-to-vehicle, vehicle-to-pole, or truck-to-vehicle accidents. One example of a response curve which meets the criteria discussed in that paper is shown in Figure 1.

The Vehicle to Vehicle Side Impact

The most common side impact accident is a "Bullet" vehicle impacting a "Target" vehicle at close to 90 degrees. Once the side door is hit by the Bullet vehicle, the door beam and outer panel deform significantly while the passenger compartment only gains a relatively small velocity change in the early stage of a crash. The side intrusion or crush increases continuously after the early penetration until the entire car reaches a common final velocity later in the crash. The responses of the Target and Bullet vehicles are functions of the impact angle and location, the impact speed, and the stiffness and weights of the vehicles.

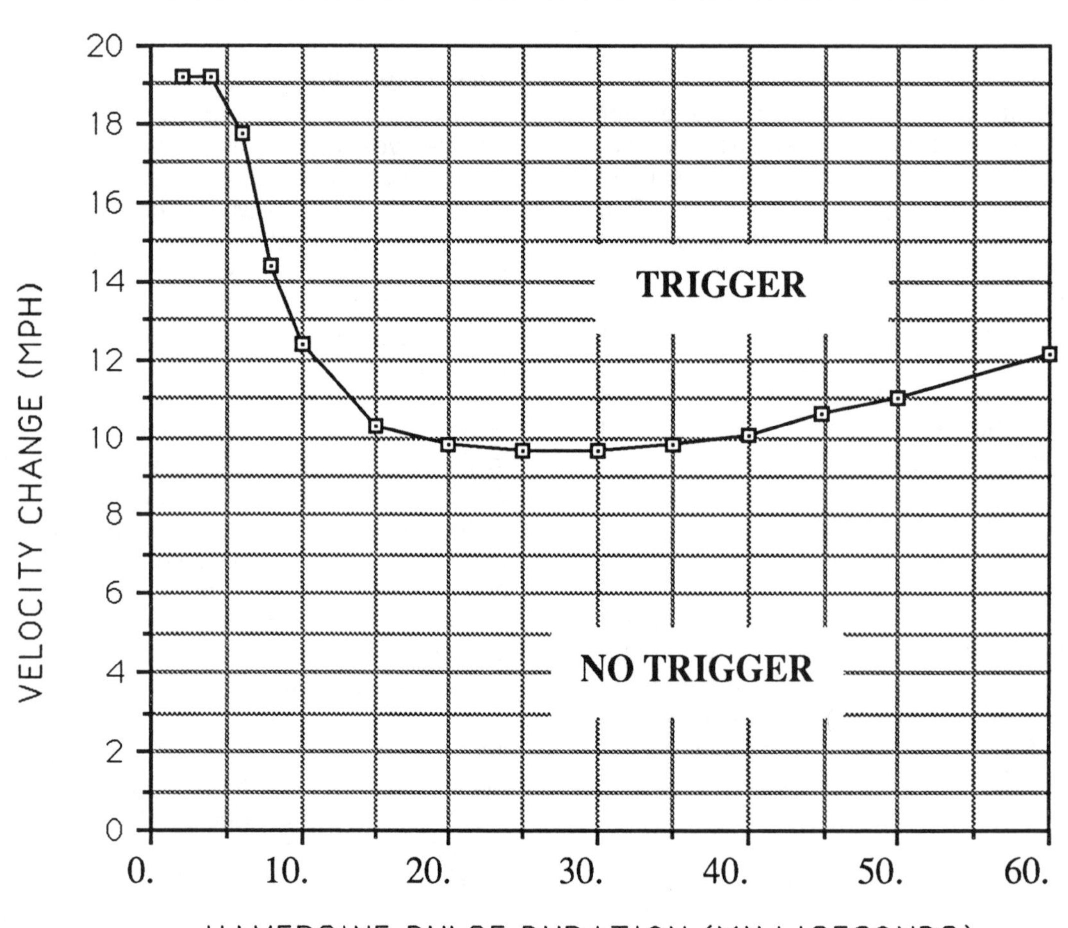

Figure 1

The horizontal axis is the duration of an idealized crash pulse (haversine) and the vertical axis is the velocity change of the crash. The velocity change required to trigger the sensor is about 19 mph (31 kph) for a 4 millisecond crash, 10 mph (16 kph) for a 20 millisecond crash and 12 mph (19 kph) for a 60 millisecond crash.

The velocity of the side door outer panel increases immediately after the impact to a maximum velocity comparable to the velocity of the Bullet vehicle, see Figure 2. The passenger compartment experiences a relatively small velocity change during this stage of the crash. As the side structure stiffens, the resistance force increases

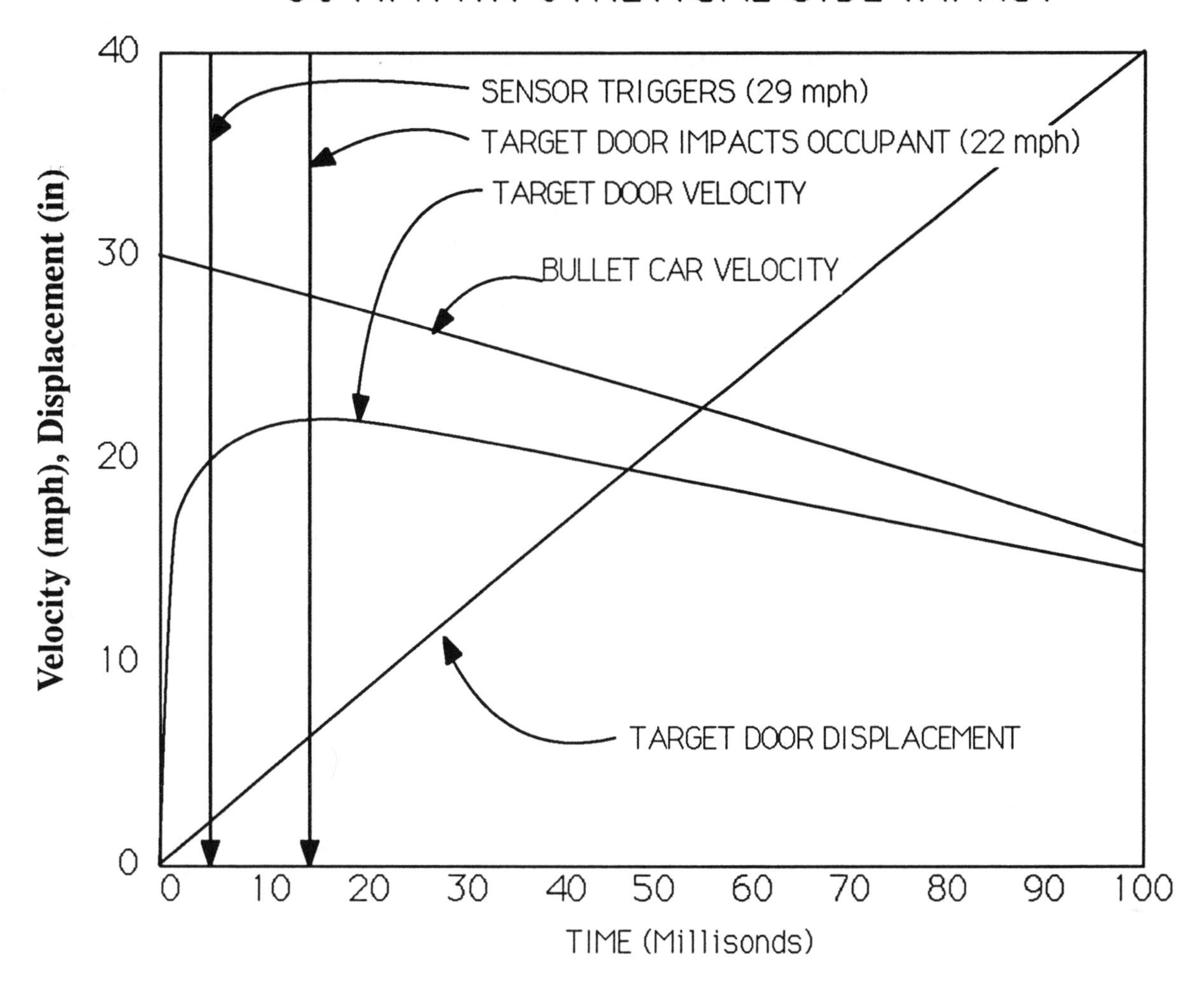

Figure 2

and starts to decelerate the side door panel until finally the side panel and the passenger compartment reach a common velocity. This final velocity is estimated to be the momentum velocity, which is the original momentum of the bullet vehicle divided by the total masses of the bullet and target vehicles, usually around one half of the bullet vehicle's initial velocity.

In Figure 2 the sensor triggers at about 4 milliseconds and experiences a 29 mph (47 kph) velocity change. If an airbag was not present, the occupant would be struck by the intruding door at 22 mph (35 kph) at about 13 milliseconds. In most vehicles the door/driver impact velocity is somewhat less and somewhat more time is available for inflating the airbag.

The time at which the occupant is struck by the inner door panel is determined by the stiffness of the vehicles, the impact condition, and the distance between the occupant and the side panel. In most of the side impact cases, the occupant is struck at a velocity significantly lower than the impact velocity but above the final vehicle velocity. In fact, the velocity, at which the target vehicle is impacted can be as high as twice the velocity that the inner door panel hits the occupant. Therefore, a crash sensor located near the outer door panel must require a higher velocity change to trigger than 16 kph in pulses in the range of 1-5 ms. If a crash sensor triggers on a 16 kph pulse, for example, and the occupant is only struck later at 10 kph there will be an undesired airbag deployment.

Crush Sensors are Not Appropriate

Since a side impact crash sensor must not falsely trigger due to hammer blows or light pounding on the side door, which can cause significant local deformations, side door deformation can not be used as the only criterion for detecting the severity of a side impact accident. A side impact displacement type sensor which responds to the crush of the side door panel, therefore, could cause frequent inadvertent sensor triggering. A side impact sensor must also trigger for other side crashes when the side door is not directly hit, but the impact is severe enough so that the occupant needs the protection of an airbag. A displacement-type sensor in these cases will not trigger until the side crush of the vehicle progresses to the location of the sensor. This will result in late triggering or no triggering of the sensor and no protection for the occupant unless the sensor covers most of the side of the vehicle. On the other hand, a velocity-type sensor will respond to the velocity change sensed in a crash and can be adjusted to a desired sensitivity to detect a side impact even though the side door is not hit directly.

Spring Mass and Ball-in-Tube Sensors

Even though spring-mass inertial sensors respond to a specified range of velocity changes, the sensitivity of these sensors increases as the pulse duration decreases. Therefore these sensors will trigger with a smaller velocity change for pulses of shorter duration than longer duration. This trend contradicts the conditions of side impact sensing as discussed above. On the other hand, viscously damped sensors, such as conventional ball-in-tube sensors, respond to the same velocity change regardless of pulse duration. These sensors also do not meet the requirements of side impact sensors, which require greater insensitivity for short, impulsive velocity changes.

Inertially Damped Sensors

In inertially damped sensors, it is the inertia of the damping fluid rather than its viscosity which provides the damping force. In these sensors, the motion of the sensing mass is opposed by a nonlinear damping force, depending on the second power of the sensing mass velocity. These sensors are naturally more sensitive to long pulses than to short pulses, but the sensitivity to very long pulses can be compensated by the bias force. Also, the sensitivity to very short pulses can be varied if a gas is used as the fluid. In this case the compressibility of the fluid can be used to cause the response curve to level out for very short pulses (<4 ms) as illustrated in Figure 1. This ability to tailor the characteristics of these sensors in the range of pulses from 1 to 50 ms makes inertially damped sensors most appropriate for side impact sensing.

A Side Impact Crash Sensor Must be Mounted On The Door

A crash sensor for sensing side impacts must be placed on the side door structure to be effective. This location is essential since it is sensing the velocity change of the portion of the vehicle which will eventually strike the occupant. Since there is insufficient signal anywhere else in the vehicle for side impacts, they can only be sensed in time with crush zone sensors. If the side door is not hit directly, the pulse propagated to the side door is delayed and stretched in its duration, compared to the pulse generated in a direct side door impact. Therefore, to be effective, a crash sensor must be more sensitive to these longer or stretched pulses.

In some installations structure is added to the vehicle to transmit the crash pulse from the vehicle door to a location within the passenger compartment where the sensor is located This effectively places the sensor on the door while it is actually mounted elsewhere.

Pressure Sensors

A side impact sensor which measures the change in pressure inside the door has recently been proposed. The fundamental assumption behind this sensor is that the change in the internal volume of the door can be used to discriminate crashes where an airbag is desired from those where it is not.

As the side door of the target vehicle is impacted, the pressure within the door increases by an amount which is a function of the cross section area of the impacting object, the point of impact, and the leakage rate from the door. This sensor uses an electronic pressure measuring element to detect any impact to the door which causes the pressure to increase within the door. The sensor assumes that the rate of change of pressure is proportional to the crash velocity change.

One problem with this sensor is that it assumes that the leakage rate from the door is not only known but that it does not change over time. Throughout its life a vehicle can undergo many events which do not require airbag deployment but which can reduce the volume within the door, create holes in the door, destroy the vapor barrier, cause the seals to leak, detach part of the inner door panel, etc. Such events will significantly change the relationship between the displacement of the outer door panel and the rate of pressure increase within the door with the result that the response of the sensor will also vary significantly over the life of the vehicle.

Another problem is that the sensor will be less sensitive to objects, such as trees, which have a smaller cross section than other objects such as a truck. The sensor will necessarily be less responsive to these impacts which are actually more severe than vehicle to vehicle impacts. This is especially the case if the impact point of the tree is at the A-pillar or B-pillar where there will be minimal damage to the door by the time sensor triggering is required. Other impacts which cause minimal damage to the door include impacts with high curbs or low poles. For example, unless multiple sensors are used, an impact by the rear seat of a vehicle may not cause significant deformation to the front door and yet the airbag could be required.

Most but not all accidents, where the point of impact is in front of the A-pillar, do not require deployment of the airbag since there is substantial rotation of the vehicle. This is not always the case, however, especially when the principle direction of force passes through the center of gravity of the target vehicle. In this case the vehicle does not rotate and there is significant deformation to the door only long after sensor triggering is required.

A further problem exists when a substantial portion of the vehicle side is involved such as when the vehicle slides sideways into a wall, highway divider or guard rail. Here clearly, the rate of change of the pressure within the door is not a good measure of the velocity of the impact. For the above reasons, the pressure change is not believed to be a good predictor of the impact velocity of the vehicle door with the occupant.

A further problem of a numerical nature must also be considered. The pressure sensor predicts accident severity, velocity change, by measuring the pressure and taking its derivative. It is not the pressure level within the door which predicts the velocity change but its rate of change. The taking of the derivative of measured data is well known to lead to significant numerical errors which further contributes to the inaccuracy of this method.

For the above reasons it is unlikely that a pressure measuring sensor will perform as well as an electronic accelerometer based system where the velocity change is determined through an integration of the measured acceleration.

Crush Sensing Safing Sensors

It may be desirable for a side impact sensing system to include safing (arming) sensors in addition to the discriminating sensors described above. In frontal impacts, velocity-type low-bias sensors located in the passenger compartment are used for safing purposes. In side impact crashes, however, the crash pulse in the passenger compartment does not provide enough information at the time that the crush zone sensor is required to trigger. Therefore, it is not possible to use a passenger-compartment safing sensor for side impact sensing system. Therefore, such sensors must also be mounted on the vehicle door unless structure is added to transmit the pulse elsewhere. Safing sensors for side impact applications can be crush sensing switches. These safing sensors should be placed in proximity to the velocity sensing sensors, and should have long contact dwells. They also must cover a sufficient portion of the vehicle side that they are always involved in the crash. This might be difficult to accomplish for reasons similar to those presented above when discussing pressure measuring sensors. Nevertheless, a combination of a velocity sensing sensor and a crush sensor significantly reduces the probability of an inadvertent deployment, by imposing a requirement that two environmental stimuli (velocity change and physical displacement) are required to initiate airbag deployment.

Anticipatory Sensors

There are several technologies which are available for use as anticipatory sensors. The most viable ones include ultrasonics, laser optics and radar, any of which could sense the approach of a vehicle, for example, providing the proper software is available to determine the nature of the impacting object and its trajectory relative to the target vehicle. Laser and radar systems are at present probably too expensive and there is concern with ultrasonics being fooled by a heavy rainfall, for example. Software systems are now available which will work with these technologies and anticipatory sensors will probably be in development within the next few years. Such a system will permit the airbag to be deployed before the impact and therefore permit the occupant to be displaced by the airbag to create the required energy absorbing space before the accident occurs.

Conclusions

To satisfy the various requirements for a side impact sensor, it was concluded that a sensor having an inertial mass is required. If this sensor is electro-mechanical, a damped sensor is desired and an inertially damped sensor is most adaptable to properly sensing side crashes. If the sensor is electronic, its response should be similar to the electro-mechanical sensor in that it should require a larger velocity change to trigger for short duration pulses (0 to 10 milliseconds) than for intermediate duration pulses (10 to 50 milliseconds). It was further concluded that sensors based on pressure measurements within the door are likely to have significant variations over the vehicle life and an improper response to many real world crashes. Finally, anticipatory sensors will probably be available in the next few years.

The most cost effective solution using available technology is the all mechanical side impact system which eliminates all of the wiring, diagnostics and backup power supply associated with electrical systems.

REFERENCES

1 Castelli, V., Breed, D.S. "Trends in Sensing Side Impacts", Society of Automotive Engineers No. 890603, 1989.

950346

A Driver-Side Airbag System Using a Mechanical Firing Microminiature Sensor

Koji Ito, Masanobu Ishikawa, and Kazunori Sakamoto
Aisin Seiki Co., Ltd.

Ichizou Shiga, Katsunobu Sakane, Yutaka Kondoh, and Masahiro Miyaji
Toyota Motor Corp.

Yasunori Iwai
Daisel Chemical Ind., Ltd.

ABSTRACT

By developing a mechanical-firing sensor using rotational inertia effect, we have completed miniaturization of the sensor and have developed a new-type mechanical-firing airbag system.

This airbag system has been confirmed to have superior occupant protection performance after conducting a variety of vehicle crash tests and sled tests.

INTRODUCTION

One of the latest trend among automobile manufacturing is to reduce weight. Under these circumstances, downsizing of an airbag assembly becomes an important issue. Accordingly, the need for miniaturization of sensors is increasing. By developing a mechanical-firing sensor in a new structure, we have completed the miniaturization of the sensor and have developed a new-type mechanical-firing airbag system.

This paper describes the structure, activation mechanism, performance, computer simulation, and evaluation of the mechanical sensor as well as important development factors for the entire system.

GENERAL DESCRIPTION OF THE MECHANICAL-FIRING AIRBAG SYSTEM

Figure 1 shows the overall schematic drawing of the system. The system is a fully-mechanical airbag system, of which all the parts are contained in the steering wheel as a pad module assembly.

The pad module assembly is attached to the frame of the steering wheel with bolts and composed of an inflator, a sensor, a bag, etc.

The sensor is located partly in the inflator and fixed so that the sensor is sealed from outside air.

The description of the system activation is given below following Figure 2.

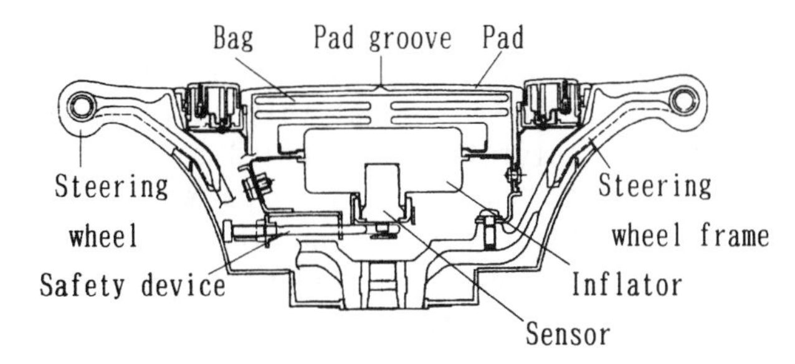

Figure 1: Schematic of the system

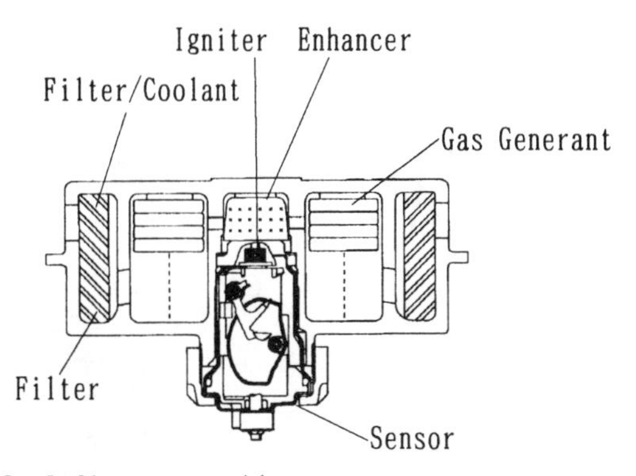

Figure 2: Inflator assembly

When the mechanical sensor detects a crash and a firing pin enters an igniter, the igniter fires and a flame from the igniter heats an enhancer and next a gas generant. As a result, a large amount of nitrogen gas from the gas generant fills the bag through a filter section.

DESIGN OF THE MECHANICAL-FIRING MICROMINIATURE SENSOR

STRUCTURE AND ACTIVATION - Figure 3 shows the structure of the sensor. The sensor is composed of a high-density weight supported by an eccentric shaft, a cam integrated with the weight, a torsion spring, and a firing pin which applies the reaction force of the torsion spring to the weight cam section and is used to trigger the igniter of the inflator.

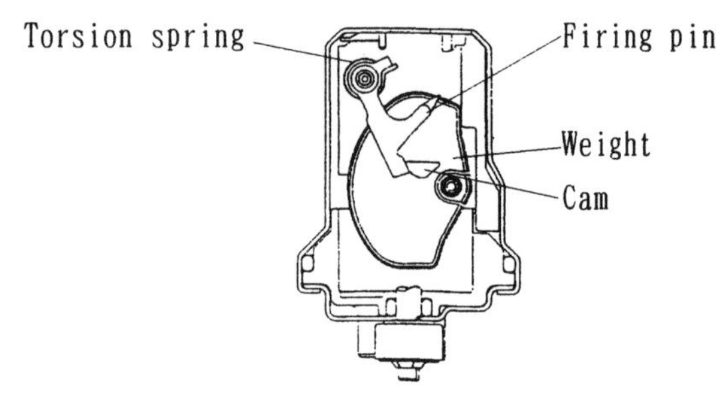

Figure 3: Structure of the sensor

Figure 4 shows the activation mechanism of the sensor. When a relatively large amount of impact acceleration is applied to the sensor due to a vehicle crash, the weight retained by the force of the torsion spring through the firing pin rotates counterclockwise against the torsion spring force. After the rotation of the weight by a predetermined angle, the firing pin is released from the weight cam and rotated counterclockwise by the torsion spring force. As a result, the pointed section of the firing pin sticks into the igniter of the inflator positioned at the end surfaace of the sensor with a certain kinetic energy , thereby exploding the igniter and thus triggering the airbag.

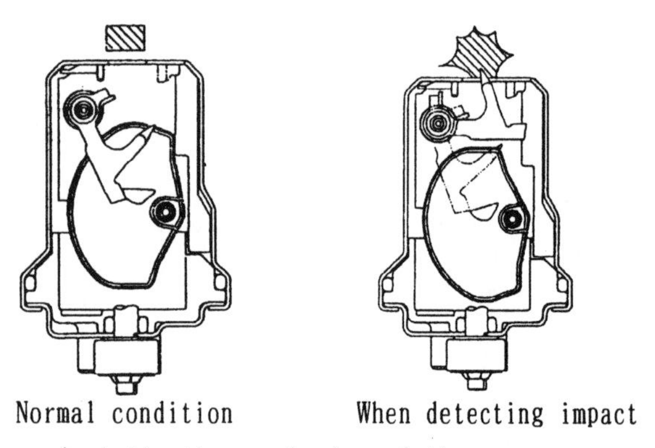

Figure 4: Activation mechanism of the sensor

Additionally, to prevent accidents due to unexpected airbag deployment before or during the vehicle assembly operation, a safety device is provided in the sensor. The safety device inhibits the sensor from activating by restricting the rotation of the weight. After the pad module assembly is installed in the steering wheel, the safety device is released so as to make the sensor operational.

PERFORMANCE REQUIREMENTS FOR THE SENSOR - The performance requirements for the sensor have been established as described below.
(1) Sensor activation conditions: The sensor must be activated within a predetermined period of time when a vehicle crash occurs at a high speed, wherein airbag deployment is required.
(2) Sensor non-activation conditions: The sensor must not be activated when a vehicle crash occurs at a low speed, during rough road running or during overload application to the steering wheel wherein airbag deployment is not required.

CRASH DISCRIMINATION REQUIREMENTS FOR THE SENSOR - The parameters for determining the crash discrimination requirements for the sensor are the force of the torsion spring, the mass, the center of gravity, and the rotation inertia the weight, and the the rotation axis positions of the firing pin and the weight, etc.

In particular, the model of the sensor is capable of meeting the above performance requirements for an airbag sensor by using the rotational inertia of the weight.

Figure 5 generally shows the sensor model and the kinetic equation for the sensor.

Next, the crash discrimination requirements for the sensor in various activation conditions are explained below.

Figure 6 shows acceleration levels and weight rotation angle in the various activation and non-activation conditions. The acceleration data shown in this paper have been

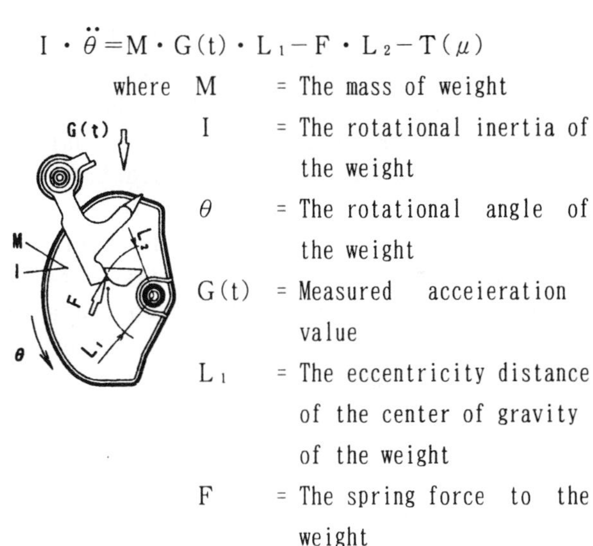

$$I \cdot \ddot{\theta} = M \cdot G(t) \cdot L_1 - F \cdot L_2 - T(\mu)$$

where M = The mass of weight

I = The rotational inertia of the weight

θ = The rotational angle of the weight

G(t) = Measured acceieration value

L_1 = The eccentricity distance of the center of gravity of the weight

F = The spring force to the weight

L_2 = The action distance of spring force

$T(\mu)$ = Friction loss

Figure 5: The sensor model and kinetic equation

obtained by simple modification of the measured acceleration data.

(1) Sensor activation conditions: Since an acceleration level higher than a threshold value and having long duration is generated during an actual vehicle crash at a high speed, the weight rotates by a certain amount against the braking force caused by spring force and the rotational inertia of the weight itself, the firing pin is then triggered.

(2) Sensor non-activation conditions: If the level of acceleration is low during an actual vehicle crash at a low speed, during rough road running, or during overload applied to the steering wheel, the weight is kept retaining by spring force and does not rotate from its initial position. If the level of acceleration is relatively high, the weight may rotate against the spring force. However, since the rotational inertia of the weight restricts abrupt weight rotation, the weight returns to its initial position after the level of acceleration drops.

Furthermore, Figure 7 shows the results of the check conducted by simulation to see if the sensor of this system conforms to the $\Delta V - t$ characteristic generally necessary for airbag sensors. Since a rightward rising curve is derived from the data plotted in the graph, it is understood that the sensor has

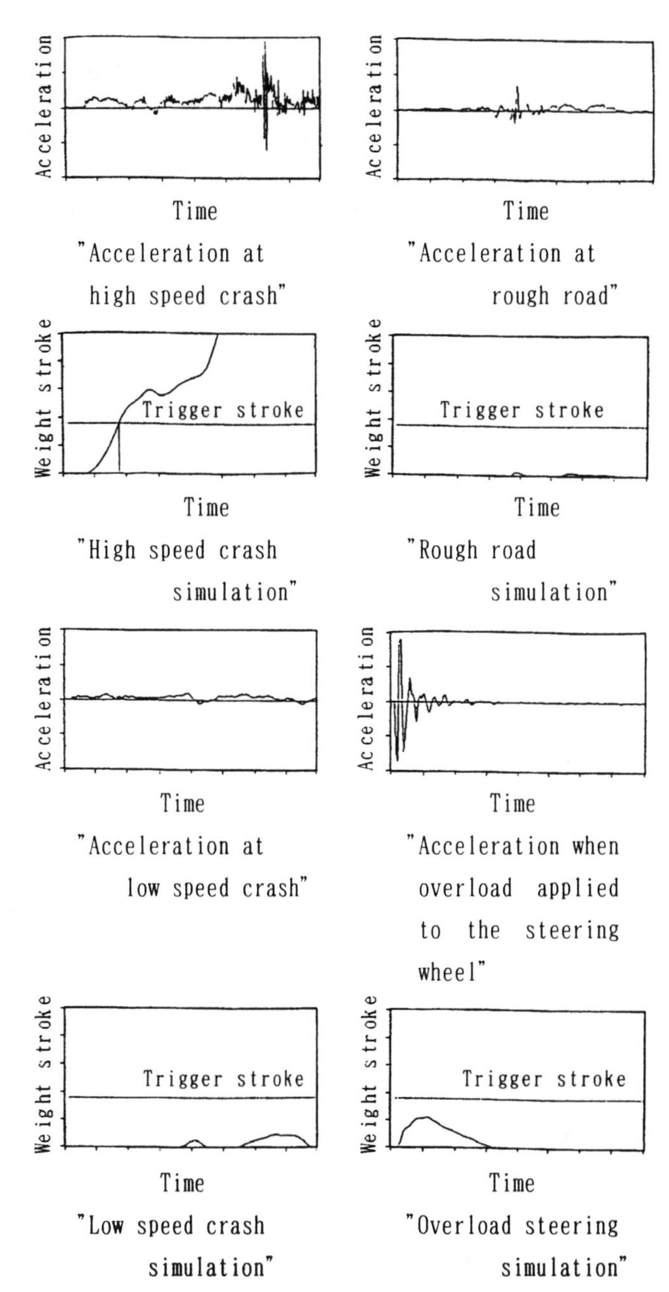

Figure 6: Acceleration and Weight stroke

satisfactory performance as an airbag sensor. [1]

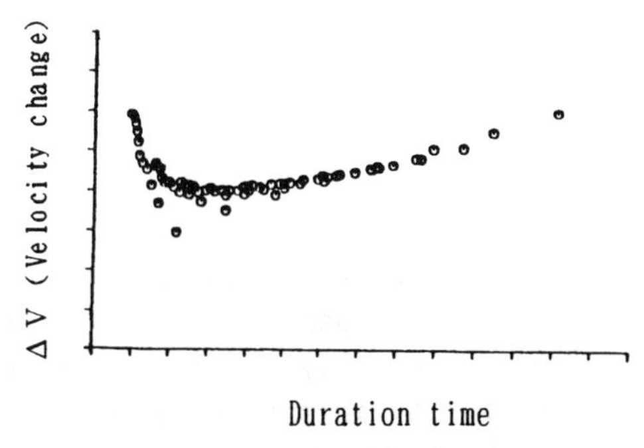

Figure 7: Detecting characteristics of the sensor

COMPARISON OF CALCULATED SIMULATION VALUES AND EXPERIMENTAL VALUES - To evaluate the validity of this computer simulation model, we checked the sensitivity performance of the simulation model and a prototype by using various sine half-waveforms and actual vehicle crash acceleration waveforms. Figure 8 shows the results of the comparison of sensing time of the simulation model and the prototype. As shown in figure, it is clear that the calculated simulation values are accurately matched with the experimental values.

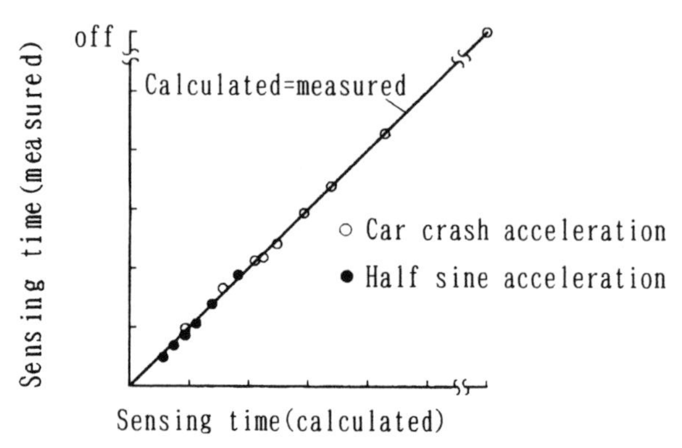

Figure 8: Comparison of sensing time

QUALITY CONTROL OF THE SENSOR - To control the quality of the sensor, we complete checks as follows.

Design Stage -

(1) By using the above-mentioned computer simulation model, the settings of the sensor are determined so that the sensor has predetermined crash decision performance even if each parameter changes within its tolerance.
(2) All of the main parts for determining the crash discrimination performance, such as the weight, the firing pin, and the torsion spring, are designed to be factors.

Manufacturing Stage-

(3) The sensor is inspected to see if it meets the performance requirements by applying several patterns of gravitational acceleration to the sensor using an impact tester.
(4) The sensor is inspected to see if the kinetic energy of the firing pin is enough to fire the igniter by measuring the initial release speed of the firing pin.

ASSEMBLIES

Important development factors for the assemblies of the sysytem are described below.

We made a pad module assembly by using unpainted single-layer injection plastics.

By evaluating and improving the inflation performance, appearance, quality, formability, light resistance, heat resistance, and scratch resistance of a variety of materials, we have developed a material capable of meeting the requirements for the pad module assembly.

The airbag assembly has also achieved the following improvements, when compared with conventional airbag assemblies.

30% lighter
12% smaller
14% fewer parts

In the performance of the system, the time between a crash and the completion of airbag deployment can be shortened by using the unpainted single-layer injection pad and the lightweight airbag.

In addition, we have confirmed that this newly-developed airbag system has sufficient occupant protection performance.

CONCLUSION

(1) We have developed a mechanical-firing microminiature sensor using rotational inertia effect.
(2) We have developed a simulation method capable of reproducing actual sensor activation. This facilitates the tuning of the sensor performance.
(3) We have improved the airbag deployment characteristic by modifying the pad and airbag.

REFERENCE

[1] Robert W. Diller, "New Sensor Develoments Leading to Sensor System Simplification," SAE paper no. 841218, (1984)

[2] Hiroaki Shinto, Kenji Ogata, Fuminori Teraoka, and Mitsuhiko Fukabori, "Development of the All-Mechanical Air Bag System," SAE paper no. 910149, (1991)

1999-01-0758

Fast Response Micro-Safing Sensor for Air Bag Systems

Masatomo Mori
Akebono Brake Industry Co., Ltd.

ABSTRACT

To introduce a new micro-safing sensor for automotive applications with excellent characteristics based on silicon micro-machining techniques to prevent malfunctions in air bag systems using an acceleration switch. This sensor is composed of an upper electrode, cantilever, and pedestal formed by silicon micro-machining. The upper electrode and cantilever contact mechanically when acceleration during collision is applied and the "ON" signal is output. To achieve a satisfactory result, we developed an optimal shape design using FEM simulation. Also, the frequency characteristic of 100Hz and threshold acceleration of 2–4 (G) are obtained by electrostatic force and chattering measure at the contact point. Other characteristics include high-speed response, compactness, and the fact it is inexpensive, making it highly suitable as a safing sensor for side air bag systems.

INTRODUCTION

Control systems such as ABS and air bags are currently installed in automotive vehicles for safety. Such equipment depends on sensors. Concerning air bag systems, most vehicles are equipped as standard with air bags for a front collision at present, while the development of a side air bag to provide protection from a side collision is being carried out and installation has begun on some vehicles. Safing sensors, whose purpose is the prevention of malfunctions, are incorporated into such air bag systems to detect an impact from collision simultaneously with the main acceleration sensor in the deployment of the air bag. However, there is the following problem concerning the acceleration sensor for the side air bag and the development of a sensor that can solve this problem is something of an urgent matter.

In a side collision the "crushable zone" is smaller than in a head-on collision; so in a side collision the deployment of the air bag must be quicker. Therefore, in side air bags analogue sensors are used because mechanical lead-switch type safing sensors, used mainly for front-collision sensors, have a slow speed of response. However, this has resulted in such sensors being expensive and subject to electromagnetic disturbance. These situations are shown in table1.

Table 1. The comparative table.

	Characteristics of present sensors		**Characteristics of new sensors**
	front collision	**side collision**	
Sensor type	mechanical	electronic	Mechanical
Response	Slow <50Hz	Fast >100Hz	Fast >100Hz
EMI	unaffected	Affected	Unaffected
cost	Inexpensive	Expensive	Inexpensive

We developed this new sensor with the features below using silicon micro-machining techniques to solve these problems.

- Compact Size & Easy Mounting (due to silicon micro-machining techniques)
- Unaffected by EMI (due to mechanical operation)
- Fast Response (because displacement is 3.5μm)

DEVICE STRUCTURE

The structure of the sensor is shown in figure 1. The 1st layer has the role of the upper electrode and stopper, and also it is possible to obtain an output signal without using special wiring to pass a substrate. Also, the following problems are solved by using the prominently shaped electrode(Figure2) as the contact point;

1. It prevents chattering of the output which occurs from the electrostatic force from the input voltage.
2. It prevents excessive electric current due to the accidental contact of the electrodes caused by the electrostatic force in the anodic bonding.

The 2nd layer receives acceleration through the cantilever, permitting displacement of the contact point on the cantilever. (Figure3,4)

Electrode material forming the above mentioned 1st and 2nd layer uses Ti/Pt. The 3rd layer has the role of stopper and pedestal. These substrate materials in 3 layers all use silicon and the shape is formed by KOH etching.

These substrates are joined by anodic bonding with sputtered glass film and then the sensor is sealed. Also formed are 3.5 μm of pole gaps with this sputtered glass film. The sensor element size is about 6 x 3 x 0.9 mm.

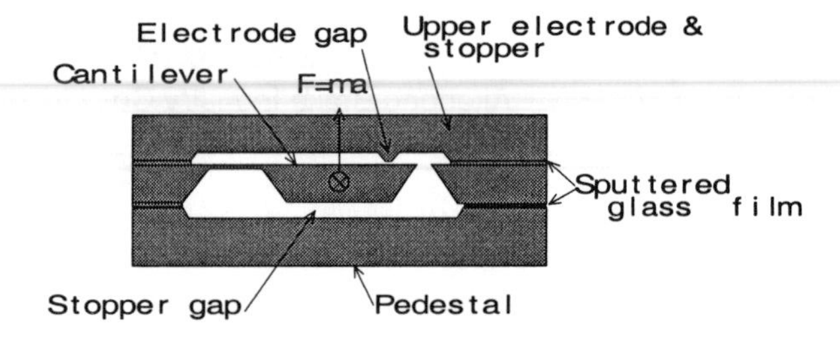

Figure 1. Structure of the sensor

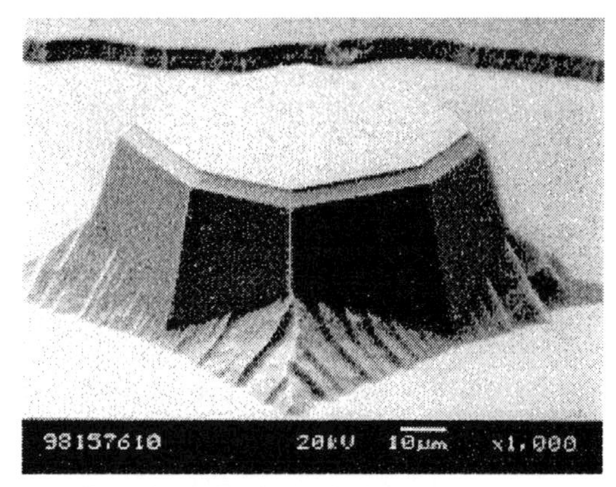

Figure 2. Prominently shaped electrode

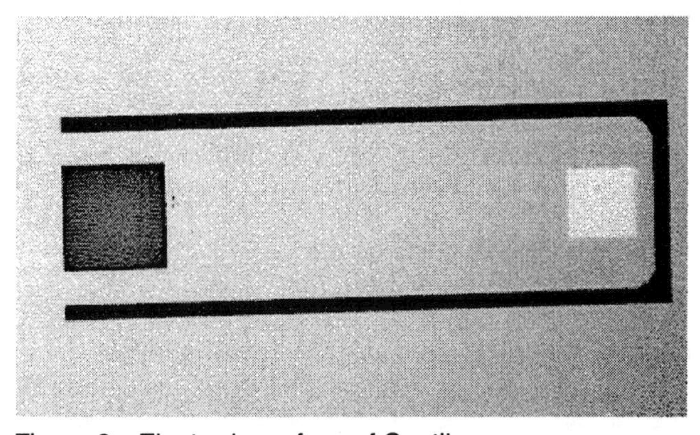

Figure 3. Electrode surface of Cantilever

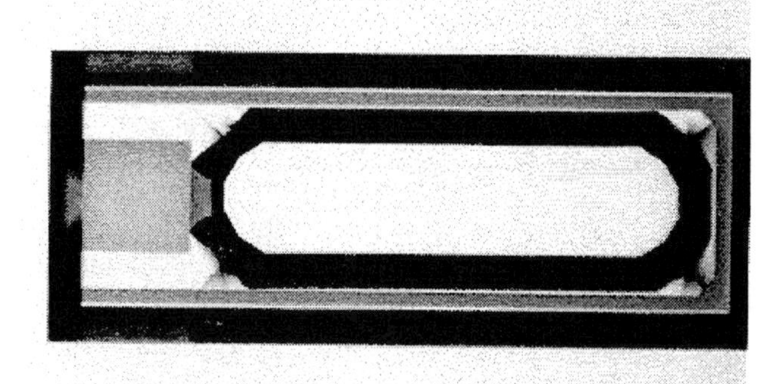

Figure 4. Etching surface of Cantilever

PRINCIPLE – When acceleration increases on the sensor shown in figure 1, the force of F=ma increases in mass. The point of contact is ON when this force is greater than the spring restoration force (kx) of the cantilever ;

$$Ma>kx$$

m :mass , a :acceleration , k :spring constant
x :displacement quantity of electrode on cantilever

The specification of the sensor is as in table 2 and the cantilever shape was then determined by FEM simulation. The simulation results are shown in table 3.

Table 2. Specification of the design

Electrode gap: x (the point of contact displacement quantity)	3.5 (μm)
Threshold acceleration	2.7 (G)
Seismic mass	0.91E-6 (kg)

Table 3. Simulation result

Spring constant of cantilever	6.9 (N/m)
Resonance frequency	131 (Hz)

CHARACTERISTICS –

The sensitivity change with the line voltage change – The sensitivity undergoes influence by the electrostatic force originated from the line voltage. Therefore, the sensitivity changes when the input voltage changes. Consequently, we made the structure that contact occurs at the prominent portion and the spring elasticity becomes dominant to the electrostatic force.

The spring elasticity Fc when cantilever does a 3.5μ displacement

$$Fc=kx$$
$$=6.9 \times 3.5E\text{-}6=2.42E\text{-}5 \text{ (N)}$$

Electrostatic force Fe when inputting 5V

$$Fe=(\varepsilon 0 \cdot \varepsilon_r \cdot S \cdot V^2) / (2 \cdot d^2)$$
$$=2.39E\text{-}7 \text{ (N)}$$

ε0 :evacuated dielectric constant
ε_r :relative dielectric constant of air
S :area of surface cantilever
V :voltage
d :displacement of the gap between the electrode and the upper part of the cantilever

Fc is dominant to Fe, so the influence of line voltage changes is adequately suppressed. The production experiment results of the sensitivity change to the line voltage are shown in figure 5, changing the gap between the electrode and the upper part of the cantilever. As for the line voltage-sensitivity change ratio, 20μ m of gaps were found to be higher.

20 μm of gap :about 0.1(G/V)
35 μm of gap :about 0.05(G/V)

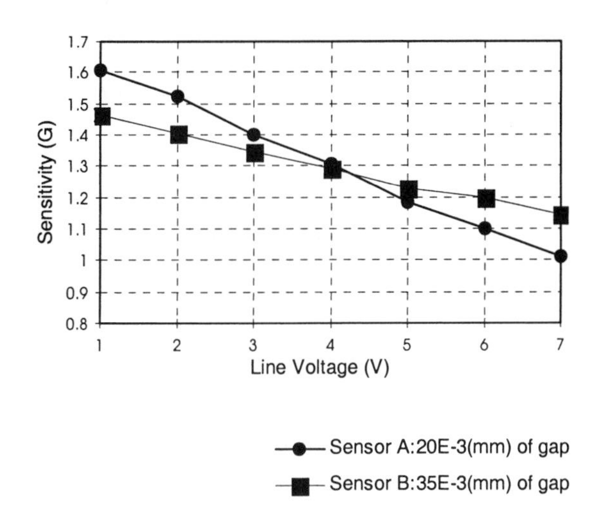

Figure 5. Line voltage-Sensitivity ratio

Chattering of the output – Chattering due to the electrostatic force can be suppressed to a degree by realizing the contact point by a prominent shape. However, when the point of contact interval approaches a minute gap, a very large electrostatic force occurs at the contact point. As a result, the chattering shown in figure 6 occurs.

Calculating this phenomenon

The area of the contact point:S=1.26E-9(m^2)
The minute distance of the point of contact part :d =2.2E-8(m)

then we get,

Fe_2=2.88E-4 (N)

It has become a value with the electrostatic force which is bigger than the spring elasticity of the cantilever.

Then, we restricted the chattering by using the damping of the cantilever, which adjusts the gaseous pressure inside the sensor. This result is shown in figure 7.

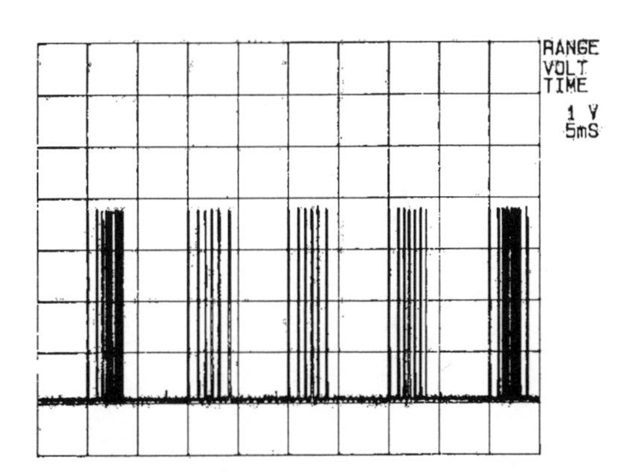

Figure 6. Pressure about 6.7E-3(Pa)
Input 100Hz/3G

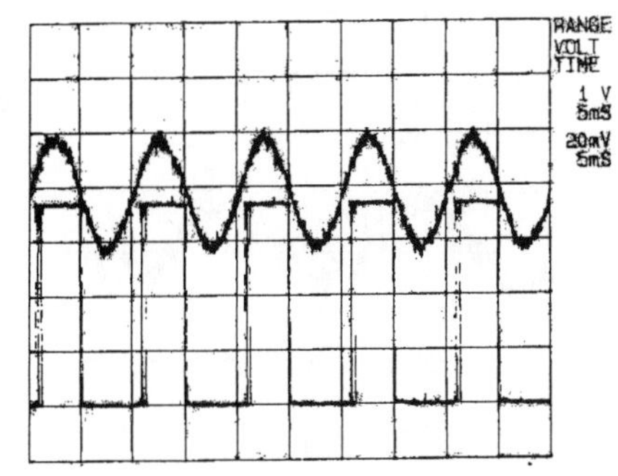

Figure 7. Pressure about 53320(Pa)
Input 100Hz/2.5G

The frequency characteristic – A measurement condition is shown in figure 8. The output of the sensor can not follow the sine wave input of equal to or more than 200 Hz and it does not reply at all at equal to or more than 600 Hz. This seems to be the reason that the resonance frequency of the sensor is about 130 Hz. Figures 9-12 shows the above result.

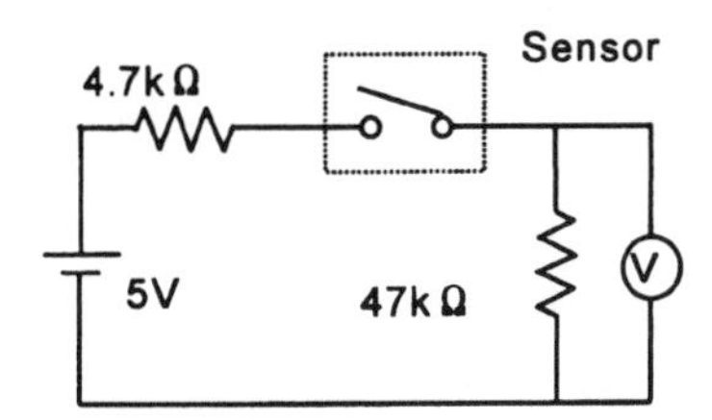

Figure 8. Measurement condition

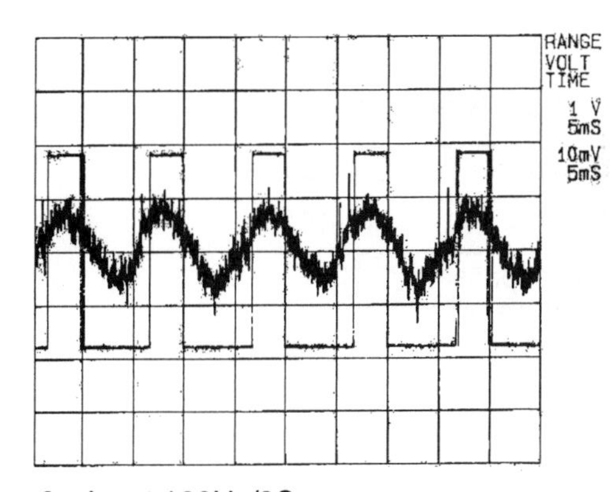

Figure 9. Input 100Hz/3G

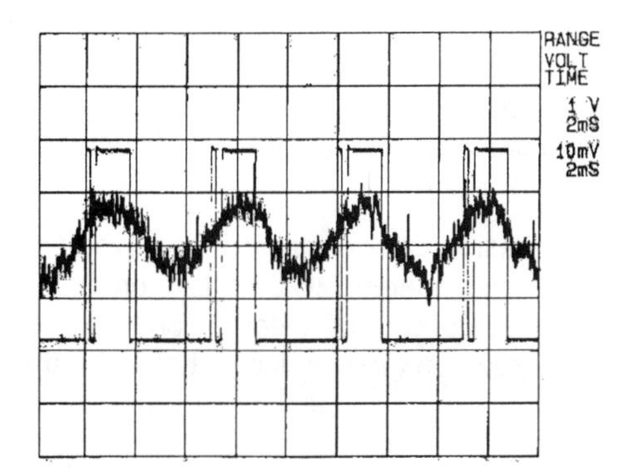

Figure 10. Input 200Hz/3G

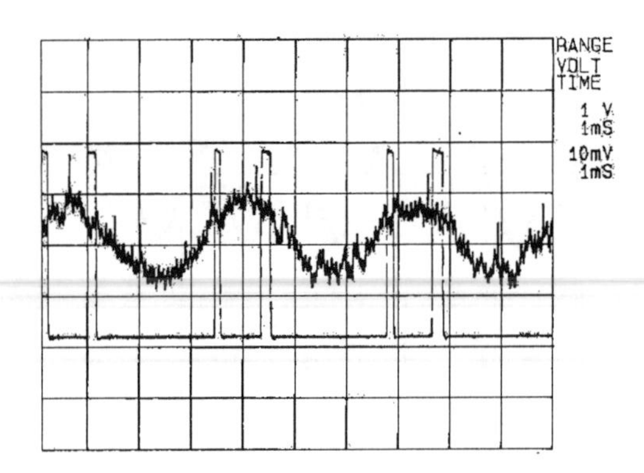

Figure 11. Input 400Hz/3G

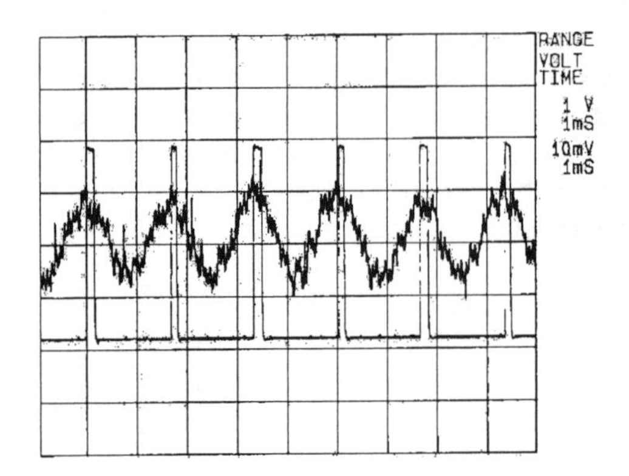

Figure 12. Input 600Hz/3G

Other characteristics – The other characteristics are shown in table 4.

Table 4. Other characteristic

Output impedance	Type. 10kΩ
Open contact	>10MΩ
Frequency response	DC ~ 100Hz
Threshold	2 ~ 4G

CONCLUSION

We made an acceleration switch using a mechanical point of contact allowing a high speed response, and could confirm a response frequency of 100 Hz and a threshold of 2-4G. In the experimental production this time the acceleration switch was at the side point of contact but next time we will make the point of contact operation in both directions of +/-.

Ultimately this design carries forward the development of a sensor with accumulated MOS FET in order to make possible the output of a large electric current necessary to activate an airbag on the identical substrate to the sensor chip.

1999-01-0765

New Aspects on Static Passenger and Child Seat Recognition and Future Dynamic Out-of-Position Detection for Airbag Control Systems

Peter Steiner
TEMIC Telefunken microelectronic GmbH, Occupant Safety Department, Ringlerstr. 17, D-85057 Ingolstadt (Germany)

Guido Wetzel
TEMIC of North America, Inc., 85 Auburn Hills, MI 48326, USA

ABSTRACT

TEMIC as an automotive electronics manufacturer has done research since several years to develop suitable sensors for occupant detection. The paper will present simulation results and technical solutions concerning this issue which is one of the key points in the future design of advanced airbags.

INTRODUCTION

Although the estimated benefits of airbags are about 3000 saved lives per year when every car is equipped with them, some residual risks with nowadays airbag technology remain. The facts show that up to mid of 1998 more then 113 deaths were reported that could have been avoided, if the airbags would have been depowered or deactivated. As an intermediate solution, it is allowed to install manual cutoff switches in cars in order to deactivate the airbag, when a child seat has to be placed on the passenger side in front of an airbag.

The new advanced airbag proposal given by the NHTSA mid of September 1998 describes goals for advanced airbag systems to automatically deactivate the airbag when risks to the passengers are detected. Therefore, new test criteria are defined which include new dummies, especially representing small persons such as babies, children and small adults. Additionally, crash tests that simulate a precrash braking situation are included in order to check situations when passengers are sitting too close to the airbag in the moment of deployment (so called „out-of-position“ situation).

GOALS OF OCCUPANT DETECTION

There are several goals, which have to be achieved in the design of occupant detection subsystems for advanced airbag application:

1st. Detection of infants in passenger side mounted child seats, mostly rear facing and too close to the airbag.

2nd.Out of position situations:

- Detection of passengers, children and adults, sitting too close to the passenger-side airbag
- Detection of passengers, children and adults, being thrown forward by non crash deceleration prior to the crash
- Detection of drivers, mostly small women, sitting too close to the driver side airbag

SIMULATION FOR SYSTEM DESIGN

In order to consider as much data as possible about the occupant's properties and its environment, powerful simulation technologies were employed in the design of occupant sensors. As shown in figure 1, the flow chart of the occupant simulation loop contains on the input side:

- CAD-data about the cars geometrical dimensions including all possible seat positions.
- Dummy data with a wide range of human body dimensions and various child seats.
- Crash data such as braking events, misuse deceleration etc. which can move the dummy without a crash.
- Sensor data including the geometrical arrangement of detection beams, weight sensitive areas etc.

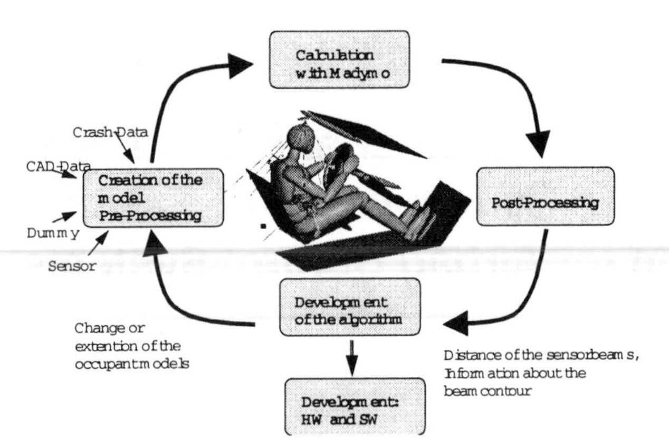

Figure 1. simulation flow chart

CHILD SEAT DETECTION – BASIC TECHNOLOGY

We employed active infrared technology combined with the triangulation principle in order to obtain contour information along a plane on the passenger side area. For child seat recognition this detection plane is positioned along the seat, as shown in figure 2.

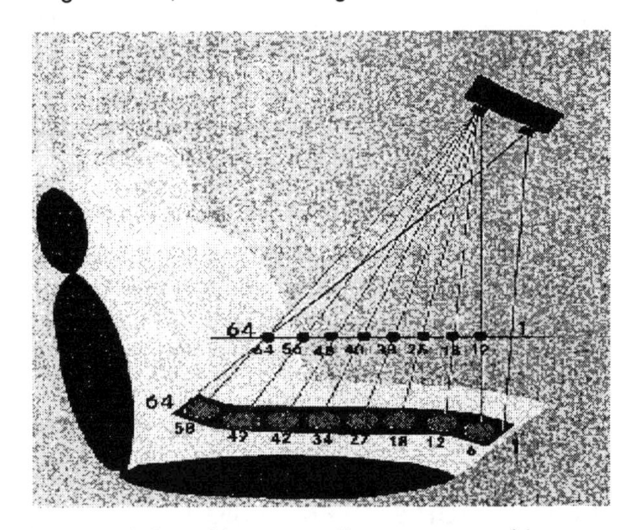

Figure 2. infrared beams on the passenger side area

The headliner mount system consists of an array of infrared LEDs that are optically projected on the seat. A CCD line observes the area along the LED spots. A shift of distance of the spot closer to the sensor gives a shift of the intensity maximum on the CCD (see Fig 3).

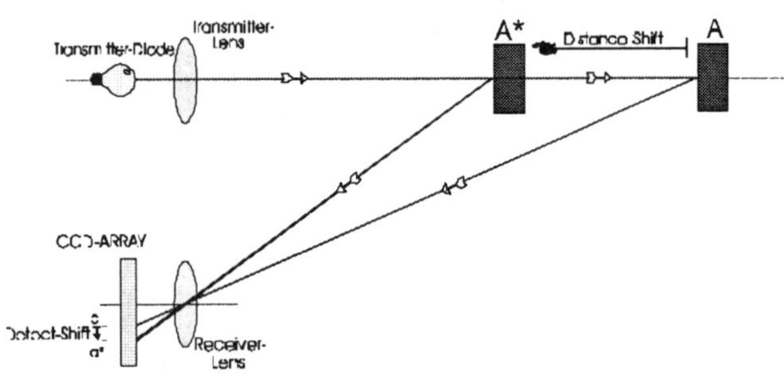

Figure 3. triangulation principle

From that information, a distance contour based on several points can be calculated. An algorithm compares the contour with characteristics of a baby in a child seat or an empty seat or an adult passenger. If it is necessary, the airbag deployment will be suppressed in a subsequent crash. The advantage of the system is that it works well with numerous child seats already in the field. There is no need for purchasing a manufacturer specific infant seat.

OUT-OF-POSITION (OOP) DETECTION

FUNCTIONAL PRINCIPLE – The identical detection principle is used for the detection of out-of-position situations. Although some out-of-position situations were already recognized by the child seat detection system, the positions of the light beams were not ideal for regular human sizes. Therefore, we inclined the detection plane more upwards so that the light beams end in the lower door area (see Fig 4 for both driver and passenger side).

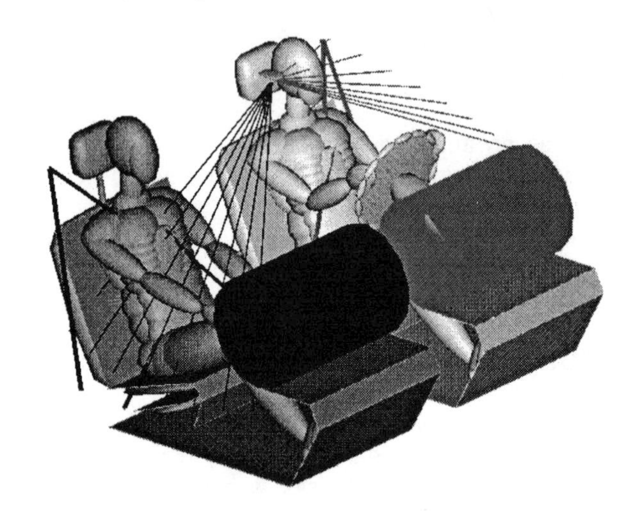

Figure 4. driver and passenger side Out of Position detection

The light beams now hit the head or upper torso of an adult occupant. During forward movement of the occupant, he passes the invisible curtain of light beams, while reducing the distance to the sensor and changing the contour. During its movement towards the airbag, the head or torso of the passenger gives a ditch in the contour. The position of this ditch represents the actual position of the occupant (Fig 5).

Passengers entering the keep-out-zone, meaning the danger area around the airbag module, can be monitored by the system.

DYNAMIC REQUIREMENTS – Madymo-simulations were performed to adjust the ideal position of the detection plane and to obtain information on the desired speed of such a system in precrash situations. Figure 6 shows the distance shift over time of each LED spot during a precrash braking event with 1 g applied. The occupant passes on his way forward one beam after the other from beam 8 (rear seat area) to beam 2 and 1 (keep out zone area).

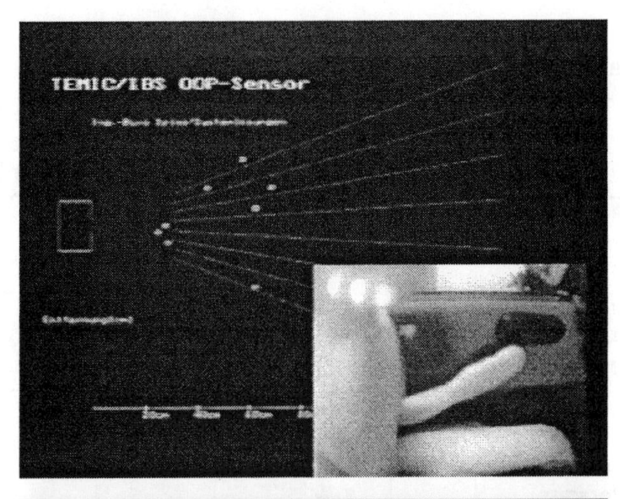

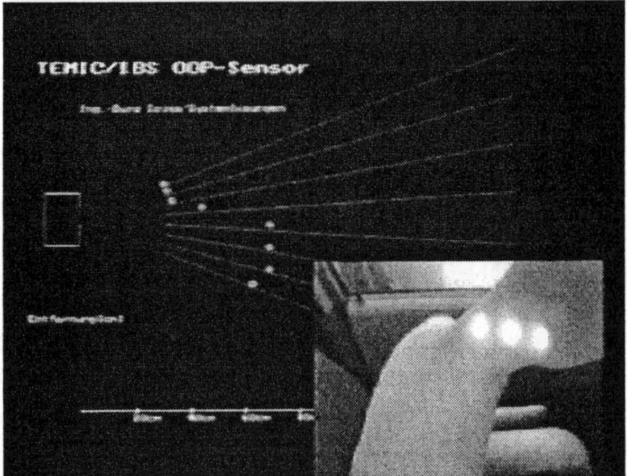

Figure 5. Occupant position represented by the sensor

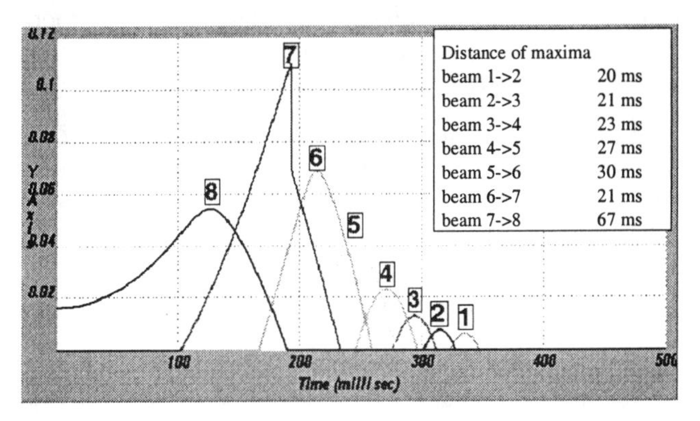

Figure 6. Sensor readout during precrash braking

When the human body enters a beam, the distance is reduced drastically. Due to the acceleration, the distance of the maxima between one beam to the other is reduced from 67 ms (beam 7→8) in the beginning of the movement down 20 ms, when the occupant is in the keep out zone. From that behavior, a read out time for the sensor of at least 20 ms can be derived. We strive for a read out time of 10 ms in our hardware design, which will be sufficient for that high dynamic situation.

The advantage of the sensor is that the working principle is based on the forward movement of an occupant. Therefore, the OOP-detector cannot be easily fooled like simple proximity sensors i.e. in the dashboard. Object movements like folding a newspaper are distinguished and filtered out by their different characteristics.

POSITION OF DETECTION PLANES

In order to find out the ideal position of the detection plane, we used again Madymo with different dummy sizes and forms. The body was also moved slightly to the left and right from the straight line towards the airbag. After several loops of optimization, we found a detection plane consisting of 8 LED's that works nearly perfectly with standard adult models (50% female to 95% male).

However, issues arose with small persons moving to the center of the car by submarining under the detection plane (Fig 7). The NPRM is dedicated to just this group of persons, since those persons are expected to be most severely hurt by an airbag-deployment.

We added a further degree of freedom to our design by using a second detection plane and restarted the optimization process. The result (Fig 7) was that the first detection plane should be elevated by some degrees and a second detection plane was introduced which is quite close to the detection plane for the child seat recognition.

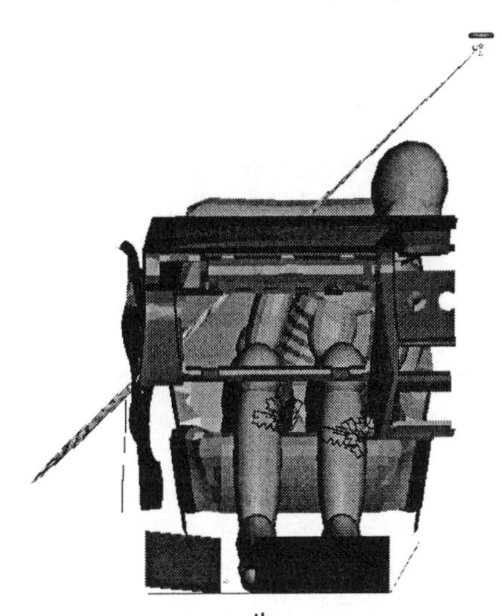

Figure 7. submarining of 5th % female

This means that the child seat detection plane could also be used for OOP-detection of small persons. As shown in figure 9, the occupant detection system on the passenger side consists of two detection planes, one plane (no.1) with multiple use either for rear facing infant seat detection or - in combination with plane no.2 - for detection of small occupants in OOP. The plane no.2 only detects the OOP-situation of tall adults.

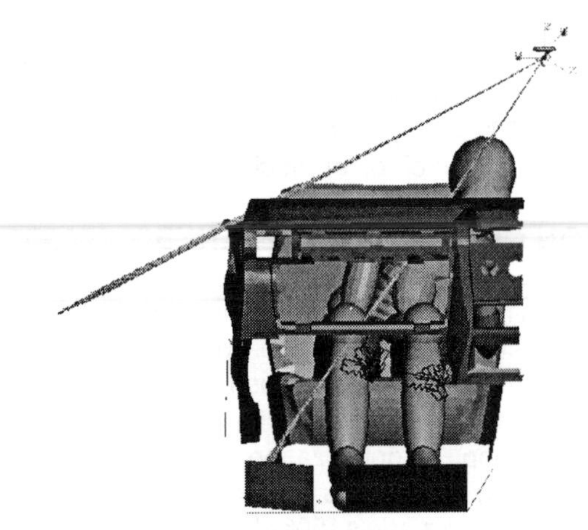

Figure 8. Madymo system simulation

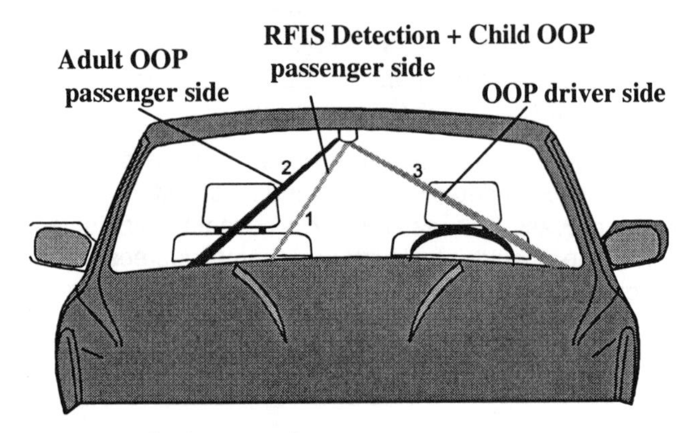

Figure 9. final system layout

Additionally, one detection plane could be added for OOP-recognition on the driver side (Fig 9). The work for this task is still under progress. We expect that one detection plane should be sufficient, since the driver is more fixed in his position relative to the steering wheel.

OCCUPANT DETECTION SOLUTIONS – MODULAR APPROACH

The final sensor system can be a fusion of different sensor technologies as shown in Tab. 1. These different sensors can be combined as needed. If for example a system is needed for rear facing infant seat detection and a basic OOP detection – the system with the single plane 1 is sufficient. Additional information can be obtained with the second or the third plane, and can be complemented by a weight sensor. If this information is needed for passenger presence detection, this information is redundant – but can also deliver additional information about the weight of the occupant to further control the power of the installed airbag.

Table 1. modular approach

Sensor Setup vs. Task	Single plane No. 1	Single plane No.2	Dual plane No.2+3	Triple plane No.1+2+3	Weight Sensor
Rear Facing Infant Seat Detection	√			√	
Passenger Presence Detection	(√)			(√)	√
OOP Passenger Side (Small Persons)	√			√	
OOP Passenger Side (Adults)		√	√	√	
OOP Driver-Side			√	√	

DISCUSSION

With the newly proposed rule, NHTSA did a good job in identifying all possible, dangerous situations in the car and passenger relating to the airbag.

It is now the objective of the OEM or the supplier to come up with a suitable solution. Our approach is the discussed optical sensor.

This sensor is easy to simulate with the means of a modern technical environment (MADYMO in this case) and is fairly easy to implement.

If we can agree to an area of 'danger' around the airbag, we can identify any 'intruding' object in this zone. With the sophisticated approach of our algorithm it is possible to 'filter' out the 'disturbing' objects.

To show the challenge, we can have a look at the different possible seating positions in a car.

All these special positions (Fig.10), regardless if forbidden by law, are actual possibilities – so the sensor system must account for these.

It was the task, to find any possible position – so we ended up with some thousands of entries in our database, and it became the challenge to reduce these data to a feasible test matrix with common denominators, which still makes sure that it covers every possibility.

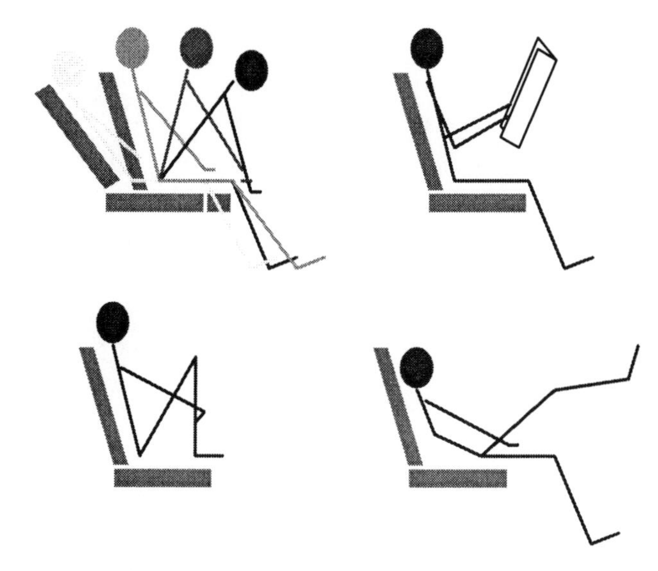

Figure 10. Seating positions

CONCLUSION

The presented system takes the following approach:

Child seats can be identified. It is a matter of the system layout whether the airbag is switched off or turned to a less dangerous stage.

The next dangerous situation is presented by the NHTSA in the precrash braking conditions. It is a possible scenario where the occupant comes close to the keep out zone and the airbag can be of harm instead of being a life saving equipment.

The presented system was laid out in simulations. We took a car interior where occupants were moved forward by the deceleration due to a major braking maneuver. The current system is able to catch movements in up to 10ms. That's what we have identified as necessary for a precrash braking of up to 1 g and is also described as a development goal in the NPRM.

With the velocity and the fast recognition time we come into a discussion of the security of the optical system. The ultimate goal is to stay within the laser class 1 – so that we do not put more danger in the car with the additional system.

ACKNOLEDGEMENTS

We like to thank the members of IBS in Seibersdorf (Germany) for their creative ideas and the cooperative style in working together on this topic.

REFERENCES

1. Federal Motor Vehicle Safety Standards; Occupant Crash Protection. 49 CFR Parts 571, 585, 587, and 595 Docket No. NHTSA 98-4405 Notice 1, RIN 2127-AG70

DEFINITIONS, ACRONYMS, ABBREVIATIONS

OOP	OUT Of Position
NHTSA	National Highway Traffic Safety Administration
NPRM	NHTSA Proposed Rule Making

960226

The BMW Seat Occupancy Monitoring System: A Step Towards "Situation Appropriate Airbag Deployment"

Klaus Kompaß
Bayerische Motoren Werke AG

Michel Witte
Interlink Electronics Europe S.A.R.L.

OVERVIEW

Future developments of airbag systems, which are now considered as standard equipment in cars, will focus on three main topics:

- new developments in airbag application fields which have not previously been fully investigated. ie: side airbags,
- cost reduction and cost effectiveness of existing systems via the introduction of uniformity in design, thus maintaining and guaranteeing quality levels and product availability,
- automatic adaptation of both existing and future airbag systems to the requirements of individual vehicles.

The aim of this presentation is to illustrate this final point. We hereby present a system which avoids unnecessary airbag deployment when the passenger seat is unoccupied.

Due to the increasing amount of cars equipped with passenger airbags, there are now a larger number of accidents during which passenger airbags are activated although nobody is actually sitting in the passenger seat. Even when one discounts those accidents which result in car "write-offs", the unnecessary deployment of the passenger airbag increases repair costs in an unjustified way. In order to avoid this type of situation, all BMW vehicles as of June 1994are equipped with a seat occupancy monitoring system (SOMS) installed in the passenger seat.

The SOMS consists of a polymer sandwich with integrated sensors whose resistance values change proportionally with applied force. The sensor mat layout is tailored to each individual seat design. Due to this flexibility of design, the sensor matrix may be mounted in three different ways:

- between the seat foam and the seat cover,
- between two layers of foam,
- between the shell of the seat and the foam.

Sensor resistance is measured electronically. The signal "passenger seat unoccupied" is obtained and the airbag-ignition-circuit is switched off only when the sensor resistance lies between two pre-defined values. Outside of these two values, and this does include if the system is damaged, the seat is always considered occupied and the airbag-ignition-circuit remains "on standby".

This presentation sets out to explain the technology on which the system is based. It also details the history of the system's conception and development as well as the different criteria it must satisfy.

INTRODUCTION

PREFACE

"Do you work in airbag development? There is nothing left to develop on an airbag, you find them in most cars nowadays!"

Often one hears this or similar responses when explaining to friends or neighbours what you do for a living. One would like, at this point, to inform them of new generator fuels, improved coatings, optimum inflation times, but one also knows that the explanation would bring only a polite smile from your "audience".

Is your neighbours reaction justified? Certainly not! It is considered by a lot of people that we are now in a phase of airbag development which is comparable to safety belt development at times when all cars were fitted with static safety belts. Similar progress as that made in the safety belt arena (inertia belts, pre-tensioners, force limiters, etc) will surely follow in the ongoing development of airbag systems. Today's buzzword is "SITUATION APPROPRIATE AIRBAG DEPLOYMENT".

SITUATION APPROPRIATE AIRBAG DEPLOYMENT

This phrase encompasses a vision of a fully automated airbag system whose activation is specifically determined by the safety needs at a precise moment:

- if a passenger is seated on the front edge of the seat the force of deployment must be reduced,
- reboard (rear-facing) baby seats must not be hit by the airbag,
- passengers who are already protected by seat belts only need the additional protection offered by an airbag in cases of higher collision speeds,
- airbag is not required to deploy if no passenger is present in the seat,
- the automatic control system reduces the effect of temperature on the combustion of the airbag fuel.

STATISTICS

In order to underline the necessity of these developments we would like to present the following facts and figures:

BMW's Accident Research Centre states that a BMW vehicle is in average, occupied by only 1.4 passengers. This means that a passenger is present in only 41% of BMW vehicles. This figure is indicative of the general situation in Germany. In the USA this figure is far lower with only 32 out of every 100 vehicles carrying a passenger.

If one supposes, theoretically, that all vehicles are equipped with airbags for both driver and passenger, as well as pyrotechnical safety belt tensioners and that the total number of vehicles is 30 million, one can safely assume there will be in the approximately 100,000 accidents each year where the collision speed lies between 25 and 40 km/h. Between these speeds the airbag must be activated and car damages can still be repaired. For your information, the safety belt tensioners would be activated at lower collision speeds, but it is not our aim to fully explain the statistics of such events. As a passenger is present in only 41% of the 100,000 collisions in which the airbag must be deployed, there are 59,000 unnecessary actuations of the passenger airbag. Assuming the not unreasonable amount of DM 3000.- (£1250, US $1900) per vehicle to repair the damage caused by the passenger airbag deployment alone, one can quite easily see the dramatic annual economical effect of some DM 177 million (£71m, US $111m) per year, unnecessary airbag activation can have.

These figures illustrate the need for measures to avoid unnecessary airbag deployment.

DEVELOPMENT OF AUTOMOTIVE SEAT OCCUPANCY MONITORING SYSTEM (SOMS)

AIRBAG TASK FORCE OF GERMAN CAR MANUFACTURERS

At the end of the 1980's, BMW had already gathered initial ideas and explored various concepts for SOM systems. At the same time BMW began with a one development.

As the basic statistical data is not influenced by vehicle make or model, it was decided, in 1990, by five leading German automotive manufacturers (Audi, Mercedes-Benz, Porsche, Volkswagen and BMW) to form a task force and to co-operate to find a solution to the problem and to develop an Seat Occupancy Monitoring System for passenger seats.

Their work concentrated on specific areas:

- Definition of the specification for such a system,
- collection of ideas for proposed solutions (it was decided to ask for solutions from as many sources as possible),
- comparison and evaluation of the proposed solutions.

REQUIREMENTS OF THE AIRBAG TASK FORCE

The primary requirement of the SOM system is to offer a fail safe system; in the event of damage or inconsistent signals, the system must act as if the passenger seat is occupied. This is in order to ensure airbag activation during a collision even if the SOM system is restricted in any way. It is, under such circumstances, preferable to activate the airbag and to save say a briefcase than to risk non-deployment of the airbag a passenger being present.

The SOM system:

- must recognise the presence of a passenger only (ie not a briefcase, box of groceries, etc), but not necessarily the position or posture of the person,
- is not required to differentiate between a baby seat and a passenger,
- must give a "passenger seat occupied" signal when a passenger is sitting in any position in the seat,
- should not influence the design of the car interior,
- should have no effect on the seat comfort.
- should guarantee a safe function throughout the lifetime of the vehicle (for this point one must also consider high mechanical and environmental constraints),
- should detect a safe weight of 12kg or more,
- should offer a "seat unoccupied" signal for weights below 12kg,
- must guarantee that the "seat unoccupied" signal is not influenced by any interference,
- must have the capability to be monitored or controlled via two data lines connected to the central "airbag release control unit",
- must be easily adapted to different car seats, interiors, etc,
- must be compatible with seat heating systems where necessary.

CHOOSING A SUITABLE SYSTEM

All offered SOM systems where evaluated based on the task force's initial guidelines. In order to decide for the most appropriated solution a cataolgue for evaluation was created.

Only the major points of the original specification have been highlighted in the following descriptions of the various systems (and the associated arguments for and against each option). We must stress that the following technologies are only those which were presented at the time of the task force. Due to increasing interest in such systems, in the future much more research time shall be spent to improve these technologies, so from time to time evaluation results may change. This report however, can only discuss the situation as it is now.

The proposed technologies were:

Capacitive measurement systems

Capacitive systems use the car's seat and body as capacitor's electrodes. The dielectric separating the electrodes is made up of either air (in the case of an unoccupied seat) or a human body. The change in capacitance caused by the change in dielectric can be measured. If the capacitance falls between two pre-defined limits the seat is considered to be occupied.

This option is attractive because of:

- easy integration into the seat,
- robustness,
- cost effectiveness,
- availability of components.

The problem with capacitive systems is that with a total capacitance of 150pF, the difference between the detection of occupied and unoccupied seats allows only 15 to 20pF. This option was, therefore, evaluated as being too sensitive to interferences.

UHF-Sensors

Seat occupancy detection can be performed in various ways using UHF sensors:

- Microwave rays can measure distances between obstacles, however, additional supplementary information is required, such as the position of the seat, in order to differentiate between a passenger sitting in the rear of the vehicle and an unoccupied front passenger seat.
- It is possible to measure dynamic movements of an obstacle located between transmitting and receiving antennae. One problem, however, remains it is impossible to differentiate between an unoccupied seat and an inactive (perhaps sleeping) passenger.
- A third possibility is to detect reductions in radiation damping between the transmitter and the receiver.

The criteria against UHF systems were the high costs involved, for example the fitting of antennae into the dashboard and the back of the seat. Furthermore, there are psychological barriers to overcome due to continuous exposure of the passenger to low level radiation.

Infra-Red-Sensors

Reflected body temperature is considered to be a measurable variant. Sensors are used to measure the reflection of, or the change in, temperature over a pre-determined period of time.

This type of system was not successful as no accurate (and regular) signal could be obtained due to differences in environmental temperature, clothing, seat heating, etc).

Optical detection systems

Again there was more than one way to use optical systems:

- Light barriers supervise a defined zone within the passenger seat area. If the light beam is interrupted the seat is deemed to be occupied. The problem with this option was the difficulty to define accurately enough the required zone to be supervised. It was not possible to obtain a reliable signal. As a result this type of system was judged unfeasible.
- Optical displacement detectors, such as those used in security systems, were evaluated. The idea was to detect the movement of the passenger, again, within a pre-determined area. The system proved to be too insensitive. In addition to this it was impossible to exclude the possibility that an inactive (sleeping) passenger could be detected as an unoccupied seat.
- Pattern recognition CCD-Chip mounted in front of the passenger seat area registers the image it "sees". This image is electronically compared to the image of an unoccupied seat. This system is very innovative, but at the time of it's presentation many negative aspects were noted; cost, complexity of the system, assembly difficulty, expensive adaptation. Furthermore, a safe detection would also be made more difficult in varying light (floodlights, etc).

Radar / Ultrasonic-Sensors

Similar to the UHF-Sensors, waves from a transmission antenna are influenced by the obstacle in their path (ie human body) when the seat is occupied. The problems which arose were almost identical to those experienced with the UHF-Sensors; expensive, difficult to integrate, plus this system does not give information when the passenger is sitting out of position.

Fibre-Optics

When the seat is occupied, fibre optic strands embedded in the seat are distorted. This distortion, or bend, alters the intensity of the light travelling through the strands, which is translated into a "seat occupied" signal. The major problems with this system is it's high fitting cost, plus the results it can achieve can be affected by the change, through time, of the tension of the seat cover.

Piezoelectric-Cable

The principle of this system is based on continuous changes of the voltage of the piezo-elements, when the seat is occupied, due to the passenger's plus the car's own vibrations during driving. This change in voltage is evaluated and indicates whether or not the seat is occupied. The trials by BMW failed due to the driving comfort experienced in modern vehicles. On a smooth road the test sample indicated an "unoccupied seat" when in fact the passenger was sitting still (eg sleeping), because the vibrations of the passenger could not be identified.

Pressure Measuring System based on Force Sensing Resistors (FSR™)

A system was offered by INTERLINK ELECTRONICS EUROPE, a Luxembourg based company. Their system could measure static forces thanks to an array of FSR™ sensors. All requirements listed by the German Airbag Task Force seemed to be satisfied by the Interlink system. As an added bonus, the costs involved corresponded favourably with the original price targets of the automotive industry.

Functional properties of the FSR

The Force Sensing Resistor (FSR) is based on a reproducible surface-effect. The technology had been patented worldwide by INTERLINK ELECTRONICS.
The typical construction of a FSR™ , as it is shown in figure 1, is based on a sandwich of two polymer films or sheets and a spacer. On one sheet a conducting pattern is deposited (screen-printing) in the form of interdigitating electrodes. On the other sheet a proprietary semiconductive polymer is deposited. Finally the two sheets are faced together so that the conducting fingers are shunted by the conducting polymer.

Structure of the force-sensing resistor

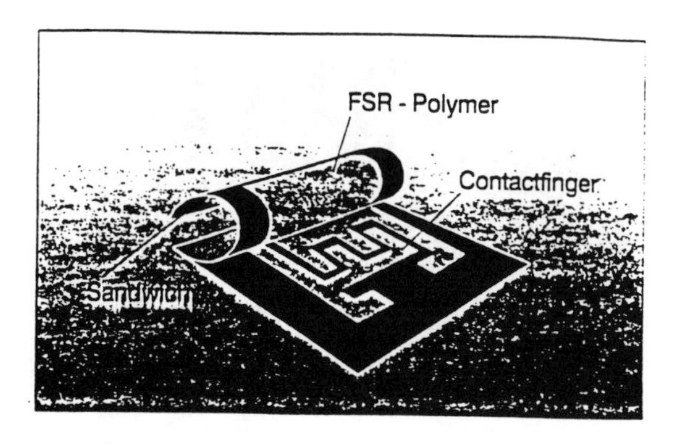

Figure 1: **BASIC FUNCTION OF THE INTERLINK FSR ARRAY**

If a force is applied upon the FSR the electric resistance of the sensor changes according to the amount of the force. Depending on the size of the used force actuator and the amount of the force, a more or less greater number of interdigitating electrodes are short circuited. As all these resistance bridges are shunted by the semiconductive polymer, the device follows the law of Ohm:

$$(1/R_{tot}) = (1/R_1) + \ldots\ldots +(1/R_N)$$

If the force with the corresponding actuator-surface and the resistance are marked up in a diagram with log-log coordinates, as it is shown in figure 2, you can see an inversely proportianal relation between the force and the resistance. The course of the shape can be adapted by INTERLINK to the specific application.

cycles are guaranteed. Extented tests proved that the used materials do not react with humidity.

Resistance/pressure curve of the FSR - element

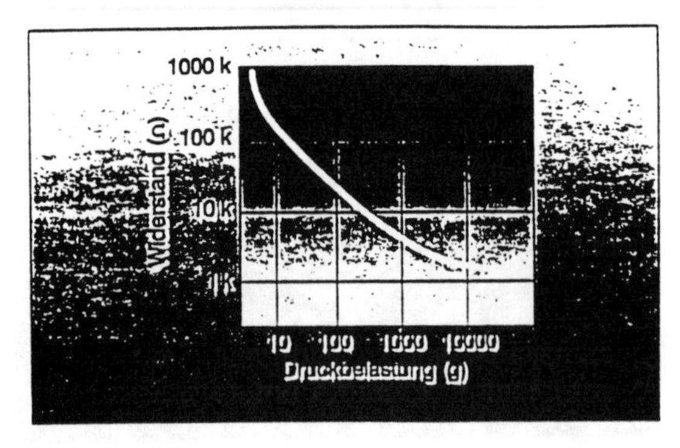

Figure 2

The FSR can be used as an intelligent switch. By defining different thresholds you can get another degree of freedom for the optimal adoption to the application.

In the case of the SOM the sensor is designed in such a way that a child with a weight of 12 kg is detected in any position in the seat. If the resistance is above a fixed threshold the seat is "unoccupied". If on the other hand the measured resistance is below a fixed threshold the system gives the message "seat occupied".
The sitting-surface had been fixed to a circle area diameter of 120 mm. By interrelating the force to the surface it is possible to avoid that a briefcase or a bag etc. causes a decrease of the resistance below a fixed threshold.

Mechanical characteristics of the FSR

The laminated thermoplastic supporting foils can be used at a temperature of -40 to 170 C. The spacer is resistant to a temperature of -40 to 150 C. At temperatures above 120 C we recommend a temperature-compensation because of the hardening of the spacer-foil.
The FSR can be designed in any possible form, that can be shaped on a two-dimensional CAD. The thickness of the FSR is of about 0.5mm. To facilitate the application of the FSR, the sensor can be furnished with an adhesive on the backside. One million of

Figure 3: **FSR-SENSORELEMENT**

One of the most important advantages of the FSR™ in comparision with membrane switches is the fact that you need only a travel of about 70 μm to release a defined signal.

ADAPTATION OF THE FSR™ TO THE SEAT

Each car manufacturer has its own conception of optimal seat-ergonomics and -comfort. In addition the costs are an important factor in the design of car-seats. So actually you can find three different types of seats:

- full seatfoam on a spring-wire- or on a spring-mat-support,
- full seatfoam on a seat-shell,
- rubber-hair-mat with cushion-foam on a spring-wire- or on a spring-mat-support.

In order not to modify the construction of a seat, the FSR™-sensor-mat has to be adopted to the seat. Beside the specifications of the task force as mentioned above, the large tolerances for the seat design and mounting must be considered.
For each seat-type Interlink developed one

specific sensor mat. The sensor-mat is designed in such a way that different kinds of seat cover can be used. For the future it would be easier and less expensive if the sensor mat could be integerated into the seat foam.

DESIGN AND LAYOUT

The criterias for the design of the sensor mat are the following:

- specifications of the task force,
- place of installation of the sensor mat,
- optical tracing of the sensor mat on the seat cover,
- stress in vertical direction and torsional stress,
- full redundancy and
- costs.

The different places of installation lead to divergent designs of the sensor mat.

Assembly between the seat foam and the seat cover

This alternative is selected if there is not enough support space in the area of the spring-wire- or of the spring-mat-support. In this case the main design criterias are:

- the seat ventilation,
- the seat ergonomics,
- the mechanical stress and
- the mounting tolerances.

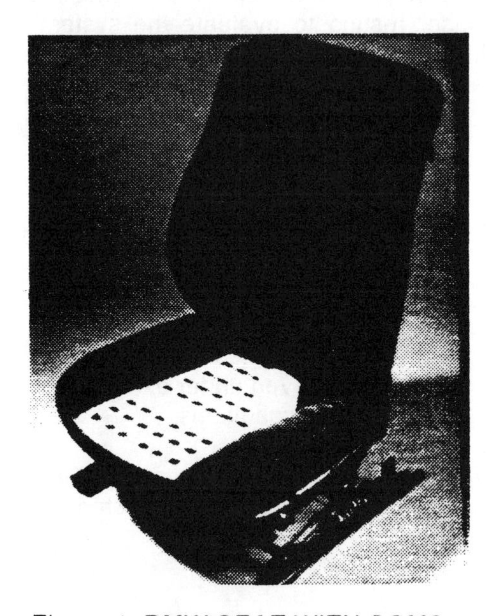

Figure 4: **BMW SEAT WITH SOMS**

For this alternative the sensor mat is designed in the shape of fishbones. This design allows an optimal adoption to the torsional stress. To ensure an optimal ventilation of the seat and to avoid optical tracing on the seat cover, the sensor mat is designed in such a way that the support foil is placed near the groove of the seat. To facilate the mounting process the mat is placed on a felt support. By using felt the sensor is uncoupled with the seat cover. The different sensor elements are designed in such a way, that there are no influences from different seat cover tensions or from initial loadings that are caused trough the mounting of the seat.

Assembly between the seat foam and the closed seat shell

This assembly alternative allows the design of a smaller sensor-mat. The tolerances of the seat foam and the seat shell are important criterias for the design of the sensor. By mounting the sensor mat directly on the seat shell there is only little mechanical stress. The problem in this case however is the curve of the shell. Therefore the sensor must be designed in such a way, that there are no initial loadings on the sensor.

Redundancy - fail safe

The FSR™-sensor-mat is designed so that each sensor element is part of the conductor. The conductors are traced all over the sensor mat and form two of the four connectors of the sensor. A diode is placed between the other two connectors. So it is possible to check the circuit and the connectors by a simple short circuit test in low-resistance direction of the diode. The failure of one or few elements of the sensor mat will have little influence on the function of the sensor mat. A failure of the circuit or a short circuit however can have fatal consequences.
For the measurement of the applied force the circuit is switched in high-resistance direction of the diode. The electronics is designed so that all failures cause low resistance and thus the signal "seat occupied".

Low-Pass-Filter

The looped conductors of the sensor mat are printed in a constant distance. The sensor mat has a defined natural capacitance. The FSR™ sensor-mat is a very primitive low pass filter. Anyway one will include on the interface connection an additional low pass filter. This will be more than enough to protect the system against EMI influences.

The mechanical traction relief

To ensure an unproblematic handling during the assembly, the conectors and the crimping of the sensor mat were molded in hotmelt. The device can support tensions up to 100 N.

ELECTRONICS

SOMS Control Unit

The raw data provided by the sensor array is not by itself sufficient to avoid unnecessary passenger airbag deployment.

In order to...

- detect any situation accurately and safely,
- ensure the system is fail safe,
- switch the occupied seat to unoccupied status,
- design and integrate the diagnostic system, etc...

a supplementary SOMS-control unit is required. This unit transmits, via a serial interface, the status of detection, as made by the sensor array, to the airbag control unit. The SOMS control unit, no larger than a matchbox, is fitted to the passenger seat, and measures, in cycles, the resistance of the array. The resistance is determined in two partial measurements in both the normal and also in the inverse direction of the diode, which is integrated into the sensor array. The measurements are evaluated by the software and transmitted to the airbag control unit as one of four possible "messages":

»seat occupied
»seat unoccupied
»system failure due to a short circuit
»system failure due to an open circuit.

If either of the latter two situations are transmitted, or data is interrupted or stopped, the airbag control unit detects an error. Any error leads to a "seat occupied" status.

In the event of a change in status from "occupied" to "unoccupied", the control unit maintains an "occupied" safety timeframe of two minutes. This is a built-in safety mechanism to ensure no false signals are transmitted if the passenger is simply changing position in the seat. The switching from "unoccupied" to "occupied" is, of course, immediate.

Integration of the Airbag Control Unit

The introduction of central airbag control units, ZAE and BAE, which are also developments of the task force, made it possible to process the information transmitted by the SOM system. The airbag control unit is able to compare different inputs and avoid incorrect reactions. In the past, based on older airbag control units, the decision to deploy the airbags was made when the seat belt was not attached. It was, therefore, possible that in an accident where the driver was wearing his seat belt, but no passenger was present (and therefore the passenger seat belt was not fastened), the driver's safety belt tensioners were activated (and the driver's airbag not deployed) but the passenger airbag was uselessly activated. This could easily lead to the confusing situation for the driver who may find it difficult to understand why his airbag did not deploy but his not present passenger's did. The new generation of airbag control units decides if in addition to the seat belt tensioner an airbag is needed and espacialy based on the SOMS information if the passenger airbag is needed.

Tests

The FSR™ sensor array, the SOMS Control Unit, as well as the complete system, have been exposed to environmental, functional and "abuse" testing to evaluate the system's suitability for daily use. The tests were designed to ensure:

»seat comfort is not compromised for the sake of the SOMS system

»seat heating must not influence the function or durability of the sensor and vice versa

»the system does not accidentally switch from "occupied" to "unoccupied" even under extremely violent driving conditions.

Crash tests have shown that, in an accident, an "unoccupied" seat is correctly detected.

Safety Philosophy

with regard to Baby Seats

A question which arises regularly in this context, and we have not, so far, addressed during this presentation, is:

»Can a SOMS detect a reboard baby seat in order to avoid airbag deployment which could be harmful for the child?

The answer is an explicit NO ! We have continuously stressed that the SOM system is required to give an "occupied" signal whenever a dubious result is transmitted and, therefore, there may be occasions when an airbag is unnecessarily deployed. A system which deactivates the airbag when a reboard baby seat is fitted to the passenger seat, must do the opposite of the SOM system. This leads to a conflict of interests when an SOM system is "asked" to detect baby seats. The development of a baby seat recognition system, is therefore, considered as high priority by most automotive manufacturers.

It is possible that within the next two years myself, or a colleague of mine taking part in this conference, will make a presentation on this development.

In the meantime, as long as no automatic baby seat recognition system is available, the passenger airbag must either be switched off or the fitting of reboard baby seats to passenger seats forbidden.

FINANCIAL CONSIDERATIONS

A rebate on fully comprehensive insurance comparable to that offered to drivers of cars fitted with ABS is not expected from insurance companies. A financial advantage for the owners of cars equipped with SOMS results from favourable insurance classification which is regularly adapted in respect to average damage repair costs. As the average repair costs will decrease for cars equipped with SOMS, one can count on an indirect rebate as a result of lower classification.

Furthermore, car owners who do not have fully comprehensive cover, and are, therefore, liable to pay any damage costs by themselves, have a direct advantage with an SOMS fitted car as repair costs are lower.

SUMMARY

In parallel with new developments in airbag technology, Situation Appropriate Airbag Deployment will remain high on the priority list of Airbag Development groups. This aim will result in an airbag system which can adapt itself to the requirements of any specific situation.

The first step in this direction is represented by the BMW Automotive Seat Occupancy Detection system shown in this presentation.

BMW is now manufacturing many of it's vehicles with a supplementary restraint system: the passenger airbag. The unpleasant side-effect of customer dissatisfaction caused by unnecessary airbag deployment represented for BMW a challenge which they have answered with the help of this innovative technology.

960500

Electronic System Design for Future Passenger Restraint Systems

Richard Vogt
Siemens AG

ABSTRACT

In comparison to a standard dual airbag system as of today, future restraint systems will introduce a variety of new and additional actuator devices which help to improve the overall restraint performance. Furthermore, sensors and sensor-subsystems will be added to improve and extend the range of crash detection and to enable an automatic adaptation of the restraint system to a given status of the vehicle, its occupants and the crash severity.

The implementation and control of these new functions require an appropriate design of the overall restraint system electronics. Two approaches are described in greater detail: the modular design based on functional building blocks, and the implementation as a distributed system including communication links.

INTRODUCTION

When frontal airbag systems found their first market introduction, the crash sensing function was usually implemented by two or more electromechanical sensors located in the front part of the vehicle ('front crash sensor' in figure 1; [1][2]. These sensors were mechanically trimmed to discriminate between must-deploy and must-not-deploy events. They contained a simple switch which, in case of a must-deploy event, closed the electronic circuit supplying ignition current to the firing squib in the airbag module. The electronic control units (ECUs) used in these systems performed mainly the two functions

- system diagnosis
- igniter energy supply

and hence were called 'diagnostic module' [3] or „DERM" (diagnosis and energy reserve module)

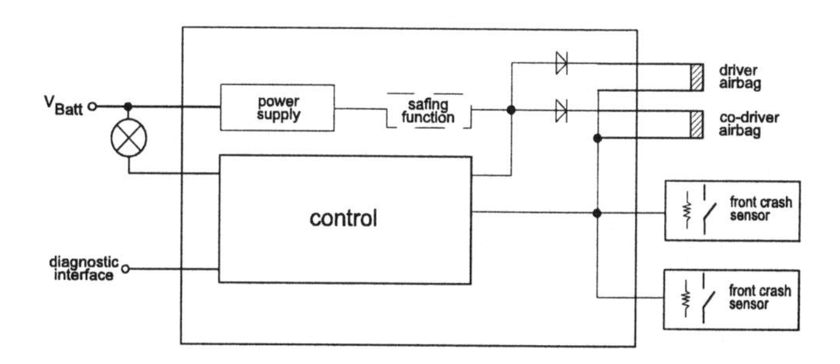

Figure 1: Diagnostic module

With improved car structures and the availability of electronic acceleration sensors and powerful microcontrollers, both at an affordable price level, the so-called 'single point sensing systems' were introduced. They no longer require the use of remote discriminating sensors in the front part of the vehicle. Instead, the deceleration of the vehicle during a crash is measured by the accelerometer and evaluated by algorithms executed by the microcontroller. Therefore, the ECU used in these systems also performs the function of

- crash sensing and discrimination

in addition to the functions listed above. Figure 2 shows an appropriate design.

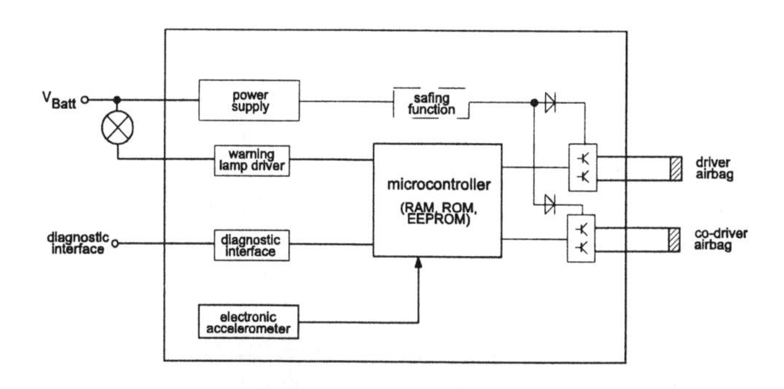

Figure 2: Single point sensing ECU

Besides cost reduction and reliability improvement, this design opens the door for improving the overall restraint system performance: the microcontroller's programmability together with the introduction of additional sensors and actuators offers the flexibility to control the vehicle's restraint devices most efficiently, with the ECU now becoming the core of a 'crash management system'.

This paper will describe new functions and features of such a crash management system and its impact on the ECU. Various strategies for designing the electronic system will then be explained.

RESTRAINT SYSTEM REQUIREMENTS

Although most of the existing restraint systems are only using one or two frontal airbags for the front seat occupants of a vehicle, further restraint devices are already available or are at least under development. The following table 1 – which may not be complete – gives an idea of the variety of restraint devices, and it indicates for which occupant in the vehicle the device may normally be applicable:

restraint device	driver	co-driver	rear seat passenger	# per vehicle
front airbag	X	X		2
belt pre-tensioner	X	X	X	4 (5)
side airbag	X	X	X	4
deployable knee bolster	X	X		2

Table 1: Restraint devices

Each 'front airbag' may be implemented with a dual stage inflator that allows a variable bag inflation characteristic, leading to an adaptive ('smart') restraint system. Furthermore, each 'side airbag' may consist of more than one bag, e. g. head bag *and* thorax bag.

ECU OUTPUTS – With respect to the ECU, each of these restraint devices has to be controlled (i. e. monitored for diagnostic purposes resp. driven for activation), which means an appropriate number of outputs has to be provided. The figures indicated in the rightmost column of table 1 sum up to 12 (13) per vehicle; in the case of dual stage inflators and dual side airbags per front seat occupant a value of 16 is reached. Even higher values are known in the literature (e. g. [4]).

In addition to those outputs connected to the restraint devices, further outputs may be needed to control the warning lamp(s) and, after an impact, to cut-off the engine's fuel supply, to unlock all doors, to place an emergency call etc.

ECU INPUTS – The complexity of the electronic system is also determined by the amount of external input lines it has to sense. The following table 2 – which again may not be complete – shows the variety of possible sensor and status signals the electronic system has to evaluate:

sensor or status	driver	co-driver	rear seat passenger	# per vehicle	speed
seat belt buckle switch	X	X	X	4 (5)	low
seat occupied	1)	X	X	3	low
rearward facing child seat		X		1	low
out-of-position	X	X		2	med.
weight	X	X		2	low
2) inflator pressure	X	X		2	low
3) inflator temperature	X	X		2	low
side impact sensing satellite	X	X	4)	2	high

1) may be useful in order to shift the automatic transmission into the parking position if the driver seat is empty

2) e. g. for compressed gas inflators

3) e. g. for adapting the performance of dual stage inflators to the inflator's initial temperature

4) depending on the side impact sensing principle [5], one sensing satellite per vehicle side may be sufficient

Table 2: Sensor and status signals

In addition to these signals, further inputs my be needed to connect diagnostic devices, pre-crash sensing devices, 'early crash' sensing devices (e. g. remote crash or crush zone sensors) for improved crash sensing performance, etc.

The impact on the ECU's complexity is not only determined by the absolute number of lines to be sensed (the figures in the table sum up to 18), but also by the amount of preprocessing required for each input signal, by the signal frequency resp. its timing characteristics (see the column 'speed' in table 2), and by the information processing necessary by the ECU's microcontroller, especially during an impact event.

DEMANDS ON THE ELECTRONIC SYSTEM DESIGN

The specification of input and output signals and their relationship is one important basis for the electronic system design. However, since these systems will be used to control a safety critical application in the automotive environment, the design will be heavily influenced by further necessities and demands, even those which are non-technical in their nature. The following table shows some of the demands that restraint system electronics have to fulfill:

demand	how	typical
flexibility	high	2 .. n outputs, 0 .. m inputs
development cycle time	short	< 2 years
volume	high	1 mio unit per year
price	low	$25 basic config.
reliability	high	detection of any failure; no inadvertant deployment
EMI robustness	high	120V/m, no malfunction

Table 3: Demands on the electronics for restraint systems

An additional important factor is that the acceptance of major design changes is critical: due to liability reasons, staying with a proven design may be preferable over the introduction of a 'revolutionary' solution. Hence 'evolutionary' improvements are more likely to be accepted.

APPROACH 1: MODULAR ECU DESIGN

A standard design approach for technical systems is to divide their functionality into smaller subsystems, so-called *modules*. Each module may then be designed to perform a dedicated function, and by combining and interconnecting the various modules the system's functionality is build up.

With respect to an ECU for restraint systems, those functional modules are, for example:

- Hardware:
  sensor, microcontroller, power supply, control unit, firing squib driver circuitry with diagnostic functions, warning lamp and other output driver, input interface circuitry, diagnostic interface
- Software:
  operating system, initialization, self-test functions, diagnostic communication interface, input and output control routines
- Crash signal processing:
  signal filtering, offset adjustment, sensor control, crash signal evaluation algorithms, deployment control, crash recording
- Mechanical design
  connector, housing, printed circuit board

Table 4: Functional modules of an ECU for restraint systems

The availability of functional modules helps to achieve the flexibility of the electronic system as demanded in table 3, above: the system is adaptable to a certain application by combining the appropriate modules.

The definition of each module shall take into consideration the following key factors:

- reusability
- standardization

Creating *reusable modules* is the key to fulfill the demand for short development cycle times, as required in table 3: once a module has been successfully developed, its design may partly or in total be used again for new applications. As a consequence, the development costs are reduced, which is a very welcome side effect. In addition, it helps to make use of already proven designs, hereby increasing the ECU's reliabilty.

Even more important is the *standardization of modules* which allows to use the same module for further applications or for other functions in the same application ('carry over' principle). Besides the positive influence on all factors already mentioned above, the multiple use of a standardized module increases the number of units, amortizing its development costs and, especially for non-software modules, lowers the piece price.

EXAMPLE OF A MODULAR DESIGN – Based on the approach described so far, functional modules for building-up an ECU for a vehicle's restraint system have been defined and implemented [6]. Some of the highlights are summarized in table 5:

- Hardware modules:
  - accelerometer subsystem
  - ASIC 1 for power supply and control functions
  - ASIC 2 for dual firing squib driver and diagnostics
  (ASIC = application specific integrated circuit)
- Software modules:
  high-level language routines for the various functions
- Crash signal processing:
  development of an 'universal' crash signal evaluation algorithm which can be trimmed to a given platform by a set of parameters
- Mechanical design
  - definition of a uniform housing assembly concept and
  - definition of printed circuit board design rules, both optimized for most efficient manufacturability
  - selection of an industry-approved connector family [7]

Table 5: Example of functional module definition

High priority was assigned to the standardization in the various areas listed above. As an example, by carefully reviewing all accessible different customer requirements the ASICs were specified and designed to cover this variety. In the field of crash signal processing, all the experience out of numerous single-point sensing applications and thousands of crash data sets was used to define an evaluation algorithm which could be adapted for use in vehicle types from subcompact up to light trucks just by trimming a small amount of parameters and storing them in the ECU's microcontroller programmable memory.

Most obvious becomes the module definition with regard to the ECU's hardware block diagram, which is shown in the following figure 3:

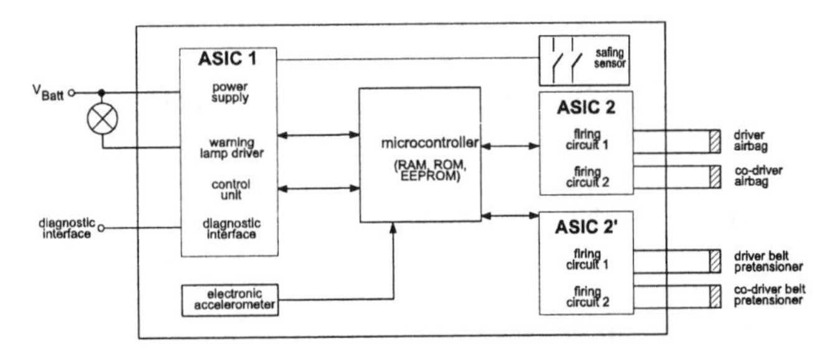

Figure 3: Example of a modular ECU design

The flexibility of this modular design becomes obvious if the customer demands, e. g., only one or two firing loops instead of the four shown in figure 3: in this case, only by deleting the second firing circuit ASIC (labeled 2') and appropriate software/hardware the ECU can be taylored to meet this demand.

EVOLUTIONARY IMPROVEMENT OF A MODULAR DESIGN – With the availablity of functional modules, design improvements and innovations can be introduced for and, at the same time, can be limited to one or more modules. Some examples, based on table 5, above, are listed below:

- Hardware module improvements:
  - new accelerometer subsystem
  - higher scale of ASIC integration (e. g. combining ASIC 1 and ASIC 2 in one chip)
  - new ASIC interconnection scheme (e. g. serial bus)
  - new microcontroller familiy
- Crash signal processing:
  - performance improvements
  - introduction of new approaches (e. g. fuzzy concepts)
- Mechanical design:
  - new assembly technologies
  - customer specific connector design

Table 6: Module improvements

By concentrating the design efforts on improving a specific module rather than redesigning the complete system the success and timely introduction of the improvement is positively influenced. Also the acceptance (and finally release) by the car manufacturer is easier to obtain.

In summary, the modular design can be seen as the most appropriate approach for building up a restraint system ECU.

APPROACH 2: DISTRIBUTED RESTRAINT SYSTEM ELECTRONICS

Although the modular design is suited to be adapted to any expansion of the restraint system, there are some practical limitations to an arbitrary ECU growth:

- ECU housing size:
  - packaging restrictions
  - connector pincount limitations
- Wiring harness:
  - number of wires
  - length of wires (cost, EMI susceptibility)
- Physical and technical requirements:
  - ECU mounting location not suitable for implementation of all functions (e. g. side impact sensing)
  - reliability aspects
- Volume and cost
  - only small volumes to be expected for a very special and complex ECU design; therefore
  - higher cost

Table 7: Limiting factors for a single ECU system

Due to these circumstances, it may be advantageous not to concentrate the complete electronic system into one single ECU, but to distribute its functionality onto several subsystems which are interconnected by a communication link. Hereby, *distributed restraint system electronics* are created.

Now, a decision has to be made on the total number of subsystems and the functions to be performed by each of these subsystems. Following factors may influence this part of the specification:

- Conditions given by the designated mounting location:
  - packaging size restrictions
  - temperature range
  - protection class requirements
- Physical requirements:
  - e. g. side impact sensing subsystem to be installed inside the vehicle's door or close to the vehicle's side structures
  - e. g. passenger presence and child seat orientation detection subsystem to be installed close to the passenger seat
- Reliability aspects:
  - e. g. reduced likelihood of common mode failures at two or more different locations resp. subsystems
- Flexibility and modularity aspects:
  - introduction of subsystems for specific functions only
  - attempt to standardize most or all of the subsystems
  - standardization of the subsystem's interconnection
- Cost

Table 8: Factors for subsystem specification

In consideration of these factors the following two alternatives for the design of a distributed restraint system electronics have been investigated in greater detail:

- Remote firing circuitry
- 'intelligent nodes'

These alternatives are described in the following two chapters.

REMOTE FIRING CIRCUITRY – The motivation for this system setup is derived by the increasing number of output stages (i. e. firing loops) demanded by advanced restraint systems (please refer to table 1). The concept is to remove all firing circuitries, including the firing squib drivers with the appropriate diagnostic functions, from the single point sensing ECU and to allocate these circuitries close to each squib. Figure 4, below, shows the principle design.

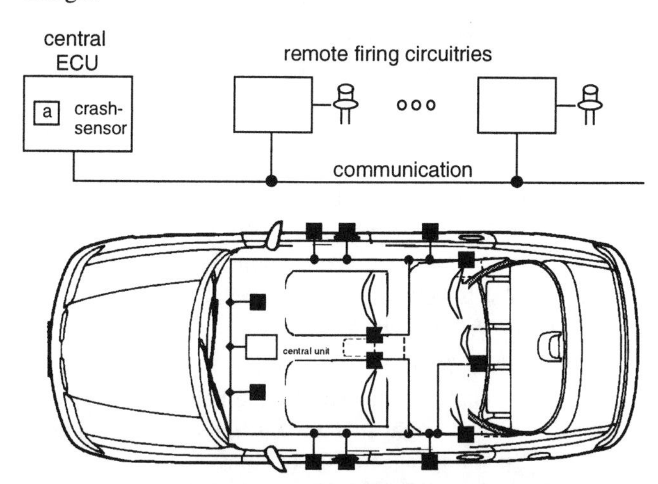

Figure 4: Distributed restraint system electronics with remote firing circuitries

An example for the design of each remote firing circuitry is given in figure 5. (The optional 'µswitch' shown in this figure indicates a micromachined switch, e. g. as described in [8], which is integrated onto a silicon chip and which is activated if a certain acceleration threshold is exceeded. Hereby this switch acts as a 'safing function' which enables the activation of the squib only if a deceleration of the vehicle is present.)

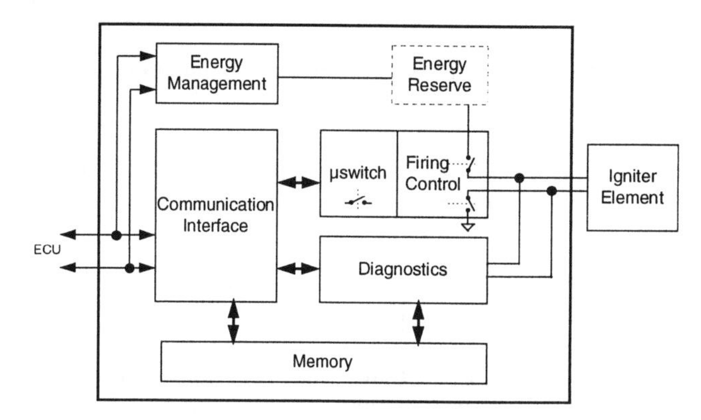

Figure 5: Remote firing circuitry block diagram

Major characteristics of this electronic system design are:

- firing controller located closely to the squib, e. g. inside the squib connector, integrated into the inflator module or even integrated into the squib itself ('smart squib')
  ⇒ short wires to the squib; reduced wiring harness failure probability; reduced EMI susceptibility
- enables the use of low energy squibs (e. g. < 1mJ)
  ⇒ simplified firing circuit design; less energy reserve demand
- use of a *bus interconnection topology* between ECU and squib drivers
  ⇒ reduced wiring effort; reduced ECU connector pincount; reduced ECU package size
- virtually 'any' number of firing controllers may be connected to the bus
  ⇒ highest flexibility; no need for ECU modification
- squib activation only by a specific bus command
  ⇒ highest safety against inadvertant deployment; no deployment under DC fault conditions
- standardization of firing controller possible
  ⇒ reduced cost, carry-over strategy

Table 9: Characteristics of a remote firing circuitry system

An important factor for the design of this system is the specification of the interconnection bus. Two examples are given:

Bus system with separate energy supply – Each firing circuitry is connected to a supply line, and the communication with the ECU is carried out via a single (or redundant) bus line. Figure 6 shows the basic topology.

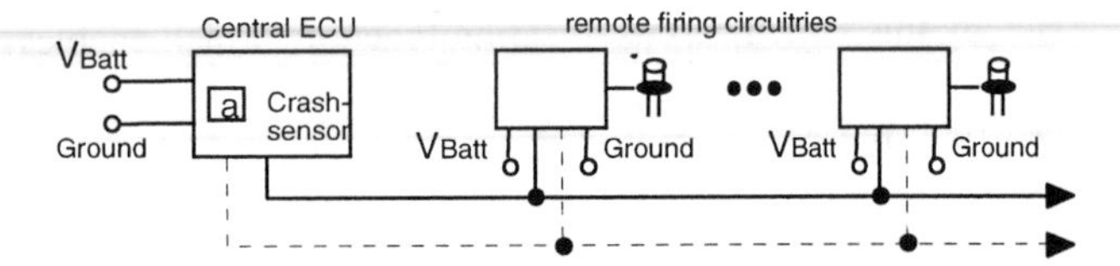

Figure 6: Separate energy supply of remote firing circuitries

Benefits of this concept are the usability of standardized bus systems (e. g. CAN). An energy reserve capacitor in the remote firing circuitry may be used as an option. However, this concept requires a minimum of three wires to each firing circuitry with appropriate supply line protection devices.

Bus system with integrated energy supply – The energy supply to each firing circuitry is done via the same bus wires as the communication. Figure 7 shows the basic topology.

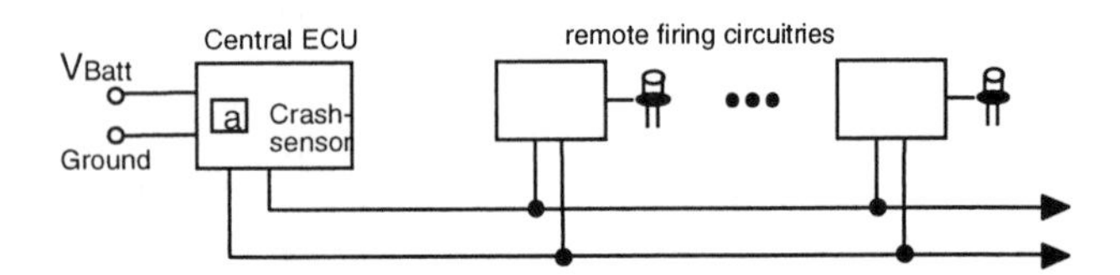

Figure 7: Energy supply via the bus line

This solution requires a specific bus communication protocol, e. g. with a DC offset voltage for energy supply purposes, and hence a certain design and circuitry effort inside the ECU. The use of an energy reserve capacitor in each firing circuitry is mandatory. However, this concept is working with two wires only, which is the same number as is used for a single 'ordinary' squib.

Summary – The major advantage of the remote firing circuitry approach is the reduction of the ECU's complexity and connector pincount as well as the flexibility of this design to work with (almost) any number of firing squibs required by the restraint system. Cost advantages are achieved especially with respect to the wiring harness. The vehicle assembly process may be optimized by integrating the firing circuitry and the associated squib into the inflator module, hereby creating a 'smart inflator' module as a preassembled and pretested production unit.

INTELLIGENT NODES – Separating dedicated functions, as e. g. the firing circuitries, from the ECU may not be sufficient if the complexity of the restraint system electronics is further increased, e. g. by additional sensor inputs in conjunction with the necessary evaluation and processing of these signals. In this case, dividing the electronic system into self-standing subsystems is a more appropriate approach. These subsystems are interconnected via a communication channel and hence are to be seen as *intelligent nodes* of the restraint system electronics.

As stated above, the definition of the subsystem's functionality is a basic part of the system specification. The following table indicates some alternative guidelines for this first part of the system design:

- Local vicinity
  Place an intelligent node close to related sensors/actuators!
- Physical necessities
  Consider any location where a sensor or actuator system is required as an intelligent node's location!
- Carry-over strategy
  Specify each intelligent node to be 'reusable' in other vehicle platforms!
- Modular expandability
  Define a set of intelligent nodes so that the system's functionality can be extended by adding further nodes!
- Dedicated functions
  Assign a certain functionality to each intelligent node!

Table 10: Guidelines for intelligent node specification

Please note that the last item listed in table 10, 'dedicated functions', may also be considered as a basis for the introduction of remote firing circuitries as described before.

Of course, further items will influence the system partitioning. As usual, cost will be a significant factor. In case of a distributed system, cost is also determined by the overhead associated with the implementation of the necessary communication between the subsystems. To keep this overhead low, the efforts (hardware, software, space, ...) spent for communication per node should be small in relation to the remaining functionality of this node. In addition it is extremly helpful to rely on a proven communication system which is accepted in the automotive industry and which is supported by hardware, software and development tools. The controller area network system (CAN) is a good example.

Based on these considerations, several approaches for a distributed restraint system electronics with intelligent nodes have been evaluated. Two examples are given below:

Modular expandable distributed system – The subsystems are defined as follows:

1. ECU #1 for all standard functions, e. g. dual frontal airbags, dual frontal belt pretensioners, left and right side airbags, interface to remote side impact sensing satellites
2. „add-on" ECU #2 for rear-seat restraint devices, e. g. dual or triple belt pretensioners, seat belt buckle switches, left and right side airbags, dual seat occupancy detection.

3. „add-on“ ECU #3 for advanced front passenger frontal airbag functionality, e. g. dual stage inflators, occupant posistion sensing

This set of ECUs offers an easy adaptability of the electronic system to various restraint system configurations, and it enables the reusability of the (standardized) subsystems from low-end to high-end applications. Figure 8 shows a system layout.

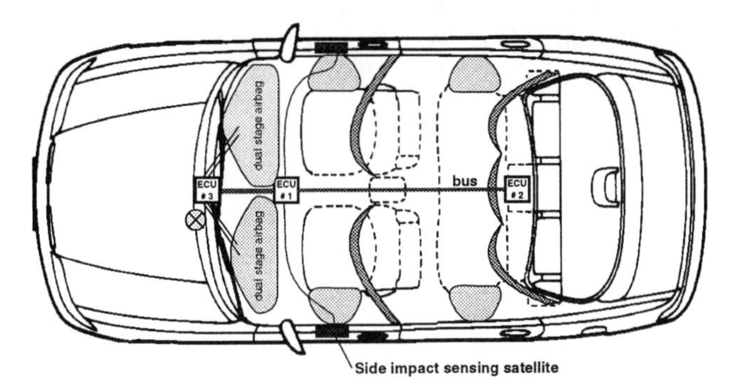

Figure 8: Modular expandable distributed system

Distributed system by expansion of existing nodes – Even „standard“ restraint systems may require a distributed electronic system. An example is a combined front and side impact protection system which – today – is using side impact sensing satellites for early side impact detection.

In consideration of the guideline „physical necessities“ in table 10, above, these side impact sensing satellites might be expanded to form the intelligent nodes of the distributed system:

1. ECU #1 (driver side): side impact sensing, evaluation and side airbag activation; front impact sensing, evaluation and driver front airbag / belt pretensioner activation
2. ECU #2 (co-driver side): equivalent to ECU #1
3. „add-on“ ECU #3 for rear-seat restraint devices, e. g. dual or triple belt pretensioners, seat belt buckle switches, left and right side airbags, dual seat occupancy detection.

Figure 9 shows the appropriate system layout.

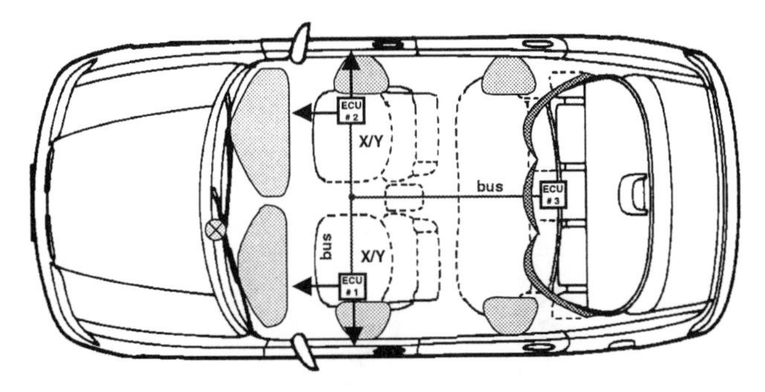

Figure 9: Expansion of existing side impact sensing nodes

Please note that this approach omits the use of a tunnel-mounted ECU. The possible lack of crash sensing performance is compensated by exchanging the acceleration signal resp. derived values hereof from the opposite node via the communication system. Furthermore, the availability of two longitudinal acceleration sensors (denoted „x“ in figure 9) enables the implementation of a safing function, and the similarity of both nodes promotes the reusability of the design.

Summary – Defining a distributed restraint system electronics based on intelligent nodes offers advantages for systems with high complexity. A careful definition of a node's functionality enables its reuse in a variety of system configurations, hereby yielding the flexibility which is needed for implementing advanced restraint systems at reasonable costs.

CONCLUSION

As has been shown, the application of standard engineering practices to the design of restraint system electronics allows handling of the increased functionality and complexity of new restraint systems. System partitioning and module resp. subsystem specification are key issues for a technically and economically successful design. The modular ECU approach as well as the distributed restraint electronic system offer the potential for easy configurability, short development times, reduced subsystem complexity and size, simplified packaging and in-vehicle assembly. These points especially support the introduction of smart restraint systems, which is a further step towards a *crash management system* with the ultimate goal of outstanding passenger safety.

ACKNOWLEDGEMENT

The author wishs to thank Michael Bischoff; John Merner and Dr. Anton Anthofer as well as all the colleagues from Siemens Automotive's Safety Electronics Subdivision for fruitful discussions and their assistance in preparing this paper.

REFERENCES

[1] White, C; Leonard W. Behr; „Inflatable Restraint Sensing and Diagnostic Strategy“; SAE-paper 901120, International Congress on Transportation Electronics 1990, Society of Automotive Engineers, Warrendale, PA, USA

[2] Breed, David S.; Thomas Sanders; Vittorio Castelli: „A Critique of Single Point Crash Sensing“; Automotive Technologies International, Denville, NJ, USA, 1991

[3] Härtl, Alfons; Richard Vogt, Anton S. Huber; „Airbag Systems – Their Permanent Monitoring and Its

Meaning to the User"; SAE-paper 901138, International Congress on Transportation Electronics 1990, Society of Automotive Engineers, Warrendale, PA, USA

[4] Koch, Michl; „X-Bag from Mercedes Benz"; Auto Motor & Sport; 2/1995, page 20

[5] Härtl, Alfons; Gerhard Mader; Lorenz Pfau; Bert Wolfram; „Physically Different Sensor Concepts for Reliable Detection of Side-Impact Collisions", International Congress on Transportation Electronics 1995, Society of Automotive Engineers, Warrendale, PA, USA

[6] Siemens Automotive; „Electronic Control Units for Central Airbag Systems"; in: Passenger Safety Systems: Electronics for Passive Protection; Siemens AG, Germany, No. J71001-A0000-A079-X-7600, 1995

[7] Siemens Components Division; „Electronics Connection System ELO for Automotive Engineering"; Siemens AG, Germany, No. A23001-G21-P001-X-7600, 1990

[8] Maciossek, A.; B. Löchel; H. J. Quenzer; B. Wagner; S. Schulze; J. Noetzel: „Galvanoplatingand Sacrificial Layers for Surface Micromachining"; in: Microelectronic Engineering; MNE '94, Davos 1994

960663

Technological Trends in Occupant Protection Systems - Recent Research Challenges from the German Point of View

Hiltmar Schubert and Karl-Friedrich Ziegahn
Fraunhofer Institute für Chemische Technologie (ICT)

ABSTRACT

Sophisticated car occupant protection systems, such as airbags and belt restraints are becoming the most frequent safety devices found today. Within the span of a few years, technology has been developed to meet market needs as well as technical requirements. But today, airbag technology is still in the basic development stage.

The vision of car occupant protection presented here includes further development of "smart protection systems", including overall safety management. Automatic restraint systems, electronics and system management will proceed to a new quality of car occupant protection. Future developments for occupant protection systems are, for example

- crash-predicting sensors
- Occupant sensing systems
- Controllable gas release for airbag inflators
- Performance depending on crash severity
- Multiple bag deployment strategies
- Alternative inflator systems
- overall safety management, including crash detection, airbag deployment, fire extinguishing system, side impact protection, new cushioning materials

concerning weight, performance and a rational cost-benefit relation.

Applied research has to challenge those requirements. The paper will give an overview of current research activities, including the German point of view of the Fraunhofer Society, a partner of automobile manufacturers for 20 years.

1. EXHAUST SYSTEM MEASUREMENTS

Airbag systems have conquered the market. Almost all automobile manufacturers offer them - either as an extra or standard accessory. The predicted growth rates are enormous as is the economic interest of the firms producing and selling these systems.

The utility of airbag systems is incontestable, as demonstrated in the reports of accident researchers, doctors and insurance companies. Just as clear , however, is the realization that an airbag system, - as valuable and important as it is - can merely be a part of a comprehensive concept of occupant security. The use of safety belts remains the most important measure for the prevention or reduction of accident injuries.

Airbag systems have passed their first crucial test: they work - in practice and in large numbers. Figuratively, one could say that airbag systems have graduated from elementary school and are now on the threshhold of high school. That is, the development is now moving toward intelligent systems. Future airbag systems require:

* **Sensors,** which trigger according to need and could possibly predict accidents
* **Gas inflator assemblies**, whose amount of gas release is dependent upon accident severity. These must be nonpoisonous, neither too hot nor too cold, create none or minimal residue, be permanently - or at least for the life of the automobile - stable. The propellants must be inexpensive, be highly efficient and be able to be produced, stored and handled well.
* **Complete systems,** consisting of driver, front-seat passenger, and side airbags, which, along with belt tighteners, are triggered according to the situation, thus avoiding unnecessary releases. A release that does not destroy the entire dashboard or windshield would be desirable. If a passenger is "out of position", he should not be unnecessarily injured by the airbags.

The **dialogue with lawmakers and media** has shown that the airbag system needs the proper conditions. Testing regulations and crash tests must be internationally harmonized. Obsolete regulations must be adapted to the current state of technology.

It also appears that the release threshold is currently set too low. Accidents with speed differences from 10 to 20 km/h require no airbag - at least not if the vehicle occupants are wearing seat belts. In this way, minor injuries caused by skin abrasions and also minor bone breaks as a result of the front-seat passenger's "wrong" position can be avoided. Airbag releases are only justified by high-speed collisions.

Here we touch upon a problem in communication with the automobile customer who must understand that an airbag release should not be triggered by every accident. Perhaps, the media, which so decisively contributed to the introduction of airbags onto the market, could help resolve this communication problem.

2. TECHNOLOGICAL CHALLENGES OF FUTURE AIRBAG DEVELOPMENT

GAS INFLATOR ASSEMBLIES - The worldwide search for alternatives for the used and proven azide continues. Solid and liquid propellents, and pressurized gases compete with each other in terms of price, efficiency, stability, and environmental impacts. Today, one cannot yet recognize a clear trend or even a winner.

SENSORICS - Electronics have developed from an airbag-triggering device to intelligent safety-management: recognition of seat occupancy, automatic shutoff if passengers are in dangerous positions or for reboard children seats. An interesting suggestion for the use of neuron networks, which indicates the real start of artificial intelligence in airbag technology, already exists. The greatest demands are placed upon side impact sensors. But still today, the ability to recognize an accident before it actually occurs appears to be only a vision.

COMPUTATIONAL PROCEDURES - The utilization of computer-supported development techniques is moving forward. They assist in designing airbag systems rapidly and efficiently for particular vehicle types and in avoiding unncessary hardware experiments.

ENVIRONMENTAL HARMONY - Airbag systems, which will be used in very large numbers in the future, must be recycled or disposed of while creating minimum environmental impact. For this, recycling strategies and reprocessing technologies are needed, which must be regarded as part of the entire recycling of old vehicles. The filling gases and combustion residues must neither contaminate the waste water nor air more than absolutely necessary; the materials used in the airbag systems should be recyclable. Environmental impacts created by the production process are to be avoided as much as possible.

RISK OF INJURY - Risks of injury by airbag systems are to be minimized.

R & D REQUIREMENTS - It appears doubtful that satisfying answers can be given to all these problems today. Although the airbag has commenced its victory march, the need for further technological developments is very great. Research and development are working hand-in-hand. Especially, in the Fraunhofer Society, which carries out contract research for industry, one can recognize clearly that the need for research in airbag technology has not decreased, but is growing continuously. We are prepared to accept the challenges and are happy to work jointly with the industry to solve the problems before us.

3. DEVELOPMENT TASKS AS ILLUSTRATED BY GAS INFLATOR ASSEMBLIES

3.1 FUNCTIONING

Gas inflators are small aggregates that produce working gas for various purposes. They differ from solid rocket propellents by the production of gases with low temperatures and exit speeds below the speed of sound.

They are basically comprised of gas generants in a pressure casing that has an opening and a triggering mechanism. The terminololgy is currently being worked out by ISO (Fig. 1).

Within motor vehicle safety systems, gas inflator systems make an ideal application because of their compact construction with the presumption of one-time use. However, because of the small size of the aggregate, in practice, one has done without control mechanisms for pressure and burning time.

Similar to solid rocket propellents, the planned function parameters have been set by design of the inflator assembly with regard to geometry, number, and substance. For this reason, the choice of the chemical composition, next to the composition of the gas, exerts a substantial influence upon the other characteristics.

One can imagine the use of gas inflator assemblies for various purposes within motor vehicle safety systems. Based upon the validated knowledge and state of technology, these comments serve to point out research potential and trends and to stimulate innovation that may go as far as to be visions of the future.

3.2 GENERAL TECHNICAL REQUIREMENTS

The following requirements should be met by a gas inflator assembly:

* Low combustion temperature (=/< 1750 K)
* Low gas-pollutant concentration (CO,NO)
* High gas exploitation (=/> 14mol/kg)
* Appropriate and reproducible pressure-time curves (short-term P max)
* Wide temperature-function range
* Resistance against thermic and mechanical environmental influences.

* Non-poisonous source materials
* Low production cost
* Long life and high reliability
* Disposal with minimal environmental impacts

Today, these requirements can be more or less fulfilled. The problem lies in finding the optimal combination for the desired function.

In general, the desire is for for future gas inflator assembly systems that are smaller, lighter and cheaper but providing equally high reliability, long life, high resistance to environmental influences, and that can be disposed of or recycled with minimal environmental impact.

Parallel to the mentioned requirements, low limits are assumed for the toxicity of gases in the vehicle passenger compartment, especially for:

CO: < 1000 ppm (in relation to a 60-liter test atmosphere)

NO: < 150 ppm (in relation to a 60-liter test atmosphere)

With regard to the dilution effect in a passenger compartment (approximately 3000 liter) compared to the new VDI proposal "Maximal Immission Concentration", these numbers appear very conservative. The hourly average value is 25 ppm. The short-term CO-peak value should not exceed 300 ppm. Since the respective circumstances vary in an "event", it is better to strive for the average values as limits. Looking at things this way, the air a smoker breathes with a maximum of 50 ppm even seems toxic.

On the one hand, these general technical requirements, demand the optimal exploitation of gas, and on the other hand, a low combustion temperature and a low CO-NOx pollutant concentration. At this point, the development of NOx plays a subordinate role in comparison to the CO problematic, as long as one maintains a balanced oxygen equilibrium.

Without the expectation of completeness, one can break down the general technical requirements according to the conceivable applications (Chart 1).

3.3 COMPOSITIONS AND CHARACTERISTICS

The categorization of gas inflator assemblies and their development trends is made primarily in terms of gas composition.

One must differentiate between gas inflator assemblies for the passenger compartment and for external use.

3.3.1. GAS INFLATOR ASSEMBLIES FOR THE PASSENGER COMPARTMENT

3.3.1.1 SOLID FUEL GAS INFLATOR ASSEMBLIES

For optimizing solid-fuel gas inflator assemblies the following options are available:

* Choice of nitrogen-containing compounds
* Choice of suitable oxidation
* Variation of the oxygen balance
* Utilization of coolants
* Utilization of catalysts

Up to now, the most important development for airbag-gas inflator assemblies is based on the carbon-free system of sodium azide/potassium nitrate which produces an ideal gas for these purposes. The creation of alkaline particles presents a disadvantage, that one, however, can reduce through the formation of slag. The production of gas inflator assemblies containing toxic sodium azide and its disposal, make up a second disadvantage. For these reasons, other systems are in development which are rich on nitrogen but include compounds containing little organic carbon. Chemical compounds, such as TAGN, NIGU, ATZ, GZT etc., which are transformed with suitable oxidators, offer an option. However, the price is an undesirable concentration of CO, CO2 and NOx ,which may be decreased by varying the temperature or by catalyst.

3.3.1.2. LIQUID FUEL INFLATOR ASSEMBLIES

Through the application of liquid fuels as gas inflators, one might have the opportunity to take advantage of carbon-free compounds which even transform themselves into working atmospheres without leaving any residues. Systems and technologies originating in the aerospace field as well as examples from defense technology development exist.

Systems consisting of hydrazine nitrate and compounds including hydroxyl ammonium nitrate (HAN) can be brought up as examples. However, one creates a different drawback, for instance, an additional transport device. Nonetheless, one should not exclude a practical application from the start.

3.3.1.3. HYBRID GAS INFLATOR ASSEMBLIES

Systems of pressurized gases (e.g. air, nitrogen, argon) in pressure-vessels that, by opening a rupturable membrane, stream into the air-bag are regarded as hybrid gas generators. A two-step pyrotechnical charge opens the membrane and heats the working atmosphere. Through this process, the cooling of the expanding gases is compensated or even overcompensated. One can design the amount of pyrotechnical charge, so that the developing carbon monoxide does not exceed the working atmosphere limit of 1000 ppm.

The greater weight created by the pressure-bomb that must meet the pressure-vessel regulation, is the disadvantage of this version.

Systems as these have already been realized.

3.3.2 GAS INFLATOR ASSEMBLIES FOR EXTERNAL USE

If one is not tied down by the CO-problematic, a realm of different systems is opened, although, based upon the present state of knowledge, one should always give preference to solid fuel gas inflators.

In this case, one will refer to developments and experiences in defense technology. Optimal adaptation for the specific application at the best possible price will have priority.

4. HANDLING AND OPERATIONAL SAFETY

The electronic equipment in modern vehicles makes possible the monitoring of the operational safety of the individual systems . In most cases, safety is dependent on the foregoing life cycle, therefore on the experienced environmental impacts /2/. The most important of these are the mechanical (shock and vibration) and thermal impacts on the gas inflator assemblies. Gas inflator assemblies are energetic systems and consequently their life is limited. Known temperature stress and aging mechanisms would make it possible to have an indicator that warned of the end of a thermally- dependent operating life.

The presence of moisture reduces the operational safety of pyrotechnical and gas inflator generants. Consequently, an air-tight isolation from the outer-atmosphere is a basic requirement.

5. DISPOSAL

In the year 2000, an estimated 70 million airbag units will be produced worldwide. Therefore, future closed-loop economies will also demand concepts for the disposal or recycling of these safety systems in ways that create as few environmental impacts as possible and/or practical. To meet these demands, technical concepts need to be developed and responsibilities for disposal allocated.

6. PROSPECTS

Gas inflator assembly systems make up an integral component of automobile safety whose significance, in respect to the risk assessment by the citizen in traffic, will increase. This indicates a corresponding market growth.

So far, praxis has shown that the already-introduced systems can increase the safety of vehicle occupants a great deal . This does not, however, exclude improvements of gas inflator assembly systems in various ways. Beyond past developments, gas inflator assemblies have additional potential and offer further application opportunities for automobile motor vehicle safety.

BIBLIOGRAPHY:

Schubert, Hiltmar, "Risks and Chances of Technology - Contributions of Applied Research to Passenger Safety," Airbag 2000 (1992); Proceedings.

Ziegahn, Karl-Friedrich, "Quali-fication of Airbag-Systems Tailored to the Automobile Environment," Airbag 2000 (1992); Proceedings.

ABOUT THE AUTHORS

Prof. Dr. Hiltmar Schubert is a senior scientist in chemistry, with special emphasis on energetic materials, including solid propellants, gas generants, and processing technology. He was the director of the ICT (Fraunhofer Institut für Chemische Technologie) until 1994 and now, after his retirement, a commissioner of the new ICT director, Prof. Dr. Peter Eyerer. Prof. Schubert's scientific career includes several hundred papers, lectures and many patents in his area of expertise. Prof. Schubert is also the president of the German Society for environmental engineering, a member of NATO's Scientific Council and a consultant for airbag technology.

Dr. Karl-Friedrich Ziegahn is a senior scientist in physics and mechanical engineering, currently serving as the general manager of ICT. His professional experience includes more than a decade on airbag qualification, especially on behalf of several automobile producers and component suppliers. Dr. Ziegahn is also the conference chairman of the "Airbag 2000" Symposium, the largest specialized airbag conference in the world.

Requirement	Free of gas-pollutants	Low on particles	Low temperature	Easily exchangeable
Application				
Passenger compartment	+	+	+	-
Motorcycle	-	+	+	-
Fire extinguisher	-	-	+	-
Mechanical equipment	-	-	-	-
Aquaplaning	-	-	-	+
Sand dissemination	-	-	-	+
External protection	-	-	+	-

Chart 1: Requirements for Gas Inflator Assemblies for a Variety of Applications

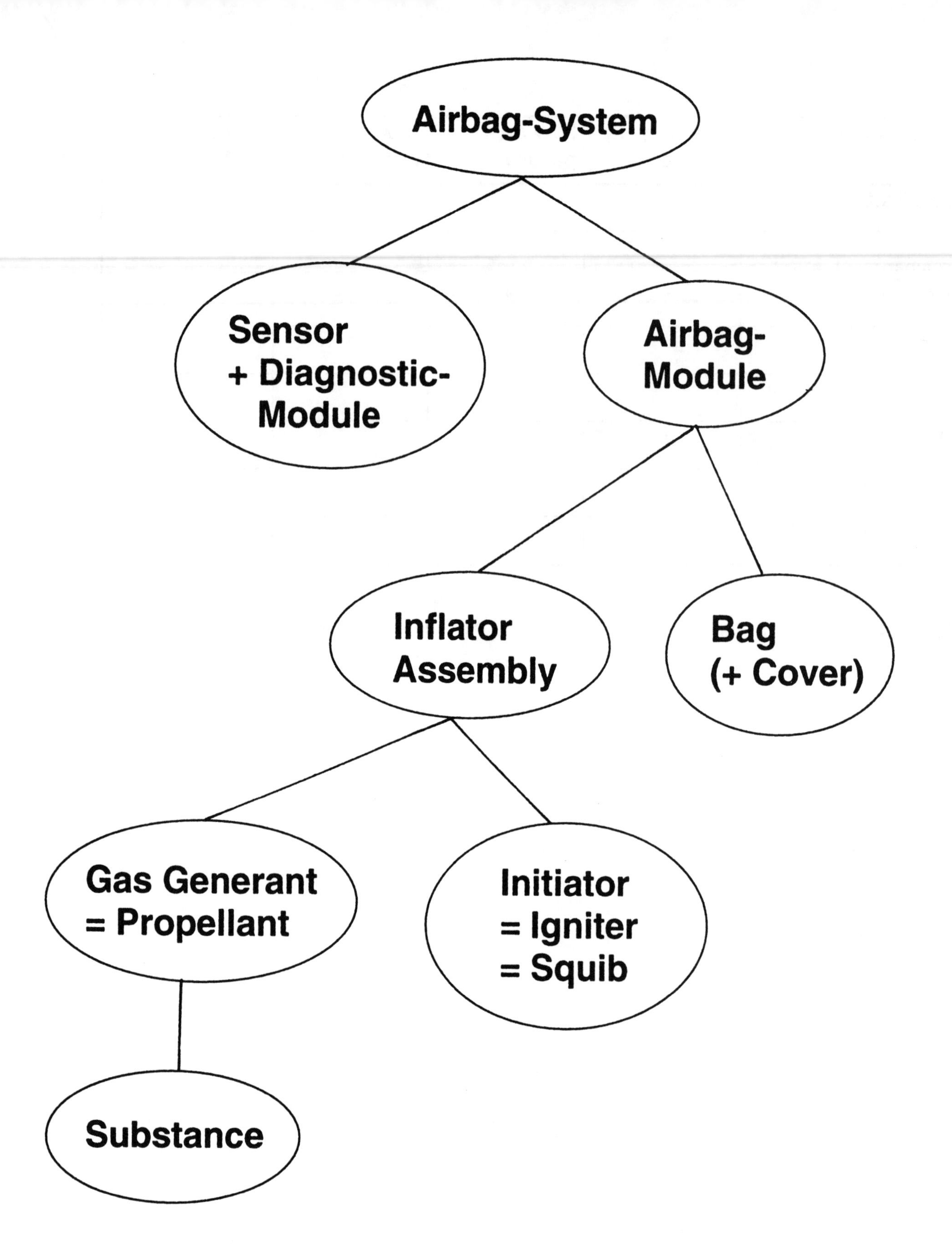

ISO-Terminology (Draft 1994)

Fig.: 1

950347

Investigation of Sensor Requirements and Expected Benefits of Predictive Crash Sensing

William R. Swihart
TRW Transportation Electronics Div.

Albert F. Lawrence IV
TRW Electronic Systems and Technology Div.

ABSTRACT

Increased motor vehicle safety standards for frontal and side impact crashes are requiring quicker sensor response to deploy air bags and safety related components. A reliable method of pre-arming crash safety systems may provide the potential for enhanced occupant protection by decreasing the time required to discriminate between real crash events and non-events. The reduced discrimination times will allow safety engineers more design flexibility in tailoring air bag inflation rates, pressures and belt pre tensioning systems to further improve their effectiveness in severe crash events and to prevent air bag induced injury due to an out of position occupant.

The paper investigates the initial system engineering requirements for a pre-arming sensor and establishes a baseline approach for further study. Prototype Forward Looking Automotive Radar sensors are used to investigate and test performance of existing radar based sensors to determine the ability of current devices to provide pre-arming capability and provides direction for further sensor development.

INTRODUCTION

Radar applications for automobiles have been widely researched for many years to provide collision warning, blind spot detection and autonomous or semi-autonomous cruise control. Deployment of radar systems with many of these features appears imminent within this decade given the development of low cost signal processing and radar transceiver functions. Cost competitiveness with existing systems, usefulness and benefit of this advanced technology will determine their success in the marketplace. Therefore, system designers and engineers have been looking at other potential functions that can enhance the usefulness and desirability of proposed systems. One approach serves to increase the utility of a forward looking radar system by providing the range and velocity information already generated by an intelligent cruise control radar to a single point electronic crash sensor module. This information can be incorporated into the air bag firing decision algorithm to decrease the time required to fire the air bag. The faster time to fire signal can significantly enhance occupant protection and reduce air bag induced injuries to occupants by tailoring the bag deployment. Tailoring of the inflation pressures [1] and energy management of inflating air bags can significantly effect the loading severity and ultimately the protection of the occupant. Loading severity is increased for those cases where occupants may be standing or may be thrown forward against the deployment door due to pre-event braking or an avoidance maneuver. Research has shown that small adults and children are more susceptible to this out-of-position condition. Faster crash sensing and optimized energy management in the occupant restraint system may improve the protection in these cases and in overall occupant protection.

FASTER TIME TO FIRE DECISIONS

Currently, single point electronic crash sensors employ solid state accelerometers and in some cases safing or arming sensors to detect events. The microprocessor in the electronic module performs calculations to discriminate between must fire and no fire events.

For frontal crash sensing, the required time to fire depends on many factors but typically extends into 10 to 20 ms range for 48.3 KPH barrier events. Most crash sensing algorithms convert the event accelerations into physical parameters and must use a significant portion of the required time to fire in order to arrive at a correct and reliable decision. Rather than improve the quality of the event accelerations to decrease the time to fire, possibly through costly structural changes to the vehicle, pre-event sensing uses information about the impending event. Current state of the art electronic crash sensors reliably provide the firing decision before the required time to fire but could be substantially improved by using range and relative velocity data from a forward looking radar sensor.

The improvement in time to discriminate between real crash events and non events was investigated by using two series of crash events for modern automobiles. Two different types of vehicles were selected for this study. The first crash series is from a typical mid-sized platform with unibody construction while the second crash series is from a typical minivan platform. The selected crash types are listed in Table 1 along with their respective required time to fire and expected time to fire using current state of the art electronic sensors. The algorithms and thresholds were modified to make appropriate use of the range and relative velocity data from a radar sensor. As can be seen from the results in Table 1, a significant improvement in time to fire can be realized with a priori knowledge of an impending event. The smallest improvement still showed a respectable ten (10) per cent improvement. This data also provides some clues as to the vehicle types and crash events that may benefit most from predictive crash sensing.

TABLE 1

Predictive Crash Sensing Performance

RTTF - Required Time to Fire, ms

TTF1 - Time to Fire With Out Predictive Crash Sensing

TTF2 - Time to Fire With Predictive Crash Sensing, ms

NF - No Fire

Mid Sized Platform, Sedan

Event	RTTF ms	TTF1 ms	TTF2 ms	Delta ms
12.9 KPH Head On Barrier	NF	NF	NF	NF
22.5 KPH Pole	95.0	88.2	52.2	36.0
25.8 KPH 30 Degree Barrier	80.0	61.5	12.6	48.9
48.3 KPH 30 Degree Barrier	55.0	53.6	13.9	39.7
48.3 KPH Head On Barrier	20.0	14.8	10.6	4.2

Minivan Platform

Event	RTTF ms	TTF1 ms	TTF2 ms	Delta ms
11.3 KPH Head On Barrier	NF	NF	NF	NF
29.0 KPH Pole	86.0	74.6	63.0	11.6
48.3 KPH 30 Degree Barrier	50.0	15.0	11.6	3.4
48.3 KPH Head On Barrier	15.0	10.8	9.5	1.3
56.4 KPH Head On Barrier	13.0	10.1	8.8	1.3

Consistently greater reduction in time to fire was demonstrated for the sedan than for the minivan. Although a detailed analysis of the individual structures has not been performed for this paper, the crash signatures of the two vehicles suggest that the stiffer and shorter structure of the minivan provides much greater energy transfer earlier in the event in which to base a firing decision. Therefore, the improvements in time to fire may not be as great for a structure of this type. Greater reductions in time to fire were observed for slower developing crashes and obliques. High speed perpendicular barrier crash events typically exhibit more energy transfer earlier in the event than that of the other crashes. Generally, the more energy available in the early stages of the event the smaller the expected improvement in time to fire. Additionally, the improvement in time to fire due to predictive crash sensing may be limited in extremely short, four (4) to eight (8) ms, time to fire requirements such as those found in short stiff vehicle structures or side impact crash events.

This data does not however indicate whether either vehicle structure or which specific crash types are the most severe in terms of occupant injury. It is also not a straightforward conclusion that reductions in time to fire will translate directly to significant improvements in occupant protection. Additional study is required to correlate the improvement in time to fire and occupant protection for specific crash types and vehicle structure. It is hypothesized by the authors that reductions in injury due to inflating airbags can be reduced for those cases where a faster time to fire results in deploying the air bag prior to the occupant coming into contact with the air bag module.

The authors have not extended the predictive crash sensing approach to consider making a firing decision only on the a priori knowledge of an impending event. The benefit of such an approach may allow initiation of the air bag prior to impact. However, this extension would require extensive research on target acquisition, tracking and destructive power of the object on the collision path.

REDUCING INJURY

Extensive research and testing have been performed to assess the injury to occupants during severe crash events. With the widespread use of airbags, much of this research has focused on the deployment loads of airbags and the resulting potential of injury to occupants that are out of position or in close proximity to the deploying air bag. Some studies have shown that loading of the airbag can influence injury [3] in real world crashes.

The design of an effective restraint system involves many factors beyond the control of the safety restraints designer relating to crash dynamics, vehicle crush characteristics, occupant size, weight, occupant position, seat belt use and many others. Some design variables available to the air bag designer are air bag inflation rate, ultimate or maximum pressure and venting. The deployment and pressurization rate of the air bag represents a compromise between not having sufficient pressure in the bag to restrain the occupant as he begins to move relative to the seat and that of having so much impulsive energy as to induce injury to an occupant in close proximity to the deploying bag. An earlier firing signal permits more time to inflate the bag and hence reduce or lower deployment loads. High and low levels of inflation aggressiveness have been considered in the referenced research. Most testing has been performed with instrumented test dummies or surrogate subjects, [4,5] that were positioned over air bag modules prior to deployment. These studies present injury values for position of subjects in relation to a deploying airbag. The injury values are based on the peak loading obtained for chest compression, chest velocity and head injury criteria (HIC). Research and test data derived from using swine as test subjects to determine the mechanism and timing of blunt injury [6] have established these injury criterion as useful in subsystem design to mitigate injury.

Other studies have developed models for the timing and mechanisms of injuries and have validated the injury criteria used in [4,5].

The reduction in injury to an occupant as a result of predictive sensing was assessed using data from [4,5]. Data is presented for chest velocity, chest compression and HIC for surrogates positioned at various longitudinal positions and at three distances from the deploying air bag. A relative reduction in injury parameter was compiled based on this data within a group of tests that contained data at different spacing. This data was then averaged to eliminate any one set of data dominating the results. As expected, the further the surrogate was from the deploying bag, the greater reduction in injury parameter thus indicating a potential reduction in injury. The results from this analysis are shown in Figure 1. These results only consider the effect of increased spacing between the subject and the depolying air bag. Although this is an important component in reducing injury, an out-of-position occupant could still be subjected to large deployment forces if positioned near the deploying bag. Additional benefit may be achieved by reducing the inflator mass flow rate and increasing inflation time to both the in- and out-of-positon occupants.

This data indicates a general trend in reducing injury and should not be taken as an absolute. The differences in surrogate position in relation to the center of the wheel caused large variations in the injury parameters. In some cases, only small reductions were noted in one or more of the parameters.

The improvement in injury values is dramatic for slight separations from the air bag module. Translating this separation to a requirement for faster air bag firing time was determined by reviewing the head movement found in a typical sedan during barrier events. Figure 2 presents the required improvement in time to fire assuming free body displacement by 25 and 50 mm at various speeds. Superimposed on this figure are film analysis data showing the time required for the head to move 25 mm at maximum head velocity. Test data is for an unbelted occupant. Test data is for barrier events for an unbelted occupant. Therefore, it is expected that improvements in time to fire due to predictive sensing in the order of 2 to 8 ms may provide a means of reducing injury.

SENSOR REQUIREMENTS

Various technologies have been investigated to provide range and relative velocity data for automotive applications such as autonomous cruise control, collision warning, collision avoidance and object warning. Most researchers have concluded that radar provides the best compromise for automotive applications due to its all weather performance [7], ability to be integrated into a vehicle and potential for low cost. Safety issues are obviated given the use of low power millimeter wave frequencies. The potential exists that the function of predictive crash sensing can be incorporated into forward looking radar sensors that are currently being developed for autonomous cruise and collision warning. A brief description of sensor operation is presented followed by a discussion of the key sensor requirements and their impact on overall effectiveness. A review of the current capabilities is discussed.

In a generic sense, any remote sensor will undergo search, target identification, trajectory computation, and threat/non threat decision as part of their operational flow. Given the dynamic nature of real world driving scenarios, the sensor needs to execute the described flow for every target representing a potential threat and needs to do so concurrently to meet response time considerations. A challenge for the system designer is to resolve the details of each step to fit general driving scenarios to maximize efficiency and/or optimize performance for given cost without generating false alarms. In this case a false alarm is defined as incorrectly identifying a target as a real threat and thus leaving the system susceptible to low speed deployment.

In the search mode, the sensor monitors its coverage space for potential threats. In a multiuse sensor, this mode is sometimes referred to as surveillance mode. This function needs to be carried out concurrently as a specific target is being tracked to insure that the entire coverage space is being monitored for safety reasons. As return energy is detected, the sensor must process this data to determine whether there is a target constituting a potential threat, whether the returns are noise, or are objects which are not a threat. This process can be challenging because real threats are immersed in clutter and because simple radar signal processing can not determine where the road is going. In some instances, tracking the target is necessary in order to estimate its trajectory to determine if it is on a collision path. Additional parameters such as speed, throttle angle, braking level, wheel slip, and steering angle could be incorporated with threat trajectory to increase confidence in the prearming decision. This additional platform information may be available on the vehicle's information network as a result of the other functions of the radar sensor.

One of the most critical requirements for the sensor is its response time, specifically the time from when a threat enters the field of view of the sensor to a determination that the threat has been confirmed and finally communicated to crash sensor processor. Required response time is a function of the worse case dynamics, distance, velocity, and acceleration of both the platform and the threat object. Figure 3 illustrates the response time needed as a function of distance and speed assuming zero deceleration. For typical range coverage and vehicle speeds a minimum time to detect a target and determine if it is on a collision path is about 1 second. In most situations when the driver becomes aware of a threat, defensive actions may increase the allowable response time. One second should be adequate for most situations. Current devices being considered are able to identify and track targets in a few milliseconds.

Another critical parameter is the spatial coverage of the sensor; this coverage directly affects how much time is available to respond taking into account the direction the threat is incident from. In order to evaluate where a target is first detected, probability of detection contours were computed assuming typical target radar cross-sections and using the sensors range, horizontal and vertical beam widths, and dynamic range. An analysis was performed to calculate the time in which the sensor had to respond assuming a target such as another vehicle was first detected when 50% of its frontal edge was within the probability of detection contour. Figure 4 and 5 illustrate the results of this analysis for a sensor with 8° and 60° total horizontal coverage respectively. In the case of an 8° coverage sensor which is typical for an autonomous cruise control or forward collision warning sensor, timely identifications and response are achievable for angles of incidence up to 15 to 20°. In future products, as the horizontal coverage of the sensor is expanded to 60°, predictive sensing warnings to angles of 35° incidence are achieved.

The existence of many pathological cases in real driving conditions complicates the signal processing and results in the majority of work in designing systems that are reliable and behave as expected. One of these cases is discriminating between a collision and a near miss. Specification of range and velocity accuracy, both systematic bias and random components, and latency are straight forward to derive given the desired response time and trajectory accuracy. While range and velocity information by themselves enable the sensor to determine that a collision may occur, inherent in the calculation is the assumption that the paths of the threat and the platform will cross. In many real driving scenarios the paths of the threat and the platform come very close but do not cross; an example of this is opposing traffic on a two lane undivided highway. By adding the additional requirement of sensing the horizontal direction of the threat from the direction of travel of the platform, distinguishing between a collision and a near miss is possible. Figure 6 is a comparison of the angular signature of two scenarios; the first is a vehicle in the path of the platform on a slight curve, and the second is a vehicle in an adjacent lane. In the first case, a collision warning would be issued based on the target angle tending toward zero. For the second example, target angle begins to diverge; this characteristic can be used to discriminate a near miss. Note that for a sensor with wide horizontal angular coverage, angular data allows early recognition of a near miss, but for typical forward looking sensors with angular coverage on the order of 8°, the object leaves the field of view approximately the same distance as when angular signature departs from that of a target which will collide.

SUMMARY

It is clear that earlier time to fire decisions can result in reduced injury to occupants and that radar sensors can be used to provide information about an impending event that can be used in the decision making process to fire the air bag. More study is required to refine the decision algorithms to provide variable thresholds as a function of threat and platform information. A first analysis of the generic sensor requirements reveals that current radar sensors envisioned for autonomous cruise control have adequate spatial coverage and range to provide this dual use function for small angle of incidence threats and for typical driving scenarios and speeds. It is further envisioned that extensive work may be required to characterize the many pathological cases encountered in real world driving conditions and that signal processing advances are required to assess the threat of specific targets. Wider field of view may also be required to extend angle of incidence of threats, although it is not clear that all of these scenarios will result in the need to fire a frontal air bag.

ACKNOWLEDGEMENTS

The authors wish to thank Thomas H. Vos, Ronald M. Muckley, David J. Bauch and Mark Carlin for their help in preparation of this paper.

---

[1] W. R. Carey, T. J. Wissing, R. G. Gehrig, G. W. Goetz, and D. A. Larson, "Energy Management in IORS", SAE Paper No. 720418, Second International Conference on Passive Restraints, May 1972.

[2] L. Patrick and G. Nyquist, "Airbag effects on the out of position Child." SAE Paper No. 720442, Second International Conference on Passive Restrainsts, Detroit, May 1972.

[3] H. Mertz, "Restraint System Performance of the 1973-76 GM Air Cushion Restraint System", SAE Paper 880400, February 1988.

[4] J. W. Melvin, J. D. Horsch, J. D. McCLeary, L. C. Wideman, J. L. Jensen and M. J. Wolanin, "Assessment of Air Bag Deployment Loads with the Small Female Hybrid III Dummy", SAE Paper 933119.

[5] John Horsch, Ian Lau, Dennis Andrzejak, David Viano, John Melvin, Jeff Pearson, David Cook, and Greg Miller, "Assessment of Air Bag Deployment Loads", SAE Paper 902324, Thirty-fourth Stapp Car Crash Conference, November, 1990.

[6] Ian V. Lau and David C. Viano, "How and When Blunt Injury Occurs - Implications to Frontal and Side Impact Protection," SAE Paper 881714, October 1989.

[7] K.W. Chang, H. Wang, G. Shreve, J. Harrison, M. Core, J. Yonaki, A. Paxton, M. Yu, C. H. Chen, G. S. Dow, B. Allen, K. Tan and P. Moffa, "Forward Looking Automotive Radar Using W-band Single-Chip Transceiver," Transactions of the IEEE.

FIGURE 1

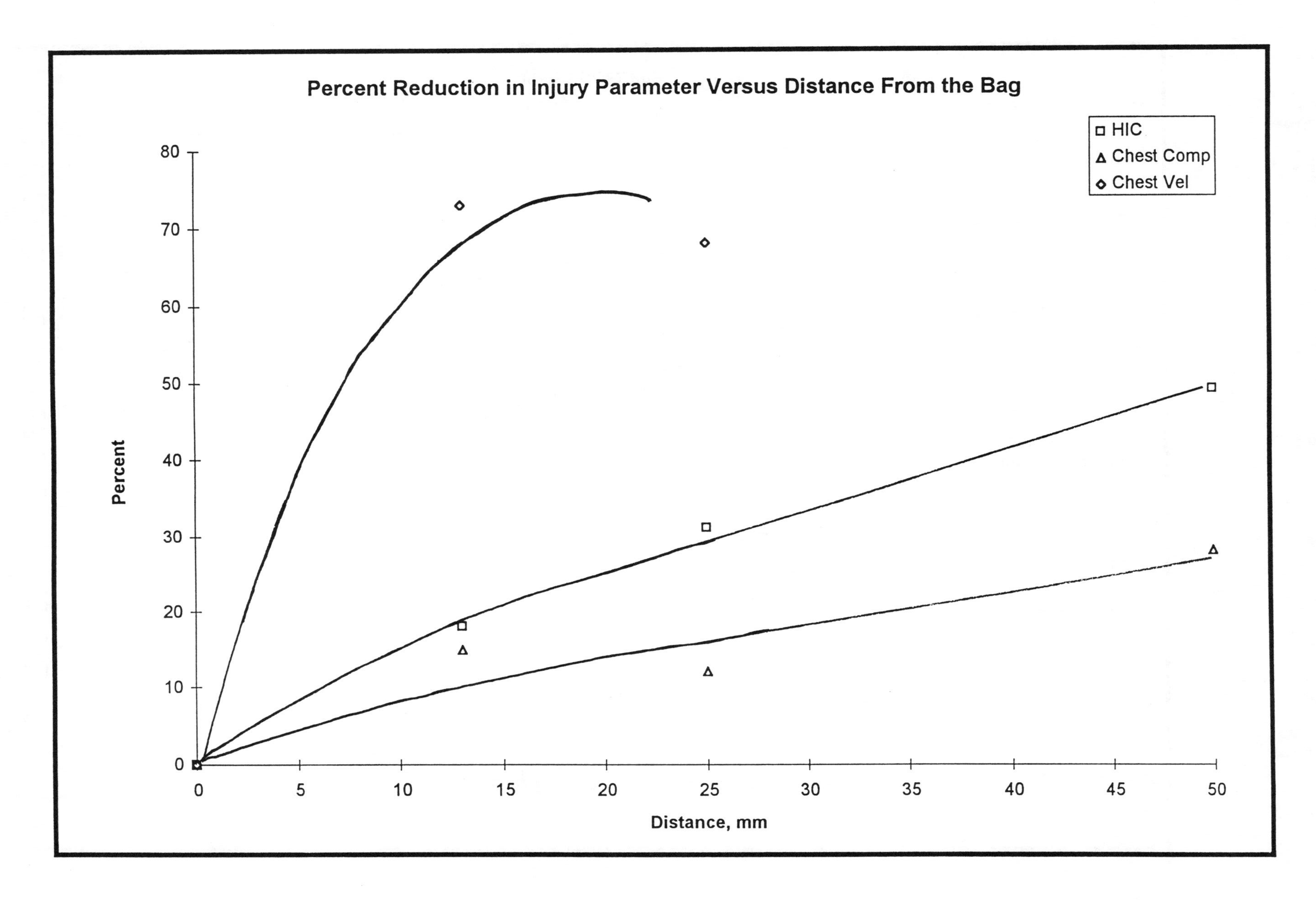

Percent Reduction in Injury Parameter Versus Distance From the Bag
HIC
Chest Comp
Chest Vel
Percent
80
70
60
50
40
30
20
10
0
0
5
10
15
20
25
30
35
40
45
50
Distance, mm

FIGURE 2

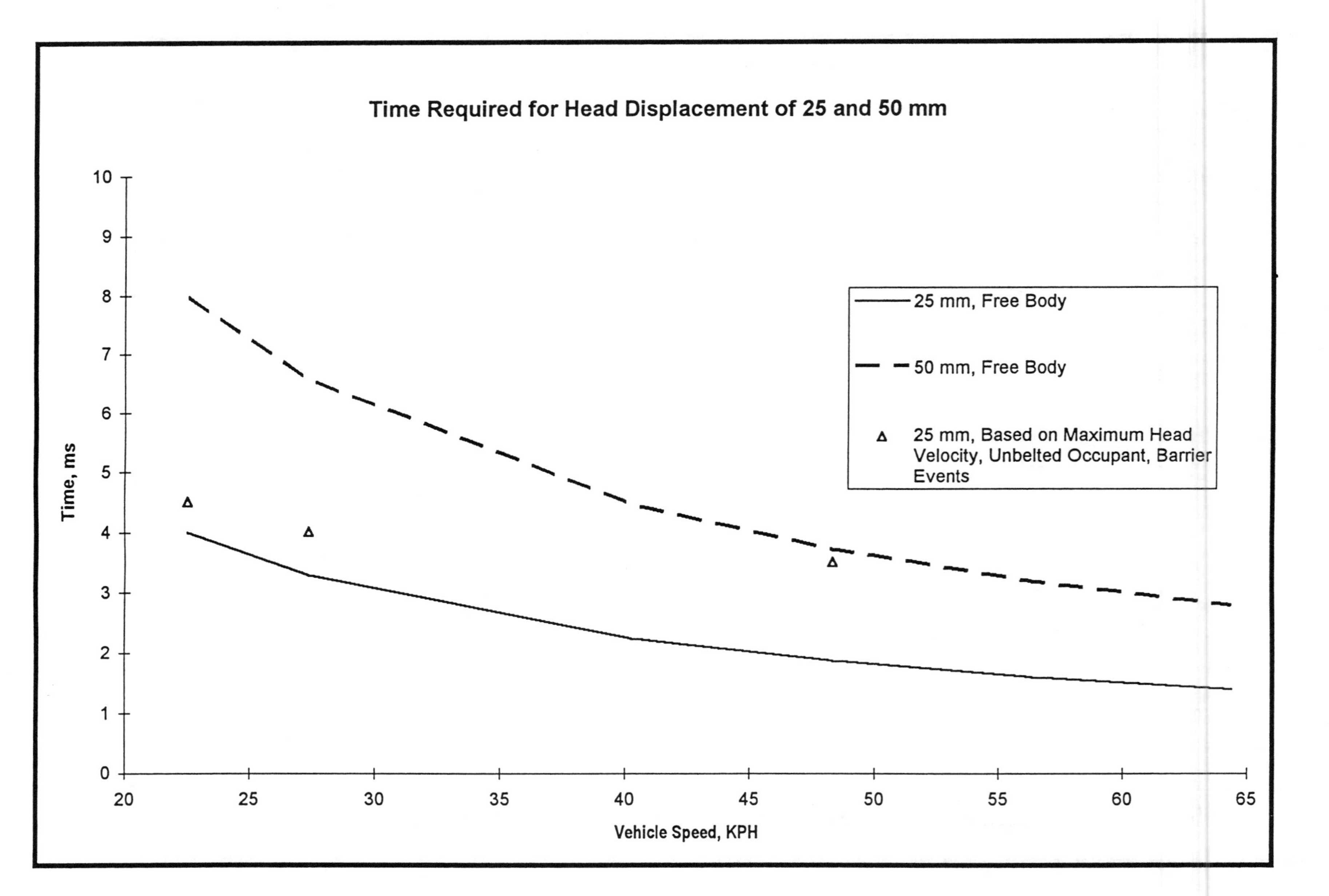
Time Required for Head Displacement of 25 and 50 mm
Time, ms
10
9
8
7
6
5
4
3
2
1
0
20
25
30
35
40
45
50
55
60
65
Vehicle Speed, KPH
25 mm, Free Body
50 mm, Free Body
25 mm, Based on Maximum Head Velocity, Unbelted Occupant, Barrier Events

FIGURE 3

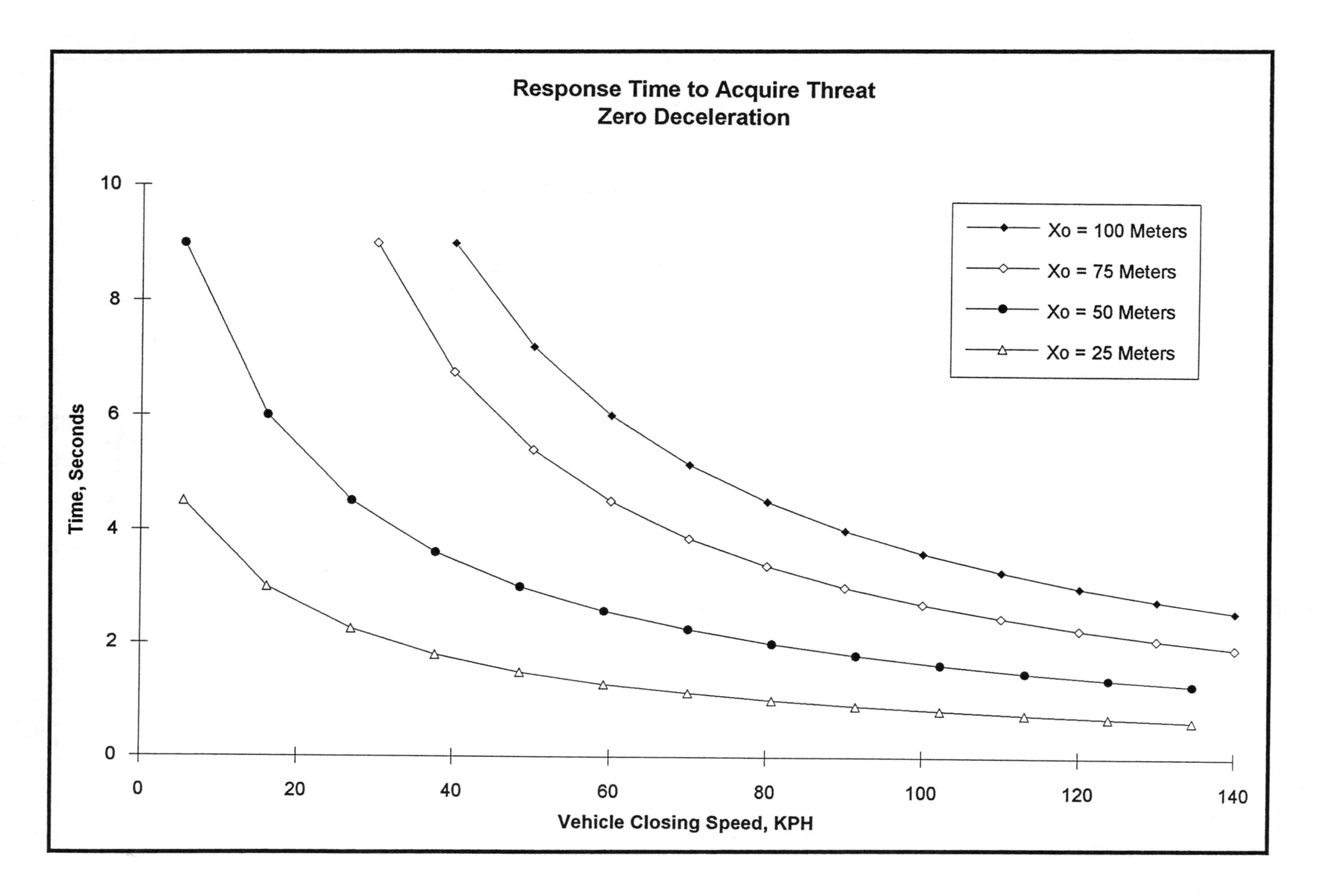
Response Time to Acquire Threat
Zero Deceleration
Xo = 100 Meters
Xo = 75 Meters
Xo = 50 Meters
Xo = 25 Meters
Time, Seconds
0
2
4
6
8
10
Vehicle Closing Speed, KPH
0
20
40
60
80
100
120
140

Figure 4

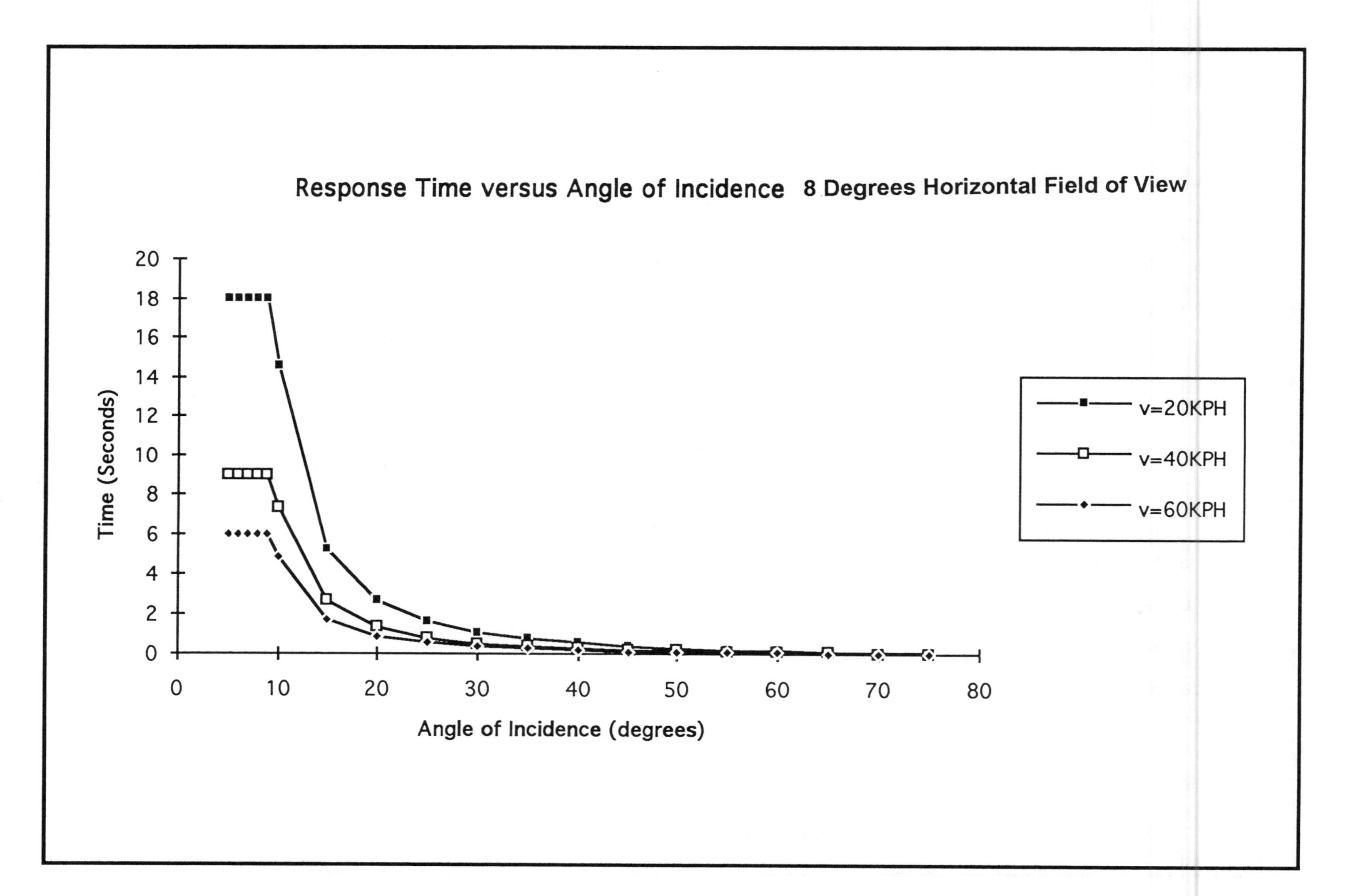

Response Time versus Angle of Incidence 8 Degrees Horizontal Field of View
Time (Seconds)
20
18
16
14
12
10
8
6
4
2
0
0
10
20
30
40
50
60
70
80
Angle of Incidence (degrees)
v=20KPH
v=40KPH
v=60KPH

Figure 5

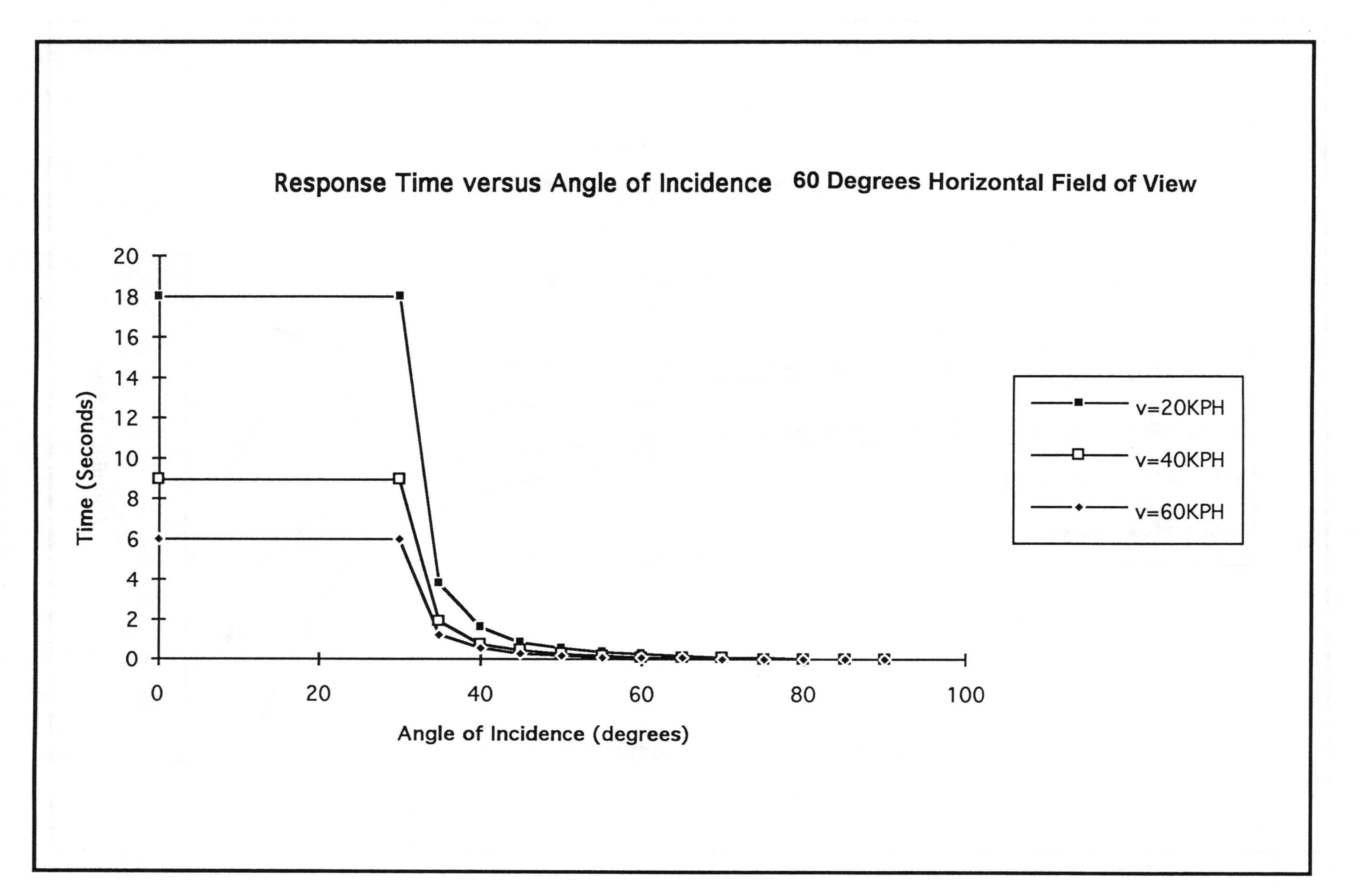
Response Time versus Angle of Incidence 60 Degrees Horizontal Field of View
Time (Seconds)
20
18
16
14
12
10
8
6
4
2
0
0
20
40
60
80
100
Angle of Incidence (degrees)
v=20KPH
v=40KPH
v=60KPH

Figure 6

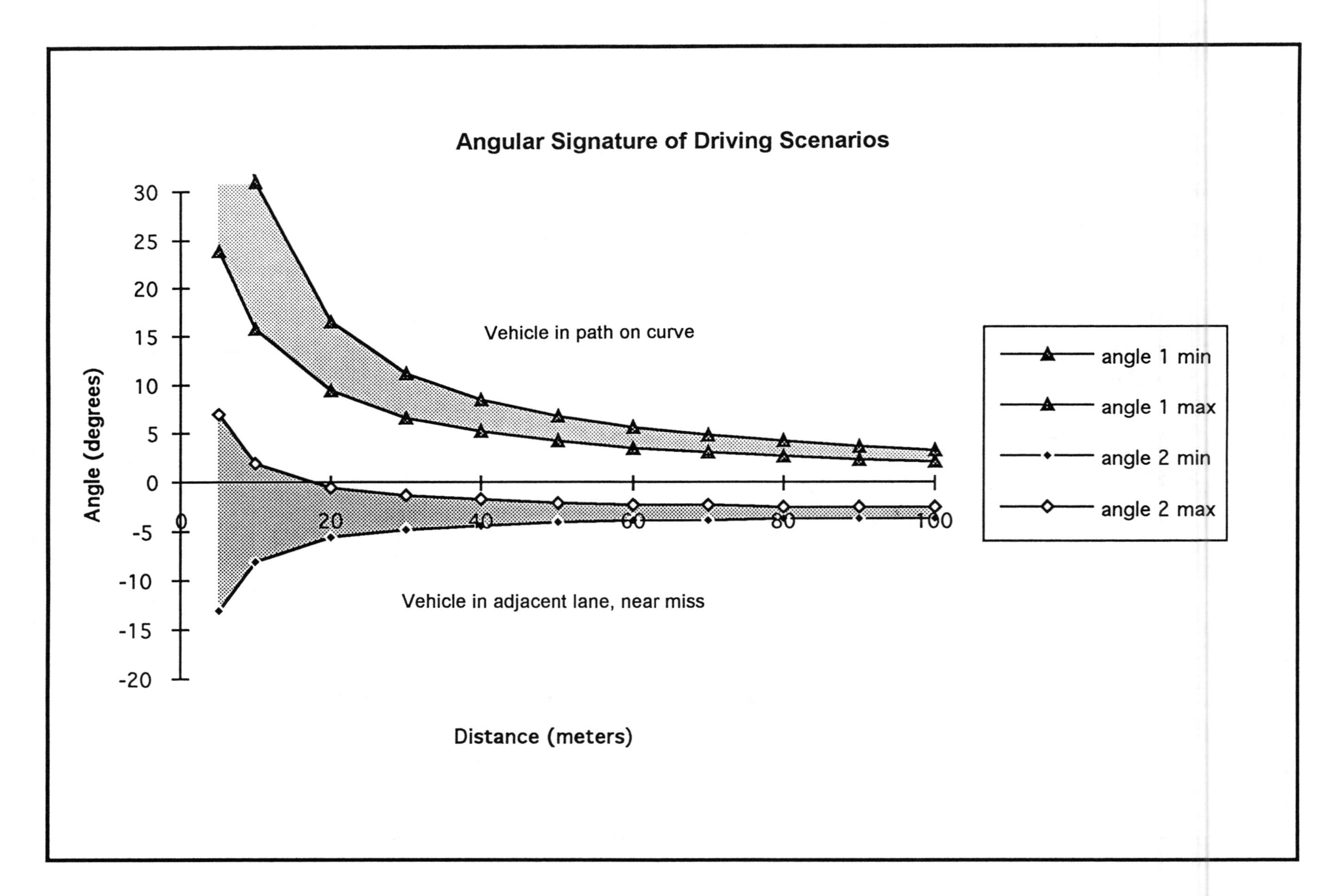

Angular Signature of Driving Scenarios
Angle (degrees)
30
25
20
15
10
5
0
-5
-10
-15
-20
0
20
40
60
80
100
Vehicle in path on curve
Vehicle in adjacent lane, near miss
Distance (meters)
angle 1 min
angle 1 max
angle 2 min
angle 2 max

950348

Physically Different Sensor Concepts for Reliable Detection of Side-Impact Collisions

Alfons Härtl, Gerhard Mader, Lorenz Pfau, and Bert Wolfram
Siemens Automotive

ABSTRACT

This paper describes new concepts to detect side impact collisions. Based on the specific system requirements for side impact detection, two physically different concepts will be described and compared to each other.

Acceleration sensing principles, applied in today's single point sensing systems, were adapted to cope with the unique requirements for side collision detection.

A more advanced and completely new concept is based on the sensing of the pressure change within the cavity of the impacted door.

Based on these sensing principles, different system configurations will be illustrated. The performance of both sensing principles will be compared on the basis of available crash and misuse test conditions.

In conclusion, it can be stated that the aforementioned sensing principles support the rigid firing requirements of a timely airbag deployment.
However, the selection of the system configuration and the physical sensing principle has to account for the individual deformation behavior of the vehicle's side structure.

INTRODUCTION

The effort within the automotive industry to improve the safety of passenger cars has been successful. Air Bags have been widely introduced to protect the occupant in frontal accidents, which will help to reduce injury numbers in head-on collisions. However, the field of supplemental restraint systems will remain challenging.

Even so, side collisions only account for 20% of all accidents, they show proportionally higher occurrence of accidents with severe and fatal injuries of approximately 50%. Air Bag systems will now be applied to protect passengers in side collisions. The first (mechanical) side air bag system has recently entered the market.

The "sensor" within the system represents a key component, which has to cope with requirements even tougher than those known for frontal impact sensors: Not only a robust distinction between fire and no fire/misuse events has to be possible, but in the case of a severe side collision, the fire decision is required within 5 ms or less.

In cooperation with different car manufactures, Siemens has developed, analyzed and proven out different sensor concepts, which have shown in several crash tests, that they absolutely satisfy the above described requirements.

This paper describes two physically different approaches.

THE PRESSURE CONCEPT

A satellite (pressure sensor plus circuitry for signal conditioning and decision making) is located inside the door cavity to monitor a dynamic pressure change.

An intruding object causes a deformation of the door and thus reduces the volume of the cavity inside the door. The rapid volume reduction leads to an adiabatic increase of the pressure within the cavity of the door. Due to the increased pressure, air starts to flow through leakage and the air mass inside the door decreases. An interesting effect observed is that the pressure amplitude (especially in the first several milliseconds) is quite insensitive to a change of the leakage size. This behavior has been proven experimentally, as well as, with a theoretical model.

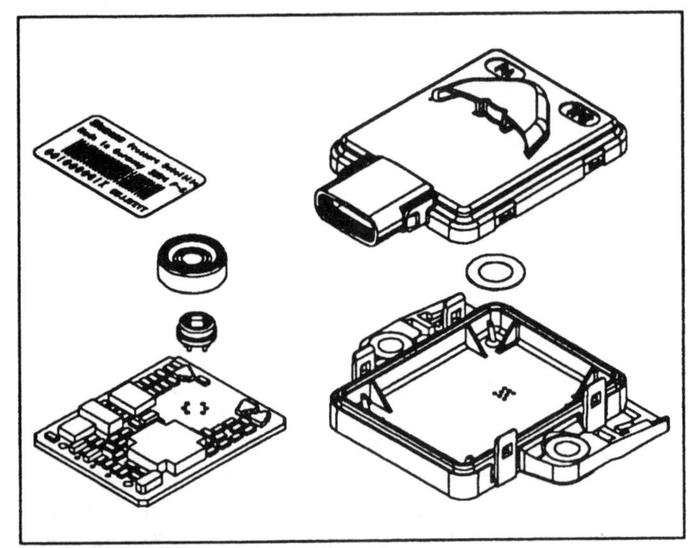

Figure 1: Pressure satellite

The pressure sensor used is a silicon micro machined component, which senses absolute pressure. The device is made up of two silicon layers with an evacuated cavity between them, to provide a pressure reference. Changes of external pressure will lead to a deflection of the silicon membrane (2nd silicon layer). Piezo resistors, implemented into the membrane, will change their value and therefore, provide information about the external pressure.

Amplification and filtering will generate the input for the decision making circuitry, where the actual pressure signal will be compared against defined firing conditions (e.g. pressure threshold and simple integral of p(t)). The measured signal is a relative value of $\Delta p/p_o$, with p_o representing the actual atmospheric pressure and Δp the dynamic pressure change. Typical pressure signals (Δp) for severe side-impacts fall in the range of 20...200 millibar (0,3...3,0 psi) with p_o=1000 mbar.
Upon recognition of a firing condition, the bag will be deployed. A remark to note is, that the comparable noise level to those pressure levels is far beyond a starting jet airplane.

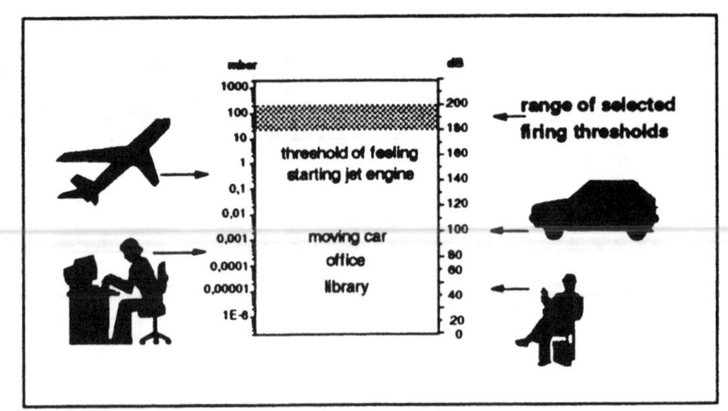

Figure 2: Pressure range, firing threshold

Another interesting fact is the possibility of a self test with this type of sensor. Due to the fact that the atmospheric pressure is contained within the output signal, high end and low end limits for the output signal can be defined and failures of the sensor (e.g. a leak of the cavity) can be detected and communicated to a warning indicator.

EXPERIMENTAL RESULTS FOR PRESSURE SENSING

During the development and prove out-phase, Siemens has worked with many different OEM's. Our pressure sensors were used in crash test on more than 30 different platforms.

Many different crash conditions have been tested (e.g. FMVSS214, EEVC, car-to-car crashes at various angles, pole impacts and truck impacts etc.). In addition to fire and nofire tests, several misuse and abuse conditions were tested (e.g. bicycle impact, foot kicks, hammer blow, frontal crash, door opening into rigid objects, door slamming and sound tests, etc.).

Figure 3 shows different fire conditions and their pressure responses. For this specific door, a fire threshold of Δp= 34mbar (p_o=1000 mbar). As shown, the signals exceeded the threshold in all cases before 5 ms.

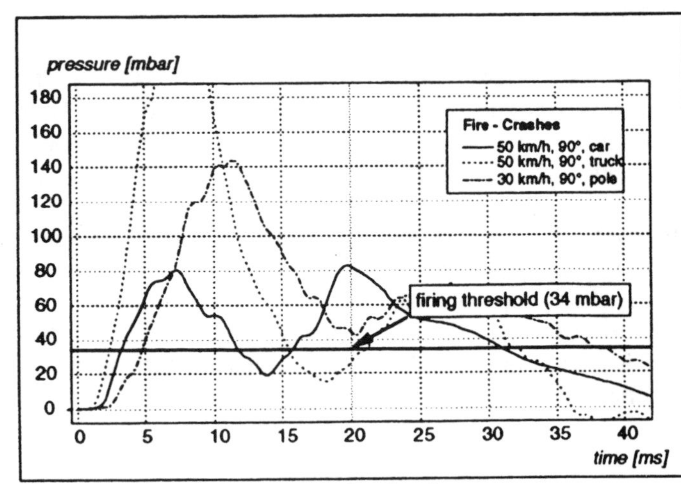

Figure 3: Example for pressure signals - fire cases

The pressure signals show all, a typical behavior for fire conditions, where a steep increase of pressure leads to a fairly high peak value before the signal is reduced, due to leakage later in the event.

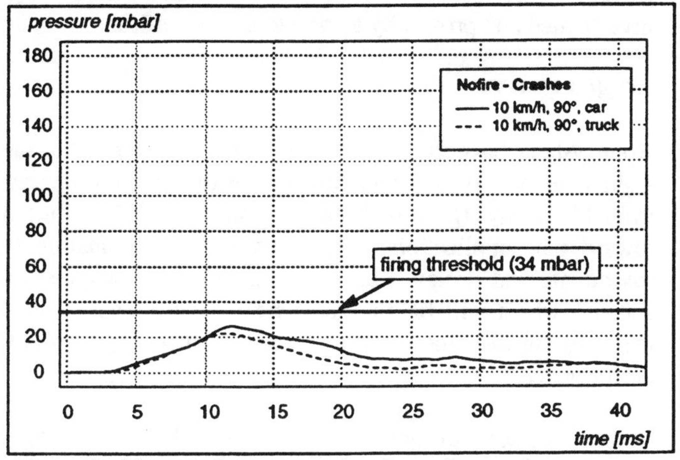

Figure 4a: Example for pressure signals - nofire-crashes

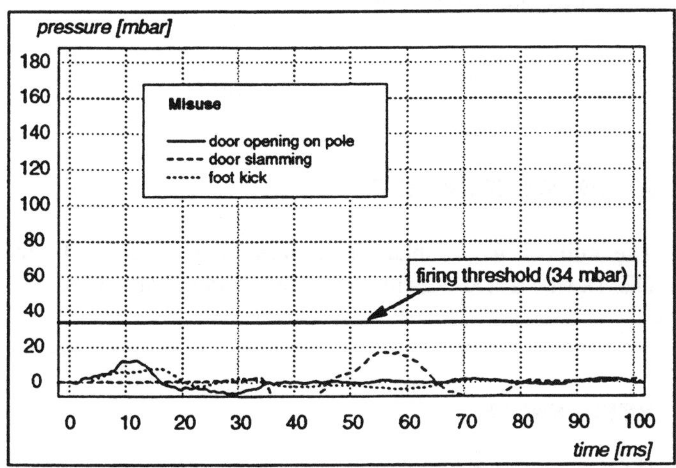

Figure 4b: Example for pressure signals - misuse tests

Figure 4 explains various misuse and nofire conditions. These events describe the low end of the fire threshold.

The comparisons of both figures 3 & 4 indicate, that a fire/nofire decision, containing a considerable safety margin, could be attempted easily with a pure threshold. High sophisticated algorithms are not necessary to distinguish these crash conditions.
More complicated crash situations might require flexible criteria for the fire decision. This can be easily implemented by the use of a simple crash algorithm.

THE ACCELERATION CONCEPT

Acceleration Sensing for side collision detection uses the well known technology of single point sensing systems. An accelerometer is the heart of a g-satellite, which is located either inside the door or close to the door structure (e.g. B-pillar, cross car beam, etc.). The signal provided by the state-of-the-art accelerometer is fed into the microcontroller. An algorithm based on physical meaningful criteria can distinguish between fire and nofire situations. In case of a fire condition, the bag will be ignited.

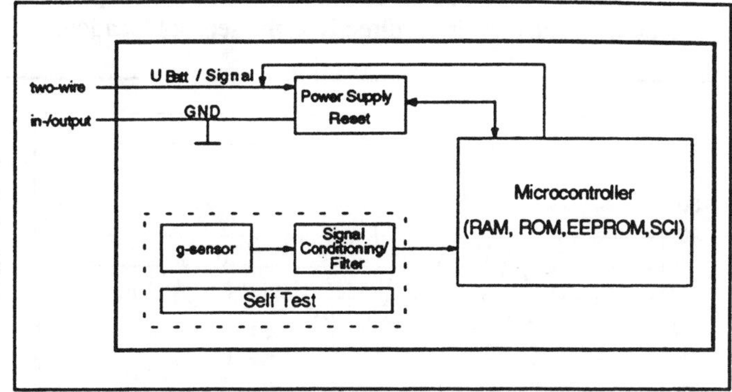

Figure 5: Block diagram of g-satellite

Depending on the location of the satellite, the g-range of the accelerometer necessary for a reliable detection of side collisions will vary. Inside the door, much higher values can be seen. The accelerometer should, therefore, provide at least a range of ± 200 g. In some cases, depending on the mounting position and door structure, a 500 g device has proven to be the right choice.

EXPERIMENTAL RESULTS FOR G-SENSING

It is common behavior to collect acceleration data from various locations during a side crash in order to determine the best feasible location. In close cooperation with car manufacturers all over the world, Siemens was able to establish a large crash data bank for side collisions. Based on the available crash data, algorithms have been developed and g-sensor satellites have participated in several hundred crash tests as well as thousands of misuse tests.

As a result of extensive testing and simulation, it can be said that positioning of the sensing device (g-sensor) is extremely critical as it relates to the crash performance.

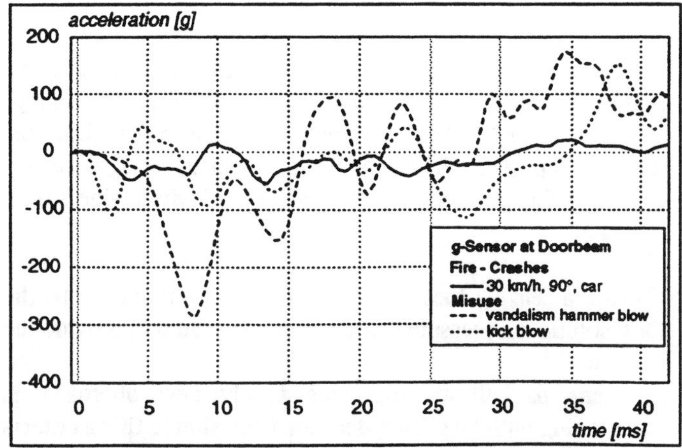

Figure 6: Example for acceleration signals - sensor position: doorbeam

Figure 6 shows fire and nofire/misuse conditions for a g-sensor located inside a door for one specific platform. Here it becomes very difficult to distinguish the abuse conditions from some fire events within the required time. A sensor located inside the door can see considerable g-forces, especially if the door is hit directly at the sensors location.

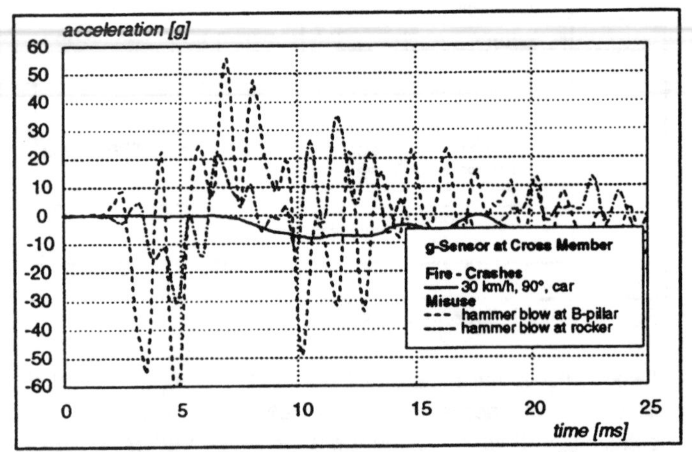

Figure 7: Example for acceleration signals - sensor position: cross member

Figure 7 reflects a scenario where the sensor is located outside the door at a cross member. The g-forces seen in the different events show much lower levels compared to the door location. A clear distinction between fire and nofire/misuse events is possible with a mainly velocity based algorithm.
If the car structure is stiff enough, the signals from an additional lateral g-sensor inside the central control unit can be combined with the satellite signals for the discrimination of fire crashes.

COMPARISON

The goal of a safety device has been clearly defined: Best crash performance has to be targeted in order to protect the passengers in all possible events.

The most feasible location for a specific platform has to be determined by testing. Packaging limitations have to be taken into account many times over.

In case a location outside the door is able to satisfy the crash requirements, the acceleration based technology (g-sensing) has proven it's reliability in single point sensing applications. Further improvements of algorithms for faster firing decisions will help this technology to be widely used within side airbag systems.

In case a sensing location inside the door turns out to be the best solution, a pressure based sensing system will provide the best fit.
Siemens has shown that this totally new approach is convincing with fast firing decisions and simple firing criteria compared to highly complex algorithms. Fire and nofire events can reliably be distinguished and packaging inside the door is easily facilitated, due to the fact, that the pressure signal is homogeneous within the door cavity.

Possible concerns of this new technology (e.g. influence of leakage, rust proofing sprayed on sensor) have been investigated and proven by testing to be insignificant.

CONCLUSION

Both systems have their advantages if they are applied in the right manner. It is fascinating to see that sensing devices available today (g-sensor) can be adapted and advanced toward new applications. It is even more fascinating to investigate and design a total new sensing system (pressure sensor), which will fill the gap where conventional systems (sensing inside the door) cannot provide satisfactory performance.

Both sensor concepts are available today and work on today's vehicle chassis without major structural modifications. By applying them to our cars, we will further improve the safety of the vehicles on our roads.

ACKNOWLEDGEMENT

The authors would like to express their gratitude to the OEM's who tested our side-impact sensors in crash tests.

REFERENCES:

[1] A. Härtl, G. Mader, L. Pfau, R. Muhr
"New Sensor Concept for Reliable Detection of Side-Impact Collisions", The 14th International Technical Conference on Enhanced Safety of Vehicles (ESV) 94-S6-0-14, (May 1994)

[2] J. Franz, U. Kippelt, "Numerical and Experimental Simulation of Different Loadcases of Side Impacts", The 14th Int. Conference on Enhanced Safety of Vehicles Proceedings 94 S6 W27, (1994)

[3] Rudolf Muhr, "Dynamische Drucksimulation bei Kompressionsversuchen, ein dynamisches System", diploma thesis, University of Regensburg, Germany, Institute of Theoretical Physics, (1994)

980557

Seat Belt Pretensioners

Helmut E. Müller and Burghard Linn

ABSTRACT

The somewhat dated FMVSS 208 regulation requires that crash dummies are belted with no slack.

Reality shows that there is 120mm slack in the belt system for a 50 percentile occupant. Test results would yield more then 20% worse injury indexes when taking slack into account.

This presentation will show how pretensioners can reduce injuries, giving a historical perspective from spring loaded systems to current pyrotechnic systems. Future developments using multi-cartridge and multi-piston systems and associated test results will be discussed.

INTRODUCTION

Pretensioned seat belts will tie the occupant to the deceleration of the car early in an accident. This will reduce the peak load experienced by the occupant by more than 20%. The forward movement is less, submarining can be avoided and a possible "out of position" situation might be corrected. Additionally occupants will be tightly kept in their seats during a rollover.

1. EVALUATION OF PRESENT PRETENSIONERS

The buckle pretensioner in fig. 1 (peak load 2000 N / 160 mm stroke) was introduced on the Opel car lines in MY 90. It is still in use on the sub-compact Corsa. Table 1 shows the drop in HIC level depending on the pretensioner force, the tightening speed and slack in the belts.

In MY93 the VW Polo was equipped with a pyrotechnically powered and mechanically triggered retractor pretensioner (fig. 2). Power and speed improved as shown in table 1. The former VW Golf and Passat models had to be retrofitted with a system mounted parallel to the wiring harness to avoid additional investments (fig. 3). 11 Million of these systems have been built, even Rolls Royce is using them.

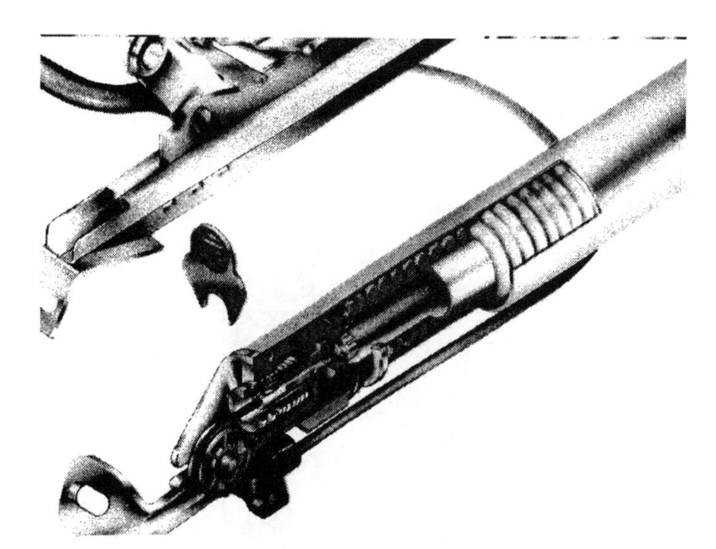

Figure 1. Opel Buckle Pretensioner

Table 1. Pretensioner "Efficiency"

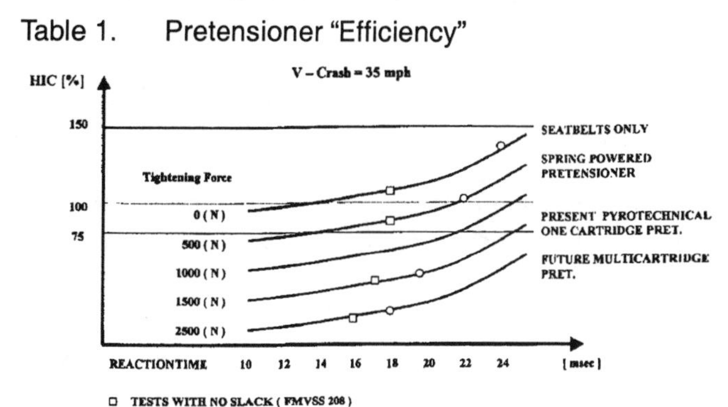

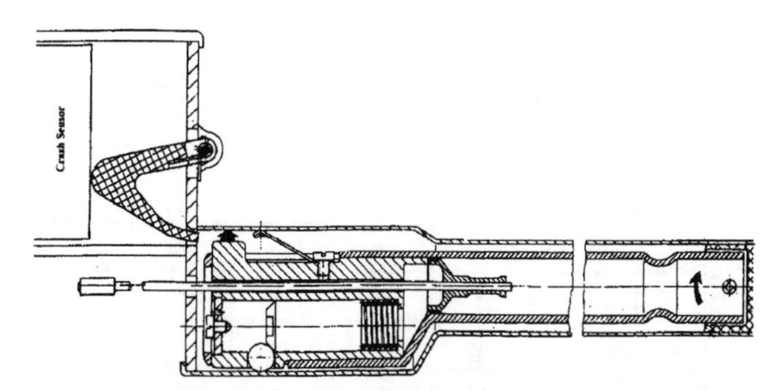

Figure 2. Mechanically Triggered Pyrotechnic Retractor Pretensioner ('93 VW Polo)

For the MY97 VW Passat a small Wankel engine, powered by 3 pyrotechnic cartridges is coupled to all 4 outboard retractors (fig. 4). The first pyrotechnic cartridge can be activated electrically or mechanically(Passat).

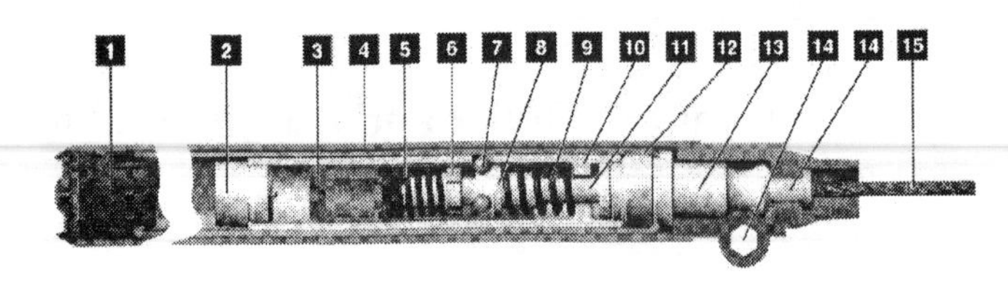

1 Bearing cap
2 Striker pin
3 Gas generator
4 Protective tube
5 Impact spring
6 Cable compressor
7 Locking balls (3x)
8 Sensor head
9 Sensor spring
10 Cylinder
11 Tubular rivet
12 Pressur cylinder
13 Sensor carrier
14 Transport packaging
15 Bowden cable

Figure 3. Mechanically Triggered Pyrotechnic Retractor Pretensioner (Former VW Golf And Passat)

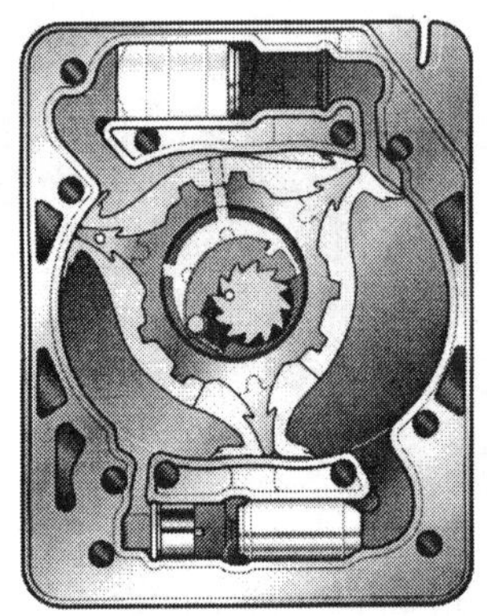

Figure 4. Wankel Retractor Pretensioner ('97 VW Passat)

2. MEASUREMENTS ON MULTI-CARTRIDGE PRETENSIONERS

Table 2 shows the gain of a multi-cartridge pretensioner over a single cartridge pretensioner and compares it to a seat belt only. The larger the slack, the more important is a high load during the full pretensioning activity.

Table 3 shows the drop in injury level in rear seats if belts are equipped with pretensioners. In addition, submarining was reduced considerably. Of course, injuries are not as severe in rear seat locations in a standard crash. But what is a standard crash in real life?

Table 2. Sled Test Results With Different Slack And Pretensioners

SLACK [mm]	PRETENSIONING	HIC [%]	HEAD ACC. (3 msec) [%]	CHEST ACC. (3 msec) [%]	BELTTIGHT. [mm]	TENSION TIME [msec]	BELT FORCE [KN]
120	NO	READ 100%			—	—	—
	ONE CARTRIDGE	50	90	80	120	12	2.0
	MULTIPLE CARTRIDGES	40	80	75	100	13	2.5
180	ONE CARTRIDGE	107	113	90	150	13	1.8
	MULTIPLE CARTRIDGES	48	85	82	185	14	2.0

SLED TEST HYBRID III, 50%

Table 3. Sled Test Results With Rear Seat Restraints w/ And w/o Pretensioners And Slack

PRETENSIONER (ONE CARTRIDGE)	SLACK [mm]	HIC [%]	HEAD ACC. (3 msec) [%]	CHEST ACC. (3 msec) [%]	PELVIS ACC. (3 msec) [%]
NO	0	READ 100%			
NO	100	128	110	108	127
YES	0	66	89	86	90
YES	100	75	86	92	108

SLED TEST HYBRID III 50%, BELT ELONGATION 12 %, ONE CARTRIDGE PRETENSIONER

3. INNOVATION IN FUTURE PRETENSIONERS

Electrically triggered cartridges are 10 times more expensive than mechanically triggered cartridges. Therefore adding one or two mechanical cartridges to a pretensioner after the first one adds little money to the system. The most simple realization of such a device as a buckle pretensioner is shown in fig. 5.

A retractor pretensioner with four pistons / four cartridges is shown in fig. 6. The pressure pulse from the active cartridge will trigger follow-on cartridges in succession. But retractor pretensioners that move the shoulder belt first, have some disadvantages in side impact crashes. The pretensioning of the belt can turn the occupant towards the impact while moving forward.

A buckle pretensioner avoids submarining and injuries to the lower body more effectively in a frontal crash. In a side crash it keeps the occupant moving with the seat, away from the intrusion.

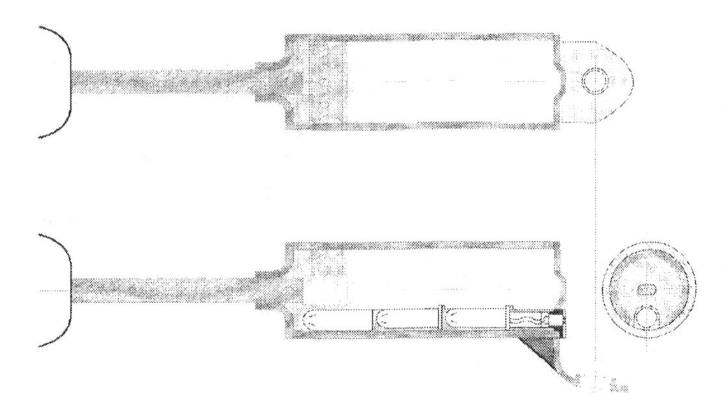

Figure 5. Three Cartridge Buckle Pretensioner

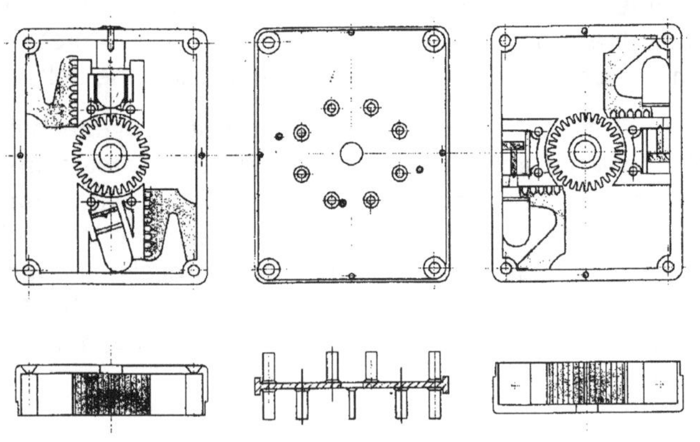

Figure 6. Retractor Pretensioner With Four Pistons And Four Cartridges

Usually belt buckles are mounted to the seat rail to better carry the crash load. With a heights adjustable seat, the seat adjustment adds to the down movement of the buckle in a crash. The reachability of the buckle might become a problem.

Fig. 7 shows a pretensioner piston integrated into a seat rail, which pulls the buckle ropes in a crash. The ropes are clamped pyrotechnically with a wedge to the piston. This allows the buckle to be mounted to the adjustable seat for best comfort while the ropes slide through the spring loaded piston.

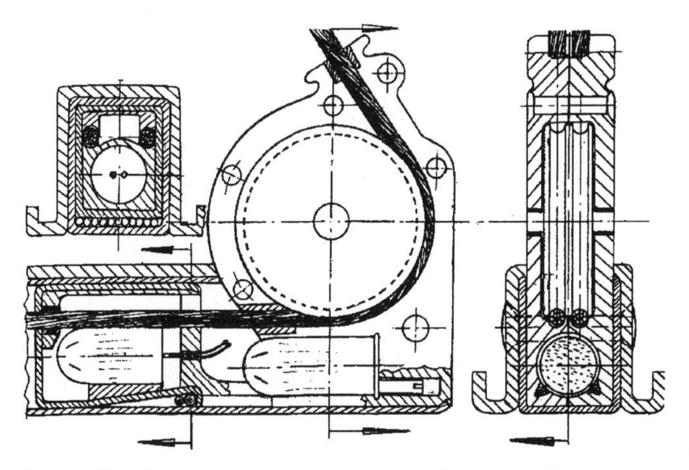

Figure 7. Pretensioner Integrated In Seat Rail For Heights Adjustable Seats

On seats w/o heights adjustment, a simple piston arrangement can be used (fig. 8). Of course, all these pretensioners are equipped with a second cartridge.

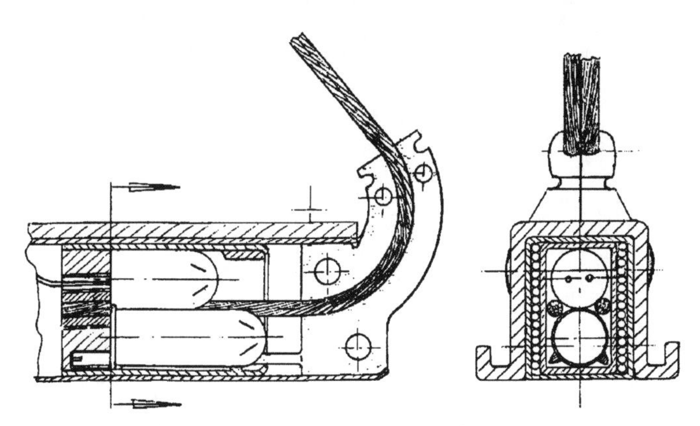

Figure 8. Pretensioner Integrated In Seat Rail With Two Cartridges (No Heights Adjustment)

If the buckle has to remain in its location even after a crash, the pretensioner might be attached to a separate lap belt retractor, which is standard on all high-end US cars for comfort reasons. Fig. 9 shows a double piston dual cartridge pretensioner which can power the lap belt retractor and the shoulder belt retractor simultaneously. This will eliminate one pretensioner and one expensive electrical trigger cartridge. This type of pretensioner lends itself for seat integrated dual retractor belt systems (fig. 10). Each retractor can be tuned in its performance through the diameter of the clutch pulley. The double piston pretensioner can also be used for buckles or belt attachments that are located close together. Table 4 shows the performance of double cartridge systems.

Load limiters are another important component to improve seat belt performance. Innovations in this area will be covered in a future paper.

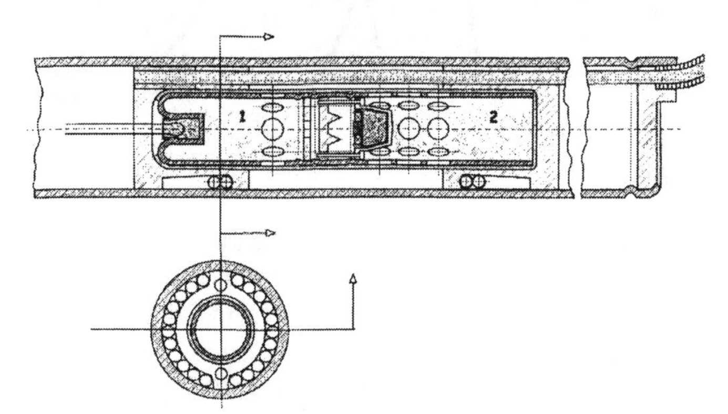

Figure 9. Double Piston, Dual Cartridge Pretensioner For Dual Retractor Belt Systems Or Double Buckle Pretensioning

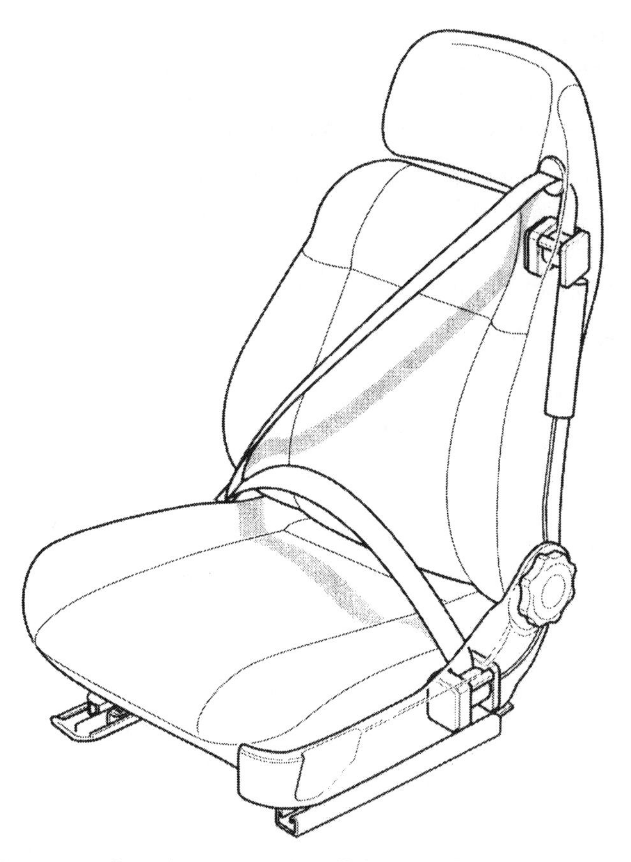

Figure 10. Seat Integrated Lap/Shoulder Belt Retractor With Double Piston Pretensioner

Table 4. Performance Of Double Cartridge Pretensioner

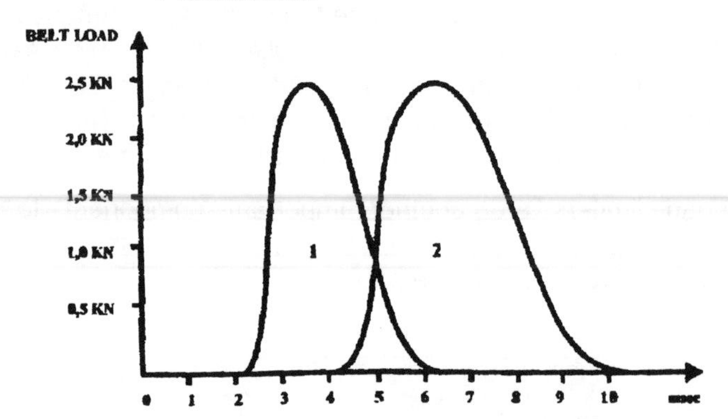

SUMMARY

- Only crashes with real slack show the advantages of seat belt pretensioners. It is urgently necessary to revise the FMVSS 208 requirements to reflect reality.
- Pretensioners should tighten the lower belt first and with greater force than the shoulder belt.
- Multi-cartridge systems improve effectiveness.
- Double piston systems decrease costs.

ABOUT THE AUTHORS

Helmut E. Müller was manager of interior design and development including passive safety at Opel for 12 years and at VW for 6 years. He is now working as an independent consultant on safety innovations.

Burghard Linn has been an engineer and manager with Opel, GM, and EDS. He now works in cooperation with H. Müller on safety innovations.

For further information, please contact:

Helmut E. Müller
Buchenweg 42
38550 Isenbüttel, Germany
Tel: 011.49.5374.5684, Fax: 011.49.5374.5185

Burghard Linn
5248 Milroy
Brighton, Michigan 48116-9727
Tel: 810.227.1223, Fax: 810.227.5179

1999-01-0080

A Method to Evaluate the Energy Capability of Seat Belt Pretensioners

Simon Xunnan He and Michael D. Wilkins
Breed Technologies

ABSTRACT

Current performance specifications of seat belt pretensioners include web retraction and belt load. These criteria may adequately represent the performance requirements of a pretensioner in a restraint system. However, by themselves, they are inadequate to evaluate potential design modifications to improve efficiency levels and thus increase energy output of the pretensioner.

This paper demonstrates a non-linear phenomenon associated with web retraction and the belt load during pretensioning. It is this non-linear behavior that promotes the insensitivity for use of web retraction in predicting the energy output of a pretensioner with design modifications. This paper proposes an energy measuring method that can be used to more accurately measure the energy output of a pretensioner and then be used to evaluate the effects on output of a pretensioner due to design modifications.

INTRODUCTION

A seat belt pretensioner is a device that tightens the seat belt in a crash event; in order to keep the occupant coupled with the seat in the early stages of a vehicle crash. It has been proven that this measure can significantly reduce injury numbers [1], [2]. Specifications for the performance of seat belt pretensioners include web retraction, belt load, and process time, etc. Web retraction is an important performance requirement for a pretensioner, because the primary function of a pretensioner is to remove belt slack. However, based on Breed's evaluation tests of several compact seat belt pretensioners, the requirement of 120 mm web retraction under the test configuration defined in the Arbeitskreis specification is very difficult to achieve [3]. This brings in a concern of efficiency of the pretensioner that must be evaluated during the development of the pretensioner.

The level of web retraction is normally used as an indicator to estimate the capacity of a pretensioner in its development. Unfortunately, this indicator is not sensitive to the change of efficiency of the pretensioner, due to the nonlinear relation between the belt force and web retraction in the normal seating configuration of a dummy, shown in Figure 1. In this nonlinearity, a small amount of increase of web retraction requires a large increase of energy.

To avoid this disadvantage, another physical quantity of belt work is introduced to measure the capacity of the pretensioner. This indicator is more sensitive to the efficiency change of the pretensioner, which makes it easier to judge the affect of design modifications to the pretensioner.

SPECIFICATIONS FOR PERFORMANCE OF SEAT BELT PRETENSIONING AND THEIR INADEQUACY

One such performance criteria for retractor pretensioners is the Arbeitskreis Specification (AK spec) [3]. This specification was developed by a cooperation of several companies that included Audi AG, Bayerische Motorenwerke AG, Mercedes Benz AG, Porsche, and Volkswagen AG. This particular specification sets performance requirements for statically testing retractor pretensioners. These performance requirements include having webbing retraction travel of greater than or equal to 120mm with at least 750 mm of webbing remaining on the spool. This includes losses due to changing from pretensioning to locking. The AK Spec also contains a requirement for maximum allowed shoulder belt force that is less than or equal to 2.5kN (562 lbs.). The whole pretensioning procedure must be completed in less than or equal to 12 milliseconds. This time is measured from the ignition signal to maximum travel. This requirement is included simply because if the pretensioning event took any longer the occupant would begin to load the belt thus reducing the performance of the pretensioner. A typical test set up for measurement of those performance data of a specific pretensioner is shown in Figure 1.

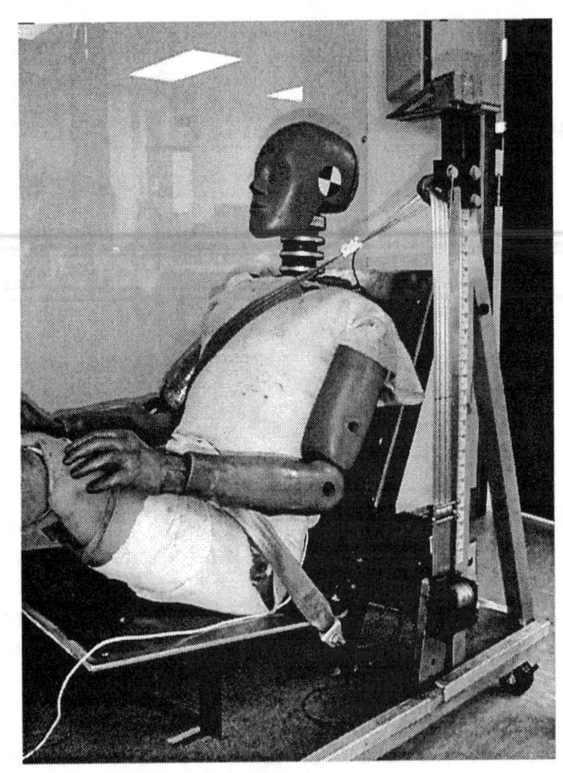

Figure 1. Test Configuration of a Pretensioner in Arbeitskreis Specification

As mentioned earlier, web retraction is an essential performance requirement for a pretensioner. This performance requirement is primarily used as a design goal in the development of a pretensioner. As a benchmark, we tested several commercially available compact pretensioners. Unfortunately, none of them met this web retraction requirement. Furthermore, in the development of our own compact pretensioners, we found that it is very difficult to use this specification to evaluate the effect of modifications on the level of web retraction.

Figure 2 shows some test results of web retraction of a specific compact pretensioner with different design modifications. Design modifications are represented by sample letters from A to K. Each sample has five tests. The dots are mean values and bars represent the data ranges.

The results indicate that the difference among the mean value of web retraction is insignificant and data spread makes it difficult to evaluate the true effect of design modifications.

The results also indicate that web retraction can be an inadequate measuring tool to evaluate potential design modifications that may lead to improved efficiency levels. Another more sensitive method of measuring the efficiency of design changes needs to be developed.

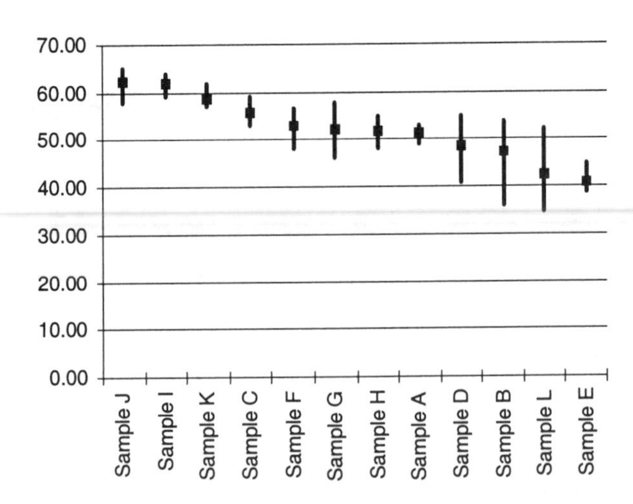

Figure 2. Test data of web retraction (mm) of a serious of modifications of a pretensioner

ENERGY DISTRIBUTION IN SEAT BELT PRETENSIONING

The process of pretensioning starts from ignition of the gas generator and ends with the belt pulling on the occupant with residual stretching. The energy source in this process is the energy stored in the gas generator and the useful work done by the energy is the work done in the belt. There is other useless work done and energy losses in between. Due to the limited energy source storage in a pretensioner, one important goal of the development of a pretensioner is to reduce these energy losses and increase the efficiency of the pretensioner.

In the entire seat belt pretensioning operation, there are basically two processes, pushing a piston by gas pressure in a closed volume and driving a mechanism to tighten the seat belt. The first process can be treated as a thermodynamic process, similar to airbag inflation. The second process is a multi-body dynamics process. The energy terms involved in the thermodynamic process are energy delivered by the gas generator E_i, energy loss by gas leaking E_l, energy left in the chamber E_s, work done on pushing the piston, and energy loss by heat transfer E_h. The energy terms involved in the multi-body dynamics include work done by the piston, energy loss E_f by friction in all sliding surfaces in the entire mechanism, and the work done by tightening the belt E_b. Notice that the work done by the piston is the same in these two processes. The energy conservation for the pretensioning can be expressed as the following equation:

$$E_i - E_l - E_s - E_h - E_f - E_b = 0 \quad (1)$$

It can also be shown in a flow chart as in Figure 3.

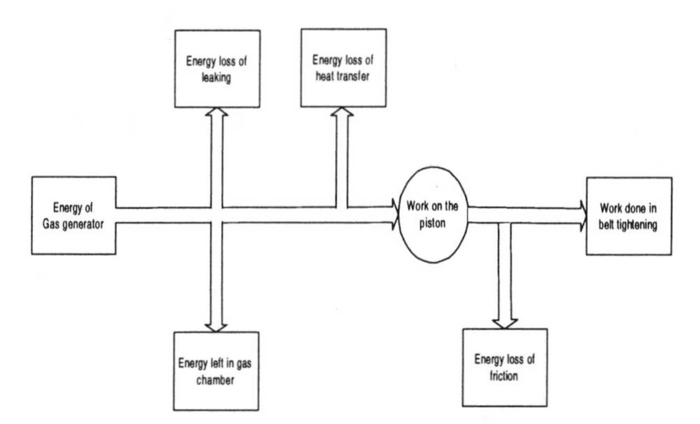

Figure 3. Energy Distribution in Belt Pretensioning

For a pyrotechnic pretensioner, the energy source comes from a gas generator, normally called a cartridge. The type and amount of propellant determines the energy delivered by the gas generator. A common way to measure how much energy a gas generator can deliver is to use a bomb test that is similar to an airbag inflator tank test. In this test, a particular gas generator is discharged into a closed vessel with a small volume, typically 10,000 mm^3, and the pressure time history is recorded. The energy conservation for this process is:

$$E_i - E_s - E_h = 0 \qquad (2)$$

The energy storage in the closed vessel E_s is:

$$E_s = \frac{C_V}{R} Vp \qquad (3)$$

in which C_V and R are gas properties, and V and p are volume and pressure time histories respectively, as shown in Figure 4.

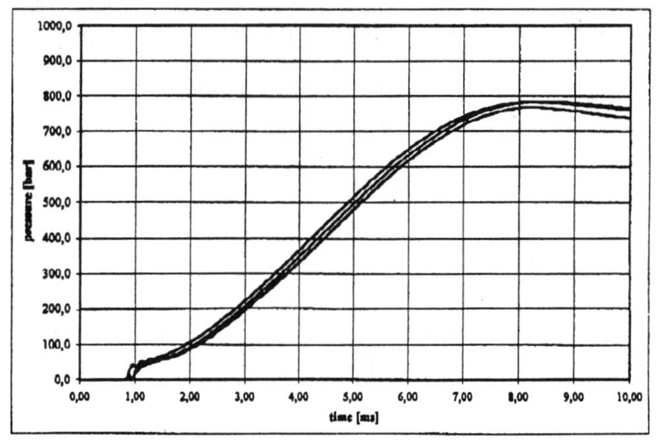

Figure 4. Bomb test pressure curve of a pretensioner gas generator

Noticing that the peak pressure occurs at very early time after ignition, the energy delivered from the gas generator can be estimated as follows by ignoring the energy loss due to heat transfer,

$$E_i = \frac{C_V}{R} Vp_{peak}$$

This relation gives an estimation of energy delivered from the gas generator.

The useful work of the energy delivered by the gas generator is the work done by tightening the belt over the occupant, represented by W_b here. This work is the integral of the belt force F_b over the distance of the web retraction, represented by d. As long as the relation between the belt force and the web retraction is known, the work W_b is found by:

$$W_b = \int_0^d F_b dx \qquad (5)$$

A METHOD OF EVALUATION OF ENERGY EFFICIENCY OF A PRETENSIONER AND APPLICATION

Several tests were run on the test apparatus shown in Figure 1 to find the reason web retraction is insensitive to design modifications. These tests were performed to find the relationship between the belt force and web retraction. The resulting belt force vs. web retraction data is plotted in Figure 5. The data reveals that the belt force vs. web retraction has a hardening non-linear relation. The higher the belt force, the higher the belt stiffness. This is especially true under high belt forces where the range of web retraction is close to the AK specification.

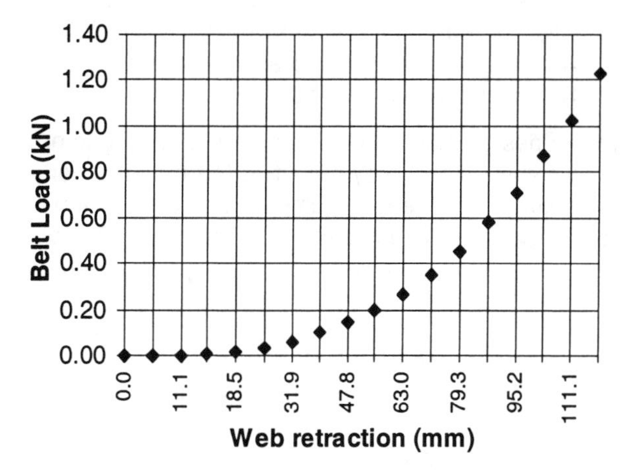

Figure 5. Test data of the relation between belt force and web retraction

There are many factors that contribute to this non-linear relation between the belt force and web retraction, such as film-spool affect, deformation of the dummy, geometry of the belt loop and so on. It is out of the scope of this paper to discuss the cause of this non-linear relation. However, it is this non-linear relation that makes the web retraction insensitive to the change of energy output of the pretensioner.

For the test data shown in Figure 5, a curve fitting can get an empirical function of the belt force F_b verses the web retraction d, as in equation. (2):

$$F_b = 0.000018 * d^{2.325} \quad (6)$$

The plot of this function and test data is shown in Figure 6 and it shows a perfect match between the function and the data from Figure 5.

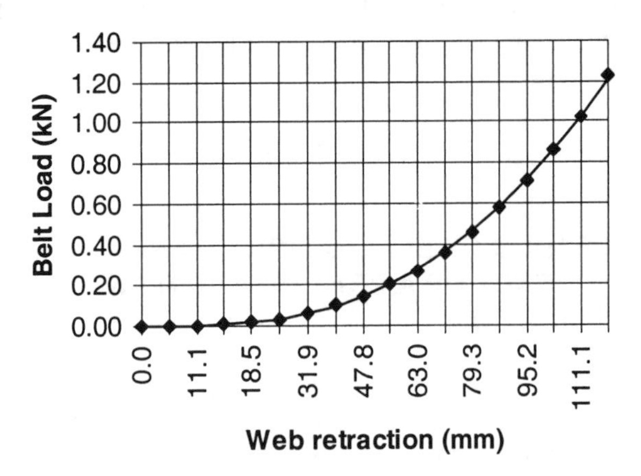

Figure 6. Curve fitting of relation between belt force and web retraction

This plot could give an explanation of the insensitivity of the web retraction to change of output energy of a pretensioner. The energy output can be measured by the work done in the seat belt. This work can be calculated as the production of the belt force and web retraction. In a belt force vs. web retraction curve as that in Figure 8, this work is simply the area under the curve.

Given a design modification with the intention to improve the efficiency, the energy output of the pretensioner will have change. For the same amount of change of the output energy, the change of the web retraction will not be the same at different web retraction level. Under the higher web retraction level, a much smaller amount of change will occur in web retraction than the change at the lower web retraction level for the same amount of energy change. The higher order of the hardening non-linearity, the larger the difference. In other words, under the same amount of energy change, the higher order of the hardening non-linearity, the smaller the change of web retraction.

This can also be explained by another plot, which is the web retraction vs. energy output or work done in the belt. In the specific case of this paper, by equations (5) and (6), the work done in the belt or the energy output can be calculated as:

$$W_b = \int_0^d F_b dx = \int_0^d 0.000018x^{2.235} dx = \frac{0.000018}{3.235} d^{3.235}$$

where W_b is in unit J and d is in unit mm. This relation can be plotted in Figure 7.

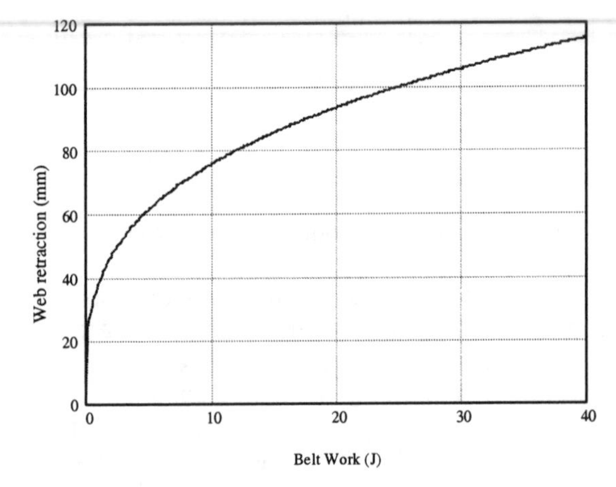

Figure 7. Work done in the belt vs. web retraction

In this example, if the web retraction is used to measure the energy output change due to design modifications, a 100% increase of the belt work change from 5(J) to 10(J) can only be identified by a 23% increase of the web retraction from 62 mm to 76 mm.

In order to be able to more clearly evaluate the change of energy output due to a design modification on the efficiency of the pretensioner, a more sensitive indicator, belt work defined in equation (5), is proposed. In this way, a 100% change of output energy of a pretensioner due to design modifications will be measured a 100% change in the belt work. By multiplying the belt force and the web retraction at the time the pretensioning finishes, the belt work can be estimated. Using this method, replacing the web retraction with the belt work in Figure 2 results in the plot in Figure 8. Note that all energy outputs are normalized to the mean value of sample K that has the highest output. The dots are mean values and bars represent the data ranges.

Compared with Figure 2, this Figure gives better indication as for the effect of each design modification on the efficiency of the pretensioner, because it spreads the differences between each test sample.

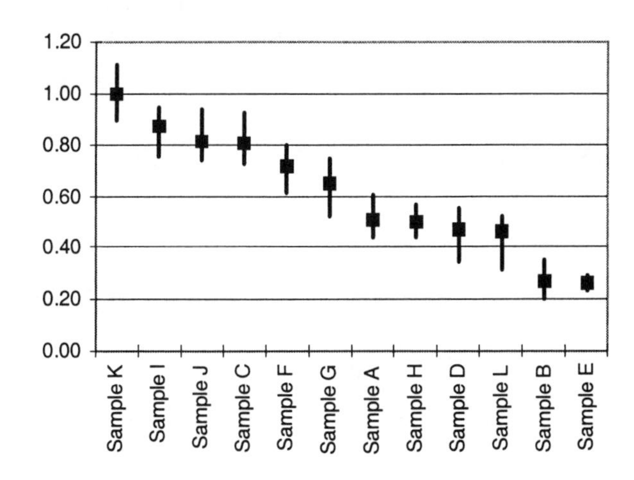

Figure 8. Test data of belt work index of a serious of modification of a pretensioner

CONCLUSION

In this paper, it demonstrated that the web retraction is not adequate to evaluate the efficiency of a pretensioner in pretensioning tests due to the hardening non-linear relation between the seat belt load and web retraction and test condition variation. A more sensitive indicator proposed in this paper is to use the work done by the belt force stretching over the web retraction. This indicator uses both web retraction and belt force, which are already measured in tests. It is useful in studying the effect of design modifications to the efficiency of a pretensioner.

This paper also discussed energy distribution within the pretensioning process. This belt work is the useful potion of the energy delivered from the gas generator. All others are energy loss and should be minimized. This new indicator itself can be used to evaluate how efficient a pretensioner is compared with the energy delivered by the gas generator that can be estimated by its tank test data.

REFERENCES

1. F. Zuppichini, "Effectiveness of a Mechanical Pretensioner on the Performance of Seat Belts," SAE Paper 905139 (FISITA), 1990 Type any references over these paragraphs.
2. J. Miller, "Occupant Performance With Constant Force Restraint Systems," SAE Paper 960502, 1996
3. Arbeitskreis Specification

970774

The Use of Magnetostrictive Sensors for Vehicle Safety Applications

Tony Gioutsos
Artistic Analytical Methods, Inc.

Hegeon Kwun
Southwest Research Institute

ABSTRACT

New sensor approaches termed magnetostrictive sensor (MsS) and nonlinear harmonics (NLH) for vehicle safety applications such as crash detection and occupant seat weight sensing are described. Both sensor approaches utilize the changes in the magnetic properties of ferromagnetic materials that occur when a stress (or a strain) is applied to the material. The MsS is a passive device suitable for vehicle crash sensing. The NLH is an active device suitable for occupant seat weight sensing. Technical features of these sensors are also discussed together with preliminary results of ongoing testing.

1.0 INTRODUCTION

Vehicle safety sensing applications have increased dramatically with the increased use of airbag technology. Of the various vehicle safety sensing applications, there are three areas which have generally produced inadequate performance, they are: 1) single point crash detection for frame/body vehicles; 2) side impact detection and; 3) occupant seat weight sensing.

In this paper, we describe two sensor approaches termed "magnetostrictive sensor (MsS)" and "nonlinear harmonics" (NLH)" for the above mentioned application areas. Both sensor approaches utilize the changes in the magnetic properties of ferromagnetic materials that occur when a stress (or a strain) is applied to the material. The MsS [1,2] is a passive device suitable for vehicle crash sensing. The NLH [3] is an active device suitable for occupant seat weight sensing

In the following, an overview of the currently proposed sensor solutions and their limitations for the three application areas mentioned above is first given in Section 2. Then a description of the MsS and NLH is given in Section 3 including their physical principles, sensor design configurations, and their general properties and technical advantages. Specific applications of these sensor approaches to crash detection and occupant seat weight sensing are then given in Section 4 followed by conclusions in Section 5.

2.0 BACKGROUND ON VEHICLE SAFETY APPLICATIONS

The use of electronic accelerometer based single point sensing for vehicle airbag deployment has increased substantially in the last few years. Their use has been slower than anticipated but still climbing. An overview of this technology is given in [4] and [5]. A natural extension of their use is for other applications including side impact detection and frontal impact detection for frame/body vehicles.

2.1 FRAME/BODY VEHICLE CRASH SENSING - For many frame/body vehicles, acceleration values near zero can be encountered until the required Time-To-Fire (TTF) for a single point module located in the passenger compartment. This leaves even the best algorithm designers with no chance of achieving the desired requirements. A robust sensing concept capable of producing information at a faster rate from the passenger compartment is desired.

2.2 SIDE IMPACT DETECTION - Similarly, side impact sensing with accelerometer based approaches has produced at best marginal results. The inherent problem with an accelerometer based approach is the variation encountered for ON/OFF crashes across the side of the door. For example, a pole crash located at the B-pillar (assume an accelerometer is mounted at the B-pillar), will produce a substantially different sensor output than a similar crash at the A-pillar. Yet, the crashes to the occupant are similar in terms of potential injury. A detailed discussion of the disadvantages of crush zone sensing is given in [5].

There are other approaches to side airbag deployment as well (Autoliv [6] and Siemens [7]). In [6], a mechanical firing pin approach is described. This sensor approach suffers from the same inherent weakness as an accelerometer based approach, namely, the large variation in sensor output based on location of the crash on the side of the vehicle. The idea behind both the accelerometer based approach and the mechanical based approach is to reduce this effect by creating an "array" sensor from a "point" sensor.

In both [5] and [8], the benefits of array sensors are described . In [8] a crush zone sensor is described. Unfortunately, this array sensor is truly an array causing other problems. In [5], frontal crash detection using a single point module approach is described as the best of both worlds (assuming an appropriate algorithm): array sensing from a point sensor. The front part of the vehicle acts as the array, the sensor simply produces similar waveforms for similar crashes (e.g. an offset pole crash on either side of the vehicle).

Therefore, it is beneficial to create an array sensor for side impact detection. By stiffening the vehicle or adding a crossbeam, the accelerometer/firing pin approaches begin to look more like a vehicle side array sensor. However, the inherent weaknesses still exists.

In [7], an air pressure approach is described that again expands on the notion of array sensing from a point. However, disadvantages of this approach include: variation over life, mounting in the door and the limited capability of the approach in sensing only those crashes impacting the door. Without dwelling on this concept, it does not provide the robustness that a magnetostrictive sensor approach can provide.

2.3 OCCUPANT WEIGHT SENSING - With the recent attention on fatalities due to airbag deployment, increased emphasis has been placed on passenger sensing. In particular, weight has been deemed the appropriate parameter for disabling the airbag. Several approaches have been proposed and are being proposed. Without getting too specific, several problems exist with current and proposed approaches including:

- Cost
- Implementation without disrupting the seat
- Temperature compensation
- Robustness to variability in seat occupancy scenarios
- Area coverage

These issues and potential solutions will be addressed further in section 4.

3.0 TECHNICAL BACKGROUND ON MsS AND NLH APPROACHES

3.1 MAGNETOSTRICTIVE SENSOR (MsS) - The MsS is a passive sensing approach which relies on a specific physical phenomenon, that exists in ferromagnetic materials, called "inverse magnetostrictive (or Villari)" effect [1,2]. The Villari effect refers to a change in magnetic induction (B) of material with application of stress, in comparison to the magnetostrictive (or Joule) effect which refers to changes in physical dimension of the material with magnetization. Being a passive device, the MsS is limited to detection of only time-varying or transient stresses (or strains) in the material such as those produced by mechanical impacts or those acoustic emission signals produced by cracking [1,2]. The MsS has a very broad frequency response, ranging from a few Hz to a few 100 kHz. The sensor can also be applied to nonferrous material such as plastics, if a thin layer of ferromagnetic material is plated or bonded to the material surface in a local area where the sensor is to be placed. In addition, the MsS requires neither a direct physical contact to the material nor a couplant for sensing.

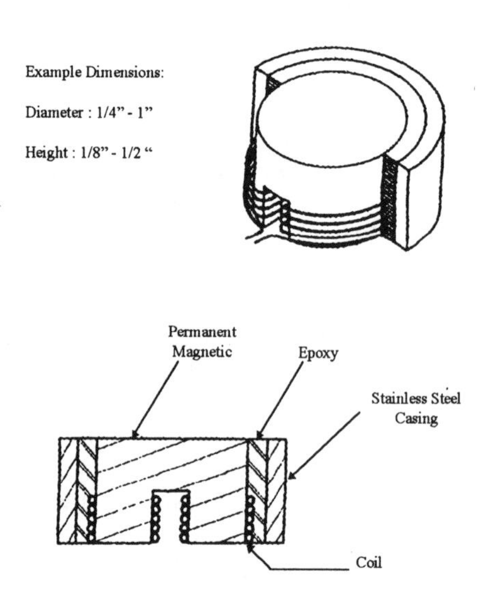

Figure 1 : Typical MsS Sensor

A typical MsS is depicted in Figure 1. It consists of an inductive coil and a permanent magnet. When stress in the material changes with time, the resulting changes in B due to the Villari effect induce an electrical voltage in the coil. The permanent magnet provides a static bias magnetic field to the material which will enhance the sensor sensitivity. The transient stresses produced by a crash or impact will propagate through the material. The attenuation of such impact stress waves in vehicles is negligibly small. Therefore, the MsS can sense the crash or impact event at a location far away from the exact impact site as long as the path for the wave to propagate is good. The MsS can survey a large area and, thus, functions like an "array" sensor from a point and is ideally suited for vehicle crash detection.

3.2 NON-LINEAR HARMONIC SENSOR (NLH) - The NLH sensor operates in a similar fashion to the MsS, but it is an active device [3]. An example NLH sensor which consists of a U-shaped ferrite core with an excitation coil and a detection coil wound on each leg is depicted in figure 2. When a ferromagnetic material such as steel is excited by a sinusoidal magnetic field (H), the corresponding magnetic induction (B) of the material is no longer sinusoidal but shows a distorted waveform. This distortion is due to nonlinear magnetic permeability and the magnetic hysteresis of the material. The distorted B-waveform contains harmonic frequencies of the applied H. In addition, the magnetic hysteresis curves of a ferromagnetic material change significantly when the material is subjected to mechanical stress or strain. The NLH sensor transmits a sinusoidal H to a ferromagnetic material and then detects the resulting B waveform.

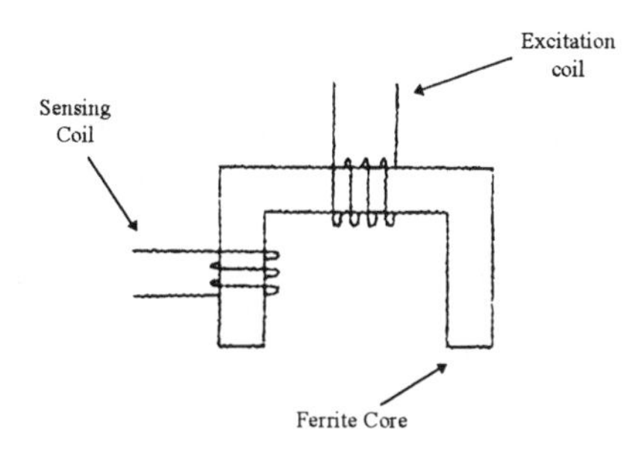

Figure 2 : Typical NLH Sensor

Typically, a coil is used for applying H and detecting B. The H is achieved by sending a sinusoidal current of a given frequency through an excitation coil (e.g. 1kHz). The resulting time-varying B induces an electrical voltage in the detection coil. The stress on the material of interest is determined by harmonic analysis of waveform. In general, the magnitude of the third harmonic of the resultant waveform is related to the stress on the material. In steel, up to about 50 percent of the yield strength, the relationship is linear (i.e. the greater the magnitude of the third harmonic the greater the stress).

The chief differences between the NLH sensor and the MsS are:

- The MsS is a passive device, whereas the NLH requires power
- The NLH sensor can sense constant stress

The NLH approach is similar to those approaches described as magnetostrictive sensors used for torque measurements [9] except that the NLH utilizes nonlinear harmonic components, whereas, the other utilizes the fundamental component.

3.3 TECHNICAL FEATURES - The two sensing approaches have the following features in common:

- Temperature insensitivity
- Non-contacting
- No moving parts
- Small and inexpensive

The MsS also has the following features:

- Passive (produces a signal without power)
- No DC offset (i.e. detects transient stresses)
- An "array" sensor from a point

The NLH also has the following features:

- Active
- Detects DC or constant stress

4.0 VEHICLE SAFETY APPLICATIONS

4.1 FRAME/BODY VEHICLE CRASH SENSING - The MsS can be used in the same fashion as an accelerometer for vehicle crash sensing [10]. Since the frontal crash sensing field is more mature, we will focus on the application of using a single point module for frame/body vehicles (e.g. trucks).

Because of the structure of frame/body vehicles, accelerometer signal values are near zero until the required TTF for passenger compartment mounting locations. This is due to the fact that the frame encounters the crash object well before the body has. By cutting a ferromagnetic hole (e.g. rubber or plastic) in the body above the frame and then facing the MsS at the frame, we are able to detect stress waveforms in the frame from the body. This will allow faster response from a single point sensor module located in the passenger compartment.

4.2 VEHICLE SIDE IMPACT DETECTION - The MsS is ideally suited for side impact detection. The device can be mounted on or near the door without necessarily being mounted on a surface. For example, one could mount the sensor on the crossbeam and face it at the door skin. Therefore, stress on the skin can be measured. Again, the "array" nature of the sensor will allow the sensor to produce waveforms even if the sensor is not contacted directly or is nearby the "hit" area. A good propagation path is all that is needed. There are other locations on or near the door that the sensor can be placed to provide a good signal with ease of mounting. For example, placing the sensor on the sill facing the base of the door will allow the door to be sensed but the sensor to be mounted within the passenger compartment. Various isolated side impacts can be detected with this arrangement. If a pole hits the A or B pillar or is on-line with the sill mounting location, the signal strength should be similar for the same impact speed. This response should be differentiated from that of an accelerometer, where depending on the mounting location (e.g. A/B pillar location or the middle of the door) substantially different waveforms will be produced.

These two features (i.e., non-contacting and array sensing from a point) make the MsS ideally suited for this application.

4.3 OCCUPANT WEIGHT SEAT SENSING

The occupant seat weight sensing problem has become a very important issue in the automotive safety community. The key to the problem is to determine the occupant's or object's weight with a very high resolution. The goal is to change the airbag deployment characteristics (e.g. do not deploy) if the weight is less than a given threshold(s). There are many issues involving this approach but basically there are three main concerns: performance, cost, and ease of implementation.

The NLH sensor is suited to this application. By placing the sensor under the seat on either the pan, springs or mounting points; stress upon the seat can be measured. For example, for seat designs which feature an array of wires holding the seat cushion in place, stress on the wires can be measured as follows. As an object is placed in the seat the foam causes stress on the wires. The stress applied to the wire by the occupant is detected by using a NLH sensor placed on the wire. The NLH output can then be converted to weight. The sensor location, design and placement will be related to the given seat, but should not affect the seat's deign. This will allow a seat weight sensor that is inexpensive, robust, and easy to implement.

5.0 CONCLUSION

We have shown that the MsS and NLH sensor have good potential for automotive safety applications. There are several Tier One suppliers testing these devices with so far excellent results. In addition, the MsS has also shown good potential for condition monitoring of combustion engines such as knock and misfire.

REFERENCES

[1] Kwun, H., "Back in Style: Magnetostrictive Sensors," *Technology Today* (Southwest Research Institute), March 1995, pp. 2-7

[2] Kwun, H. and Teller, C.M., Patent #5456113, "Nondestructive Evaluation of Steel Cables and Ropes Using Magnetostrictive Induced Ultrasonic Waves and Magnetostrictively Detected Acoustic Emissions"; also Patent #5457994 and other pending patents. Assignee: Southwest Research Institute

[3] Kwun H. and Burkhardt G. L., "Nondestructive Measurement of Stress in Ferromagnetic Steels Using Harmonic Analysis of Induced Voltage", *Non-Destructive Testing International*, June 1987, pp. 167-171

[4] Gioutsos, T. and Gillis, Ed, "Testing Techniques for Electronic Single Point Sensing Systems," *SAE International Congress and Exposition*, Paper # 940803, 1994

[5] Gioutsos, T. and Gillis, Ed " Tradeoffs and Testing for Frontal Crash Sensing Systems," *SAE Worldwide Pass. Car Conf. and Expo.*, Paper # 932911,1993

[6] Dahlen, M. "Side Airbag Systems: Seat-mounted vs. Door-mounted", *SAE Side Impact Protection Toptec*, 1994

[7] Hartl, A. Mader, G. Pfau, L. and Wolfram, B., "Physically Different Sensor Concepts for Reliable Detection of Side-Impact Collisions", ," *SAE International Congress and Exposition*, 1995

[8] Breed, D., et al. ,"Performance of a Crush Sensor For Use With Automobile Air Bag Systems," *SAE International Congress and Expo.*, Paper # 920122, 1992

[9] Klauber, R., et. al., "Miniature Magnetostrictive Misfire Sensor," Paper # 920236

[10] Gioutsos, T., Patent # 5,580,084, "System and Method for Controlling Vehicle Safety Device", Assignee : Artistic Analytical Methods, Inc.

1999-01-1327

Further Results on the Use of Magnetostrictive Sensors for Vehicle Crash Detection

Tony Gioutsos and Michael Murray
BREED Technologies, Inc.

ABSTRACT

This paper expands on an earlier paper on the use of Magnetostrictive Sensors (MsS) for vehicle crash detection sensing. Analysis of vehicle crash data has shown promising results especially for use in side impact detection. Topics concerned with sensor system implementation are discussed, as well as the use of magnetostrictive sensors for frontal crash severity measurement and pedestrian impact applications.

INTRODUCTION

Airbag systems are increasingly being used to protect occupants during side impact crash events. The requirements for an effective sensing system are implementation without significant design changes, robustness with respect to variability of crash types, low cost, and immunity to abuse and environmental conditions. The goal of a sensing system is to provide "array" sensing from a "point". This type of sensing system attempts to monitor a wide area using a single sensor.

Currently, accelerometers are being used for the majority of side impact detection systems. However, they have shown limitations in their array sensing capabilities. This paper explores the use of magnetostrictive sensors (MsS) for side impact crash detection. The paper begins with a brief description of the MsS and its operation. Discussion then focuses on issues in the use of MsS for vehicle crash sensing. Following this we delve into crash data analysis and system design. Finally a brief description of using MsS for Crash Severity and pedestrian impact detection.

MAGNETOSTRICTIVE SENSING

BACKGROUND – As detailed in [1], the MsS is a passive sensing approach which relies on a physical phenomenon that exists in ferromagnetic materials, called the "inverse magnetostrictive (or Villari)" effect. The Villari effect refers to a change in magnetic induction (B) of material with application of stress, in comparison to the magnetostrictive (or Joule) effect, which refers to changes in physical dimension of the material with magnetization.

The MsS is a passive device; meaning that it requires no power source. Being a passive device, the MsS is limited to detection of only time varying or transient stresses (or strains) in the material such as those produced by mechanical impacts or those acoustic emission signals produced by cracking. The MsS has a very broad frequency response, ranging from a few Hz to a few 100 kHz. The sensor can also be applied to nonferrous material such as plastics, if a thin layer of ferromagnetic material is plated or bonded to the material surface in a local area where the sensor is to be placed. In addition, the MsS requires neither a direct physical contact to the material nor a couplet for sensing.

SENSOR DESCRIPTION AND OPERATION – As Figure 1 depicts the MsS consists of an inductive coil and a permanent magnet. When stress in a material changes with time, the resulting changes in B due to the Villari effect induce an electrical voltage in the coil. The permanent magnet provides a static bias magnetic field to the material, which will enhance the sensor sensitivity. The transient stresses produced by a crash or impact will propagate through the material. The attenuation of such impact stress waves in vehicles is negligibly small. Therefore, the MsS can sense the crash or impact event at a location far away from the exact impact site as long as the path for the wave to propagate is good. The MsS can survey a large area and, thus, functions like an "array" sensor from a point and is ideally suited for vehicle crash detection, especially crush zone sensing like side impact detection.

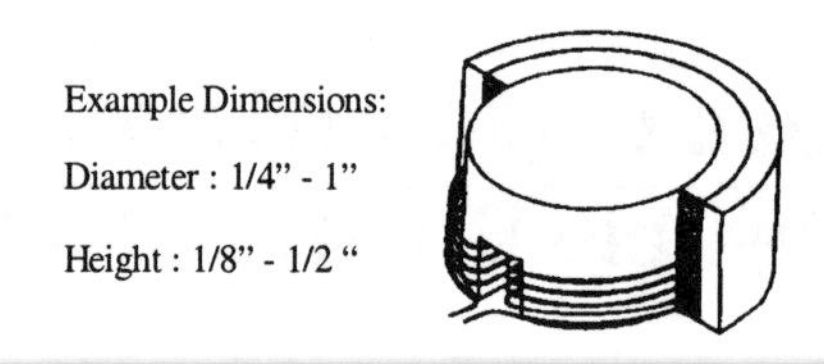

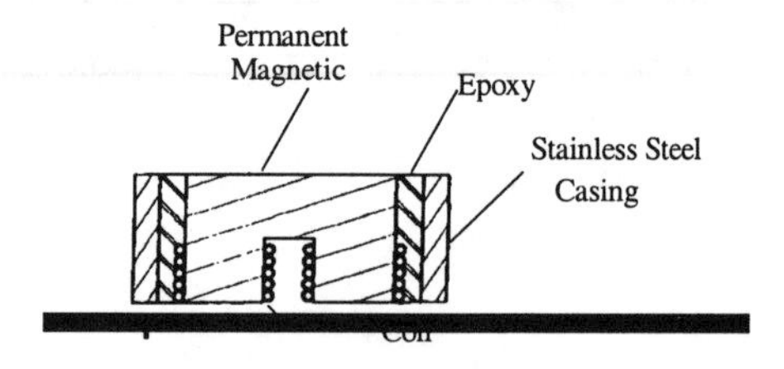

Figure 1. Diagram of Magnetostrictive Sensor (MsS)

MSS ISSUES IN CRASH SENSING

Since the publishing of [1] many factors concerned with crash sensing have been discovered and analyzed. These issues are types of stress waves, stress vs. area, displacement sensing, mounting, voltage levels, non-metallic performance and frequency content.

STRESS WAVES – In this paper, we will not give a detailed description of acoustic emissions (stress waves), except to state their relevance to vehicle crash sensing. Basically, there are two types: flexural and longitudinal. A general statement about these waves is that flexural waves are low frequency in content and longitudinal waves are higher frequency in content, which also means they travel faster. Flexural waves are also more similar to an acceleration waveform than are longitudinal waves. The MsS senses both types of waves. Because of this, the information content in a MsS waveform is much richer than that of an acceleration waveform. This leads to faster decision times and more robust performance.

STRESS VS. AREA – Stress can be applied in various ways. For example, for side impact crash sensing the stress can be applied across the entire door (e.g., 214 or car-to-car), a large portion of the door (e.g., pole), or in a small area (e.g., abuse). As the stress is applied, it disperses or transmits throughout the structure as long as there is a good mechanical coupling. So in a 214-type crash, the stress waves basically are transmitting throughout the door area in a symmetric fashion. This implies that the voltage levels out of a MsS will be constant no matter where it is placed near or on the door.

In a pole crash, unless the pole impacts directly on the sensor (higher signal), the voltage levels will be approximately constant because the dispersion is small relative to the stress per unit area. However, in abuse cases the stress per unit area is small and dispersion becomes a large issue. The voltage near the stress point will be large, but only for a short time. But since the stress per unit area is small, the voltage levels drop off significantly as measured by MsS further from the point of stress.

Using the knowledge of stress wave type and stress per unit area versus crash type will lead to some very important conclusions for sensor placement as will be discussed.

DISPLACEMENT SENSING – The magnetostrictive concept can be used as a displacement sensor, if desired. For example, by placing the sensor on a surface without adequate mounting and stressing the surface, the sensor will "move" off of the surface causing a large change in the magnetic field; thereby causing a large voltage in the coil. This voltage is independent from the voltage induced by the acoustic emissions. In general, this voltage varies in vehicle crashes just like accelerometer waveforms. It is location specific (i.e., if the stress is applied nearby the signal is much greater). This fact is one that leads to further conclusions about sensor configuration.

MOUNTING – In [1], using the sensor across a gap was discussed in detail. The gap approach is simply another way to use the sensor as a displacement sensor as discussed above. However, it is our determination that this type of use of the sensor will produce wildly variant signals. Figure 2 depicts the output voltage (peak-to-peak) for a MsS mounted with various torque values (the sensor was designed with two flanges to screw into the sensing surface). A standardized drop test was used to produce the stress (no damage was done to the material). Note that the peak voltage drops significantly as the torque is increased till about 6 lbs., where it then remains relativitely constant. This is due to the displacement effect discussed above. Therefore, to reduce variation we recommend a torque mounting of 10-ft lbs.

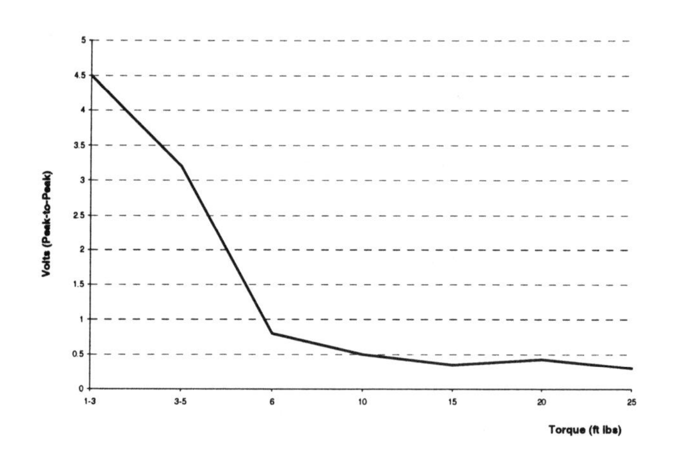

Figure 2. MsS Volts (Peak-to-Peak) for Various Torque

VOLTAGE LEVELS – The current experimental MsS design has produced low voltage levels in vehicle crash testing. This design has a small number of coil windings (about fifty) and is in the process of being redesigned to increase voltage and also allow simplified manufacturing.

The voltage levels vary depending on location and use but once mounted with approximately 10-ft lbs of torque, they produce around 10 millivolts peak-to-peak. Noise levels for this device are on the order of microvolts, so there is no problem with separating crash signals from noise. However, as will be discussed later, higher output values will allow for better use in a sensor system configuration.

NON-METALLIC PERFORMANCE – As discussed in [1], the sensor can be used to detect acoustic emissions for non-ferromagnetic materials by bonding or plating. As will be seen later, we have collected excellent data for a vehicle with plastic doors. The cross member and interior portion of the door were ferromagnetic and simply bonded to the outside plastic skin. The signal strength was on the order of other data we have seen.

FREQUENCY RESPONSE – As stated previously, the frequency response of this sensor is high. We have collected data with a 10 kHz sample rate. Figure 3 depicts a spectrum of a typical crash. Note that the content exists until 50 kHz, the Nyquist frequency as would be expected. The fall off is due to anti-aliasing filtering.

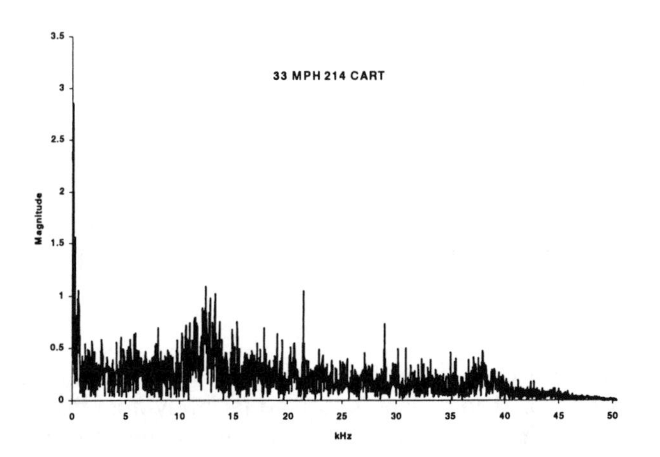

Figure 3. Frequency Spectrum for Typical MsS Waveform

VEHICLE CRASH DATA

The following sets of data are presented as an overview of the MsS performance during crash testing.

COMPACT VEHICLE WITH PLASTIC DOOR – The first vehicle was a compact vehicle with plastic doorskin. The MsS's used were an earlier design without flanges. Because of this, the displacement effect was not removed. However, this is basically irrelevant because as we shall see the displacement part of the waveform (i.e., flexural waves) was low frequency and could be filtered out.

Figure 4 depicts the location of three MsS sensors and their placement near the door. Sensor 13 was placed on the inner B-pillar near the floor. Sensor 5 was placed on the door face near the "H" point. Sensor 8 was placed on the door face near the hinge.

Figure 5 depicts two accelerometer waveforms taken from the B-pillar. The crashes are a 20 mph pole test at the H point (a side airbag deploy crash with a required time-to-fire (TTF) of 12 msec) and a 10 mph 90° 214 cart (a no deploy crash). Note that the energy content is very similar (energy is related to average peak-to-peak values). This is not to say an advanced algorithm similar to one discussed in [2] could not separate these two waveforms. However, this might not be the case if the pole impact is moved slightly further from the B-pillar.

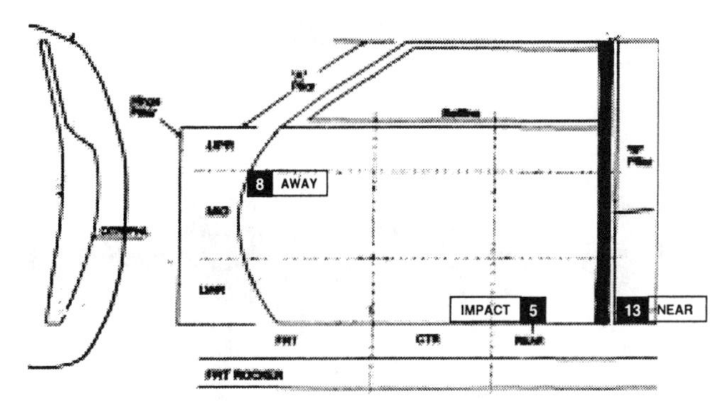

Figure 4. Sensor Location Map for Compact Vehicle

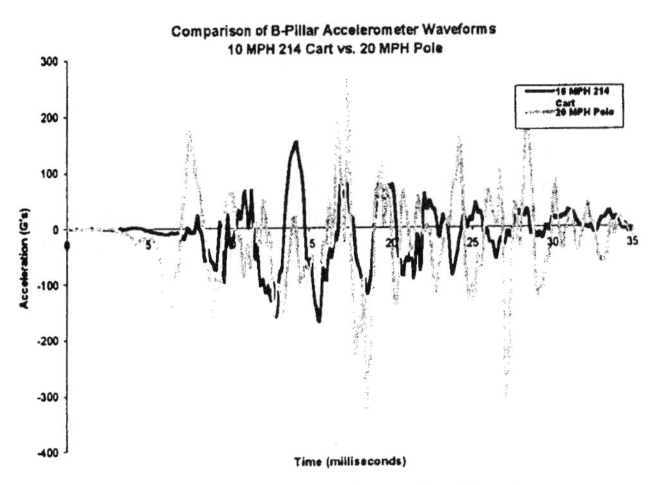

Figure 5. Accelerometer Waveforms for Side Impact Crashes

Figures 6 – 8 depict the MsS response. Note that all three produce substantial differences between deploy and no-deploy crashes no matter where the location, by the required TTF. Since sensor #5 is right at the impact point, one expects this difference. In Figure 6, we have high passed filtered the data to remove low frequency displacement information. Note the huge separation in energy well before the 12 msec TTF.

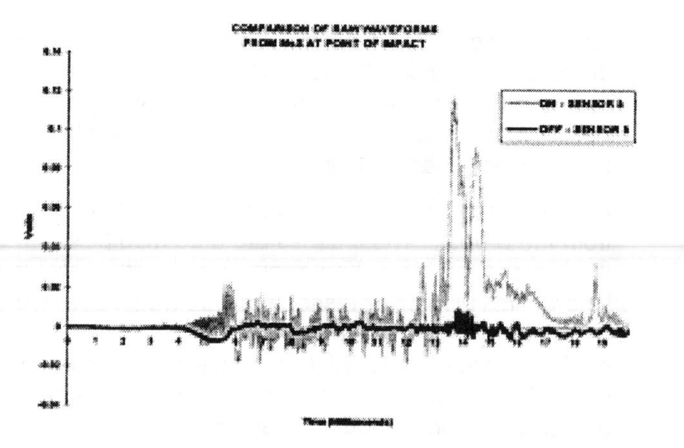

(a) MsS Waveforms (Sensor At Impact Point)

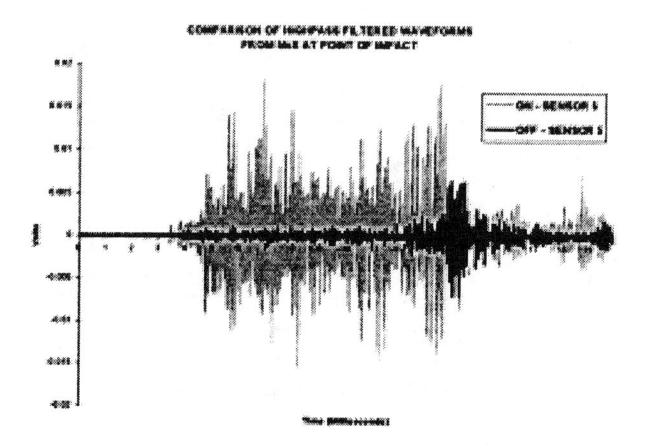

(b) MsS Filtered Waveforms (Sensor At Impact Point)

Figure 6.

Figure 7 depicts the waveforms from the B-pillar location. Note that the low frequency content is negligible, as it should be since this location did not exhibit any displacement changes in either crash. Note again, even without the sensor located at the pole location, there is an easy separation of the crashes in terms of energy.

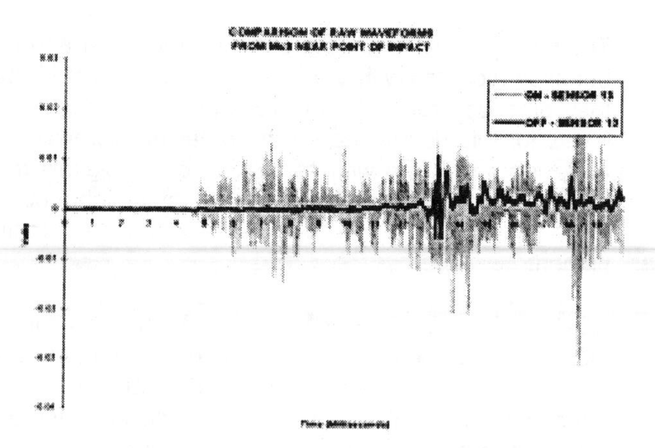

(a) MsS Waveforms (Sensor Near Impact Point)

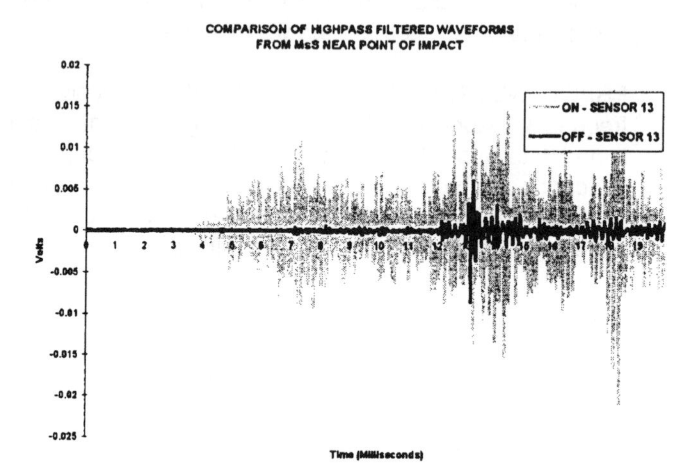

(b) MsS Filtered Waveforms (Sensor Near Impact Point)

Figure 7.

Figure 8 depicts the MsS response from a worst case scenario. The sensor is located far away from the pole impact point but is directly in-line with the no-deploy 214 cart hit! After removing the low frequency content, it is evident that the pole crash has substantially more energy than the 214-cart no-deploy test. This demonstrates the array sensing capabilities of the sensor.

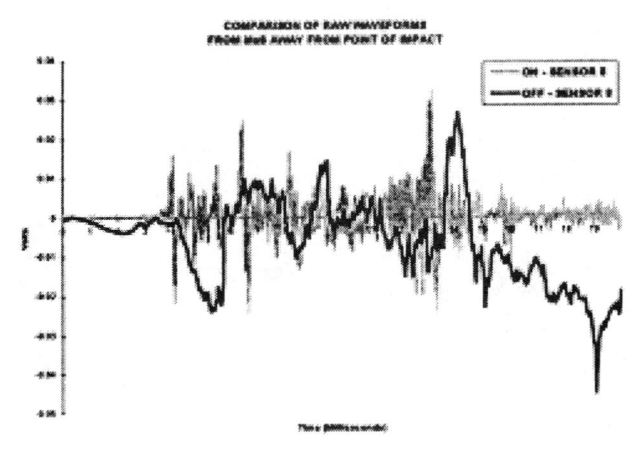

(a) MsS Waveforms (Sensor Away From Impact Point)

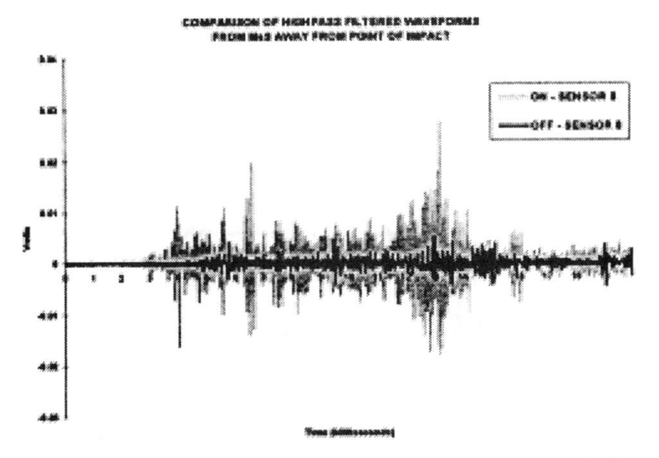

(b) MsS Filtered Waveforms (Away From Impact Point)

Figure 8.

Figure 9 is a plot of the three sensor locations for the pole crash. Note that after removing the low frequency content (Figure 9(b)), that the signal energy is almost exactly the same! This is an extremely impressive feature of the MsS: the sensor produces the "same" energy content for an isolated crash like a pole hit. A similar depiction of accelerometer waveforms is not available, but would obviously not show nearly similar energy content.

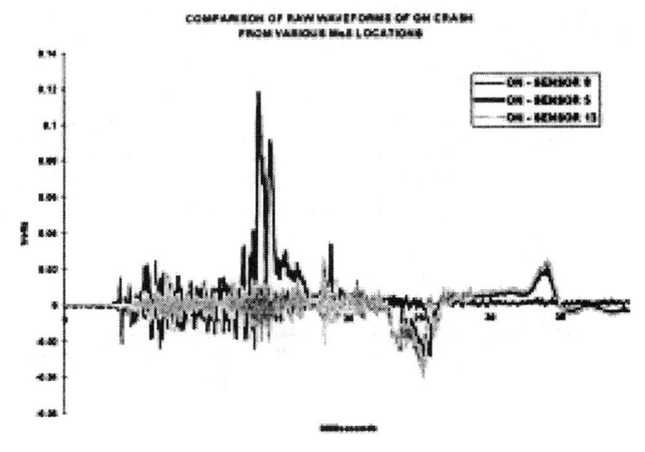

(a) MsS Waveforms for 20 MPH Pole (All Sensors)

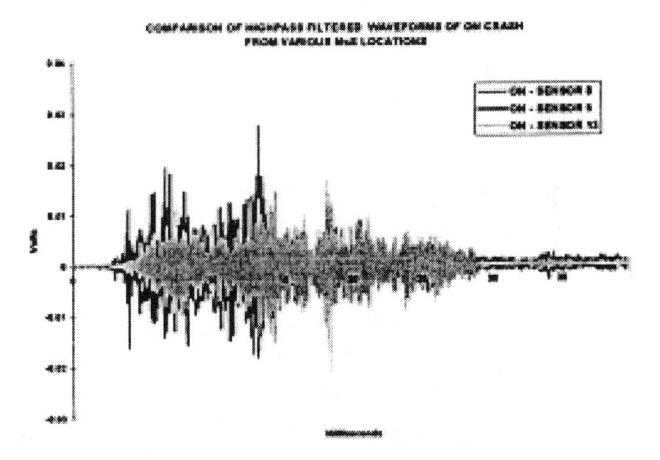

(b) MsS Filtered Waveforms for 20 MPH Pole (All Sensors)

Figure 9.

MINIVAN SIDE IMPACT DETECTION – Figure 10 depicts acceleration waveforms at a B-pillar location for a deploy pole (required TTF of 6 msec) crash and a door slam. Note the almost exact waveform until approximately 8 msec. A simple door slam can confuse the accelerometer-based system. However, Figure 11 depicts the MsS B-pillar response with a 3 kHz low pass filtered waveform (data acquisition constraints). Note the ease of separation because the doorslam produces basically no signal. This is exactly what should happen since a doorslam has minimal change in stress on the door and surrounding areas. This points out another feature of the MsS: its general immunity to abuse events.

Figure 10. Accelerometer Waveforms (Pole vs. Doorslam)

Figure 11. MsS Waveforms (Pole vs. Doorslam)

INSIDE DOOR MOUNTING – Figures 12 and 13 depict waveforms from an inner door mounting (near speaker) for a very large SUV and a minivan respectively. Figure 12 contains two waveforms and a "band" representing the maximum value over all the abuse events listed. Note again the ease of separation. Figure 13 shows the similarity between waveforms for the same tests (33.1 mph 214 test) and again the ease of energy separation from a threshold no deploy event. The inner door location is our preferred location due to fast response versus limited abuse problems (e.g., hammer blow right on sensor is difficult to do from inside the vehicle).

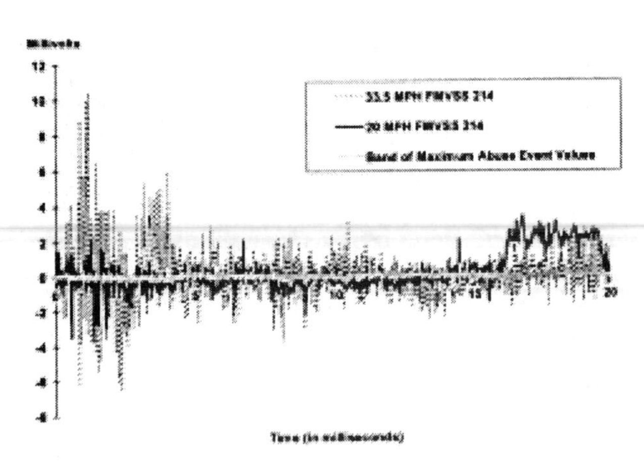

Figure 12. MsS Waveforms - SUV (Sensor on Inner Door)

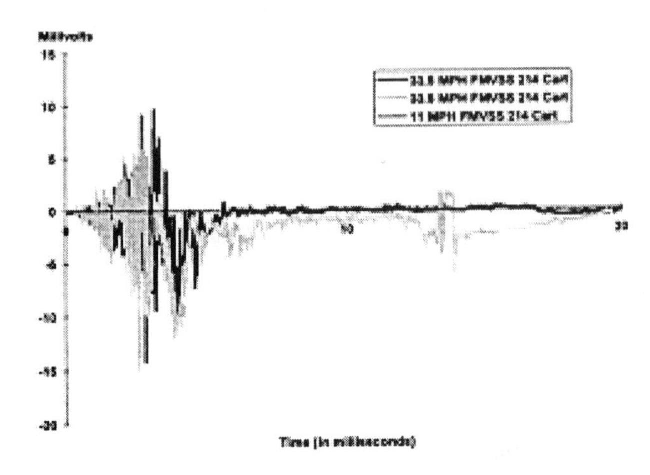

Figure 13. MsS Waveforms - Minivan (Sensor on Inner Door)

PEDESTRIAN IMPACT – The sensor has also begun to be investigated for use in a pedestrian impact detection system. Two similar tests were run; a simulated legform was sent into a vehicle at 40 Kph. A summary of the results is incomplete, however, by observing Figure 14, one can note again the array sensing capabilities of the sensor. The three waveforms are outputs of three MsS's placed on the inner bumper at the center, right center and far right portion. The simulated legform impacted at the center of the bumper for this test. The three waveforms have approximately the same signal strength (although different delays due to acoustic emissions propagation), and different frequency content. This difference in frequency content is currently being investigated as a potential way of determining impact point.

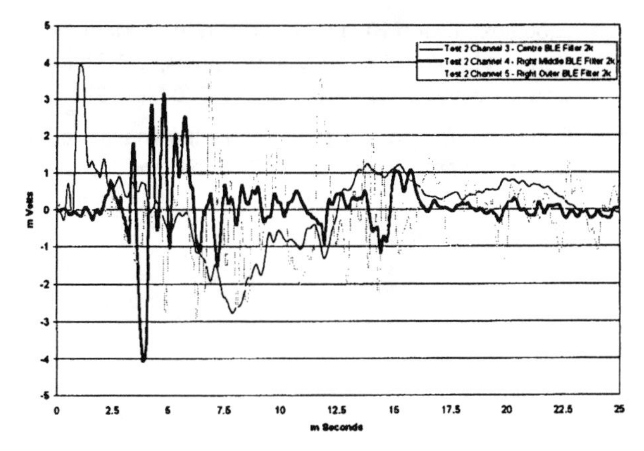

Figure 14. MsS Waveforms from Pedestrian Impact Testing

CRASH SEVERITY – In [2], frontal crash severity problems and issues are presented. A proposed system is also discussed, which uses a Ball-In-Tube(s) as a frontal sensor(s). An accelerometer frontal sensor(s) could also be used. But what about MsS? The sensors could be used and are currently being investigated. Location placements now considered include bumper frame rails and the inner hood. The analysis is currently continuing.

MSS ALGORITHM

Figure 15 depicts a flow diagram for a MsS vehicle crash detection algorithm. The MsS provides data to a band pass filter (implemented in hardware or software). Then using proprietary waveform time estimating means, we calculate energy over a "time" segment. This segment avoids short time abuse events (e.g., hammer hit on sensor < 2 msec) and long time no deploy events. If the band pass filtered and time segmented energy is greater than a threshold, a deploy decision is reached.

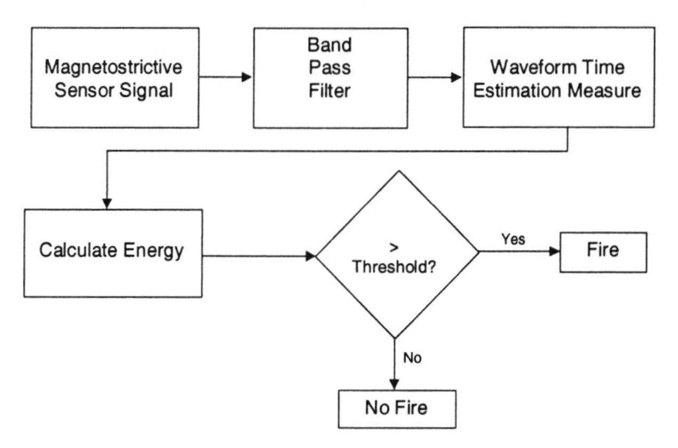

Figure 15. MsS Algorithm Flow Diagram

MSS IMPLEMENTATION

The MsS can be implemented in three ways for side impact detection. All three implementations should also have a safing sensor (i.e., accelerometer) before firing the airbag (this currently is done for accelerometer based side impact systems where the side safing accelerometer is a lateral accelerometer in the single point module – SDM).

The first implementation is similar to accelerometer based side impact sensing systems; that is to use a satellite electronics module. Basically, the MsS is configured into a "satellite" of appropriate electronics to process the signal, determine a deployment and transmit a signal to the SDM. The MsS, in essence, replaces the accelerometer. The cost of this "satellite" should be less than an accelerometer satellite due to sensor cost and the passivity of the MsS.

The second configuration uses a "central" side impact unit. Figure 16 depicts this scenario. Two stand-alone MsS's are input to one set of electronics that is used to determine firing for both sides of the vehicle. The MsS is rugged enough to stand-alone and preliminary testing indicates immunity of the MsS to EMI. This approach could incorporate an accelerometer safing sensor in the central electronic unit. It should be obvious that the cost of this system (even with an accelerometer safing sensor) will be substantially lower.

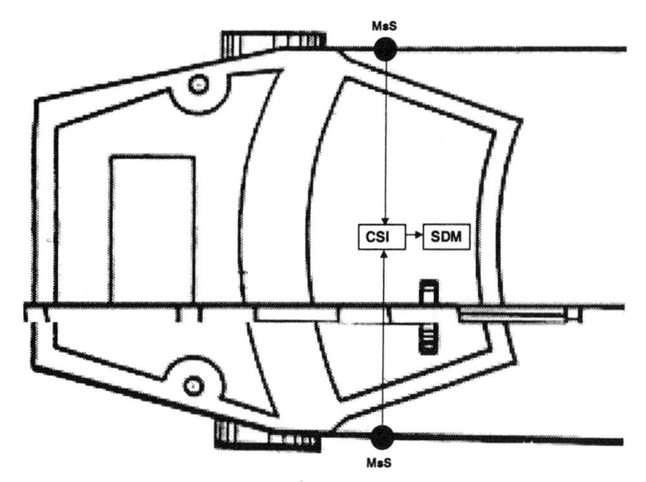

Figure 16. MsS Implementation Using Central Side Impact Unit

The final implementation would be to directly connect two stand alone MsS's to the SDM. All processing, safing etc. can then be handled directly from the SDM. This approach is extremely cost effective.

CONCLUSION

In summary, the signal processed output from the MsS showed the ability to readily discriminate between ON and OFF crash impacts. The results from the testing and algorithm reveal that the MsS can provide an excellent solution to the side impact-sensing problem. Although the voltage levels were small for the crashes, the waves for the OFF crash were even smaller and allowed for easy discrimination. Also, the ability to discriminate using any of the sensor locations appears to solve a major problem of the other sensing approaches. Finally, implementation of the sensor into a side impact detection system can be extremely cost effective relative to an accelerometer based side impact detection "satellite".

REFERENCES

1. Gioutsos, T. and Kwun, H., "The Use of Magnetostrictive Sensors for Vehicle Safety Applications," SAE International Congress and Exposition, Paper # 970774, 1997
2. Gioutsos, T. and Tabar, D., " Determination of Vehicle Crash Severity Using a Ball-In-Tube and Accelerometer System (BASS)", SAE International Congress and Exposition, paper #99PC-302, 1999

980555

The X-By-Wire Concept: Time-Triggered Information Exchange and Fail Silence Support by new System Services

Elmar Dilger, Thomas Führer, Bernd Müller
Robert Bosch GmbH, Stuttgart, Germany

Stefan Poledna
Robert Bosch AG, Vienna, Austria

ABSTRACT

This paper presents the conceptual model and the fundamental mechanisms for software development in the context of the Brite-EuRam project *Safety Related Fault Tolerant Systems in Vehicles* (nick-named X-By-Wire). The objective of the X-By-Wire project is to achieve a framework for the introduction of safety related fault tolerant electronic systems without mechanical backup in vehicles.

To achieve the required level of fault-tolerance, an X-By-Wire system must be designed as a distributed system comprising a number of fault-tolerant units connected by a reliable real-time communication system.

For the communication system, the time-triggered TTP/C real-time communication protocol was selected. TTP/C provides fault-tolerance message transfer, state synchronization, reliable detection of node failures, a global time base, and a distributed membership service.

Redundancy is used for masking failures of individual processor nodes and hardware peripherals. To keep the required number of processors low, the processor nodes must exhibit fail-silent behavior, i.e., produce either correct results or none at all.

At the logical design level, the so called XBW-model was developed to combine the basic mechanisms for system design (BASEMENT and DFR). For the detailed software design of embedded distributed fault-tolerant hard real-time systems, the framework of the DFR meta object model is used. The model supports systematic error detection strategies for achieving fail-silence behavior at the node level and fault tolerance strategies for achieving fail-operational behavior at the system level. This eases the application software development for distributed fault-tolerant real-time systems considerably.

Within the X-By-Wire project, a prototype of a steer-by-wire system is developed. This prototype will demonstrate the application of the principles, model, and mechanisms developed by the X-By-Wire partners.

1. INTRODUCTION

It is the objective of the Brite-EuRam III-project "Safety Related Fault Tolerant Systems in Vehicles" to define a framework for the introduction of safety related fault tolerant electronic systems without mechanical backup in vehicles (so-called "X-By-Wire systems"). The "x" in "X-By-Wire" represents any safety related application such as steering, braking, powertrain or suspension control. These applications will greatly increase overall vehicle safety by liberating the driver from routine tasks and assisting the driver to cope with critical situations.

The competitiveness of the European vehicle industry strongly depends on increasing safety standards. It is obvious that in the mid term this will only be possible by the replacement of mechanical backup systems by highly dependable distributed fault tolerant mechatronic systems onboard a car. This goal can only be achieved by introducing such systems based on a properly designed system architecture.

Similar considerations exist in the US within the SAE particularly in the *SAE Vehicle Network for Multiplexing and Data Communications Standards Committee*. In 1993 this committee published a document [SAE94a] on the requirements of safety critical control applications onboard vehicles. In this document the topics of temporal performance, dependability and implementation constraints of safety critical automotive networks are established and a typical benchmark problem of safety critical application is defined. Furthermore, three classes for vehicle communications have been identified, covering body electronics (SAE class A, B) and system electronics (SAE class C). In a companion document [SAE94b] the SAE came to the conclusion that none of the surveyed protocols (J1850, CAN, VAN, AUTOLAN, etc.) satisfies the requirements of distributed safety critical applications onboard vehicles.

The consortium of the Brite-Euram III Project "Safety Related Fault Tolerant Systems in Vehicles" consists of the following partners: Daimler-Benz Research, Fiat Research Centre, Ford

Europe, Volvo, Bosch, Magneti Marelli, Mecel, University of Chalmers, and the Vienna University of Technology.

The system architecture presented in this paper uses a strictly time-triggered, distributed computer architecture where fault-tolerance mechanisms are based on the fail silence property. High level system services of the real-time operating system provide the necessary support for the system and application programmer. These ideas will be proven by the prototype implementation of a "steer-by-wire" demonstrator.

2. REQUIREMENTS ON CLASS C REAL-TIME COMMUNICATION SYSTEMS

A real-time communication system intended for data exchange of safety-critical applications in the automotive environment has to meet requirements that exceed the constraints of communication systems common in other application domains. The following section gives an outline of fundamental requirements for communication systems to be suitable for safety-critical applications like steer-by-wire. It follows the taxonomy given in [SAE94a].

Regularity of Information Transfer: Most safety critical applications are control-oriented, performing their task in a loop consisting of sampling inputs, processing these inputs and outputting the calculated setpoints.

Due to the regularity inherent to control applications a large amount of data exchange takes place periodically. We call data elements belonging to this type "repetitive". Repetitive data elements carry state information, i.e., they contain real-time images of state variables in the environment.

Besides the periodic control application, a real-time system must be able to cope with events that are outside its sphere of control, e.g., failure events. These so-called chance events result in a demand for irregular communication. If there is a minimum interarrival period between two successive chance events of the same type, the event and thus the data element to be disseminated via the bus can be transformed into a quasi-periodic one [Mok83].

Minimal Message Latency: The latency of information exchange is defined as the interval between the point in time when a data element is produced by the sending task and the point in time when it is used by the receiving task. The maximum allowable latency is determined by the application's needs, it can be as stringent as 1 *ms* or less. The actual latency achieved by the communication system depends on various aspects like the bandwidth of the communication medium, the protocol's logical structure, and, most important, the medium access method. Thus minimizing the latency jitter or, ideally, maintaining a constant latency is of utmost importance.

Fault-tolerance: In safety-critical applications it is not sufficient to detect an error. The system must exhibit a fail-operational behavior, i.e., the system must be able to maintain its function after the occurrence of an error. This behavior, which must be guaranteed up to a pre-defined number of failures, can only be achieved if the system is fault-tolerant. Class C requirements for short response times - even in the presence of faults - can only be guaranteed by active redundancy. Redundant systems require replica determinism to maintain state synchronism [Pol96a]. This means that actively redundant components have to make equivalent decisions at about the same point in time.

Robustness: The communication system will be the most vital part of the distributed system, yet probably the component most exposed to electromagnetic interference. A design that is utmost tolerant to electromagnetic interference, and also that can recover from a "blackout" with a minimal latency is therefore required.

Error detection: When distributed control is applied, several nodes will participate in common control functions. Failures in one node (beyond the pre-defined number of failures) will require the remaining nodes to take appropriate action, for example to switch over to a mode with degraded performance, to use alternative sensors or actuators, etc. A consensus of which nodes and functions that are operational, a membership agreement is thus necessary, at the node level and at the level of control functions. It is important that changes in this membership are detected unanimously and timely.

Acknowledgment and Atomic Transmission: In many applications it is important for the sender to know whether its message was received or not. This information has to be provided by a proper acknowledgment scheme. Moreover, some applications require that a message is received by either all recipients or by none, i.e., atomic transmissions. In this case the sender should be aware of correct or incorrect reception.

Testing: A real-time architecture should support a constructive testing method, which means that each subsystem is tested independently. If it passes this test an integration with other subsystems must not have any side-effect.

Configurability/Composability: In many engineering disciplines, systems are built by integrating a set of well-specified and tested subsystems. It is important that the properties that have been established at the subsystem level are maintained during the system integration. Such a constructive approach to system building is only possible if the architecture supports composability.

3. STATE OF THE ART EVENT-TRIGGERED (ET) COMMUNICATION SYSTEMS

If a communication system transports event messages, the temporal control is external to the communication system, i.e., it is in the sphere of control of the host computers to decide when a message must be sent [Kop93b]. If, for example, a number of nodes decide to send a message to a particular receiving node at the same point in real time, then there is a problem: If the communication system has dedicated channels between any two nodes, all messages will arrive simultaneously at the receiver and overload the receiver. On the other hand, if the communication system uses a single channel that serializes the traffic, then a conflict for the access to this channel is unavoidable. Different single channel ET protocols resolve such an access conflict by different techniques: by a random access technique (Ethernet), by some predefined order relation (token protocol), by message priority (CAN), etc. This does not solve the fundamental problem, that the temporal control at the Communication-Network Interface (CNI) is not defined by an ET protocol. Temporal control in an ET system is thus a global issue, depending on the behavior of all nodes in the system. From the point of view of temporal behavior, ET systems are not composable.

In a number of proposals the (sometimes fuzzy) notion of real-time network management is suggested to solve this fundamental problem of node co-ordination in the temporal domain. The following quote taken from the minutes of a meeting of the SAE Multiplexing Committee, March 2, 1995, about the SAE J1850 single channel communication protocol for automotive applications paints a vivid picture of this issue (emphasis added):

"SAE J1850 is a complete document. However, its content is not sufficient to guarantee that devices designed and built to its requirements will communicate as intended to perform some operational function. The reason for this is that SAE J1850 does not provide a network management framework to manage total network traffic. SAE J1850, and its companion J2178, establish how a device, A, can report a parameter, X, to another device B, during normal vehicle operation. However, neither of these documents provides a framework for agreement on why and ***when*** A shall report X. Agreement on this is necessary so that devices which communicate to cooperatively accomplish some function know what to expect from the other devices. Without this agreement, the ***interoperability*** of designs cannot be assured."

4. THE TIME TRIGGERED COMMUNICATION PROTOCOL TTP/C FOR CLASS C

Time Division Multiple Access (TDMA): This method of bus access is based on asserting each node a certain time slice within a period in which it has exclusive sending rights. Each node knows about the time slices of the others ensuring no node will send at the exclusive sending slot of an other node. The advantage of this technique is, that each node has the same priority for its bus access and latency can easily be calculated.

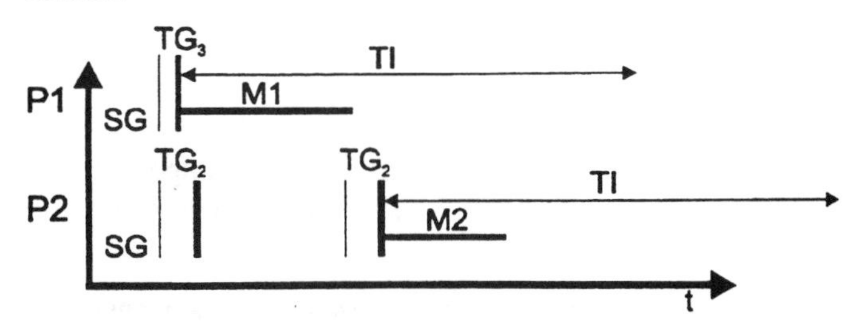

Fig.4.1. Minislotting

Some protocols use the occurrence of an event to start a TDMA period, e.g. Minislotting [ARI91]. Fig. 4.1 illustrates one protocol period, where two nodes try to access the bus due to an event starting the protocol machine. Both nodes have to wait for the synchronization gap (SG) to start their personal terminal gap (TG_i). This terminal gap corresponds to the priority of the node. In this figure, node P2 will be allowed to access the bus as soon as node P1 has finished sending its message (M1), its terminal gap has passed and the bus is free. After sending a message, the transit interval (TI) is started. Within this interval no further sending is allowed. The condition SG > Max $\{Tg_i\}$ ensures that a new period can only be started, if every node has sent its message within the current period.

The most consequent way to realize time division multiple access of distributed nodes within a safety critical real-time system is to base the protocol execution on a strictly time triggered activation. This means, that the protocol execution is initiated by the progression of the *global time*. Each node in this real-time network refers to this global time and the point in time when a message will be sent is known *a priori* to all receivers. A protocol which operates according to this description is the *Time Triggered Protocol* for *Class C* applications (TTP/C), which has been developed at the Technical University of Vienna, Department of Real Time Systems [Kop93a, Kru97, Kop97].

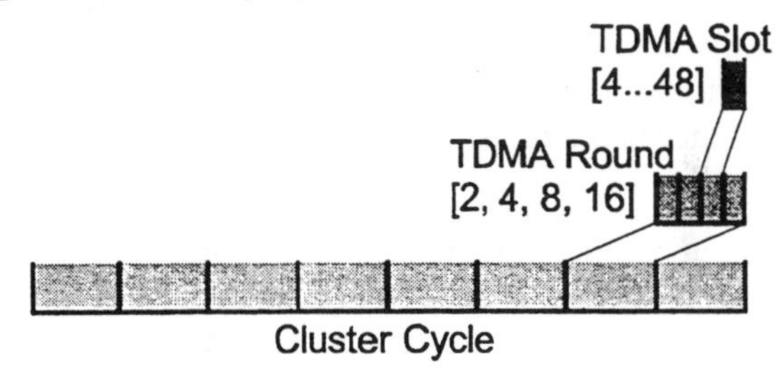

Fig. 4.2. TDMA Cycle

Fig. 4.2 illustrates how TTP/C's media access is organized. The sequence of TDMA slots in which each node sends one message forms a TDMA round. After finishing one TDMA round another one will be started. The temporal access pattern is the same but possibly different messages are sent. The number of different TDMA rounds determines the length of a cluster cycle (TDMA cycle). After a cluster cycle is finished, the transmission pattern starts over again at the beginning of the cluster cycle.

TTP/C–Overview of the Protocol Services: TTP is an integrated time-triggered protocol for real-time systems with TDMA-media access. The host using the TTP/C protocol machine only needs to provide the data to be sent in the Controller Network Interface (CNI) and has to read the received date out of the CNI. This interface is only passed by state information, data and some parameters needed for protocol configuration (see Fig. 4.3.).

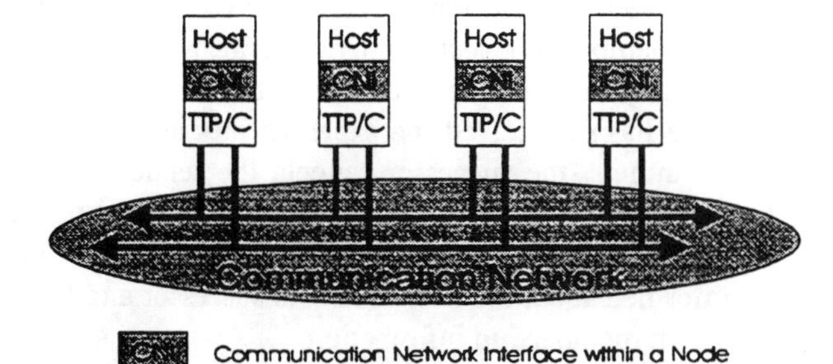

Fig. 4.3. Scheme of a TTP/C network

The TTP/C Controller operates autonomously without any control signals from the host computer. The necessary control information is stored in the controller in a personalized message descriptor list (MEDL):

- for each message, the instant when the message has to be sent and the address of the CNI-data field where the data must be fetched.
- the instant when a particular message has to arrive and the address of the CNI-data field where the received data must be stored.
- additional information for the protocol operation.

The length of the MEDL corresponds to the cluster cycle described above.

Each TTP/C controller has access to the replicated bus (channel 1 and 2). If the global time reaches an instant that is contained in the MEDL, the actions specified in the MEDL are carried out. So a message will be sent on both channels within the sending slot. Because of the protocol structure and the media access method used, a *minimal message latency* can be achieved without additional overhead. In addition, latency jitter in the range of microseconds can be guaranteed.

To support *fault-tolerance*, two TTP/C controllers can be operated in parallel. The replicas act as one fault tolerant unit (FTU). If a message is sent by an FTU, the same message will be sent physically on both channels by the first replica at a predefined sending slot and again by the second replica in a later sending slot.

State synchronization is guaranteed by including the controller state (C-State) into the CRC-calculation (normal frames) or by explicitly sending the C-State (initialization frames). The C-State is a data structure that contains the internal state of the controller which consists of the time field, the MEDL field and the membership vector.

Robustness of a TTP/C network is guaranteed by cabling according to the CAN-specification, Manchester coding and appropriate hardware units, e.g. high speed CAN-drivers with fault tolerance properties tolerating bus failures, such as shortcuts of CAN-HIGH with mass.

With TTP/C it is possible to support *error detection* at the receiver's side since the point in time when a message should arrive is known. If this message is not received at the expected point in time, all receivers detect this transmission error. This *a priori* knowledge is also used to implement the membership service. Every node of a TTP/C network has its own local membership vector providing membership information for any node in the network. If the successor of the sender node hasn't received the expected message on any of the two channels, the successor cancels the sender from its local membership vector. As the membership vector is part of the CRC calculation of the next message, the original sender will be informed about the decision of its successor and has to adapt its own membership information.

To *configure* the protocol on the one hand and to guarantee *composability* on the other, TTP/C offers the CNI Control Fields. Composability is achieved by temporal encapsulation of the components within the TTP/C network. Message transmission is carried out by the TTP/C controllers completely autonomous, no control flow passes the CNI. Therefore, each TTP/C controller acts as a temporal firewall [Kop98] which guarantees that the communication pattern is kept unchanged, regardless of varying system conditions such as load. It is therefore possible to test individual nodes in the network before the complete system is available. Furthermore, it is guaranteed that a node, once tested, can be integrated in the system without further effort.

Global Time Base: The most critical part of TTP/C is the provision of a global time base. As any action of the protocol depends on the progression of the global time, synchronization mechanisms have to be included in the protocol services. The global time, which is based on *macro ticks*, is corrected according to the result of a synchronization algorithm. The macro tick is generated by the controller's hardware. Each macro tick is generated by a number of micro ticks which can be varied to enable clock correction.

The clock synchronization algorithm is extremely robust, it can withstand up to f arbitrary, even malicious, faults given that the number t of TTP/C controllers satisfies the condition $t > 3f$. During operation, TTP/C synchronizes the local clocks to generate a common time base of known precision. The algorithm is based on the *a priori* knowledge of the message's sending time (information in the MEDL). The local node knows the expected arrival time and starts time measurement before this expected time by a hardware mechanism. At the end of this measurement window, the message should have reached the node. Due to the difference of expected and actual arrival, the time difference can be calculated and added to a push-down stack of depth four [Kop97]. The fault-tolerant average algorithm [Kop87] will be activated at any node of the network at the same instant (MEDL).

5. FAIL-SILENCE PROPERTY

To mask the failure of individual components it is necessary to employ redundancy. The number of redundant components, denoted t, which is required to tolerate f faults depends on the behavior that the individual components exhibit in the case of failures:

$t > 3f$ is necessary to tolerate failures of components that behave arbitrarily. This failure semantics is often called *Byzantine* failures.

$t > 2f$ is necessary to tolerate *consistent* failures. Consistency requires that all non-failed components have the same perception of the failure behavior of the failed component.

$t > f$ is necessary to tolerate failures of fail-silent components. A fail-silent component produces either correct results—in the value and time domain—or it produces no results at all.

Given these results, it is obvious that the least number of components is necessary to implement fault-tolerance based on fail-silence. This, however, is of utmost importance, since the number of components translates directly into system cost, size, weight, energy consumption and heat dissipation. Furthermore, the construction of fault-tolerant systems out of fail-silent components is relatively easy, compared to components that fail consistently or byzantine. There are no voters or majority building devices necessary. It is sufficient to pick the result that is delivered by any of the fail-silent components. It is therefore advantageous to build system architectures that are based on the principle of fail-silence.

Fail-silent computers: To build a fail-silent computer architecture, each and every node must therefore deliver either results which are correct in both the value and the time domain or no results at all. To achieve a sufficiently high coverage for the fail silence assumption [Pow92] it is necessary to employ extensive error detection strategies at the node level. There is a broad variety of error detection strategies. Some of them are hardware based (e.g., watchdogs or some signature checking methodologies), others are software based. The architecture described here supports both types of error detection mechanisms. At the software level it supports error detection strategies which can be applied systematically[1]: double execution, double execution with reference check, validity checks (for messages, history-states, and resources), assertion checking, and signature checks. Experimental results have indeed shown that it is possible to achieve a high error detection coverage by combining some of these strategies [KFA+95]. The individual mechanisms and implementations will be described in the following.

Mechanical fail-silence: For any physical realization of electronic instructions, actuators are needed. In a *steer-by-wire* system for instance actuators are necessary to steer the wheels and to give feedback to the driver. A failing actuator (for instance a motor driving a gear) may block the whole system in a way, that not even a second, redundant actuator is able to provide enough force to guarantee the specified service. Thus, the fail silence property is of particular importance when studying fault-tolerance of actuators.

One way of achieving fail silence is to use two or more non-self-locking motors and non-self-locking gears. For the same reasons as for the sensors, it is preferable to have actuators based on different physical principles, for instance electromotors and hydraulic actuators. However, this may be in contradiction to environmental considerations that demand a purely dry, non-hydraulic system. A detailed dependability analysis may decide on this. Focusing on electromotors, for dependability and reliability reasons brushless motors should be used, as these do not suffer from the consequences of wear and have much smaller failure rates than motors with brushes.

[1]By systematically applicable it is meant that the error detection strategy can be applied to a piece of software without any knowledge of the application domain [Pol96a].

6. SYSTEM SOFTWARE ARCHITECTURE

The design of the system software architecture was guided by a number of objectives and constraints: it should provide systematic, i.e., application independent, mechanisms for error detection, robustness, and fault tolerance; it should support composability, reusability, testability, and reduction of complexity; and it must make economic use of hardware resources like run-time, memory, and communication bandwidth.

At the logical level of software design, the partners tried to combine the most important mechanisms of the already existing approaches of BASEMENT [HLS+96] and DFR [PT98a] in the so called XBW-model. For the detailed design of distributed, fault tolerant software this abstract model is not sufficient as it is not detailed and specified enough.

The software workgroup and the prototype workgroup of the X-By-Wire project decided to use Bosch's *DFR model* for detailed design. The DFR model is a meta object model specifically designed for distributed, fault-tolerant, hard real-time systems. The DFR model defines a semantic framework for the construction of application-specific object models. It comprises three separate, orthogonal domains: the *value domain*, the *time domain*, and the *distribution domain*.

Value domain: this domain is concerned with the functional behavior of objects without any consideration of synchronization, temporal events, or distribution issues. In the DFR model, the value domain contains all the imperative code written by the application developer(s).

The value domain provides three essential object types[2]. The *subsystem* is the basic packaging construct which serves as the unit of distribution, replication, composability, and as focal point for the application of error-detection and fault-tolerance strategies. A subsystem encapsulates the functional characteristics of a real-time application object. It provides *processes* and *messages*. The process is the smallest schedulable unit of a DFR system and gives a time-invariant description of an independently active thread of control. Messages are the basic communication mechanism between different processes. Messages are implemented as state messages [Kop93b], i.e., there is no synchronization relation between the sending and receiving processes; thus, messages support transparency of timing, scheduling, and distribution.

Time domain: this domain is concerned with the temporal behavior of the objects of a concurrent software system located on a single processor node. It addresses the dynamic interaction of autonomously active objects and the events triggering changes in the system state. In the DFR model, the application developers describe the time domain purely declaratively.

The most important object types of the time domain are *operating modes* and *tasks*. An operating mode encapsulates the timing characteristics for a specific functional mode of a

[2] The DFR model provides a number of other object types not described here. For a more detailed discussion, the interested reader is referred to [PT98a].

processor node. Its timing behavior is defined by a set of tasks and their triggers. In the DFR model, the task is the unit of scheduling. It comprises a set of processes which are executed in a fixed sequence after a specific trigger (time or event).

Distribution domain: this domain is concerned with the association of software components to specific processor nodes in a distributed system and with the temporal pattern of communication between the nodes. As for the time domain, the application developers describe the distribution domain purely declaratively.

The central object type of the distribution domain is the *TTP-bus* which connects a number of processor nodes called *p-nodes*. A TTP-bus can be operated in one or more *bus-modes* each characterized by the timing of the messages transmitted over the bus. In the DFR model, redundancy is introduced by allocating a specific subsystem to more than one p-node.

These three domains were designed such that objects in different domains are as independent from each other as possible; thus, the application developer can design and test the software components in isolation without having to consider variable characteristics of the environment the components will be used in. In particular, objects of the value domain are independent from the other two domains and can be used in different temporal and distributional settings without any change of the source code.

7. SYSTEMATIC MECHANISMS FOR ERROR-DETECTION AND FAULT-TOLERANCE

The DFR meta object model supports systematic mechanisms for node-level error detection and system-level fault-tolerance as they are necessary in safety critical systems. The overriding goal in the design of these mechanisms was to decouple the application software as far as possible from the fault-tolerance considerations, i.e., a clear separation between functional and non-functional behavior. For instance, the implementation of a process should be unconcerned with the degree of replication.

Node-level error detection strategies: these strategies - double execution, double execution with reference check, validity checks, and signature checks - were designed to support fail-silent behavior of processor nodes. The mechanisms supported provide high coverage for the fail-silence assumption. If the implementation of the value domain satisfies certain constraints, different error detection mechanisms can be used for specific subsystems without modification of the source code.

When double execution is specified for a given subsystem, each of its processes is executed twice per activation; a difference in the results of the two executions indicates that a transient fault of short duration occurred.

Double execution with reference check extends the error detection coverage to transient faults with longer duration and permanent faults. In this case, for each of the subsystem's processes an additional execution with reference data is performed between the first and second execution. A difference between the result calculated by the reference execution and the expected result indicates a fault of longer duration.

Validity checks provide detection of faults resulting in the corruption of data. They are implemented by storing redundant information like check sums and checking the consistency of this information with the data's value on every access.

Assertions are used to check the correctness of a node´s state with respect to application specific criteria. For messages and resources, the application programmer can specify invariants which have to be satisfied during all executions. Pre- and postconditions are used to specify correctness criteria for entry and exit of a particular process.

Signature checking provides detection of control flow errors. When signature checking is specified for a subsystem, all tasks containing processes of the subsystem are checked concerning the order of process execution.

Distributed fault-tolerance: Tolerance of permanent faults requires redundancy which is achieved by replication of physical components. Replication of hardware components takes place at the granularity of nodes and their peripherals - called smallest replaceable units (SRUs). In the DFR model, software components are replicated at the granularity level of subsystems, i.e., replicated and non-replicated subsystems may reside on the same SRU. This differentiation between the granularity level of replicated software and hardware components is advantageous for economic reasons by making it possible to replicate only critical software parts. This allows the application designer to select the necessary level of redundancy without incurring the high hardware costs for the complete replication of SRUs.

Replicated subsystems must show identical - replica deterministic [Pol96a] - behavior in the value domain as well as in the time domain. The DFR model supports enforcement of replica determinism in both domains.

Non-determinism in the value domain arises due to replicated sensors which return slightly different readings. Agreement on these diverging values is supported in the DFR model by the *RDA-messages* (replica deterministic agreed messages) mechanism [PT98b]. It provides a framework for processes to exchange the individual sensor readings and to attain agreement on them. This framework supports the development of subsystems that can be used with replication degrees from 1 to *n* without any changes in the source code.

Replica non-determinism in the time domain arises due to slight variations in the processing speed of nodes hosting replicated subsystems. This variations in the processing speed are caused by divergences of oscillators, different CPU types for processing nodes, differences in the software, and others. To ensure replica determinism in the time domain, it is sufficient to guarantee that all replicated processes receive and send messages in the same order. This is ensured by the *timed message* [Pol97] mechanism. Timed messages are associated with an agreed upon validity time. Message delivery to the application processes is based on this validity time which guarantees identical message delivery for all replicas.

Closely related to replica determinism enforcement is the membership service. The TTP bus provides a consistent and timely view on the group of correct nodes that is attached to the bus. This node-level membership is translated by the operating system to a subsystem-level membership. This membership service indicates the replication degree for a given subsystem. It is thus possible to query the status of certain functions.

8. TOOL SUPPORT

The DFR model supports static system construction. This means that the application's function and structure is completely defined during design-time and does not change during run-time. The DFR model is semantically rich and explicitly provides complete information on the static system structure, i.e., all the application objects and their associations.

This information is used by a tool to adapt the independent, reusable software components from the value, time, and distribution domains to a specific system. For instance, if two subsystems are assigned to different processing nodes, the communication between them must take place via global messages on the TTP bus. The necessary mechanisms to relay this information to and from the bus must be provided.

The DFR model depends heavily on the adaptation of the source code since the software components of the different domains have minimal coupling. Based on the information contained in the DFR model, a suitable tool can thus adapt the source code to a specific system, relieving the user from any manual source code changes which would make software reuse impossible.

In addition to its use in source code adaptation, the *a priori* knowledge available in the DFR model facilitates source code analysis, configuration of the run-time system and various optimizations. All these aspects can be handled automatically by a tool, providing improved robustness and high resource efficiency without compromising on composability and reusability.

The DFR model and its associated mechanisms are mapped to the xERCOS[3] kernel by a tool named Off-Line Tool (xOLT). The xOLT analyzes the (generic) source code of all components of a single processor node, checks the system for local and global consistency, and generates highly optimized code tailored to a specific implementation of the xERCOS kernel.

The xOLT supports incremental and distributed software development, and it provides access to all the design information pertinent to a specific application. In addition, the OLT generates the complete code for the interface between the processor node and the TTP communication controller.

[3] Embedded Real-time Control Operating System for X-By-Wire systems.

9. PROTOTYPE

Overview of the non-mobile prototype: The partners of the X-By-Wire project agreed upon to build a non-mobile prototype, intended to demonstrate the general ideas for a steering system. Fig 9.1. is a scheme of the steering-prototype without mechanical backup. The TTP/C communication bus is the backbone of the demonstrator as it connects the three electronic main parts: steering-wheel actuator, steering control unit and steering actuator.

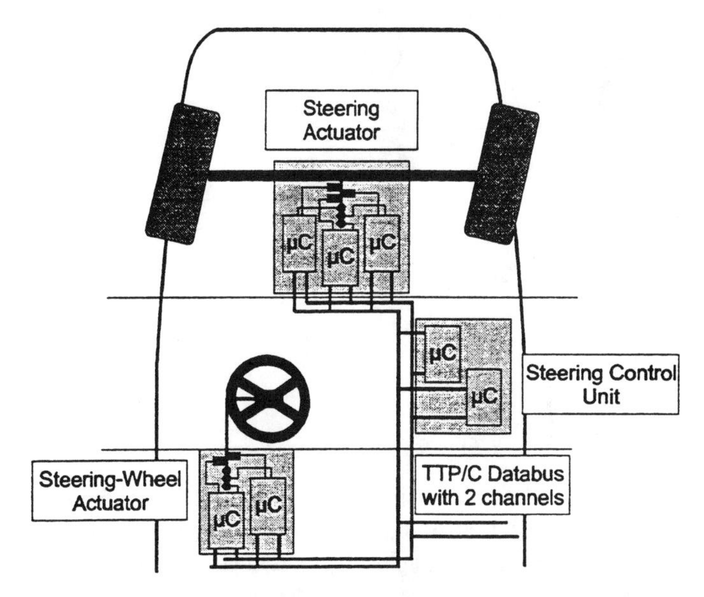

Fig 9.1. X-By-Wire Prototype

Two replicated nodes form the fault-tolerant *steering-wheel actuator*. They get the commands from the driver by angle and torque measurement and receive the feedback of the road-wheels to map this feedback to the steering-wheel.

The *steering control unit* also consists of two nodes. They receive the commands from the driver (provided by the steering-wheel actuator) and the feedback from the road-wheels. Furthermore, the steering control unit performs the overall control- and comfort functionality. This function is intended to assist the driver in critical situations, e.g., the car is starting to drift, it also helps the driver during parking maneuvers. Additional information from other systems, e.g., ABS can be taken into account as well. The output is sent by means of the TTP/C-bus to the steering actuator as well as to the steering-wheel actuator.

At the front of the prototype, three microcontrollers form the *steering actuator*. They are performing a simple control-loop by driving the electro-motors to reach the desired angle of the road wheels. They also control the current angle and torque by reading their angle and torque sensors. The feedback-angle and feedback-torque is communicated to the rest of the system, e.g., to the steering-wheel actuator to give the driver a feedback from the road-wheels.

Detailed description of the steering actuator: This section describes the most important parts of the steering actuator. First of all, a basic notation about the measurement and processing of the current steering angle and steering torque at the road-wheels has to be defined. As the steering actuator is placed at the stub shaft of a commercial car, the angle of interest is the mean steer angle β as described in Fig. 9.2. [Gil92]. The necessary angles at the wheels to drive the car into a curve with radius r under β are marked as β_l (left steer angle) and β_r (right steer angle) and are given by the construction of the used steering axle of the demonstrator.

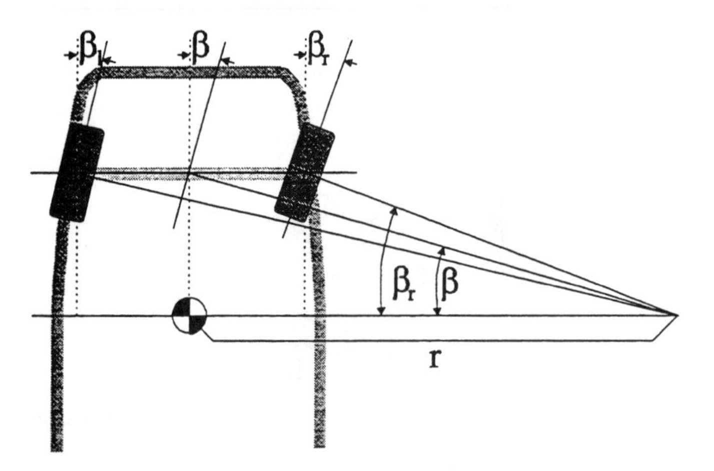

Fig. 9.2. Mean Steer Angle

In the following, a combination of the definitions of the value domain (see section 6) and of the distribution domain of the steering actuator is graphically presented in Fig. 9.3. The object model description (value domain) of the DFR meta object model describes the functional encapsulation into subsystems (container boxes) with its processes (gray shaded boxes) and which messages are used in the system (arrows).

The distribution domain describes the mapping of the subsystems to the *p-nodes* (see section 6) and which messages are global ones. The mapping of the subsystems influences which messages could be treated as interprocess-messages. This kind of messages (gray shaded arrows) will only be sent and received by subsystems of the same physical node. All other messages have to be sent on the TTP/C bus for common use. While the dashed arrows represent RDA-messages (see section 7, RDA-message mechanism) and are exchanged between replicas with the same functionality, the filled arrows correspond to simple global messages to be exchanged within the distributed real-time system.

On the base of the graphical notation explained above, Fig. 9.3. represents the encapsulation of the steering-actuator. The functionality is splitted into the subsystems *S_Torque_Sensor* and *S_Angle_Sensor*, measuring the current mean steer angle (see Fig. 9.2.) and mean steer torque, the subsystem for the overall control loop *S_Control* and the subsystem *S_Actuator*, which drives the motors according to the settings of *S_Control*. The subsystem *S_Diagnostics* is the node local representation of diagnosis functionality, which could be initiated by the steering control unit (not shown in Fig. 9.3.).

The figure also represents the RDA-messages *S_Torque_RDA* and *S_Angle_RDA* and the interprocess-messages *S_Actuator_Setpoint* and *S_D_Command*.

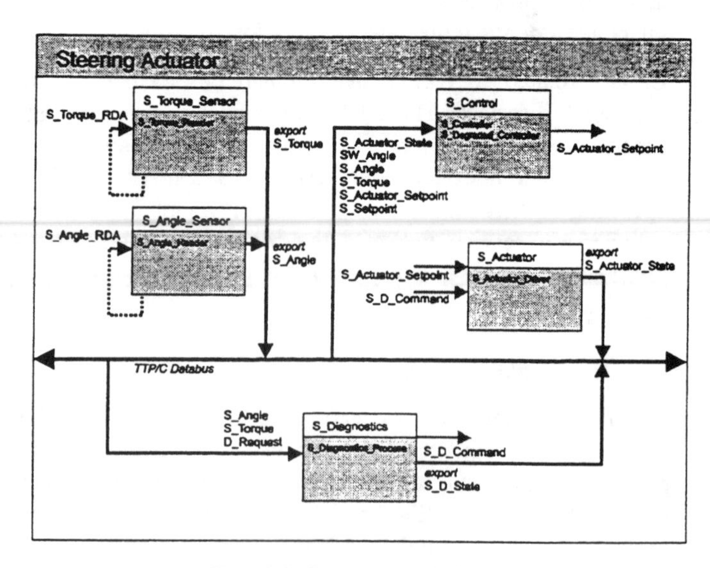

Fig. 9.3. Steering Subsystem

While the necessity of the RDA-messages should be clear in the context (see section 7), the interprocess-message *S_Actuator_Setpoint* holds the calculated actuator-setting of *S_Control* and *S_D_Command* contains a command for diagnosis, e.g., a setpoint to start fault injection. All other messages are global ones and are used according the DFR-meta object model definitions of the prototype.

10. CONCLUSION

The paper presents the technologies developed by the partners of the Brite-EuRam III-project "Safety Related Fault Tolerant Systems in Vehicles" (X-By-Wire). These technologies provide a framework for the introduction of safety related systems without mechanical backup. The focus is on time-triggered information exchange, the fail-silence approach, and the support of distribution by new system services.

The requirements on class C real-time communication systems preclude the use of the wide spread event-driven communication protocols. TTP/C offers the necessary temporary encapsulation by its strictly time-triggered processing, its reliable message transfer based on a globally synchronized time base, and its efficient membership service.

The fail silence property guarantees for each component, that it provides correct functionality and sends correct results at a pre-defined point in time. Otherwise the component remains silent.

The DFR meta object model provides the conceptual framework and systematic mechanisms for node-level error detection and distributed fault-tolerance. It was the basis for the development of the xERCOS operating system for embedded real-time control applications and the accompanying off-line tool xOLT.

The steer-by-wire prototype developed in the context of the X-By-Wire project will demonstrate the application of these technologies and the feasibility of safety related fault-tolerant electronic systems without mechanical backup in vehicles.

ACKNOWLEDGMENTS

Thanks to all partners of the X-By-Wire project for the efficient cooperation, many ideas, solutions and work already done. Thanks to Prof. Dr. Hermann Kopetz and his staff for the detailed information about TTP/C. Some details were taken out of the "Specification of the Basic TTP/C Protocol" which was available for the project partners.

REFERENCES

[ARI91] ARINC (1991). *Multi-Transmitter Data Bus ARINC 629 - Part 1*: Technical Description. Aeronautical Radio Inc., Annapolis, Maryland 21401.

[Gil92] Gillespie, T. D. *Fundamentals of Vehicle Dynamics.* ISBN 1-56091-199-9

[HLS+96] Hansson, H. A., Lawson, H. W., Strömberg, M., Larson, S. (1996). BASEMENT: A Distributed Real-Time Architecture for Vehicle Applications. In *Real Time Systems*. Kluwer Academic Publishers, Boston.

[KFA+95] Karlsson, J., Folkesson, P., Arlat, J., Crouzet, Y. and Leber, G. Integration and Comparision of Three Physical Fault Injection Techniques. In *Predictably Dependable Computing Systems*. B. Randell, J.-C. Laprie, H. Kopetz, and B. Littlewood (eds). Springer, 1995, pages 309–327.

[Kop87] Kopetz, H., & Ochsenreiter, W. (1987). *Clock Synchronisation in Distributed Real-Time Systems.* IEEE Trans. Computers. Vol. 36(8). (pp. 933-940).

[Kop93a] Kopetz, H., & Gruensteidl, G. (1993). *TTP - A Time-Triggered Protocol for Fault-Tolerant Real-Time Systems.* Proc. 23rd IEEE International Symposium on Fault-Tolerant Computing (FTCS-23). Toulouse, France. IEEE Press. (pp. 524-532), appeared also in a revised version in IEEE Computer. Vol. 24 (1). (pp. 22-26).

[Kop93b] Kopetz, H. (1993). *Should Responsive Systems be Event-Triggered or Time-Triggered?* IEICE Trans. on Information and Systems Japan (Special Issue on Responsive Computer Systems). Vol. E76-D(11). (pp. 1325-1332).

[Kop97] Kopetz, H. *Real-Time Systems: Design Principles for Distributed Embedded Applications.* Kluwer Academic Publishers. 1997.

[Kop98] Kopetz, H. *Component-Based Design of large Distributed Real-Time Systems.* To appear in Control Engineering Practice.

[Kru97] Krug, M. *A Prototype Implementation of a TTP/C Controller*, SAE Technical Paper Series, 970296, February 1997.

[Mok83] Mok, A.K.: *Fundamental Design Problems of Distributed Systems for the Hard Realtime Environment.* Ph.D. dissertation, MIT, 1983.

[Pol96a] Poledna, S. *Fault-Tolerant Real-Time Systems: The Problem of Replica Determinism.* Kluwer Academic Publishers. 1996.

[Pol97] Poledna, S. *Deterministic Operation of Dissimilar Replicated Task Sets in Fault-Tolerant Distributed Real-Time Systems.* In Proccedings of the Sixth IFIP International Working Conference on Dependable Computing for Critical Applications (DCCA-6). Springer Lecture Notes Series. Grainau, Germany. Mar. 1997.

[Pow92] Powell, D. *Failure Mode Assumptions and Assumption Coverage.* In Proceedings of the 22th International Symposium on Fault-Tolerant Computing. Computer Society Press of the IEEE. Boston, Massachusetts. Jul. 1992, pages 386–395.

[PT98a] Poledna, S. and Tanzer, C. *DFR Objects: A Meta Object Model for Distributed Fault-Tolerant Hard Real-Time Systems. Submitted to IEEE International Symposium on Object-Oriented Real-Time Distributed Computing.* 1998.

[PT98b] Poledna, S. and Tanzer, C. *Software Support for Fault-Tolerance. Submitted to International Symposium on Fault-Tolerant Computing.* 1998.

[SAE94a] J2056 I.R. Class C Multiplexing, Part 1 JUN93 Applications Requirements, Society of Automotive Engineers, Warrendale, PA, 1994

[SAE94b] J2056 I.R. Class C Multiplexing, Part 2 APR93 Survey of Known Protocols, Society of Automotive Engineers, Warrendale, PA, 1994.

980560

Dynamic Traffic Light, Vehicle Signalling Display

Hamid Kashefy

ABSTRACT

The present invention is a meaningful traffic light for motor vehicles. By flashing light(s) of colours orange, green or red, it indicates one ***predetermined unit*** variation in speed or in frequency of pulses received from speed sensor of the vehicle on which the traffic light is installed. Also, different configurations of said lights symbolise stationary state, or different levels of speed of the vehicle.

When difference between the *current speed* and the speed at which the latest flashing happens is equal or greater than one *unit*, a flash of a light arises.

INTRODUCTION

Would it be safer or less expensive if instead of the usual traffic lights only intensity coded red lights were installed at every intersection?

What functions should Dynamic Traffic Light perform in order to prevent traffic crashes?

As a matter of fact, driving is a potentially dangerous activity which requires dynamic coordination. The so called Centre High Mounted Stop Lamp (CHMSL), and the other required red lights which are also referred to as ***Stop Lamps*** in traffic safety literatures, do not uniquely signify the intention of a driver for stopping his/her vehicle and, therefore, *they are not necessarily stop lamps.* Dynamic Traffic Light, which also consists of ***Advanced Stop Sign System***, takes the place of the CHMSL on motor vehicles since the brake lighting system that is presently in use on motor vehicles is too primitive and too static.

It is primitive because it has a single and simple function which does not precisely reflect the state of motion of leading vehicles. If a lead vehicle is stopped or it has low speed, there is no indication from a distance to a following driver who may be travelling at high speed, that the lead vehicle is actually stopped or it has low speed; unless the following driver advances perilously closer to the lead vehicle and then react.

If a lead driver partially releases the accelerator pedal and does not brake at all, speed of his/her vehicle is reduced. This decrease in speed which may be insignificant to the lead driver, may have significant consequences to a following driver, ***but it is not informed to the following driver by any light.***

The brake lighting system which is presently in use on motor vehicles is static because though it denotes deceleration when a lead driver uses the brake, it does not indicate to the following drivers that how fast the speed of the lead vehicle is reduced. When a vehicle is travelling at any speed, whether the driver of this vehicle just taps the brake pedal or (s)he depresses the brake pedal rigorously for swift deceleration, the red brake lights are simply turned on. Moreover, the brake lights are intensity coded at least with the presence light and, in effect, the brake lights may not be detected easily by following driver(s).

DYNAMIC TRAFFIC LIGHT, VEHICLE SIGNALLING DISPLAY

"Colour-coded systems employing green tail lamps, amber turn lamps and red stop lamps have frequently been shown to substantially decrease reaction times of following drivers. ... Mortimer (1970), also reports that a colour coded rear lighting and signalling system produced fewer errors (13 compared to 33) and fewer missed signals (9 compared to 50) and was subjectively judged to be superior to a conventional all-red system. ... The most recent research poses question about the need to provide additional information (with rear lighting system of motor vehicles), such as speed (e.g., Jollife, et al., 1971), deceleration (e.g., Voevodsky, 1974), or accelerator position (e.g., Mortimer and Sturgis, 1976).

People are notoriously poor at estimating deceleration (or acceleration) rates of vehicles ahead (e.g., Olson, et al., 1961; (Mortimer, et al., 1974b). Therefore, a rear signal indicating the level of deceleration has intuitive appeal.

Most of the studies reviewed are in agreement that spatial separation of functions is beneficial. ... However, since separating the presence and stop lamps would require separate lamp assemblies, *and since available data strongly suggest that a colour coded system would be more effective than an all-red system,* ***serious consideration should be given to the desirability and feasibility of a completely colour-coded system"*** (1)*.

* Numbers in parentheses designate references at end of paper.

Dynamic Traffic light is also an illustration of the feasibility of implementation of a sophisticated and, therefore, meaningful colour coded system which maintains the integrity of the required signal systems.

"Lave (1987) ran a number of cross-section regressions on state data, disaggregated by highway type. The dependent variable was the fatality rate, the explanatory variables measured both the limit effect and the coordination effect. If the desired consequences of speed laws is to limit dangerously high speeds, then we should observe a positive relationship between the average speed of travel and the fatality rate. If the desired consequences of speed laws is to coordinate the traffic flow, then we should observe a positive relationship between speed-variation (the standard deviation of the observed speed distribution) and the fatality rate. The regression results were a surprise to the conventional wisdom: average speed, the variable designed to measure the effect of going too fast, was not statistically significant; but ***speed variation***, the variable designed to measure the effect of uncoordinated speeds, was significant. ***That is, it was easy to detect the fatality-inducing consequences of uncoordinated flows, but impossible to detect any fatality-inducing consequence of going too fast***" (2). Therefore, drivers are not able indeed to immediately perceive speed variations of leading vehicles and react on time. Otherwise, most of the accidents would not happen even at high speeds.

"The precise purpose of the brake application (panic stop, ordinary stop, deceleration, disengagement of cruise control) *requires the following driver to be aware of the traffic environment ahead*" (3).

"In fact, one of the few indications in the literature of a link between driver performance measures and crash involvement rates **relates to *driver information processing abilities***" (4). As used in this description, the term ***Highway Information*** means the information that drivers need while driving, in order to improve their performance and reduce their crash involvement rates.

A number of situations which may lead to traffic crashes are identified. In this invention, amber*, red and green lights are used to implement ***a simple language of light***, in order to dynamically coordinate the speed variations of vehicles. At every moment, Dynamic Traffic Light of leading vehicles help following driver(s), who just look at leading vehicles ***on any lane*** from a distance, to realize the following situations:

a. Whether a lead vehicle is in motion or its is stopped regardless of braking;
b. If the lead vehicle is moving, whether it is travelling with low speed or it is travelling at some higher speeds (ie: the range or level of speed of the leading vehicles);
c. Whether leading vehicles are travelling with a constant speed or they are accelerating or decelerating;
d. If leading vehicles are accelerating or decelerating, how fast the speed of these vehicles is changing (ie: what is the magnitude of their acceleration or deceleration), and whether a vehicle is stopping;
e. Whether a head on vehicle is accelerating. If so, how fast its speed is increasing (ie: what is the magnitude of acceleration).

In drawings which illustrate an overall view of this invention, Fig. 1 depicts the front view of a lamps' casing for Dynamic Traffic Light, and portrays the configuration of amber, red and green lights.

Fig. 2 illustrates top-side view of the lamps casing of the traffic light (shown in Fig 1), and the casing for its electronic circuits which is depicted as a platform for the lamps casing at the moment of installation behind the rear window.

Fig. 3 represents an optional small casing containing an optional pulse amplifier/transmitter to be installed near the fuse box of vehicle, in order to amplify and transmit the speed sensor pulses to Dynamic Traffic Light in the rear of vehicle. Said device allows transmission of speed sensor pulses without extending an additional wire to the rear of vehicle.

FUNCTIONS OF DYNAMIC TRAFFIC LIGHT – As used in this description, the term ***Unit of Speed*** means a predetermined speed (such as 4 km/h, 3 km/h, 2 km/h or X km/h, where X can be any other desired number) for the automatic occurrence of flashes of the green, amber or red lights. Also the term ***reference speed*** means the speed at which the latest flashing of any light occurs, or the maximum speed after the latest flash of any light arises.

Fig. 4 illustrates all of the well defined functions of Dynamic Traffic Light for providing the ***unique Highway Information*** as follows:

1. If speed is greater than 31 km/h, green lights LG1 and LG2 and/or LG3 remain on to inform following drivers about the forward motion of lead vehicle Fig. 4-a.
2. If speed that is greater than 31 km/h is reduced without brake application, orange (or yellow) lights LO1, LO2 and LO3 momentarily flash on while all green lights which are on flash off, per every unit decrease in reference speed, Fig. 4-d. ***Flashes of orange lights of a lead vehicle would arise as advanced warning signals, in order to increase the expectancy of following driver(s) about the usage of brake by lead driver.***
   A spontaneous reaction of a following driver to the repellent flashes of orange lights of a lead vehicle would be decreasing speed by reducing pressure on the accelerator pedal. Swift deceleration without

---

* Amber is used as orange or yellow colours.

applying brake would keep orange lights on almost steadily, and keep the green lights almost off.

3. If while brake is not applied speed is reduced to between 32 km/h and 7 km/h, only the amber light LO3 stays on while all of the green lights are turned off. This indicates the low speed of leading vehicle to following drivers. If a vehicle with the low speed (that is greater than 7km/h) still decelerates, then orange lights LO1 and LO2 flash on per every unit decrease in reference speed while LO3 remains on, Fig. 4-e.
4. When a vehicle is stopped or in advance, when speed of the vehicle is above 0 km/h and less than 8 km/h, ***regardless of braking***, red light LR5 (Fig. 4-g) and two or more extra red lights turn on in the shape of a downward triangle and remain on, in order to indicate *the total stop of the vehicle* (Fig. 4-j). *The Advanced Stop Sign (or* Stop Light) of Dynamic Traffic Light could provide ***more time*** for a following driver to stop his/her vehicle.
5. If at any speed, a driver has to brake intensely to totally block wheels of his/her vehicle *while the vehicle is still in motion*, then the Stop Sign glows and does not flash as explained in function No. 4 above, for as long as the wheels of said vehicle do not spin and its speed sensor does not generate pulses. This urges following driver(s) for total stop or for changing lane if possible, *before that following vehicles approach the troubled lead vehicle unsafely.*

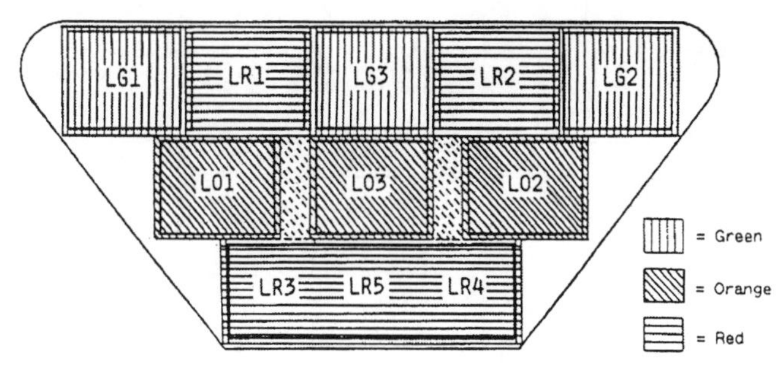

Figure 1. Front view of lamps casing of Dynamic Traffic Light

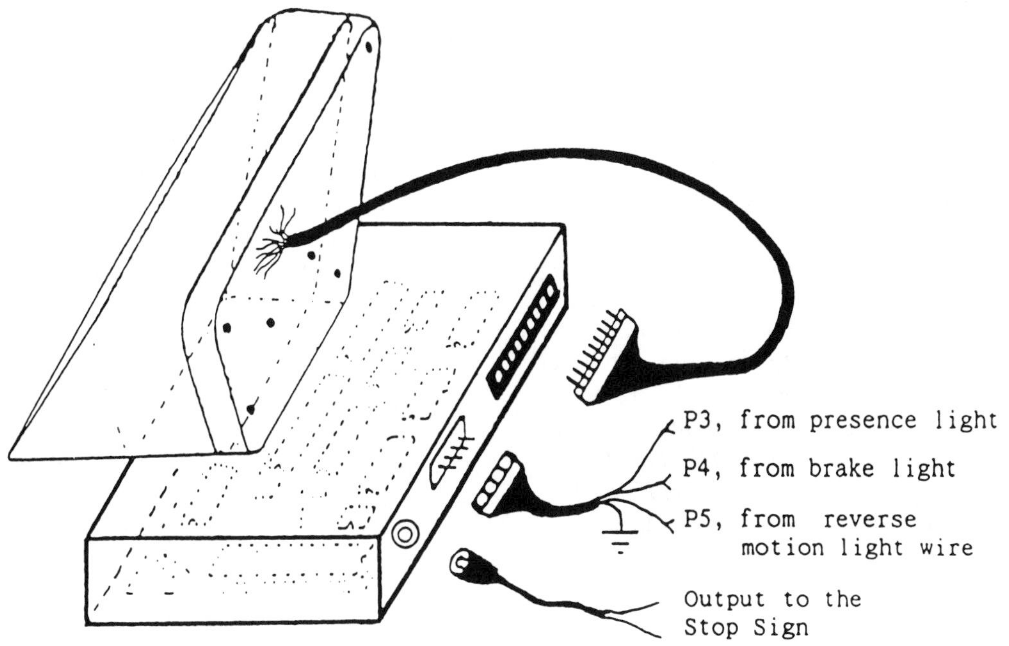

Figure 2. Top-side view of Dynamic Traffic Light with its input and output connectors

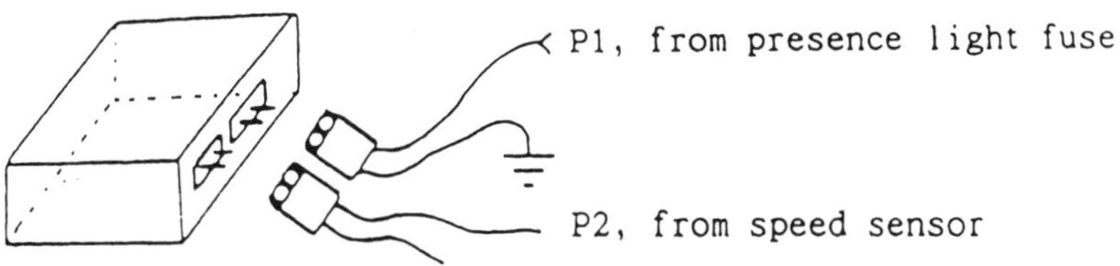

Figure 3. Optional Transmitter allows pulse transmission through the presence light wire.

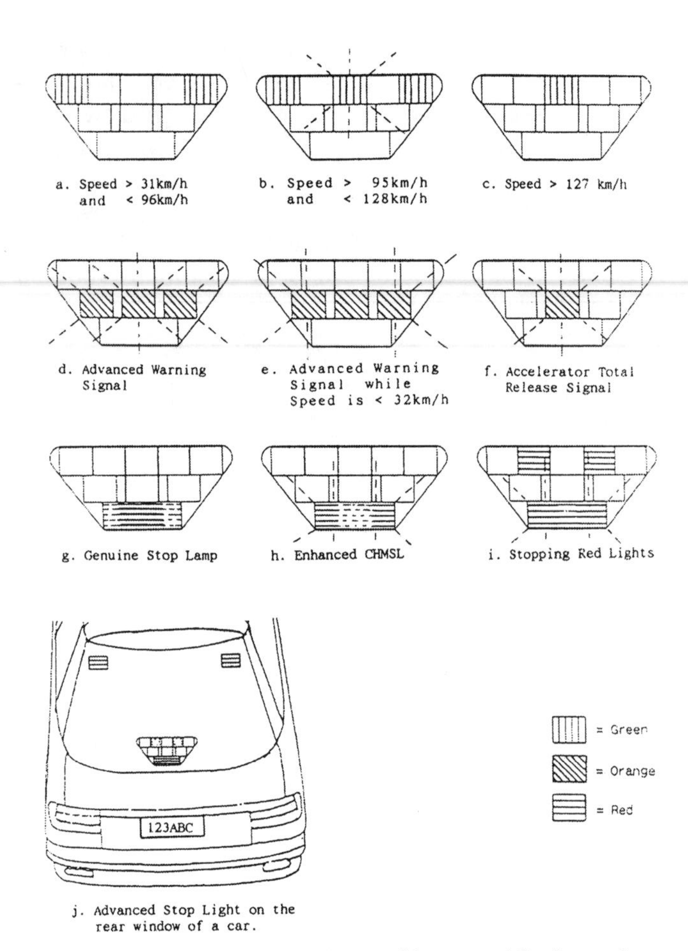

Figure 4. Speed Level Indicator Signs and flashes of Dynamic Traffic Light which arise only when speed varies by a predetermined unit.

6. When brake is applied and/or when the Stop Sign is activated, green and orange lights are turned off. If at any speed that is greater than 7 km/h brake is applied, red lights LR3 and LR4 are momentarily turned on without any delay (except the rise-time of light bulbs) to function as the required CHMSL, Fig. 4-h; the function of the required CHMSL is culminated in the first moment, since after the first fraction of a second the two ordinary red lights are perceived anyways. Then per every *unit* decrease in *reference speed*, the red lights LR3 and LR4 flash on to behave as an enhanced CHMSL which also indicates the rate of deceleration. *The faster speed is reduced, the faster red lights LR3 and LR4 flash on, and the brighter said red lights glow. Indeed, when red lights are on and do not flash, they denote total stop.* ***Thus flashes of the red lights would be a reverse complement to the Stop Sign.*** Flashes, specially of the red light, would be occurred the most efficiently if fast-rise light bulbs or LEDs are used.

EXTENDED FUNCTIONS – In order to employ the red and green lights the best, the electronic circuits needed to implement the above six functions are expanded (at low cost), in order to evolve and increase the functions of the red and green lights by the following functions which are numbered as continuation of the above functions:

7. The same method that is used to keep at least one orange light on is used to turn on the so called *Stopping Red Lights* LR1 and LR2, when speed is in the range defined in function No. 3. In that case, LR3 and LR4 continue to deliver their information as explained in function No. 6. This informs the following driver(s) that ***the vehicle, on which some red lights flash and some red lights of its traffic light remain on without flashing, is stopping,*** Fig. 4-i.
   A fine detail about the stopping red lights is that, while speed is under 8 km/h and the Stop Sign is activated, the *Stopping Red Lights* are deactivated and prevented from flashing if driver taps the brake pedal. This preserves the integrity of the Stop Sign, since no red light should flash when the Stop Sign is activated.
8. If every 1 km/h increase in the speed that is mentioned in function No. 1 does not take more than a predetermined time interval, such as ~0.5 s (ie: if acceleration is more than ~0.55 m/s2), then green light LG3 flashes on momentarily per each unit increment in the *reference speed* while green lights LG1 and LG2 remain on Fig. 4-b. In addition to said restriction, when speed that is lower than 32 km/h is increased, green light is prevented from flashing; because when a vehicle is accelerated from rest position on slippery road, false flashes of green light could be produced as tires might lose friction with ground and spin while vehicle is not in motion. False flashes of green light might never occur on vehicles equipped with a Traction Control System.
   Different levels of speed are ***coded*** by the three green lights as low speed of vehicle is coded by the orange light LO3:
9. If speed is increased to between 95 km/h and 128 km/h, green light LG3 is turned on and remains on as LG1 and LG2 are left on, in order to inform following drivers from a distance that speed of lead vehicle is in said range, Fig. 4-b. If the mentioned speed is increased, then green light LG3 momentarily flashes off (while LG1 and LG2 remain on), at least per each *unit* increase in the *reference speed*. This happens with the restriction mentioned in 8.

10. If speed is increased to 128 km/h or more, green lights LG1 and LG2 turn off and only LG3 remains on to indicate that speed of vehicle is 128 km/h or higher, Fig. 4-c. If this speed is increased, green light LG3 and (LG1, LG2) flash off and on interchangeably with said restriction, at least per every *unit* increase in the *reference speed.*

    Usually, a vehicle should accelerate to overtake other vehicles on the road. Naturally, a vehicle is the most dangerous to pedestrians who intend to cross the street, when the vehicle is being accelerated. "Almost two out of three deaths in 1993 occurred in places classified as rural. In urban areas, nearly one fourth of the victims were pedestrians; in rural areas, the victims were mostly occupants of motor vehicles. More than one half of all death occurred in night accidents. ... Of fatal accidents in rural areas, head on collisions between motor vehicles are the most prominent type" (6).

    Drivers of head on vehicles and pedestrians may also benefit from the Dynamic Traffic Light:

11. The flashes of green light, which would occur per each *unit* increase in the *reference speed* of vehicles, could be shown (without any restrictions) by a light of a colour other than green, orange or red (e.g., violet) in front of the vehicle on which the Dynamic Traffic Light is installed. Said light would inform *the driver of a head on vehicle or pedestrians* that the vehicle, on which a light flashes in front of it, is accelerating at the rate that the light flashes. The faster the vehicle accelerates, the faster said light would flash and the brighter it would glow.

    The same method that is used in this invention to produce the immediate first flash of CHMSLs LR3 and LR4 (function No. 6) can be used to perform the following:

12. As soon as accelerator pedal is totally released by a lead driver, before that the lead driver depresses the brake pedal, orange light LO3 of Dynamic Traffic Light of his/her vehicle could be flashed once, Fig. 4-f. As any other flashes of orange light (function No. 2), the occurrence of the Accelerator Total Release Signal would only mean one *unit* decrease in speed. ***This would increase the expectancy of following driver that brake might soon be used by the lead driver.*** However, if it is accepted that more often the release of the accelerator pedal is not followed by braking, and that such crafty flashes of the orange light LO3 would defile those flashes of the orange lights that would occur per each unit decrease in speed, then function No. 12 may be omitted, in order to allow more salient flashes of the orange lights to occur.

    Dynamic Traffic Light would always keep at least one light on, ***and only lights of the same colour could be on at a time.*** If vehicle is set for reverse motion, only the Stop Sign would remain on.

POSSIBLE VARIATIONS ON THE LIGHTS OF THE THREE COLOURS – Dynamic Traffic Light can have any other number of amber, green and red lights, and said lights can have any other geometric arrangements or signal parameters such as size and intensity. Each light should have a separate housing in the lamps' casing. As a result, the lights of the same colour, which might be illuminated simultaneously, could be distinguished from each other from a fairly far distance when lights with appropriate signal parameters are used. "A small lamp must emit less candlepower, to minimize glare, and need also emit less candlepower to reach adequate daytime brightness levels, compared to a lamp of greater area" (5). A lamps' casing which is approximately 1.5 times larger than the casing illustrated in Fig. 1 may be suitable for cars. The lamps' casing can have any other desired size and/or geometric shape, such as the lamps' casing shown in Fig. 5 and Fig. 6.

On some vehicles, instead of the genuine Stop Lamp LR5 (Fig. 4-g), a stop sign of a conventional shape can be illuminated, as it is depicted in Figures 4-j and 5.

An appropriate configuration of the lights of the three colours of Dynamic Traffic Light could be replaced for the rear lighting system that is currently in use on motor vehicles. Since with such colour coded system the red brake light would not be intensity coded any more, it would not need to have the maximum intensity to be perceived easier even in daylight. Rather, the flashes of the enhanced brake lighting system and *the change of colour to red*, and not the intensity of the red lights would clearly indicate braking. Hence, the *Stopping Red Lights* and the *Advanced Stop Light*, which would be independent from the brake lights, could have the maximum intensity to be perceived with even greater accuracy and speed of interpretation, from a fairly far distance.

However said lights should be configured, multiple lights of the same colour are needed because:

a) When speed is increased, consecutive flashes of green light would turn the green light almost off. Hence, at least one green light should flash on or remain on while other green light(s) flash(es) off.
b) With the existence of three lights of green colour, it also becomes feasible to code different levels of speed, as stated earlier.
c) While at least one orange remains on to indicate the low speed of a vehicle, at least one orange light should flash on per each *unit* decrease in speed of that vehicle.
d) At least three red lights are required; one as enhanced flashy CHMSL which is initially turned on with no delay, one as the *Stopping Red Light*, and one as the genuine Stop Lamp (ie: as *Advanced Stop Light*). A Stop Sign of a particular shape would require the configuration of more than one red light. Since the enhanced flashy CHMSL accomplishes the

function of the required CHMSL, the need for an additional red light as the required CHMSL is eliminated.

e) Since eventually a lamp can burn out, it is wiser to have at least two lights of the same colour. Then if, for example, LO1 (or LG1) is not functional any more, LO2 (or LG2) functions alone until the driver may perceive the problem and replace the burnt out light bulb.

Each of the above functions are offered as a new and unique solution to prevent or relief one of the identified situation in driving. As a result, naturally the complexity and the cost of the rear lighting system of motor vehicles would be increased. The Dynamic Traffic Light with all of its mentioned functions would be the *optimum innovation* in the rear lighting system of motor vehicles. Yet, for reducing its cost, the traffic light could be reduced to feature any subset of the mentioned functions. For example, it could portray the functions No. 1 to No. 6, or it could only feature the *Advanced Stop Light* (functions No. 4 and 5). Also, the Dynamic Traffic Light could easily embody the required speedometer as digital meter for combining their cost.

ABOUT THE *UNIT OF SPEED* – "A reasonable value for constant deceleration for good tires on dry level pavement is 5 m/s^2, even though a few exceptional vehicles under ideal conditions can even exceed the value of deceleration due to gravity of 9.8 m/s^2" (4). That is, with 5 m/s^2 as maximum constant deceleration by brake application, every second speed is reduced maximum by 18 km/h. Therefore, it appears that when *Unit of Speed* is chosen to be 2 km/h, 9 to 10 flashes per second would be reasonable as maximum rate of flashing by red light of Dynamic Traffic Light, for precise reflection of the rate of deceleration of a lead vehicle.

The repulsing and momentous flashes of orange light should arise well before that flashes of red lights occur. Nevertheless, since insignificant speed variations always occur, the integrity of flashes of the orange lights should be maintained by not allowing such flashes to arise when only 2 km/h (which is chosen as *Unit of Speed* for flashes of the red lights) is reduced from speed. Hence, when brake is not applied and speed is reduced, Dynamic Traffic Light would dynamically set the *Unit of Speed* to 3 km/h. The unit would be reset to 2 km/h when brake is applied.

The conditional flashes of the green light, which may tempt a driver to accelerate hastily, are also supposed to contribute in road safety while these flashes would stimulate the traffic to move ahead. It seems that on the average ~2m/s^2 is a reasonable presumption for magnitude of acceleration of a fairly fast going passenger car. In that case, every second speed is increased by ~7.2 km/h. Therefore, with 4 km/h as Unit of Speed for flashes of green light, the latter would flash 2 (or maximum 3) times per second to reflect the magnitude of acceleration of a vehicle.

MORE ABOUT THE STOP SIGN – The red light LR5 is already implemented as a genuine Stop Lamp in Dynamic Traffic Light. Because "*more than three fourths* of the struck vehicles in rear-end collisions are stopped" (3), then it would be ideal if the already required CHMSL be used only as a genuine Stop Lamp. But because the CHMSL is already in use and it is turned on every time that brake pedal is depressed, the Stop Lamp should be uniquely configured with few red lights, in order to be distinguished from an ordinary CHMSL (or from the Stopping red lights).

On many vehicles such as mini vans with large and vertical rear window, on all buses and on many trucks, the extra red lights could be alternatively arranged in the shape of a standard Stop Sign. The latter might even have greater effect when it is installed on vehicles because it would be illuminated by light bulbs. Instead of turning on the whole area of the Stop Light, only its edges could be illuminated, Fig. 5. The Advanced Stop Light could also be installed on the sides of the school buses for increasing the safety of these special vehicles. The Stop Light on the sides of the other long vehicles might be useful too.

Since usually the rear window of cars is small or inclined, the proposed Stop Sign could be configured as in Fig. 4-j on the rear window of cars, so that the Stop Sign be detected from a fairly far distance.

The fact that drivers are obsessed by the quality and the amount of the information that is provided by the intensity coded brake lighting system, and the absence of the idea of the *Advanced Stop Sign System* for motor vehicles has led many drivers to request flashy CHMSL for clear perception of a stationary vehicle. The most peculiar problem with the idea of flashing the required CHMSL as an indication that a vehicle is stopped is that flashing is indeed an activity or action. Associating the state of stillness of a vehicle with the activity of a light is not logical. That should be the reason that the flashes of the CHMSL are speculated to be so annoying and undesirable as an indication that a vehicle is stopped.

Simultaneous flashes of many CHMSLs might increase tension until lead drivers accelerate. *On the other hand, both the movement of vehicle and the flashes of Dynamic Traffic Light, that would occur when speed variations happen, are activities or actions.*

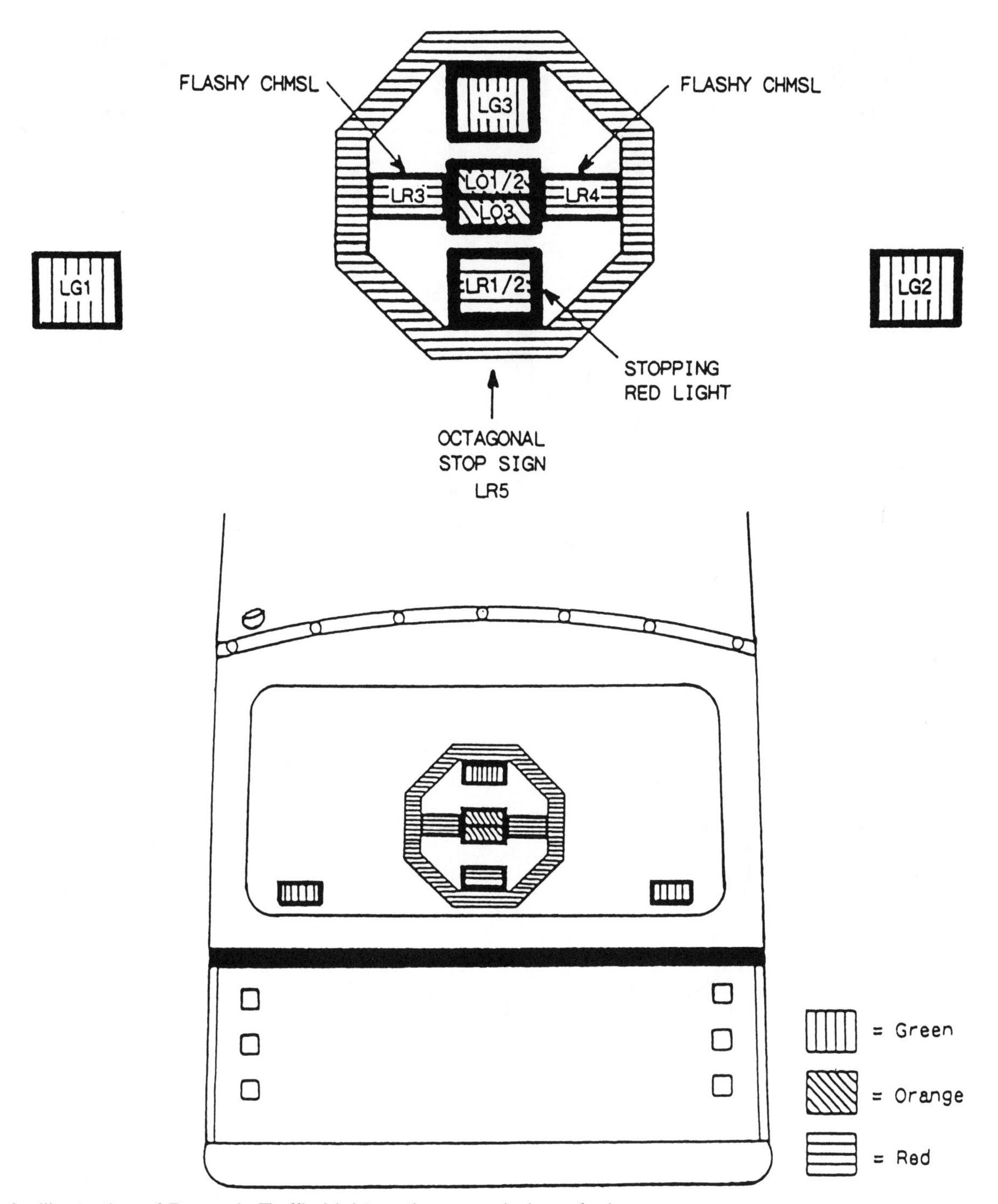

Figure 5. An illustration of Dynamic Traffic Light on the rear window of a bus

A self-evident but important fact to emphasize is that, the large percentage of struck vehicles which are stopped do not suddenly physically appear in front of following vehicles. Evidently, the struck vehicles exist as moving and leading vehicles on the active roads before coming to stop and be struck. But first their speed variations and then their stop is not obvious to the following drivers who get involved in the collisions.

"The Lead Vehicle Stationary (LVS, which is known as the first crash subtype) does not involve simply a 'too slow' reaction of the following driver to a sudden crash threat" (7). Rather, it involves a "recognition error" or a delayed recognition. The *Advanced Stop Light* of a lead vehicle would send an unambiguous message to following drivers that the lead vehicle is stopped; since a particular configuration of red lights would remain on (with above average intensity) as Stop Sign while no red light would flash. This uncovers the usefulness of the flashes of the red lights also as a reverse complement for the Stop Sign.

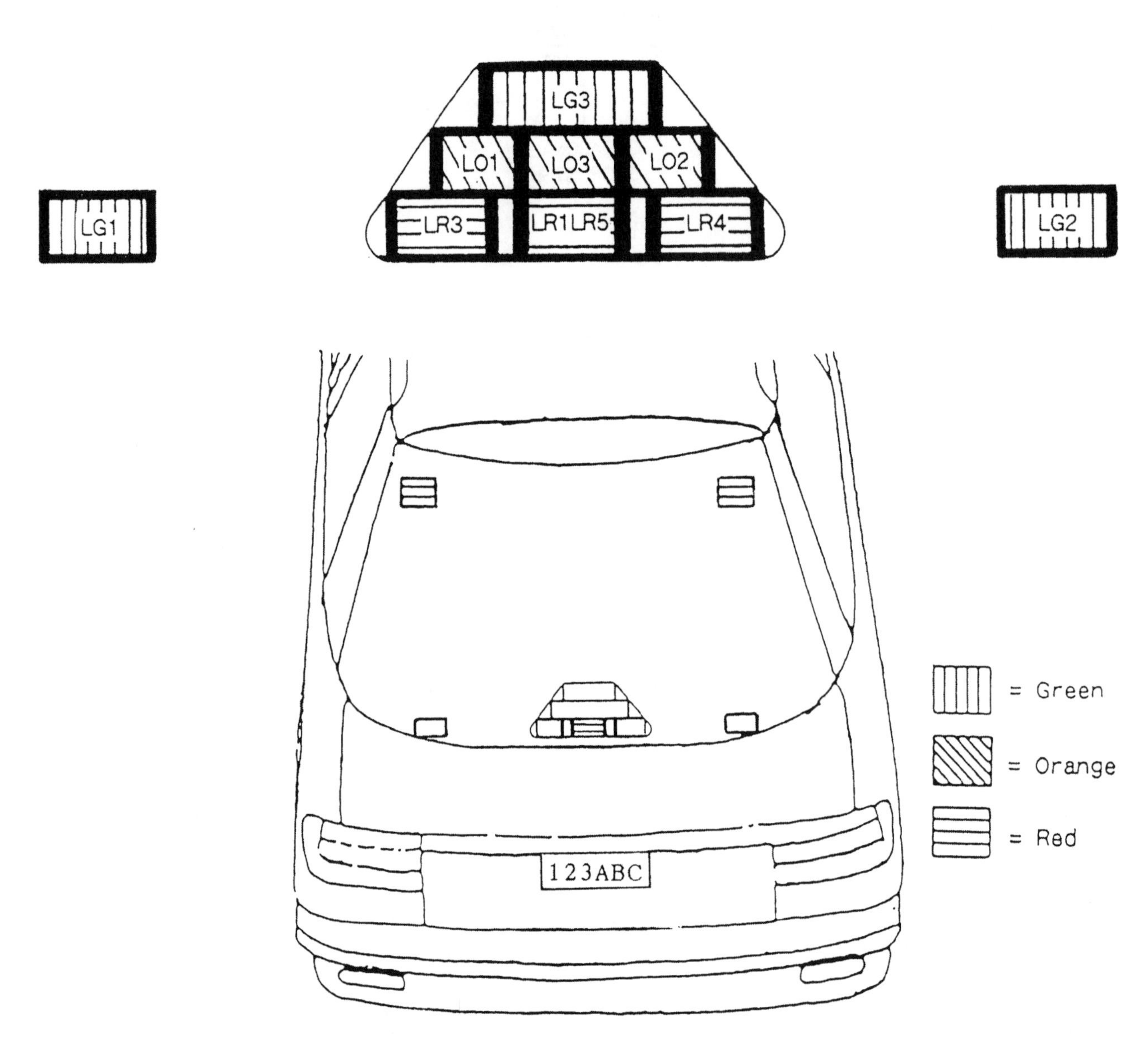

Figure 6. An illustration of possibility of different arrangements of the lights of Dynamic Traffic Light, on the rear window of a car.

MORE ABOUT THE FLASHES OF THE RED LIGHTS –

The greatest difference between the functions of the Dynamic Traffic Light and the flashes used in the studies by NHTSA is that, in those studies the new concept of the Unit of Speed, the *new Advanced Warning Signals* prior to braking, the *Enhanced CHMSL*, the *new Stopping Red Lights*, the *new Advanced Stop Sign System*, and the *new Speed Level Indicator Signs* which are presented here did not exist. Rather, Few frequency levels were chosen to generate crafty flashes of the red lights.

"... Evans and Rothery [1974] investigated the ability to judge the sign of relative motion in a car following situation. ...***A driver could perceive correctly that a lead car was coming closer, but realize too late that the closing speed was much greater than thought***" (4).

The occurrence of flashes of the red lights per each unit decrease in speed would allow a following driver to observe the precise rate of decrease in speed of the lead vehicle, in order to depress the brake pedal accordingly. Every driver already knows how hardly (s)he should depress the brake pedal of his/her own vehicle, in order to obtain the intended reduction in his/her speed.

"Not all Lead Vehicle Moving (LVM) crashes are precipitated by rapid deceleration of the lead vehicle. Many involve slow decelerations (e.g., *typical slowing before a turn*) or simply a speed differential between the lead and following vehicles. ... The provision of an effective warning affords the driver the opportunity to take corrective action" (7). The latter requires the functions of the orange lights.

MORE ABOUT THE FLASHES OF THE ORANGE LIGHTS – "It appears that the following driver does not attempt to, or is unable to, maintain a desired spacing by acceleration or decelerating when the actual spacing becomes larger or smaller than desired. Rather, when the vehicles move apart, the driver accelerates, and when they approach, the driver decelerates" (4). "The driver of a passenger vehicle who is following a truck which enters an upgrade section is not given any overt signal that the truck speed is dropping substantially below that of his own vehicle" (5). Also, a lead driver may reduce the speed of his/her vehicle first by partial lifting of accelerator pedal, and then by applying brake. In the case of vehicles with manual transmission, drivers may reduce their speed first by shifting to lower gears and then by braking. Currently, such reductions in speed are not informed to following drivers by any light.

"A second major casual factor associated with rear-end crashes was *following too closely*. ... an increased delay before perceiving the need for braking, followed by a faster movement of the foot to the pedal and subsequent large deceleration, will increase the risk of being struck in the rear" (7), or will lead a following driver to get involved in other kinds of accidents resulted from losing the control of his/her vehicle (e.g., going out of the road, and possibly colliding with a fixed object).

The effects of functions of Dynamic Traffic Light are beyond reducing reaction time of drivers, because they would actually provide ***more time*** for following drivers to react to the discrepancies between their vehicles and the flow of traffic ahead of them. With the ***repulsing*** flashes of the orange light, a following driver could maintain a desired spacing with a lead vehicle whose speed might be reduced without brake application. As a result, even if reaction time of following driver is increased by fatigue for example, (s)he would have ***more time*** to react to the possible sudden use of brake by lead driver. A vehicle with 100 km/h speed travels 27.7m/s. Hence, one second (or more) or 27.7m distance (or more) can be deterministic in saving or taking lives.

The safety of Motorcyclists could also be greatly increased by *Advanced Warning System* of Dynamic Traffic Light, because Motorcyclists often reduce their speed first by reducing gas and perhaps then by applying brake. This creates great dangers for them since when they reduce speed without braking, no light indicates their deceleration. Thus following drivers react with some delay.

MORE ABOUT THE GREEN LIGHTS – If age of the driver is considered to be an extreme case, then with the two mentioned restrictions for the occurrence of flashes of the green light(s), an old driver would feel more confident to increase speed when (s)he observes the flashes of the green light(s) on a lead vehicle (and possibly on the vehicle(s) leading the lead vehicle); and a very young driver would be discouraged to increase speed when (s)he does not notice the flashes of the green light on leading vehicles. The restricted flashes of the green light(s) would guarantee that the increment in speed of a leading vehicle is meaningful and following driver could increase the speed of his/her vehicle more confidently.

If speed of a lead vehicle is reduced rapidly and changing to another lane is the best reaction for the following driver, then the speed level indicator signs of Dynamic Traffic Lights of leading vehicles on the other lanes would accelerate the following driver's decision making for choosing a more appropriate lane.

"Denton [1976] finds that a subject's selection of a target speed is highly influenced by the subject's previous speed. After simulated driving at about 70 mph for three minutes, subjects underestimated a simulated 30 mph by between 5 to 15 mph; the perception that the speed is lower than actual persisted for at least 4 minutes" (4). *Speed level indicator signs would be specially useful when a driver reaches a slower highway or road after travelling on a fast highway.*

WHAT SHOULD DRIVERS KNOW IN ORDER TO RESPOND PROPERLY TO THE FUNCTIONS OF DYNAMIC TRAFFIC LIGHT? – Since the meaning of the green, orange and red lights of traffic lights are already ***universally known***, drivers should only be informed that:

1. The triangular red sign shown in Fig. 4-j or Fig. 6 represents the Stop Sign. Also when red light is on and does not flash, it implies total stop.
2. The faster orange or red lights of a lead vehicle flash, the hastier the following driver should reduce the speed of his/her vehicle. *Drivers should also be informed that flashes of the orange light of a lead vehicle will arise only when its speed is reduced without brake application, and it does not necessarily imply braking.*
3. When orange light of Dynamic Traffic Light of a lead vehicle remains on, the speed of the lead vehicle is very low (under 32 km/h).

If the extended functions of the green and red lights should be implemented too, then drivers should also be informed that:

1. The faster the green light(s) of a lead vehicle (and possibly the green light(s) of the vehicle leading the lead vehicle) flash(es), the more confidently following driver can increase speed on the lane on which (s)he observes the flashes of the green light(s) of the lead vehicle. Drivers should also be informed that flashes of a light in front of a vehicle indicate the acceleration of that vehicle.
2. Instead of studying the mentioned functions, drivers can gradually ***acquire*** that the three green lights are indeed used to code different levels of speed.
3. Besides having knowledge of the function of the *Advanced Stop Sign System,* drivers can also soon acquire that if a red light of Dynamic Traffic Light of a lead vehicle remains on steadily, while the flashy red light of that vehicle flashes, it denotes that the lead vehicle is stopping.

DYNAMIC TRAFFIC LIGHT VERSUS OTHER RELEVANT SAFETY FEATURES – Collision Warning Radar System (CWRS) – There are at least two disadvantages with CWRS which is going to be installed on future cars; as drivers cannot continuously keep their eyes on the lights of the mentioned radar system *on the dash of their own car,* they should be dependent on the beep sound of the radar system of their car. And since there are usually irritations on the road while driving, the beep sound of the radar system either adds to the mentioned irritations for drivers with good hearing, or it can be missed by drivers with low hearing.

Unlike CWRS which warns drivers when they have already reached the frontier of danger on the lane on which they travel, *a great capability of Dynamic Traffic Light* is that it allows a following driver to simultaneously monitor:

a) the state of motion of leading vehicles on all lanes;
b) the decrease in speed or distance of leading vehicles on all lanes;

from as far as following drivers' eyes can see the meaningful flashes and the speed level indicator signs of Dynamic Traffic Lights of leading vehicles. Hence, following driver would have the opportunity to predict and prevent hazards in car following situations.

A sleepy driver who is awakened by an inter_vehicle sensing device has even less time to realize the actual situation on the road, in order to react correctly. Dynamic Traffic Light, which would help to keep drivers attentive and awake in the first place, would display the actual situation on the road so that the lax driver could react more precisely as soon as (s)he is awakened.

Automatic Braking System, which is under test to be used with the mention radar system on future cars, may add to the frustrations resulted from the surprises that a driver may encounter in driving when a running car is automatically braked, specially on slippery roads.

Advanced Brake Warning System (ABWS) – The ABWS activates the CHMSL or all of the required red lights when accelerator pedal is lifted at a rate greater than 0.3 m/s. "If ABWS activates all of the required red lights then in a tailgating situation, lead driver can easily deceive following driver by a false alarm, when (s)he taps the brake pedal" (3).

Usually in the first fraction of a second after braking, the CHMSL is perceived rather than other required red lights; this is a proven fact since CHMSL has reduced rear-end collisions. If following driver is aware that the ABWS can only activate the CHMSL, then in the first fraction of a second after that lead driver uses the brake, following driver perceives the CHMSL but not the other required intensity coded red lights. *In that first moment,* following driver has two choices:

a) (s)he should not promptly brake until (s)he is sure that the other required red lights are activated too. In that case, the effect of the conventional CHMSL is neutralized and the so called ABWS does not function as a genuine ABWS since following driver does not brake on time. On one hand the ABWS should report that brake is used, and on the other hand it should report that brake is not used but gas pedal is released. Therefore, following driver needs to obtain additional information from other required red lights to be sure whether lead driver only releases the accelerator pedal, or (s)he also uses the brake.
b) (s)he should promptly brake. Hence, when lead driver just taps the brake pedal, following driver is deceived to brake as CHMSL is activated.

Even if all of the above facts about the unclear function of ABWS are ignored, *the fact that braking by following driver is not always the best reaction to the release of accelerator pedal by lead driver (particulary on wet, icy, snowy and gravelly pavements, specially at curves) cannot be disregarded.* It seems logical to presume that since drivers have always reacted to red light by braking, they continue to react by braking to a red light that is turned on by ABWS, even when the activation of ABWS is not resulted from braking. The latter prevents traffic to advance and on slippery roads, the unnecessary braking itself can add to the risk of getting involved in an accident which falls in a category other than those accidents that are prevented by ABWS. Red light should be exclusively used to denote braking, and of course to implement the *Advanced Stop Light.*

Anti-lock Braking System (ABS) – The questionable ABS is installed on many motor vehicles with the intent to stop a running vehicle on time to prevent traffic crashes. However, besides other deficiencies of this relatively expensive braking system (such as stopping the vehicle later than conventional brakes on snow or gravel), *if a driver reacts too late to a stimulus (ie: to a lead vehicle) and does not brake on time*, no computer can support the brake to stop the vehicle as if brake was applied significantly sooner. Even sophisticated new cars can get involved in sever accidents and, therefore, new cars are equipped with air bags which have disadvantages too.

Two separate recent studies, conducted for Transport Canada and for General Motors, have shown that ***people would drive more aggressively if they know that they have a safety feature in their car***. In an analysis of more than one million cars, the American Highway Loss Data Institute found no pattern of decline in insurance claims because of anti-lock brakes. Hence, the problem of losing the control of vehicle should be hunted in the information processing abilities, ***amount of time that a driver is given to react***, and in the reaction time of drivers, rather than in the conventional brakes.

THE FUNCTIONS OF DYNAMIC TRAFFIC LIGHT ARE NOT OBSCURE OR VEXATIOUS – Said functions (Fig. 4-a to 4-j) may seem complicated when they are considered all together. But the fact is that maximum three lights of the same colour can be turned on at the same time, and each function performed by the lights of the same colour is well defined and its meaning is self evident.

Whether the functions of Dynamic Traffic Light are clear or not should be judged based on each of the individual functions performed by lights of a particular colour. For example, considering that ***the meaning of the three colours of traffic lights are universally known***, what else could a Centre High Mounted Green Light whose colour changes to orange mean except warning the following driver(s) to be more attentive to the variations in the flow of traffic ahead?

If with the interference of the ABWS the red light is used for two purposes (ie: to denote braking, and to report that brake is not used but accelerator pedal is lifted), then the message that red lights should deliver is ambiguous. But when few lights of different colours are used to implement few well defined and unique functions, not only there would be no vagueness about the meaning of the function of each of those lights, but also said lights would provide adequate and clear data for following drivers. Thus, the possible concurrent flashes of the traffic light on many lead vehicles would not be annoying for following drivers who would instead enjoy driving with the ***Highway Information*** and confidence.

Not only observation of the possible flashes of Dynamic Traffic Light of all leading vehicles would not be confusing for a following driver, but also (s)he would predict the changes in the motion of his/her leading vehicle when (s)he observes the flashes of the orange lights of Dynamic Traffic Light of the vehicles ahead of his/her leading vehicle. If a lead driver does not reduce speed subsequent to the occurrence of flashes of the orange lights on the vehicle(s) ahead of him/her (e.g., because of influence of alcohol), then following driver, who would be most possibly attracted by flashes of the orange light of Dynamic Traffic Light of the vehicle(s) leading the lead vehicle, would have the opportunity to understand the mishandling of the situation by his/her lead driver. That would naturally increase the expectancy of the following driver that brake might soon be used by the diver leading the lead driver.

Tension and aggression in driving may be usually resulted from the frustrations related to consecutive unexpected situations that a driver encounters during driving. If the unique ***Highway Information*** that could be generated by the Dynamic Traffic Light already existed on the active roads, then following drivers would not experience most of the unexpected situations that they have been suffering from them. The Dynamic Traffic Light which provides more time for following drivers to react, would reduce the utilization of brake for emergency situations, *and consequently, the average life time of brakes would increase too.*

"The Tri-Level statistics portray rear-end crashes as resulting largely from driver inattention and other forms of delayed recognition (i.e., ***conscious*** driver does not properly perceive, comprehend, and/or react to vehicle in his or her forward travel path). *There is little involvement of vehicle factors, indirect human causes (e.g., alcohol)*, or environmental factors. This pattern is true for both LVS and LVM crash subtypes -- specially the LVS crashes. ... (ie: the driver 'looked but did not see' the crash threat)" (7).

THEORY OF OPERATION

Fig. 7 illustrates a complete block diagram of this invention. Dynamic Traffic Light operates by monitoring the speed of the rotor shaft in transmission. Rotational speed is monitored by keeping track of fluctuations of frequency of pulses received from speed sensor during predetermined equal time intervals.

After that the sensor's pulses are amplified, they are transmitted to the receiver (pulse detector) in Dynamic Traffic Light in the rear of vehicle. A 10 Hz timebase is used to generate equal time intervals of 0.1s. Every timebase pulse generates three sequential pulses: a decision pulse, followed by an update pulse, followed by a clear (or reset) pulse.

A first counting section counts the pulses detected by pulse detector during each of the equal time intervals. In any vehicle, at the end of any time interval, the number that is present in the first counter should be the ***current speed*** of vehicle per km/h. Therefore, the frequency of pulses generated by the speed sensor of that vehicle should be (vehicle speed * 10)* Hz;

As used in this description, the word "*memory*" means a mean for storing a number, and the term "***current speed***" means (the number of pulses generated by speed sensor

* More about this in the appendix.

and counted by the first counter during the latest time interval) km/h. Also as used in this description, the term "***reference speed***" means the number of pulses that was counted by the first counter during a previous time interval, and was stored in the memory at the end of that time interval if:

i) at the end of said time interval, Dynamic Traffic Light had generated a pulse to indicate that the *current speed* was increased or decreased at least by one *Unit of Speed*, relevant to the value which was stored in *the memory*; or

ii) at the end of said time interval, the *current speed* had increased at least by 1 km/h.

Referring to Fig. 7, output of the first counter is fed to a first magnitude comparator, to the *memory* and to a subtracter. Output of said *memory* is fed to said first comparator, to said subtracter and to a speed level determinator. The output of said speed level determinator section, which is at VDD level, is amplified to VCC level by a voltage amplifier section whose outputs drive transistors and trigger a number of timers, in order to coordinate the operations of a number of relays. The latter powers a specific set of lamps, in order to produce the appropriate flashing of a particular light or, to implement the signs which indicate different levels of speed, Fig. 4.

Output of the *memory* can also be digitally displayed so that the required speedometer be incorporated in the Dynamic Traffic Light as a digital meter.

Said first magnitude comparator realizes whether speed is increased or decreased (at least by 1 km/h) by comparing the *current speed* with the *reference speed*. A second magnitude comparator realizes whether *current speed* is altered at least by one *Unit of Speed* or not, relevant to the *reference speed*.

At the end of each time interval, the decision maker gates section is fed by the two magnitude comparators, by a first timer, and by the decision and update pulses. This enables the gates section to generate appropriate pulses at the end of a time interval.

FOR CREATION OF THE FLASHES OF THE ORANGE OR RED LIGHTS – Said decision maker gates instruct the ALUs (or subtracter) to subtract the current value of said first counter (ie: the *current speed*) from the value that is present in the memory (ie: *reference speed*) or vice versa, whichever subtraction that generates a positive number as result, which can be zero.

If at the end of a time interval said first comparator realizes that the *current speed* is ***decreased*** at least by 1 km/h, and if simultaneously said second comparator is realizing that the result of subtraction performed by said subtracter is equal or greater than one *Unit of Speed*, then the decision maker gates section generates a pulse so named **pulse A**, in order to indicate that speed is decreased at least by one *Unit*. **pulse A** is then amplified to trigger timers which drive their corresponding relays for flashing orange or red lights. The latter flashes if while brake is applied **pulse A** is generated. Otherwise, the occurrence of **pulse A** results in flashes of the orange lights.

After that **pulse A** is generated, the decision maker gates section decides to update said *memory* to the *current speed*. When brake is not used, the second set of the inputs of said second magnitude comparator is set to binary number 3 to represent one *Unit of Speed*. When brake is applied, *Unit of Speed* is reduced to 2km/h.

FOR CREATION OF THE FLASHES OF THE GREEN LIGHTS – The same method that is used to generate the **pulse A** can be used to generate a pulse when speed is increased by one *unit*. However, a second counter can be used to count the number of time intervals by the end of which speed is *increased* at least by 1 km/h. This allows to impose a restriction for the occurrence of flashes of the green light(s). Referring to Fig. 7, if at the end of a time interval said first comparator realizes that the *current speed* is ***increased***, then the decision maker gates section concurrently does the following:

i) It advances a second counter by one, in order to count the number of time intervals by the end of which speed is ***increased*** at least by 1 km/h. When count reaches one *Unit of Speed*, the QB output of said second counter (so named **pulse B**) is amplified to trigger timers which drive their corresponding relays to flash at least one green light. Concurrently, **pulse B** resets said second counter to zero through a diode, so that said second counter resumes counting the mentioned time intervals and flash the green light(s) if count reaches one unit again. There is, however, a restriction for the second counter to be advanced: If the magnitude of acceleration is less than ~0.55 m/s2 (ie: if every ~0.5s speed is not increased at least by 1km/h), then before that said second counter counts up to one unit, said decision maker gates reset it to zero. Hence, when acceleration is not significant, green lights do not flash even though speed is increased by one *unit*.

ii) It triggers said first timer to delay (for ~0.5 s) resetting the second counter to zero. When magnitude of acceleration is equal or greater than ~0.55 m/s2, said first timer is triggered (or reset) per every 1 km/h increase in speed so that its output remains activated for a limited time without interruption. This delays resetting said second counter at the end of a number of consecutive time intervals, for as long as output of said first timer lasts. Thus the second counter counts up to one *Unit of Speed* to produce the **pulse B** if output of the first timer lasts long enough.

iii) It updates the value stored in memory to the current value of said first counter (ie: to current speed) to assure that a flash of a light occurs when the current speed is altered at least by one unit.

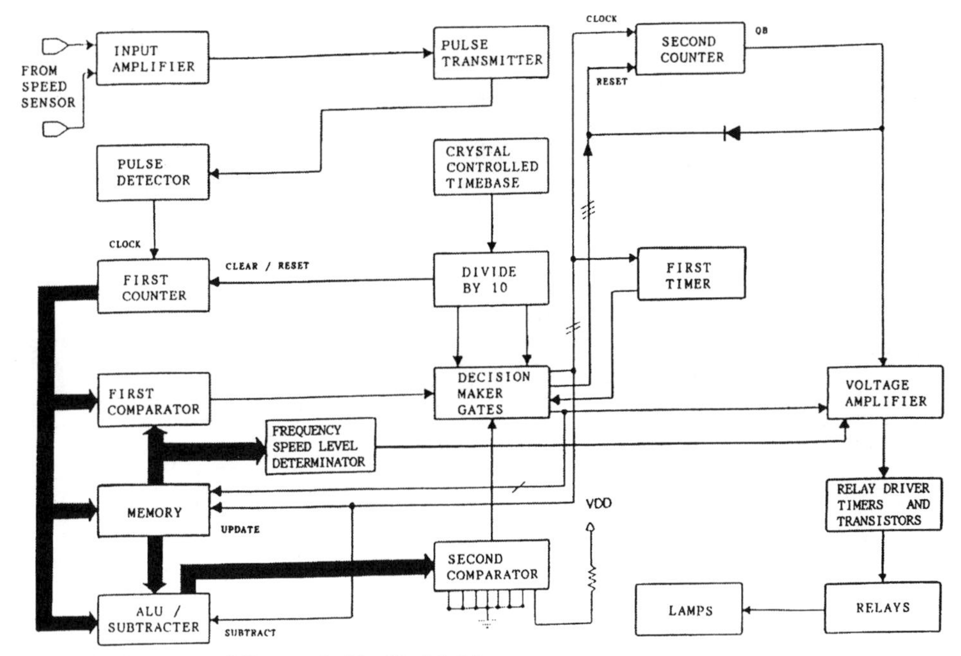

Figure 7. A complete block diagram of Dynamic Traffic Light

Whether **pulse A** or **pulse B** is generated or not, at the end of each time interval the clear pulse resets said first counter to zero, in order to restart counting the next sample period.

ALTERNATIVE METHOD – Instead of using said subtracter and said second comparator to generate the **pulse A**, it is possible to use the same method that is used to generate the **pulse B** to produce the **pulse A**; meaning that similar to the second counter, a third counter can be used to count the number of time intervals by the end of which speed is *reduced* at least by 1 km/h.

However it is more economical to generate the **pulse A** by using a third counter rather than by using said subtracter and said second comparator, the sensors should produce twice as many pulses to be counted during shorter timebase of 0.05s (or 20 Hz). Otherwise, Whether magnitude of deceleration is ~2.5 m/s^2 or more, with 2 km/h as *Unit of Speed*, maximum 5 flashes of the red light is generated per second. As a result, besides the fact that orange or red lights may flash when speed is actually reduced by more than one *Unit of Speed*, maximum five flashes of red light per second is too little to reflect the maximum rate of decrease in speed of a vehicle when brake is applied. In contrast, when said subtracter and said second magnitude comparators are used with the convenient 10 Hz timebase, red lights can be flashed maximum 9 to 10 times per second, in order to precisely reflect the maximum rate of decrease in speed of a vehicle.

MORE ABOUT THE PULSE TRANSMISSION – One trivial but inconvenient method for transmission of the speed sensor pulses is to extend a wire from the speed sensor (which is usually installed in front of car) to the pulse amplifier in the rear of car. Nevertheless, I have discovered that the two most commonly used micro-chips, namely NE555 timer and LM339 operational amplifier, can together function as a remarkable transmitter and receiver with astounding bandwidth of just under 20 mega-hertz.

While these transmitter and receiver are immune to statics, pulses could be transmitted ***one by one*** from the amplifier/transmitter to the Dynamic Traffic Light, through the presence light wire. The latter would also power both the transmitter which would be installed in front of vehicle, and the Dynamic Traffic Light which would be installed in the rear of vehicle.

CIRCUIT DESCRIPTION

BRIEF DESCRIPTION OF THE RELATED DRAWINGS – In drawings which illustrate the circuitries needed to implement the Dynamic Traffic Light with its transmitter and receiver, Fig. 16 illustrates the wiring of lamps contained in the lamps casing that is portrayed in Fig. 1, Fig. 5 or Fig. 6.

The optional speed sensor's pulse amplifier/transmitter, Fig. 13, is contained in the small casing illustrated in Fig. 3. Its power supply, which generates VCC1, obtains its **P1** input from presence light fuse or from the wire from which electricity flows through said fuse.

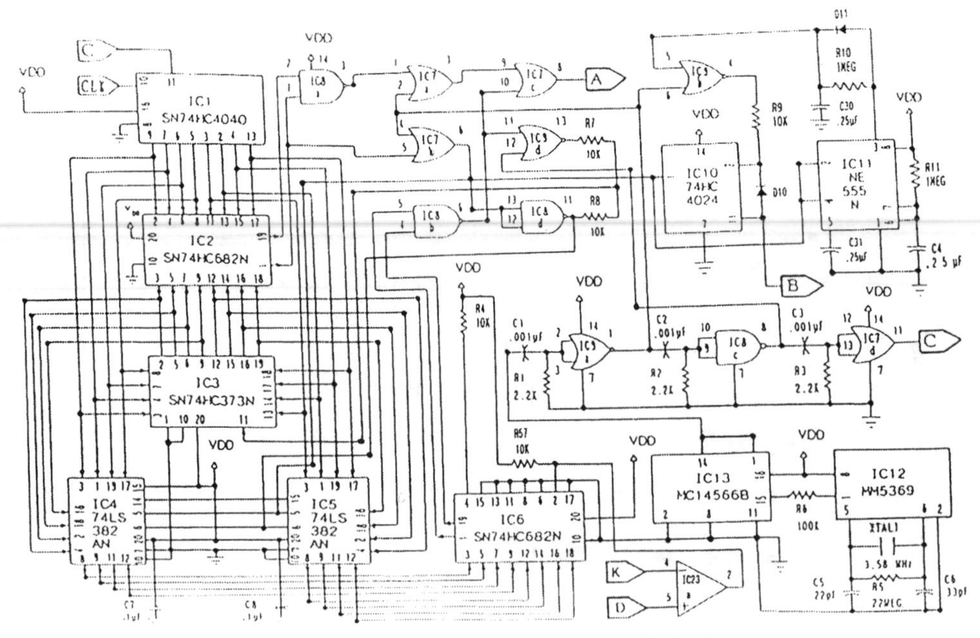

Figure 8. The heart of Dynamic Traffic Light. It generates the Pulse A and the Pulse B.

The casing illustrated in Fig. 2 (which should be preferably installed in the rear of vehicle) contains every other circuitry (on the main board) as follows:

The circuitry illustrated in Fig. 8 which is the heart of this invention. It produces **pulse A** or **pulse B** to indicate one unit decrease or increase in frequency (or speed) respectively. **pulse A** and **pulse B** are amplified by voltage comparator/amplifiers in Fig. 9 to trigger timers in Fig. 10 which drive relays in Fig. 11. The P4 and P5 inputs in Figure 11 are directed to affect the operation of the circuitries in Fig. 11 and in Fig.8.

Fig. 9 also illustrates the speed level determinator having a set of eight diodes, few voltage comparators and a logic gate at the output of IC3.

Fig. 15 illustrates the power supply that obtains its **P3** input from presence light wire in the rear of vehicle, and generates VCC2, VCC3, VCC4 and VDD for the different sections of Dynamic Traffic Light.

Fig. 14-a illustrates the pulse detector (receiver) that is associated with the pulse amplifier/transmitter depicted in Fig. 13. If instead of said transmitter and pulse detector a twisted pair wire should be used to direct the speed sensor pulses to a pulse amplifier in Dynamic Traffic Light in the rear of vehicle, then the circuitry illustrated in Fig. 14-b should be used to amplify the speed sensor pulses for advancing the first counter (IC1) in Fig. 8. In that case circuitries illustrated in Figure 13 (ie: the casing in Fig. 3), in Fig. 14-a and the ~5 m wire in Fig. 15 should be omitted.

The input and output pins of all of the unused op-amps of LM339 should be grounded.

BASIC OPERATIONS – Referring to Fig. 8, IC1 a SN74HC4040 12 stage binary counter is advanced by output pin 2 of IC24-a (or by output pin 13 of IC23-d) to count the incoming pulses from sensor during time intervals of 0.1s. Only the first 8 least significant bits of this counter are needed, in order to monitor up to 255 km/h. The 60Hz signal is generated by IC12, an MM5369, 17 stage programmable oscillator/divider and its support components. The divider (IC13) consists of one 14556 which contains three independent function block. Here, IC13 is used only as a divide by 10 counter to generate 10 Hz timebase. The output of divider is fed to three pulse generators comprising: IC9-a, C1 and R1; IC8-c, C2 and R2; and IC7-d, C3 and R3; in order to generate the three sequential pulses; namely the decision, the update and the clear pulses.

Said first counter (IC1) feeds the first 8 bit magnitude comparator (IC2), an 8 bit transparent latch or the memory (IC3), and two 4 bit arithmetic logic units (IC4 and IC5) which are cascaded to form an 8 bits ALU which is used as subtracter. Whether the binary number represented by said first counter should be reduced from the value latched by IC3 or vice versa is determined by select inputs pin 5, 6 and 7 of said subtracter IC4-IC5.

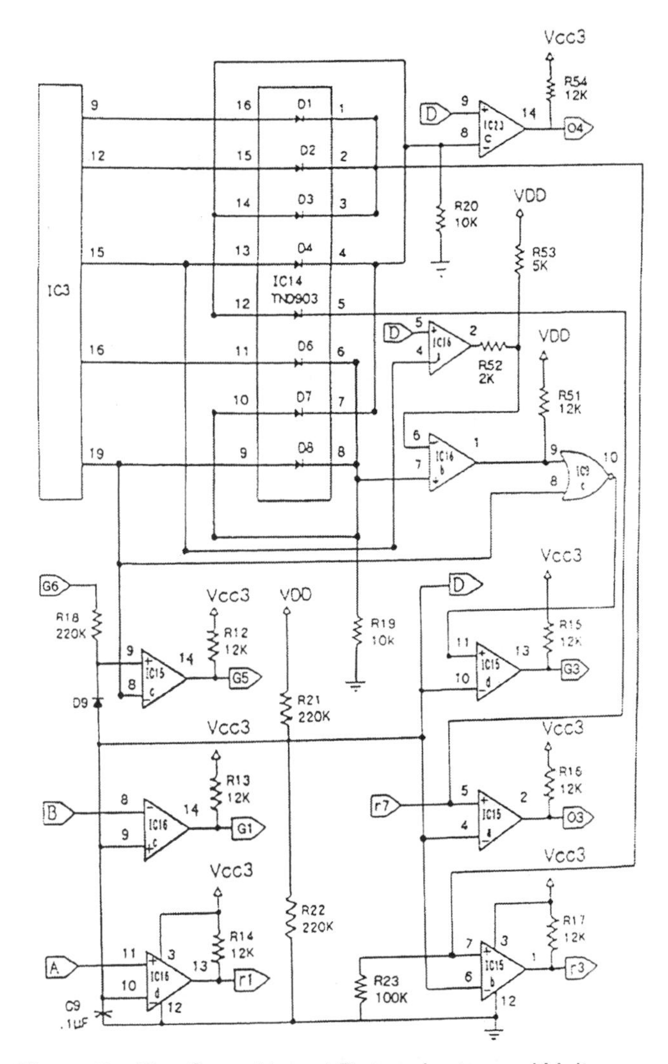

Figure 9. The Speed Level Determinator and Voltage Amplifier sections are shown here.

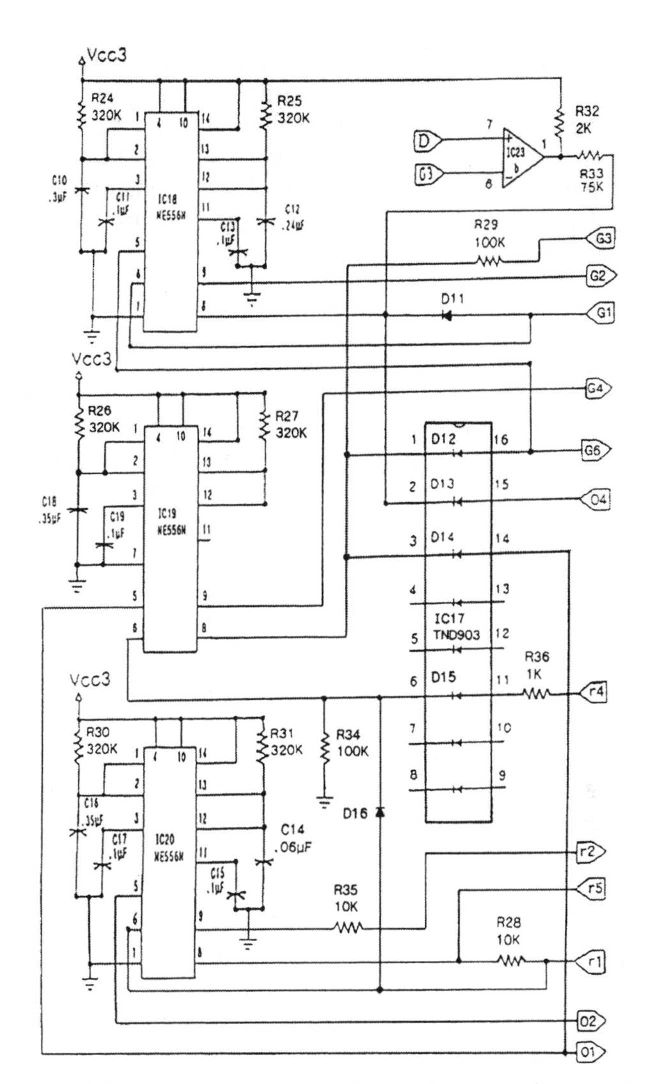

Figure 10. The Relay Driver Timers consist of a set of three NE556 timers.

In order to perform subtraction, pin 7 of ALUs should be grounded and their pin 5 and pin 6 should be in opposite state. At the end of a time interval, the decision pulse, output pin 1 of IC9-a, pulls low input pin 2 and pin 4 of the two OR gates IC7-a and IC7-b respectively. If at this moment, the value represented by said first counter (IC1) is greater than the value latched at the output of the memory (IC3), then pin 1 of IC2 goes low and output pin 6 of IC7-b pulls low input pin 5 of IC4 and IC5. Simultaneously, IC8-d inverts output of IC7-b to feed input pin 6 of IC3 and IC4. Consequently, output of IC3 (the latched value of said first counter) is subtracted from the current value of said first counter. Otherwise, if pin 1 of IC2 is high, the current value of said first counter is subtracted from its previously latched value.

The result of subtraction is fed to the first set of the inputs of IC6, the second 8 bits magnitude comparator. The second set of the inputs of said second magnitude comparator is set to binary number 3 to represent one unit of speed.

Only when the current speed is less than the reference speed, both outputs of IC2 (pin 1 and 19) go high to pull low output pin 3 of IC8-a. When brake is applied output k (from Fig. 11) is fed to pin 4 of IC23-a (Fig. 8), in order to pull low input pin 2 of IC6. This changes the unit of speed from 3 km/h to 2 km/h when brake is applied.

In Fig. 9, a 2.5 volts reference voltage is generated by dropping the regulated supply voltage (VDD) across two identical resistors R21 and R22, filtering it and feeding it to nine op-amps configured as voltage comparators/amplifiers.

CREATION OF THE TWO PULSES '**A**' AND '**B**' – The decision maker gats comprise the following: IC7-a, IC7-c, IC9-d, IC8-a and IC8-b; IC7-b and IC8-d; and IC9-b. At the end of a time interval, if the results of subtraction is greater than or equal to one unit of speed, then both outputs pin 1 and pin 19 of IC6 go high to indicate that speed (or the frequency of pulses received from the sensor) is altered at least by one unit. Consequently, input at pin 10 of IC7-c and at pin 11 of IC9-d are pulled low. If pin 1 of IC7-a is already pulled low, then **pulse A** is generated at output pin 8 of IC7-c. Otherwise, if pin 5 of IC7-b is already pulled low then not only the low at output pin 6 of IC7-b together with output pin 11 of IC8-d) instruct the ALUs for subtraction, but also concurrently, the low at output pin 6 of IC7-b does the following:

i) It triggers IC11, a 555 timer, by pulling low pin 2 of this IC. IC9-b resets IC10, a 7 stage binary counter, by applying the inverted version of the decision pulse to pin 2 of IC10. The high output of the 555 timer (IC11) feeds pin 5 of IC9-b to prevent this IC from repeatedly resetting IC10. R10 and C30 prevent the occurrence of a low pulse at output pin 3 of IC11, if this IC is reset while it is already triggered to produce a high pulse.

ii) It advances IC10 to count the number of time intervals at the end of which the number of pulses counted by IC1 is increased at least by 1, relevant to the value that is latched by IC3. The reset pin 2 of the second counter IC10 remains low, and it can count up while the output of the 555 timer (IC11) remains high. If the output of this timer remains activated long enough, then said second counter (IC10) succeeds to count at least up to one unit of speed. Then the QB output at pin 11 of IC10 is taken as **pulse B**. The latter feeds pin 2 of IC10 through diode D10 so that, the second counter resets itself to zero to restart counting the mentioned time intervals.

iii) Since the low at output pin 6 of IC7-b occurs when at the end of a time interval the number of pulses counted by IC1 is increased at least by one, this low is inverted by IC8-d to update the memory to the current value of counter.

If at the end of a time interval **pulse A** is generated, the update pulse output pin 8 of IC8-c pulls low input pin 12 of IC9-d. This creates a high pulse at pin 13 of this IC to update the latch (IC3) to the current value of the first counter.

Instead of using the second counter, the same method that is used to generate **pulse A** can be used to generate **pulse B**; an OR gate can be fed by pin 6 of IC7-b and by pin 6 of IC8-b, in order to generate **pulse B**.

Referring to Fig. 11 and Fig. 16, as long as relays are not activated green lights LG1, LG2 and LG3 receive VCC4 from output 6 of relay 3, and output 9 of relay 2 provides ground for LG1 and LG2 to keep them on, as in Fig. 4-a.

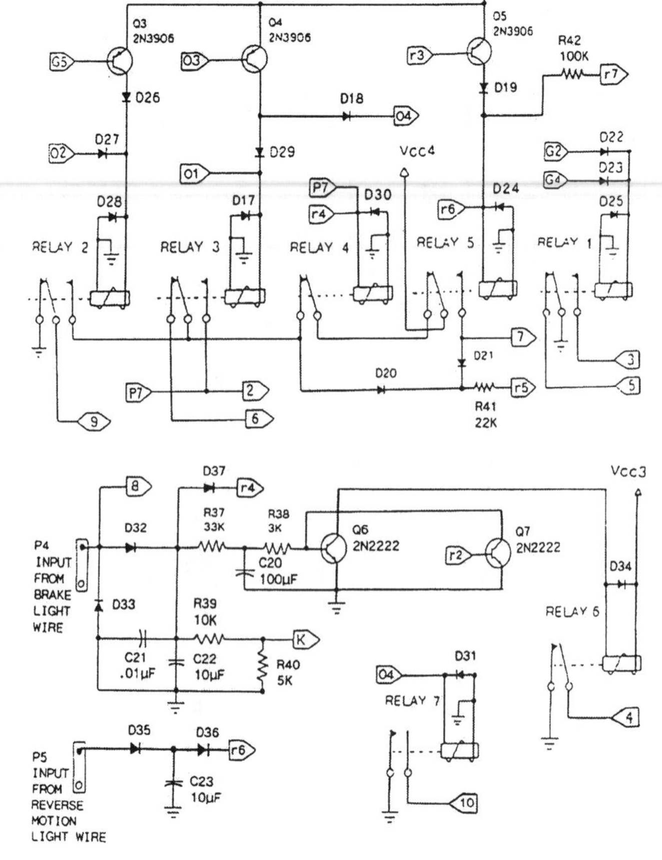

Figure 11. The Relay Driver Transistors and a set of seven relays. Transistors Q6 and Q7, and relay 7 contribute in implementation of the Enhanced CHMSL.

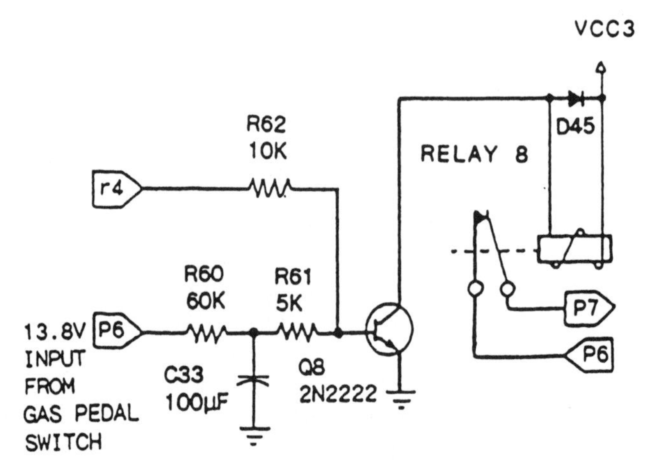

Figure 12. The circuitry to generate the Accelerator Total Release Signal.

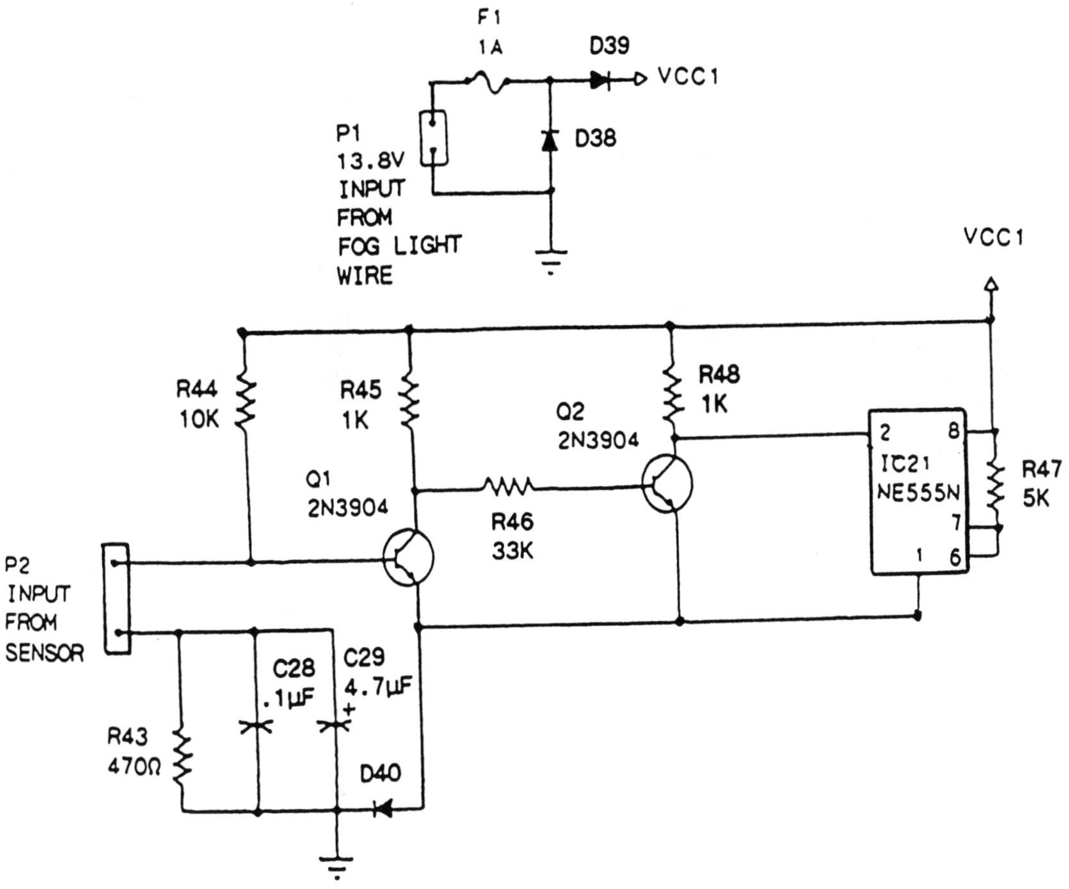

Figure 13. The Imput Amplifier section and the optional transmitter are shown here. The transmitter consists of a NE555 timer (IC21), and a resistor (R47).

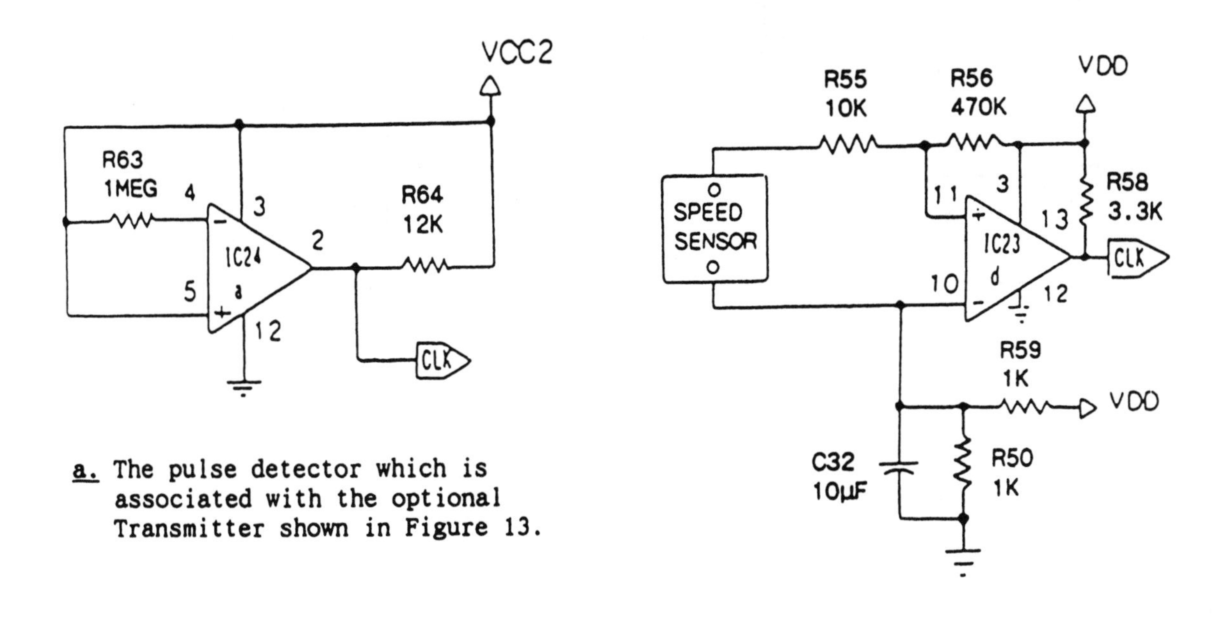

a. The pulse detector which is associated with the optional Transmitter shown in Figure 13.

b. The pulse amplifier which can be replaced for the circuitry depicted in Figures 13 and 15.

Figure 14. The circuitry which receives the speed sensor's pulses.

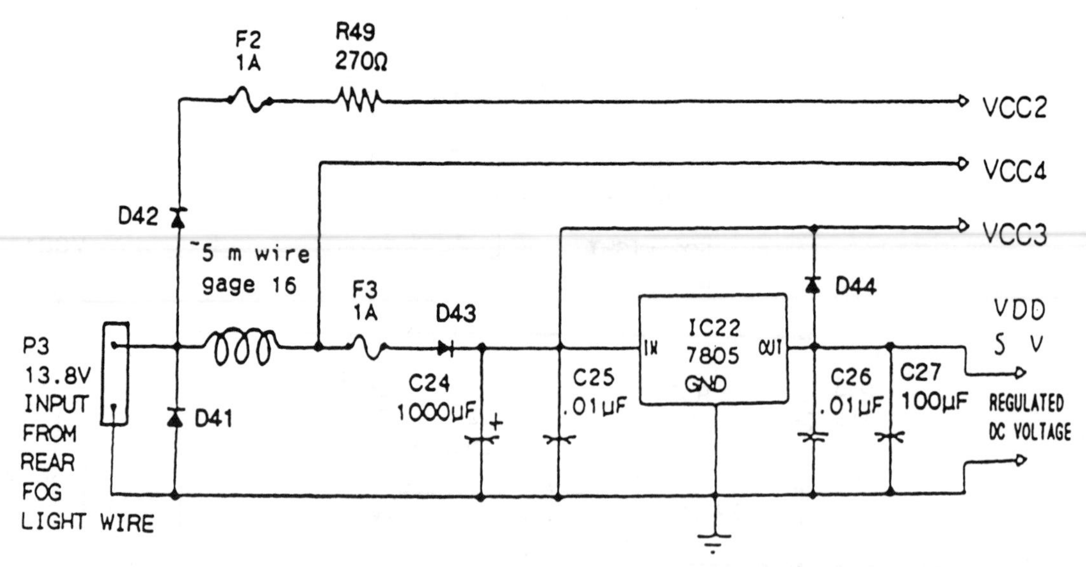

Figure 15. The power supply which generates: VDD to mainly power the circuitry in Figure 8; VCC2 to power the pulse detector; VCC3 to power the Relay Driver Timers/Transistors and the Voltage Amplifier Section; and VCC4 to power the Lamps section through the Relays section.

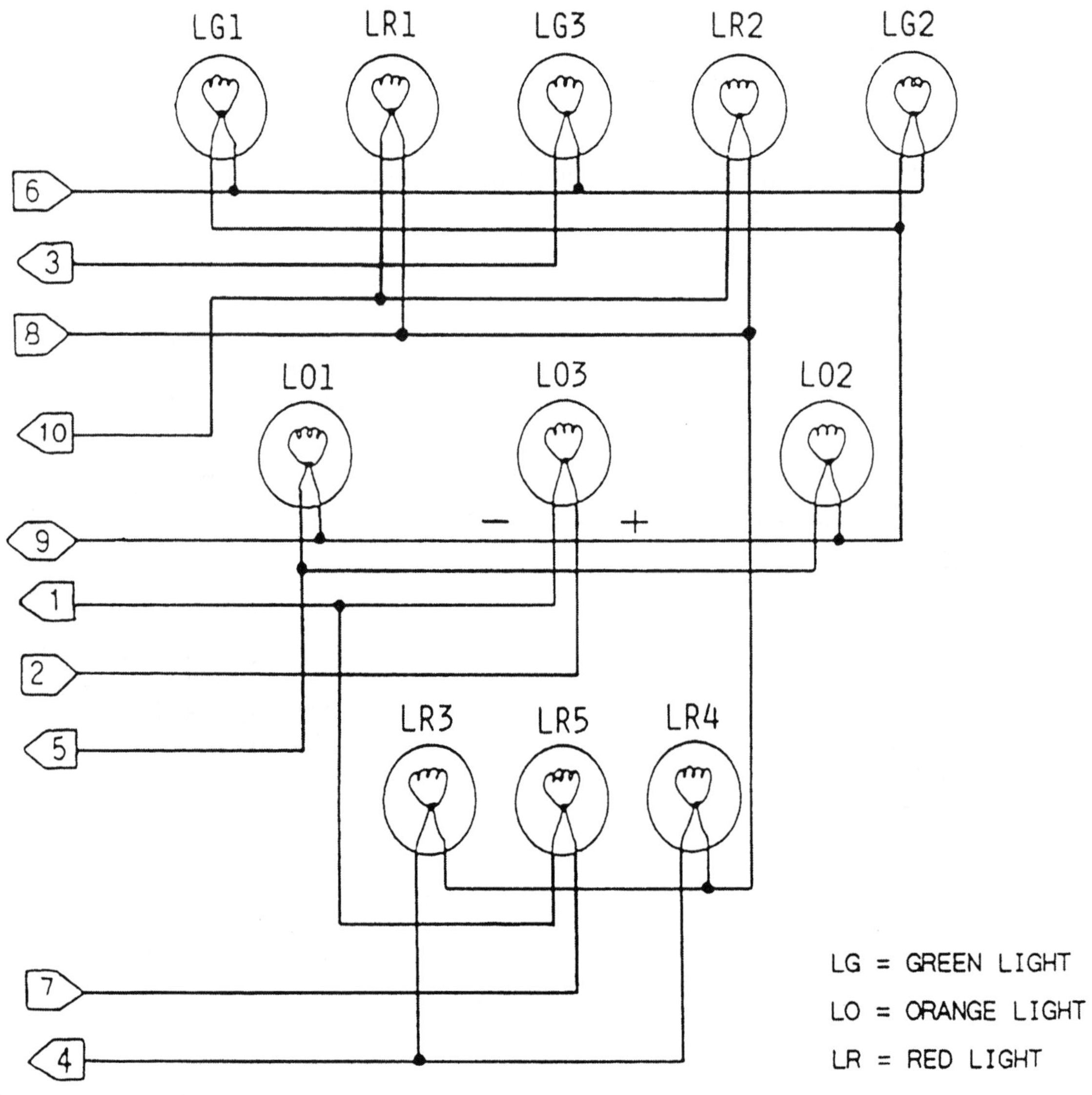

Figure 16. The connection of eleven light bulbs which should be contained in the lamps casing illustrated in Figure 17.

Referring to Fig. 9 and Fig. 10, when speed is increased and **pulse B** is generated, output (G1) pin 14 of IC16-c is pulled low to pull low inputs pin 6 and 8 of timer IC18 in order to generate G2 and G6 at its output pins 9 and 5 respectively. G2 energizes relay 1 during a time determined by R25 and C12, for providing ground 3 for green light LG3 to flash on this light. IC16-a and IC16-b form an AND gate which together with diodes D6 and D8, and with IC9-c recognize binary number that is equal to or greater than 96; when speed is greater than 95 km/h, input at pin 11 of comparator IC15-d goes low to pull low output (G3) at pin 13 of this IC. G3 pulls low input pin 2 of IC19 to create G4 (inverted version of G3) to close relay 1 and to keep LG3 on as illustrated in Fig. 4-b.

The G3 (output pin 13 of IC15-d), O1 (output pin 5 of IC19) and G6 (output pin 5 of IC18) are fed through resistor R29, diodes D14 and D12 respectively, to input pin 8 of IC19, in order to pull low output (G4) pin 9 of IC19. This results in release of relay 1 for flashing off lamp LG3.

If the speed that is greater than 95 km/h is increased and **pulse B** is generated, G6 pulls low G4 to release relay 1 and flash off LG3 (as in Fig. 4-b and Fig. 4-a interchangeably). Since G2 should not close relay 1 when G4 releases this relay to flash off LG3, G3 disables G2 by creating a voltage at the output of IC23-b (Fig. 10) which feeds pin 8 of timer IC18.

If speed is greater than 127 km/h, the high voltage at pin 19 of IC3 (Fig. 9) pulls low G5 output pin 14 of IC15-c which drives transistor Q3 to energize relay 2 (Fig. 11). Consequently, output 9 of relay 2 (ground) is disconnected from lamps LG1 and LG2 to turn them off, and only LG3 is kept on by relay 1, Fig. 4-c. If this high speed is increased and **pulse B** is generated, output pin 5 of IC18 (G6) pulls high output (G5) pin 14 of comparator IC15-c during a time determined by R24 and C10. Thus relay 2 is shortly released by transistor Q3, while output at pin 3 of IC19 (G4) is pulled low by G6 to releases relay 1. As output 9 of relay 2 provides ground for LG1 and LG2 to flash on these lights, output 3 of relay 1 is disconnected from LG3 to flash off this light (Fig. 4-a and Fig. 4-c, interchangeably).

IMPLEMENTATION OF THE FUNCTIONS OF THE ORANGE LIGHTS – The latter perform three functions:

Flashes of the Orange Lights – When speed is decreased and **pulse A** is generated, output pin 13 of IC16-d (r1) is pulled low to trigger the two timers of IC20. The latter generate the two pulses O2 and r2 at its pin 5 and pin 9 respectively. Pulse r1 also triggers a timer of IC19 to generate O1 at its pin 5. When speed is reduced and brake is not applied, the following happen:

a) O1 closes relay 3, for disconnecting output 6 of this relay from all of the green lights and provide output 2 of this relay for the orange light LO3;
b) O1 feeds pin 8 of IC19 through diode D14 in order to pull low G4 and to release relay 1 if it was energized by G4. As a result, ground 3 is disconnected from green light LG3, and ground 5 is provided for orange lights LO1 and LO2; and
c) O2 activates relay 2 to disconnect ground 9 from green lights LG1 and LG2, for providing output 9 of this relay (ie: VCC4) for LO1 and LO2.

Hence, when speed is reduced without brake application, every time **pulse A** is generated all green lights flash off while all orange lights which are off flash on, Fig. 4-d.

Low Speed Indicator Sign – When speed is reduced to under 32 km/h, the following happen:

a) Pin 5 of IC14 goes low, to pull low (O3) output pin 2 of IC15-a. O3 drives transistor Q4 which activates relay 3 to disconnect output 6 of this relay (VCC4) and to turn off all green lights. Output 2 of relay 3 provides VCC4 for orange light LO3 which is grounded by 1.
b) Voltage at output pin 4 and pin 7 of IC14 is inverted/amplified by IC23-c (Fig. 9) to create O4 (inverted version of O3) which, through diode D13, feeds input pin 8 of the timer IC20. Thus O4 does not allow pulse G2 to be produced, in order to prevent green light LG3 from flashing at low speed.

Accelerator Total Release Signal – A normally closed spst switch should be connected to the accelerator pedal, with the same method that a switch is installed with brake pedal for conduction of electricity to the brake light when brake is applied. As soon as accelerator pedal is totally released, the switch provides 13.8 V voltage as **P6** input for the circuit illustrated in Fig. 12. This voltage provides (with no delay) output **P7** of relay 8 to feed the orange light LO3 which is grounded. Resistor R60 and capacitor C33 create a time delay of approximately 1s before that the **P6** drives transistor Q8 for activation of relay 8 and disconnecting voltage **P7** from orange light LO3. The result is that as soon as gas pedal is released, the orange light LO3 flashes once with no delay (except the rising time of the bulb), as in Fig. 4-f. When gas pedal is released and voltage **P6** is disconnected, relay 8 is released with a short delay, after that C33 is discharged. As a result, if gas pedal is repeatedly depressed and totally released (during short time intervals) LO3 flashes only once with the first release of accelerator pedal. If while LO3 is being flashed on by **P7** brake is applied, r4 drives Q8 to close relay 8 sooner and interrupt such flash of orange light so that only red lights flash.

IMPLEMENTATION OF THE FUNCTIONS OF THE RED LIGHTS – The latter performs basically three functions:

Flashes of the Red Lights – When brake is applied and **pulse A** is generated, the following happen:

i) Just before creation of pulse A, relay 4 is activated by r4 (Fig. 11), and VCC4 is disconnected from input of relay 2 and relay 3 which power orange and green lamps. Instead, output 8 provides voltage for red lights (LR1, LR2) and (LR3, LR4). Since LR3 and LR4 are already grounded by output 1 of relay 6, they glow with no delay.

ii) As circuit in Fig. 12, resistor R37 and capacitor C20 (Fig. 11) create a delay before that the P4 voltage (from brake light wire) drives transistor Q6 for activating relay 6 and disconnecting ground 1 from LR3 and LR4. Then, as soon as pulse A is generated, the pulse r2 drives transistor Q7 for pulling low base of Q6. Consequently, relay 6 is released for as long as pulse r2 lasts (ie: for ~0.1 s), and red lights LR3 and LR4 are flashed on.

 The advantages of flashing the red lights with this method (rather than by activating relay 6 directly by the pulse r2) is that The function of the required CHMSL is culminated in the first moment by the LR3/LR4 (as they are turned on with no delay). Also the elements of the light bulbs for LR3 and LR4 are warmed up with no delay (just before that pulse A is generated) during the first flashing, to be turned on more brightly during their second and subsequent flashes by r2. Moreover, it may take less time to flash the red lights by repeatedly releasing a relay that is already activated, rather than by energizing the relay. Since r2 can be generated up to ten times per second, it should not last more than ~0.09 s. Because this time is not enough to turn on a 12 volts lamp brightly, one way to turn on a bulb brightly with car electricity during ~0.1 s is to choose ~10 v fast rising bulb for LR3 and LR4. As the magnitude of deceleration increases, the intensity of red lights LR3 and LR4 increase.

iii) Voltage r4 feeds pin 6 of IC19 to prevent this timer from being triggered by r1. This alleviate the fast production of the pulse r2 as r1 triggers two timers instead of three timers.

Operation of the Stopping Red Lights – If speed is reduced to less than 32 km/h, Q4 which closes relay 3 also closes relay 7 (by O4) (Fig. 11) to provide ground (output 10 of relay 7) for red lights LR1/LR2. Thus the latter are turned on as Stopping Red Lights.

Operation of the Stop Light – When speed is less than 8 km/h or when speed sensor does not generate pulses because of hard braking, output pin 1, 2 and 3 of IC14 go low to pull low r3 output pin 1 of IC15-b. Pulse r3 drives transistor Q5 to energize relay 5, in order to disconnect VCC4 input from relay 4 and turn off all the orange and green lights. At the same time, output 7 of relay 5 powers red light(s) LR5. This voltage can also be directed to power two or more extra red lights (LR5) which together constitute the Stop Sign. Also when car is set for reverse motion, r6 (Fig. 11) activates relay 5 to turn on the Stop Sign, and to prevent the activation of other lights.

The voltages (r3 and P6) which energize the relay 5 also provide output r7 to be fed to input pin 5 of IC15-a through resistor R42. This prevents the Q4 from energizing relay 7 in order to keep LR1/LR2 off when speed is under 8km/h. The voltage r5 in Fig. 11, is fed to pin 8 of IC20, in order to prevent the operation of relay 6 when brake is not applied or when Stop Sign is on.

IMPLEMENTATION OF THE OPTIONAL TRANSMITTER AND RECEIVER – Referring to Fig. 3 and Fig. 13, the speed sensor can be connected to P2 input of pulse amplifier and transmitter, via a twisted pair cable and a 0.1" female Molex connector. One side of the sensor is AC coupled to ground through C28 and C29 capacitors and the other side is passed on to the input amplifier, which is composed of Q1, Q2 and associated bias resistor. The speed sensor voltage is biased slightly positive to ensure that Q1 turns on reliably. The amplified signals are then fed to input Pin 2 of IC21, a 555 timer which operates as a pulse transmitter. Neither pin 6 nor pin 7 of the 555 timer is connected to a capacitor and hence, it is not set in astable or monostable mode. It would equally function as a transmitter if it was set in its monostable or astable mode instead of the configuration shown in Fig. 13. The timer functions as a fast transmitter when no capacitor is used at its pin 7 and 6.

The pulse detector, IC24-a in Fig 14-a, is one of the comparators of LM339 (quad voltage comparator), and is set as a non-inverting voltage comparator. *The pulse transmitter transmits the amplified pulses through presence light fuse (ie: through P1 input of its power supply* Fig. 13) to pulse detector in Dynamic Traffic Light (in the rear of vehicle) which also obtains its power (ie: VCC2) from presence light wire (P3 input, Fig. 15). There are, however, restrictions for pulse transmission with this method:

i) VCC1, which is branched off from the beginning of presence light wire to power the amplifier and transmitter, should not be filtered and should directly or through a rectifier diode power the amplifier and transmitter.

ii) VCC2, which is branched off from a point further from the beginning of the line of electricity (ie: from presence light wire in the rear of vehicle) to power pulse detector IC24-a, should not be filtered and should directly or only through a diode and/or a protection resistor power the pulse detector.

iii) P3, the voltage which is branched off further from the beginning of the line of electricity at the beginning of which P1 is branched off, must traverse through a wire ~4.5 m or longer, or through a resistor (~470W) to power circuitries other than said pulse detector, Fig. 15.

iv) If said transmitter should be used in a communication environment, it can be repeatedly used throughout the line of electricity for regenerating the received pulses by the comparator. In that case, the regenerator (555 timers) should be powered through a coil and through a rectifier diode, and said regenerators should be grounded by a diode.

CONSTRUCTION OF THE LAMPS CASING – The lamps casing of the Dynamic Traffic Light may be constructed basically from three layers of the geometric shapes illustrated in Fig. 1, Fig. 5 or Fig. 6. In drawings which illustrate the embodiment of the lamps casing that is pictured in Fig. 1 and 5, Fig. 17-a is bottom/side view of layer 1 and Fig. 17-b and 17-c are top view of layers 2 and 3 respectively.

Fig. 17-d illustrates the three layers at the time of assembly.

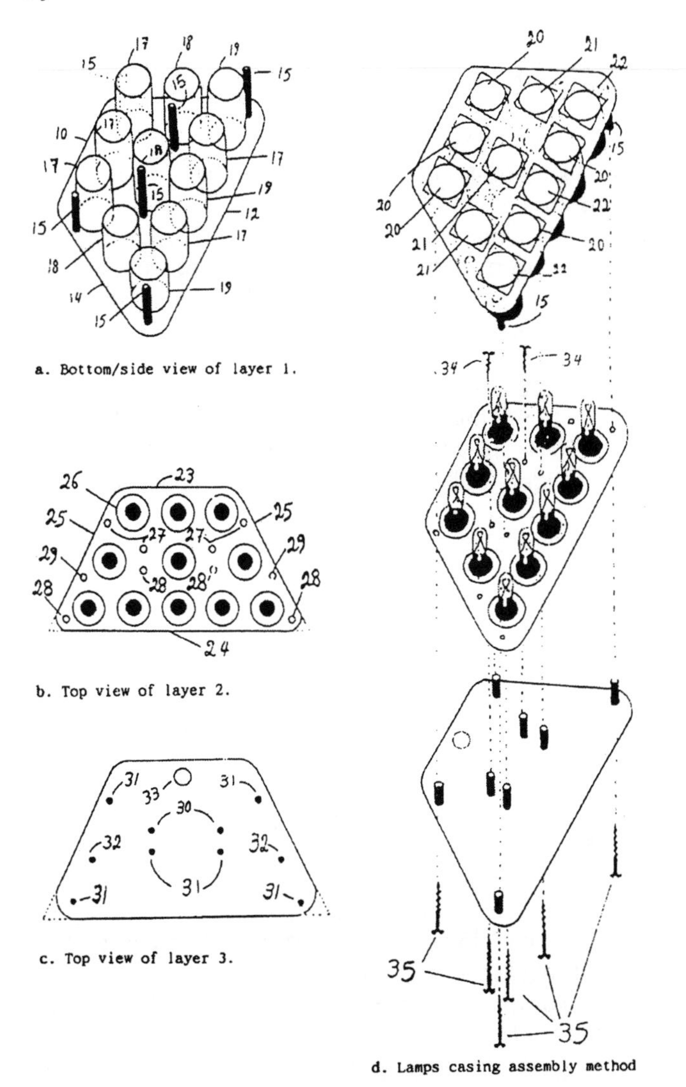

a. Bottom/side view of layer 1.

b. Top view of layer 2.

c. Top view of layer 3.

d. Lamps casing assembly method

Figure 17. The three layers of the lamps casing

PARTS LIST

All **RESISTORS** are 1/4-watt, 5%
R1-R3 --- 2,200 ohms
R4, R7-R9, R19, R20, R28, R35, R39, R44, R55, R57, R62 --- 10,000 ohms
R5 --- 22 megohms
R6, R23, R29, R34, R42 --- 100,000 ohms
R10, R11, R63 --- 1 megohms
R12-R17, R51, R54, R64 --- 12,000 ohms
R18, R21, R22 --- 220,000 ohms
R24-R27, R30, R31 --- 320,000 ohms
R32, R52 --- 2000 ohms
R33 --- 75,000 ohms
R36, R45, R48, R50, R59 --- 1000 ohms
R37, R46 --- 33,000 ohms
R38 --- 3000 ohms
R40, R47, R53, R61 --- 5000 ohms
R41 --- 22,000 ohms
R43 --- 470 ohms
R49 --- 270 ohms
R56 --- 470,000 ohms
R58 --- 3,300 ohms
R60 --- 60,000 ohms

CAPACITORS

Capacitors C20, C22, C23, C24, C27, C29, C32 and C33 are electrolytic, all others are disc. Capacitors C10, C12, C14, C16 and C18 must be +/-5%. C1-C3 --- 0.001 μF
C4, C12, C30, C31 --- 0.25 μF
C5 --- 22 pF
C6 --- 33 pF
C7-C9, C11, C13, C15, C17, C19, C28 --- 0.1 μF
C22, C23, C32 --- 10 μF
C10 --- 0.3 μF
C29 --- 4.7 μF
C14 --- 0.06 μF
C16, C18 --- 0.35 μF
C20, C27, C33 --- 100 μF
C21, C25, C26 --- 0.01 μF
C24 --- 1000 μF

SEMICONDUCTORS

IC1 --- SN74HC4040 asynchronous 12-bit binary counter
IC2, IC6 --- SN74HC682N 8-bit magnitude comparators
IC3---SN74HC373N octal D-type transparent latch with 3-state output
IC4, IC5 --- SN74LS382AN 4-bit arithmetic logic units
IC7 --- SN74HC32N quadruple 2-input positive-or gates
IC8 --- SN74HC00N quadruple 2-input positive-nand gates
IC9 --- SN74HC02N quadruple 2-input positive-nor gates
IC10 --- SN74HC4024 asynchronous 7-bit binary counter
IC11, IC21 --- NE555N timer
IC12 --- MM5369 17-stage oscillator/divider
IC13 --- MC14566B industrial time base generator
IC14, IC17 --- TND903 silicon diode arrays
IC15, IC16, IC23, IC24 --- LM339N quad comparators
IC18, IC19, IC20 --- NE556N dual timers
IC22 --- 7805 voltage regulator

DIODES

D1-D16, D18-D21, D26, D29, D32, D33, D35-D44, D46 --- 1N4004 rectifier diode
D17, D22-D25, D27, D28, D30, D31, D34, D45 --- 1N4148 silicon diode
Note: TND903 micro-chip which is array of diodes can be used instead of 1N4004 rectifier diodes (as IC14 and IC17 are used), except for D11, D16, D40, D43 and D44.

TRANSISTORS

all used with an appropriate heat sink.
Q1, Q2 --- 2N3904 NPN transistors
Q3, Q4, Q5 --- 2N3906 PNP transistors
Q6-Q8 --- 2N2222 NPN transistors,

RELAYS

For employing bulbs with higher wattage, form C relays for cars can be used such as (FORD F57B-14B192-AA) 12 v, 10/20 A, 75W coil. But in order to avoid resonance, mercury to mercury form C relays can be used (such as HGSR51111N00 from CP clair).

MISCELLANEOUS

* F1, F2, F3 --- 1A slow blow fuse
* XTAL1 --- 3.58-MHZ colour-burst crystal
* A wire ~4.5 m long, gage 16

Also for the construction of the casing in Fig 1:
* 9 holders for miniature (wedge or two pin) lamps
* LO1, LO2, LR1, LR2, LR5 --- 12.8 volts miniature wedge bulbs, GE lamp No. 912
* LG1, LG2, LG3 --- 12.8 volts miniature wedge bulbs, GE lamp No. 922 OR 912
* LO3 --- 12.8 v miniature wedge bulb, GE lamp No. 921
* LR3, LR4 --- 10 v fast-rise bulbs with at least 20 candle-power as ideal bulbs, or 2 pin 10.5 v halogen GE lamp No. 794
* Two additional red lights for the stop sign, GE lamp No. 921
* Three layers of the shape of trapezoid (made of bake lite, phenolic G10, plastic material to resist high temperature) for assembly of the lamps casing
* A multi-conductor cable with 4 wires gage 18
* A 0.1" 4-pin Molex connector for P3, P4, P5 and ground
* A 4 poles female connector
* A 4 poles male connector
* A multi-conductor cable with 10 wires gage 18
* A 0.1" 10-pin Molex connector
* A 10 poles male connector
* A 10 poles female connector
* A 0.1" 2-pin Molex connector for P1, P6 and ground
* A 0.1" 2-pin Molex connector for P2
* Connectors to deliver output 7 of relay 5 to the additional red lights

CONCLUSIONS

1. The intensity coded brake lighting system that is in use on motor vehicles is not informative enough to following drivers and, therefore, it is insufficiently appropriate for a goal as important to prevent traffic crashes. "The precise purpose of the brake application (panic stop, ordinary stop, deceleration, disengagement of cruise control) ***requires the following driver to be aware of the traffic environment ahead***" (3).
2. The provision of the Dynamic Traffic Light, which consists in the new concept of *Unit of Speed*, the *new Advanced Warning System*, the *Enhanced CHMSL*, the *new Stopping Red Lights*, the *new Advance Stop Sign System* and the *new Speed Level Indicator Signs*, affords the drivers the opportunity to take corrective action when the discrepancies between their vehicles and the flow of traffic ahead occur. Contrary to the other safety features which may encourage drivers to drive more aggressively, both the change of colours from green and the logical flashes of the lights of Dynamic Traffic Light would ***enlighten*** following drivers as to the presence of dangers related to the speed variations of leading vehicles.
3. Contrary to a system such as ABWS, Dynamic Traffic Light would help traffic to move ahead since not only by providing the *Highway Information* it would have a great potential to prevent traffic crashes, but also the restricted flashes of the green light(s) would signal following drivers to feel more confident about going faster.
4. *When the relatively very inexpensive Dynamic Traffic Light could provide the Highway Information* ***to alert and prevent*** *drivers from falling into hazardous situations, a conventional brake with its simple and standard operation would be adequate to stop a vehicle on time, or at least to reduce the severity of collisions.* Hence, the use of the relatively expensive safety features such as ABS brakes, with their own deficiencies and with their unusual operation, would become less critical in vehicles. Consequently, the Dynamic Traffic Light could help to reduce the cost of motor vehicles while the existence of lights of the three colours could enhance the appearance and increase the attraction of motor vehicles. Moreover, the required speedometer could be easily incorporated (as a digital meter) in this traffic light, for merging the cost of said speedometer with the cost of the Dynamic Traffic Light.
5. If the ***unique highway information*** that could be created by Dynamic Traffic Light already existed for drivers, then most probably a great number of sever accidents would not occur. Since "small reduction in reaction time promise large reductions in crash rates" (4), without Dynamic Traffic Light, *which provides even* ***more time*** *for following drivers to react*, vehicles are prone to accidents.
6. The unique and robust functions of Dynamic Traffic Light are literally the most natural method for allowing following drivers to maintain the control of their vehicles, and reduce their crash involvement rates. *Said functions, which preserve the integrity of the required signals on motor vehicles, should be treated as axiomatic for road safety.* Dynamic Traffic Light should be considered for safety regulatory priorities, and be expressly permitted to be used freely as an optional auxiliary device for motor vehicles.

Evidently, if the use of the required CHMSL was not delayed for more than a decade by costly research (after that its usefulness had been confirmed by the common sense and wisdom), then the CHMSL would have prevented a great amount of the loss which was resulted from the vehicles rear-end crashes during the more than a decade.

In view of the considerable amount of information regarding the usefulness of colour coded systems, and with the existence of the meaningful Dynamic Traffic Light, spending on further sophisticated research for rear lighting system of motor vehicles will be *a double loss*. Dynamic Traffic Light would be the most beneficial if it is used sooner rather than latter, after that its usefulness is confirmed by the common sense and wisdom.

In light of the previous research and the available data, isn't it the time that turning on but not flashing a conventional configuration of lights as the ***Advanced Stop Light on the rear window of motor vehicles***, when vehicles are (*almost stopped*) or stopped regardless of braking, or when vehicles are in motion and their wheels are blocked by hard braking, should be considered as an absolutely necessary function for rear lighting system of motor vehicles?

ACKNOWLEDGEMENT

I would like to thank Mr. Edward T. Kuczynski of Woodbridge Group and the SAE staff who made it possible for me to present this paper in 1998 SAE Congress. I also would like to thank Mr. James G. White of Transport Canada and Mr. Michael Perel of the U.S. National Highway Traffic Safety Administration, who provided me with some reports which gave me a broader perspective on the issue of rear signalling of motor vehicles. Said reports also enabled me to refer to more evidences in support of my ideas.

REFERENCES

1. SAE technical paper series #830565, International Congress & Exposition, February 28 - March 4, 1983.
2. P. Rothe, 1990, Challenging the Old Order.
3. U.S. National Highway Traffic Safety Administration (NHTSA), Docket No. 96-41, Notice 1, 1996.
4. Evans, 1991, Traffic Safety and the Driver.
5. Rudolf G. Mortimer, Car and Truck Rear Lighting and Signalling: The application of research findings. Symposium on psychological aspects of driver behaviour, Netherlands, 1971 (II.1.B).
6. Accident Facts, 1994 edition, National Safety Council.
7. Assessment of IVHS Countermeasures for Collision Avoidance: Rear-End Crashes. DOT HS 807 995, final report, May 1993.

CONTACT

Hamid Kashefy
P.O. Box 554,
Station Place du Parc
Montreal, PQ H2W 2P1
Canada

APPENDIX

STANDARDIZATION OF THE UNIT OF SPEED

On different cars, the signals should occur when speed is altered at least by one standard unit. Moreover, the speed level indicator signs on all cars should indicate the same speed level at the same speed. For this reason, the frequency of pulses that are generated by speed sensor should be exactly 10 times than the speed of the vehicle on which the sensor is installed; for standard operation of Dynamic Traffic Light, one pulse should be generated per 0.1 s at 1 km/h of speed. Thus the following are defined:

u = speed defined as 1 km/h
p = number of pulses that speed sensor should generate per rotation of rotor shaft in transmission (ie: per rotation of transmission) at 1km/h
p' = number of pulses that sensor should generate per rotation of speedometer cable (ie: number of poles of the speed sensor)
S = RPM of speedometer cable at 1km/h
t = transmission revolution per .1s
n = number of pulses that sensor should generate per rotation of transmission during .1s at 1km/h, and that is defined to be 1.
T = transmission RPM
d = differential ratio of vehicle
D" = diameter of wheel (wheel size)
c = πD" circumference of wheel
w = wheel RPM

And from mechanics:

$$w = T / d \text{ or } T = w * d, \qquad \text{(Eq. A1)}$$

and

$$u = w * c \text{ or } w = u / c \qquad \text{(Eq. A2)}$$

We want c per centimetre as we have speed per km/h. But since wheel size is normally given per inch, we change c to centimetre:

$$c = (\pi D\text{" in} * 2.54 \text{ cm/in}) = 2.54\pi D\text{" cm.} \qquad \text{(Eq. A3)}$$

Also we want u as centimetre per minute:

$$\Longrightarrow u = (1\text{km/h} * 1000\text{m/km}) * (100\text{cm/1m} * 1\text{h/60min}) = 1666.67 \text{ cm/min} \qquad \text{(Eq. A4)}$$

From equations (2), (3) and (4) we conclude that:

$$w = 1666.67\text{cm/min} / (2.54 * \pi D\text{"cm}) = 208.86 / D\text{"} \qquad \text{(Eq. A5)}$$

which is wheel revolution per minute at 1 km/h.

Also from (1) and (5) ==>

$$T = (208.86 / D\text{"}) * d \qquad \text{(Eq. A6)}$$

and t should be obtained as it was defined:

$$t = (T * 1\text{min/60s}) * (1\text{s}/10) = T / 600 \qquad \text{(Eq. A7)}$$

Also n should be obtained as it is defined:

$$n = t * p \qquad \text{(Eq. A8)}$$

From (7) and (8) ==>

$$n = (T / 600) * p \qquad \text{(Eq. A9)}$$

and n is defined to be constant number 1 ==>

$$1 = (T / 600) * p \qquad \text{(Eq. A10)}$$

$$\Longrightarrow T = 600 / p \qquad \text{(Eq. A11)}$$

From equations (6) and (11)

$$\Longrightarrow (208.86*d*p) / (600*D\text{"}) = 1 \qquad \text{(Eq. A12)}$$

$$\mathbf{\Longrightarrow p = (2.87 * D\text{"}) / d} \qquad \text{(Eq. A13)}$$

Therefore, if for example, wheels of a car have 15" in diameter and differential ratio of that car is 2.93, then the number of pulses that sensor should generate at 1km/h per revolution of transmission must be

$$p = (2.87 * 15\text{"}) / 2.93 = 14.79, \qquad \text{(Eq. A14)}$$

in order for the sensor to generate 1 pulse per each 0.1 s at 1 km/h of speed.

With the above example, If speed sensor is mounted directly on the transmission, then S may be equal to T and the above calculation may be final; so the number can be rounded to 15. In order to determine the error involved per each 1 km/h of speed when we round the value obtained for variable p, we replace values for variables d, D", and p in the equation (12)

$$\Longrightarrow (208.86 * 2.93 * 15) / (600 * 15\text{"}) = 1.02 \qquad \text{(Eq. A15)}$$

$$\Longrightarrow 1.02 - 1 = 0.02 \text{ km/h error} \quad \text{(Eq. A16)}$$

per 1km/h of speed.

Since the error is always too small we can ignore it. Thus designing an appropriate speed sensor for different vehicles, or adjustment in gears ratio guarantees the standard operations of Dynamic Traffic Light.

Production of 1 pulse per 0.1 s at 1 km/h can be accomplished in at least three ways when speed sensor (or speedometer cable) should be rotated (ie: when S <> T):

I) A particular speed sensor can be chosen to determine a fixed value for p'. For example, a 20 pole magnetic resistance element sensors (MRE) (or photocoupler sensors) can be used to have p' >= 20.

It naturally follows from above calculations that

$$p = (S * p') / T \quad \text{(Eq. A17)}$$

$$\Longrightarrow S = (p * T) / p' \quad \text{(Eq. A18)}$$

$$\Longrightarrow (S / T) = (p / p') \quad \text{(Eq. A19)}$$

In the above example, with p = 14.69 and p' = 20 and with equations (19) we conclude that:

$$S / T = 20 / 14.7 = 1.36 \quad \text{(Eq. A20)}$$

Therefore, it is revealed that per one rotation of the transmission, the speed sensor (or speedometer cable) should rotate 1.36 times in order for the 20 pole sensor to generate 1 pulse per 0.1 s at 1 km/h of speed. So the number of teeth of the gears that should rotate the speedometer should be adjusted to obtain the desired value for S. The 1.36 is not a rounded value and, therefore, the error involved in recognizing 1 km/h of speed is almost zero when said gears ratio is adjusted.

II) Instead of adjustments to the gears ratio for rotating the speedometer cable, an appropriate speed sensor can be designed to obtain a desired value for p'. In that case, at 1 km/h, S or T/S is known in every vehicle and p' can be found from the equation (19).

With the above example for d and D", from equations (13) and (19) it is concluded that p' = 19.99. we round this value to 20 to represent the number of poles of the required speed sensor. In order to reveal the error involved for measuring each 1 km/h of speed after we round the value that we obtained for the variable p', we first replace 20 for p' in the equation (19) to get a value for p:

$$\Longrightarrow p = 19.99 * 0.73 = 14.59 \quad \text{(Eq. A21)}$$

Replacing p, d and D" in the equation (12), the result should be 1:

$$(208.86 * 2.93 * 14.59) / (600 * 15'') = 0.99 \quad \text{(Eq. A22)}$$

==> there is 0.01 km/h error per 1 km/h.

III) The third way to assure standard operations of Dynamic Traffic Light is that the differential ratio and/or wheel size should be adapted to a unique speed sensor such as the mentioned 20 pole MRE sensor. Then from the obtained equation (13) and (17), we conclude that:

$$d = (2.87 * D'' / p') * (T / S) \quad \text{(Eq. A23)}$$

Similarly, if d is already determined, then wheel size (D") can be calculated in order to generate pulses with the desired frequency.

980556

Driving Factors and Future Developments of Airbag Technology

Karl-Friedrich Ziegahn
Fraunhofer Institut für Chemische Technologie (ICT)

The Author:

Dr. Ziegahn is head of general management of Fraunhofer ICT and deputy director of the environmental engineering division. He serves as chairman of the **airbag 2000+-symposium** and is a technical expert for airbag R&D.

Abstract:

Airbag systems have undergone an unbelievable success story during the past years. The story really began in 1969 when the US Federal Government first proposed that all cars be equipped with inflatable restraints. It took nearly 20 years to get the market break-through and in the year 2000 one expects a market volume of round about 70 million airbag units worldwide. Challenges have not only been set to the developers and engineers but also to policy makers, marketing experts, biomedical scientists and the consumer associations before the real success story could happen.
Today the smart car occupant protection system is again one step before another quantum jump. Cost, efficiency and performance drive nearly all car manufacturers and their system supplier towards new frontiers of airbag technology.

Implications of the current systems like the risk of fatal injuries, danger to children and the 'out-of-position' problem challenge again engineers and scientists. The gas generator technology has reached a technological level which allows solutions for a smart, tailorable system configuration. Occupant protection is not only based on it, as it could only be a part - a supplementary part - of the whole protection management. On the other hand, gas generator technology provides a full range of other application, from fire extinguishing systems to the opening and closing of separation devices. Gas generators of the future will be inexpensive, easy to handle, non-toxic and tailorable to specific applications.

Airbag system configuration has to be adapted to different requirements, e. g. to avoid that the 'helping hand' becomes unintentionally dangerous to the person who has to be helped. Future systems will be less aggressive and tailored to specific crash scenarios. But all overall airbag technology improvements in cars occupant protection. This contribution will highlighting some of the trends and ideas at the turn of the century.

The Airbag Success Story

Although the basic principles of airbag technologies had been developed some twenty or thirty years ago the broad market introduction along with fundamental cost reductions and remarkable improvements took place within the last decade or even within the last five years. The world market for airbag systems is assumed to reach 70 million units in the year 2000 (Fig. 1)/1/.
There is no doubt among experts that this innovative driver and passenger protection system is contributing to traffic safety in an effective manner. In Europe, even without any legislative requirement, the consumers obviously like this feature and honour it when buying a new car.
But airbag technology is still far ahead to be perfect. It has not yet reached its final state of development. Recent observations, as also reported in public media, show a further need for developments towards perfection, both from economical as from safety point of view.

Challenges for Airbag Developers

Challenges for airbag developers include

- mainly inexpensive and non-toxic gas generators (inflators)
- intelligent sensor systems and electronic management
- tailored threshold levels and deployment characteristics
- minimized risk of injuries
- advanced numerical simulation and design
- consideration of environmental issues

At the turn of the century we expect smart, tailored airbag systems as a part of the integrated car occupant protection strategy, avoiding unnecessary risks, especially for out-of-position passengers, children and at low impact speeds. The production costs have to be further cut down to increase efficiency and to enable a broad market introduction. The consumer wants a system high in reliability but low in costs and he or she wants to be sure that airbags do not 'kill our kids' as FORTUNE headlined in August 1996/2/.

But there are also the requirements of the car manufacturer. Concerning, for example, the easy adaptation to different car concepts, the handling during assembly and maintenance of the car and last but not least the environmental safe recovery or final disposal as a part of the car recycling strategy.

The Dynamics of the Airbag Market

The airbag market has undergone some remarkable changes within the last five years. Joint ventures, mergers and the entrance of new players did change traditional supply structures. Only system provider are still partners of the automobile manufacturer and component manufacturers had to decide about their most liked position (level) on the market.

Research and Development Tasks

Not surprisingly R & D still plays an important role in airbag technology. The time intervals between innovations shorten to less than one year, although most of the experts thought some years ago that most of the development has been done by then. Cost, performance and new application potentials are the main driving factors which lead to new airbag systems marked by a broad variety of technical solutions (azide, non-azide, hybride, liquid fuel etc.) Sensor technology is aiming on previewing sensors, neural (networks allowing the implementation of artificial intelligence and occupant sensing.
R & D of airbag technology is much more connected to market development than in most other technical areas. Corporate R & D of the main players is challenged by time constraints as well as by cost implications. The more fundamental developments undergo risks of delays or cancellations as the market requires fast solutions. Therefore there is a need and a potential for external R & D offered by universities and contract research organizations like Germany's FRAUNHOFER. It does not only develop new alternative gas generator substances but runs also most of the independent qualification tests and service life evaluations /3/. Current research activities include formulation of substances for gas generants, processing in pilot plants, production technology evaluation of environmental impact, recycling, recovery, and disposal concepts, assessment of gas toxicity, numerical simulation of airbag-module strength, investigation of combustion phenomena as well as the whole set of testing.

Gas Generator Technology versus Airbag Technology

A very interesting field is the investigation other application areas for airbag technology. As an alternative for pressurized gas storage high energetical substances offer a wide variety of applications as for example *in general*:

- to provide energy 'in situ'
- to initiate fast movements
- to deliver non-inflammable gas for fire extinguishing systems
- to open and close separation units
- to produce foams and blow forms
- to blow out particles and liquid doplets
- to be applied for other safety systems, e. g. for avalanche rescue purposes.

All in all the gas generator technology as a part of airbag technology can be seen as a very promising development field. Both areas are currently overlapping to a large extent but one can imagine that due to different requirements they may expand in different directions:

The very general advantages of chemical gas generators compared to pressurized gas are:

- Size is smaller
- Weight is lower
- Maintenance needs less efforts
- Gas exhaust characteristics can be tailored
- No pressurized vessels require no special surveillance and accordingly are free from appropriate regulations

Even if there are other constraints for high energy substances it can be expected that within specific applications as the examples show the use of gas generators will increase. This requires a tailored modification of combustion velocity and generated gas volume, a task which is not trivial to overcome (Fig. 3).

The gas generator propellant is being determined by its chemical composition and its shape. The selection of basic compounds and the addition of functional substances like cooling agents, slag formers, stabilizers or catalysts provide a great variety of potential solutions but require chemical experience and know-how. A detailed discussion of the chemistry will be given by Schubert e. a. /4/.

Airbag System Configuration

Airbags are being designed to fulfill two requirements:

- as a primary safety device to protect the unbelted occupant (according to FMVSS 208)
- as a supplementary safety device to diminish the risk of injuries for the belt wearing occupant (main application in Europe)

Also the 'out-of-position' occupant problem has to be considered.

These design requirements are partly contradictionary. The protection of unbelted occupants requires a larger volume and a higher deployment rate of airbags than those of supplementary system.

This leads to more 'aggressive' airbags which may cause serious injuries to out-of-position occupants. The dilemmas is well known and many experts ask for other crash test conditions which are adapted to belted occupants. In that case the energy of airbags may be reduced and airbag induced injuries may be minimized. Bigi and Bosio /5/list some design improvements which could help to reduce out-of-position occupant injuries:

- reduction of bag fill rate as much as possible
- optimization of the bag folding pattern
- retreatment of the module in deeper regions of the dashboard or steering wheel
- optimization of cover design

Besides of these design requirements airbags should not be deployed when the seat is not occupied or a rear facing infant carrier is placed in the seat. Tailorable gas filling devices should at least allow to reduce airbag deployment energy

- in case of out-of-position occupants
- at minor impact speeds
- at lower crash severities
- when occupants are belted
- for all small and light weighted occupants
- if seats are in the most forward position

Future trends in airbag configuration will lead to a more flexible, situation related system. Looking far ahead possible developments are

- **the Multi-bag system**
  Several small bags are surrounding the occupants. This will reduce the need to fill one large volume; multiple bags allow a specific protection against different impact directions at different times considering the actual size and weight as well as the position of the occupant.

- **Centralized versus decentralized gas generation**
  A decentralized gas generation requires many inflator units all around the vehicle, especially in the case of multi-bag systems. It could be economically wise to investigate the potential of a centralized gas generation unit combined with a controllable gas generator which provides the filling gas depending on the occupant/crash situation.

- **Refill of bags during a 'long-term'-crash**

  The first large crash impact which triggers the airbags may be followed by a series of other impacts (e. g. hit a car, leaving the road and crash against a tree). In this case it could be helpful to have a second airbag available or to refill the one first used. The potential of refilling a deployed airbag again should therefore be investigated whereby concern should be given to the occupants position.

- **Previewing sensors and artificial intelligence**

  Another possibility of reducing airbags aggressivity is the time elongation between triggering decision time and the catch of the occupant. Today's and future cars have more and more different sensor systems on board, as for example speedometer, triaxial accelerometers, car dynamic control, skid sensors, anti-spin control (traction control), ABS, temperature, distance control, GPS. It should be possible to monitor simultaneously all these and more data and by using artificial intelligence to evaluate very precisely the actual movement of the car, this for setting the smart occupant protection into some sort of 'alert status', evaluating different crash scenarios which are related to this specific speed, movement, angle of the car and probability of occurrence. With the assumed increase of computer performance this would probably be the easiest way to improve the protection system.

- **Increased car body strength and stiffness by gas generators**

  Folded honey comb structures for example can be expanded by a filling gas leading to a larger impact absorption volume. Also hollow tubes and bars of the car body can be filled with gas generated by high energy substances and can increase the stiffness of the structure. Even the separation of parts of the automobile by using cutting explosives along predesigned border lines is thinkable, if this will lead to a minimization of the overall crash impact on the occupants.

Conclusions

Looking 20 years ahead in occupant protection engineers and designers have to fantasize more or less. Some of the ideas described above may seem as not feasible today or tomorrow. Airbags and the gas generator technology are only one but important piece of the occupants protection puzzle. Legislation and consumers' behavior, road design and traffic management are other important influences which determine survivability of the future road traffic. Nevertheless several ten thousands of fatal injuries annually in USA, Europe and Asia require more than only traditional thinking about enhancing vehicles safety. Research and development are challenged by those figures to optimize car occupants protection.

References

/1/	Akzo	Automobil Produktion, Februar 1995
/2/	R. Norton	„Why airbags are killing kids"
/3/	K.-F. Ziegahn Th. Hirth T. Reichert	Environmental Qualification of Airbag Systems, SAE Conference Proc., 1996, Detroit
/4/	H. Schubert H. Schmid N. Eisenreich	Various Gasgenerator-Design for Car Occupant Protection 4. Airbag Symposium of ICT, Proc., 1998, Karlsruhe
/5/	D. Bigi A.C.Bosio	Tailorable Restraint Systems: The Future in Frontal Protection 3. Airbag Symposium of ICT, Proc., 1996, Karlsruhe
/6/	H. Schubert K.-F. Ziegahn	Technological Trends in Occupant Protection Systems - Recent Research Challenges from the German Point of View SAE Conference Proc., 1996, Detroit

Appendix

Figure 1:

70 Mio
60 Mio
35 Mio
28 Mio

1994 1995 1998 2000

World Market / Million Airbags

Figure 2:

Today

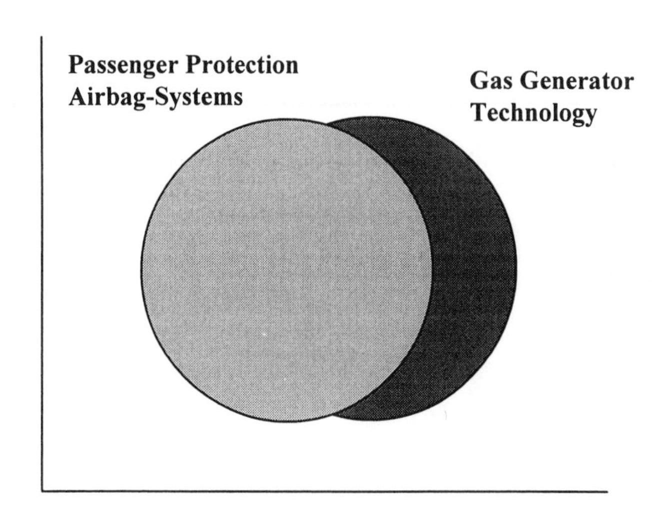

Tomorrow

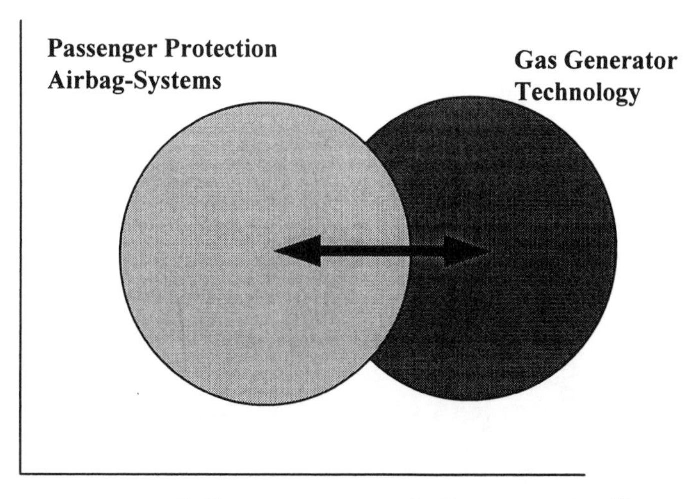

New applications for gas generators and other passenger protection concepts will result in a decrease of overlapping of technical demands.

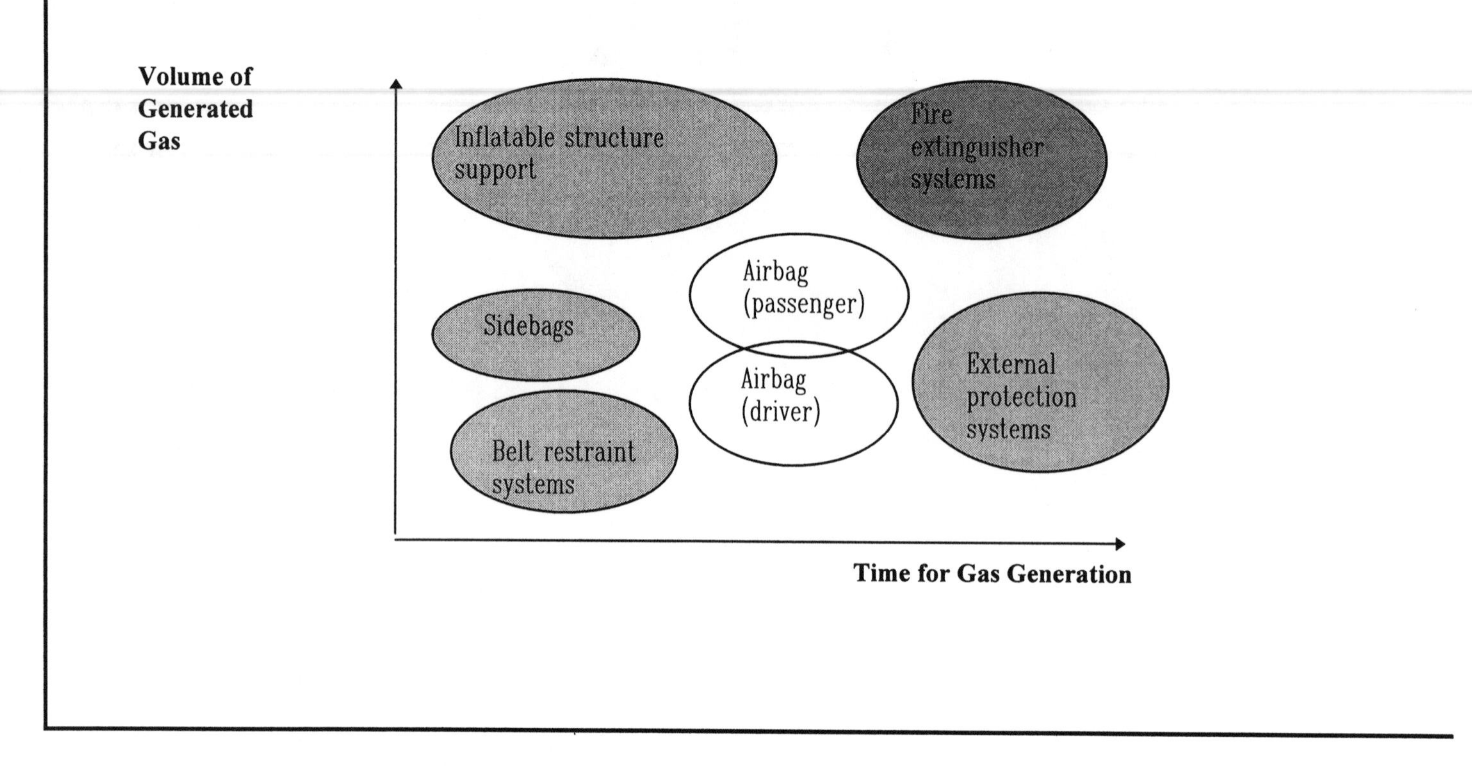

Figure 3: New applications of gas generators require tailored modification of combustion characteristics.

980646

An Innovative Approach to Adaptive Airbag Modules

Shawn Ryan
Advanced Development Group, Delphi Interior & Lighting Systems

ABSTRACT

An airbag module with adaptive capability can be achieved by several methods. Separate inflators, a single hybrid inflator with multiple heaters, or a dual-stage pyrotechnic inflator can all be used to control the inflation level of an airbag. In addition, Pyrotechnically Actuated Venting (PAV) offers an innovative method for regulating airbag inflation energy. PAV allows a controlled amount of gas to be vented out of the module before it enters the cushion. This paper discusses several methods for achieving a variable output airbag module. Static and sled tests were conducted to evaluate PAV in comparison to other adaptive restraint modules.

INTRODUCTION

The airbag is an important part of a vehicle's occupant restraint system. In combination with the knee bolsters, seat belts, and a steering column, the airbag manages the energy of an occupant during a frontal collision. According to the Special Crash Investigation (SCI) database, airbags have saved over 2500 lives. Through November 1997, SCI has attributed 87 deaths to airbags. Forty-nine of these deaths have been infants and children who were riding in the front passenger seat. Most of the 35 drivers whose deaths are associated with airbags have been smaller adults. In many of these instances the occupant was unbelted, or improperly belted.

There are two primary modes by which an airbag can cause injury to the occupant. The first is punch-out force. Punch-out injuries occur when the occupant is very close to the module. During punch-out, the cushion volume is small, the pressure is high, and the occupant inhibits the cushion from unfolding. Punch-out loading can cause injuries to the occupant's chest. The second mode is membrane force. Membrane force occurs as the cushion is expanding and interacts with the occupant before it has achieved its designed shape. Membrane loading can cause injuries to the occupant's head and neck.

One method of reducing the risk of injury is to reduce the power at which the airbag deploys. Recent legislative changes have been implemented to allow less aggressive airbags. The risk of injury from punch-out and membrane loading is reduced with less aggressive airbags. Less aggressive airbags may provide less restraint to normally seated, unbelted occupants in high severity crash scenarios.

The current amendment to the Federal Motor Vehicle Safety Standard 208 which allows the use of less aggressive airbags expires on September 1, 2001. The National Highway and Traffic Safety Administration has indicated future regulations will restore high severity crash testing and add lower severity testing with smaller occupants. Once high severity, unbelted test requirements are restored, adaptive airbag modules will maintain the reduced risk of injury initially achieved through depowering.

DISCUSSION

While there are many possible approaches to achieve variable output airbag modules, only a few are considered in this paper. This paper discusses separate inflators; dual stage pyrotechnic inflators with separate combustion chambers; dual heater hybrid inflators; and PAV. All of these systems require the development of additional sensing technology not included in this discussion.

SEPARATE INFLATORS

One method of achieving a variable output airbag module is by using a primary and a secondary inflator in a module. The inflators may use sodium azide, hybrid, or non-azide technology. The primary inflator would provide restraint for low level deployments. The secondary inflator would provide additional gas for higher severity deployments. For a high level deployment, both inflators would deploy simultaneously. For a low level deployment, only the primary inflator would deploy. An offset time between the two inflators may be introduced to achieve additional levels of restraint. The total energy and mass input to the cushion is the same with an offset deployment as a high level deployment, however more of the energy is input to the cushion later in the event.

Dual inflator systems offer some unique advantages. Different types of inflators using different technologies can be used in the same module providing flexibility in the design. The primary and secondary inflators may be tuned independently for greater control of the high and low level tank curves. In many cases, a validated design can be used for the primary inflator. Communization between the primary and secondary inflator can reduce manufacturing costs.

Dual inflator systems also have some drawbacks. Dual inflator systems are larger, have more mass, and cost more than a single inflator using the same technology. When only the primary inflator is required for restraint, disposal of the live secondary inflator becomes an issue. Finally, since there are two separate combustion processes, the variability doubles over a single inflator for a given output.

DUAL CHAMBERED PYROTECHNIC INFLATORS

The simplest embodiment dual stage pyrotechnic inflator has a single body with separate combustion chambers. Pyrotechnic inflators produce gas and heat from the combustion of a gas generant. For a high level deployment, both chambers are ignited simultaneously. For a low level deployment, only the primary chamber is ignited. The primary and secondary chambers produce energy independently. Because the primary and secondary inflators in a separate inflator system also produce energy independently, it is expected that a dual chambered pyrotechnic inflator could provide similar performance to a separate inflator system using the same technology. The total energy and mass input to the cushion would be the same with an offset deployment as a high level deployment, however, more of the energy would be input to the cushion later in the event.

There are also some drawbacks to dual chambered pyrotechnic inflators. Dual level pyrotechnic inflators are less expensive, smaller, and lower mass than separate primary and secondary inflators with the same technology. Current designs indicate dual level pyrotechnic inflators using advanced non-azide technology can be made smaller than other dual level systems.

When only the primary stage is required for restraint, disposal of the live secondary stage becomes an issue. Since there are two separate combustion processes, the variability doubles over a single output inflator for a given output.

DUAL HEATER HYBRIDS

Another method of achieving a variable output airbag module is by using a primary and a secondary heater in a hybrid inflator. Hybrid inflators consist of stored gas and a pyrotechnic material. The pyrotechnic material is ignited to produce heat and some additional gas. The heat expands the stored gas to fill the cushion. For high level deployment, both heaters are ignited simultaneously. For low-level deployment, only the primary heater is ignited. The dual heater hybrid systems differ from the separate inflator systems and dual chamber pyrotechnic inflators in that the total mass produced by the dual heater hybrid inflator remains relatively constant.

To achieve intermediate restraint levels, an offset time between the ignition of the two heaters could be introduced. Although the total energy produced by the two heaters would be the same as a high level deployment, the time interval that the heaters produce the energy would lengthen. Because the mass would flow out of the inflator, the heat transferred to the gas would change. In general, the heat transferred to the gas would decrease as the offset time increased. Initial testing has shown the longer the offset time, the lower the peak tank pressure. The exact relationship between peak tank pressure and offset time will depend on the inflator design.

As with dual chamber pyrotechnic inflators, dual heater hybrids can be made smaller, less expensive, and with less mass than separate primary and secondary inflators using the same technology. By varying the offset time, the total heat energy input to the cushion could be adjusted between the high and low levels.

A dual heater hybrid inflator presents some challenges. Dual heater hybrids must perform under a variety of internal operating pressures. This increases design considerations and makes it more difficult to control variability. In particular, the effect of offset time on restraint depends heavily on the internal design of the inflator. The ability to independently tailor the high and low level tank curves is also limited by the inflator design geometry.

PYROTECHNICALLY ACTUATED VENTING (PAV)

PAV controls the inflation by allowing a portion of the gas produced by the inflator to be vented out of the module. The maximum amount of gas vented out of the module is directly related to the portion of inflator ports aligned with a vent slot in the module. This amount can be tailored by design geometry to meet module performance requirements. It is important to understand that the average pressure in the cushion does not drive the mass flow out of the vent slot. Instead, the mass flow out of the module is driven by localized pressure from the inflator ports. The localized pressure forces a large percentage of gas from the inflator out of a relatively small opening. For a normally open system, a mechanism is incorporated into the module such that when actuated it closes the vent slot. A normally closed system would open the vent slot when the mechanism is actuated. A normally open PAV system was used in this study.

In order to respond to information from a crash severity sensor, the mechanism must move within milliseconds. Also, the mechanism must have sufficient power to overcome resistance due to impingement of gas from the inflator. An initiator can actuate the mechanism. An initiator is a device that converts an electrical signal into a pyrotechnic event. For a high output, the inflator would be deployed and the mechanism would be actuated at the same time. The mechanism can close the vent in less than two milliseconds, causing all of the gas produced by the inflator to enter the cushion. When only the inflator is deployed, a portion of the gas escapes through the vent slot and the result is low level deployment.

By altering the offset time between the inflator and initiator deployment, the inflation level for a PAV module can be controlled. Figure [1] shows the average pressure curves in a 60 L closed tank for a passenger PAV system with different offset times. Figure [2] shows the average pressure curves in a 60 L closed tank for a driver PAV system with different offset times. As the offset time increases, the tank peak pressure decreases from the high level to the low level. In offset deployments, the tank pressure follows the low level pressure curve until the mechanism is actuated.

Figure [1]: Pyrotechnically Actuated Venting
Average 60 L Tank Pressure
Passenger Hybrid Inflator

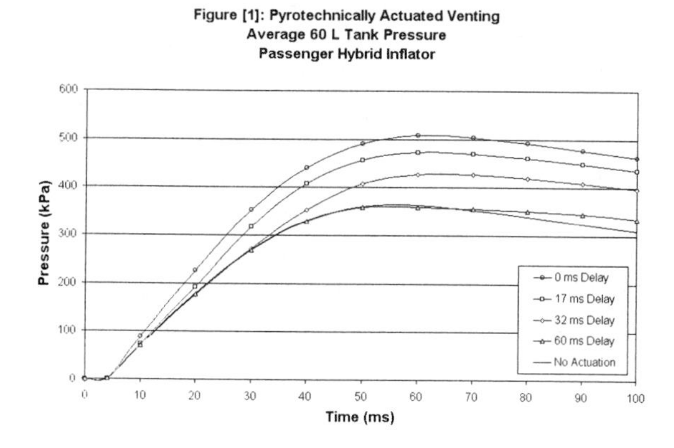

Figure [2]: Pyrotechnically Actuated Venting
Average 60 L Tank Pressure
Driver Non-Azide Inflator

Pressure (kPa)
Time (ms)
0 ms Delay
9 ms Delay
18 ms Delay
47 ms Delay
No Actuation

PAV modules have several advantages. In a PAV system, the offset time directly affects the total energy input to the cushion. Since PAV works on a module level, it can be applied to single output inflators that are already validated. From Figure [1] and Figure [2], there is little difference between the low level and a long offset time. This allows the initiator to always be fired to eliminate disposal concerns. The tank testing indicated PAV did not increase the variability of the system.

Although inflator design is simplified with PAV, the module design considerations increase. The design must ensure that the mechanism operates with the same reliability as other adaptive modules. It is difficult to apply PAV to designs where the ports are not located on the circumference of the inflator, and/or the inflator ports are not localized. The ability to tune the low level is limited to scaling the magnitude of the high level tank pressure curve.

TESTING

In order to compare the performance of variable output airbag modules, several tests were conducted. For these tests, a generic mid-sized car environment was simulated. The passenger adaptive modules utilized separate hybrid inflators, a dual heater hybrid inflator, and a single output hybrid inflator with a PAV mechanism. The passenger adaptive modules were compared to a baseline module, which used a single output sodium azide inflator. The driver adaptive modules utilized separate hybrid inflators, a dual heater hybrid inflator, and a single output non-azide inflator with a PAV mechanism. The driver adaptive modules were also compared to a baseline module, which used a single output sodium azide inflator.

One sled test series focused on evaluating the high level performance of the adaptive modules. For the purposes of this evaluation it was assumed that the high level performance should be designed to restrain an unbelted 50th percentile male Anthropomorphic Test Device (ATD) in a high severity crash. The sled pulse approximated a 30 mph barrier impact for a mid sized vehicle. This test simulated the government requirement for U.S. vehicles before changes were made to allow depowering. For the separate inflators, both inflators were deployed at the same time; for the dual heater hybrid, both heaters were ignited at the same time; and for PAV, the inflator and the mechanism were actuated at the same time.

In each test series, the ATD was instrumented to provide the following data: Head Injury Criteria (HIC), neck flexion, neck extension, neck shear forward, neck shear rearward, neck axial tension, neck axial compression, chest acceleration, chest deflection, chest viscous criteria, and femur loads. This data, along with film and kinematics analysis was reviewed to evaluate each adaptive module. In this test series, the neck axial compression data subset best illustrated the overall differences between each module. Other data subsets when considered individually may not reflect the same trend. In this series, the modules that provided less restraint allowed the passenger ATD to penetrate far enough into the cushion to make contact with the windshield. The cushions that allowed more contact with the windshield resulted in higher neck axial compression. Film and kinematics analysis showed all of the passenger adaptive modules were allowed more penetration than the baseline. This could be adjusted by using higher output inflators in each adaptive module and/or tailoring the cushion venting. The passenger ATD neck axial compression numbers for this test series are shown in Figure [3]. The data is graphed in terms of percentage of Injury Assessment Reference Value (IARV). A value below 100% IARV is desirable.

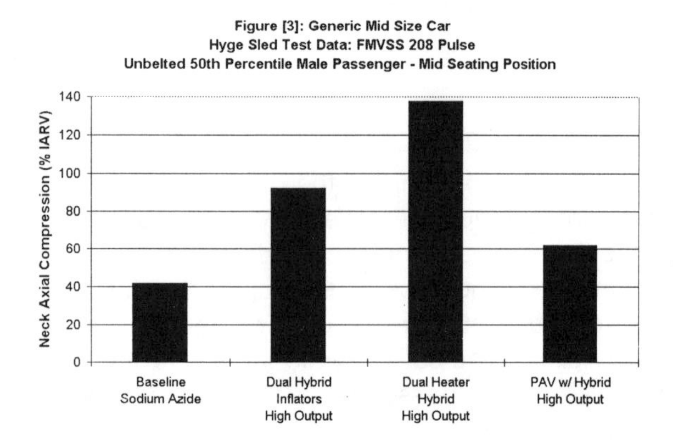

Figure [3]: Generic Mid Size Car
Hyge Sled Test Data: FMVSS 208 Pulse
Unbelted 50th Percentile Male Passenger - Mid Seating Position

The driver ATD data indicates the separate inflator system had the best correlation to the baseline performance. Film and kinematics analysis showed dual heater hybrid and PAV driver modules allowed more penetration into the cushion by the ATD than the baseline. This can be adjusted by tuning the inflator outputs or module characteristics such as cushion venting. As with the passenger side, the systems that provided less restraint allowed the driver ATD to make contact with the windshield, thus increasing neck axial compression. The driver ATD neck axial compression numbers for this test series are shown in Figure [4].

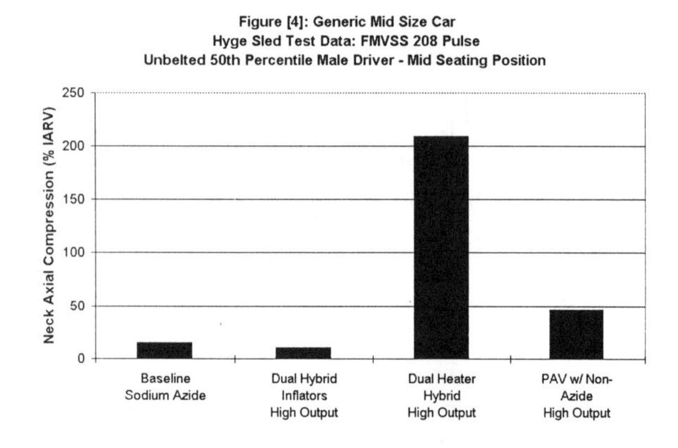

Figure [4]: Generic Mid Size Car
Hyge Sled Test Data: FMVSS 208 Pulse
Unbelted 50th Percentile Male Driver - Mid Seating Position

Another sled test series evaluated the low level restraint of the adaptive modules. For this evaluation, the generic AAMA pulse was used with unbelted 5th percentile female ATDs seated full forward. For the separate inflators, only the primary inflator was deployed; for the dual heater hybrid, only the primary heater was ignited; and for PAV, only the inflator was deployed.

The passenger ATD data indicates good correlation between all of the low output adaptive modules. In this test series, the baseline modules provided a stiffer cushion. This resulted in higher chest and neck numbers for the baseline. In this series, the neck extension data subset best illustrates the difference in performance for each system. The passenger ATD neck extension numbers are shown in Figure [5]. The higher neck extension from the baseline system is due to higher membrane loading from the cushion under the ATD's chin.

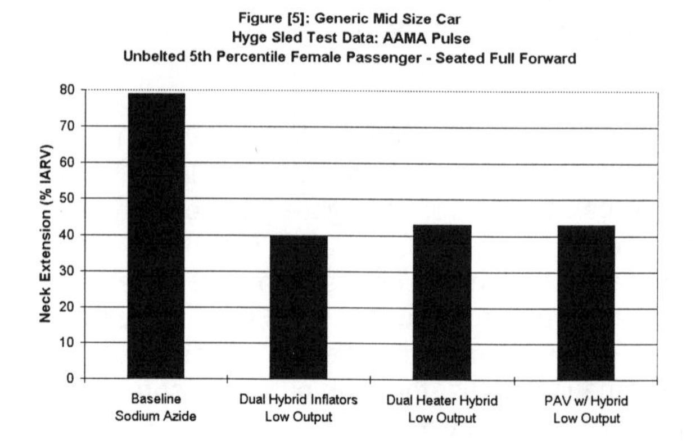

Figure [5]: Generic Mid Size Car
Hyge Sled Test Data: AAMA Pulse
Unbelted 5th Percentile Female Passenger - Seated Full Forward

The neck axial tension data subset provided the best illustration of overall driver module performance in this test series. The driver neck axial tension numbers are shown in Figure [6]. As with neck extension, axial tension is often related to membrane loading from the cushion under the ATD's chin. The stiffer baseline cushion resulted in greater membrane force than the PAV and the separate inflator driver modules. Film and kinematics analysis showed the driver dual heater hybrid modules allowed the driver ATD to penetrate far enough into the cushion to trap the cushion under the ATD's chin. This resulted in a higher neck axial tension than the baseline. The low level performance of the driver dual heater hybrid system can be adjusted by tuning the inflator output and/or modifying the module.

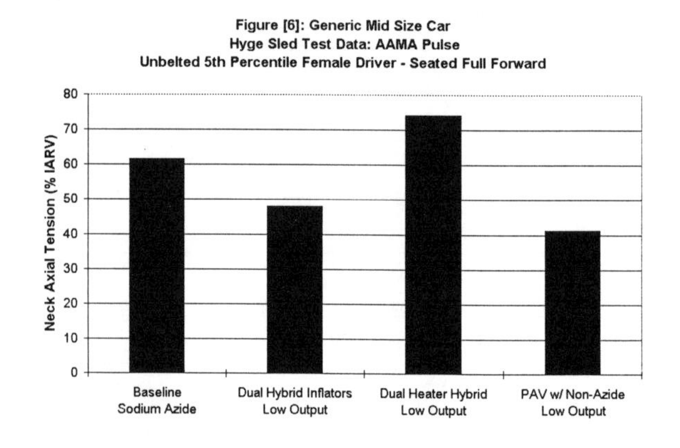

Figure [6]: Generic Mid Size Car
Hyge Sled Test Data: AAMA Pulse
Unbelted 5th Percentile Female Driver - Seated Full Forward

A test series was conducted to evaluate Out Of Position (OOP) occupant performance for each module. The tests were conducted on static fixtures that simulated an OOP 5th percentile female. The passenger ATD was positioned for maximum neck interaction. The driver ATD was positioned for high neck interaction with some chest loading.

The passenger ATD data demonstrates decreased membrane loading for all low outputs over the baseline. All adaptive passenger modules showed decreased membrane loading at low level than at high level. Again, the neck extension data subset most clearly demonstrates the overall performance for each system. Passenger ATD neck extension data is shown in Figure [7].

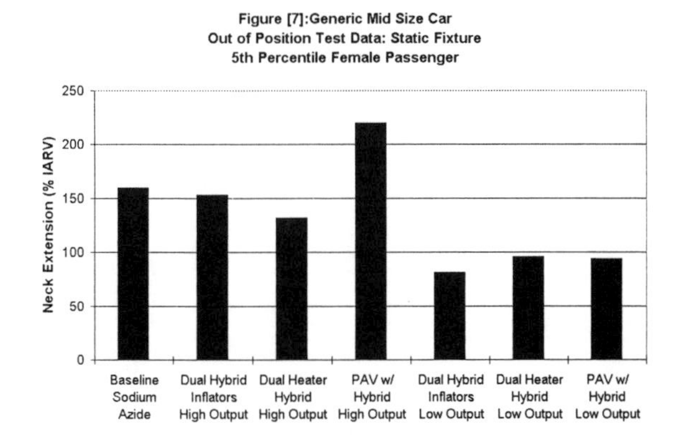

All adaptive driver modules showed decreased membrane loading at low level than at high level. The low level separate inflator system did not demonstrate and improvement over the baseline system. The neck axial tension data subset best illustrates the performance of each system. Driver ATD neck axial tension data is shown in Figure [8].

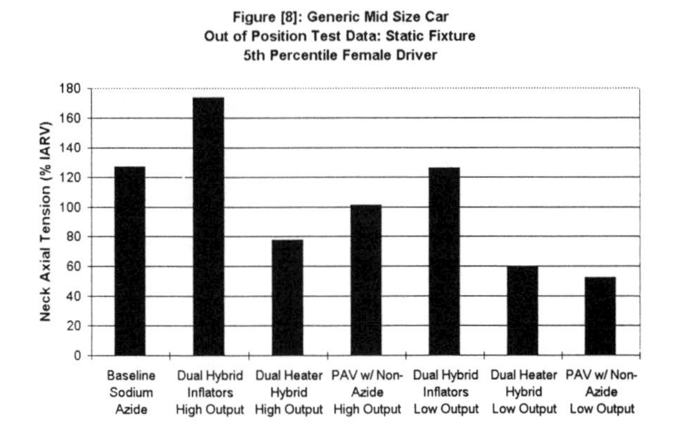

CONCLUSION

Several methods for creating a variable output airbag module were compared. Each method has potential advantages and trade-offs over other systems. With an offset deployment, separate inflator and dual chambered pyrotechnic systems would provide a maximum amount of energy to the cushion late in the event. Normally open PAV systems provide energy to the cushion that decreases from the full level to the low level as offset time increases. The effect of offset deployments in a dual heater hybrid depends on the inflator design.

Tank testing proved Pyrotechnically Actuated Venting can alter the energy delivered to the cushion. The sled test series with the unbelted 50th percentile male ATD indicated that most of the adaptive modules required additional tuning to provide restraint equivalent to the baseline in a high severity scenario. The sled test series with the unbelted 5th percentile female ATD demonstrated a properly tuned low level is capable of reducing the risk of membrane loading to a small occupant in a low severity scenario. The OOP testing indicated low levels also decrease the risk of injury over high levels when the occupant is in close proximity to the airbag. The OOP testing confirmed that by altering the energy delivered to the cushion, PAV is capable of altering occupant performance.

ACKNOWLEDGMENTS

Many people contributed their insight and assisted in gathering data. In particular, Allen Starner, Bob Neiderman, James Webber, and Alex Damman were instrumental in the completion of this paper.

ABBREVIATIONS

AAMA -- American Automotive Manufacturers Association

ATD -- Anthropomorphic Test Device

IARV -- Injury Assessment Reference Value

OOP -- Out Of Position

PAV -- Pyrotechnically Actuated Venting

SCI -- Special Crash Investigations database

Striking

A DOUBLE BLOW FOR SAFETY

MAURICE GLOVER ON A REVOLUTIONARY AIRBAG SYSTEM DUE TO BE REVEALED NEXT YEAR ON THE RENAULT LAGUNA

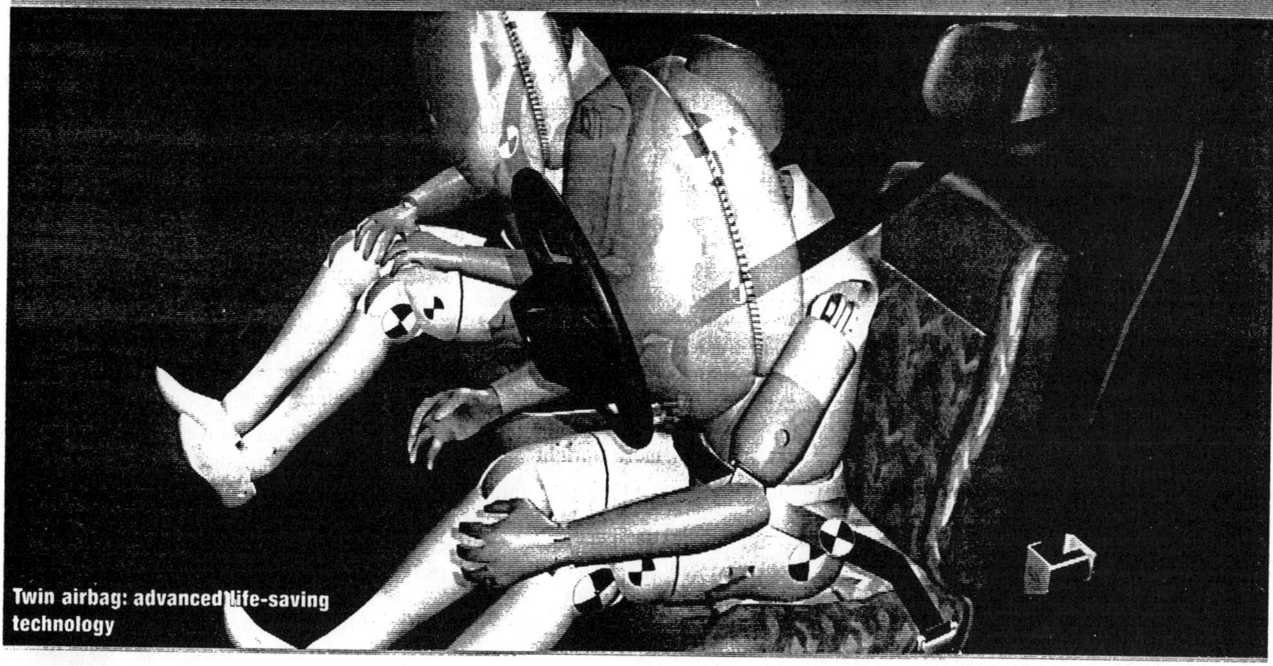

Twin airbag: advanced life-saving technology

A revolutionary double-action airbag which promises to help shield passengers from injuries in high-speed car crashes is among an array of advanced electronic equipment due to be revealed next year on the new Renault Laguna.

The ground-breaking safety system will be a key feature of P5, the platform that has been designed to carry most of the French company's next-generation of upper-range passenger car models.

As the first P5-based product, the Laguna is set to give a debut to the twin airbag, and it will be the company's star attraction at the Paris Motor Show, according to engineers based at the Aubevoye Technical Centre in northern France.

When operated in conjunction with the latest version of adaptive safety belts, the new system is claimed to prevent injuries to the driver and front seat passenger at impact speeds of up to 65 km/h (40 mph)

"We have devised a protection system with the purpose of building on the excellent results which have been by our Mégane model in the Euro NCAP test procedure," says safety department chief Christian Steyer.

Based on smart computer technology, the new system uses two airbags in one package and varies the level of restraint it offers according to the type of impact shock it senses.

The equipment, developed jointly with Swedish safety

Levels of restraint vary according to speed of impact

The system is an extension of ESP which brakes wheels independently when sensors detect a skid

systems supplier Autoliv, is designed to operate at three levels. The first involves restraint by seatbelt only, as at present, and covers "mild crash" impacts at low speeds. But accidents at speeds above 35 km/h bring the second level of protection into action and operate a seatbelt pre-tensioner at the same time as triggering the deployment of a 35-litre capacity airbag.

Severe impacts at speeds of more than 45 km/h fire the third level of protection, which involves a further seatbelt pre-tensioning action backed by the deployment of the second airbag, which has a 60-litre capacity.

"We know we are the first to introduce this lifesaving technology, but we would expect it to be made widely available very soon after it appears," says Steyer. 'We do not have any argument with this probability.

"The system was designed for future top-level Renault vehicles, but it will be introduced on the Laguna because it shares the P5 platform. The idea of the double airbag is the result of an 18-month development programme with Autoliv, and we are very pleased with the way it works.

"It is significant that we have also been able to reduce the level of pressure exerted by the action which pre-tensions the seatbelts, which also works to cut down the risk of chest injuries. In our new system, this stress is distributed much more evenly."

Farid Bendjellal, deputy engineer of vehicle customer specifications, explains: "Our intention in carrying out the second seatbelt pre-tensioning action is to pull the body back to a more upright position. In taking it closer to the seat backrest, we are effectively providing an additional 5 cm of knee room, and as everybody knows, this extra space can be a crucial factor in the avoidance of injuries in an accident situation.

"What this development programme has achieved is the expansion of the area in which seatbelts and airbags are effective. That is obviously a key point for Renault, which has made safety a key issue, but we are all hoping it will prove to be good for everyone else as well."

Bendjellal says the system will be produced by Autoliv in France but will also be supplied by companies in Germany and Britain when contracts have been formalised.

By 2002, the P5 will build on the double airbag development with the introduction of airbags incorporated in part of the straps of seatbelts used by rear seat passengers.

"After a great amount of research, our conclusion is that the bag-in-the-belt system is the best way forward in raising safety levels for occupants of rear seats," says Bendjellal. "Because the airbag is located in the lap-strap rather than on a diagonal belt, it is perfectly positioned in front of the occupant and offers the same level of protection as a front airbag."

According to a Renault UK spokesman, dynamic driving control and automatic tyre pressure monitoring are also likely to be included in the equipment offered as standard in the next version of the company's mainstream mid-size car for business and family use.

"The features will bolster our view that the new range is likely to be perceived more as a scaled down upper sector car rather than an upper-medium model that is directly comparable with the Ford Mondeo or the Vauxhall Vectra," he says.

"Although specification levels have yet to be set for the UK market, there is no doubt that this new car will be the most technically advanced vehicle in the sector. We are already working to position it above the area presently occupied by many German products and we aim to turn the tables on those manufacturers as far as high technology is concerned.

"Unlike our rivals, we will not baffle motorists with science by confronting them with so many buttons on the dashboard that they never get to use half of them. Our technology will be simpler to understand. The company is dedicating itself to making every advanced feature more user-friendly."

The comments came after Renault Group research chief Christian Balle previewed a selection of forthcoming technologies. "Both dynamic driving control and tyre pressure monitoring will be available for the Laguna, but features like trajectory control assistance, which vibrates the steering wheel if the car drifts off line and an accident prevention system we call 'stop and go' are still another five years away from the production line," says Balle.

An extension of ESP, which brakes wheels independently and controls engine speed when sensors detect the onset of a skid, the patented Renault system is claimed to reflect further progress in the control of vehicle understeer.

"This is particularly effective when excessive speed is applied during cornering and it helps prevent the vehicle drifting toward the outside of the bend," says Balle.

"Initially, it corrects the yaw rate by braking one wheel and checks the engine speed, but if severe understeering is detected, all four wheels are braked and engine torque is reduced. This increases the driver's ability to make use of the road and control the understeer."

Michelin has worked alongside Renault to develop the tyre pressure monitor, which is able to detect the slow leaks which are believed to lead to 80 per cent of all punctures. The equipment illuminates dashboard indicators to warn of a drop in pressure or a puncture and is able to display the pressure in each tyre. AE

980907

Accurate Predictive Algorithm for Air Bag Expansion by Fusing the Conventional Predictive Algorithm and Proximity Sensor

Nao Kitada and Kajiro Watanabe
College of Engineering Hosei University

ABSTRACT

The airbag systems in the first generation had been developed and are equipped in real automobiles. This paper is aimed at describing a new airbag scheme that might be categorized in the 1.5 generation.

The airbag system always needs some delay time between the triggering and complete expanding. The existence of the delay time is the main cause of difficulty for accurate airbag triggering. The predictive airbag expanding algorithm that compensates the delay time was proposed and the validity was examined in the first generation development.

Development of the 1.5 generation airbag systems with the higher performance are our next problem. Airbag equipped in automobiles must receive driver's body at the optimal timing when collision by which the effect of airbag is extremely improved. The more accurate predictive airbag system is required.

The algorithm combines the signal from the conventional acceleration sensor and that from a proximity sensor which provides the preview information of collision, and provides the accurate timing. The application of Kalman filter to the signal from the acceleration sensor yields noise suppressed acceleration of the movement of driver. The predictive algorithm based only on the signal from the acceleration sensor can predict driver's motion **only after collision**. The assist of the proximity sensor to the conventional predictive algorithm provides the prediction before and after collision.

Here we theoretically consider how the use of proximity sensor improves the prediction accuracy. We carry out simulations to show the validity of the new method.

Key words: airbag , predictive algorithm , proximity sensor , sensor fusion

INTRODUCTION

Airbag system in the first generation [4)5)6)7)8)] is now a prerequisite component for the safe automobile driving and the most new automobiles have been equipped.

Difficulty in developing airbag algorithm is the existence of the delay time between the triggering time and the airbag expanding time [9)]. Driver's body must arrive to the surface of airbag at the time when the expansion is completed, otherwise airbag can't exhibit the effect. Thus it must be triggered by predicting when the driver's body arrives to the position of the surface of expanded airbag.

The use of a proximity sensor that can preview the collision situation helps prediction of driver's motion toward to front window during the time before collision. I.e., the preview signal from the proximity sensor and the present automobile velocity can predict the jerk, acceleration, velocity and displacement of driver's motion due to collision beforehand. On the other hand ,the conventional predictive algorithm can predict driver's motion **after collision.**

Combination of the preview from the proximity sensor and the prediction from the acceleration sensor gives overall prediction which yields more accurate trigger timing. Here we describe how to effectively combine the signals from a proximity sensor and the acceleration sensor.

SYSTEM DESCRIPTION

Figure 1 shows a situation when an automobile collides to an object. A mass-spring-damper system in the inside of automobile in Figure 1 shows a dynamic model of driver's motion when collision. Figure 1(a) shows the situation when a bumper collides to the object, i.e. at a moment when an automobile collides to the object. Figure1(b) shows a situation when automobile's body collides to the object where we let t=0 be collision time.

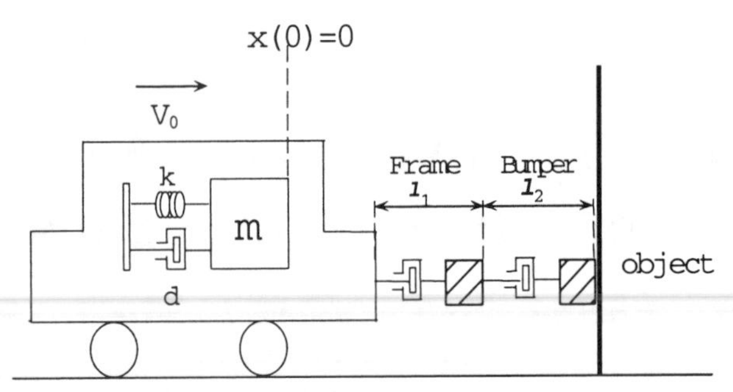

Figure 1(a).Collision of Object and Bumper

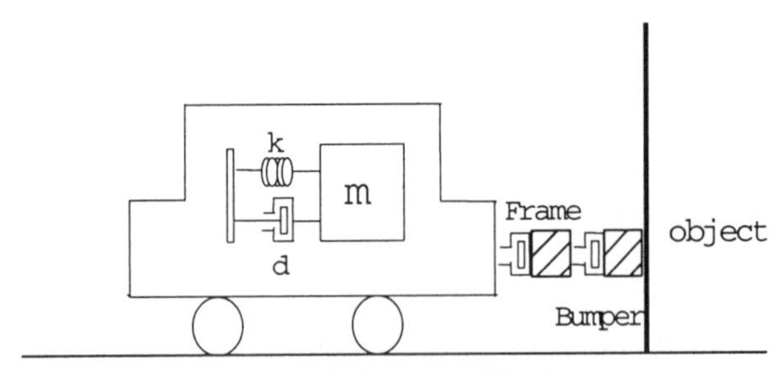

Figure 1(b).Collision of Object and Body

Define the variables and constants for the system in Figure 1.

- m : mass of driver's body
- k: stiffness constant of seat belt
- d: damping coefficient between driver's body and seat
- l_1 : distance from automobile body to frame
- l_2 : distance from frame to bumper
- V_0 : running velocity of automobile before collision
- $x(t)$: displacement of driver's body
- $v(t)$: velocity of driver's motion

Assume for the system in Figure 1.

(A1) Acceleration upon collision changes linearly(constant jerk model).

(A2) Two sensors ,i.e., a proximity sensor and an acceleration sensor are used.

(A3) Moving part of the driver's body at collision is described by only mass model.

Assumptions (A1) and (A3) are required to make the model of driver's motion simple by which the algorithm itself is also very simple so that the necessary calculation time is very short. Also we cite Assumption (A3), because the stiffness constant of the seat belt is very small, and the collision occurs in the very short time and thus any driver cannot resist. Therefore the stiffness constant and damping coefficient between driver's body and seat can be neglected.

AIRBAG EXPANSION PREDICTIVE ALGORITHM

ESTIMATION OF THE ACCELERATION AND JERK OF DRIVER'S BODY BY THE KALMAN FILTERING – Here we describe how to estimate the acceleration and jerk from the measured acceleration toward the driving direction. The acceleration is very much high frequency noise corrupted. The Kalman filtering is used to estimate the noise suppressed acceleration and the jerk of the driver's motion.

Define the variables and constants for the Kalman filtering.

- a_m : measured acceleration
- $\hat{a}$: estimated acceleration
- $\hat{a}'$: estimated jerk
- n: measurement noise
- R: variance of measurement noise
- ω'_a: system noise
- Q: covariance matrix of system noise

and define the matrices and vectors.

$$A=\begin{bmatrix}0&1\\0&0\end{bmatrix}\quad B=\begin{bmatrix}0\\1\end{bmatrix}\quad C=[1\quad 0]\quad a_e=\begin{bmatrix}\hat{a}\\\hat{a}'\end{bmatrix}\quad G_k=\begin{bmatrix}k_1\\k_2\end{bmatrix}\quad Q=\begin{bmatrix}0&0\\0&Q\end{bmatrix} \qquad \text{(Eq. 1)}$$

Then from Assumption(A1), the equation of driver's motion can be described by;

$$\frac{d}{dt}a_e = Aa_e + B\omega'_a$$
$$a_m = Ca_e + n \qquad \text{(Eq. 2)}$$

Then Kalman filter is given by the following equation;

$$\frac{d}{dt}a_e = (A-G_kC)\,a_e + G_k a_m \qquad \text{(Eq. 3)}$$

where the Kalman gain vector Gk is obtained by solving P for the following Riccati equation;

$$PA^T + AP - \frac{1}{R}PC^TCP = -Q\ ,\quad G_k = \frac{1}{R}PC^T \qquad \text{(Eq. 4)}$$

By solving the Riccati equation, the Kalman gains are finally determined as follows;

$$k_1 = \sqrt{\frac{2\sqrt{RQ}}{R}} \quad k_2 = \sqrt{\frac{Q}{R}} \qquad \text{(Eq. 5)}$$

Eq.(3) with the coefficients in eq.(5) provides an optimal acceleration and jerk.

ESTIMATION OF VELOCITY AND DISPLACEMENT OF DRIVER'S MOTION – The equation of driver's motion is given by driver's dynamic model in Figure 1. In spite of Assumption (A3) that neglects the stiffness and the damper, we first consider a general model that does not neglect these factors. The driver's motion of equation is given by

$$\frac{d}{dt}\begin{bmatrix} v \\ x \end{bmatrix} = \begin{bmatrix} -\frac{d}{m} & -\frac{k}{m} \\ 1 & 0 \end{bmatrix}\begin{bmatrix} v \\ x \end{bmatrix} + \begin{bmatrix} 1 \\ 0 \end{bmatrix}\hat{a}\ ,\quad \begin{bmatrix} v(0) \\ x(0) \end{bmatrix} = \begin{bmatrix} V_0 \\ 0 \end{bmatrix} \qquad \text{(Eq. 6)}$$

Solution of eq.(6) provides the driver's displacement $x(t)$ and velocity $v(t)$.

A set of simultaneous equation of eq.(2) and eq.(6) is a general equation that relates the jerk, acceleration, velocity and displacement. Solution of the simultaneous equation at the time $t+T$ from the time t under the Assumption (A3) provides the following state transition equation.

$$\begin{bmatrix} \hat{x}(t+T) \\ \hat{v}(t+T) \\ \hat{a}(t+T) \\ \hat{a}'(t+T) \end{bmatrix} = \begin{bmatrix} 1 & T & \frac{1}{2!}T^2 & \frac{1}{3!}T^3 \\ 0 & 1 & T & \frac{1}{2!}T^2 \\ 0 & 0 & 1 & T \\ 0 & 0 & 0 & 1 \end{bmatrix}\begin{bmatrix} \hat{x}(t) \\ \hat{v}(t) \\ \hat{a}(t) \\ \hat{a}'(t) \end{bmatrix} \qquad \text{(Eq. 7)}$$

Eq.(7) shows the predication relation of driver's motion. I.e., from the present state $\left[\hat{x}(t)\ \ \hat{v}(t)\ \ \hat{a}(t)\ \ \hat{a}'(t)\right]^T$, we can calculate the T future state $\left[\hat{x}(t+T)\ \ \hat{v}(t+T)\ \ \hat{a}(t+T)\ \ \hat{a}'(t+T)\right]^T$.

ESTIMATION OF ACCELERATION FROM RUNNING VELOCITY – We can estimate the optimal trigger timing by the conventional predictive algorithm based on the relation of eq.(7). In this paper, the air bag triggering is judged not only by conventional predictive algorithm but also by the use proximity sensor information. First here we describe how to preview the collision situation by the proximity sensor.

We estimate the collision acceleration from the preview sensor and the running speed of automobile. The time when the collision occurs will be described in the following section. Let be the time interval during when the collision is continuing(time when the object arrives to the frame) and $\bar{a}$ be the average acceleration. Then we have

$$\begin{cases} l_1 + l_2 = V_0\lambda - \frac{1}{2}\bar{a}\lambda^2 \\ 0 = V_0 - \bar{a}\lambda \end{cases} \qquad \text{(Eq. 8)}$$

From eq.(8), the average acceleration(constant) is estimated as following;

$$\bar{a} = \frac{V_0^2}{2(l_1 + l_2)} \qquad \text{(Eq. 9)}$$

We can obtain the average jerk, velocity and displacement by differentiating or incomplete single integrating and/or incomplete double integrating eq (9), respectively.

Then the jerk is zero(constant), the velocity and the displacement is calculated by solving following equation.

$$\begin{aligned} \frac{dv}{dt} &= -\frac{1}{T_d}v + \bar{a}\ ,\quad v(0)=0 \\ \frac{dx}{dt} &= -\frac{1}{T_d}x + v\ ,\quad x(0)=0 \end{aligned} \qquad \text{(Eq. 10)}$$

PREDICTION OF COLLISION TIME BY USING PROXIMITY SENSOR – Statistics of collisions, tangency and rear-end collisions by automobiles themselves, collisions and tangency to other objects and a rising to median, account for 94.8% [10] of total automobile accidents. Automobiles and these objects includes metal or are metal themselves. Thus we employ a metal detector as the proximity sensor.

The distance to the object to be collided is measured by a proximity sensor that can detect metal if the metal is within l. Prediction time t_f to collide to objects just after catching by the proximity sensor is calculated by following equation.

$$t_f = \frac{l}{V_0} \qquad \text{(Eq. 11)}$$

From eq.(11), we can preview if we are going to collide.

PREDICTIVE ALGORITHM – Figure 2 summarizes and shows the proposed predictive algorithm for air bag expansion explained in 3.1 , 3.2 , 3.3 and 3.4.

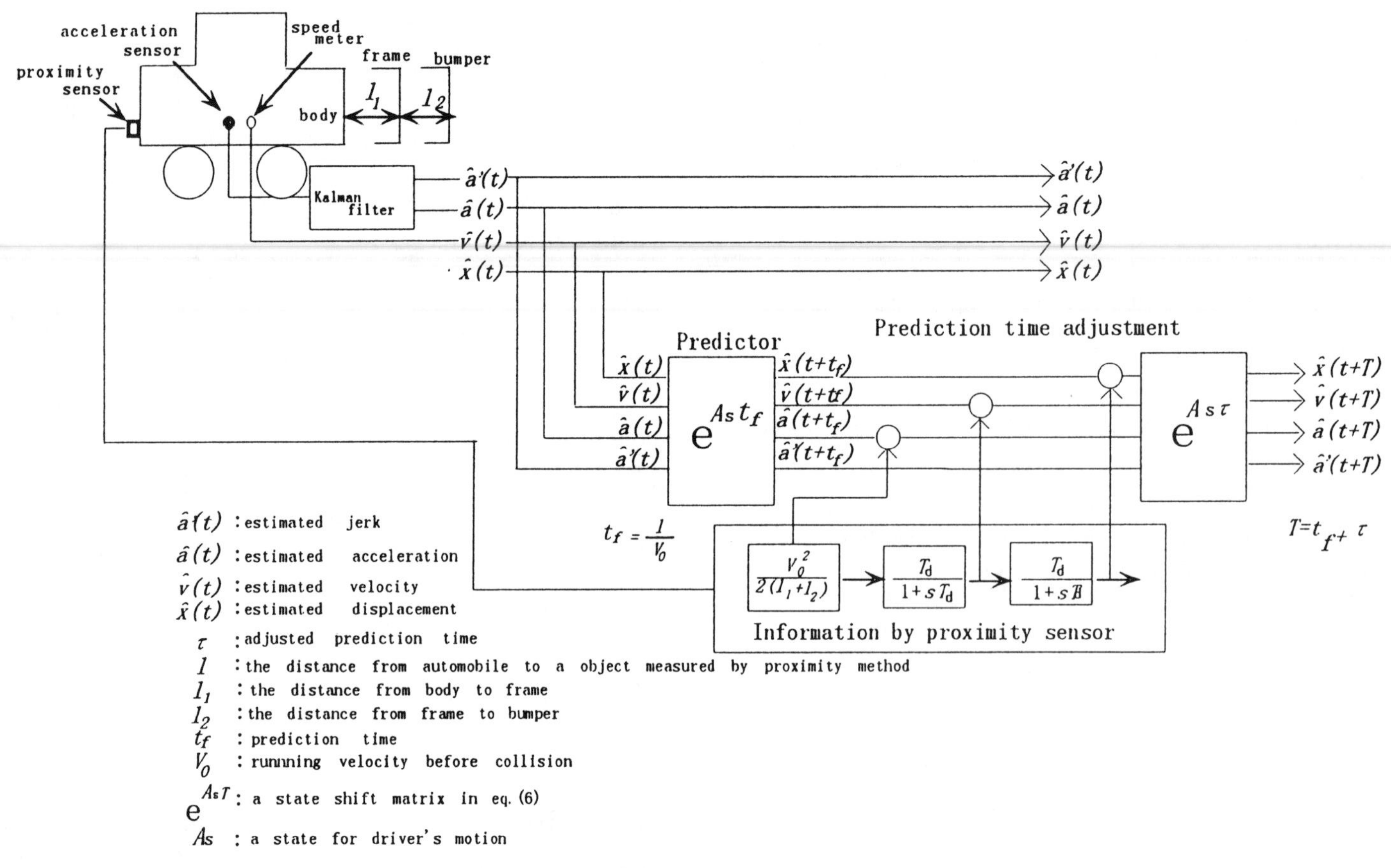

Figure 2. A Basic Scheme of the Proposed Method

AIR BAG TRIGGER JUDGMENT – The distance from driver's standard position to air bag expanding position is 0.15[m] and the delay time for air bag expansion is 30[ms]. Thus the judgment procedure is as follows;

1. Reset the output of the acceleration sensor immediately after the proximity sensor catch the objects to eliminate an excessive acceleration.
2. If {(the proximity sensor catches the object) or (accreditation level is higher than a certain level)} and $x(t+T)$=0.15[m] then air bag triggerotherwise keep air bag trigger = OFF.

In the algorithm above, if the proximity sensor fail to catch the object, the algorithm proceeds the conventional prediction procedure as shown in Figure 3

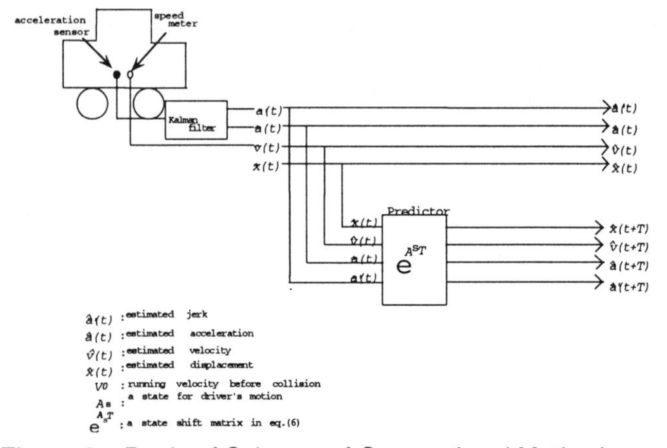

Figure 3. Basic of Scheme of Conventional Method

SIMULATIONS

SIMULATIONS UNDER THE REAL COLLISION DATA – It is very expensive to obtain the collision acceleration data by the crash experiments of real automobiles. Therefore, to verify the proposed predictive algorithm, we use the real experimental data that are refereed in literature [11]. The details of the experimental situation are shown in Figure 4. The experiments used three automobiles to obtain the data. The front automobile pulls the experimental automobile to a wall to situate a head-on collision. In order to acquire the data from the experimental automobile ,a measurement automobile is connected by a cable. In the experimental automobile, two acceleration sensors are installed to measure the acceleration of automobile itself and inside of dummy head.

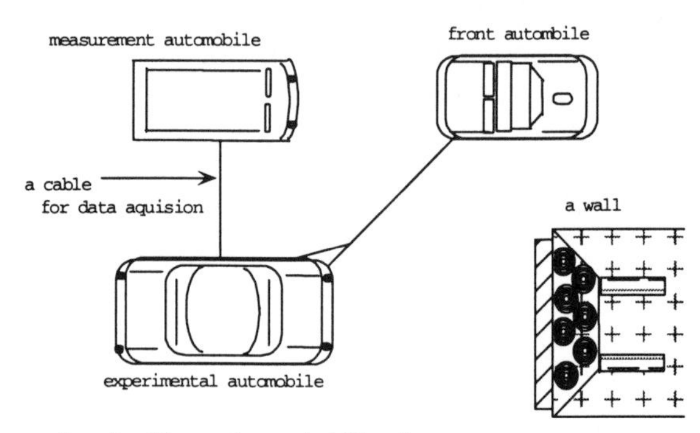

Figure 4. An Experimental Device

Figure 5 [11] shows the real deceleration data for the head-on collision. By using the data, we carried out simulations by the conventional predictive algorithm in Figure 3 and by the proposed predictive algorithm in Figure 2.

Values of the variables and constants for the simulations are as follows;

(a): V_0 =33.8[km/h]

(b): V_0 =43.5[km/h]

τ =10[ms]

t_f =20[ms]

T=30[ms]

l_1 =0.25[m]

l_2 =0.30[m]

R = 1

Q = 10

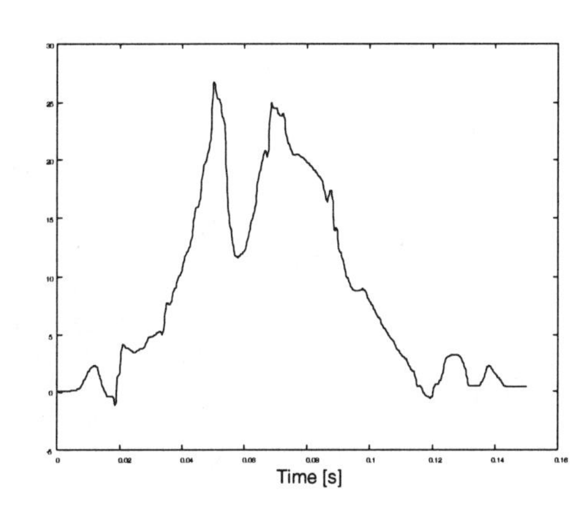

(a) V_0 =33.8[km/h]

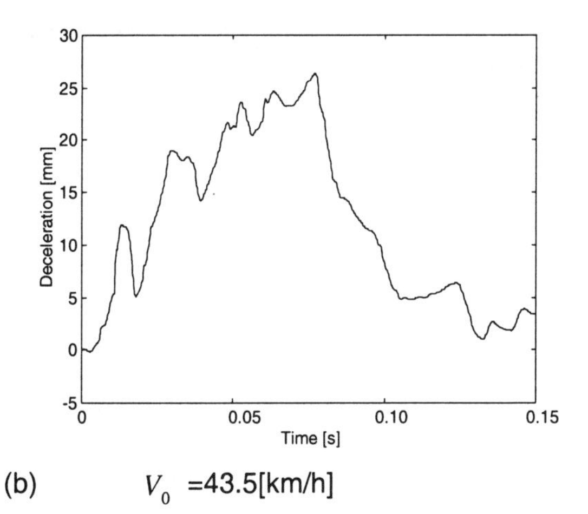

(b) V_0 =43.5[km/h]

Figure 5. Real Deceleration Data [11]

SIMULATIONS BY USING THE CONVENTIONAL ALGORITHM – Figure 6(a), (b) show the predictive and actual displacement of driver's body by the simulations via the conventional predictive algorithm in Figure 3. The displacement is accurately predicted, but it should be noted that the prediction is carried out after collision t = 0. This leads to inaccurate prediction just after the collision. Especially, it becomes serious when the automobile runs is the very fast speed.

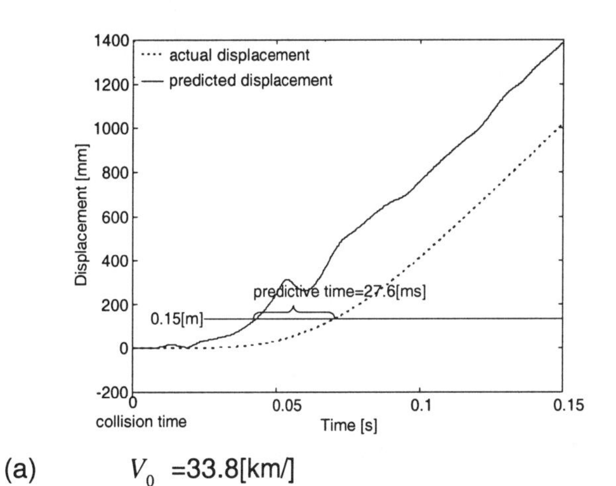

(a) V_0 =33.8[km/]

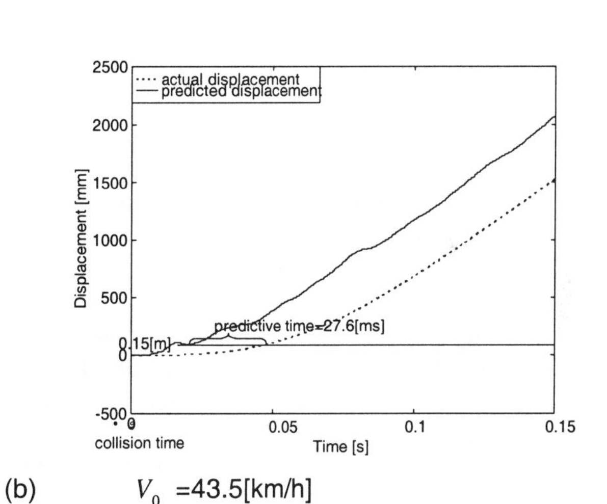

(b) V_0 =43.5[km/h]

Figure 6. Simulation by Using the Conventional Predicted Algorithm

SIMULATIONS BY USING PROPOSED PREDICTIVE ALGORITHM – Figure 7 shows the predictive and actual displacement of driver's body simulations by using the proposed predictive algorithm in Figure 2. Both cases in Figure 7(a), (b) show a good prediction. In these cases , because the preview function of proximity sensor , the displacements are predicted before collision.

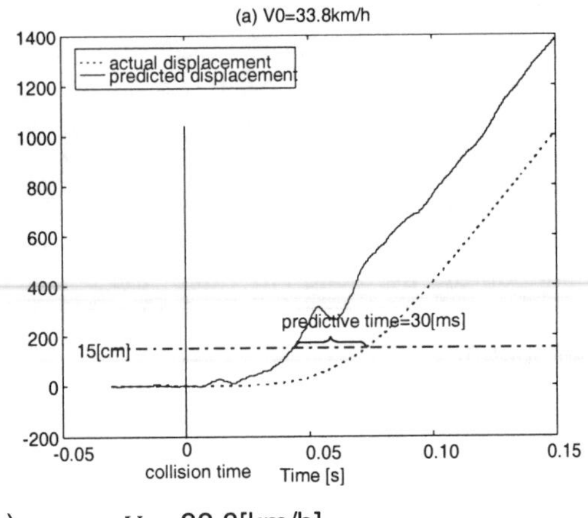

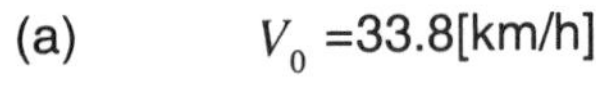
(a) V_0 =33.8[km/h]

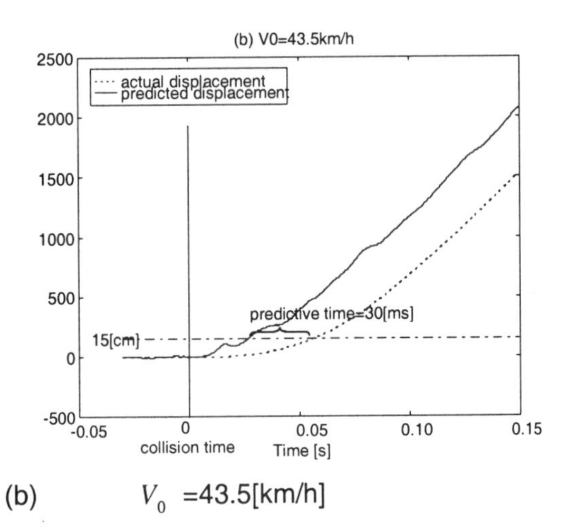

(b) V_0 =43.5[km/h]

Figure 7. Simulations by Using Proposed Predicted Algorithm

CONSIDERATION OF SIMULATIONS – The simulations by using the conventional predictive algorithm can provide the prediction of 30[ms] future **after collision** as shown in Figure 6. Thus the accurate predictions were not obtained immediately after collision. The simulations by the newly proposed predictive algorithm can predict 30[ms] future from before collision as shown in Figure 7. Being compared with the conventional predictive algorithm, the proposed predictive algorithm can provide more accurate prediction. Thus the new prediction algorithm gives better trigger timing. The validity of the proposed accurate predictive algorithm for air bag expansion is shown.

CONCLUSIONS

In this paper, we propose an accurate predictive algorithm for air bag expansion by fusing the conventional predictive algorithm and proximity sensor.

The predictive algorithm can preview the future collision before collision by using the proximity sensor and can estimate optimal trigger timing by the calculated collision data beforehand. We combine the preview information that can catch the future collision before collision by the proximity sensor and the prediction of driver's motion after collision by the acceleration sensor.

The simulations using real data show the validity of the proposed algorithm.

REFERENCES

1. Messer Schmitt Patent Japan-49-55031 (1974)
2. Bosch Patent Japan-63-503531 (1998)
3. Kanto-Seiki Patent Japan-4-146851 (1992)
4. Susan Enouen, Dennis A. Guenther, Roger A. Saul, Thomas F. MacLaughlin,
5. "Comparison of Models Simulating Occupant Response with Air Bags", SAE Trans.
6. Section 3, 840451, pp.343358 (1984)
7. Russel Brantman ,David Breed, "Use of Computer Simulation in Evaluating Airbag System Performance ",SAE Trans. Section 4, 851188,pp.10721081 (1985)
8. Naoki Suzuki ,Shigemitu Inoue, Ryoji Nakahama, "Determination of Airbag Sensor Threshold Level by Graphic Method ",SAE Trans. Section 6, 890193, pp.307316 (1989)
9. Dietrich E. Bergfried, Bernhard Mattes, Martin Rutz, "Electronic Crash Sensors for Restraint System", SAE Trans. Section 6, 901136,pp.12521260 (1990)
10. Robert W. Diller, "Electronic Sensing of Automobile Crashes of Airbag Deployment", SAE Trans.,910276 ,pp.6568 (1991)
11. Kajiro Watanabe ,Yasushi Umezawa ,"Optimal Timing to Trigger an Airbag" SAE Trans.,930242 , (1993)
12. "Automotive Engineers' Hand Book", Society of Automotive Engineers of Japan
13. "Automotive Industries Hand Book", Nissan Automotive joint-stock cooperation

970167

Roof Airbags

Helmut E. Mueller
INOVA

ABSTRACT

1. In case of a side impact crash an airbag installation which is part of the door seal improves HIC-values considerably and covers FMVSS 201 requirements in this area.

2. A clamp-in-formed headliner contains roof-structure-supports and acts as noise dampening to the sheet metal roof.

Also, it carries
- side airbags along the roof rail
- airbags between the front passengers
- airbags between the front and rear passengers

All generators and supply tubes are mounted to this module.

Altogether, it is a solution for an all-around-head-airbag system.

INTRODUCTION

After introducing thorax airbag systems there is a need for head impact protection. Also the FMVSS 201 requires a passive surface energy absorbtion to specific survival values.

Two investigations, which are not finished yet, could be a new milestone.

1. HEAD AIRBAG AS PART OF THE DOOR SEAL.

Fig. 1 and 2 show the seal schematically and an adapted enclosure for a bag and a 'tube' generator.
The bag is added to the seal from the instrument panel to the 'B'-pillar (Fig. 3).

The safety results were sufficient (Fig. 4), but the noise level of the included generator was to high.

We decided to install a seperate gas generator at the end of the tube in the area of the lower 'A'-pillar and / or 'C'-pillar.

The tube served as a quick pressure transfer agent. The result showed a respond which was three milliseconds faster with the tube than without one.

The Pressure-Transfer-Tube is a 'MUST'.

Fig. 5 and 6 show the area of protection.
Some real crashtests were achieved and the results were evaluated by MADYMO method (Fig 7 and 8).

As the bag is mounted in the 'stiff' area of the flange, its influence to the upcoming FMVSS 201-law was tested.

The drop in HIC-Figures was 1350 to 1050. The packaging is shown in Fig. 9.

To come down to HIC 750, hardfoam (Fig. 10) was added.

The rear passengers and the FMVSS 201 require the protection of the rear flange as well.

Fig. 11 and 12 show sections on the the 'A'-pillar and grip handle area. This bag is part of the front door seal.

The bag moves on between 'B'-pillar and formed headliner without mounting (Fig. 14).

In a four door car the bag continous with the upper seal as shown in Fig. 12. Front and rear door seal is one assembly. A two door car requires the connection to the window flange without seal (Fig. 13).

Schematical layout Fig. 15.

2. 'ALL AROUND' HEAD AIRBAG

In some accidents it will be advantageous to have more than a front and side airbag for protection of the head. An airbag between the front passengers would prevent them from 'banging' their heads together.

In fact most rear passengers are not belted, therefore a roof airbag in front of the rear passengers would protect both front and rear passengers.

Now there is a solution:
Presently between roof and formed headliner, there are the roof reinforcement brackets and a glued-on sheet of felt or carboard for noise dampening.

The layout of the module headliner is such that leaf spring type roof supports are glued into or onto the headliner itself. With cams at the ends of the roof brackets, the formed headliner is clamped to the sheet metal. The cams are covered by the grip handle mountings (Fig. 18 and 19).

In this manner, the module headliner is trim part, roof reinforcement and noise dampening - all in one. It saves money, weight, assembly time and improves head clearance.

The bags are fastened to the roof brackets and are packaged invisibly behind the headliner (Fig. 16). Typical sections show the package in the grip handle-2 door window- 'A'-pillar- and roof-rail area (Fig 20,21,23,24). Even a bag behind the sunvisor against overroll is possible. The headliner itself is precut and its ends of cloth or vinyl are covered in a notch (Fig.22).

Between the roof rail and the headliner, there is enough space to locate the power supply with the module (Fig. 21).

For sunroof versions there is enough space between the sunvisors for a bag (Fig 17).

The roof rail bag has to fire as quickly as possible.

The bags between the passengers have to be as slow as the todays frontal bags. This covers all risks against 'out of position'.

SUMMARY

For an optimal protection in an average passenger car concerning bags one needs:

A: In case of a frontal crash
- driver and passenger frontal bags
- head bags for rear passengers
- knee bags for front and rear

B: In case of a sside crash
- thorax bags
- roof rail bags
- in-between-passenger bags

The complete system can be realised for less money than car manufacturers paid for the frontal bags 3 years ago.

ABOUT THE AUTHOR

My experience is based on being head of interior design and development for 18 years.

For further information, please consult:

Helmut E. Mueller, Buchenweg 42
38550 Isenbuettel, Germany
Ph: 0049/5374/5684
0049/172/4256628
Fax: 0049/5374/5185

1 head airbag for side impacts

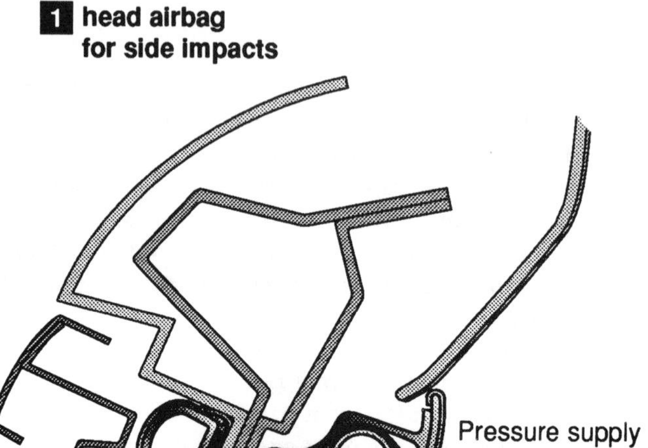

2 section

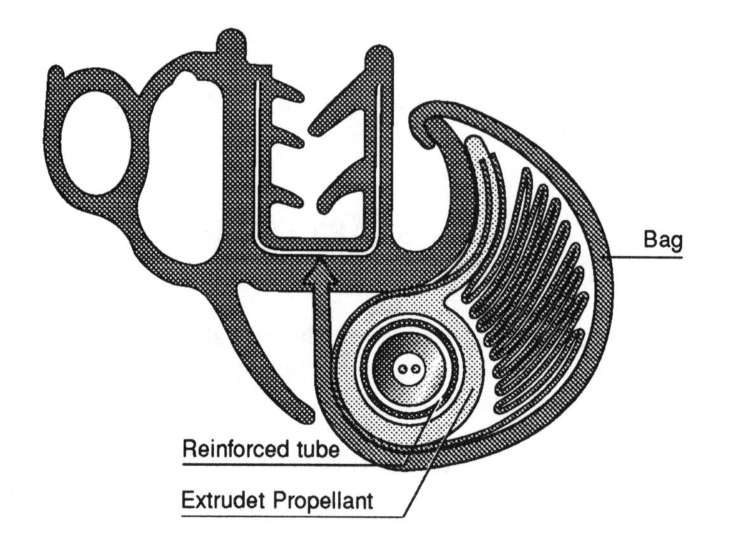

3 head airbag for side impacts

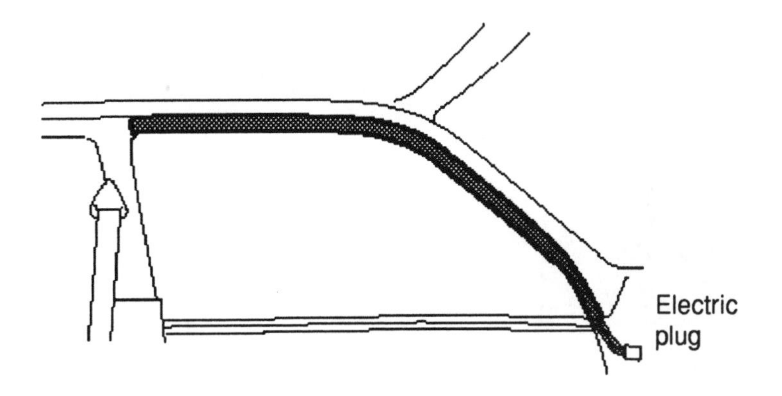

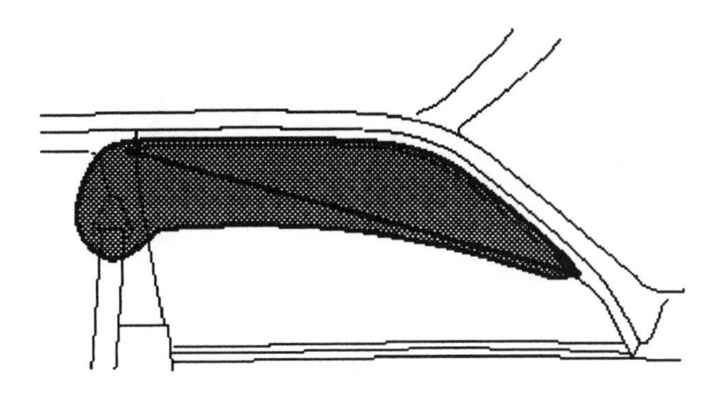

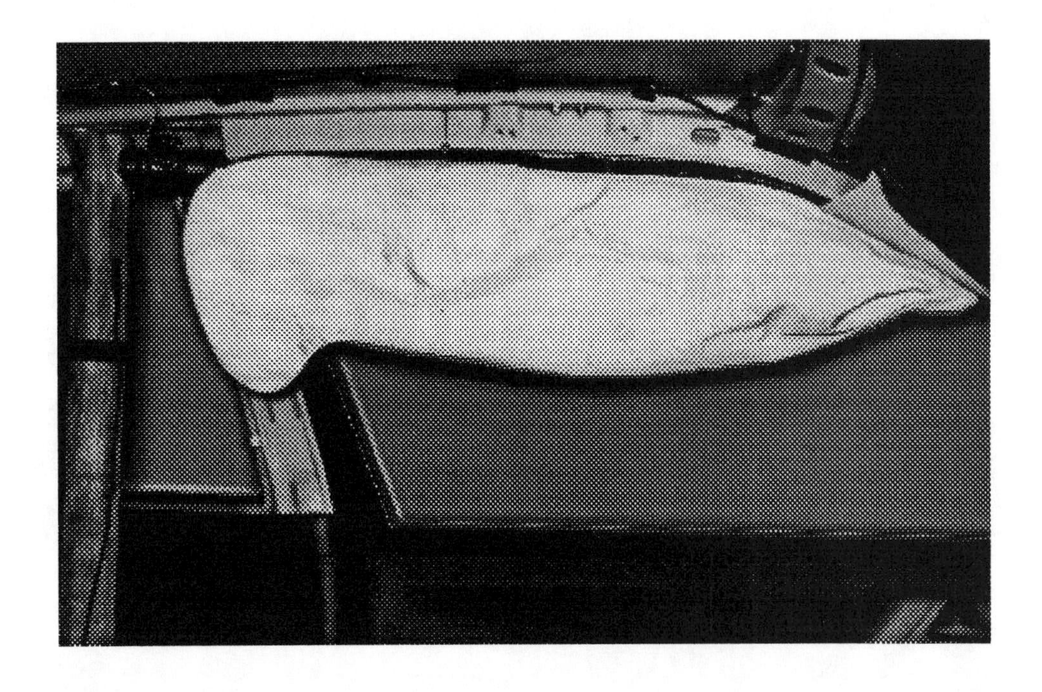

4 positive tests

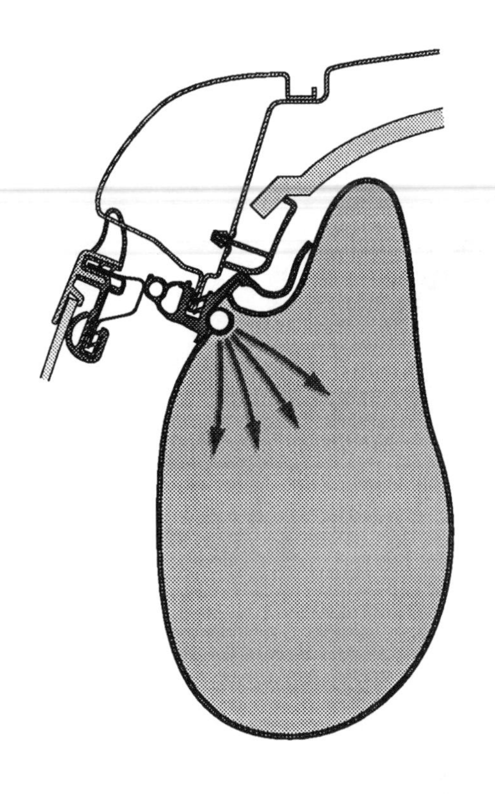

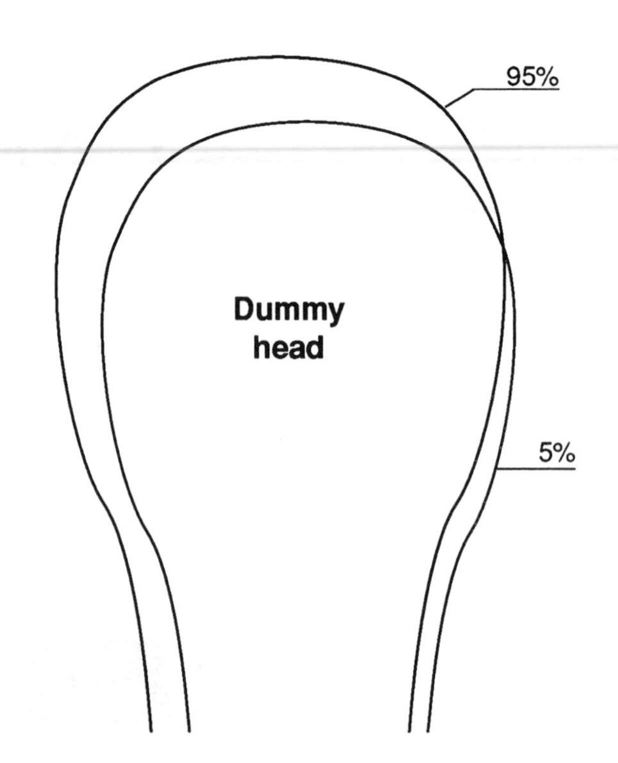

5 protection area

6 head airbags – schematic

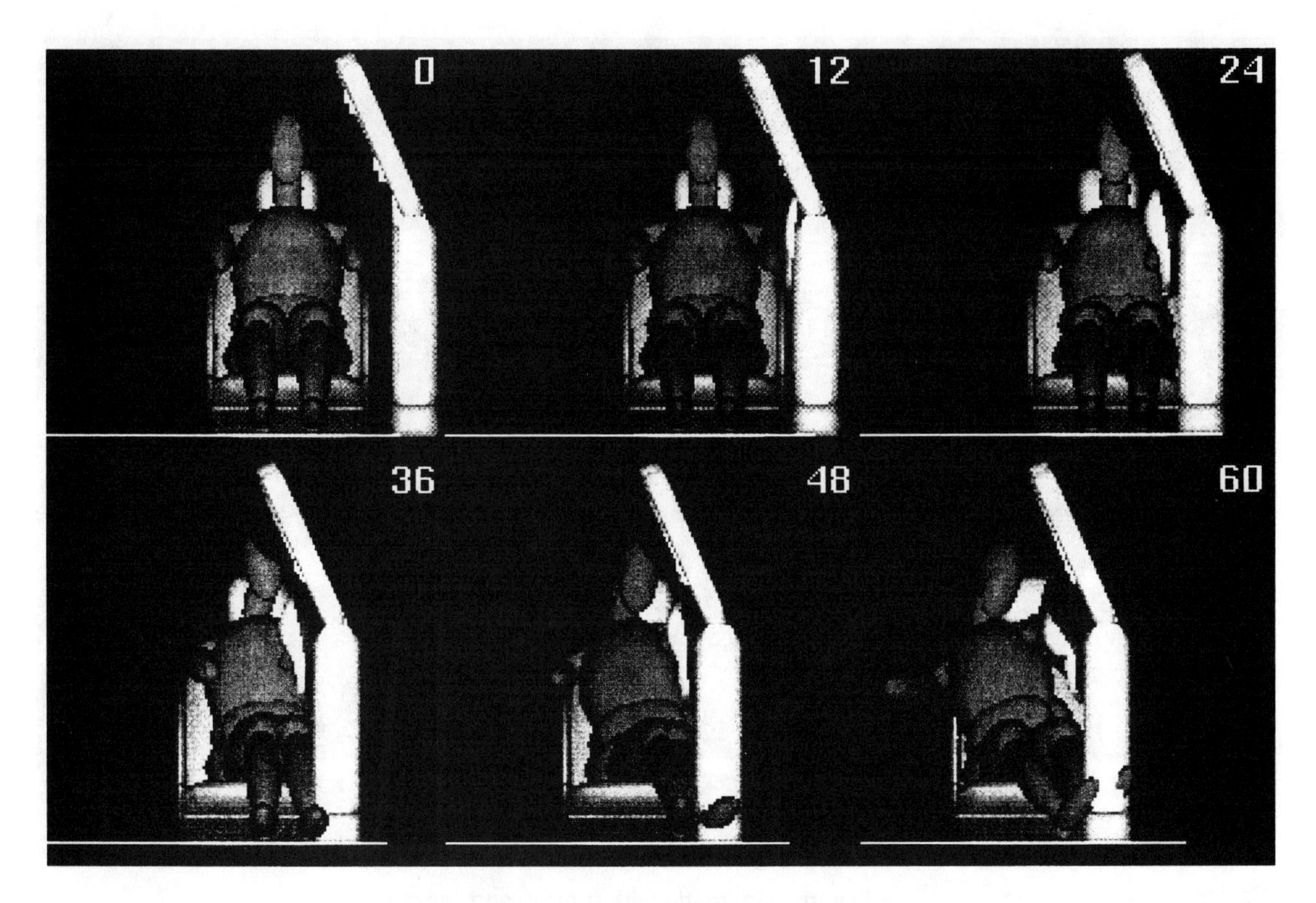

7 Madymo evaluation

- Dummy cinematic with thorax and "seal"–bag at "post"–impact
- 95% dummy/max. volume at 20m sec.
- Seat rearmost location

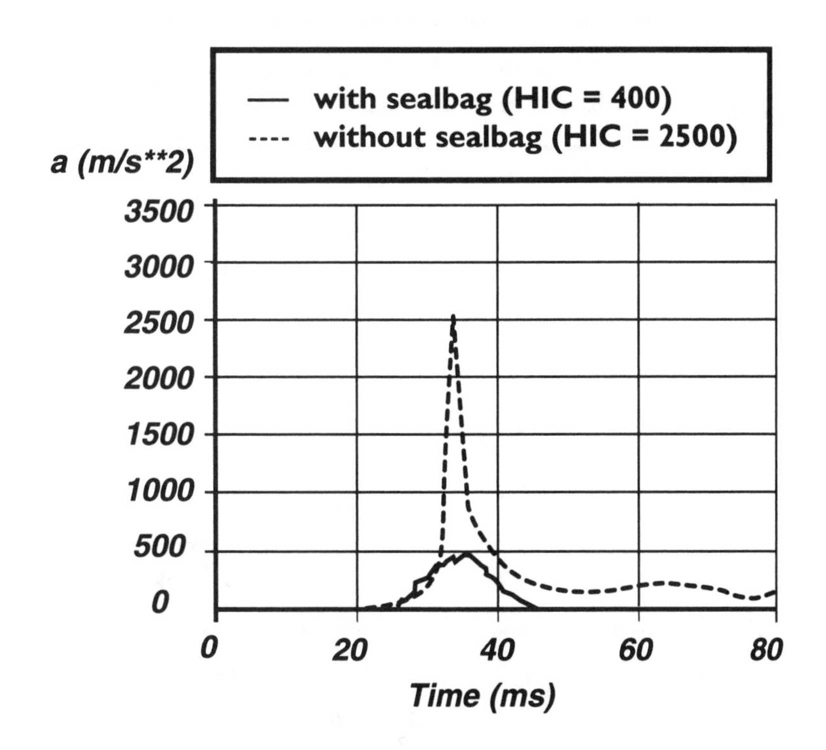

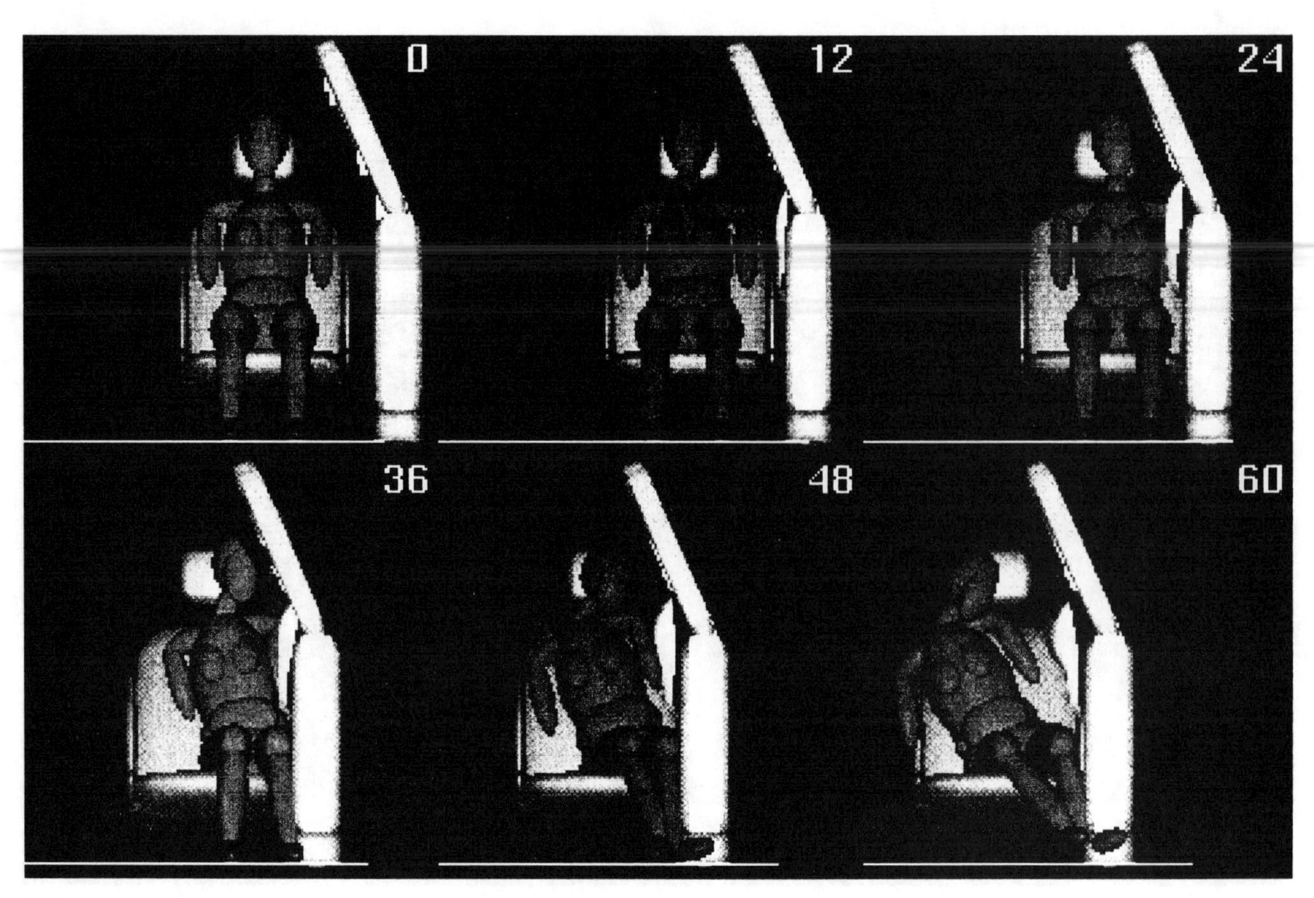

8 Madymo evaluation

- Dummy cinematic with thorax and "seal"–bag at "post" – impact
- 5% women/max. volume at 20m sec.
- seat foremost location

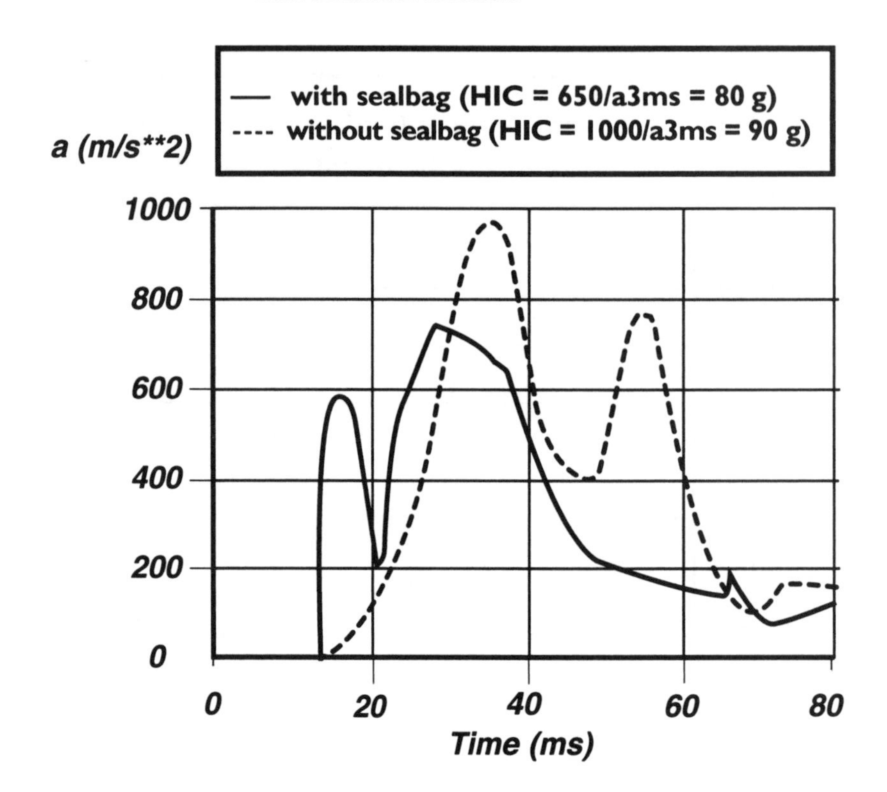

9 HIC 1350 to HIC 1050

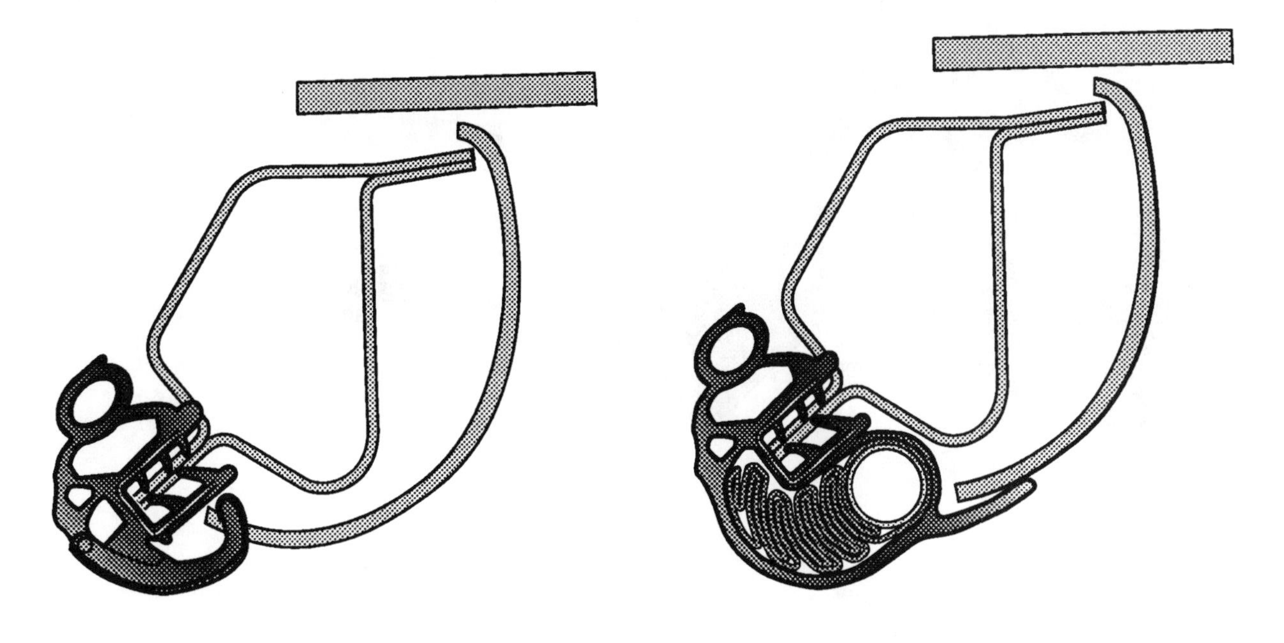

10 HIC 750

11 A-pillar section

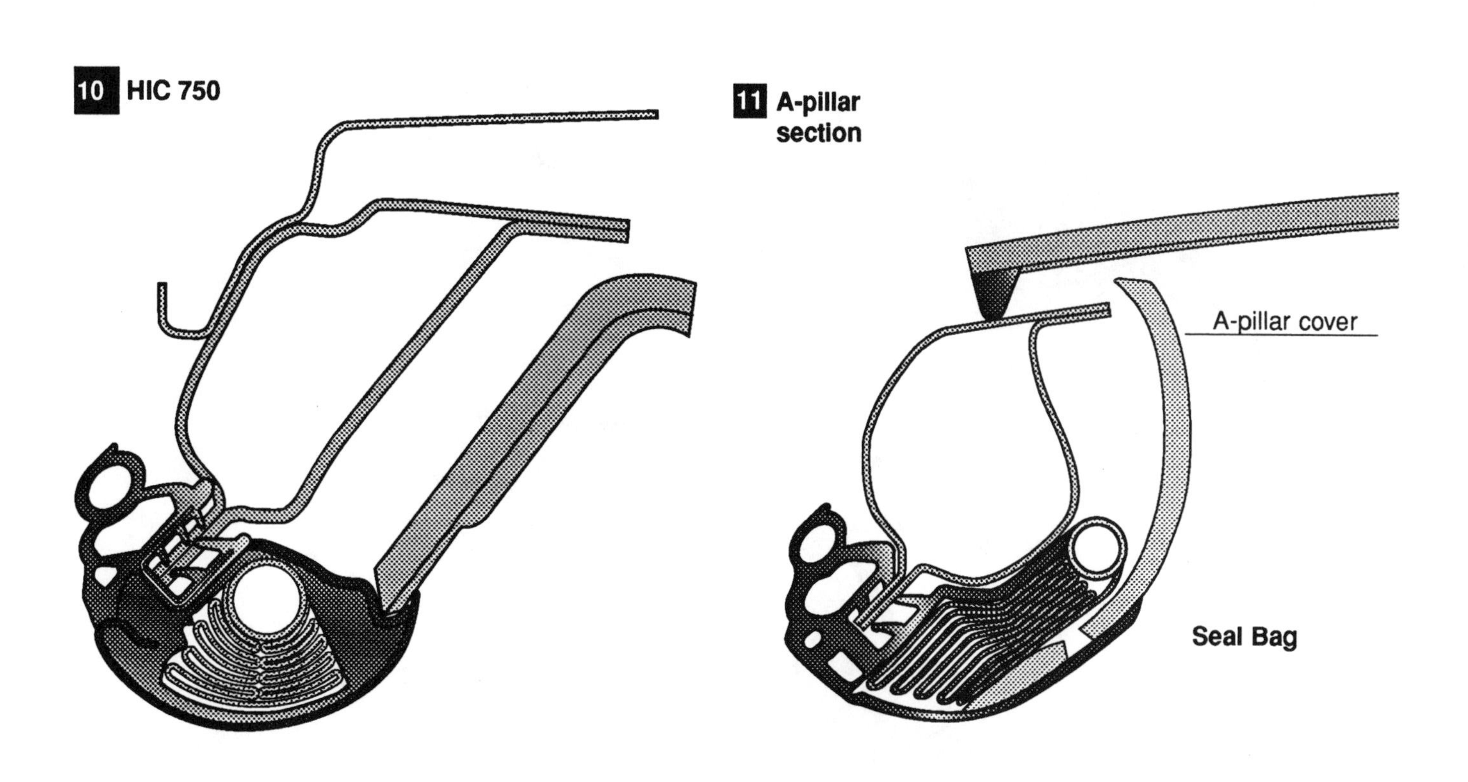

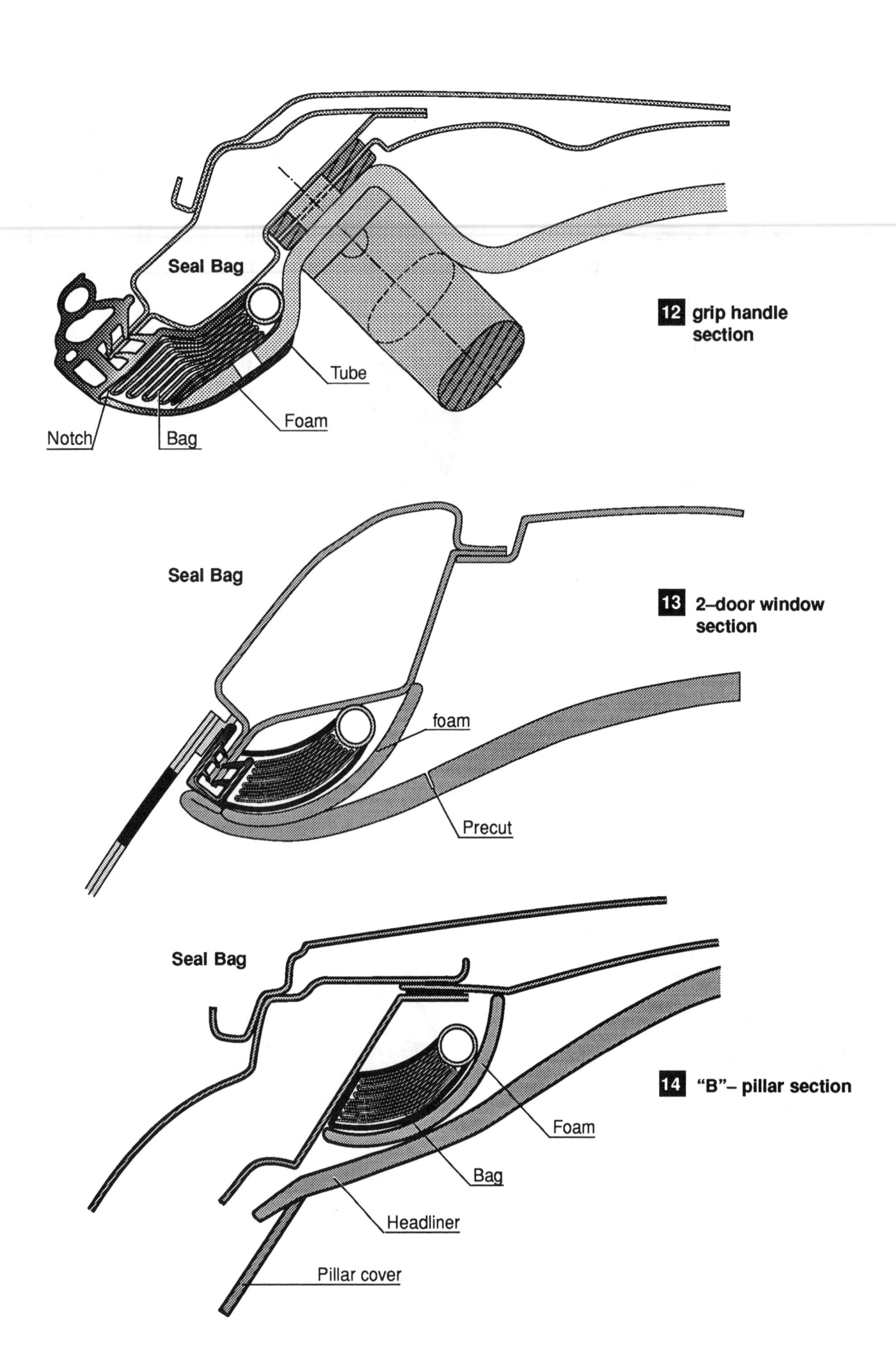
Seal Bag
Tube
Foam
Notch
Bag
12 grip handle section
Seal Bag
foam
Precut
13 2–door window section
Seal Bag
Foam
Bag
Headliner
Pillar cover
14 "B"– pillar section

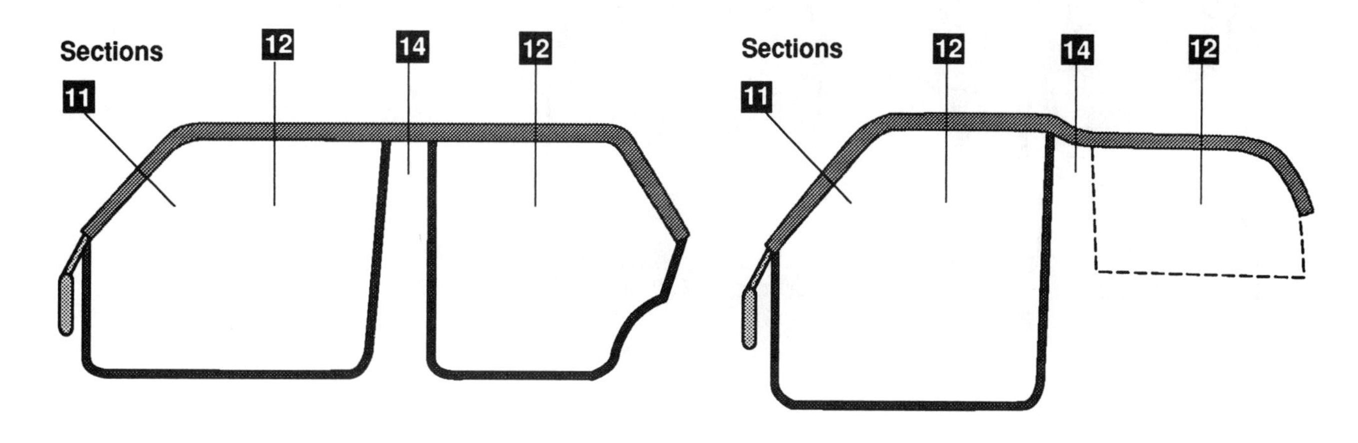

15 schematic layout seat bag

16 Layout of module headliner with "ROOF" bags

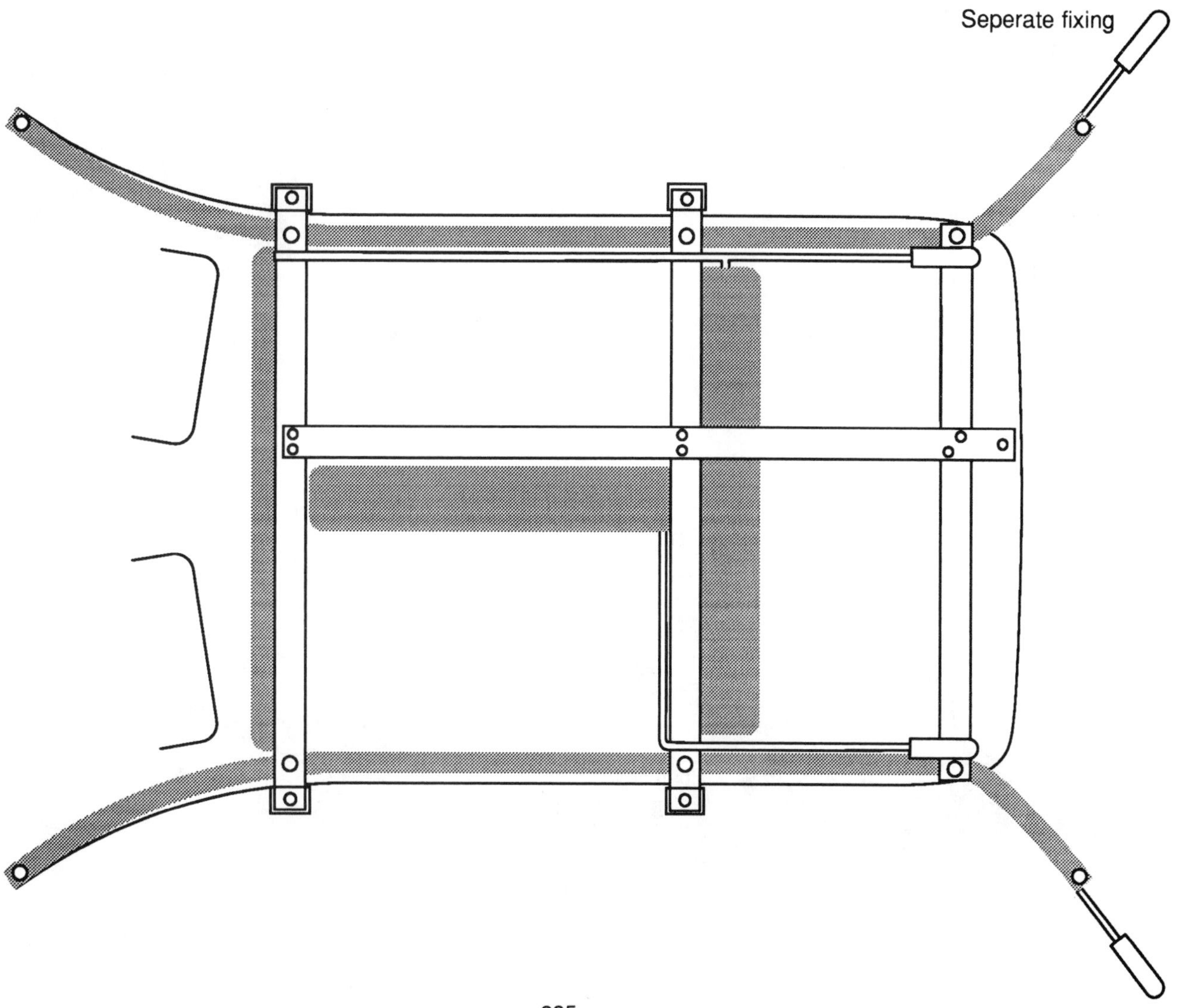

17 **roof bags**

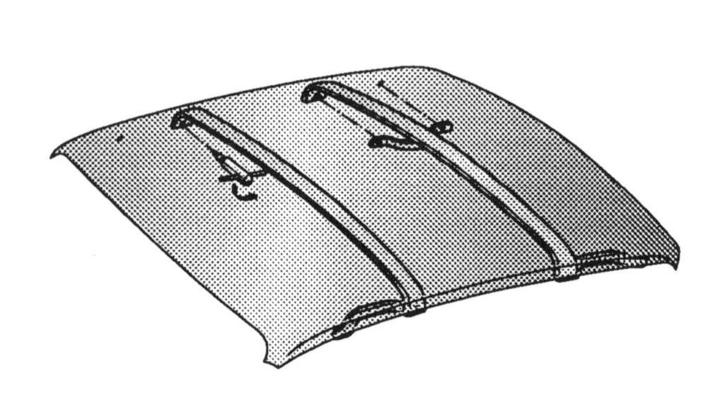

18 **formed headliner leafspring reinforced**

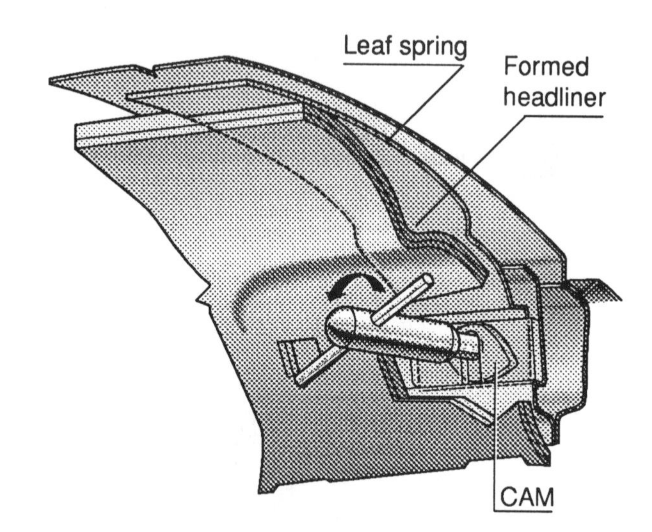

19 **section**

20 **section**

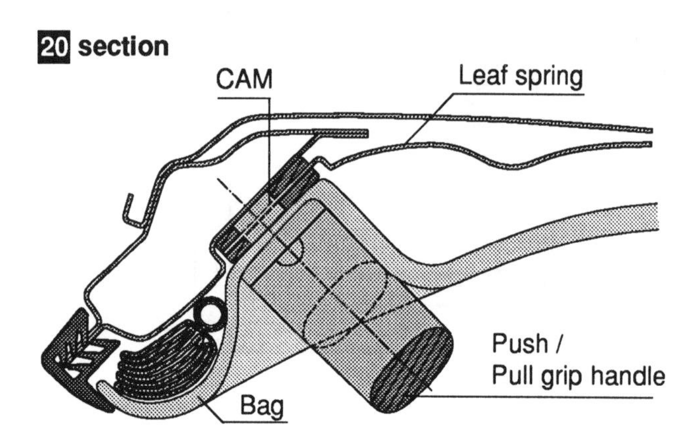

21 **2-door window/ section**

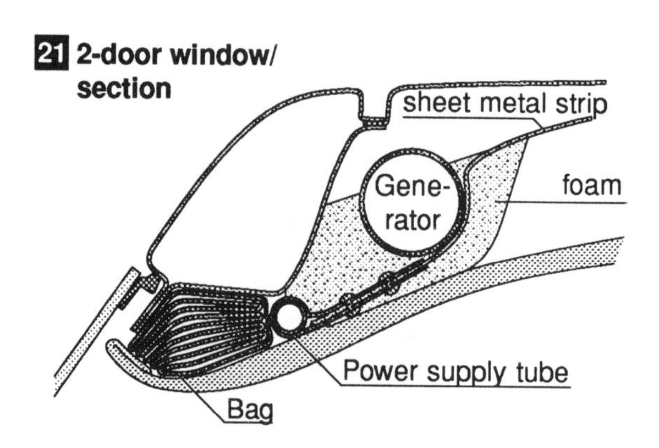

22 Bag beetween rear and front passengers fixed to leaf spring

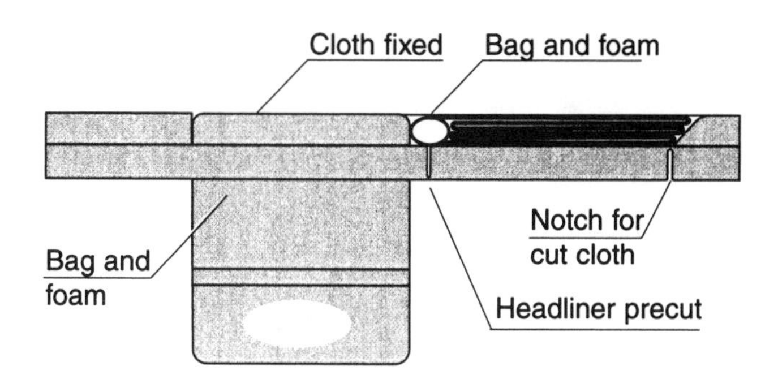

23 "A" – pillar section

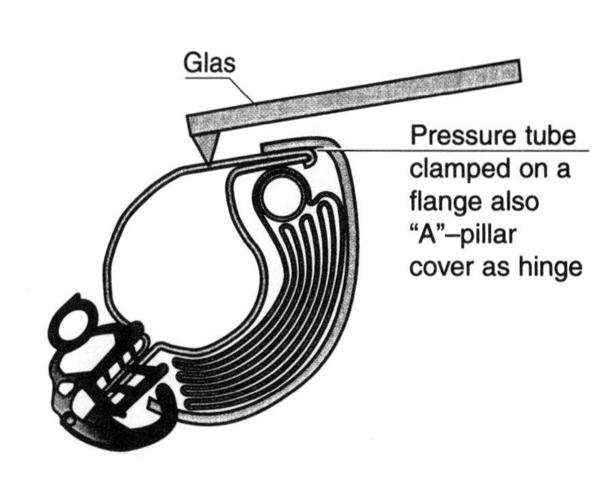

24 Roof-rail section

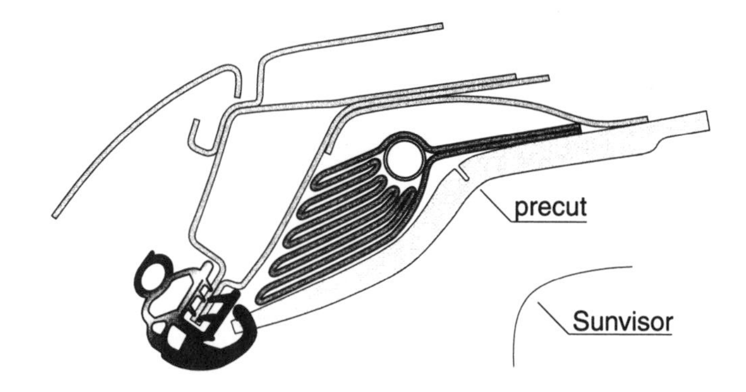

25 Sunvisor "X" section

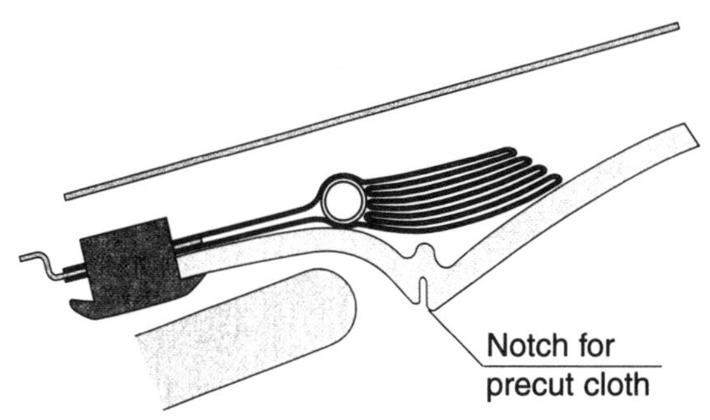

Side impact and sensing

*by **Kevin Jost**, Associate Editor*

Each year in North America, about 8000 automobile deaths and 27,000 serious injuries are attributed to side impacts, according to TRW officials. One way to reduce these numbers is to equip a vehicle with side-impact air bags. On the heels of the introduction of the first production side-impact air bag, Volvo's seat-mounted unit, electronics companies such as TRW, Robert Bosch, and Siemens Automotive are helping to develop more sophisticated systems with improved sensing capabilities.

When designing for side-impact air-bag sensing, the first challenge was to develop a sensor which reacted in about 2.5-5 ms, says the Executive Vice President & General Manager of the TRW Automotive Electronics Group, Philippe Lemaitre. It was thought that predictive sensing was needed, but sensors have been developed to the point at which this is not the case (see sidebar). The second sensing challenge was the variation of structure collapse depending on side-impact intrusion location. This was accommodated at TRW with the system's computer algorithm. The third hurdle was the variation of lateral stiffness among vehicles and the resulting effect on crash-signal transmission. If a vehicle's structure is stiff laterally, meaning the side-impact energy transmits well to the floorpan or crossmember beams, then a single, central electronic box with two crossmember-mounted peripheral sensors is sufficient for side-impact detection. If the vehicle is not stiff laterally, structure intrusion can occur without the central unit or a crossmember sensor detecting any acceleration. In this case, the side sensors need to be in the crush zone (i.e., the door inner panel or the B-pillar) to respond within the required time. TRW has developed systems for both possibilities.

According to James Chamberlin, Vice President of TRW's Technology, Planning & Quality, Occupant Restraint System Group, it was a major engineering challenge to develop a side-impact air-bag system that is triggered fast enough, protects the appropriate areas of the human body, keeps the head inside the vehicle, and positions the occupant to dissipate the energy. But TRW and other companies have developed systems, and the product plans of car companies worldwide call for introductions in the 1996-1999 time frame.

Automakers and suppliers are developing side-impact air bags that deploy from the seat cushion, seat back, door, and roof rail/B-pillar. A year ago, product plans called for predominantly small,

Side-impact crash and air-bag deployment simulation using computer-aided design (courtesy of Siemens Automotive).

Next-generation side-impact sensors

To meet the greater requirements of side-impact sensing, engineers at Robert Bosch have developed a sensor with a triggering time of less than 5 ms. Air-bag deployment within 10-12 ms after triggering would be possible with a new inflator. The company expects to begin producing the sensor in 1996 for a MY 1997 European vehicle, and in MY 1998 for a U.S. model, for door-mounted side-impact air bags.

Prototype testing of these satellite accelerometers has shown promising results at B-pillar, crossmember, or door vehicle locations. These accelerometers can be connected to a central sensing unit which deploys not only the driver and passenger frontal air bags, but also the side-impact air bags. The final trigger decision and command are made in this single electronic control unit. A high-speed digital communication link is used between the peripheral acceleration sensors and the central unit for fast response.

Bosch also offers a stand-alone air-bag triggering unit which deploys just the side air bags and is independent from the frontal air-bag system. Benefits of the stand-alone triggering unit are less wiring, components can be sourced independently, development time to add an air bag with a stand-alone system is less, and the wiring is more direct and can go directly to the side inflator—but total electronics content is greater.

Siemens Automotive has introduced an alternative to traditional side-impact sensors, which it says will enhance the effectiveness of side-impact air bags. In comparing the behavior of its absolute pressure sensor with that of traditional accelerometer sensing concepts for vehicle impact detection and air-bag deployment, company engineers claim the pressure sensor is up to three times faster, firing the air bag less than 5 ms after impact.

The pressure sensor is mounted inside the vehicle door cavity, where it monitors dynamic pressure changes to the air within the cavity. The micromachined unit is made of two silicon layers with an evacuated cavity between them which provides pressure refer-

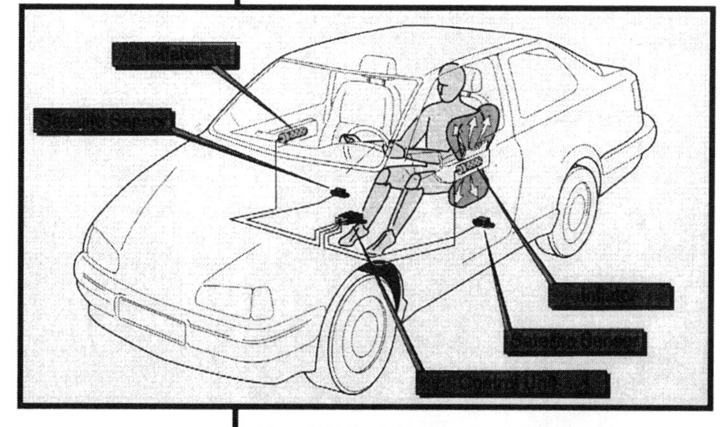

Bosch side air-bag system with central sensing unit.

door-mounted bags of 9-11 L for thorax protection, says Mr. Chamberlin, but now, the predominant position is seat-back mounted. There are advantages to both: seat modules are relatively small and door bags are always deployed toward the occupant. There are out-of-position concerns with seat-mounted bags if an occupant is leaning against the door—the bag deploys from slightly behind and not completely from the side.

About 47% of deaths in side impacts are due to head injury, which occurs when the occupant's head strikes the hood of the intruding vehicle, or the window sill or B-pillar of the occupant's car. (Some injury of the inboard side of the head has been found to be caused by impacting the other occupant or, in the case of an unbelted occupant, the opposite B-pillar.) Mr. Chamberlin says the best way to prevent injury is not to provide a head air bag, because the side glass is gone by the time the head reaches it in a severe impact greater than 50 km/h, and there is nothing for the air bag to react against. He believes the best way to control head injury is to control the upper torso. If the upper torso is controlled, then the head can be kept away from anything it could strike, and head injuries are reduced dramatically—up to 65%.

An air-bag system is designed for each specific vehicle. Much of the side-impact air-bag development is being done for Europe. European vehicles have B-pillars which are farther forward than their North American counterparts, so a B-pillar-mounted system may be desired. TRW has developed and tested headliner-mounted air-bag systems primarily for Asian customers' smaller vehicles, but the trend is toward seat- or door-mounted systems. □

Reader Feedback

Please circle all the numbers on the enclosed Reader Service Card which pertain to this article.

Circle 271	This article is directly related to my work
Circle 272	Technically, a very useful article
Circle 273	Read for information only
Circle 274	Would like more articles in this area
Circle 275	Did not find article useful

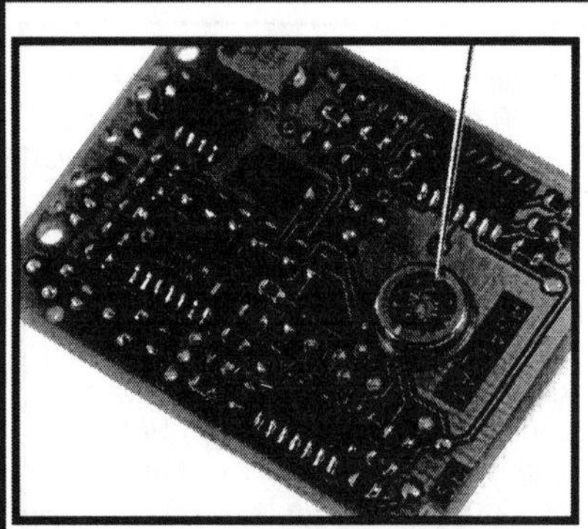

Siemens side-impact ECU featuring the company's patented absolute pressure sensor (located by needle in photo), which triggers side-air-bag deployment in under 5 ms.

ences. Changes of external pressure lead to a deflection of the second, silicon-membrane layer. Resistors incorporated into the membrane provide pressure-change information to a microcontroller. Typical dynamic pressure changes for severe side-impact collisions fall in the 2-20 kPa range at an atmospheric pressure of 100 kPa.

During product testing, the sensor was evaluated in crashes of more than 30 vehicle platforms, at various angles, and in collisions with poles and trucks. The sensor concept showed an extremely high immunity against simulated misuse and abuse conditions including bicycle impact, foot kicks, hammer blows, frontal crashes, and door slams. The Siemens system will be introduced in the European market by model year 1997. □

934217

Side Impact Airbag Technology

by R F Else, BSc, CEng, MIMechE, *Autoliv Ltd*

SYNOPSIS

Airbag technology is now widely employed for the increased protection of occupants in frontal impacts. Electrolux Autoliv has developed a related technology & test methodology for side impact protection. The use of such a product has shown significant benefits over the sole use of foam structures in reducing occupant injury criteria.

1 INTRODUCTION

Side impact crashes are a significant cause of injury to vehicle occupants and although the occurrence of such accidents is lower than frontal impacts (see fig 1) the severity of the resulting injuries is higher. This is primarily because the energy absorption potential of the side structure of the vehicle is much lower than the frontal structure (typically by a factor of 5). On the struck side of the vehicle the occupant is located very close to the sites of structural intrusion and this leads to injury being at a rate double that of occupants located on the non-struck side.

The situation is being addressed by legal frameworks in both the USA & Europe - although the test methodologies differ in each territory (mainly to reflect the typical vehicle population).

Various technical solutions are being developed to comply with such legislation. It is important to note that the effectiveness of such measures should be evident in real world injury reduction rates - rather than statutory test results to a narrow test envelope. In particular the occupant population involved in side impacts has a high median age and the ability of the human anatomy to withstand injury significantly reduces with age. See Fig.2.

2 OCCUPANT PROTECTION STRATEGIES

The root causes of injuries to occupants involved in side impact crashes can be broadly split into two areas:

- The behaviour of the vehicle structure when it is struck
- The interaction of the structure with struck side occupant

It has been shown that the vehicle structural behaviour in a side impact event needs careful control. There are two related key objectives in the development of such a structure. The first is to limit the intrusion of the side components of the vehicle body (by increasing the energy absorption of the structure to reduce both the intrusion and the impact velocity to the occupant - the latter being a critical characteristic). The second is to ensure the intruding structure interacts with occupant in a way that limits peak loading to the occupants' anatomy. In particular the door structure should remain substantially vertical as it intrudes. It is vital that the vehicle is designed such that the structural intrusion and the energy transference to the occupant is closely controlled and the crash worthiness of the structure is optimised before the additional protection for the occupant interface is developed.

There are two basic strategies for managing the energy transfer to the occupant from the structural intrusion:

- The sole use of energy absorbing foams and similar materials

- The use of an airbag system in conjunction with energy absorbing foams.

If the vehicle side impact crash worthiness is of a high level, then the legal side impact legislation tests can typically be met by using only foam structures which are placed between the occupant and the structural components of the vehicle sides. Whilst this may satisfy a legislative requirement it does not offer the optimum solution to reduce occupant injuries to the lowest practical levels which should result in the best 'real world' performance. For several years Electrolux Autoliv have been researching & developing side impact airbag systems which are now being productionised for the first vehicle applications. These systems offer a measure of protection significantly above that of conventional foam structures.

3 SIDE IMPACT AIRBAGS

3.1 System Configuration

The airbag system that is the subject of the development programme is used in conjunction with energy absorbing foams. Basically the airbag system functions to provide protection to the chest whilst padding provides protection to the pelvis region. The airbag is used to spread the load from the intruding structure over the widest possible area of the chest, but it also moves the occupant laterally away from the intruding structure reducing the deformation based injury criteria. Not only does this reduce the peak biomechamic loading on the occupant but it also results in a reduced head excursion trajectory (typically 50-70mm) to minimise adverse contacts to this vulnerable body segment. Note that the distribution on injuries to body segments (in the range A.I.S 3-6) is biased to the chest/pelvis/abdomen - these form typical 56% of all side impact injuries. See Fig.3. Figure 4 shows a schematic arrangement of the airbag system with a door mounted module.

3.2 System components

The airbag system that Electrolux Autoliv have developed is similar to a frontal impact system in that it consists of:-

- A sensor/initiator system

- An inflation means (gas generators)

- A bag - typically 12L

It compliments the foam padding installed at pelvis level in the vehicle side/door structure. The strategy for the system components is as follows:-

3.2.1 Sensors

There are a number of potential sensor types to provide system initiation - these are discussed below:

(a) Conventional occupant compartment systems, similar to frontal crash devices are not appropriate as they will respond too late in a side impact event. The occupant will have been injured by the intruding side structure before the vehicle will have experienced any significant lateral accelerations.

(b) By mounting the sensor or accelerometer in the side structure will ensure that the relevant accelerations are sensed earlier in the crash event, but even at very sampling rates crash discrimination will be difficult. The ability to filter out minor localised impact to the vehicle, (eg swinging the door open forcibly against a post for instance), without initiating deployment is also difficult.

(c) The use of membrane switch type devices which requires the closure of several 'serial' contacts to provide a 'fire' signal was also investigated. Such an arrangement is feasible but still requires an electrical power supply and a system diagnostic function - to check for discontinuities in the wiring.

(d) Predictive systems such as short range radar devices scanning the area immediately to the side of the vehicle are complex, expensive and at an early development stage. They are not considered to be a cost effective solution to the problem.

(e) The idealised sensor device should be:

- Responsive to the depth of intrusion into the vehicle
- Responsive to the speed of the intrusion into the vehicle
- Require no electrical power or diagnostics
- Able to directly initiate deployment of the airbag system with a minimum time delay - typically within 2ms
- Unresponsive to kicking or hammering on the door

Electrolux Autoliv have developed a sensor that fulfils the idealised objectives highlighted in (c) above. The device is a mechanical 'percussion cap' sensor. A pyrotechnic charge is ignited when a striker pin exceeds a critical depth/speed characteristic - the latter being optimised for the vehicle side impact structural collapse mechanism. The initial pyrotechnic flame is used to directly ignite the gas generators in the side impact airbag module by allowing it to be 'conducted' down shock tubes between the sensor and the module. This latter effect takes place within 1ms.

The location of the sensor and its rigidised mountings form part of the vehicle structure/airbag development programme.

The pyrotechnic sensor configuration is shown in figures 5 & 5(a).

3.2.2 Inflators

Because a side impact event occurs so rapidly it is vital that the airbag can be deployed extremely quickly. To achieve this the effects of inflation gas flow paths need to be minimised and the actual burning process of the gas generation material needs to occur rapidly.

To achieve these goals a system has been developed that uses two micro gas generators at either end of the module - minimising gas flow path lengths (these are also shown in figure 5).

The pyrotechnic charge in the gas generators is a double base material which incorporates stabilisers to enhance its long term temperature ageing characteristics. The material (up to 4g) is ignited directly by the flame front travelling down the shock tubes. Using this system the airbag can be fully inflated in around 8ms - still allowing there to be typically a 100mm clearance between the door surface & the occupants chest.

3.2.3 Airbags

Initial development work used airbags of around 8L which operated without vents. As the system was refined & optimised the bag volume has increased to around 12L (to reduce chest deflection at high door contact velocities) and now incorporates vents. A parametric study has been conducted to determine the relationship between the airbag characteristics and the resulting injury criteria - primarily by matching the force/deflection characteristics of the airbag to those of the occupant. The airbag must be softer than the chest - or the chest deflection will be increased instead of being reduced.

4 TEST RESULTS

Initial test work was undertaken using a dynamic sled facility which replicated a vehicle side impact event by striking a seated dummy with a moving door structure. The velocity/time history of the door closely simulates an actual crash test for the first 20ms. (Door velocities were used in the range 10-12 m/s corresponding to typical side impact tests on actual vehicles).

The development work has concentrated on the Bio Sid dummy which is the most biofedelic device available for such tests. The work has centred around the reduction in deflection based criteria - as proposed in the European legislation, rather than the acceleration based criteria stipulated in the American regulations. (Side impact injury mechanisms differ from those in frontal impacts - primarily because the loading rate is higher).

As a comparison to the airbag performance, tests were also run using foam structures alone - to determine the benefit derived from the use of airbag technology. (Note that the airbag was used in conjunction with an open cell foam padding at the pelvis level - 75mm thick with a density of 30-40Kg/m^3).

The force deflection characteristics of the airbag with different initial pressures and vent areas were determined and matched with the same characteristics for the Bio Sid dummy. The test results indicated that a 12L airbag with an initial inflation pressure of 40KPa was the optimum configuration - with a vent area of $300mm^2$. This work was undertaken using a drop test rig with an impactor shaped to the same form as the chest of the dummy - the impactor being dropped at a suitable speed onto an inflated airbag.

The use of a 'characterised' airbag system in conjunction with pelvis level padding resulted in a very significant reduction in chest deflection (-40%) & viscous response (-60%) when compared to purely foam structures - at both test velocities (10m/s & 12 m/s). Tests to the American criteria for T.T.I. showed a significant benefit (-45%) when using the airbag - particularly at a 12m/s test speed. Typical test results are shown in figures 6,7 & 8.

The test work has also indicated that there is a relationship between injury levels and the impacting door velocity. Chest deflection is propotional to door velocity to the power of 2, whilst VC is proportional in the same way to the power of 4. This highlights the importance of good vehicle structural design to reduce the velocity of the side structure intrusion.

5 PRODUCTIONISATION

The design concepts described above are now being taken to a production stage. Actual crash testing has substantiated the benefits shown during the sled test phases. The embodiments of the design will be vehicle dependant. The airbag module for the first application will be seat mounted - deploying between the occupant and the door structure from a position on the outer edge of the seat. The sensor mechanism is also housed in the seat and is strategically located in a position to sense the early intrusion of the vehicle structure.

It is envisaged that such systems will show significant 'real world' injury reductions by reducing the severity of the interaction of the intruding vehicle structure with the typical vulnerable individual likely to be involved in such a crash.

6 ACKNOWLEDGEMENTS

This paper is based upon the work undertaken by **Yngve Håland of Electrolux Autoliv in Sweden & Bengt Pipkorn of the Department of Injury Prevention at Chalmers University of Technology at Gothenburg, Sweden.**

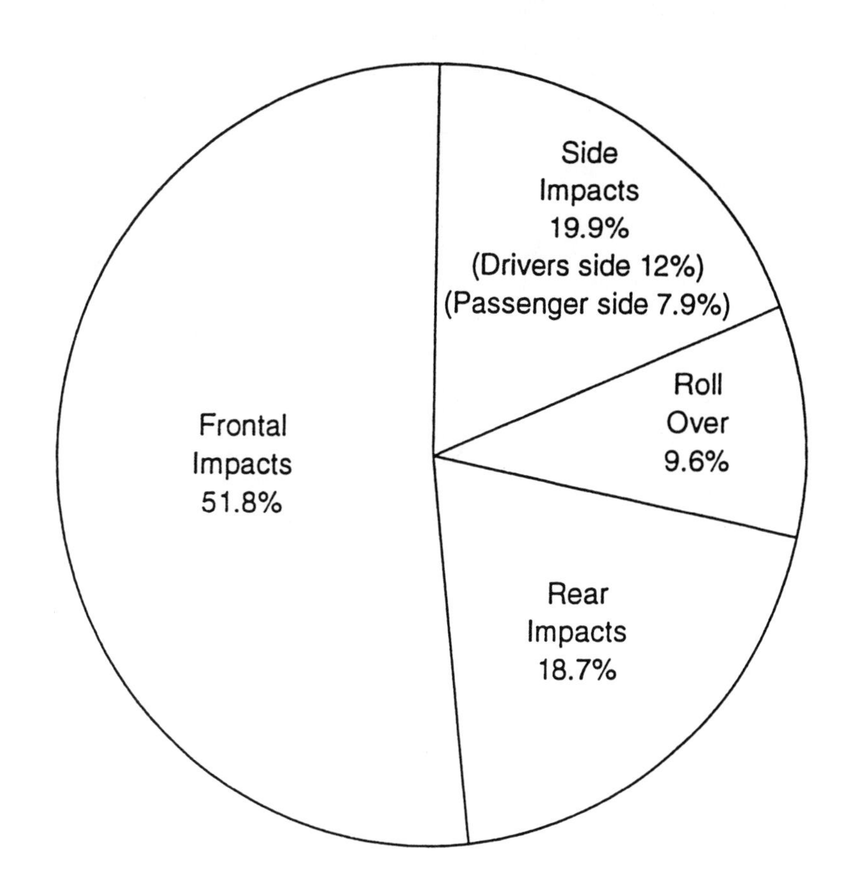

Fig 1 - Distribution of Impact Directions

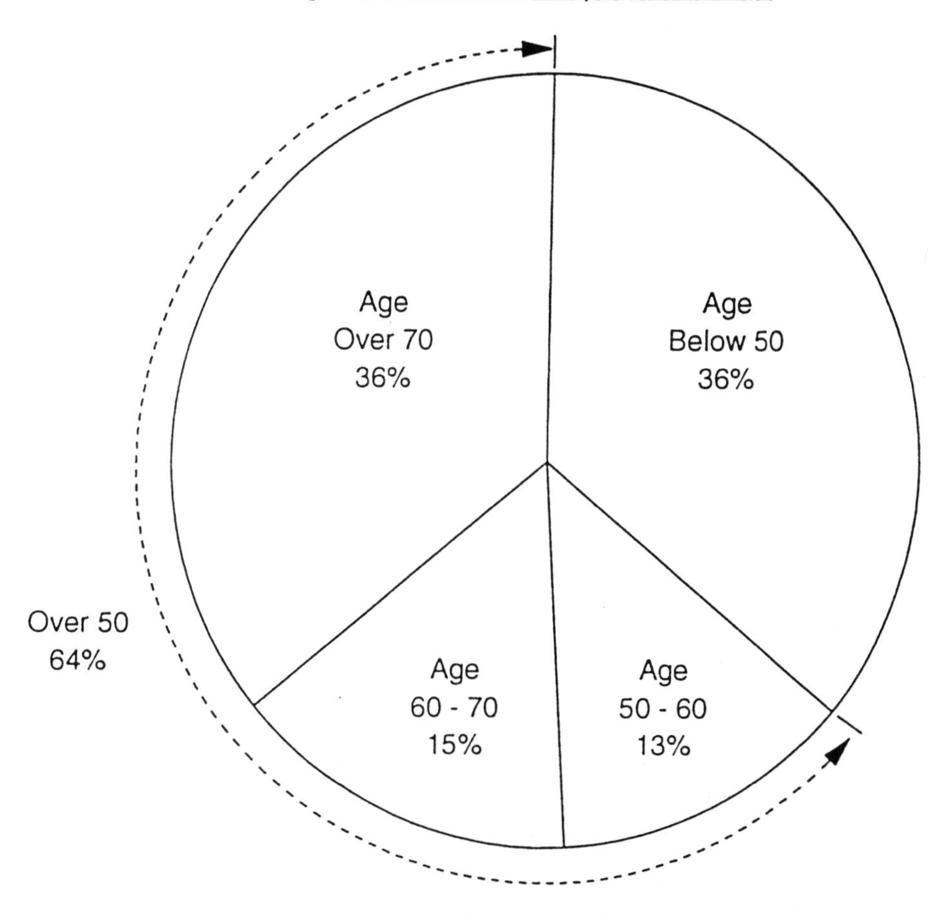

Fig 2 - Driver Age Distribution in side Impacts

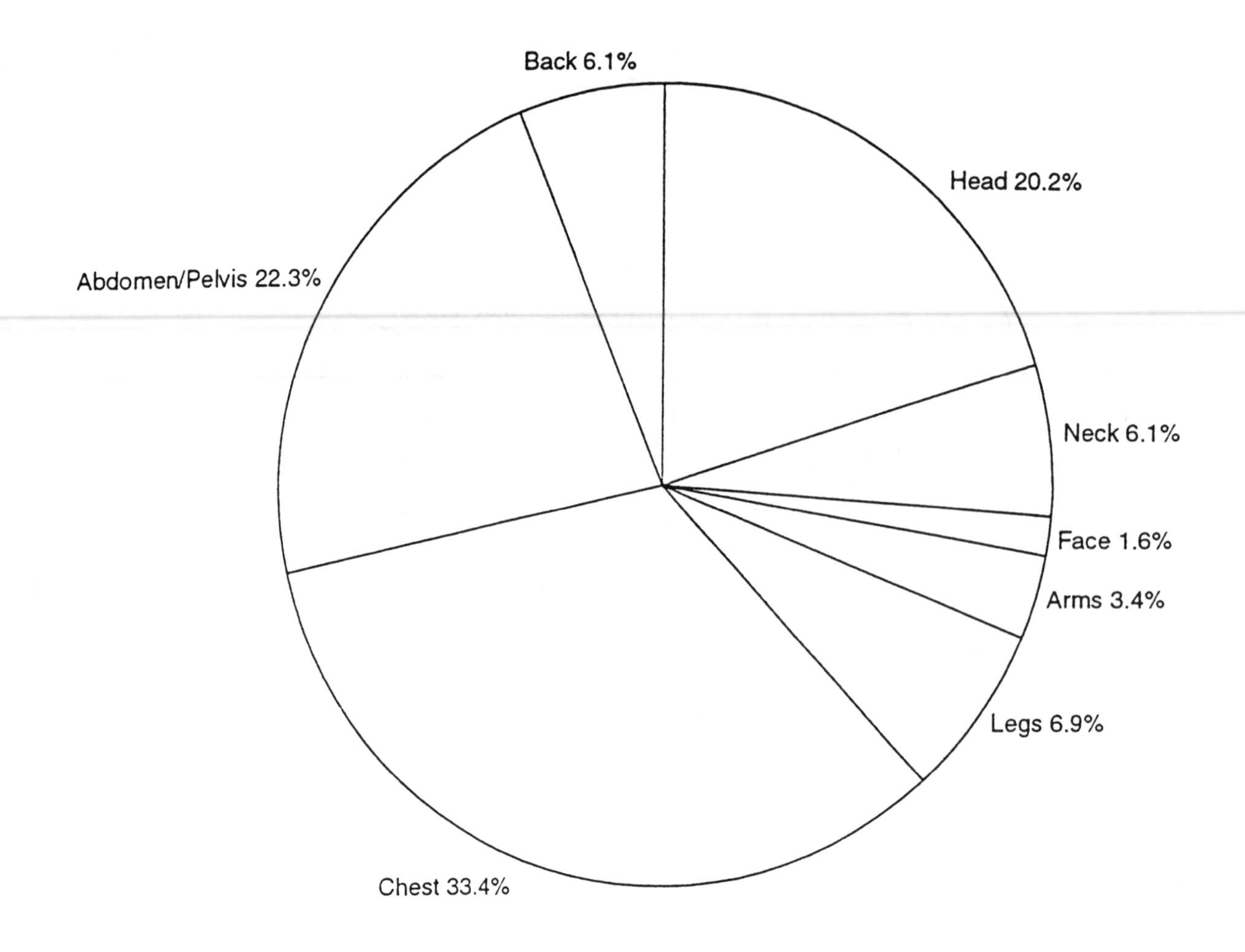

Fig. 3 Distribution of Injuries (AIS 3-6) in side impacts

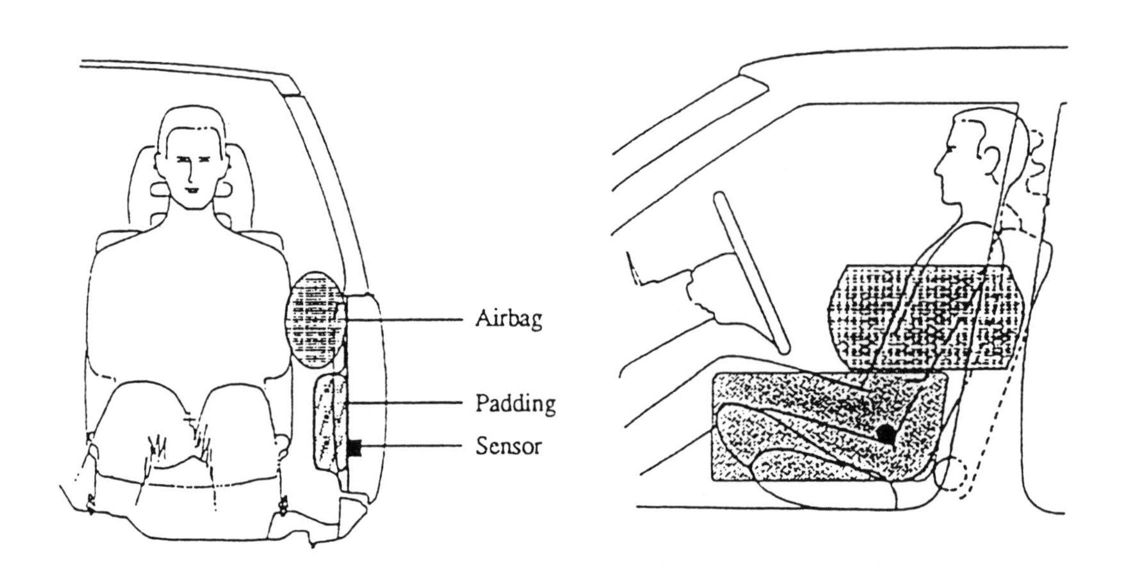

Fig. 4 Airbag & Padding System Schematic

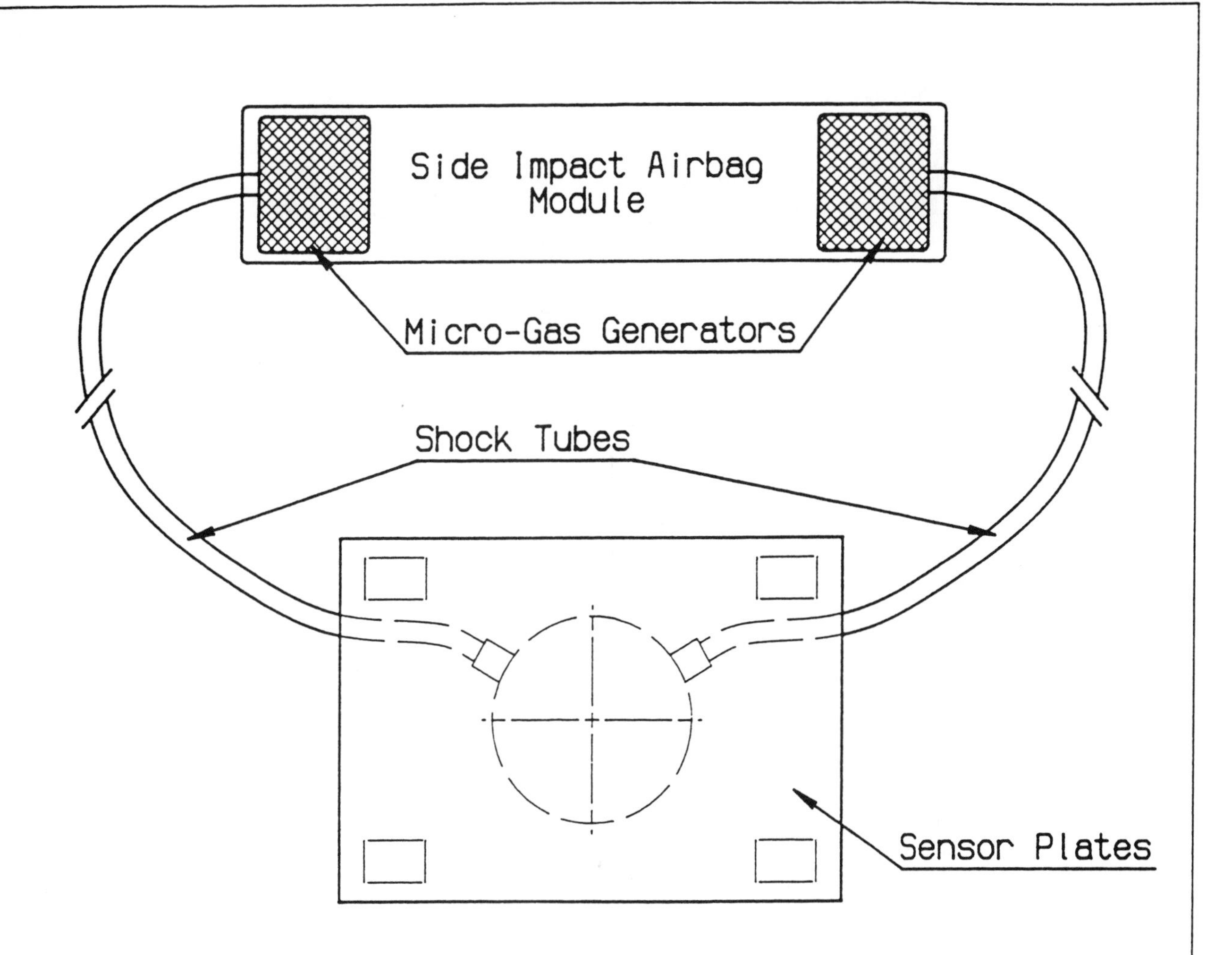

Note :- Sensor is typically mounted in the vehicle structure at a relatively low level in a region that will be crushed in the early part of a side impact event.

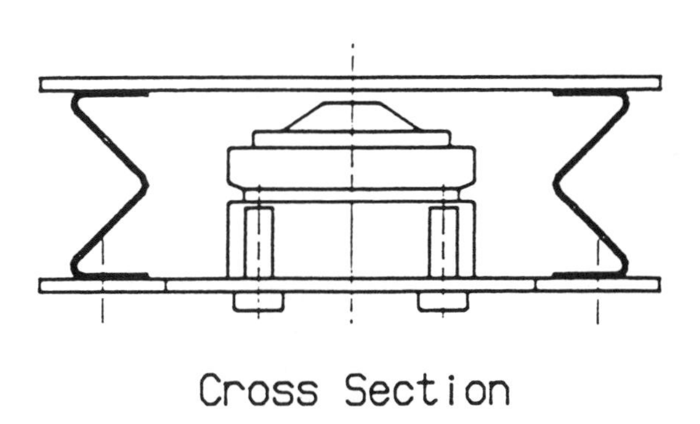

Cross Section

Fig. 5 Sensor configuration schematic

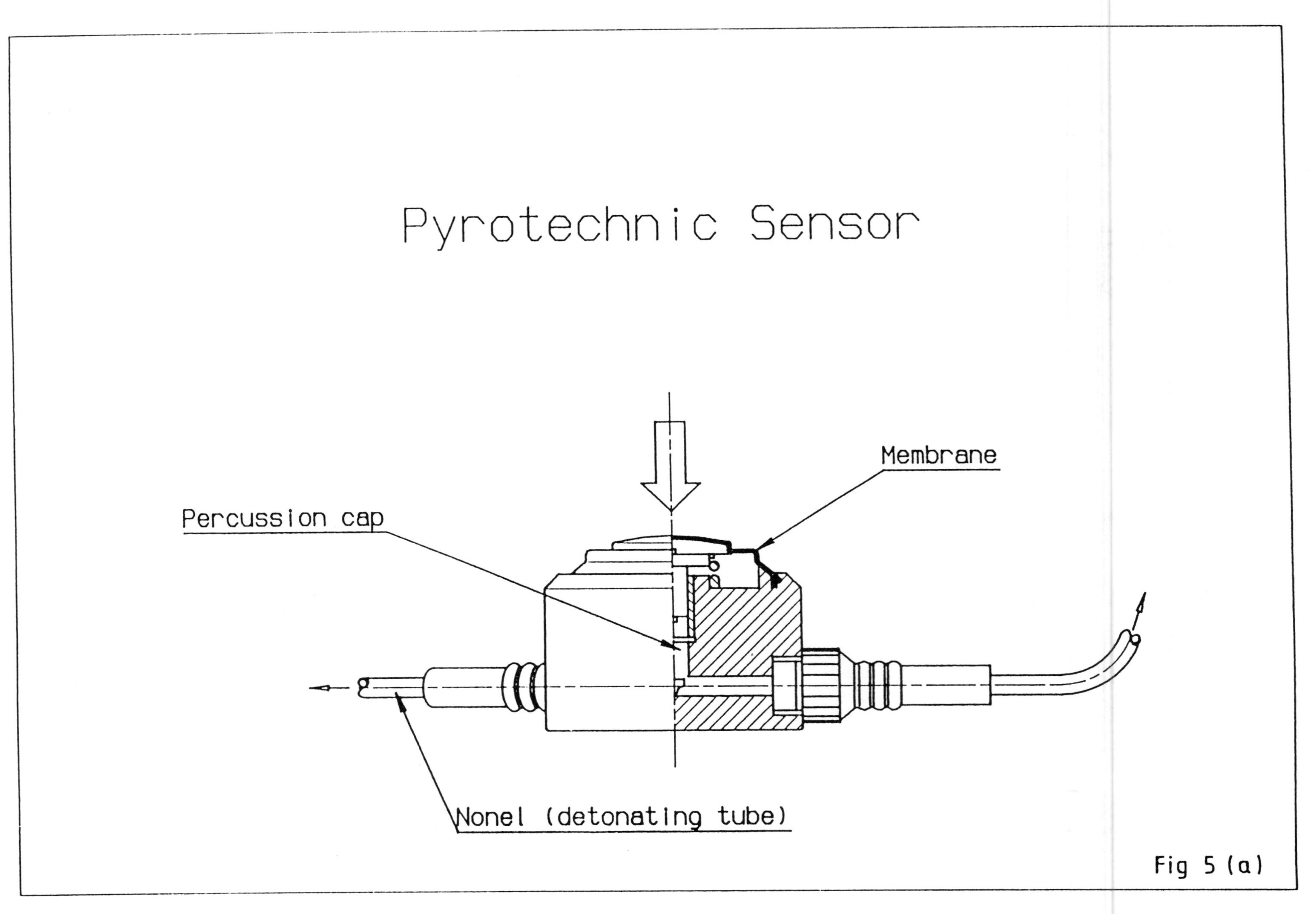
Pyrotechnic Sensor
Membrane
Percussion cap
Nonel (detonating tube)

Fig 5 (a)

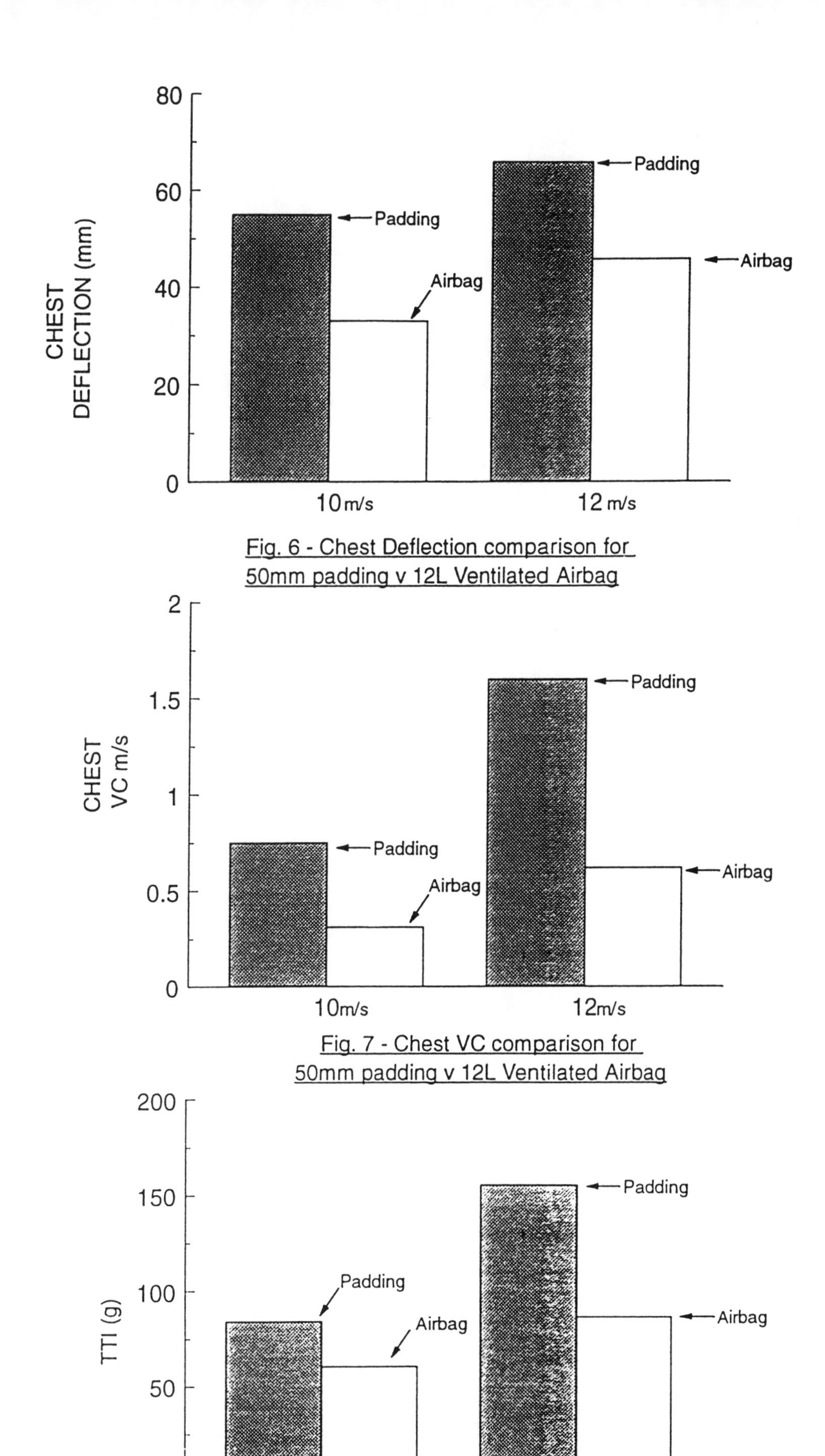

Fig. 6 - Chest Deflection comparison for 50mm padding v 12L Ventilated Airbag

Fig. 7 - Chest VC comparison for 50mm padding v 12L Ventilated Airbag

Fig. 8 - TTI Comparison for 50mm padding v 12L Ventilated Airbag

SAFING SENSOR REQUIREMENTS FOR USE WITH ELECTRONIC CRASH SENSING FOR AIRBAG DEPLOYMENTS.

David F. Gallup, Robert J. Bolender

TRW Transportation Electronics Division

ABSTRACT

This paper offers insight into the role safing sensors will take with electronic discriminators over the course of the next several years. While electronic discrimination offers the potential of lower overall system price and fast reaction times, it is also susceptible to a new set of failure modes such as electronic errors, EMI, and other electrical failures. Additionally, older requirements such as extreme chemical compatibility and long closure duration are relaxed for electronic discrimination. The paper analyzes the requirements for safing sensors for use with electronic discriminators. It evaluates the existing technology and concludes that while existing state of the art has proven to meet the needs of mechanical sensing systems very well, current sensors require improvement to work with electronic discriminators. It concludes with a brief discussion of TRW's development of a new mechanical safing sensor tailored to meet the needs of electronic discrimination.

INTRODUCTION

Electronic crash sensing is an emerging technology for controlling active automotive safety systems such as airbags and seatbelt pretensioners. The transistion from electromechanical sensing systems that currently dominate the market, to electronic, algorithm based crash discrimination, necessitates reexamination of the requirements placed upon the rest of the safety system. Electronic discrimination eliminates some of mechanical systems' failure modes but introduces new failure modes to the system. These failure modes and their effects result in a need to reexamine safing sensor requirements and technologies.

AIR BAG SYSTEMS

Electromechanical air bag systems have been in use since the early 1970's. High reliability was the focus of most systems. They tended to be high price, required routing of wires to the front of the vehicle (the best crash sensing location), and had separate diagnostic modules. Current mechanical systems are priced less and include one of the crash sensors inside the diagnostic package; however, the cost is still relatively high and routing of wires to the discriminating sensors is still a common requirement.

Air bag systems utilizing an electronic discriminator have been in use since the early 1980's. Electronic sensing shows potential for easier installation due to a centralized, single point sensing location, lower overall system cost, and more advanced diagnostic functions. Difficulties include a lack of data from field use and validation of electronic sensing methodology (algorithm methods). With mechanical systems, it is possible to run a crash test at 16 MPH, at 30 MPH, and interpolate performance to all the crashes in-between. Mechanical sensors are robust, allowing them to be placed in the engine compartment where crash sensing is easier. Discrimination of crash events becomes more difficult as the sensing location moves further back in the vehicle. Algorithms are required to properly discriminate between crash events from a poor location. This results in increased complexity which makes interpolation of intermediate crash performance and extrapolation to unusual events more difficult.

Electronic discrimination brings a new set of failure modes that need consideration. Among these are items listed below:

ALGORITHM ERRORS are a concern for many automotive OEMs. Algorithms that work on multiple car lines and are independent of mounting location are difficult to write, difficult to test, and have limited field experience to draw upon. Algorithms have been developed to meet all anticipated events; but, there are many unknowns concerning the real world performance of crash impact algorithms. In addition, the sensing location is critical, there is electrical noise that affects the low level signal from the accelerometer, and algorithms must be forgiving to manufacturing variation in the production of accelerometers.

ELECTRICAL SINGLE POINT FAILURES within the electronic sensing package can be a concern. Most of these are eliminated by design at minimal cost. Combinations of failure modes cannot be completely eliminated, leaving a need for protection.

SYSTEM SHORTS to ground or to a current source are failure modes that deserve consideration. Mechanical systems usually place the firing squib in between the discriminating and safing sensors. This isolation by mechanical switches provides a high degree of protection. In a single-point sensing unit, it is advantageous to place a safing sensor on the opposite side of the squib from the discriminator. This circuit provides additional protection against mechanical or electrical shorts in the system.

ELECTRO MAGNETIC INTERFERENCE (EMI) is a significant performance concern. Most systems have been designed to minimize the effects upon the electronics; however, EMI is capable of causing significant changes to the signal from the accelerometer. These changes directly affect the algorithm. A method of mitigation is to place two accelerometers in parallel; but, in opposite directions. Such two accelerometer systems would be more inherently immune to EMI since EMI would cause a common drift of the accelerometer signal and it is easy to design in compensation for this. It would not be completely immune; however, it would be much less susceptible to detrimental EMI affects. Cost is a significant drawback in two accelerometer systems since accelerometer cost is a significant portion of total material costs.

SAFING SENSOR UTILIZATION

The use of a mechanical safing sensor greatly enhances the reliability of electronic discrimination systems. A mechanical switch provides immunity to the above listed issues of EMI, accelerometer or algorithm error, single point electrical failure, and provides more protection from electrical shorts. It also provides an additional level of security during the introduction of electronic discrimination. The demands for mechanical safing sensors designed for use with electronic discriminators will be different from those for use with mechanical systems. Price, reusability, fast response time, small size, ease of installation, and reliability are the main considerations while chemical compatibility and long duration are not as significant.

As airbag systems have become more common and their use has spread from the initial market of luxury cars, price has become an increasingly important factor. Some European automotive manufacturers have rated price as more important than reliability and North American manufacturers are not far behind. The Asian market is still placing a higher value upon reliability; however, there is a strong drive to lower price.

The reuseability of the sensing system after an airbag deployment is becoming more important as a result of the drive to lower system cost. Automotive manufacturers from all nations have expressed concern regarding the capability of mechanical sensors to repeatedly switch the firing current. Concern exists that mechanical sensors may not reliably perform this function. As a result, they have expressed the desire for the safing sensor to always close before the algorithm so that it only has to pass the firing current and is not doing the switching. Diagnosis of which sensor performed the switching action during the deployment has been designed into some systems so that if the safing sensor did not do the switching, then the module could be used again. The module would be replaced if the safing sensor switched the firing current. As a result, a safing sensor capable of switching the firing current multiple times would be an added value to automotive manufacturers by eliminating the requirement for diagnosis and necessity of replacing modules after use. The current carrying capacity required of the safing sensor will increase with time as multiple airbags and seat belt pretensioners are phased in.

Reaction times for safing sensors used with electronic discriminators need to be faster than those used in mechanical systems. Algorithms do not have to wait for mass movement to close contacts before firing. Also, they can be more sophisticated in evaluating the crash signal to determine fire or no fire. Hence, electronic discrimination can inherently be very fast in crash discrimination. A production vehicle with extremely fast discrimination performed by an algorithm required and demonstrated a reaction time of 4.2 milliseconds in a 55 KPH, 0 degree crash. Typical reaction times for current safing sensors are approximately 10 to 15 milliseconds for these types of events with a good crash sensing location. In the vehicle described above, the reaction time was required in order to allow time to fill a large passenger side airbag. This places strong demands upon the safing sensor if it is to close before the algorithm. Reaction times of electronic discriminator algorithms will continue to improve, creating a need for improved safing sensor speed.

Sensing system package size is very important to automotive manufacturers. This places obvious demands upon the safing sensor to be as small as possible. While space has always been a concern, the practice of mounting safing sensors on separate brackets will not continue and increasing customer expectations will continue to drive this issue in the future. One method for conserving space is to raise the sensor off the PCB to allow components to be placed underneath the sensor. This method can be employed with relatively low cost as long as crash pulse transmission is maintained and the safing sensor does not determine overall package height.

Ease of installation can be another major consideration. Many mechanical sensors today are adapted to work with electronic sensing by adding brackets to permit soldering to the PCB. Cost and piece count are reduced by the eliminating this bracket and designing a sensor that can be soldered directly to the PCB.

Closure duration is an important consideration in distributed mechanical systems because this assures that there will be overlap between the discriminating and safing sensor closures so that the airbag will fire. Mechanical sensors will open and close multiple times during a crash and it is an important part of modeling to be sure that there is a time period during which both mechanical systems. Drawbacks for use as a safing sensor with electronic sensing include size and reaction time. They are relatively large for use with electronic discrimination and the fact that the mass rolls forward to the contact means that rotational inertia, while less than that of a ball, must still be overcome in addition to translational inertia. This slows down the sensor.

DOWNSIZED ROLAMITE sensors are a more advanced version of the standard Rolamite sensors. Reduced size and cost are among the primary advances in the technology. The fundamental principle of operation is unchanged and as such they are very low friction and highly reliable. The primary disadvantage of this sensor for use as a safing sensor with an electronic discriminator is the reaction time. The mass must overcome rotational inertia. For some vehicles this is not a problem; however, it may be a limitation with others.

REED SWITCH sensors are relatively new compared to the above technologies. They consist of a tube shaped magnetic mass, restrained by a coil spring, moving over a hollow shaft. The hollow shaft contains a reed switch that closes when the magnetic field of the mass travels over it. Smaller size and lower price are primary advantages of reed switch sensors and they are being increasingly employed in safing sensor applications due to their lower cost, particularly among European automotive manufacturers. They are also fast because there is no rotational inertia to overcome. Primary disadvantages for their use as safing sensors with electronic discrimination include performance at temperature and reusability. Because of the magnets, the sensors can take permanent threshold shifts after exposure to high temperature. The contact closure can become noisy after high current usage. Reed switches can also have the distance between the reeds change in response large impacts such as drop, and this degrades their performance.

ROTARY sensors operate with an eccentric mass that rotates on a shaft and is held back by a spring. During a crash, the mass rotates forward to close the contacts. These have seen less use than the Rolamite and ball in tube sensors; but, have been in production for some time. They offer good contact quality due to a high degree of wiping and long closure times are achievable. Primary disadvantages include a slow reaction time and larger size.

WEIGAND sensors consist of a moving magnet in a tube. The moving magnet is restrained by magnets at both ends of the tube. An acceleration pulse will send the magnet down the tube and past a wiegand coil that produces a voltage pulse instead of a mechanical closure. These sensors are in development and are not in production. Primary disadvantages include additional electronics to record the voltage pulse and close a switch, size and cost.

ANALYSIS OF SAFING SENSOR TECHNOLOGIES

Currently, there is no safing sensor that is ideally suited for use with electronic discrimination. None of the described sensors meets the combined requirements of low cost, reusability, fast reaction time, small size, easy installation, and reliability. Many are well suited for use with mechanical systems; however, they are not optimized for use with electronic discrimination. As such, TRW has decided that advancements to the state of the art in mechanical safing sensors is required. Therefore, with guidance from many automotive manufacturers, TRW is developing a new sensor tailored for use with electronic discriminators.

COIL SPRING sensors are the product of this development and consist of a tubular mass moving on a shaft and restrained by a coil spring. Movement of the mass down the shaft permits the contacts to close. The sensor meets the new requirements by design. The mass only needs to translate down the shaft and has no rotational inertia to overcome to close the contacts. In this respect it is very similar to the reed switch design. It has mechanical sensors are closed before the required time to fire. Electronic discrimination relaxes this constraint significantly since the electronics will latch closed for a significant period after it is determined that an airbag deployment is required. This eliminates the need to be concerned about the discriminating and safing sensors opening and closing out of phase.

The environment in which the safing sensors will be contained is also less harsh than many current mechanical systems which were designed to be capable of mounting in the engine compartment. The passenger compartment will see smaller temperature ranges and fewer harsh chemicals. The sealing requirements of the sensor are therefore less demanding.

High reliability must obviously be maintained. In the drive for lower cost systems, it cannot be forgotten that peoples' lives may depend on a properly functioning airbag system. Electronic discriminator systems must, as a minimum, maintain the high reliability standards established by mechanical systems. OEMs often feel more comfortable with mechanical systems because they are more 'intuitive.' Features which enhance the reliability of the overall system and decrease the risks of new technology must be utilized to assure the acceptance of the technology in the market.

CURRENT SAFING SENSOR TECHNOLOGIES

BALL IN TUBE sensors operate on the principle of a tightly toleranced ball, nominally restrained by a magnet, which can travel down a tube to bridge two contacts. The tube that it travels down is also tightly toleranced so that the nominal gap between the two parts is very close. This provides a 'gas damping' effect that helps filter out high frequency shocks. With a high degree of gas damping the device behaves as a velocimeter and the sensor is used in the role of a discriminator. It should be noted that the gas damping is not required for discrimination. When this technology is employed near the passenger compartment, as would be required for use with an electronic discriminator, the damping is largely eliminated to meet the more difficult reaction time requirements. With the damping eliminated, the sensor behaves like a spring mass system. This sensor is appealing because the fundamental operating mechanism is simple and it is easy to put together. It has seen widespread use in mechanical systems. Drawbacks for use as a safing sensor with electronic sensing include reaction time and size. It is not possible to predict whether or not the ball will roll, slide, or 'float' down the tube. Rotation of the ball due to rolling slows sensor reaction time significantly by requiring energy to overcome rotational and well as translational inertia. Ball in tube sensors are also relatively large, requiring significant space.

PISTON IN TUBE sensors are newly developed sensors that are descendants of the ball in tube sensors. However, instead of a ball, restrained by a magnet, sliding in a tube, it is a ducted piston, restrained by a spring, sliding in a Teflon coated tube. Bimetallic contacts provide temperature compensation. Advantages of this sensor over the ball in tube sensor include less particle sensitivity, a more repeatable mechanism, and a faster reaction time. The last two improvements are due to the fact that the moving mass can no longer rotate. Rotational inertia is not an issue and this improves repeatability. Drawbacks for use as a safing sensor with an electronic sensing module include size and cost. It is still a relatively large sensor; and forming the ducts in the mass, stamping the bimetallic contacts, and coating the tube with Teflon are expensive operations.

ROLAMITE sensors operate upon the principle of a thin metal band wrapped around a rolling mass. The band is a spring that holds the mass away from a contact. The sensors are normally hermetically sealed and have negligible friction that results in a very repeatable sensor. They have a high degree of reliability and have consequently seen widespread use in contacts that have a controlled, short distance between them that also permit a very fast reaction time. Special efforts have been taken to enhance the current carrying capacity of the sensor through materials choices and careful design consideration. This can be an option for those customers desiring special high current carrying safing sensors in addition to the fast reaction time. Other goals were achieved by employing the latest design techniques of team development, the QFD process, and including the vendors in the design team. This has resulted in a design with very few parts, simple assembly, and easily manufactured parts since vendors and process engineers are part of the design team.

System cost has been reduced by designing the sensor to be soldered directly to the PCB without the addition of brackets, fasteners, or additional components. At no additional cost, diagnosis is available that will assure sensor presence and an intact firing path to and from the sensor. There are redundant connections including redundant contact welds to assure reliability targets. Traceability is being designed in at the front end to assure that the level of component traceability required by automotive manufacturers can be met with little or no additional cost. No magnets are employed to assure that there is minimal temperature dependence.

CONCLUSION

The advancement of electronic discrimination for crash sensing has resulted in a new set of failure modes and sensing requirements. Electronic discriminators offer the potential of being very fast and low cost, but require additional protection against new failure modes such as EMI. Overall system costs need to be driven down further while maintaining reliability. Additionally, some constraints have been relaxed such as closure duration and chemical compatibility. These new requirements demand a safing sensor that is lower cost, reusable, fast, easily installed, and reliable. While state of the art safing sensors have proven to meet the needs of mechanical sensing systems very well, they are not optimized to work with electronic discriminators. As a result, TRW has worked with many automotive manufacturers to understand and develop a new sensor that is tailored to meet the new requirements. The result of this development is the coil spring sensor that will be ready for production in January of 1994.

973240

A Remotely Mounted Crash Detection System

David B. Rich, Walter K. Kosiak, Gregory J. Manlove, and Dwight L. Schwarz
Delco Electronics Corp.

ABSTRACT

This paper describes a crash detection system developed for use in automotive passive restraint applications. The system is unique in that the small, rugged module performs all of the functions necessary to determine when airbag deployment is required. Because it is designed to be mounted near the impact zone, the system is well suited to three difficult crash discrimination situations: front impacts in body-on-frame vehicles such as light trucks, side impacts in any automobile, and crash severity discrimination for future adaptive restraint systems.

INTRODUCTION

The increasing usage of safety belts and the addition of supplemental frontal airbag restraints have brought about a significant decline in frontal impact injuries and fatalities [1,2]. Increased attention is now also being directed towards improving the level of occupant protection during side impacts. For both frontal and side airbag restraints, the crash sensor is a key system component.

FRONTAL IMPACTS - In frontal impacts the front structure, or crush zone, of the vehicle decelerates early in the crash event, intruding into the engine compartment and absorbing crash energy. Thus, the full distance between the occupant and the vehicle steering wheel or instrument panel is available for an airbag system to provide cushioning. Traditionally, the trigger time requirements for a frontal crash sensing system are defined such that an unbelted occupant has moved less than 125 mm by the time the airbag is fully inflated. Typically, this means that the sensor and algorithm must initiate airbag deployment within 20-50 milliseconds for a variety of frontal impacts.

Most modern vehicles use accelerometer-based crash sensors mounted in the passenger compartment to sense and discriminate frontal impacts. Typically, this location provides timely discrimination of impact events. However, for some body-on-frame vehicle structures such as light trucks, the isolation between the passenger compartment and the frame effectively delays the crash deceleration signal from reaching a sensor mounted in the passenger compartment, particularly for pole and angled barrier impacts. At the time that an airbag trigger is required, crash acceleration and velocity signals measured in the passenger compartment may be too small to make reliable discrimination decisions.

SIDE IMPACTS - Vehicle manufacturers are investigating structural enhancements, energy-absorbing padding, and side airbags to improve occupant protection during a side impact collision. However, the requirements of a side airbag sensing system are substantially different than those for frontal sensing systems. In the case of side impacts, the space between the exterior and interior of the vehicle and the space between the interior and the occupant are much smaller than in the case of frontal impacts. As a result, intrusion into the passenger compartment begins very early in the event, and the allowed occupant displacement is much less. Therefore, the trigger times required to initiate the airbag inflation are very short -- often less than five milliseconds.

CRASH SEVERITY SENSING - The next generation of adaptive airbag restraint technologies, such as variable-level inflation and dynamic occupant position-based deployment, will require an indication of the severity of the crash as quickly as possible. By combining information from a central passenger compartment sensor and a forward satellite sensor located in the crush zone, the airbag control system will have an early measurement of the severity of the crash event.

We have developed a crash detection system that reliably discriminates frontal and side impacts for a wide variety of vehicles. With some changes, this system can be extended to provide crash severity sensing for future adaptive airbag systems. In the following sections, we will discuss the overall sensing system design and the design of the key components: the custom IC, the acceleration sensor, and the package.

SENSING SYSTEM DESIGN

Electronic crash sensors for restraint systems have been in high volume production for many years and are continuing to become more and more sophisticated [3,4]. The performance requirements for crash detection systems in frontal and side impacts, as well as for frontal crash severity sensing, are very demanding. To reliably discriminate crashes in these situations, many vehicles require satellite sensors mounted in the crush zone.

Figure 1 shows typical signal plots for a 13.4 m/s (30 mph) frontal barrier impact for a light truck vehicle. Note that the acceleration signal amplitudes seen at the radiator support location are at least an order of magnitude greater than the amplitudes seen in the tunnel. Similarly, signal plots of a 15.2 m/s (34 mph) side impact barrier test for a midsize car are shown in Figure 2. The acceleration signal seen in the B-pillar location is much higher in magnitude than that seen in the tunnel location. For both the front and side impacts, another important characteristic to note is that the acceleration velocity signals at the satellite sensor location are evident much earlier in the crash event than the signals in the passenger compartment location. This early detection is critical when occupant protection relies on fast airbag deployment. The figures also show the required airbag trigger times for each of the crash tests. Clearly, these data demonstrate that locating sensors in the crush zone can provide much higher signal amplitudes much earlier in the crash event, for both side and frontal collisions.

To provide reliable discrimination in these situations, our sensing system is designed to be mounted in the crush zone. Figure 3 shows the vehicle configuration of a sensing system combining front and side satellite sensors with a central sensor located in the passenger compartment. These sensors interface with the central control module to communicate crash discrimination information as well as satellite state-of-health status. We chose to implement this interface as a unidirectional, 2-wire current loop that provides power to the satellite and information back to the central control module. In many instances, the sensor in the passenger compartment is used to improve system reliability against single-point system faults.

A block diagram of the satellite sensor is shown in Figure 4. The system is partitioned into a custom IC and a sensor, plus a few discrete components.

CUSTOM IC DESIGN

The custom integrated circuit is implemented in a 1.2u CMOS process with twin-poly capacitors, high-voltage devices, and bipolar NPN transistors. This IC processes the input sensor signal, determines if an impact of sufficient severity has occurred to require deployment of

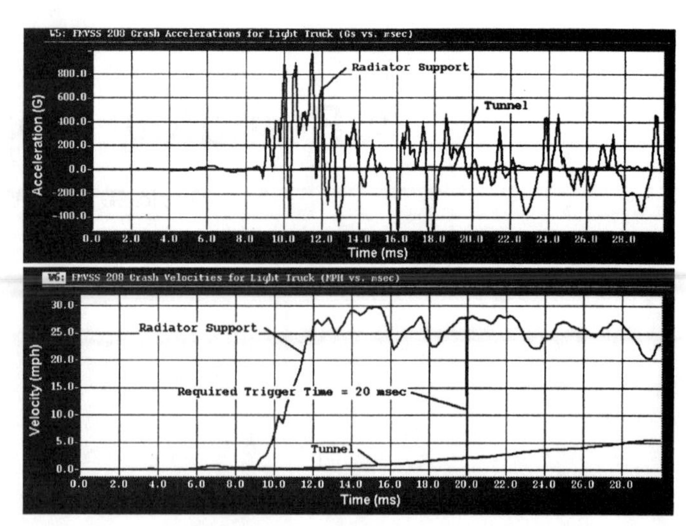

Figure 1: Acceleration (top) and velocity (bottom) crash data recorded during a 13.4 m/s (30 mph) frontal barrier impact of a light truck. Data are shown from two locations: the radiator support, and the tunnel inside the passenger compartment.

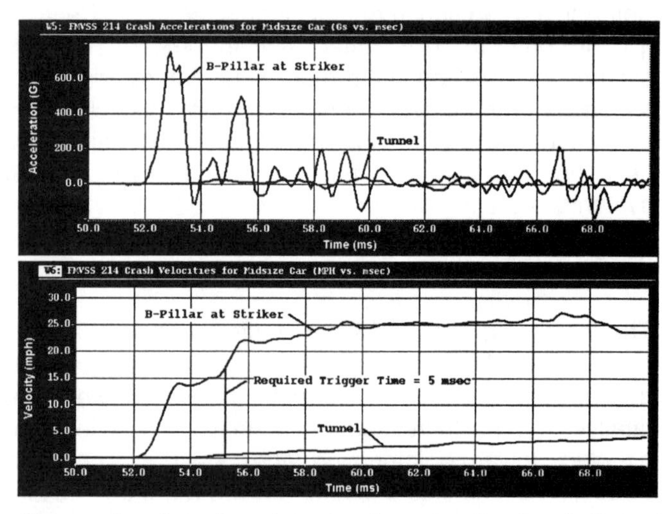

Figure 2: Acceleration (top) and velocity (bottom) crash data recorded during a 15.2 m/s (34 mph) side impact barrier test for a midsize car. Data are shown from two locations: the B-pillar at striker, and the tunnel inside the passenger compartment.

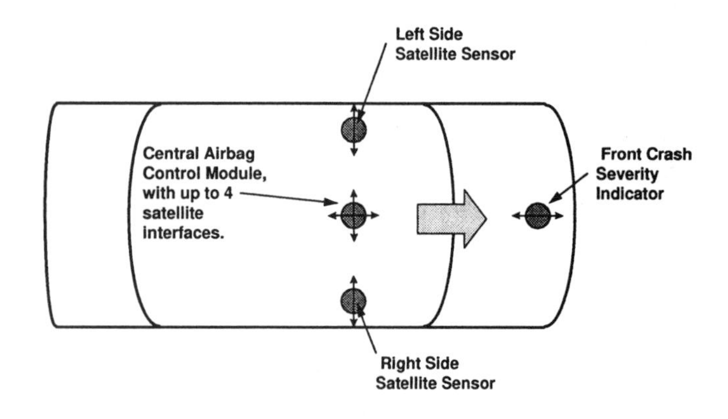

Figure 3: Vehicle configuration of the sensing system using front and side satellite sensors.

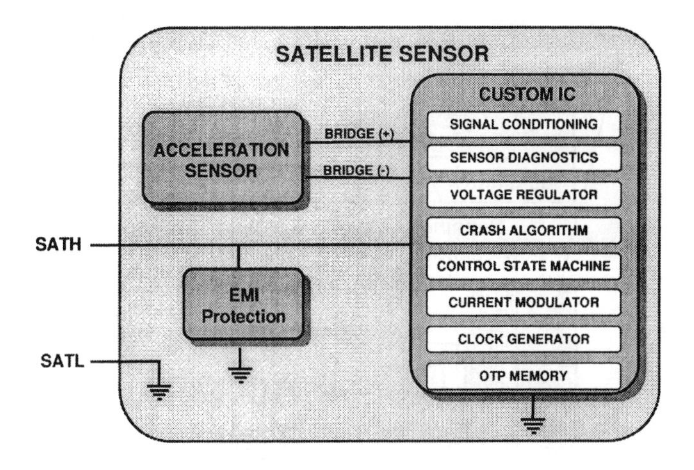

Figure 4: Block diagram of the satellite sensor.

the air-bag, and communicates with the central airbag control module.

Some of the features of the custom integrated circuit include:

- Accurate integrated oscillator and regulator
- Acceleration sensor interface
- State machine
- Crash sensing algorithm
- Two-wire serial communication
- Non-volatile memory to calibrate for various vehicle and sensor characteristics

OSCILLATOR AND REGULATOR - The regulator is an LDO with a p-channel pass transistor. It requires one external capacitor and regulates 5 volts to within +/- 5% from the battery input. The battery line can vary in amplitude from 6-40 V. The oscillator is an on-chip capacitive design using a temperature compensated current source. The absolute value of the current source is modified with the non-volatile circuitry to adjust the center frequency to 4 MHz. The oscillator spec is +/- 5% over the entire temperature range of -40 to 120 °C.

SENSOR INTERFACE - The acceleration sensor interface must amplify the input signal from approximately 1.8 μV-s^2/m (17.7 μV/G) to 408 μV-s^2/m (4 mV/G); this requires a nominal gain of 225. As the sensor has process variations which affect the sensitivity, the interface circuit can compensate the gain from 180-280.

The sensor has a sensitivity with a negative temperature coefficient of 0.22 %/°C. A positive gain coefficient of the same magnitude has been incorporated into the interface circuit. The acceleration accuracy of the module over the entire temperature range is +/- 8%, including hysteresis, over the lifetime of the part.

A key feature to this sensor interface circuit is the output offset. Because the output acceleration from this circuit is directly applied to the crash detection algorithm, offsets directly translate to calculation errors. Figure 5 shows a block diagram of the active offset cancellation circuitry and the input gain stage.

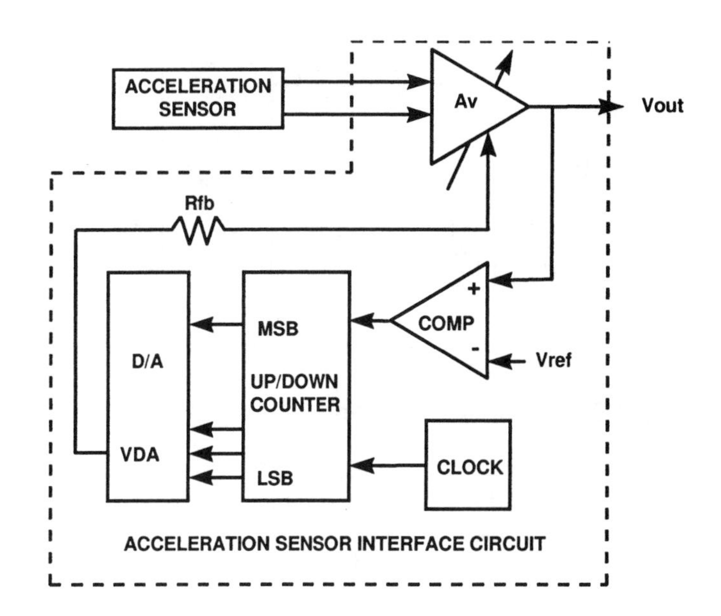

Figure 5: Block diagram of the acceleration sensor interface circuit.

The sensor has a differential output which is converted to a single-ended output at Vout, and the gain is adjusted in this stage. The Vout signal is applied to the positive terminal of a comparator and compared to Vref. When Vout is above Vref, the up/down counter counts up. This increments the digital-to-analog converter (D/A) output, which is applied through a resistor to the negative terminal of the gain stage. The up/down counter counts up until the Vout signal drops below the Vref limit. At this point, the output is regulated to within one D/A count of Vref. The Vout signal will toggle one least significant bit (LSB) around this point. This is typically less than 10 mV of error, which is negligible in the operation of the algorithm.

STATE MACHINE - The digital state machine is used to control the operation of the custom integrated circuit and the communication to the central controller. There is a power-up sequence including self test to assure operation of the circuit. The part is then in the normal operation state. During this state, an OK pulse is transmitted to the central control box every second. If a fault is detected within the satellite sensor, the part transitions to the fault detected state. If the sensor interface circuit detects an acceleration signal greater than a specified level, the part transitions to the discriminate state. During this time, if a crash is detected, the deploy signal is sent to the central controller; otherwise the part returns to the normal operation state.

ALGORITHM - The block diagram of the crash-sensing algorithm of the custom integrated circuit is shown in Figure 6. Analog circuitry within the IC calculates four crash parameters that are used to discriminate crash

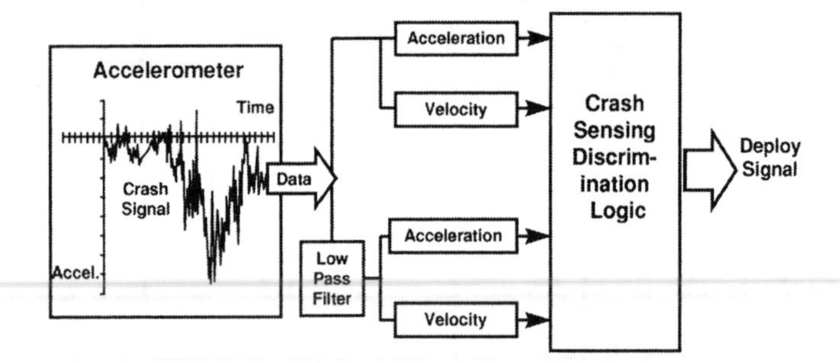

Figure 6: Diagram of the crash sensing algorithm.

severity. All four of the parameters are compared against thresholds that can be tailored to meet an individual vehicle's crash sensing performance requirements. The thresholds are adjustable with 8-bit D/A's and are stored non-volatile memory during manufacture of the satellite sensor.

2-WIRE SERIAL COMMUNICATION - The custom integrated circuit has a low nominal operating current, typically less than 5 mA. The communication with the central controller is achieved through current modulation of the module. The custom IC generates an additional current of approximately 38 mA which the central controller interprets as a logical one where the normal current of 5 mA is interpreted as a logical zero.

The custom IC communicates the ID of the unit, the state-of-health, and the deploy message to the central controller. The controller is updated once every second for state-of-health and virtually immediately upon detection of a crash event that requires airbag deployment.

NON-VOLATILE MEMORY - The custom integrated circuit contains UV-EPROM to trim the sensor interface circuit and oscillator, as well as to specify the vehicle-dependent algorithm parameters and identification message. UV-EPROM is incorporated in a latch which can be adjusted until the desired result is achieved; then the value is programmed. The non-volatile latch can only be programmed during the manufacturing calibration. Voltage margining of the programming depth and a parity chain are incorporated in the latch design to assure that valid data are maintained during the lifetime of the part.

SENSOR DESIGN AND FABRICATION

Piezoresistive accelerometers have been under development for many years [5,6]. Since 1993, millions of these sensors have been used in frontal impact detection systems [6,7].

This piezoresistive accelerometer is produced using a combination of bulk and surface micro-machining processes. It has a dynamic range of ±5000 m/s^2 (±500 G) and uses a patented "compensating beam" technique

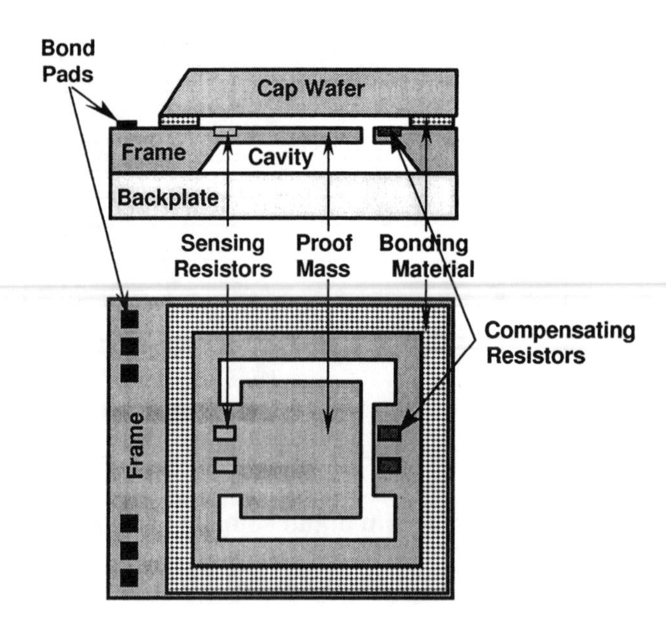

Figure 7: Diagram of the accelerometer.

to reduce errors caused by internal and external stresses. The cell, which is completely sealed at the die level, is mounted directly to the ceramic substrate.

DESIGN - The sensing element of the device consists of a thin paddle connected to a frame in a cantilever fashion, as shown in Figure 7. The paddle acts as a proof mass and is free to respond to accelerations. As the beam flexes in response to acceleration, the resistance values of the piezoresistors implanted in the beam change.

The resistors are arranged in a Wheatstone bridge circuit, which translates the small changes in resistance into a differential voltage. By arranging the compensating and sensing resistors on opposite sides of the Wheatstone bridge, the device rejects unwanted error strains and passes the desired strains due to acceleration. Performance of the device is summarized in Table 1

FABRICATION PROCESS - Summary diagrams of the fabrication process are shown in Figure 8 [8]. Starting with a p-type silicon substrate, an n-type epitaxial layer is grown to a well-controlled thickness on the top surface.

Parameter	Value	Units
Dynamic Range	±5000	m/s^2
Sensitivity	0.390	mV·s^2/m
Sensitivity Accuracy	±8	% FS
Bāndwidth (-3 db)	0.1-1500	Hz
Bias Error	±1.0	%FS
Noise, 0.1-1500 Hz	±0.5	% FS
Temperature Range	-40 to 120	°C

Table 1: Performance of the accelerometer, including the signal conditioning and amplification.

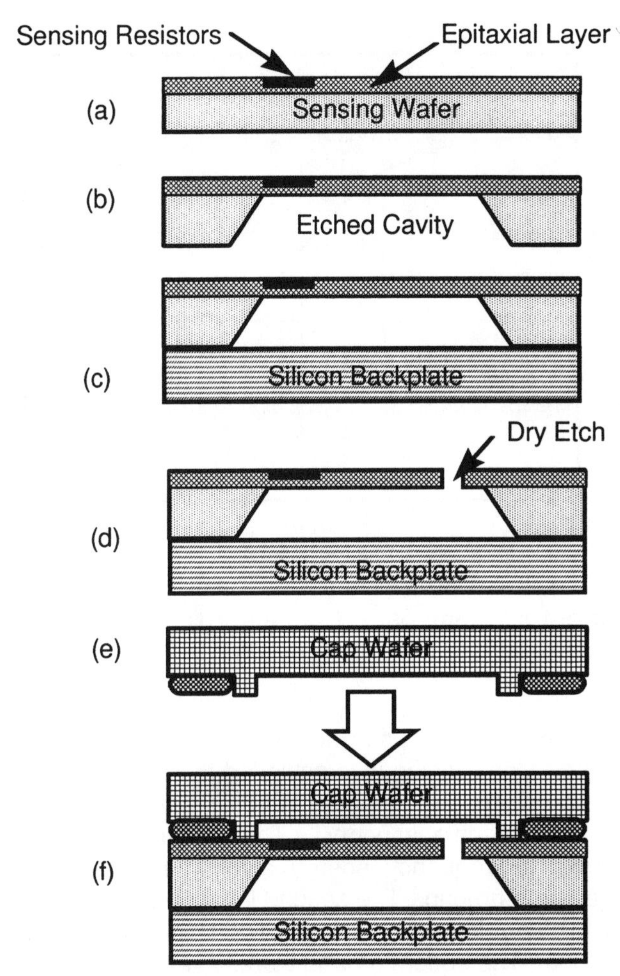

Figure 8: Diagram of the sensor fabrication process sequence.

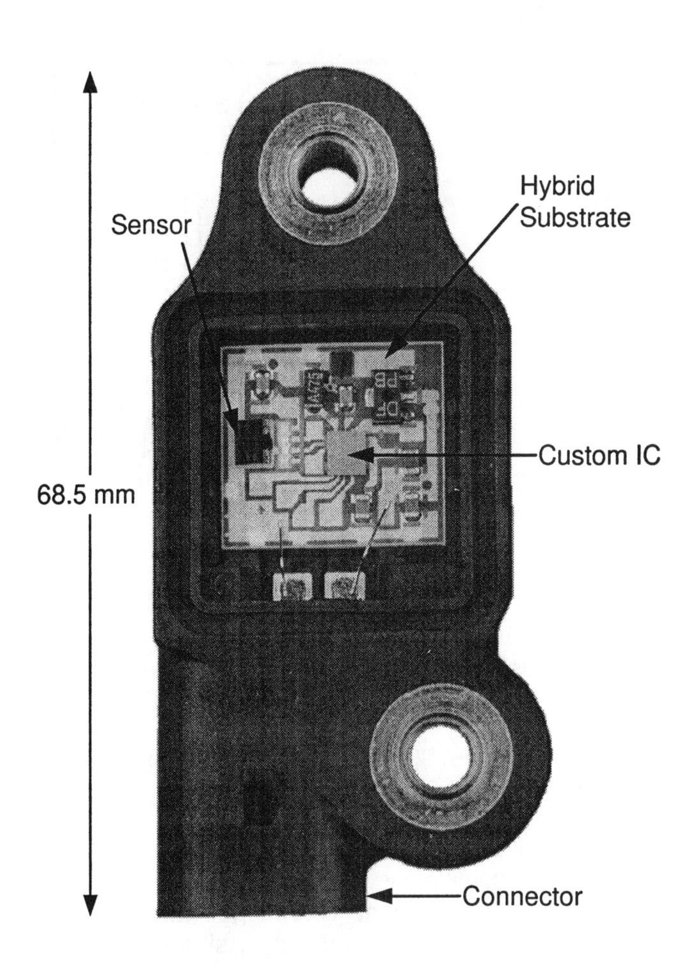

Figure 9: Photograph of the module with the cover removed.

Standard integrated circuit processes are then used to create piezoresistors and the sensor circuit (a). An electro-chemical etch process is used to etch a cavity into the back side of the wafer, forming a thin diaphragm of the epitaxial layer (b). At this point, the device is essentially a pressure sensor. Millions of high-reliability pressure sensors are produced each year using this process [7].

A silicon backplate wafer is bonded to the back side of the sensing wafer using a low-temperature fusion bonding process, forming a two-wafer stack (c). Then, a dry etch process is used to cut through the diaphragm from the front side, forming a cantilevered beam in a paddle shape (d). The sensing element is now free to move.

A separate cap wafer is made using wet etch processes, and is bonded to the two-wafer stack using a screen printed glass frit process, (e) and (f). This combination of front- and back-side caps not only protects the accelerometer during the harsh wafer dicing and assembly operations, but also allows the device to be directly mounted to the ceramic substrate and covered with standard circuit passivation materials.

PACKAGE DESIGN

The package was designed to meet the demands of both the passenger compartment and under-hood automotive environments. Particular attention was paid to balancing the requirements of size, rigidity, environmental protection, mounting flexibility, error proofing, and electro-magnetic interference (EMI) protection in a cost-effective way.

Several factors contribute to the compact size of the module, shown in Figure 9. Hybrid surface-mount technology, unpackaged ASIC's and aluminum wirebond interconnects are used to produce a circuit area of less than 3.1 cm^2. A connector and two bushings are integrated into the housing body. The overall dimensions are 68.5 mm by 33 mm by 14 mm, excluding the connector latch and locating pin.

The rigidity of the module is important to faithfully transmit the crash signal to the acceleration sensor. Because the package size is small and the ceramic substrate is stiff, the first mode of resonance is greater than 3.2 kHz, as shown in Figure 11. When this signal is filtered by the two-pole 800 Hz filter within the

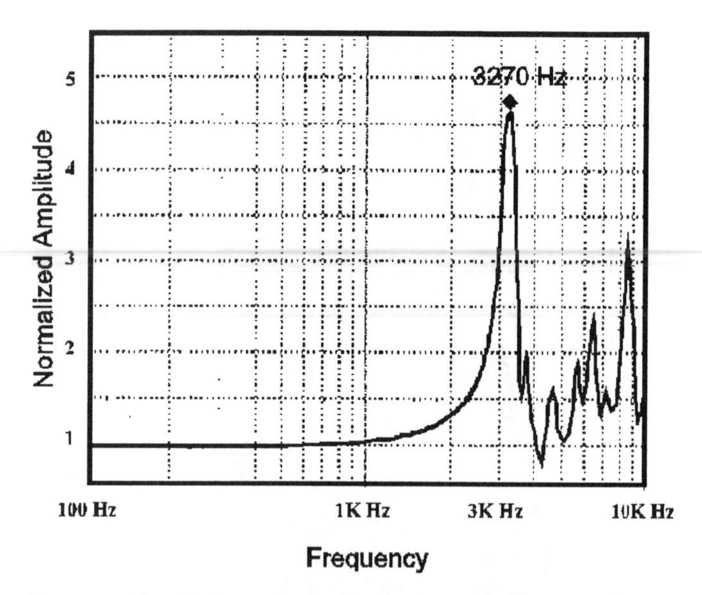

Figure 11: Frequency response of the package, measured using a reference accelerometer in the location of the sensing cell.

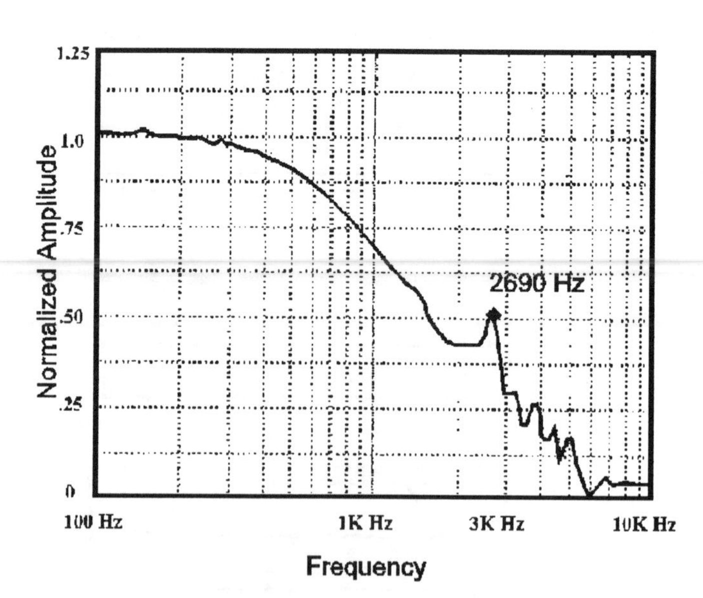

Figure 10: Composite frequency response of the package, sensor and filter circuitry.

conditioning IC, any unwanted high frequency response is further reduced (see Figure 10).

Environmental protection is achieved by passivating the circuitry and completely sealing the exterior of the module. A very compliant silicone gel passivates the circuit, while a cover attached with a silicone adhesive seals the exterior of the module. The connector is made weather tight through the use of a three-rib compliant seal. The module has successfully passed all applicable automotive testing, including powered temperature cycle, fluids compatibility, 96 hour salt fog, powered dunk, and water submersion testing.

To reduce assembly labor and part count at the vehicle assembly plant, an alternative mounting scheme was developed. In addition to a standard through-hole bushing design, an optional configuration using threaded bushings and pre-attached screws is available. Using this threaded bushing configuration, the module can be installed in a single operation by placing the head portion of the screws through keyhole slots and tightening the screws.

Due to the many possible algorithm calibrations, vehicle designs, and mounting locations, error proofing the installation is critical. In addition to the 63 possible electrical ID's, there are 17 possible mechanical keys that fit within the connector shroud. Each unique key insures that only a mating connector with a matching key configuration may be connected.

Many of these advantages were achieved through the use of a plastic housing. One disadvantage of using a plastic housing is the lack of EMI shielding when compared to a metal case. A combination of conditioning the input/output signals and the use of a ground plane was used to overcome this potential problem. Conditioning was achieved through the use of a pi network consisting of two 2200 pF capacitors and a 600 Ω ferrite chip impeder. This network effectively shunts high frequency energy coming in through the connector to ground. The second level of protection, the ground plane, is formed using a two-layer hybrid circuit technique. The bottom ground plane layer and the upper active circuit are separated by approximately 25 μm of dielectric material. This combination allows the module to withstand 150 V/m of radiated and conducted EMI.

CONCLUSION

A remotely mounted crash detection system has been developed that meets the sensing performance requirements of automotive restraints systems and the durability requirements for mounting near the impact zone. The system is well-suited for frontal impact, side impact, and future crash severity discrimination applications on a wide variety of vehicles and is currently being fitted on high-volume production vehicles.

REFERENCES

1. C.J. Kahane, "Fatality Reduction by Air Bags: Analysis of Accident Data Through Early 1996." DOT HS808470. Washington, DC: U.S. Department of Transportation. 1996.
2. S.A. Ferguson, A.K. Lund, M.A. Greene, "Driver Fatalities in 1985-1994 Air Bag Cars", Insurance Institute for Highway Safety, Arlington, VA. 1995.
3. T.D. Hendrix, J.P. Kelley, W.L. Piper, "Mechanical Versus Accelerometer Based Sensing for Supplemental Inflatatable Restraint Systems", SAE 901121; 1990.
4. R. Vogt, "Electronic System Design for Future Passenger Restraint Systems", SAE 960500; 1996.
5. L. Roylance and J. Angell, "A Batch-Fabricated Silicon Accelerometer," *IEEE Trans. Elect. Dev.*, ED-26, (1979) 1911-1917.

6. W. Yun and R. Howe, "Recent Developments In Silicon Microaccelerometers," *Sensors*, vol. 9 (1992) 31-41.
7. D. Sparks and R. Brown, "Buying micromachined sensor in high volume," *Sensors*, vol. 12 (1995) 53-56.
8. D. Sparks, D. Rich, C. Gerhart, and J. Frazee, "A Bi-Directional Accelerometer and Flow Sensor Made Using a Peizoresistive Cantilever", ATA 6th European Congress on Lightweight and Small Cars: the Answer to Future Needs, Cernobbio, Italy July 2-3 1997, vol. 2, p 1119-1125. ATA-97A2IV40.

Saab's security & safety priority

THE all new Saab 900 and Saab 9000 have built in safety which place Saab amongst the safest passenger vehicles.

In a recent study by the Monash University Accident Research Centre, the Saab was placed second for safety in research conducted on 220,000 accidents in Victoria and New South Wales.

The new Saab range has improvements which further enhance passenger safety.

Saab itself has subjected the 900 series to 19 different crash test methods: 13 at high speed to assess the survival space, restraint systems and occupant injury criteria - and six at relatively low speed (8 to 25 km/h) to assess the performance of the self-repairing bumpers, easy-repair pumper supports and airbag sensor testing.

In Australia the combination of parts pricing and ease of repair has lead to several leading insurance companies reducing the new vehicle's rating by some four points over the previous model. This represents a saving of 14% in the insurance premium.

Passive safety

Passengers are protected by an integral safety cage that surrounds the interior space. The cage has been refined based on data gathered from accidents and advances in structural technology. By optimising each element of the integral steel pressings, paying particular attention to the characteristics of the joints, Saab has developed a weight efficient structure that is claimed to withstand high crash loads and not collapse.

The crash characteristics of the front and rear body sections have been optimised to absorb crash energy by spreading the impact loads as widely as possible in all types of accidents.

An example of this approach is the special reinforcement running from the base of each A-pillar to the front of each sill to prevent distortion of the door frame by a front wheel as it moves back under the influence of a severe offset fontal impact. Saab engineers have calculated that this element is equivalent to extending the front crumple zone and hence the overall car length.

Occupant protection from side impact has been a feature of Saab since 1972 when door reinforcements were fitted. The new Saabs now have a network of pillars, members and reinforcements in the side elements of the safety cage to link these beams to the main structure when a side impact occurs.

Special emphasis has been placed on the role played by the roof structure in providing body stiffness and crash protection in roll-over accidents. High quality steel with a gauge thickness of between 1.2 and 1.5 mm is used together with considerable reinforcement and optimisation of the roof pillar design. As a result the load required to collapse the front corner of the roof is almost three times the weight of the car.

Seating safety

The Saab 900 has restraint systems for each of the five seat positions, all provided with full inertia-reel three-point lap-sash seat belts.

In the front, both of the height-adjustable seat belts incorporate spring-loaded pretensioning devices within the inertia reel mechanisms under the B-pillar trim. In the first stages of a severe impact the system is triggered to tighten the belt and take up any slack to prevent excessive forward movement. Both inboard and outboard lower belt anchorages are

Continued next page >

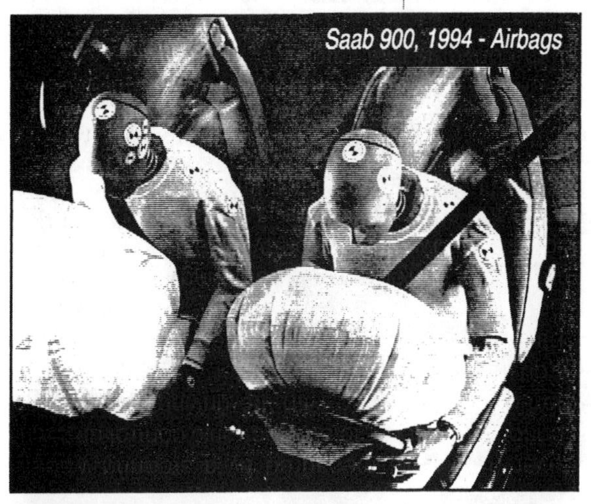

Saab 900, 1994 - Airbags

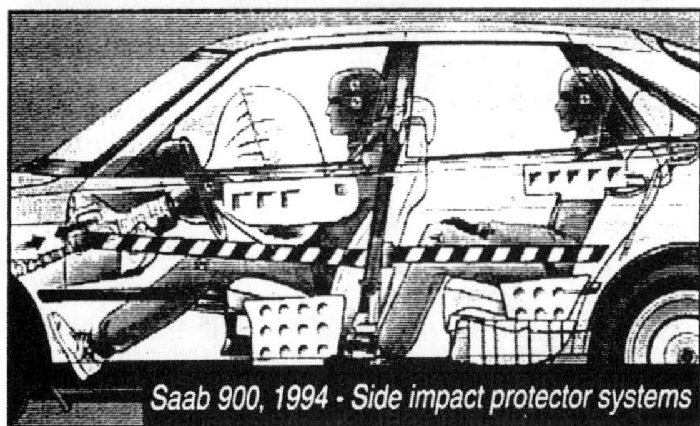

Saab 900, 1994 - Side impact protector systems

Saab 900, 1994 - Safeseat

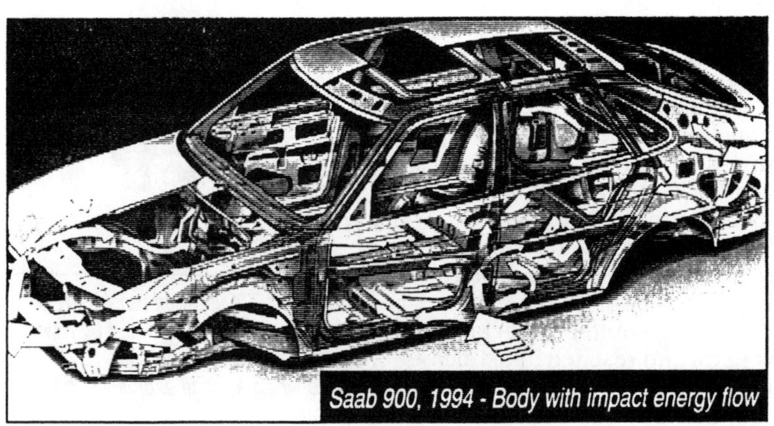

Saab 900, 1994 - Body with impact energy flow

In a rear-end collison , the movement of the occupant's head must be arrested in time. All Saabs have sturdy head rests

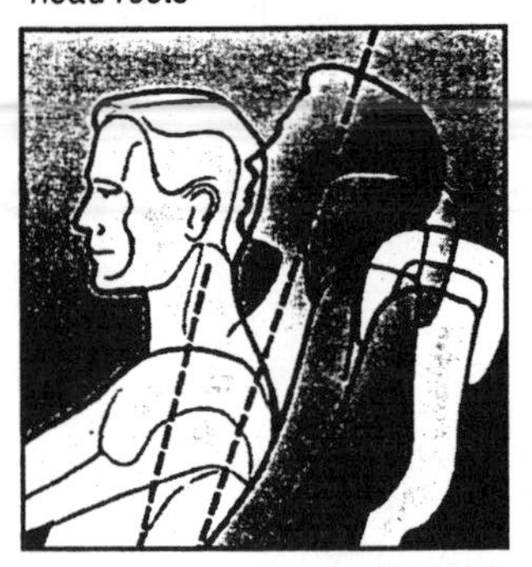

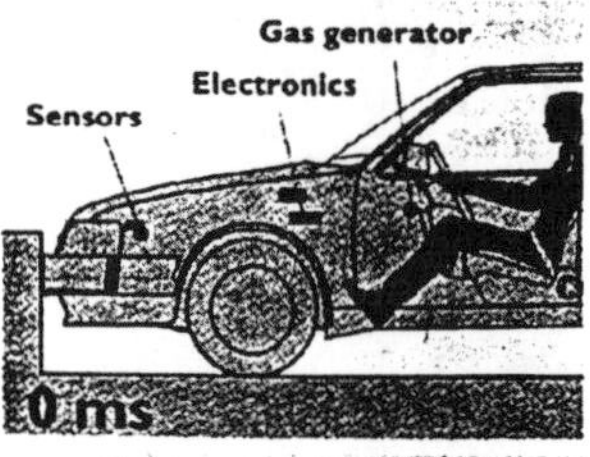

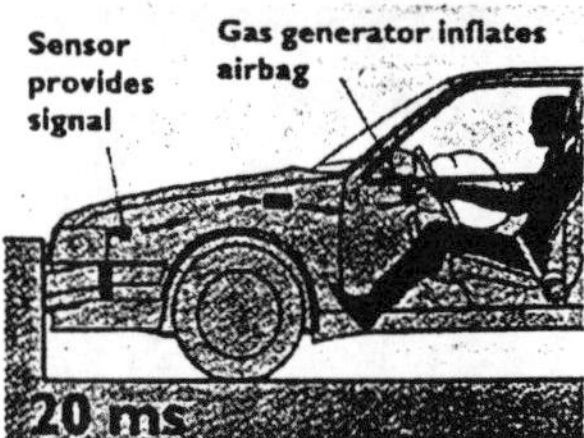

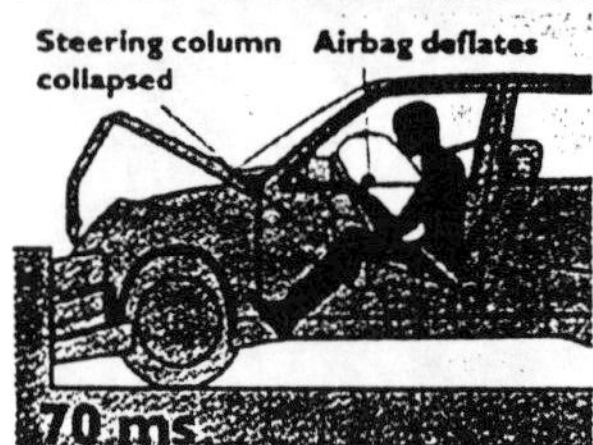

In principle, the airbag inflates before the driver has time to move forwards upon impact, since the car body's crumple zones are first compressed. The entire sequence is over in one tenth of a second.

< *from previous page*

mounted on the seat frame, to ensure that the lap restraint runs across the best anatomical position to minimise internal injuries. The steering wheel is split into collapsible sections and connected by joints so that there is minimal intrusion even during a severe fontal impact.

The location of the ignition key has been retained on the transmission tunnel, this is considered a safer position, while conventional ignition locations infront of the driver have the potential of creating knee injury in a frontal impact.

Safeseat

In the event of a crash, a unique combination of various safety systems, known as the Saab Safeseat, help protect rear seat passengers. The rear seats now have full three-point lap and diagonal inertia-reel belt systems for all three seat positions. The rear seat can still be folded in a 40/100 split.

A patented structural full-width cross beam is attached to the top of the larger part of the backrest. This beam is latched to special reinforced brackets on the tops of the rear wheel housings and incorporated mountings for three head restraints. As well as providing the upper anchorage for the seat belt systems, the cross-beam also acts as a load restraint to protect passengers from being hit by shifting objects under hard braking or during a sever frontal impact.

When large objects are carried the right-hand rear seat rest can be released from the cross-beam, while allowing two passengers to be properly restrained and protected. If larger loads need to be carried instead of passengers, then the cross-beam can be unlatched and lowered with the other part of the backrest to provide an extended boot floor. The cross-beam hinges up to form a load restraining lip immediately behind the front seats. This also means that the rear seat head restraints can stay in position and need not be removed when folding the seat flush with the floor.

Another feature of the seating system is a dual purpose cross beam under the front edge of the rear seat. This element is shaped to prevent the occupants submerging under the lap belt section of the three-point harness and at the same time contributes substantially to door intrusion resistance in a side impact. To spread the localised loads more evenly, the lower trailing edges of the rear doors, as well as the front doors, are injected with rigid structural foam to create a continuous load path, dissipating forces away from the cross-beam ends.

Auxiliary features of the Safeseat system include dual optional integrated drop-down child seats and factory fitted anchorages in the rear passenger compartment for installing older types of rearward-facing child seats.

Airbags

A full 'US-sized' airbag is included in the standard specification of all new Saab 900s, mounted in the steering wheel hub. A full-size passenger-side airbag is offered as an option.

Vehicle security

The security system offers additional security through a door locking mechanism which has a deadlock setting when the door is locked with the key.

This system will allow the door to be opened only by the original key.

Even if a would-be thief broke a window and reached in to release the door handle or locking button, that door, or any of the car doors could not be unlocked.

High speed tests

30 mph barrier test
50 km/h barrier impact, to verify steering wheel intrusion
35 mph barrier impact – nominal laden condition
35 mph barrier impact – full laden condition
55 km/h, 50 per cent offset impact – 15 degree barrier test
30 degree barrier impact - 30 mph (left and right)
Pole impact
Side impact (latest US standard) – 27 degrees 33.5 mph
Side impact (earlier US standard) - 90 degrees 20 mph sled test for fuel integrity
Side impact (EEC proposal) – 90 degree 50 km/h
Rear impact - 30 mph – nominal laden condition
Rear impact - 35 mph – overladen condition
Roll-over test – 30 mph

Low speed tests

Pole impact
30 degree barrier impact (left and right)
Front under-ride impact
50 per cent offset front impact
50 per cent offset rear impact

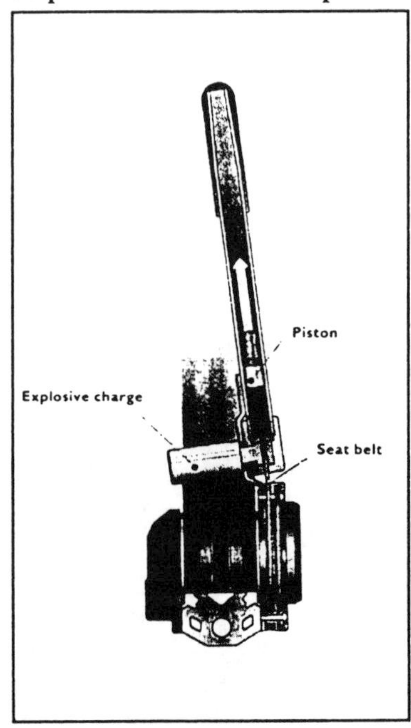

Saab 900, 1994 seatbelt tensioner, on collison, the belt is pulled taut by a wire coupled to a piston which shoots forward when a small explosive charge detonates

2000-01-0822

Future Electrical Steering Systems: Realizations with Safety Requirements

Werner Harter, Wolfgang Pfeiffer
Robert Bosch GmbH

Peter Dominke, Gerhard Ruck
ZF Lenksysteme GmbH

Peter Blessing
Heilbronn University of practice science

ABSTRACT

Additional future requirements for automobiles such as improved vehicle dynamics control, enhanced comfort, increased safety and compact packaging are met by modern electrical steering systems.

Based on these requirements the new functionality is realized by various additional electrical components for measuring, signal processing and actuator control. However, the reliability of these new systems has to meet the standard of today's automotive steering products. To achieve the demands of the respective components (e.g. sensors, bus systems, electronic control units, power units, actuators) the systems have to be fault-tolerant and/or fail-silent. The realization of the derived safety structures requires both expertise and experience in design and mass production of safety relevant electrical systems. Beside system safety and system availability the redundant electrical systems also have to meet economic and market requirements.

Within this scope the paper discusses three different realizations of electrical steering systems

1. Electrical power steering system (mechanical system with electrical boosting)
2. Steer-by-wire system with hydraulic back-up and
3. Full steer-by-wire system

The paper presents solutions for these systems and discusses the various advantages and disadvantages, respectively. Furthermore strategies for failure detection, failure localization and failure treatment are presented. Finally the various specifications for the components used are discussed.

INTRODUCTION

In this paper the technical solutions and safety aspects of various electrical steering systems are described, starting from conventional steering systems.

For car manufacturers and end customers the use of electrical steering systems offers many advantages concerning flexibility, enhancement of familiar steering functions and the introduction of innovative steering functions.

New steering functions which are even coupled with automatic steering interventions, call for an adaption of regulations concerning the approval of steering equipment.

Development and production of the next generations of electrical steering systems up to purely electrical steering systems create high safety demands on components and systems.

Reliable and safe electrical steering systems can be realized by using appropriate safety techniques for these new systems and their components combined with the know-how of safety relevant vehicle systems.

At the same time the transition to purely electrical steering systems will take place step by step via systems with mechanical or hydraulic backup.

CURRENT STEERING SYSTEMS AND THEIR SAFETY ASPECTS

Because steering systems besides braking systems rank among the most demanding vehicle systems regarding safety engineering, the control of safety is a basic prerequisite for introducing a new steering system into the market.

Comparing safety aspects of conventional and new steering systems demonstrates that with every generation of steering systems new steering functions are added, whose safety must be ensured by appropriate system and constructional design.

MECHANICAL STEERING SYSTEM – Conventional steering systems are based on the mechanical steering of which two variants are in use. The components of a rack-and-pinion steering (Fig. 1) and ball-and-nut-steering, which is used for higher steering forces, are purely mechanical: Steering wheel, steering column, steering pinion, rack, ball-and-nut gear, tie rod. Additional components are various universal joints and bearings.

Figure 1. Mechanical steering

With steering systems the loss of the steering control is considered as a substantial safety-relevant effect. This could be caused with a mechanical steering system by break or lockup due to design errors, material failures or production defects.

In practice, however, these types of faults are excluded, because on the one hand mechanical components are designed with sufficient safety margins to be safe and on the other hand production defects can be practically neglected because of the large experience of steering systems manufacturers with mechanical systems and the applied high production and quality standards. These assumptions are confirmed by field experience with mechanical steering systems, too.

HYDRAULIC POWER STEERING SYSTEM – The hydraulic power steering system (Fig. 2) today is the most used steering system. It is based on the components of the mechanical steering system, in addition there is a hydraulic system, usually consisting of hydro pump with V-belt drive, hydraulic lines, oil reservoir and steering valve.

The essential new function of this power steering is the hydraulic support of the steering movement, so that the driver's steering-wheel effort is reduced. Therefore in the event of failure, the loss of steering boost arises as a new safety aspect in comparison to purely manual steering. This can be caused by a leakage of the hydraulic system or by a hydro pump failure.

Figure 2. Hydraulic power steering

Since by design the manual steering system is further available, in case of a failure the steering function is further available and the driver can adapt himself by the usually slowly rising steering-wheel effort in good time to the missing steering boost.

In the regulation ECE R79 concerning the approval of vehicles with regard to steering equipment the maximum permitted steering control effort with a failure in the steering equipment is stipulated. In the context of the vehicle approval the appropriate investigations for the failure modes of the hydraulic power steering are performed and the compliance with the prescribed steering efforts is checked.

NEW STEERING SYSTEMS AND THEIR BENEFITS

New steering systems use more and more electrical components. This is mainly because of two reasons: Energy saving and installation simplification by modular design.

ELECTRO-HYDRAULIC POWER STEERING SYSTEM – The electro-hydraulic power steering system (Fig. 3) is based on the rack-and-pinion hydraulic power steering and its essential new feature is an electrically driven hydraulic pump, which substitutes the steering pump driven so far by the vehicle engine.

Additional new components compared to the hydraulic power steering are: Electric motor, electronic control unit and an optional sensor for steering velocity. The pressure supply unit integrates electric motor and electronic as well as hydraulic pump and oil reservoir.

The decoupling of the pump drive from the vehicle engine allows a free selection of the installation location in the vehicle and in the consequence the delivery of a fully functional and checked axle steering module by the steering system manufacturer.

The electronically controlled electric drive results in an energy saving up to 75% depending on load distribution and control strategy. Additionally a variable steering boost is possible functionally depending on steering- and vehicle-velocity and other parameters.

Figure 3. Electrohydraulic power steering

As to safety the electro-hydraulic and the hydraulic power steering have the same failure effect: Failures of new additional components only affect steering boost, and the mechanical rack-and-pinion steering is still available. Omission of the V-belt drive, by which the steering pump has been operated so far, improves the steering boost reliability even further.

ELECTRIC POWER STEERING SYSTEM – The electric power steering system (Fig. 4) combines a mechanical steering system with an electronically controlled electric motor to a dry power steering. The hydraulic system, which so far delivered the steering boost, is substituted by an electrical system. For this, a torque sensor measures the steering wheel torque and an electronic control unit calculates the necessary servo torque. This is delivered by an electric motor in such a way that the desired torque curve at the steering wheel is created.

Depending on the necessary steering forces the electric motor engages by a worm gear at the steering column or at the pinion and for high forces directly at the rack by a ball-and-nut gear. In figure 4 the pinion-solution is represented, which is intended for middle class vehicles.

The components involved in the electrical power steering are besides the mechanical steering components: Electric motor, electronic control unit, power electronics, steering wheel torque sensor and CAN data bus to other systems.

The electrical power steering system offers large benefits compared to the hydraulic power steering. Apart from about 80% lower energy consumption the omission of the hydraulic fluid increases the environmental compatibility. The electrical power steering is delivered to the car manufacturer as a complete system module ready-to-install. The adaptation of the servo power assistance to certain vehicle types as well as the modification of the control strategy dependent on different parameters and vehicle sizes are easily and rapidly feasible.

Figure 4. Electrical power steering

From the safety point of view as with the other power steering systems due to failures in electrical components, again the steering boost can be impaired, here by faults of components of the electrical servo system. The steering system's unintentional self activity as well as too strong steering boosts are to be concerned as new potential safety critical effects, which must be avoided by appropriate countermeasures.

FUTURE STEERING SYSTEMS

The main feature of future steering systems is the missing direct mechanical link between steering wheel and steered wheels. With such a steer-by-wire steering system (Fig. 5) the missing steering column's function must be reproduced in both directions of action. In forward direction the angle set by the driver at the steering wheel is measured by a steering angle sensor and transferred with the suitable steering ratio to the wheels. In reverse direction the steering torque occurring at the wheels is picked up via a torque sensor and attenuated respectively, modified fed back to the driver as a counter torque on the steering wheel.

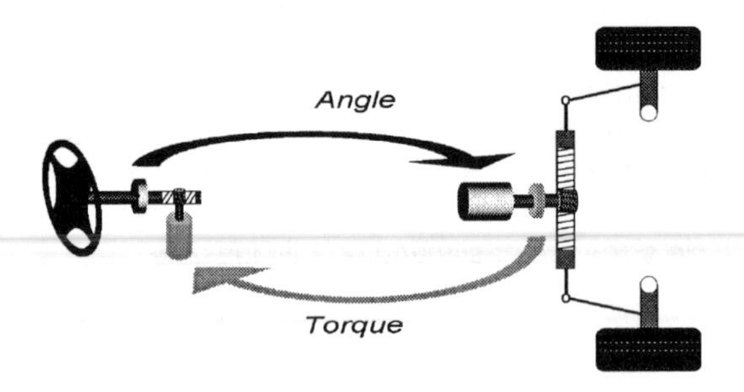

Figure 5. Principle illustration steer-by-wire

First, steering wheel module and steering module are implemented with familiar components of mechanical and electrical steering systems, like: Steering wheel, gearbox, electrical motors, rack. The operational principle is, however, in principle open for more futuristic designs like sidestick operation on the driver's side and single wheel steering on the wheel side.

While in systems with mechanical connection in the case of electrical errors only the steering boost is concerned, corresponding measures must be taken with steer-by-wire systems that in case of any electrical failure steering control is always guaranteed.

ADVANTAGES OF STEER-BY-WIRE SYSTEMS – Steer-by-wire is a universal actuator for automatic steering intervention. – For vehicle dynamic steering intervention a steering angle actuator is needed which does not affect the steering wheel while rapidly correcting the vehicle wheels. On the other hand, a torque actuator will be needed for automatic lateral guidance interference and future steering systems of autonomous driving, thus imparting a superimposed torque onto the steering wheel and letting the driver with that know the intended direction, evaluated by the lateral guidance control system.

Steer-by-wire meets both requirements ideally.

Along with "drive by wire" and "brake by wire" it provides the condition to materialize vehicle dynamics and comfort oriented automatic controls in one system.

Design advantages for the automaker – The rigid steering column curbs the design freedom for the engine compartment. On either side space has to be provided (left-hand or right-hand driving). Steer-by-wire implies that no steering column impairs the good usage of engine compartment.

Further advantages –

- Improvement of steering comfort. Road disturbance is first evaluated and assessed by the system; only useful information is fed to the driver.
- Improvement of steering comfort. Road disturbance is first evaluated and assessed by the system; only useful information is fed to the driver.
- Steering wheel return torques and steering ratio can be adjusted variably by software; thus rendering the steering system adaptable to any target group and driving situation without any need to redesign.
- Reduction of injury risk in an accident by missing steering column.
- Steering behaviour (understeering, oversteering, self-steering properties) can be matched by software.

REGULATIONS VERSUS TECHNOLOGICAL EVOLUTION

Approval of steering systems is in compliance with ECE R79 regulation. It comprises types and functions of former and recent steering systems, based on the mechanical connection between steering wheel and vehicle wheels, resulting in a nearly constant predetermined relation between steering wheel and vehicle wheels position.

Recent development of steering systems has primarily been focused on improving power steering, economy of energy and feasible installation.

Meanwhile the special committee of steering systems within the "Fachausschuss Kfz-Technik" has submitted a suggestion to amend ECE R79, to be dealt with on an international level. This suggestion is to extend the range of this regulation on the one hand, enabling purely electrical steering systems and electrical steering systems with a hydraulic backup to be included. On the other hand, new functions of future steer-by-wire systems, such as vehicle dynamics steering intervention, variable steering ratio and automatic lateral guidance are to be considered.

SAFETY REQUIREMENTS OF ELECTRICAL STEERING SYSTEMS

Electrical steering systems applying electric and electronic components within steering technology for the first time will duly be submitted to strict safety requirements.

CLASSIFICATION OF THE SYSTEMS – In order to assess safety critical aspects of electrical steering systems DIN standards V 19250, V 19251, and IEC 61508 can be considered.

Criteria to be assessed in case of failure are:

- Extent of damage
- Frequency of driving situations, in which the failure reveals effects
- Failure control
- Failure probability

This means classification into the highest possible safety level for the automobile. The probability of a steering failure will thus have to be minimised to the smallest possible value.

FAIL-SAFE QUALITY AND FAILURE TOLERANCE –

Along with familiar components of mechanical steering systems electrical steering systems entail electric and electronic components which are new for steering systems.

Various sensors are required for steering angle, steering torque and motor positions providing the interface between the mechanical and the electrical subsystem.

Electric motors are used as actuators to be triggered by means of appropriate power electronics. The core is an electronic control unit (ECU) competent for controlling and safety monitoring the whole steering system. The ECU communicates with other systems via a serial data bus. The energy for the electrical steering system is supplied by the electrical system of the vehicle.

While mechanical components properly have a history of proven reliability, the new electric and electronic components will explicitly be demanded to be highly reliable and capable to be monitored.

Due to the high safety level for electrical steering systems the demand is made that a component failure is to be coped with. This is expressed in terms of ′Fail-safe-quality′ and ′Failure-tolerance-quality′ within safety technology terms, (DIN V19250, NTG 3004).

Fail-safe behaviour means that a safe condition must be maintained, i.e. steerability must be sustained even when electrical faults occur.

Failure tolerance means that a possible maximum system functionality must be sustained even when a component fails to work.

Concerning steerability the fail-safe demand and the failure-tolerance requirement can be considered equally significant.

While electrical power steering and the steer-by-wire system with hydraulic backup get into a safe system condition when the electrical system is switched off due to electrical failures the safety structure of the purely electrical steer-by-wire system must be additionally equipped in terms of failure tolerance, when failures in the vehicle electrical system or in electric components occur.

DEMANDS ON MANUFACTURERS OF ELECTRICAL STEERING SYSTEMS

High safety requirements of current and future electrical steering systems with increasing integration of functions and systems in the vehicle result in higher requirements for suppliers of these electrical steering systems.

The following essential conditions must be met:

- Competence and experience with safety relevant systems and components.
- Know-how in development and production of electrical/electronic vehicle systems.
- Experience in development and implementation of effective safety architectures.
- Availability of reliable, monitored components and the required technologies.

ZF Lenksysteme GmbH can use the experience and the know-how of Robert Bosch GmbH and ZF Friedrichshafen AG for the development of such systems.

IMPLEMENTATION OF SAFETY REQUIREMENTS

In order to meet the safety demands for steering systems also for electrical steering systems, high quality standards in development and manufacturing will have to be maintained and safety technical principles for components and systems will have to be applied.

For example:

- Careful design with correct selection of materials and dimensioning. Right from the start of design, potential failure mechanisms must be analyzed with regard to their probability of occurrence, mechanisms of detection, and their effects.
- Failure rugged operation principle of components. On the one hand, this demand must be met during various operational phases of the electrical steering system and, on the other hand, has to be met in terms of environmental conditions and aging of components.
- Application of homogeneous and diversity redundancy. In order to achieve the required fail-safe quality even under economic restrictions a well aligned realization of homogeneous redundancy (using operational principles of the same kind) and diversity redundancy (using operational principles of a different kind) becomes necessary . The latter facilitates control of common mode failure effects within the system, meaning multiple failures based on a common cause.
- Detection and location of failures by means of intelligent monitoring of components. Apart from well proven procedures such as CRC-checks for data transfer and effective electronics self-tests, as well as monitoring plausibility and gradients of physical variables, model based methods are applied. A statement is made about the correctness of a processed variable using dynamic comparing calculation with other processed variables.
- Redundancy management with "graceful degradation" in case of failure. A modular division of partial functions permits the partial switch-off of the functionality concerned after locating a failure, thus limiting the functional restriction to the required extent.
- Realization of independent switch-off paths for defective subsystems. Even the occurrence of any simple fault must result in a positive switch-off of the

defective subsystem to ensure the fail-safe quality of the system.

- Intensive failure simulation and testing. Particularly potential failure causes of new electrical and electronic components of electrical steering systems require a detailed evidence about the efficiency of utilized failure detection and locating mechanisms during development on the test bench and in the vehicle.
- Reliable and controlled production process. The introduction of electrical steering systems into the market does not only call for a high quality manufacturing standard of mechanical components, but likewise dependable production processes for microelectronics, power electronics and mechatronic components.
- Use of methodical analysis techniques. From the initial draftphase unto the final supply, quality and safety of components and systems will have to be checked continuously by systematic methods, such as FMEA, fault tree analysis etc.

SYSTEM STRUCTURES OF SAFE ELECTRICAL STEERING SYSTEMS

SYSTEM STRUCTURE OF ELECTRICAL POWER STEERING – Functional description – In an electrical power steering system the steering torque initiated by the driver (Fig. 6) is measured by a steering wheel torque sensor and is fed into an electronic control unit. The latter then calculates along with the driving speed a reference torque for the steering motor, which, however, can optionally also depend on the steering angle and steering angle velocity. By means of the calculated reference torque the currents of the steering motor are actuated. Figure 6 shows the pinion-type realization, where at the pinion the electrical torque is superimposed to the torque initiated by the driver. In further versions both torques can be superimposed either on the steering column or on the rack. In case of a failing electrical component of this steering system the non-boosted mechanical intervention by the driver is maintained.

Safety features – The system's fail-safe behaviour concerning electrical faults is accomplished by detecting and evaluating all electrical failures. In case of major electrical faults the electrical power steering system is switched off.

Sensor failures or failures in the electronic control unit might be considered as an example, resulting in an unintentional self-activity of the steering or in a too strong steering boost . Risks of that kind are avoided by an effective monitoring strategy where failures are detected on time and the power steering system is switched-off. One detection method for this constitutes checking sensor signals and motor currents for plausible system conditions on an second path.

SYSTEM STRUCTURE OF STEER-BY-WIRE SYSTEM WITH HYDRAULIC BACKUP – Functional description – The steer-by-wire system with hydraulic backup is shown in Fig. 7. It represents an advantageous combination of the functional possibilities of electrical steer-by-wire systems with the high reliability of proven hydraulic elements. The system consists of components at the steering wheel and at the vehicle wheel level, an electronic control unit and a hydraulic backup. Steering wheel motor and sensors for the steering wheel angle and the steering wheel torque are arranged at the steering wheel. These components identify the driver's desire and reproduce the return forces to the steering wheel which are transferred to the steering wheel by conventional steering systems. These feedback forces are important to gain a safe feeling while driving.

At the vehicle wheels side the system consists of an electric motor directing the mechanically coupled wheels via a gear and a rack, and of sensors to measure angles and torques.

The electronic control unit registers periodically all sensor values, processes them via efficient control algorithms and supplies the control signals to actuate the motors. Via a serial data bus, the electronic control unit communicates with a vehicle guidance unit which coordinates the superior steering interventions, e. g. to improve vehicle dynamics. This unit at the same time constitutes an interface to the driver information system, and to additional control units for engine and brakes. The control unit in Fig. 7 is presented as a central control unit. It can also be divided into two modules arranged close to the steering wheel and steering motor, and connected to a data bus system for communication.

A closed hydraulic unit, consisting of a hydraulic pump at the steering wheel and a plunger on the vehicle wheel level, constitutes the backup. Both sides of these components are connected with each other by hydraulic lines. During normal operation the plunger is bypassed. In case of failure, the fail-safe switching valve actuated by the electronic control unit will close the bypass. Thus, via the hydraulic backup, the steering actuator can be operated by means of the steering wheel. Without electric current, the switching valve must be closed. In case of failure of the 42V vehicle electrical system thus the hydraulic bypass is automatically closed and the backup safely activated.

If the steering wheel motor can still be controlled during backup operation it can be adequately actuated to support power steering. The increased pressure needed to operate the hydraulic backup is provided by means of a small pressure reservoir with check valve. This pressure accumulator compensates the leakage which occurs during the vehicle life time. The pressure within the backup level is continuously monitored by a pressure or displacement sensor.

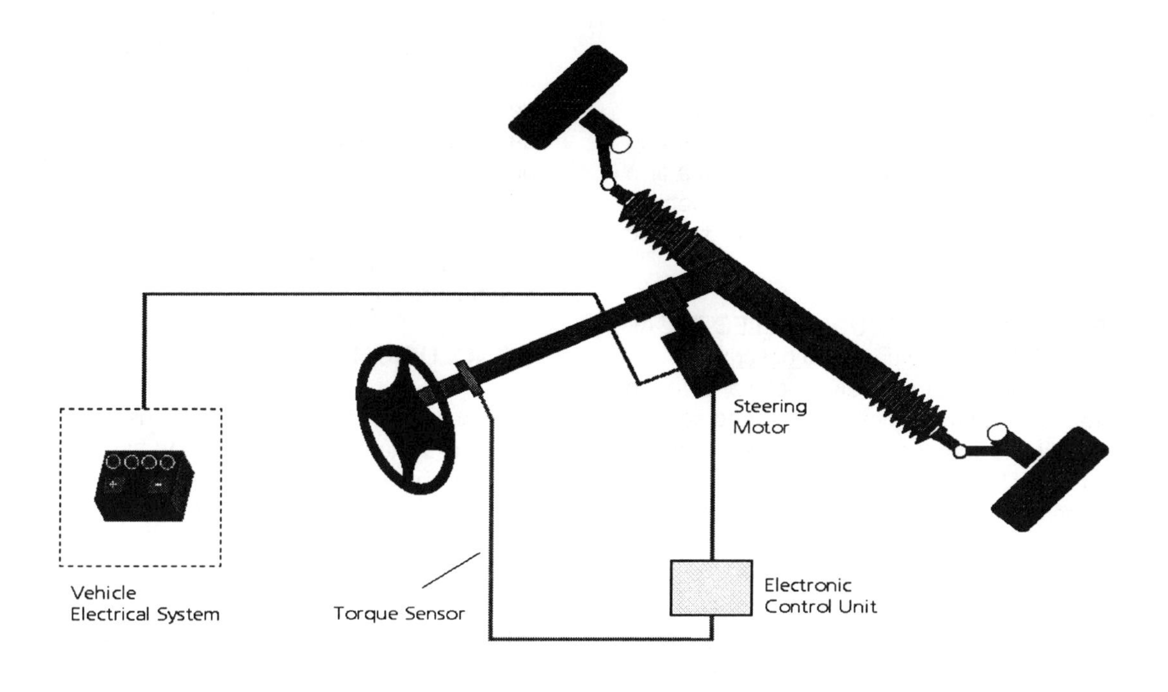

Figure 6. System structure of electrical power steering

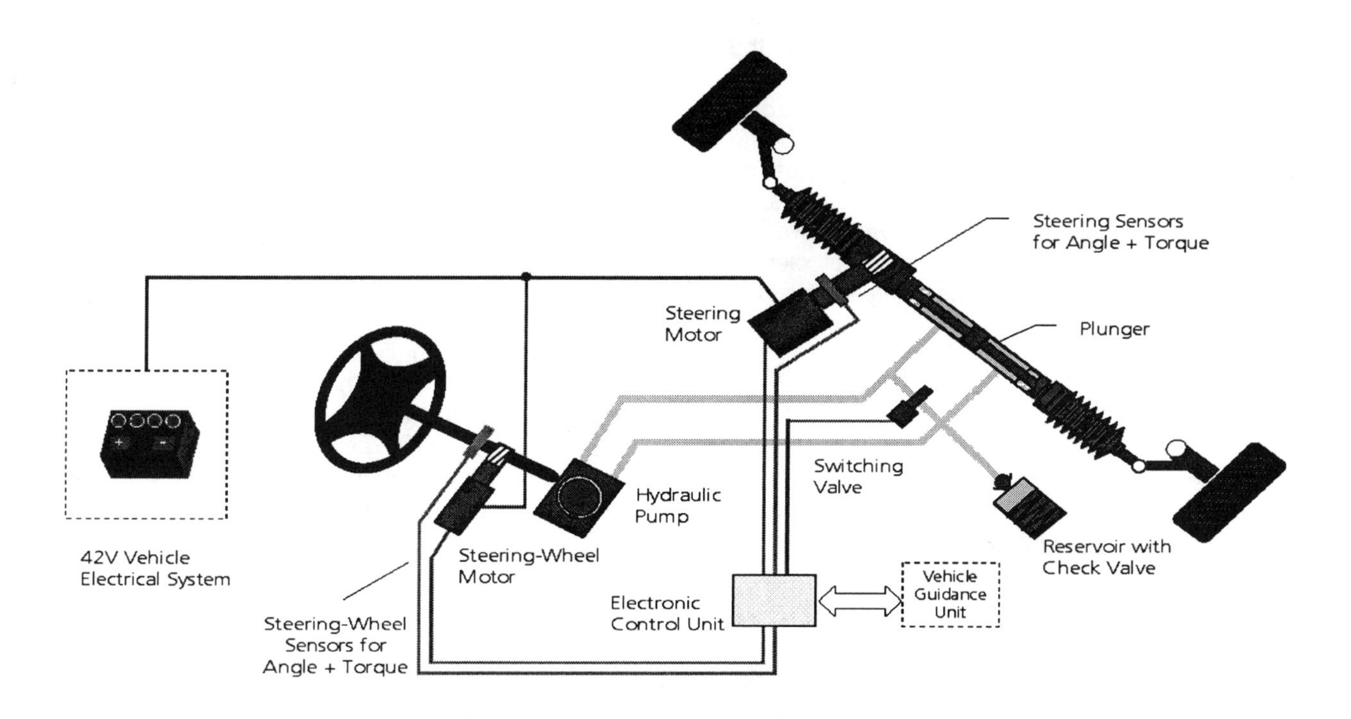

Figure 7. System structure of steer-by-wire system with hydraulic backup

Safety features – The system's fail-safe behaviour concerning electrical faults is accomplished by detecting and evaluating all electrical failures. According to the respective importance of the fault the functionality of the system is reduced. In case of major electrical faults the electrical steering system is completely switched off and the switching valve is safely actuated, establishing a firm hydraulic link between steering wheel and the vehicle wheels. On the hydraulic backup level vehicle dynamic intervention is no longer possible.

SYSTEM STRUCTURE OF THE PURELY ELECTRICAL STEER-BY-WIRE SYSTEM – Functional description – Fig. 8 shows the structure of a purely electrical steering system. The reduced safety by omitting steering column and hydraulic backup is compensated by higher demands on the safety structure of electrical and electronic components. Again, the system consists of components at the steering wheel, on the vehicle wheel level, and it comprises a control unit and a 42V vehicle electrical system. In this case this must be implemented as a safe 42V vehicle electrical system containing additional elements for the diagnosis of charge condition, as well as for the disconnection of batteries.

The steering wheel motor and sensors indicating steering wheel angle and steering wheel torque are arranged at the steering wheel. These components identify the driver's desire and reproduce the return forces transferred to the steering wheel. For a safe acquisition of steering wheel position two redundant steering angle sensors are used. Power stage and power supply for the steering wheel motor are likewise redundant. In order to exert a return force on the steering wheel in case of a defective steering motor a torsion spring is available to generate the return torque. Optionally a second steering wheel motor can be used in order to redundantly generate the return torque. On the vehicle wheel level the system is equipped with a redundant set of electric motors and redundant sensors measuring angles and torques.

The electronic control unit is designed fail-safe in terms of redundant power supply, signal processing and power actuation.

Sensor values are identified periodically and redundantly, further processed via matched control algorithms and the calculated actuation signals are supplied to the two steering motors as well as the steering wheel motor.

As to the link between the electronic control unit and the vehicle guidance unit as well as dividing the functions of these components to the decentralized units the explanations are in accordance with what has been described earlier referring to the steer-by-wire system with hydraulic backup.

Safety features – Failure tolerance is required in these areas: sensors, electronics, actuators, vehicle electrical system and data transmission. This is accomplished by appropriate redundant structures.

The fail-safe behaviour against electric faults is to be ensured by a complete detection and locating of all electric failures. Locating a defective channel during signal detection or signal processing requires majority decisions. The needed redundancy is achieved either by hardware components or by including additional processing variables of the same kind. The defective channel is then switched-off consequently. In spite of electrical faults both steerability and vehicle dynamic interventions are ensured on account of the redundant system structure.

CONCLUSION

This contribution presented various types of electrical steering systems and their safety aspects. The electro-hydraulic power steering does no longer operate the hydraulic pump via a V-belt drive from the internal combustion engine. Rather, an electric motor is used, yielding energy savings and flexibility of installation. Electrical power steering pursues this trend and offers additional advantages since no hydraulic system is required.

A steer-by-wire system with hydraulic backup and a purely electrical system were discussed. It had been stated that redundant fail-safe structures for electric and electronic components are to be established due to the fact that no mechanical or hydraulic connection between steering wheel and vehicle wheels are available.

Future innovative steering functions, such as vehicle dynamic interventions, collision avoidance, individual wheel steering, tracking assistance, automatic lateral guidance, and finally autonomous driving functions will be implemented in a system compound of various vehicle systems. Future steering systems will thus have to be integrated into a system compound, in terms of interfaces and functions. The steer-by-wire principle becomes absolutely necessary when those innovative functions are to be achieved.

The transition to purely electrical steering systems will proceed step by step, both for safety reasons and acceptance by the customer. The path will lead from electrical power steering via a steer-by-wire system with a hydraulic or mechanical backup towards purely electrical steer-by-wire systems.

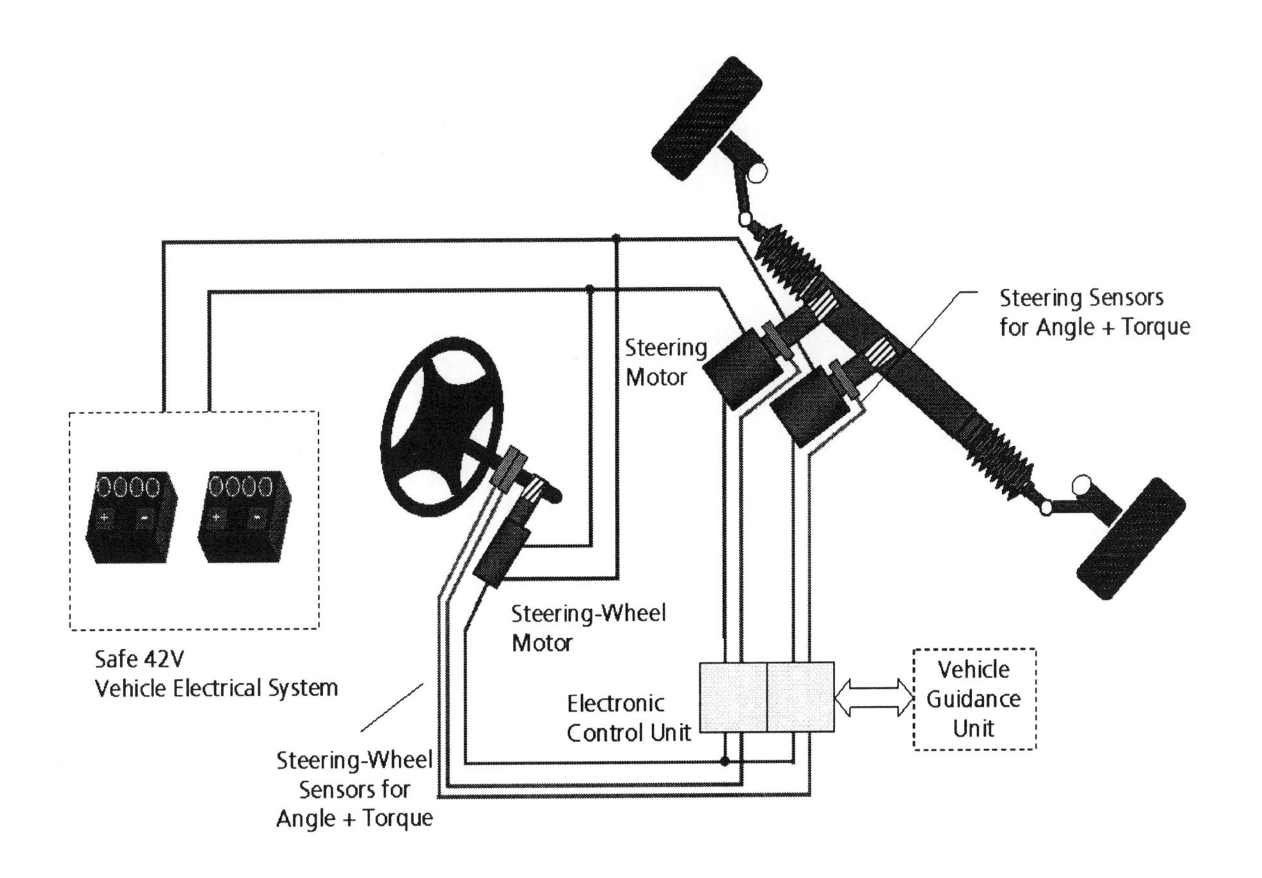

Figure 8. System structure of purely electrical steer-by-wire system

REFERENCES

1. Branneby, P.; Palmgren, B.; Isaksson, A.; Pettersson, T.; Franzen, S. (SAAB): „Improved Active and Passive Safety by Using Active Lateral Dynamic Control and an Unconventional Steering Unit". 13th Int. Technical Conference on Experimental Safety Vehicles, Paris, 1991, pp. 224-230.
2. Daimler-Chrysler: Forschung und Technologie: Steer-by-wire, Neuartige Assistenzsysteme, Mobiler Arbeitsplatz, Internetsite http://www.daimlerchrysler.de/investor/annual98/fue1_g.htm 3/99.
3. Leventon, W.: „An Automobile Oracle Speaks on the Next Millennium", PD&D-Net, June 1997.
4. Junker, H.: „Evolution in der Lenkungstechnologie". Automobil-Industrie 1/92, S. 17-21, 1992.
5. Ackermann, J.;Bünte, T.; Sienel, W.; Jeebe, H.; Naab, K. (BMW): „Fahrsicherheit durch robuste Lenkregelung". Automatisierungstechnik 44 (1996) 5, pp. 219-225.
6. Krämer, W.; Hackl, M.: „Potential Functions and Benefits of Electronic Steering Assistance". XXVI. FISITA Congress, Praha, June 17-12, 1996, Paper B0304.

DEFINITIONS, ACRONYMS, ABBREVIATIONS

- ECE R79 — Regulation No. 79. Uniform provisions concerning the approval of vehicles with regard to steering equipment.
- DIN V 19250 — Leittechnik: Grundlegende Sicherheitsbetrachtungen für MSR-Schutzeinrichtungen
- DIN V 19251 — Leittechnik: MSR Schutzeinrichtungen. Anforderungen und Maßnahmen zur gesicherten Funktion
- IEC 61508 — Functional safety of electrical/electronic/programmable electronic safety-related systems
- NTG 3004 — Zuverlässigkeitsbegriffe im Hinblick auf komplexe Software und Hardware

974113

A safety analysis methodology and its automotive application

I R KENDALL
Jaguar Cars Limited, Coventry, UK
K M HOBLEY
University of Leeds, UK

This paper will introduce the PASSPORT methodology for safety analysis, which is recommended by the "MISRA Guidelines" published in 1994 as the UK motor industry's recommended approach to safety-related electronic systems and software. It is a powerful methodology for maximising the benefit of FMEA and Fault Tree type analyses, in order to provide a clearer understanding of the hazard and integrity implications of systems. PASSPORT has its origin as an EC project, part of the DRIVE II programme (Project No V2058). It is a joint paper by Leeds University, one of the prime movers in generating the ideas behind the PASSPORT method and Jaguar Cars, who are actively applying the techniques on real automotive projects. Both are members of the MISRA (Motor Industry Software Reliability Association) consortium responsible for the Guidelines and are now partners in the new EC Framework IV project COMPASS. COMPASS has been set up to create computer based tools to support the PASSPORT methodology and to widen its acceptance in the automotive industry, and we will describe how both theory and practice are being used to feed into this new project.

1. Introduction

Recent developments in automotive electronics have seen systems aimed at improving road safety, transport efficiency and environmental quality. It is, however, clear that systems with the power to advise on, or to control, road safety, transport efficiency or environmental quality, also have the potential to give rise to new safety concerns. Experience shows that, unless positive action is taken, then at some stage in the life of a system a failure to perform as expected will indeed occur. Random faults may occur for many reasons (e.g. component wear and communication breakdown), and systematic faults may be made in the software or in the overall design of the system. The undesirable effects of component wear, communication breakdown and software faults can be overcome by using a suitable design, but this is only effective if the design is correct. No responsible developer will deliberately produce an incorrect design, nevertheless design faults in both hardware and software do occur, and the reason for them can usually be traced to an incomplete understanding of the system due to its complexity. In the future, systems are likely to be even more complex than today.

In order to assure the functional safety of a system, the development process must include safety engineering activities that run in parallel with the normal development life-cycle. The UK motor industry consortium, MISRA, was established to make recommendation on what these activities should be, and this resulted in the MISRA "Development Guidelines for Vehicle Based Software" published in November 1994 [MISRA 94]. In compiling these guidelines a study was carried out to survey a number of different approaches to safety engineering, and this concluded that the work carried out as part of the European Framework III research project, PASSPORT (V2057/8) [PASSPORT 95] was highly appropriate for automotive application. Since then, the new European ESPRIT project COMPASS (EP 22816) has been formed with the explicit aim of producing a software tool for implementing the PASSPORT methodology.

2. PASSPORT Prospective System Safety Analysis

[PASSPORT 95] refers to a generic process known as Prospective System Safety Analysis, or PSSA. ("Prospective" to indicate that it should be performed before the real system exists.) The framework for PSSA divides the safety analysis task into two separate phases:

1. A Preliminary Safety Analysis (PSA) is performed as part of the feasibility study when the system concept is being proposed. The objective is to discover whether there are any safety hazards associated with the system, and if so, to identify the top-level safety requirements and the safety integrity levels associated with them.
2. A Detailed Safety Analysis (DSA) is performed in parallel with the system design. The objective is to analyse the design for confirmation that all the safety requirements have been implemented to the required integrity level, including any new ones introduced as a result of the architecture chosen for the system.

Figure 1 shows the flow of information into and out of the PSSA, and between the PSA and the DSA. COMPASS will implement support for both PSA and DSA.

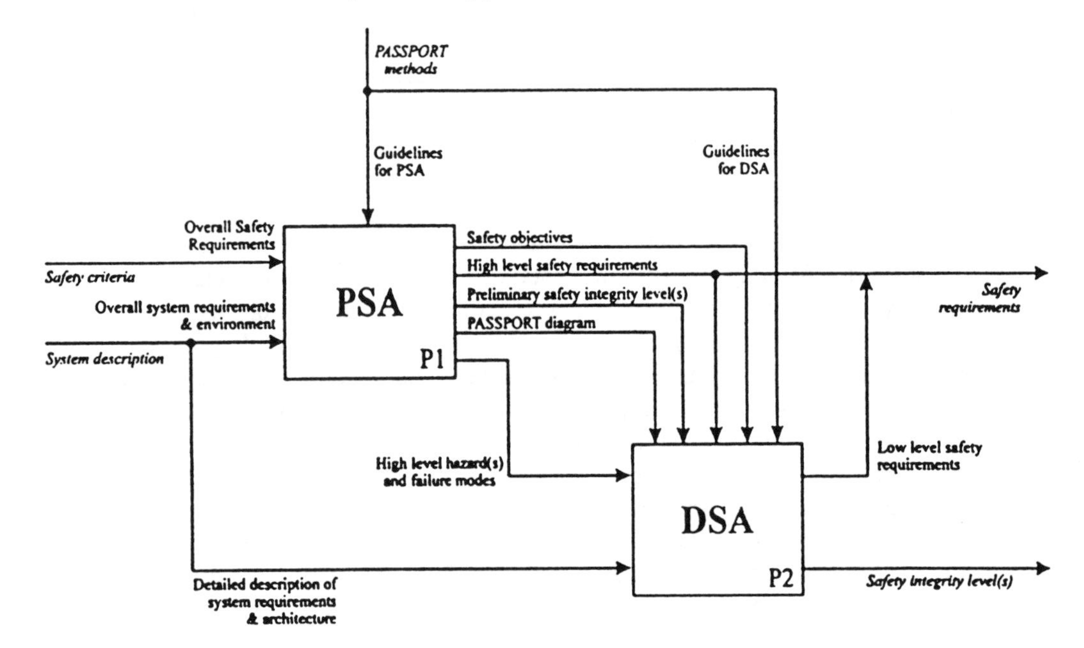

Figure 1 : Prospective System Safety Analysis

The methodology for both the PSA and the DSA follows the same basic pattern:
a. Produce a model of the system.
b. Check the model for completeness and consistency.
c. Undertake hazard analysis.
Whilst the techniques for performing task (c), e.g. Failure Mode and Effects Analysis (FMEA) and Fault Tree Analysis (FTA) are well known, it was discovered that there were no suitable overall modelling techniques available upon which one could perform such a hazard analysis.

3. Preliminary Safety Analysis

The aim of a PSA is to identify how the proposed system, or Target of Evaluation (TOE), interacts with its environment, and then to discover whether any of these interactions could result in a hazardous situation in the case of a failure of one or more parts of the TOE. The first task is therefore to produce a model that clearly shows the relationship between the TOE and its environment.

3.1 The Passport Diagram

A safety hazard will be caused by *what* the TOE might do to its environment, it is not necessary to know during this phase *how* it might happen. Indeed, since many design decisions will not have been taken, the 'how' may still be unknown. For this reason the model devised, the PASSPORT Diagram, is a functional model.
The basic building blocks of a PASSPORT Diagram (see Figure 2) are the *Nucleus of the TOE*; a set of *terminators* which either take input from, or provide output to, the operating environment of the TOE; the *information sets* that pass between the terminators and the nucleus of the TOE; and the *flow* of these information sets. Whilst the diagram is based on the data flow diagrams used to model informatic systems, the types of terminator and information sets include anything and everything that might effect the safety of the TOE (e.g. movement of items, algorithms used and the development process itself).

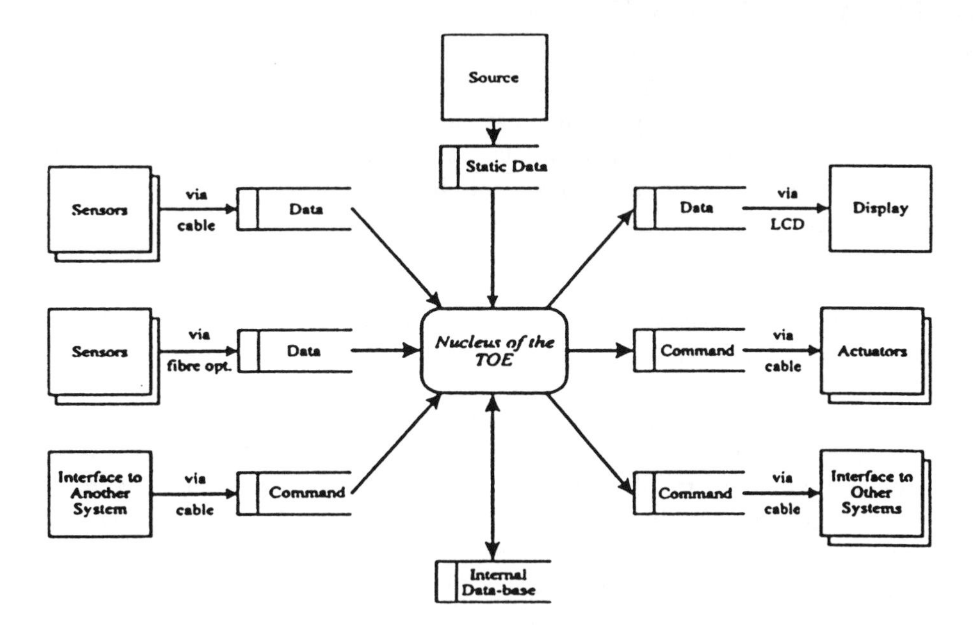

Figure 2 :An Example PASSPORT Diagram

3.1.1 Completeness and Consistency Checks

The PASSPORT Diagram model can be checked for completeness by confirming that each of the top-level system requirements can be achieved. Since a PSA should be performed with a team of people with a wide range of expertise, the act of creating the PASSPORT Diagram model can help to highlight misunderstandings between engineers at a very early stage in the life-cycle.

The (self-)consistency of the PASSPORT Diagram model can be checked by applying the maxim "what goes in must come out, and what comes out must have gone in". The checks ensure that there are no unnecessary inputs to the TOE, and that all the data is present for the system to function as intended.

3.2 Preliminary Hazard Analysis

Once the PASSPORT Diagram has been shown to be complete and consistent then the only way that the TOE can effect its environment, for good or ill, is contained within it. Each element of the PASSPORT Diagram is then systematically analysed and the question "what if ....?" asked (e.g. what if the actuator failed to operate; operated with no command, etc?, or what if the information was corrupted; failed to arrive, etc?). The effectiveness of this task can be increased by the use of checklists and guidewords to remind the analysis team of all the modes of operation that the system may undertake. By this means a preliminary hazard list can be built up for the system.

Very few, if any, real systems start with a blank sheet of paper. They are usually built from a number of existing sub-systems with some new items to provide the additional functionality. There is therefore normally sufficient information, even at the concept stage of the life-cycle, to perform an analysis; and to discover how each of the hazards in the hazard list might occur by building up a tree of preliminary events that could lead to the final undesirable event. By analysing the leaves of each tree it is then possible to identify the top-level safety requirements necessary to reduce the risk of the hazard.

3.3 Safety Integrity Levels

[MISRA 94] shows an overview of the thought process required to assign an integrity level to a hazard (see Figure 3). This uses the novel concept of "controllability" [DRIVE 92] to assess the risk associated with a system hazard, which relies on the notion of the ability of someone finding themselves in a hazardous situation to mitigate the possible effects, i.e. to control it. Such an approach, although still highly subjective, overcomes many of the situational factors that normally make assessing automotive hazards difficult. The process considers both the type of control that has been lost, and whether any other control features remain that might help to alleviate the situation in time. Each hazard is placed into one of five controllability classes (uncontrollable, difficult to control, debilitating, distracting and nuisance). These five controllability categories are then mapped directly onto five safety integrity levels.

The degree of care that will be taken to implement each of the safety requirements will depend on the importance of each associated hazard. The concept of a safety integrity level arises naturally from the fact that some activities are perceived as being more hazardous than others. Their use is desirable because the costs associated with high integrity levels can be very great. A balance must therefore be struck between using too low a level, which will increase risk, and using too high a level, which will result in unnecessary costs.

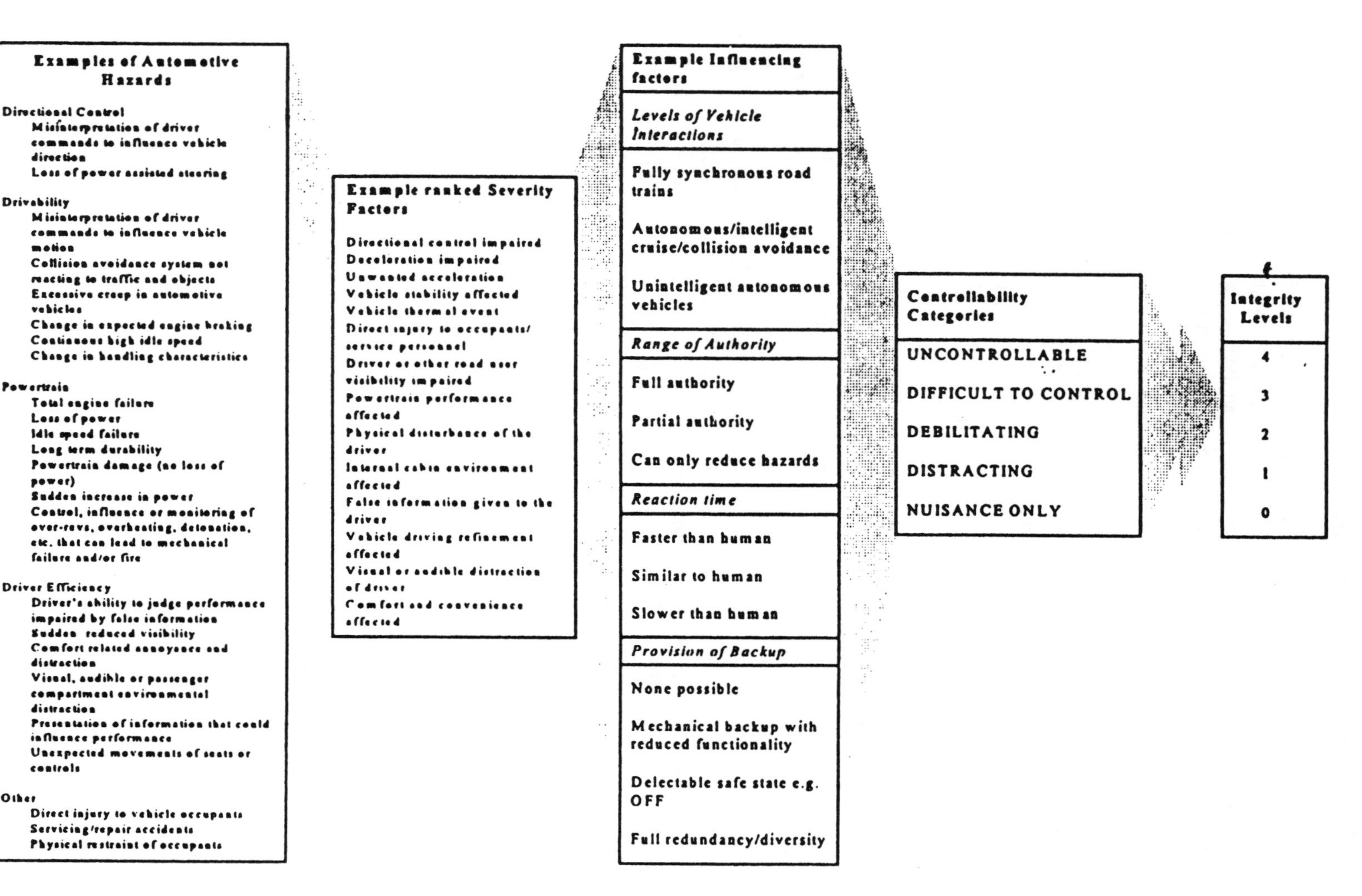

Figure 3 : Guide to Assigning Integrity Levels (from [MISRA 94])

4. Detailed Safety Analysis

The second phase of the PSSA is an analysis of the detailed design of the system. The aim is to confirm the findings of the PSA, check whether the design chosen needs some additional safety requirements and to perform a hazard analysis on all the critical items. However, before these tasks can begin it is necessary to ensure that we have a complete and consistent model of the system.

By this stage in the life-cycle there are normally two models of the system, though their distinction may not be clear. One model is the functional model, normally used by the software engineers, which describes what the system is to do. The other model is the physical model, normally used by the electrical and electronic engineers, which describes how the system is to be implemented. It is necessary to ensure that these two separate models are self-consistent, but no technique to do this could be found. The PASSPORT Cross model was therefore devised.

4.1 The Passport Cross

The PASSPORT Cross model is based upon the Business Systems Planning system, developed by IBM [IBM] to provide a means of performing consistency checks between two different representations of a system, but it has been extended to cover all aspects of DSA.

The main PASSPORT Cross consists of four matrices sharing common axes (see Figure 4). There are two *connection* matrices and two *projection* matrices. A recognised functional modelling methodology, such as Hately-Pirbhai [H-P 88], can be used to describe the *Functional Elements* (FE) and *Information Sets* (IS) upon which they operate. The first connection matrix, FE-IS, is completed by indicating which Information Sets are being used by each Functional Element.

The physical model, e.g. as created using existing circuit schematic tools, can be used to describe the *Physical Elements* (PE) and *Communication Facilities* (CF) that join them. The second connection matrix, CF-PE, is completed by indicating which Communication Facilities permit which Physical Elements to communicate.

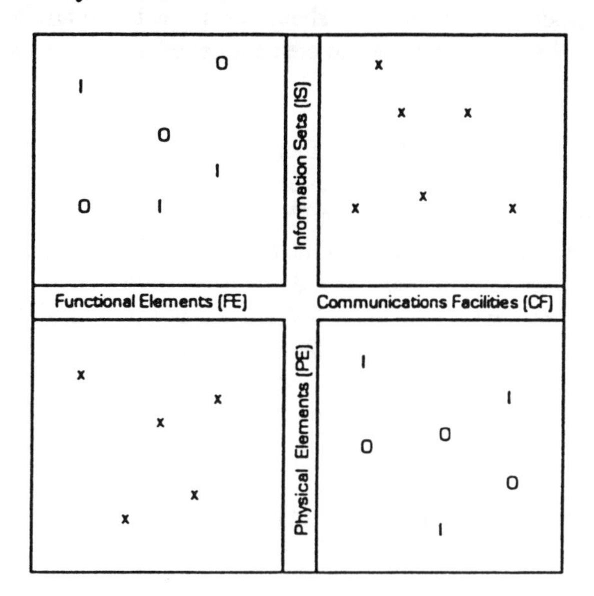

Figure 4 : The PASSPORT Cross

The matrices CF-IS and FE-PE are referred to as projection matrices. They relate the elements of the functional model to the elements of the physical model. The matrix CF-IS identifies which Communication Facilities are being used to transmit each Information Set, and the matrix FE-PE identifies which Physical Elements are being used to implement each Functional Element. It is this ability to relate the physical model to the functional model that lies at the heart of the power of the PASSPORT DSA method.

4.1.1 Consistency Checks

There are a number of different consistency checks that may be performed on the PASSPORT Cross model. They may be grouped together into two basic types: *Intra-matrix* (those checks that are performed upon each matrix independently to confirm that it is well-formed) and *Inter-matrix* (those checks which are performed across the set of matrices as a whole to confirm that they fully relate to each other). The full set of consistency rules can be found in [PASSPORT 95].

4.1.2 Levels of Decomposition

In complex systems, which might use a particular Physical Element to perform more than one function, e.g. to support re-use, or in which a single function might be performed by more than one Physical Element, e.g. to provide redundancy, the models will exhibit different levels of decomposition.

The PASSPORT methodology includes some special extensions to allow these apparent "mismatches" in levels of decomposition to be dealt with in an equally robust manner. (Achieved by introducing two new matrices, CF-FE and IS-PE, known as *flow* matrices.)

4.2 Hazard Analysis

There are two forms of hazard analysis that must be performed during a Detail Safety Analysis of the design, FMEA and FTA.
Ideally every failure mode of every element should be subjected to an FMEA. However, this is impractical and it is therefore necessary to create a list of *sensitive elements* for which it is prudent to perform an FMEA. The PASSPORT Cross model enables such sensitive elements to be identified.
One type of sensitive element is a Physical Element that implements a safety requirement; this can be readily identified using the traceability facilities of the PASSPORT Cross model. Another type of sensitive element would be a Physical Element that implements a number of functions; this can be identified by looking at the rows of the FE-PE matrix.
An FTA should be performed upon each hazard identified during the FMEA. Each FTA will identify a list of *common events* and *weak links* in the physical model that have potential safety consequences, and for which care should be taken during the remainder of the development life-cycle. The PASSPORT Cross model can assist the FTA process as it naturally provides information on the inter-connections between various elements.

5. Example Application of the Techniques

Jaguar Cars have been applying the PASSPORT methodology on its recent electronic throttle development projects. PSA in particular has proved to be an extremely useful technique for facilitating progress in the early stages of a project, when there is uncertainty about what the critical issues are. The ability to condense the view of the system into a PASSPORT Diagram, to identify the hazards and to elicit the key safety requirements, has allowed more focus to be attained earlier in the vehicle programme than would have previously been the case. The PSA process will establish the correct level of detail appropriate for moving the project forward. Previously a much more time consuming FMEA activity would have been started, inevitably covering more detail than required, most of which would likely to be incorrect once the design is underway, and with unclear objectives. An example of the PASSPORT diagram for an electronic throttle system is shown Figure 5. The PSA identified the main safety requirements, such as the need for redundant sensors, each of which is traceable back to a hazard; to confirm that the worst hazard for electronic throttle is "unintended acceleration"; and to recommend an integrity level for the system of 3.
As yet it has been impossible to carry out a DSA as defined in the PASSPORT methodology for the electronic throttle project. The main reason for this is the fact that the physical and functional designs for the system are maintained by completely disjoint methods, and across several internal and external organisations - the hardware as a series of engineering drawings, schematics and diagrams; and the software partly in Hately-Pirbhai [H-P 88] structured design, and partly in natural language descriptions. However, as one would expect, FMEA [HANDBOOK] and FTA [A-M 93] have been performed in the usual way using existing standalone FMEA and FTA software packages. Although it would be possible to perform the PASSPORT DSA on paper, collecting the information from the disjoint sources mentioned, this would require a significant amount of effort and work duplication, and would undoubtedly result in very large, unwieldy documents. To attempt such an activity on paper may also prove unbeneficial due to the high risk of human error.

This experience has been confirmed by The University of Leeds, who have attempted full DSA on some transport telematic systems on behalf of other EC projects.

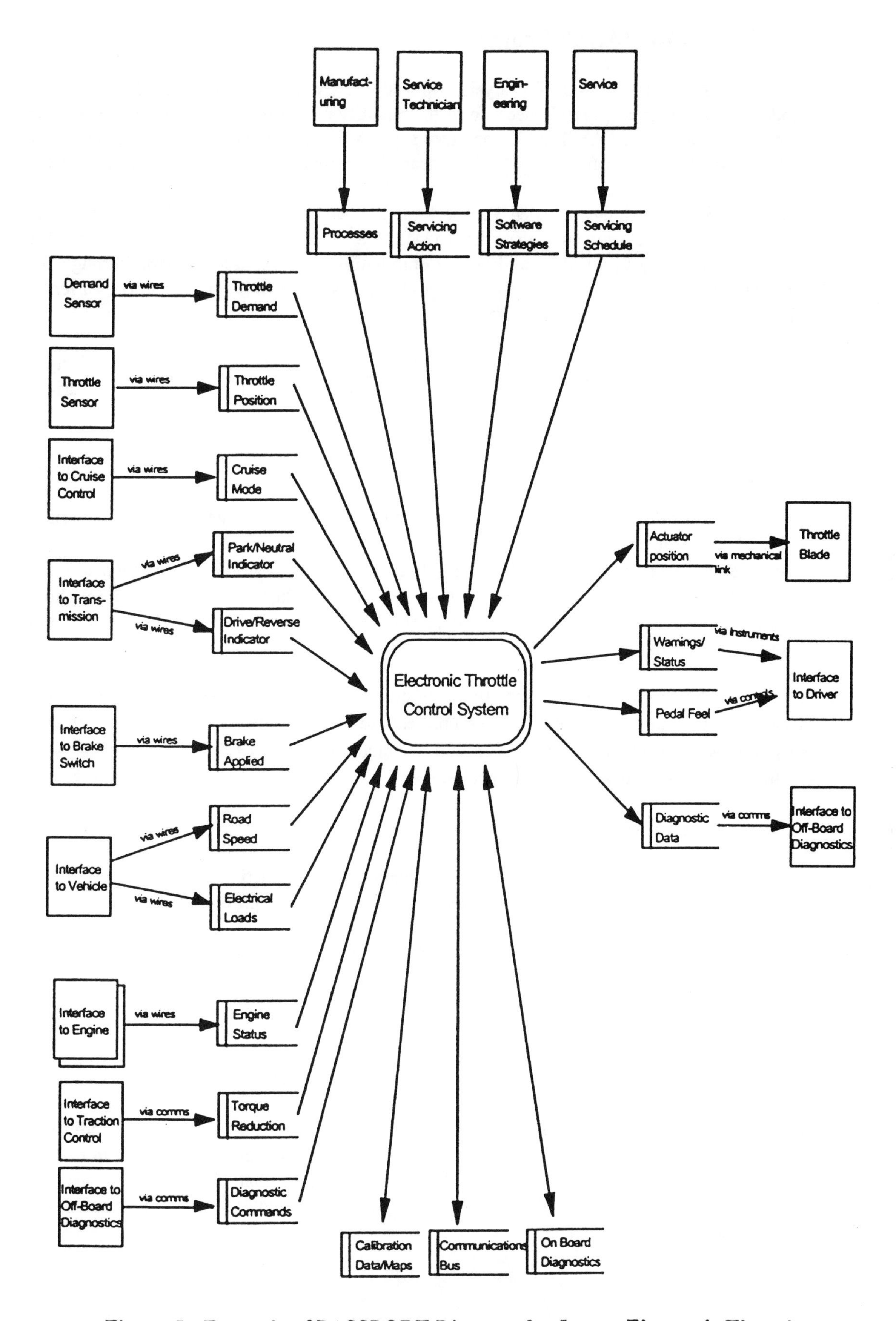

Figure 5 : Example of PASSPORT Diagram for Jaguar Electronic Throttle

6. The "COMPASS" Project

Carrying out a full PASSPORT analysis, especially DSA, is a complex problem, even for modestly sized systems, mainly because of the quantity of information generated. Hence, to take full advantage of PASSPORT, computer support for the methodology is essential. The COMPASS project, part funded by the European Commission under Framework IV ESPRIT (EP 22816), has been established to develop such a computer based tool for PASSPORT. (COMPASS = **COM**puter assisted **P**assport for the **A**rchitecture and **S**afety of **S**ystems.) Initial studies have estimated that the COMPASS tool has the potential to save the automotive industry up to £600 million, by eliminating a key source of error (through the physical versus functional model consistency checks), and by reducing the effort currently expended on FMEA and related analysis activities.

The tool will use a two-stage process to guide its users through the analysis of system designs in a way that allows critical components or functions to be identified, under various hazardous scenarios. This new computer based tool will :

- enable Europe's motor industry to realise efficiency gains over traditional methods by targeting analysis at critical parts;
- improve car safety, helping car makers and their suppliers work together through the rule-based modelling systems;
- offer engineers of all disciplines the ability to perform more effective safety analysis;
- provide "right first time" results by working from the concept stage throughout development. This helps avoid the costly re-engineering of systems when unexpected risks are found late in the development cycle.

The project has been underway since the beginning of 1997, and at the time of writing this paper, is at the point of refining user requirements to an initial design. An α-prototype will duly follow in 1998, and the project will conclude in June 1999 with β-version user evaluation. Those involved in the project include:

Project Manager	Lotus - UK
Users	Jaguar Cars Ltd. (luxury car manufacturer) - UK Lotus Engineering (automotive design consultancy) - UK TÜV Rheinland (assessment authority) - GERMANY The University of Leeds (academic research) - UK
Developers	LSC Group Ltd. - UK Athens Technology Centre SA - GREECE CYTEC Datensysteme GmbH - GERMANY

The developers will be responsible for further developing the β-version into an industrial strength, commercial tool after the end of the funded project, which they will seek to market across the European Union and beyond. It should be pointed out that, although the project has an automotive bias, it is strongly believed that it will have many uses in other application domains and industry sectors.

The main "user needs" for COMPASS can be summarised as :

- To manage the quantity of information required for safety analysis.
- To manage the quantity of information generated during safety analysis.
- To act as a diagramming aid.
- To assist uniformity of approach and facilitate effective assessment.
- Where possible, to act as an "expert system" or check-listing aid for the safety analysis process.
- To improve efficiency during safety analysis activities.
- To assist in the promotion of a "safety culture".

The tool will be designed to work on Windows NT client platforms, supporting standalone and multi-user operation, including UNIX or Windows server. It will be based on "industry standards" wherever possible (e.g. a STEP compliant database), and the intention is to allow it to interface with existing hardware and software design tool, to allow import and export of data.

7. Conclusions

The project PASSPORT has produced a systematic methodology suitable for the safety analysis of vehicle electrical/electronic based systems, and in order to do this two new modelling techniques, the PASSPORT Diagram and the PASSPORT Cross, have been devised. The methodology has been successfully applied, at least in part, to a number of real projects, thus demonstrating its potential benefit for application. However, full implementation of DSA on even moderate sized systems, is impractical without computer tool support. The COMPASS project will provide this computer support for the PASSPORT methodology, including information storage and retrieval, diagramming and modelling aids, automated consistency checking, and "expert system" style advice for the analysis. This should enable the methodology to be practical, even on large scale projects, thus leaving the design and development teams to concentrate on the identification of any potential hazards, and the design measures taken to deal them.

8. References

[MISRA 94] MISRA, Development Guidelines for Vehicle Based Software, The Motor Industry Research Association (MIRA), ISBN 0 9524156 0 7, 1994.

[PASSPORT 95] PASSPORT, Framework for Prospective System Safety Analysis, Deliverable N° 9, DRIVE II Project PASSPORT (V2058), 1995.

[DRIVE 92] DRIVE Safely, Towards a European Standard : The Development of Safe Road Transport Informatic Systems (Draft 2), DRIVE Project V1051, 1992.

[IBM] IBM, Business System Planning - Information System Guidelines, IBM Document No. GE/20/02572.

[H-P 88] Strategies for Real-Time System Specification by Derek Hatley and Imtiaz Pirbhai. Dorset House Publishing, New York, 1988.

[HANDBOOK] Failure Mode and Effects Analysis Handbook. Automotive Safety and Engineering Standards. Ford Motor Company, 1995.

[A-M 93] Reliability and Risk Assessment. J.D. Andrews and T.R. Moss. Longman Scientific and Technical. 1993.

9. Acknowledgement

The work described in this paper was funded by the European Commission DG XIII Telematics Programmes (DRIVE Safely, and PASSPORT), by the European Commission ESPRIT Programme (COMPASS), and by the UK Department of Trade and Industry SafeIT programme (MISRA).

916028

Proposal for a Guideline for Safety Related Electronics in Road Transport Systems (Drive Project V1051)

Winfried Asmuth, G. Heuser, H. Trier, and J. Sonntag
TÜV Rheinland

Today more and more safety related vehicle subsystems are equipped with programmable electronics. Current research programs—in Europe DRIVE and PROMETHEUS—are thinking about high tech systems such as automatic distance control, steering by wire, etc. which have all a very high level of risk. The project V 1051 of the DRIVE programme is intended to ensure a high safety standard in road transport systems. The main task is to specify a guideline for developing and production process as well as for the certification procedure of hard- and software of porgammable electronic systems.

This presentation is mainly concerned with measures against random hardware failures as well as with methods of analysis used to detect weak points in the design of Road Transport Informatics (RTI) systems (Figure 1).

DRIVE SAFELY: Meeting with Industry

System Architecture and Hardware Aspects

- Introduction
- Approach to attain safety
- Measures against failures
  - Measures to control failures
  - Various safety architectures
  - Evaluation of measures
- Analysis of Failures
- Summary

Figure 1. Structure of the Report

Firstly the general approach to attain safety of RTI systems will be explained, followed by measures against failures. The main emphasis is on the measures to control random failures. We look at both the various architectures for different safe situations and at an approach to assign measures to integrity levels. Following that, methods to analyses and to detect weakpoints in the design will be considered. Finally there is a short summary of the presentation.

Approach to attain safety (Figure 2)

In order to ensure safe operation of safety critical electronic systems, it is necessary to recognize the various possible causes of failures and to ensure that adequate precautions are taken against each.

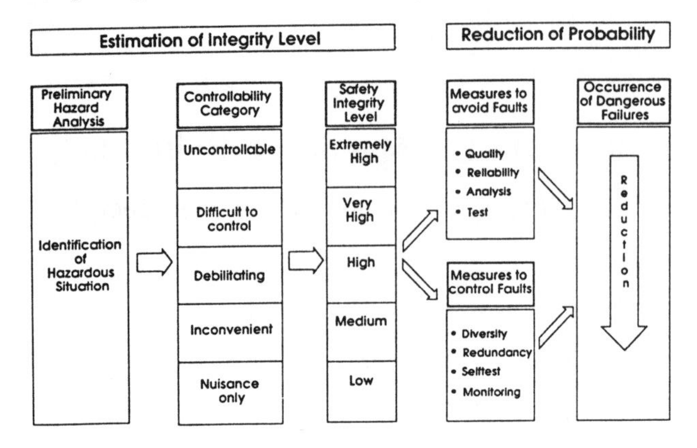

Figure 2. Approach to Attain Safety

The main safety objective is to avoid hazardous malfunction of the electronic system with sufficient probability for a particular application. This goal could be attained, if the following approach is applied:

In a preliminary hazard analysis the hazards must be identified and evaluated.

The degree of probability to control the dangerous situation by the road user places each hazard into one of five controllability categories from nuisance to uncontrollable.

Assigned to these categories are safety integrity levels which are ranged from low to extremely high.

According to the integrity level a combination of measures with suitable effectiveness must be taken to avoid and to control faults.

By quality measures which are applied during the development process, faults can be avoided.

By structural and fault detection measures, faults can be controlled during the operation phase.

The controlling measures are necessary, especially at higher integrity levels, because the reliability of electronic systems is often limited. Both the measures to avoid and to control faults increase safety by reducing the probability of the occurrence of dangerous failures.

Measurement Failures (Figure 3)

Precaution must be taken against both random hardware failures and systematic failures in hard- and software. Suitable measures must be taken to avoid faults during the development and production phase and to control faults during the operation phase of the system.

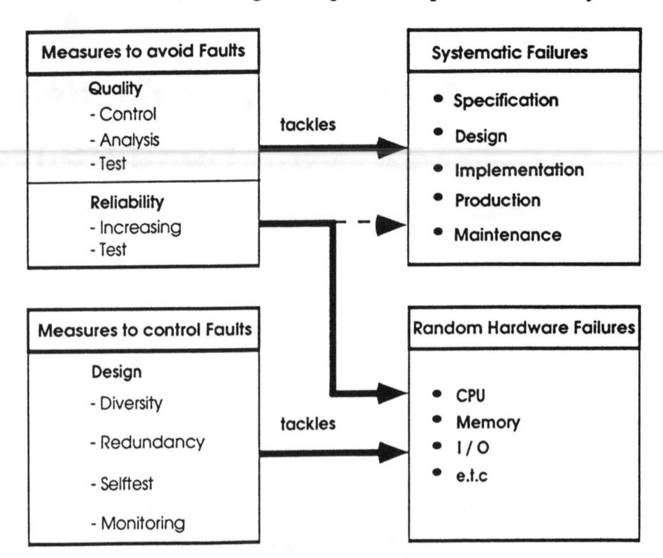

Figure 3. Measures Against Failures

Software faults always result in systematic failures. They can only be avoided by quality control measures during the development process and controlled by diverse redundancy during operation.

Hardware failures can be caused by both systematic faults made during the development and productions processes, as well as by random faults in components. Systematic hardware faults can be avoided through quality measures. The occurrence of random hardware failures can be reduced by improving reliability, but this is often limited, especially with complex programmable electronic systems. Therefore additional design measures are necessary to control these random failure.

The measures against failures can be summarized into three safety elements:

- Quality---> tackles mainly systematic failures
- Reliability---> tackles mainly random failures
- Design---> tackles also mainly random failures

Measures to control faults (Figure 4)

If safety cannot sufficiently be attained by quality and reliability measures, then measures are necessary to control failures during the operation phase.

The first basic safety requirement on RTI systems for all levels of integrity is, that a single fault may never lead to a dangerous situation. To attain this goal faults must be detected and an appropriate protection measure must be carried out. The second requirement is, if a fault is undetected, it should never lead to a dangerous situation even in combination with one or more other faults.

These safety requirements can mainly be fulfilled with the aid of two methods: The first method uses comparison of results of redundant systems. The second method uses test procedures to detect faults in systems.

If quality + reliability < safety
then measures are necessary to control faults.

The objective of the control measures is to ensure that faults never lead to a dangerous situation.

To achieve this faults must be detected and an appropriate protection measure must be carried out.

Fault detection measures

- System structure (Comparision of the result)
- Test routines to recognize faults
- Inspection, maintenance

Figure 4. Measures to Control Faults

In addition to these technical measures, periodical inspection and preventive maintenance should be carried out to detect faults.

Faults detection measures (Figure 5)

In the first place the designer has to specify a system architecture which is related to the integrity level and the predicted safe state of the process in the event of a failure.

System structures

- One channel with selftest and monitoring
- Two channel with comparison of the result
- Three channel with voter
- etc.

Test routines

- CPU-, RAM-, ROM-, I/O-Test
- plausibility check of Input/Output data
- watchdog
- etc.

Organizational Measures

- Periodical Inspection
- Maintenance
- etc.

Figure 5. Fault Detection Measures

The following structure are feasible:

- One channel with selftest and monitoring
- Two channel with comparison of the result
- Three channel with voter etc.

Secondly, he has to select adequate test methods to recognize faults in each channel. Possible test routines are:

- CPU-, RAM-, ROM-, I/O-Test
- Plausibility Check of Input/Output data
- Watchdog

Thirdly, he has to determine organizational measures to detect failures during the operation phase of the system. Inspection and maintenance should be performed periodically.

Architecture of Safety-Related Systems (Figure 6)

Modes of safe design

We have to distinguish between fail safe, fail soft and fail operational design.

Modes of safe Design	Required safety function	Modes of operation	Possible System Structures
Fail safe	Emergency cut out (operation terminated)	- Protection/ Monitoring system - Control system	- One channel with fault detection measures - Two channel with fail safe comparator
Fail soft	Emergency operation (Limp home)	- Safety Monitor with Limp home function - Control system	System with redundancy for "Limp home"
Fail operational	Continue operation with full function	Control system	Two out of three voting system

Figure 6. Architecture of Safety-Related Systems

Required safety function

In the first case the process can be transferred into the safe state after the detection of dangerous failures, for example by switching off the energy supply (**Emergency cut out**).

In the second case the operation must be continued by degraded functional capabilities or performance in case of failures (**Emergency operation**).

In the third case the operation must be continued by full functionality of the process (**Continue operation**).

Modes of operation

A safety related system can be used as a protection system or can be part of a control system.

Protection systems get several inputs from the monitored process and their sole purpose is to transfer the process into a *safe state* (Emergency cut out) if they recognize a potentially dangerous state of that process.

Control systems use information from the process to influence process operation. In this case we have to distinguish between the following safety functional requirements in case of failures:

- Emergency cut out
- Emergency operation (limp home)
- Continue operation with full functions

Possible System Structures

Fail safe can be attained by:

- One-channel system with selftest and external safety monitor
- Two-channel system with fail safe comparator

Fail soft can be attained by one-channel system with redundancy for "limp home."

Fail operational can be attained by a two out of three voting system.

Competition between Reliability and Safety (Figure 7)

Safety and reliability can be found in competition in certain applications.

Modes of safe Design	Relation between Reliability and Safety	Safety Requirements on the availability of the	
		system function	safety function
Fail safe	Safety and reliability in competition System reliability is limited by safety	none	depends on the integrity level
Fail soft	Safety and reliability in competition System reliability is limited by safety	conditional required for the "Limp home" system	depends on the integrity level
Fail operational	Reliability and safety have the same objectives	equal to the safety function	depends on the integrity level

Figure 7. Competition Between Reliability and Safety

By **fail safe** and **fail soft** design reliability of the system will be reduced, because parts of the system or their functions are no longer available. In these cases safety and reliability are in completion; system reliability is limited by safety. Only if the part of the system under consideration is designed to **fail operational**, are safety and reliability achieved by the same measures. In this case reliability and safety have the same objectives.

Safety requirements on the availability of the system function must be made only if the system is designed "**fail operational**" and conditional for the Limp home function if the system is designed "**fail soft.**" Availability of safety function is required for all modes of safe design, but the degree of availability depends on the integrity level.

Required Effectiveness of Measures (Figure 8)

A set of measures is necessary to avoid and to control hardware failures of (programmable) electronic systems. A combination of measures with suitable effectiveness must be selected according to the estimated integrity level.

Three classes are proposed:

- little effectiveness
- average effectiveness
- great effectiveness.

The effectiveness classes are established for each integrity level. The required effectiveness of the measures to avoid systematic and to control random hardware failures ranges from little to great. Measures to control systematic failures are only required at higher integrity levels. Lower integrity levels only require measures with little effectiveness.

Hardware Measures to		Safety Integrity Level				
avoid (Development)	control (Operation)	Low	Medium	High	Very High	Extremely High
Systematic failures		little	little to average	average	average	great
	Systematic failures	--	--	--	average	great
	Random failures	little	little to average	average	great	great

Figure 8. Required Effectiveness of Measures

These measures are able to tackle failures with relatively high probability of occurrence. In contrast, higher integrity levels require measures with great effectiveness which are also able to tackle failures with low probability of occurrences.

Assignment of Structural Measures to Integrity Levels (Figure 9)

Possible measures to avoid and to control failures are listed and evaluated in the draft of the proposed standard. The list should be an aid for the developer to find suitable measures for a particular application; however other adequate measures with the same effectiveness could be applied. Different combinations of measures are possible for a safety integrity level, where the selected combination for a particular application depends strongly on the considered process. The total

Structural Measures	Safety Integrity Level				
	Low	Medium	High	Very High	Extremely High
One-channel structure					
- with selftest	X				
- with selftest and monitoring		X			
Two-channel (fail safe)					
- homogenous			X	X+	
- diverse				X	X+
Three-channel (fail operational)			X	X+	X+

X required
X+ required with additional high value measures

Figure 9. Assignments of Structural Measures to Integrity Levels

effectiveness of all measures should be examined carefully to ensure that the safety requirements are fulfilled.

An example is given for structured measures (see Figure 9).

Failures Analysis (Figure 10)

To prevent dangerous situations of a system in case of failures all thinkable modes of component or sub-system failures and their effect must be considered. Many standardized analysis methods exist.

Two of these are:

- Fault Tree Analysis (FTA)
- Failure Mode Effect Analysis (FMEA).

To prevent dangerous situation failures of items must be considered.

Useful analysis methods are:

- Fault Tree Analysis (FTA)
- Failure Mode and Effect Analysis (FMEA)

FTA is applied to identify combination of failures and to identify safety-critical items (top-down approach).

FMEA is a qualitative evaluation of failures, especially to detect weakpoints in design, construction and production of systems (bottom-up approach).

Both are applied to complement each other during safety analysis.

Figure 10. Failure Analysis

The objective of the FTA is to identify all possible combination of component or sub-system failures, which could lead to the undesired event. The FTA is a top-down method and can be started before a detailed design document is produced. This method can support the hazard analysis process to identify safety critical items. The output of this analysis is a clear reviewable diagram(s).

The objective of the FMEA is the qualitative evaluation of component or sub-system failures of a system. The FMEA is a bottom up approach, whereby the various failure modes and effects of components or sub-systems will be analyzed, especially to detect weakpoints during the design, construction and production phase of a system.

The methods are not applied in competition, on the contrary the will be applied to complement each other. During safety analysis the FTA is used to identify safety critical components/sub-systems and the FMEA is used to detect weakpoints of these items.

FMEA of Complex Electronic Devices (Figure 11)

The FMEA will be performed to give an answer of the question: "What happens if a sub-system/component fails in a particular failure mode"?

FMEA should answer the questions:

What happens it a sub-system/component fails in a particular failure mode.

General approach

- Failure mode of items must be identified
- Failure effect must be analysed
- Failure with very low probability must be excluded

Approach for complex electronic system

- It is economically unjustifiable to evaluate all failures of complex electronic design.
- Therefore FMEA should preferably be performed on module level of complex electronic unit and on component level of actuators, sensors, electric and simple electronic circuits.
- The electronic unit itself must be constructed according to the prescribed measure against failures.

Figure 11. FMEA of Complex Electronic Devices

Firstly all failure modes of components/sub-system must be identified. Secondly the effect of the failure must be analyzed. To design safety devices it is necessary to excluded failures with very low probability.

Another problem occurs where applying the FMEA on complex (programmable) electronic systems, because it is impossible or economically unjustifiable to evaluate the very high number of failures and their resulting effects. An applicable approach is to perform the FMEA on module level of complex electronic systems and of component level for actuators, sensors, electric and simple electronic circuits. The electronic unit itself must be constructed according to the prescribed measure against failures.

Consideration of Multiple Failures (Figure 12)

The standard FMEA regards only the consideration of a single failure. In standards on road traffic signal system, electric signalling systems for railroads, burner control systems, press control devices etc. multiple failures are considered according to a particular fault flow chart.

For RTI system an appropriate fault flow chart is proposed. The chart begins with the "1st failure" (e. g. emitter-collector of any transistor short circuit). Verification that after each *"1st failure" no hazardous situation may occur* must be made. If so, what else happens after the "1st failure"?

- The automatically detected "1st failure" immediately leads to the "safe situation." If all first failures run

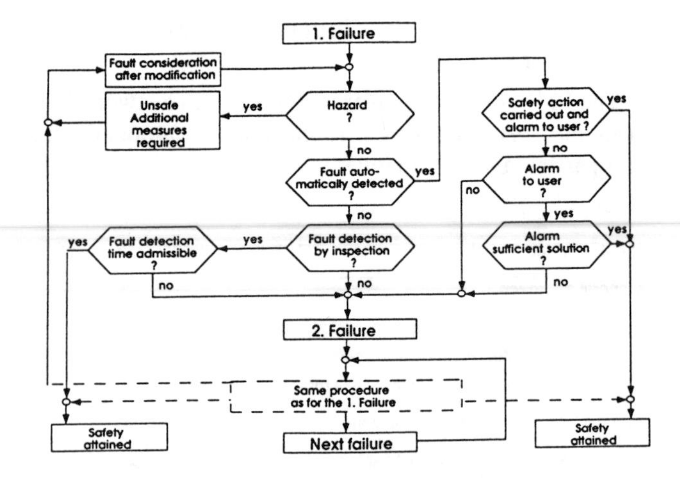

Figure 12. Consideration of Multiple Failures

If safety cannot be achieved sufficiently by quality and reliability measures,
then measures are necessary to control failures.

This is usual if electronic systems are applied.

The control measures must be able to detect faults and to carry out a safety action with sufficient probability

The different safe situation must be considered when designing the system. These are:

- Emergency cut out (fail safe design)
- Emergency operation (fail soft design)
- Continual operation (fail operational design)

Figure 13. Summary

this course, the safety device is fail safe to the highest possible level. Only relay switching circuits or small discrete electronic systems can be designed this way.

- The automatically detected "1st failure" leads to an alarm signal for the user. If this solution is not sufficient further failures must be considered or the system must be modified in order to ensure a switch over to the "Safe Situation."
- The "1st failure" will be detected during the next system inspection (automatic check or manual test). In general it is not assumed that during the interval between failure appearance and failure detection any "2nd failure" will occur and thus, in combination with the "1st failure" create a hazardous situation. It is often a question of probability calculation whether or not this assumption is permitted. If this solution is not sufficient further failures must be considered or the system must be modified in order to ensure a switch over to the "Safe Situation."
- The "1st failure" will not be detected at all (not detected by routine inspections).

In the last mentioned case, or if the alarm signal and inspection is not a sufficient solution, the "1st failure" has to be combined with any "2nd failure" and the flow chart is followed for the same routine. According to the level required, this may be repeated infinitely.

CONVENIENCE SYSTEMS

2000-01-0171

Automotive Application of Biometric Systems and Fingerprint

Jan Lichtermann and Rod Pettit
Robert Bosch GmbH

ABSTRACT

Until now, the use of biometric systems has not been in the public eye. The high cost of sensors and processing has meant that biometrics was previously restricted to high security access, financial transaction and law enforcement applications. However, as a result of improvements in technology, biometric sensor price and reliability have achieved levels where biometrics is being seriously considered for automotive systems.

This paper introduces the field of biometrics, the key terms and processes. Fingerprint Technology and Identification by Fingerprint are discussed, as are the use and applicability of biometrics in automotive applications, including Personal Profiling, Keyless Engine Start and vehicle access authorization.

The key findings of investigations over the last years are discussed.

INTRODUCTION

Automotive access and engine start security systems have, until now, been achieved by the classical access method of validation by something the user possesses, such as an ignition key. Bank auto-teller machines have improved the security level with the requirement that a customer must also provide something known, such as a PIN number. Keys, transponders and smart-cards can be stolen, and, in some cases, copied. PIN numbers or other forms of knowledge, such as passwords, can be forgotten or observed and are also inconvenient to remember and use.

Entry to a vehicle is currently granted through mechanical key-lock systems, remote keys (either infrared or UHF) or more recently introduced transponder based keyless entry-systems [1]. Vehicle immobilization is based either on mechanical key-lock systems (antiquated), transponder technology (standard) or transponder based keyless entry (new). In all these cases however, the driver has to carry something that provides authentication.

Biometrics provides the advantage that access is based on who the user is, and not on what is possessed or known. This implies that the driver himself becomes the key. Convenience is improved, as a key or transponder are no longer required. The driver cannot be identified by current systems, unless each driver uses a unique key or transponder. In contrast, biometric systems can uniquely identify drivers. The knowledge of who is actually driving a vehicle facilitates automatic reinstatement of a driver's preferences.

WHAT IS BIOMETRICS?

Biometrics uses the measurement of biological characteristics and a *Biometric* is any human physiological or behavioral characteristic that is universal, unique, permanent and collectable [2]. An alternate definition for *Biometrics* is given as "Automatically recognizing a person using distinguishing traits." [3]. The identification of individuals using a biometric is known as *Biometric Identification*. There is currently no distinct definition for the terms Biometrics and Biometric Identification and both terms are used synonymous throughout literature.

Identification is the process of associating an identity with an individual (Who am I?). Measured biometric characteristics are compared with a list of previously measured characteristics of individuals. When a match is found, the individual's identity is deduced to be that of the matching individual found in the list.

Authentication is the process of verifying that a person is who he claims to be (Am I who I say I am?). It requires that the person being subjected to scrutiny (*Claimant*) supply information about who he claims to be. This may be in the form of a unique number or index that is used to select the person's known characteristics out of an existing list of previously measured characteristics of individuals, or it may be in the form of simultaneously supplying a previously validated version of the person's known characteristics. The known characteristics are then compared against those measured and a decision is made as to whether the person's measured characteristics match those known to be the true characteristics of the claimed identity.

The *Identification Process* requires that a database of the unique biometric characteristics (also called *Features*) of people be stored inside or available to the system, while

Authentication differs in that it can be achieved by supplying the system with only the features of the person being tested at the time and no search through a database is required. The result is that Authentication is faster than Identification and that a person's biometric features can be carried by the person (perhaps in an encrypted smart card) and do not have to be stored centrally.

OVERVIEW OF BIOMETRIC TECHNOLOGIES

THE PRINCIPLE OF A BIOMETRIC SYSTEM – Without any assistance, people can recognize other people. Analyzing faces and voices is the most common method of accomplishing this identification. Other forms of identification such as hair, skin and eye color, height, dress and body shape also give us clues about the person and it is usually a combination of attributes that we use to make the final decision and although we may think it is a clear-cut decision, there is always the possibility that we are deceived by a well disguised impostor.

Current security technologies using keys, transponders and passwords provide a definite and hard decision about whether or not access should be permitted. Biometrics is fundamentally different. It involves predicting the likelihood that the actual biometric measurement just made is associated with a particular individual. The actual Yes/No decision is usually controlled by a number of thresholds that can be used to alter the performance of accurate impostor detection versus the discomfort of denying access to a valid user.

Due to cost and system complexity issues, automated biometric identification systems are typically limited to working with only one biometric characteristic. However, *Multimodal Systems* [4], that use two or more biometrics simultaneously, do exist and by weighting the results obtained from the differing methods, these systems can reduce the instances of valid user rejection without compromising impostor detection.

SYSTEM FUNCTIONAL BLOCKS – Biometric identification systems are usually accomplished using the key functional blocks shown in Figure 1.

The relevant biometric data is collected using a biometric sensor during the *Data Acquisition* phase. The quality of the captured data is of great importance and some form of automated data quality assessment is generally necessary. The results of this assessment can be used to adjust sensor parameters, provide feedback to the user about improved positioning or as a parameter to tune the performance of the processes that follow.

The purpose of the *Signal Pre-processing* phase is to normalize the data and apply filters to remove the distortion introduced by noise, manufacturing tolerances and environmental conditions such as ambient noise in a Voice Recognition System.

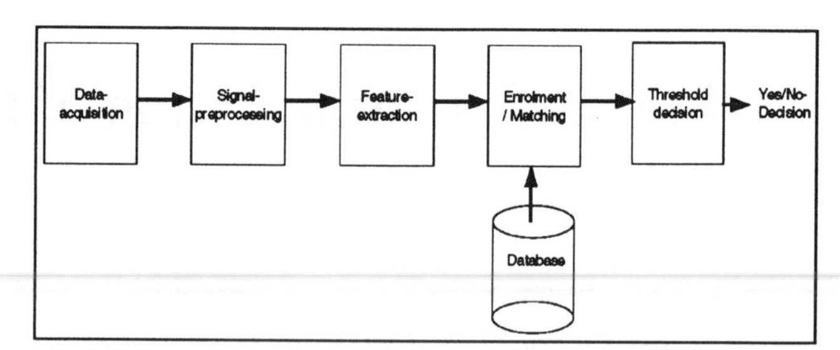

Figure 1. Key functional blocks of a biometric identification system.

The *Feature Extraction* phase reduces the enormous amount of data captured by the sensor into the much smaller amounts of information (Features) that permit the differentiation of people. This reduction is necessary for two reasons. The first is to reduce the amount of memory space required to remember known people and the second is to reduce the complexity required for the following Matching phase. For example, humans can recognize the letter 'i' independent of the font or size. Simplistically, the only relevant information needed for comparison is that the letter has a small vertical line with a dot on top.

To introduce a new person to a biometric system is called *Enrolment* and the set of features extracted for specific people are called *Feature Templates* which are stored together with other information, such as name and access privileges, into the database of the system. During enrolment, it is usual to repeat the Data Acquisition to Feature Extraction steps multiple times, to make sure that typical and relevant features are stored. Some systems use a quality assessment to select the best out of several acquisitions and more sophisticated systems may even combine the information from several acquisitions into one enhanced feature template.

Authentication and identification processes both involve comparing the recently extracted features of a Claimant, with the Feature Templates of enrolled users. The process of comparison is called *Matching* and is a non-trivial task, as it is normal for there to be no exact correlation between extracted features and stored templates. This non-exact correlation is due to a number of factors. Sensors may have a limited field of view and it cannot be guaranteed that the same area of the Claimant is exposed to the sensor each time. It is also most likely that the area is translated and rotated relative to the area sampled during enrolment and that features are distorted due to the elastic nature of skin. Features can also be hidden or altered by clothing, dirt, injury and environmental conditions.

Hence, a comparison provides a *Matching Score* that indicates the number of coincident features found between those of a Claimant and those found in an enrolment template. This matching score must then be further translated into a Yes/No decision.

To quantify the performance of a biometric system, two failure possibilities need to be considered:

- The system might decide to allow access to an impostor, because his features are close to those of an enrolled person. This is quantified by the *False Acceptance Rate (FAR)*.
- The system might decide to reject a valid enrolled user and this is quantified by the *False Rejection Rate (FRR)*.

FAR and FRR values, which are interdependent, can usually be tuned to alter system performance. This interdependent behavior is typically plotted as a hyperbolic curve of FAR versus FRR and a working point for the system selected. At the two extremes, a system can be tuned for high security, but have a high chance of rejection of valid users, or tuned for very low rejection of valid users, but have decreased success in impostor detection. The point where FAR and FRR are the same, is known as the *Equal Error Rate (EER)* and this value is often used when comparing the performance of systems. Debate on the worth of FAR, FRR & EER values highlights the strong need for the development of standardized test methods for biometrics.

THE DIFFERENT BIOMETRIC TYPES – The list of human characteristics currently being used for biometrics is varied and continuing to grow as more research is undertaken. Some of the techniques have already proved themselves as reliable and form the basis of commercial systems, while other methods have still to demonstrate their worth or be commercially viable.

A short list of the most well known methods includes DNA, Ear shape, Fingerprints, Face, Hand and Finger Geometry, Infra-red Facial and Hand Vein Thermograms, Iris, Keystroke Dynamics, Odor, Retina, Signature & Voice.

For reasons of user acceptance, reliability, robustness and physical implementation difficulties, only a small number of methods are currently feasible for automotive use. These methods are fingerprint, speaker and face recognition.

Hand Recognition, for example, while in common use for building access control applications, is not suitable in the automotive environment due to the physical space required and the ergonomic complexity associated with hand placement.

Speaker Recognition offers a non-intrusive and natural interface. However voice is a behavioral biometric and is affected by health and emotions. To extract features that are both unique and invariant is very difficult and the ambient noise associated with automotive applications makes the task even more complex. Furthermore, reproduction of an earlier recorded voice can potentially be used to circumvent a speaker authentication system.

Face Recognition can be used without any action or knowledge of the user. Video cameras with a very high dynamic range are required to be able to distinguish facial features in conditions from strong sunlight to total darkness. High computing power and very complex algorithms are needed to both find and recognize a face within a scene which has randomly moving objects in the background and continuously changing illumination conditions; such as occur when driving.

The uniqueness of a fingerprint is not questioned and with over one hundred years of history [2], fingerprint is the most widely researched and accepted biometric method. Due to its wide use and long period of acceptance, a diverse range of sensor technologies has been developed and the associated cost reduced. The automotive industry can leverage the investment and experience currently coming from the PC and telecommunications mass markets which are currently driving fingerprint technology. As sensors only need to be large enough to sample the tip of a finger, they are small and easily located in a position accessible to an automotive user. However, its use as a law enforcement tool [5] means that fingerprint has a negative stigma to overcome.

Of these three methods, only fingerprint is viable for the next generation of vehicles. While speaker and face recognition systems are currently offered commercially, their poor performance in the automotive environment currently prohibits their use and further development is required before reliable systems can be realized. Video cameras are also still very expensive when compared to fingerprint sensors.

FINGERPRINT TECHNOLOGY

WHERE DO FINGERPRINTS COME FROM? – Fingerprints are composed of *Ridges* and *Valleys*. These contours emanate from subcutaneous structures and stresses within the epidermis, which is the border between living and dead cells in skin. The formation of the contour pattern itself is determined during fetal development and from then on remains permanent and unique for each person and each finger. The new cells from these structures continually drift to the surface of the finger and become callous during this transition. Hence fingerprints are robust and continually renewed, even when subjected to abrasion.

THE FEATURES OF A FINGERPRINT – For either authentication or identification, a decision has to be made whether two given fingerprints match. The comparison method selected depends on what are considered to be the features of a fingerprint. The comparison of ridges is most often.

If the microscopic structure of the ridge flow is examined, local ridge characteristics (called *Minutiae* - minute details) can be seen. To date, as many as 150 different types of Minutiae have been found and although other approaches are also used, such as counting the ridges between reference points and the distance between skin pores, today's fingerprint systems are usually based on two minutiae types called *Ridge Endings* (the point where a ridge abruptly stops) and *Ridge Bifurcations* (the point where a single ridge divides) (Figure 2). When detailing a Minutiae for comparison purposes, it is described by its *Type* (Ending or Bifurcation), *Position* (X & Y coordinates) and *Orientation* (Direction of flow of the ridge at the minutiae). A fingerprint is described by the total of all its minutiae and the uniqueness of a fingerprint is based on the probability that no two fingerprints will have the same configuration of minutiae.

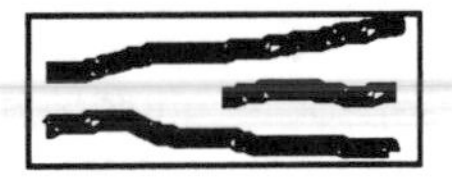

ridge ending **ridge bifurcation**

Figure 2. The most popular minutiae types.

FINGERPRINT IMAGE ACQUISITION – Varying physical principles are used by fingerprint sensors, but the end result is the same. A two-dimensional, grayscale image representation of the fingerprint is created.

Optical sensors are the oldest and most familiar type of sensor. A light source illuminates the fingertip, which is placed on one surface of a prism. The differences in refractive index between ridges that touch the surface and valleys that do not, alters the reflected light which is conducted through a lens system to a CCD-element.

Capacitance based silicon sensors have recently been developed by many companies. These sensors are usually large silicon chips (15 x 15 mm) with an array of capacitive electrodes which typically provide a resolution of 500 DPI. Each electrode forms a capacitance together with the fingertip surface just above the area of the electrode. The distance between the skin and an electrode differs between ridges, that directly touch the sensor surface, and valleys, that are separated by an air gap from the sensor surface. This difference in distance causes a difference in capacitance, which is measured by the sensor and finally results in an image describing the contours of the fingertip.

Some capacitive sensor manufacturers use a DC electric field to measure capacitance, while others use an AC field. These sensors have the advantage that they are flat and that they can easily be manufactured using standard silicon processes. Due to the requirement for close contact between the sensor electrodes and the finger, the silicon sensor surface is only protected by a thin coating and hence these sensors are susceptible to mechanical and Electrostatic Discharge (ESD) damage.

A *Thermal* silicon sensor, which consists of a 320 x 40 pixel array of electrodes, senses differences in heat. As fingerprint valleys are insulated from the sensor surface by air, the thermal conductance differs from ridges which make direct contact with the sensor surface. Using the natural heat of the finger, the sensor is able to detect the contours of the finger by the different thermal energy transferred to the sensor.

The thermal sensor differs from the other types in that a finger must be dragged across the sensor surface and the resulting multiple small image segments are then combined to create a single complete image of the fingerprint. The sensor's main advantages are a significantly smaller silicon area compared to capacitive sensors and the possibility of a thicker protective layer between the silicon surface and the finger.

Ultrasonic sensors emit a sound-wave towards the fingertip and due to changes in acoustic impedance, some of the energy is reflected back towards the sensor at the interface of materials of different density. The time difference between emission and receipt of an echo is proportional to the distance the sound-wave had to cover to reach these interfaces and thus the contour of the fingerprint can be determined. Bulky commercial sensors, with up to 500 DPI resolution, exist and it is still unclear if cost effective sensors can be realized.

FEATURE EXTRACTION – The output of a fingerprint sensor is a two-dimensional, grayscale image representation of the fingerprint that has been presented to the sensor.

A binary image is then computed, in which the Ridges are represented by ones and Valleys represented by zeroes.

Ridges are usually more than one pixel wide and this complicates the search for line ending and bifurcation minutiae. Therefore a much simpler image is computed with Ridge Lines represented by a one pixel width line. The resultant image is known as the *Skeleton*.

Using the Skeleton, line end and fork detection is relatively simple. Initially the start of all Ridge lines need to be found. Once a ridge line is detected, a tracking algorithm sequentially enumerates all pixels along the ridge line. The result is a list of minutiae with X and Y co-ordinates, type (either line ending or bifurcation) and ridge orientation.

MATCHING

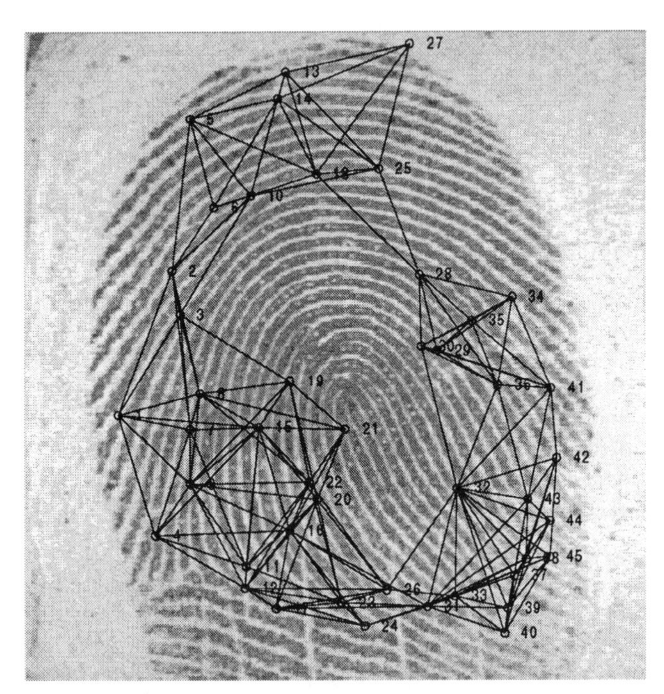

Figure 3. Minutiae Features.

Given two minutiae feature lists, matching determines whether these fingerprints belong to the same finger. The matching strategy depends on the choice of features and Jain [6] gives a good summary of publications dedicated to matching.

Minutiae matching is often referred to as *Point Pattern Matching* and at first glance appears a trivial task, however a lot of issues have to be considered to achieve satisfactory performance.

Even using the same finger, no two images will be the same. Finger translation, rotation and distortion all lead to the minutiae lists never being identical and thus it is not possible to compare the absolute coordinates contained in minutiae lists. The effect of translation and rotation can be reduced by controlling finger position with a mechanical guide which provides hard limits and tactile feedback to the user.

A comparison of minutiae is only practical in the overlapping portion of two templates. It is likely that some features will be present in the first list, but missing in the second and vice versa. As a consequence, the sensor cannot be too small. The sensor needs to be large enough that a reliable decision can be made based on the features found in an overlapping area. The area size depends on the quality of the finger-guide and the typical feature density.

AUTOMOTIVE APPLICATION OF BIOMETRICS

When first thinking about implementing biometrics in automotive applications, the main benefit appears to be improved security. As will be shown later, this is only partially true. Improvements in user convenience and efficiency are the main driver for biometrics. For the application of biometrics in the automotive field, three areas can be identified.

Currently, the biometric identification market is driven by the PC (password replacement), E-commerce (digital signature release), the mobile telephone (PIN replacement), banking (automatic teller machines) and building access markets. When biometrics is in widespread everyday use, there will be the expectation that the same technology will be available in automotive applications.

UNIQUE CHARACTERISTICS OF BIOMETRIC SYSTEMS – The use of a biometric system for security related authorization has consequences which must be understood from the outset and methods developed to handle them:

- Personal features are unique and inseparable from the person; so it is not possible to lend them to another. This implies that a vehicle with a biometric system cannot simply be loaned, unless the person is enrolled into the system. This is a characteristic of all biometric systems; independent of the biometric method.
- Any new user must first be enrolled into the system and added to a database of known users. An administrator is required to authorize the biometric system to learn the characteristics of a new user. It is also reasonable to enable the passing-on of the right to enroll users from an initial administrator to others. As a result, multiple users may have the right to enroll new users.
- An effective interface for the administration of users and access rights must be accomplished. As authorization is uniquely linked to a specific person, it is possible to set user-specific usage restrictions. Examples could be maximum speed or an expiration date.

Furthermore, any automotive application must allow for the following important scenarios:

- For motor vehicle rentals, access permission must be given at the sales counter. Together with the access rights discussed above, this offers new opportunities to ensure that a vehicle can only be driven by those customers enrolled at the sales counter and only for the duration of the contract.
- Lending the vehicle to a friend or colleague.
- Hotel valet parking.
- Vehicle workshop servicing.
- The possibility of driving the vehicle when the enrolled driver cannot, as in an emergency.
- Sale of the vehicle.
- Handling of the vehicle during manufacture and transport.

If we accept that higher security through biometrics is not our goal, then while biometric methods are faster and more convenient, any system must also include a non-biometric, non-person-specific bypass method for vehicle entry and driving authorization. One such method would be to provide a mechanical or remote key for vehicle entry and a transponder-based immobilizer as used in today's solutions.

It should be emphasized that the bypass solution is only needed for special situations. Normally, a user will operate a vehicle with the biometric system and thus profit from the advantages already outlined.

PERSONAL PROFILE – Existing seat memories offer the possibility of storing seat and mirror position for different drivers and to recall these settings whenever the same driver uses the vehicle. The user interface is typically through numbered or colored pushbuttons which are assigned to different drivers. The maximum number of drivers is limited by the number of buttons and the assignment of buttons needs coordination between the drivers. Furthermore, each driver must remember his assigned button.

As biometric systems can uniquely identify drivers, automated Personal Profile systems becomes viable.

The person specific pushbuttons are replaced by a biometric sensor, which is used for recalling, as well as storing, the user's specific settings. The maximum number of drivers is no longer limited by the number of pushbuttons, only by memory capacity. User comfort is increased, as user specific buttons (either numbered or colored) do not need to be remembered or coordinated. All the driver has to do when he enters the vehicle is to put his finger on the sensor and adjustment of the accessories can start immediately.

Other, not so obvious, settings can also be remembered, such as driving style and suspension tuning parameters, navigation system destination settings, telephone numbers, e-mail or billing account details for a Telematics system, as well as driver status information such as the length of time a person has driven versus the amount of rest.

IMMOBILIZATION AND ENGINE START – Challenge-response based transponder immobilizers, together with encrypted communication to engine management systems, have all but eradicated instances of vehicle theft due to 'hot wiring' and component exchange. The only additional security benefit provided by biometrics is in overcoming the unauthorized use of a key or transponder card, as biometric features cannot be stolen or copied. Also convenience is improved as a key or transponder card is no longer required to operate the engine.

As the biometric system is a replacement for today's ignition key, it has to allow for the usual power-on sequence. i.e. Off, Accessories, Ignition and Start. For safety and regulatory reasons, the system may only allow engine start if the driver simultaneously presses the clutch or break pedal while placing his finger on the sensor. The logistics of electrical supply, engine start and steering lock operation needs to be carefully considered.

Once again, convenience is improved. When a driver wants to start the engine, he only has to put his finger on the sensor while pressing the break or clutch pedal. This is faster and simpler than fiddling the key into the ignition lock. Figure 4 shows a study where the fingerprint sensor is integrated into the gear-stick; with biometric system authorization and engine start in one action.

As the driver is uniquely identified by the biometric system, specific rights can be assigned for each driver. This may be useful when lending a vehicle for a limited period of time and particularly for motor vehicle rentals. Once the agreed period of time is expired, heavy limitations can be put on the vehicle operation, such as speed or distance.

VEHICLE ENTRY – Usually, vehicle access is granted by either a mechanical or remote key and although transponder based Passive Entry systems [1] are just being introduced into the market, all of these systems require that the driver carries something. Once an ignition key is no longer required for engine start, it makes sense to also use a biometric system for vehicle entry. This completely eliminates the necessity for a key or transponder. A driver parking at the beach will no longer have to find a good hiding place for the keys.

A fingerprint sensor near the door handle requires the driver only to touch the sensor to gain access. System options can ensure that valuables stored in the glove box or trunk during valet parking and vehicle workshop servicing are kept secure.

Due to the hostile environment, the use of fingerprint sensors on the external surfaces of a vehicle is a technical challenge and although inconvenient, mechanical covers may ultimately be required to protect the sensor.

The current state of the art in fingerprint sensor technology does not offer a realistic solution for the problem of external mounting and each of the existing sensor technologies must be further developed in order to support this application.

DEMONSTRATION VEHICLE – To prove the benefits and functionality of biometrics, Bosch has fitted a demonstration vehicle. Work initially centered around a PC based proof-of-concept and included fingerprint based engine start and Personal Profile control of seat & mirror positions. Using the PC based solution first provided the benefits of permitting rapid implementation of system concepts, the use of universal tools for algorithm development and a graphical interface for visualization and demonstration purposes. It also permitted the development of a platform independent system which could later be ported to an embedded system once issues such as processor performance and cost requirements were more clearly known. Current activities include the addition of door access control and the development of embedded electronics solutions to reduce operating time, package size and power consumption.

Figure 4. A fingerprint sensor integrated in gear-stick.

CONCLUSION

Advances in semiconductor technologies, as well as data processing techniques, mean that the implementation of biometric applications, which require a large amount of processing power, can now be implemented at a reasonable cost. Many sensor suppliers are also developing ASIC based system solutions that implement all of the functional blocks needed for a biometric identification system.

Over the next few years, publicly visible biometric systems will become commonplace. Both automated and personal bank teller applications using fingerprint, speaker and iris recognition already exist and will expand. Computer network access and log-on applications involving fingerprint, speaker and face recognition are just beginning to secure acceptance and are expected to be widespread.

Biometric will appear in automotive applications over the next few years. Existing sensor technology was developed for the office environment and needs a lot of improvement to be reliable in the automotive environment, but sensor manufacturers are achieving promising progress.

As biometric applications become an everyday occurrence, public understanding and acceptance will develop to a point where automotive customers will not only accept, but also expect, the convenience that systems utilizing biometric solutions can bring.

REFERENCES

1. Schmitz S., Roser Ch., - A New State-of-the-Art Keyless Entry System, SAE Technical Paper Series, 980381, 1998
2. Jain A.K., Bolle R., Pankanti S. - Personal Identification in Networked Society, Kluwer Academic Publishers, 1999.
3. Biometric Consortium – An introduction to biometrics, www.biometrics.org
4. Hong L. - Automatic Personal Identification Using Fingerprints, Dissertation, Department of Computer Science, Michigan State University,
5. Lee H.C. and Gaensslen R.E.- Advances in Fingerprint Technology, CRC Series in Forensic and Police Science, CRC Press LLC 1994
6. Jain A.K., Hong L., Pankanti S. – An Identity-Authentication System Using Fingerprints, Proceedings of the IEEE, Vol. 85, No. 9, September 1997

CONTACT

Jan Lichtermann graduated from the University of Kaiserslautern, Germany, with a degree in Computer Science and has worked as a scientist in VLSI and computer architecture design at the University of Kaiserslautern. After receiving his Dr.-Ing. in 1998, he joined the Corporate R&D division of Robert Bosch GmbH where he is currently responsible for the advanced development of automotive biometric and security systems. Jan.Lichtermann@de.bosch.com

Rod Pettit graduated from Swinburne University, Melbourne, Australia, with a degree in Electronics Engineering and a diploma in Computer Systems Engineering. He joined Robert Bosch Australia in 1981 and is currently responsible for advanced product and technology development.

2000-01-0130

An Integrated Automobile Keyless Operation System

Tricia Liu, William Liew and Herbert Everss
*e*Drive, Inc.

ABSTRACT

An integrated automobile keyless operation system is developed. The system consolidates the ignition key and lock/switch, security and keyless entry and other functions into one unit. The system uses a small handheld portable wireless remote controller to replace the ignition key, the door key and the trunk key. A receiver replaces the ignition switch and functions as controlling switches for door-locks and/or trunk-lock. The system incorporates the latest advancement in wireless technologies and digital-signal processing. By consolidating components, the design cuts manufacturing cost, improves reliability and offers outstanding convenience and enhanced security.

INTRODUCTION

Due to historical reasons, ignition switch/lock, security/immobilizer device, remote starting, keyless entry and keyless trunk release were introduced to automobiles separately and at different time. These separated devices presents such problems as higher manufacturing cost, complicated assembly, inconvenient to use and lower reliability. With the technology advancements in data processing, IC chip and power switches, using electrical/electronic devices to replace the ignition switch/lock and consolidate these separated devices will be seen in next generation vehicles.

A car key has been the symbol of operating a vehicle since automotive infancy a century ago. Technically speaking, the major problem with the conventional ignition switch is that ignition switches are easily accessible, and therefore do not provide sufficient anti-theft protection. In addition, in cold or hot weather it is an unpleasant experience to enter a car to start the engine, the heater or air conditioner. Stepping into a hot or cold car, using a specific key and inserting the key, are antiquated.

The electronic anti-theft devices provide some extra protection for motor vehicles, but all are extra add-on devices to make up for the weakness in the ignition switch. A major drawback of these devices is that a driver must carry and operate an anti-theft device and an ignition key. Manufacturers and service-providers have to manufacture, install anti-theft devices and ignition switches separately.

Remote starting devices cannot function as independent ignition driving devices. They use complicated circuits and are expensive to manufacture. In addition, they do not provide any secure technologies to differentiate millions of cars and have serious technical flaws dealing with safety and security issues. Two vehicle-starting systems installed in one car also add extra manufacturing cost.

The objectives of the work described in this paper include:

1. to use an electronic device to replace the conventional ignition lock and ignition key and to provide a reliable, highly secure, user-friendly and low-cost integrated keyless operation system which is consistent with the existing industrial standards and avoids changing significantly the existing vehicle circuits and allows simple "plug-in" when the invented system is installed on an existing motor vehicle;
2. to avoid having to detect engine-running conditions and avoid using a feed-back-control circuit to de-energize the motor starter and therefore overcome the problems of lower reliability and higher manufacturing cost associated with the existing remote-starting devices;
3. and to use an integrated-circuit (IC) chip, such as a microprocessor, digital signal processor (DSP), application-specific-integrated-circuit (ASIC) or programmable-logic-array to reduce the cost of the keyless motor vehicle starting system and to make its development much more efficient and its reliability much higher.

DESCRIPTION OF THE KEYLESS OPERATION SYSTEM

THE DESIGN – The system includes two major parts. One is a small handheld, remote controller and the other is a receiving device. The remote controller replaces the conventional ignition key, the remote controllers for remote keyless entry, remote trunk release and remote starting. The receiver replaces the lock of conventional ignition system. The remote controller is approximately 1"x 1/2"x 1 1/2". Push-buttons are provided for locking/unlocking doors, starting/stopping engine and releasing trunk locks, etc. Differently coded electromagnetic signals are transmitted through the air when one of the buttons is pressed. The receiving device decodes the signal transmitted by the remote controller and takes action based on which button is pressed.

Figure 1.

The remote controller includes a transmitter integrated-circuit (IC) chip, a radio-frequency (RF) modulator, a battery and normal-off button switches: a power-code switch and a starter-code switch, door-unlock switch and trunk release switch. The transmitter IC chip performs the functions of generating a power-code, a starter-code, a door-unlock code and a trunk-release code. The power-code switch is used to turn on and off the power-code generating function of the transmitter IC chip. The starter-code switch is used to turn on and off the starter-code generating function of the transmitter IC chip. Similarly, the door-unlock code and the trunk release code are used to actuate the door-unlocking and trunk-release mechanism, respectively. The IC chip design is open to incorporate more desired functions, for instance, remotely open/close windows and sunroofs, into the remote controller.

As shown in Fig. 2, the receiving/controlling board has an RF demodulator, a receiver integrated-circuit (IC) chip, an ACC-circuit switch, a starting-circuit switch, a first ignition-circuit switch, a second ignition-circuit switch, a door-unlocking circuit and a trunk-release circuit. The receiver IC chip performs the functions of detecting the power-code, starter-code, door-unlocking code and trunk-release code, and generating output signals to control the ACC-circuit switch, the starting-circuit switch, the first ignition-circuit switch and the second ignition-circuit switch, door-unlocking switch and trunk-release switch. The receiver IC chip also maintains an OFF-state, an ACC-state, a START-state, a RUN-state, a door-unlocking state and a trunk-release state. As the examples of controlling procedures, when the starter-code from the remote controller is properly detected by the receiver IC chip, the receiver IC chip switches its state to the START-state. When the starter-code from the remote controller disappears, the receiver IC chip switches its state from the START-state to the RUN-state. Whenever the power-code switch on the remote controller is pressed once and the power-code is detected by the receiver IC chip, the receiver IC chip switches its state from the OFF-state to the ACC-state, or from the ACC-state to the OFF-state, or from the RUN-state to the OFF-state. In the OFF-state, the ACC-circuit switch, starting-circuit switch, first ignition-circuit switch and second ignition-circuit switch are all turned off. In the ACC-state, the ACC-circuit switch is turned on and the starting-circuit switch, first ignition-circuit switch and second ignition-circuit switch are turned off. In the START-state, the starting-circuit switch and the first ignition-circuit switch are turned on while the ACC-switch and the second ignition-circuit switch are turned off. In the RUN-state, the ACC-circuit switch, the first ignition-circuit switch and the second ignition-circuit switch are turned on while the starting-circuit switch is turned off.

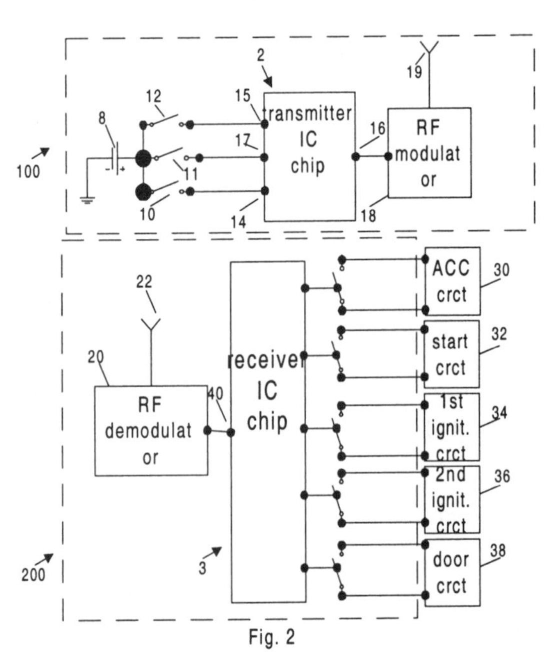

Fig. 2

OPERATION PROCEDURE – To start a motor vehicle equipped with the keyless starting system in this invention, the starter-code switch on the remote controller is pressed and held until the engine of the vehicle is started. To stall the engine of the motor vehicle when the engine is running, the driver of the vehicle needs to press the power-code switch momentarily. To turn on the ACC (power supply for radio, air conditioner, fan, etc.), the driver needs to press the power-code momentarily. To unlock the car driver door when approach the vehicle, the driver needs to press the door-unlocking button. Similarly, to release trunk, the driver needs to press the trunk-release button. All the operation can be performed by using one remote controller.

Fig. 3 illustrates a logic flow chart of the software loaded into the transmitter IC chip 2 to implement the functions of generating the power-code and the starter-code. The software implements a process to perform these tasks. Starting from the beginning of the process, the port 14 input signal at step 3-1 is checked. If the port 14 input signal at step 3-1 is high, then the port 15 input signal is checked at step 3-2. If the port 15 input signal at step 3-2 is high, the port 16 outputs all 0's at step 3-3. If the port 15 input signal at step 3-2 is low, the port 16 at step 3-4 outputs the power-code, which is a series of 0 and 1's encoded according to a first rule. If the port 14 input signal at step 3-1 is low, the port 15 input signal at step 3-5 is checked. If the port 15 input signal at step 3-5 is high, the port 16 at step 3-6 outputs the starter code, which is a series of 0 and 1's encoded according to a second rule. After completing the tasks at steps 3-3, 3-4 or 3-6, the transmitter IC chip 2 goes back to the beginning of the process at step 3-1 and starts another cycle of the process.

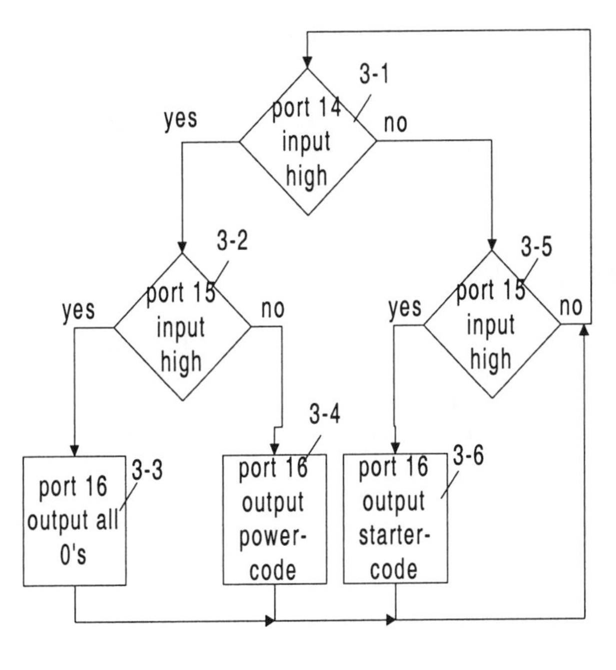

Fig. 3

Fig. 4 illustrates a logic flow chart of the software loaded into the receiver IC chip 3 to implement the functions of detecting the power-code and the starter-code, generating output signals to control the ACC-circuit switch, the starting-circuit switch, the first ignition-circuit switch and the second ignition-circuit switch, and maintaining a state-machine with the OFF-state, the ACC-state, the START-state and the RUN-state. The software implements a process to perform these tasks. Starting from the beginning of the process, an initial state of the OFF-state is set at step 7-0. Then the receiver IC chip tries to detect the power-code transmitted from the portable wireless transmitter 100 at step 7-1. At step 7-2 it is checked whether the power-code is detected or not. If it is found at step 7-2 that the power-code is detected, then it is checked at step 7-3 whether the current state is the OFF-state. If it is found at step 7-3 that the current state is the OFF-state, the state of the receiver IC chip 3 is switched to the ACC-state at step 7-4, and the port 41 outputs a high voltage and the ports 42, 43 and 44 output low voltages at step 7-5. If it is found at step 7-3 that the current state is not the OFF-state, then it is checked at step 7-6 whether the current state is the ACC-state. If it is found at step 7-6 that the current state is the ACC-state, the state of the receiver IC chip 3 is switched to the OFF-state at step 7-7, and the ports 41, 42, 43 and 44 output low voltages at step 7-8. If it is found at step 7-6 that the current state is not the ACC-state, then it is checked at step 7-9 whether the current state is the RUN-state. If it is found at step 7-9 that the current state is the RUN-state, the state of the receiver IC chip 3 is switched to the OFF-state at step 7-7, and the ports 41, 42, 43 and 44 output low voltages at step 7-8. If it is found at step 7-9 that the current state is not the RUN-state, the receiver IC chip 3 goes back to step 7-1 to start another cycle of the process. If it is found at step 7-2 that the power-code is not detected, then the receiver IC chip tries to detect the starter-code at step 7-10. At step 7-11 it is checked whether the starter-code is detected or not. If it is found at step 7-11 that the starter-code is detected, then the state of the receiver IC chip 3 is switched to the START-state at step 7-12, and the ports 42 and 43 output high voltages and the ports 41 and 44 output low voltages at step 7-13. If it is found at step 7-11 that the starter-code is not detected, then it is checked at step 7-14 whether the current state of the receiver IC chip 3 is the START-state. If it is found at step 7-14 that the current state is the START-state, the state of the receiver IC chip 3 is switched to the RUN-state at step 7-15 and the ports 41, 43 and 44 output high voltages and the port 42 outputs a low voltage at step 7-16. If it is found at step 7-14 that the current state is not the START-state, the receiver IC chip 3 goes back to step 7-1 to start another cycle of the process. After completing the tasks at steps 7-5, 7-8, 7-13 or 7-16, the transmitter IC chip 2 goes back to step 7-1 and starts another cycle of the process.

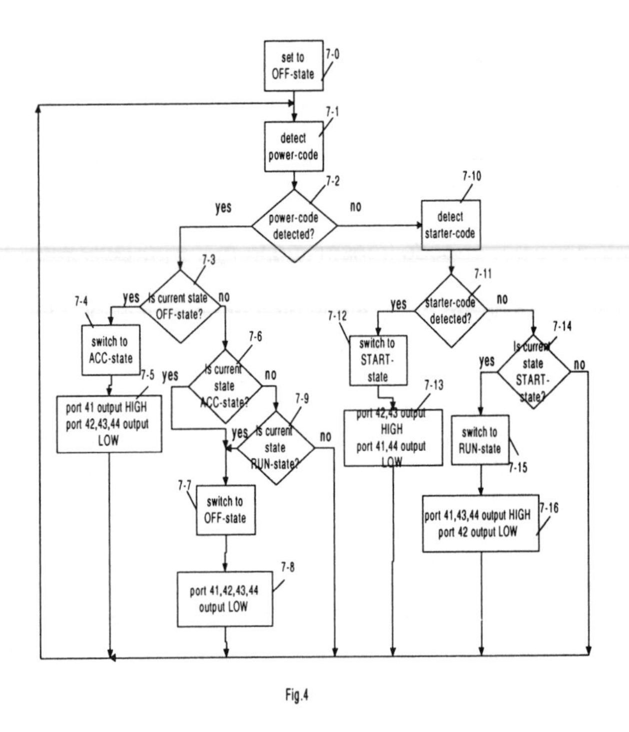

Fig.4

SAFETY ISSUES

- Descrambling-Proof As described above, the system uses the most advanced digital signal processing and wireless communications technologies to make it immune to descrambling and interfering signals. Code hopping technology is also incorporated.
- It won't start a vehicle when the transmission lever is not engaged in the "P" position. The system has this safety feature like all the conventional ignition locks do. Signals indicating status of the transmission lever can also be drawn and used to prevent undesired starting and stopping.
- The engine will be turned off if it is started for more than ten minutes without driving. This will prevent such problems as carbon monoxide presence in an enclosed garage when the vehicle is left started and forgotten.
- The engine will be turned off if the car doors are tampered. This will prevent the cars being driven away by unauthorized persons. The car doors will be automatically locked when the engine is started.
- The system deploys a special electronic enclosure technology to prevent accidental operation of the remote. The technology also makes the remote child-proof.

In a word, the system matches or exceeds the safety level offered by the combination of conventional ignition lock/switch, security, keyless entry and keyless starting devices.

TESTING – Road tests were performed on country roads for more than 4500 miles since March 1998 at the average speed of 45 miles/hour. Start/drive/stop tests were performed for more than 1000 times under various conditions including radio-frequency interference areas. Tests are under the ambient temperature from –10°C to 40°C. Road condition: with occasional pot-holes, moderate curves and ups-and-downs.

SPECIFICATIONS – The frequency used is the unlicensed band, such as 315, 433 and 418 MHz. Operation Voltage: 6~20 volt; Receiver Standby Current: < 5 mA; Remote-Controller Standby Current: 0 mA; Remote controller battery life span: 10 years assuming 15 times of button-presses per day, 365 days per year. Operation Temperature(behind cluster panel): –40°C to 80°C.

INSTALLATION – Both OEM and aftermarket installation can be done in a simple hardness plug-in fashion. Final packaging design can be done according to specific make and model to achieve the OEM installation time to be less than that of conventional ignition switch/lock. It is designed such that aftermarket installation can be achieved within 45 minutes.

COMPATIBILITY WITH THE IDB (INFORMATION DATA BUS) – When used with IDB, the system actually will enjoy its full security strength and advantages, some of which can not be realized by using the dash-board ignition lock. For instance, when used with IDB, the system can fundamentally solve the problem of hot wiring, while dashboard ignition lock and the existing security devices can not. OEM installation when used with IDB will be much simplified.

CONCLUSION

An integrated keyless operation system has been developed. The system consolidates the functions of conventional ignition switch/lock/key, anti-theft device, keyless entry, keyless trunk release and remote starting into one simple unit. Drivers never need to operate different separate systems. The system also meets the auto-manufacturers' demands for "decontenting" and parts consolidation. It not only reduces the cost for components, it also simplifies assembly. It offers higher reliability than the discrete systems. The system is open to incorporate with other passive keyless entry technologies without encountering battery consumption problem of the remote controller. The system is suitable for recent volume application for its outstanding performance/cost ratio.

REFERENCES

1. Schmitz, Stephan et al., "A New State-of-the-Art Keyless Entry System", SAE International Congress & Exposition, February 1998, Detroit, MI, USA, Session: Cae - Computer-Aided-Engineering (B&C);
2. Liu, Chunyan Tricia, "Keyless Motor Vehicle Starting System with Anti-Theft Feature", U.S. Patent Number 5,689,142, November 18, 1997.

CONTACT

Ms. C. Tricia Liu, Chief Operating Officer of eDrive, Inc., 12 Mountain Ave., Montville, NJ 07045, Tel: 973-257-1912, email: tricia@edriveinc.com;

Dr. William Liew, Chief Technical Officer of eDrive, Inc., 12 Mountain Ave., Montville, NJ 07045, Tel: 973-257-1912, also a former senior engineer of AT&T/Lucent Bell Labs. email:william@edriveinc.com;

Mr. Herbert Everss, advisor of eDrive, Inc., also a former senior executive of Siemens Automotive and Mannesmann VDO. 2950 W. Square Lake Rd.., Suite #100, Troy, MI, 48098, Tel: 248-641-1446;

980381

A New State-of-the-Art Keyless Entry System

Stephan Schmitz and Christopher Roser
Robert Bosch GmbH, Stuttgart, Germany

ABSTRACT

A new keyless entry system, for automotive security and comfort applications, has been developed. The system utilises a single chip transponder for vehicle immobilisation, keyless entry and remote control functionality. The transponder system can be embedded into a key fob or an ISO smart card. System security and communication speed is provided by an advanced anti-collision protocol in combination with sophisticated challenge response and uni-directional protocols. The system can be activated by using either a push-button transponder, where long range access to the vehicle is provided, or by touching an actuator near the door or trunk. Due to the inductive coupling between the transponder and the vehicle mounted antennas, the vehicle door or trunk opens on successful verification as if there were no locks. Additionally, inside the vehicle, the transponder can be used as an immobiliser.

INTRODUCTION

Vehicle immobiliser systems based on transponder technology are currently used as standard equipment for vehicles. With the increase of immobiliser equipped vehicles, the theft of vehicles has decreased. Due to the demanding security requirements, the immobiliser is gaining more and more importance. However, next generation vehicle immobilisers will offer additional functionality like keyless entry and remote control. A keyless entry system with the combined functionality of a remote control, keyless entry and a immobiliser will be included in next generation vehicle lock systems.

The new keyless entry system described here was developed for automotive security and driver comfort and will replace existing immobilisers and UHF remote controls. To achieve this objective, a vehicle based interrogation system is activated by pulling the door handle and starts a medium range, bi-directional signal transmission (up to 1.5 m) between the vehicle and the transponder carried by the user. Additionally, a special protocol provides long range access to the vehicle (more than 10 m), whenever the user presses a button located on the transponder.

Interrogation systems are commonly used in access building applications, production line control, shoplifting prevention, ski ticketing, automatic fare collection and electronic ID cards etc. However, the practical operation of such a system in vehicle applications requires an enormous development effort, which has to meet the following requirements:

The communication area of the system has to be optimised for vehicle applications. Special antennas have to be developed to provide the required operating range.

A high data rate in combination with a fast anti collision and communication protocol is used to minimise the communication time.

To avoid the utilisation of approach detection sensors a fast electronically operated latch with mechanical backup (a High Integrated Security latch, HIS-latch) is used to minimise the reaction time of the latch.

The stand-by current of the keyless entry system should be minimised to avoid a flat vehicle battery scenario, if the vehicle was not used for a longer period of time.

A sophisticated cryptographic algorithm has to be developed to guarantee the security of the system for driving permission, vehicle access and remote control.

A user initiated communication protocol has to be developed in order to allow the combination of remote control with keyless entry functionality.

To provide maximum handling comfort for the end user a ISO smart card package for keyless entry functionality has to be achieved. If in addition a remote control functionality is desired, the package of a key fob of key head is required.

For each packaging realisation (key fob, key head or ISO

smart card) the utilisation of 3 different applications (i.e. 2 vehicles and building access control) should be achieved.

The battery life time of a ISO smart card and a key fob or key head realisation has to be more than 3 years.

The paper firstly gives a brief description of the operating principles and functionality. Then we discuss the communication protocol and the cryptographic algorithm. The next part presents physical properties such as the operating range and the overall timing of a keyless entry system. Possible multiple applications will be discussed in the last section.

KEYLESS ENTRY SYSTEM DESCRIPTION

Figure 1 illustrates a smart card based keyless entry system and the components involved. As soon as the user pulls the door handle, an antenna electronics is activated and a LF (125 kHz) signal transmission between the vehicle and the portable transponder is initiated. All transponders within the communication area are identified and one transponder is selected for the security check. On successful verification of the transponder, the antenna electronics provides an opening signal to the HIS-latch. The control unit adjusts the driver's seat and positions the mirrors according to the drivers unique preferences.

Depending on the packaging form factor, the transponder communicates via UHF (key head or key fob packaging) or via LF (ISO smart card packaging). The integration of a SAW resonator is currently not possible in the smart card solution. The vehicle sends data to the transponder always via LF. To allow for a flat transponder battery, a passive bi-directional communication can be established with both packaging scenarios.

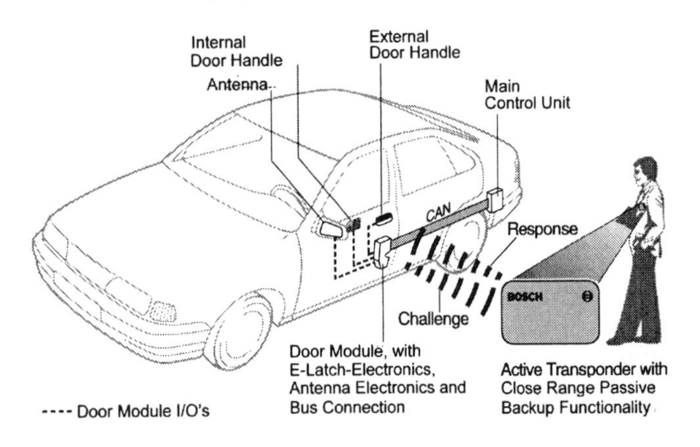

Fig. 1: Realisation of a smart card based keyless entry system

By utilising the key head/fob packaged transponder with remote control functionality, the user can have access to vehicle locking or unlocking applications from a greater distance. By pushing a transponder mounted button, a unidirectional UHF sequence is sent which wakes up the vehicle based UHF receiver. The UHF receiver demodulates the UHF transmission and sends the bit pattern to the antenna electronics, which validates the code and determines which action is being requested (i.e. locking or unlocking the doors or opening the trunk).

In order to provide higher comfort for the end user, a special memory organisation provides the user the ability to use his transponder for multiple vehicles. The unique structure also allows up to three different applications (e.g. vehicles, garage door opening, building access control). For each of these applications, two different kinds of security protocols exist. Whenever the security should be at a maximum, mutual authentication is carried out. Maximum security is required for driving permission. In vehicle access applications, where the security is limited by the breaking of a window, a single challenge response protocol is utilised.

The keyless entry system includes the following components:

TRANSPONDER

As illustrated in Fig. 2, the transponder consists of a hardwired core logic with interfaces for external components such as the coil, battery, UHF transmitter and push buttons. The form factor independent packaging features of the transponder enables embedding of the transponder into an ISO smart card or into a key head/fob.

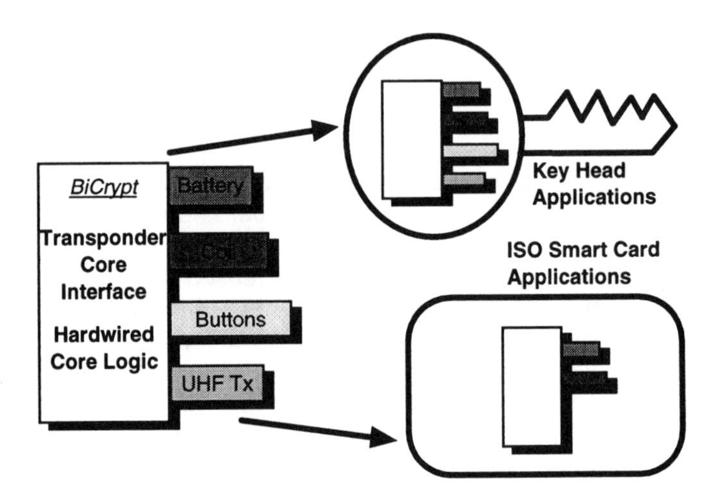

Fig. 2: Transponder schematic overview and packaging features

LF- ANTENNA

A wound loop antenna is built into the driver's door mirror as shown in Figure 1 for generating a rotating magnetic field as described by Hirano et. al., 1988.

UHF RECEIVER

The UHF receiver wakes up when an appropriate UHF sequence is received from the transponder due to an intentional activation by the user. On validation via the uni-directional protocol, the vehicle's door is opened. The UHF receiver is connected to the control unit.

DOOR HANDLE

In order to use existing vehicle components, the door handle initiates the keyless entry system. An additional push button integrated into the door handle activates the lock process after a successful verification.

DOOR MODULE

The door module combines the antenna electronics, bus connection and HIS-latch electronics. The door module is connected to the door handle and the actuators. An interface connects the door module with the body work bus.

CONTROL UNIT

The control unit is connected to the body work bus (e.g. CAN) and positions the driver's seat, mirror, etc.. The actuators for global locking or unlocking and trunk opening are furthermore controlled by the control unit.

COMMUNICATION PROTOCOL

As described above, immobiliser applications, vehicle access applications and remote control applications require different security levels. For each of these applications a different communication protocol exists. However, all protocols share common features. Especially for vehicle immobiliser and vehicle access applications, the protocols are quite similar due to their bi-directional character. The uni-directional protocol lacks any anti-collision feature and is therefore treated separately.

BI-DIRECTIONAL PROTOCOL

This protocol consists of a wake-up command, a special sequence to wake up the transponder in the field. An activation command then sets the transponder into receive mode. The following data field, the public object identification code (POIC), defines the kind of application (e.g. one specific vehicle). By receiving the POIC and a successful verification (see figure 3) the transponder is set into Activate mode. For each application, 8 time slots exist and each corresponds to a specific group number.

As soon as the transponders have received their specific POIC, they answer with a beacon in a specific time slot defined by their internal group number. The door module's logic is now able to select one of the transponders in the communication area with a Select command. A next command decides whether a single Challenge Response (vehicle access applications) or a Mutual Authentication (vehicle driving application) protocol is executed.

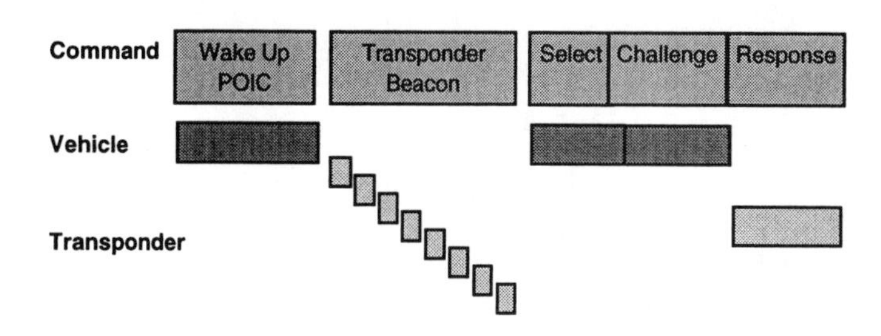

Fig. 3: Schematic overview of the bi-directional time slot protocol

For each protocol, a different kind of secret object identification code (SOIC) is stored in the antenna electronics. The SOIC is diversified into a secret key for each transponder by concatenation of the transponder's unique manufacturer code and the SOIC. After receipt of the challenge, the transponders use this unique secret key for enciphering. The antenna electronics is therefore able to uniquely identify the transponder by the response received.

As seen in figure 3, the bi-directional protocol is restricted to a limited number of transponders (8 shown here). However, the number of transponders allowed can easily be increased by allowing more than one transponder to answer in the eighth time slot. A further separation of transponders can be achieved by an second anti-collision protocol as presented in figure 4.

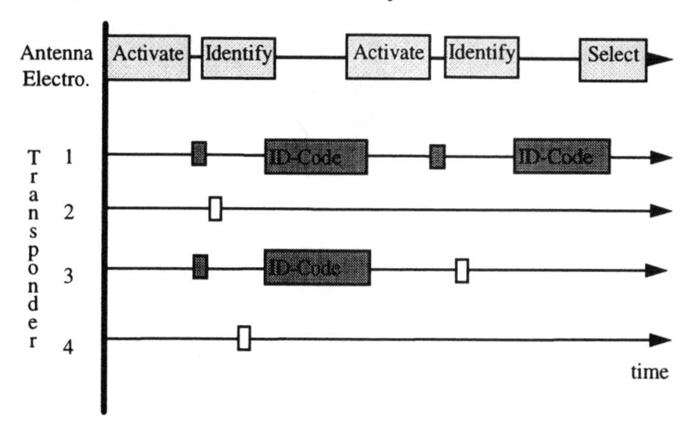

Fig. 4: Overview of the anti-collision protocol for multiple transponders in one time slot

An Activate command and a random number are sent to the transponders in the eight time slot. These transponders combine the random number received with their unique manufacturer code to randomly define one of

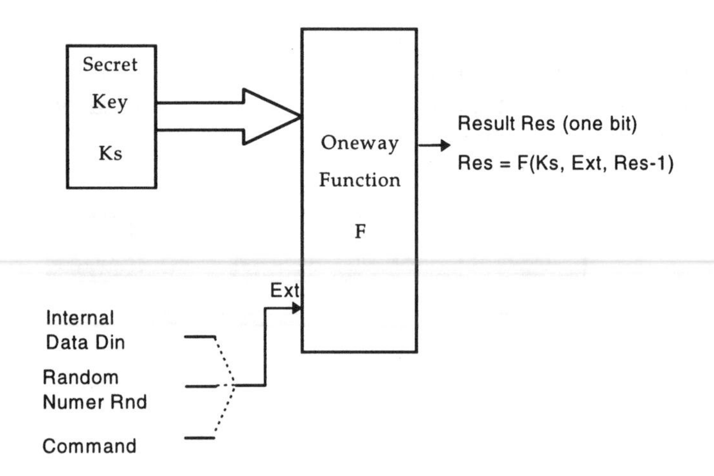

eight time slots. Each of these transponders will respond with a beacon in the defined time slot. An Identify command is sent by the antenna electronics directly after the receipt of the first transponder beacon and sets all other transponders into a Halt mode. Transponders, which have already responded, will reply with their unique manufacturer code.

In the case where two or more transponders answer at the same time (see fig. 4) the procedure is repeated with a new random number. The selection of a single transponder can therefore be achieved.

POIC	Group No.	Counter	Challenge	Response	Commands

UNI-DIRECTIONAL PROTOCOL

The uni-directional protocol is used whenever the user presses the transponder mounted remote control button to gain access to the car from a greater distance. Instead of a rolling code, a special kind of Challenge Response protocol is used. An internal counter is increased, if the transponder mounted button is pressed. This counter value is also transmitted. Whenever the counter has a higher value than that previously stored in the antenna electronics, the sequence is accepted. On verification of the response, the driver is authorised to have access to the vehicle. Figure 5 presents an overview of the uni-directional protocol sequence.

Fig. 5: Schematic overview of the uni-directional protocol

In each transponder, two secret keys, one group number and the counter value are stored in memory.

ALGORITHM DESCRIPTION

A one way cryptographic function specifically designed for easy and compact implementation in hardware and software is used. The algorithm is based on a one-bit-output-per-clock internal structure. Figure 6 represents the algorithm procedure.

Fig. 6: Algorithm structure

At each clock cycle, the Secret Key (Ks) and one external bit (Ext) are input to the one way function (F) to calculate the next bit of the result (Res). The external bit can be either internal data, a random number or other data like received bits of the commands.

The drawing below (figure 7) presents how the algorithm is used to perform the mutual authentication between the transponder and the antenna electronics.

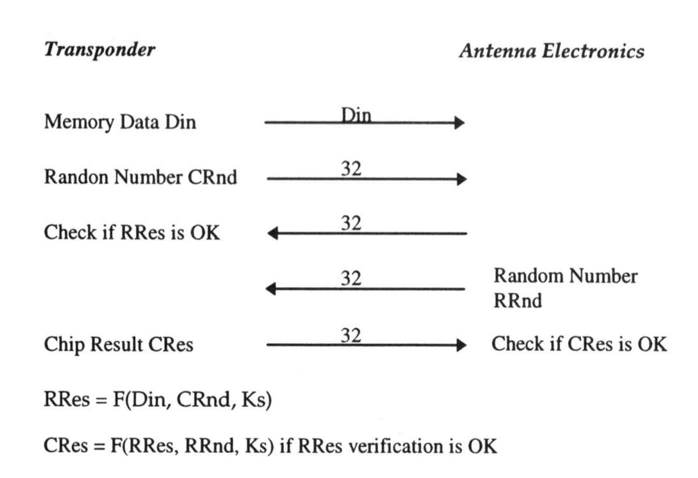

Fig. 7: Mutual authentication

The transponder and antenna electronics use the same secret key Ks during authentication. Secret keys contained in the chips can be diversified using a common mother key and unique personalised data.

Step 1 : The algorithm is initialised by reading transponder internal data Din. Transponder and antenna electronics are synchronised in a random state determined by Din and secret key Ks.

Step 2 : The transponder sends to the antenna electronics a «question» (random number CRnd) which is used as an external input to the algorithm. CRnd places the algorithm into a random internal state.

Step 3 : The antenna electronics sends the « answer » to the question (Rres) which is compared by the transponder with the result calculated internally. If the comparison is successful, the transponder considers that the antenna electronics is authenticated.

Step 4 : The antenna electronics sends to the transponder a «question» (random number RRnd) which is used as an external input to the automate. RRnd places the algorithm into a random internal state.

Step 5 : If the antenna electronics authentication was successful, the transponder sends the « answer » to the question (result CRes) which is compared by the antenna electronics with the result calculated internally. If the comparison is successful, the antenna electronics considers that the transponder is authenticated.

PHYSICAL PROPERTIES

OPERATING RANGE

The operating range is a critical issue for any keyless entry system. Two requirements conflict. On the one hand, the range should be maximised to provide as much comfort as possible, while on the other, the range should be minimised to ensure anti-theft security. If the communication area is too large, a thief could gain access to the car, while the driver is either approaching or leaving the vehicle.

To discover the best transponder opening positions, many situations were evaluated and this resulted in us determining that the optimum range is between 1.5 and 2.0 m. However, the range should be shaped more rectangularly than quadratically, with the long side in parallel with the vehicle (see figure 8).

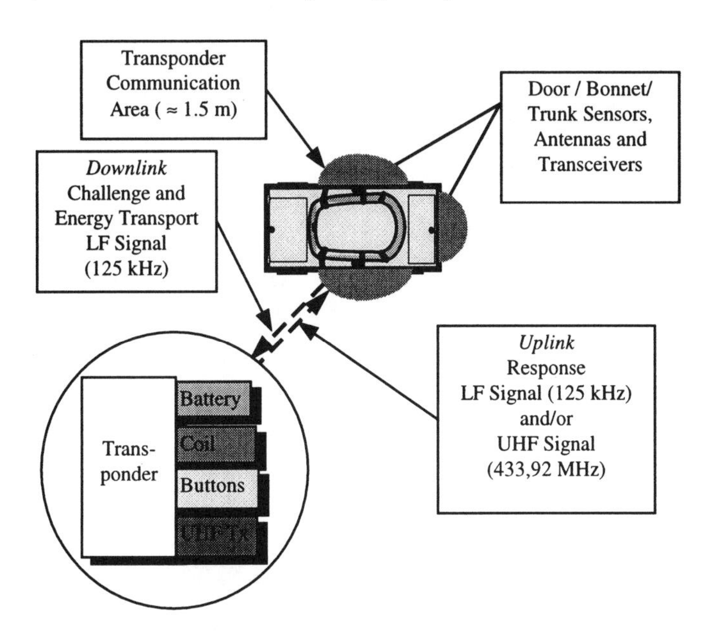

Fig. 8: Schematic overview of keyless entry system communication features

To have the best control of communication range, we use inductive coupling between the transponder and the antenna. For very small distances compared to the wavelength, the magnetic field generated by small loop antennas decreases with the cube of the distance. It is therefore well suited for data transmission, where the range has to be controlled.

Another advantage of inductive coupling, is the possibility to transfer energy at close distances via the magnetic field. This is necessary in a flat transponder battery scenario. Ideal places for mounting an antenna are the driver and passenger side mirrors and the rear bumper.

Depending on the packaging form factor, the communication area and the data transmission frequency from the transponder to the vehicle changes. Due to the limited thickness of a smart card, no SAW resonator can be embedded and the data transmission is restricted to a bi-directional LF communication only. However, utilisation of a key head/fob provides enough space to install a SAW resonator and the communication from the transponder to the antenna is established using UHF. The UHF emitter can furthermore be used for remote control functions.

Simulations and tests have shown that the operating range depends strongly on the coil area and the receiver stand-by current. Tests with a smart card implementation of a keyless entry system have resulted in a communication range of 1.5 m. Using a key fob implementation provides a communication distance of more than 1.5 m. The main reason for the increase is the higher stand-by current.

In the case of a smart card with a small battery (30 mAh), the life time should be at least three years. This puts enormous pressure on the stand-by and the reading and writing current consumption for the smart card embedded transponder in order to assure the required life time and communication distance of 1.5 m. Simulations of a transponder with a stand-by current of 700 nA show that the required communication area can be achieved. Transponder read/write cycles will also not consume more than 4 mA. By assuming 100 applications per day a 30 mAh smart card battery will last the required 3 years.

When utilising key head/fob packaging, the higher battery capacity allows a higher stand-by current of 3 µA. The range can therefore be increased to more than 1.5 m. A battery capacity of 120 mAh still provides a small key head package and a life time of 3 years.

OVERALL TIMING

The overall timing for the vehicles' door opening is a critical issue. Mechanical door locks, as used today, provide a door handle which has to be pulled to open the door. Tests with random users have revealed an opening time from as fast as 60 ms to as long as 300 ms. By adjusting an electronically driven HIS-latch to a total delay time (communication and HIS-latch) of 160 ms, most user felt that this delay (pulling the door handle and door opening) is just acceptable. Therefore it is essential for any system to have fast communication and opening times.

By using current existing HIS-latch units in combination with the fast protocol an opening time of nearly 130 ms can be realised. An overall communication time of less than 160 ms can therefore be assured and customer requirements fulfilled.

MULTIPLE APPLICATIONS

The fast growth of smart cards and the utilisation of multiple remote control systems (i.e. vehicle access and garage opener) are inconvenient for the end user. Specifically in the field of smart card applications, a multi application card would be highly beneficial. With the command structure utilising POICs, the keyless entry system could therefore be used for three different kinds of applications (i.e. vehicle, building access and garage opening), all within a single key fob/head.

The ASIC developed, also combines the functionality of a vehicle keyless entry system and stored value applications such as ticketing and cash less money transfer.

CONCLUSION

We have described a newly developed multifunctional keyless entry system. The system combines the functionality of immobiliser and remote control. Moreover, the comfort for the end user is increased due to the feature of keyless entry and multiple application utilisation.

To satisfy automotive keyless entry requirements, an extremely fast anti collision protocol was implemented and used in combination with a fast HIS-latch to assure a door opening time of less than 130 ms. An innovative cryptographic algorithm ensures the same security for vehicle immobiliser application as in todays' banking systems.

The form factor independent packaging concept (key head/fob or smart card), gives the choice to the end user. Both systems fulfil the customer requirements in operating range and in a life time of more than 3 years.

REFERENCES

1. Balanis, C. A., Antenna Theory, 2nd ed., Wiley, New York, 1997.
2. Hirano M., Takeuchi, M. Tomoda, T., Nakano, K.-I., "Keyless Entry System With Radio Transponder", IEEE Trans. Ind. Electr., Vol. 35, No2, 208-216,1988.

94C059

Automotive Entry and Security Systems

Keith W. Banks
TRW Transportation Electronics Division

ABSTRACT

Increasing consumer demand for vehicle and personal security is strongly influencing the development of vehicle entry and security systems. Vehicle theft is a major problem in both the U.S. and Europe, costing billions of dollars annually.

Automotive suppliers are working on new, innovative electronic entry and security systems to protect vehicles from inadvertent entry and theft. These technologies include passive entry systems, vehicle immobilizers and "rolling code" security. With the continuing increase in automotive electronics, these systems will be integrated into overall vehicle electronics as cost effective theft deterrents for consumers.

As much as vehicle quality and value were the benchmark priorities in the 1980s, safety and security are topping motorists' concerns in the 1990s. A heightened awareness of increasing crime rates is causing people to take extra precautions to protect themselves and their possessions.

The traditional lock and key is expected to soon give way to a new generation of OEM vehicle entry and security systems as the demand for more protection and convenience continues to grow. These electronic products are better equipped to combat the auto theft epidemic that plagues many urban centers particularly in the United States and Europe.

The U.S. leads the world in auto theft (see figure 1). In 1992, 1.6 million vehicles were stolen, 2.9 million had contents or accessories taken and an estimated 25,000 motorists were subjected to carjackings, according to the National Insurance Crime Bureau (NICB). The cost to motorists and insurance companies is staggering - nearly $8 billion annually.

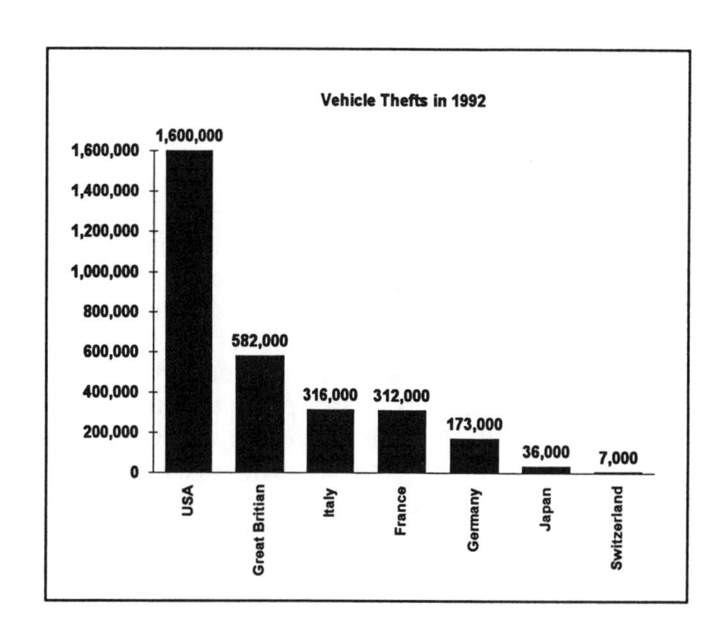

Figure 1
Vehicle Thefts in 1992

Europe is also paying a heavy price. Since the borders to Eastern Europe were opened, Germany, France and Great Britain have witnessed a sharp increase in auto thefts. Last year, the German insurance industry paid out 1.5 billion Deutsche marks ($1.0 Billion) and British insurers 650 million pounds ($1.0 Billion).

The enormity of the problem has forced governments and insurance industries to take action. In France, insurance companies issued a regulation called "SRA" that uses a star-rating system to award policy holders for the level of theft deterrent equipment they install. The more stars, the greater the insurance discount. In Great Britain, insurance companies offer discounts for certain types of theft deterrent systems, and they are mandated on high-risk vehicles.

Germany is feeling additional pressure to act because auto theft recovery rates have dropped from

65% in 1991 to 23% in 1992. Last year, auto makers agreed to German insurance companies regulations requiring electronic immobilization equipment to be installed in all new cars by January 1995, or the policy-holder will receive only 90% of the stolen car's insured value. With the possibility of French and British insurance companies enacting a similar regulation, Europe is currently a hotbed of development for immobilizer technology.

In the U.S., twelve states currently mandate insurance discounts for designated equipment, with criteria and discounts for these devices differing between states and insurance companies. Nationwide, discounts range from 5% off the comprehensive portion of an insurance premium for devices that must be activated by the driver and up to 35% for automatically activated devices. The disparity between "passive" and "active" systems is due to the fact that drivers often fail to engage their security system when they leave the vehicle. Discounts are also available on specified vehicle recovery systems.

As examples of actual savings, consider the policies of two large auto insurers. State Farm Insurance provides a 10% discount for certain types of theft deterrent equipment on the comprehensive portion of its insurance, which averages $206.00 annually. Thus, a 10% discount saves the policy holder $20.60 a year. At AAA, where comprehensive has three components - theft, fire and vandalism - a 10% policy discount applies only to one-third of the total charge. So, using the $206.00 policy model, the discount would be $8.66 annually.

What all of this means to auto manufacturers and suppliers is a booming vehicle entry and security industry. For example, in Britain, 51% of British-made vehicles were equipped with standard car alarms last year, according to the Society of Motor Manufacturers and Traders (SMMT). Immobilizers are available as standard or at point of sale on 96% of British cars - up 36% from 1991.

In the U.S., sales of OEM and after market security products totaled $542 million last year and are predicted to reach $724 million by 1997, according to the Electronic Industries Association (EIA) (see figure 2). In recent years, after market business has dominated the U.S. auto security industry. In 1993, 15% of the people who purchased a vehicle and 10% of those who leased one had an after market security system installed, according to a national survey by CNW Marketing/Research. Further, 1994 installation rates of after market devices are expected to rise to 16% for new vehicles and 13% for leased.

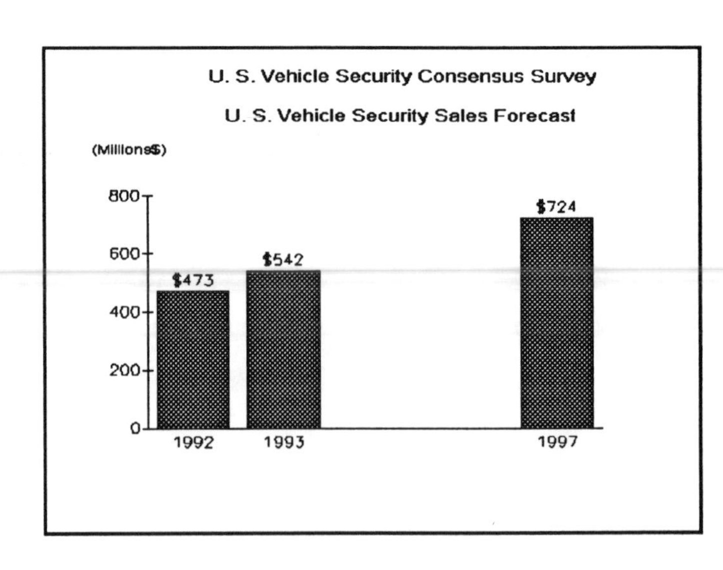

Figure 2
U.S. Vehicle Security Consumer Survey
U.S. Vehicle Security Sales Forecast

However, the OEM market is changing. While only 1% of the U.S. and import trucks came with factory-installed anti-theft systems in 1992, auto makers raised that to 4% in 1993, according to Automotive News. Import auto makers also increased the number of vehicles with anti-theft systems sold in the U.S. from 17% in 1992 to 19% last year. The percentage of North American-built cars remained constant with nearly 8% in 1992 and 1993.

In France, sports cars and convertibles are overwhelming favorites for theft. In Great Britain, they are Cosworths and hot hatchbacks. In the U.S., six of the top 10 stolen vehicles in 1991 were sport utility models, according to the NHTSA. The NICB says the reason is partly due to the high demand for these exports in Central and South America, where road conditions are poor. When motorists know that their vehicle is a higher theft risk, one might think they are more likely to purchase a security system. However, that isn't always the case. In the U.S., the Ford Mustang was the car most often stolen in 1990 and 1991, but 61% of Mustang owners reported they don't use an theft deterrent device, according to a survey by Industrial Marketing Research (IMR). In addition, the GMC Jimmy sport utility ranked second on the list in 1991, but 71% didn't have a security system either.

DEMOGRAPHICS

Who thinks security systems are important? According to a national survey by CNW Marketing/Research, more women than men find security to be important (see figure 3). The greatest percentage of women who think that security is important are women 56 to 65 years old. Single women's interest in security declines from older to younger age groups. People aged 18-25 report the least interest in theft and security of all the age groups.

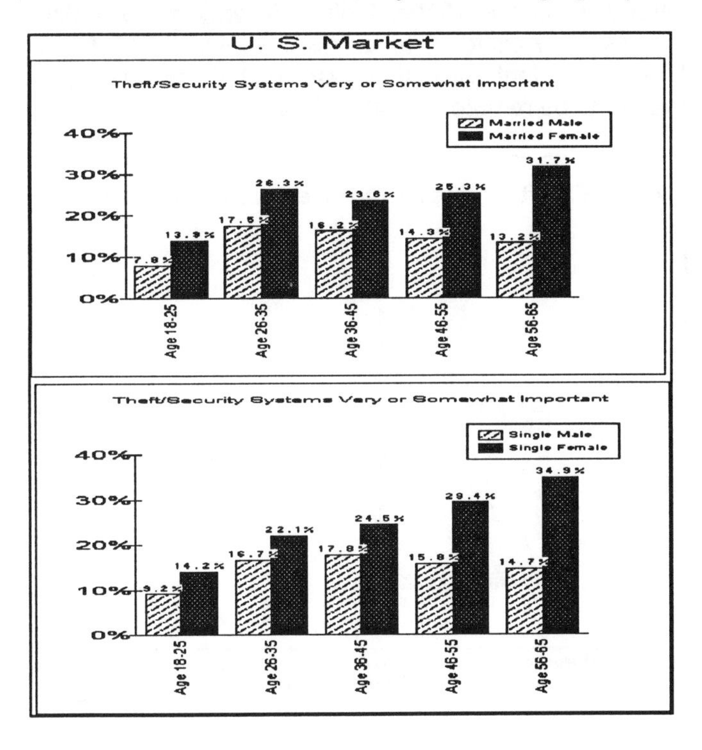

Figure 3
U.S Market

At age 25, however, interest in security systems apparently changes. As people start to earn a better living and can afford more expensive cars, they are more interested in security systems to protect those vehicles. According to a survey by Simmons Research, the greatest percentage of car owners with a security system - 26% - are between the ages of 25 and 34 (see figure 4).

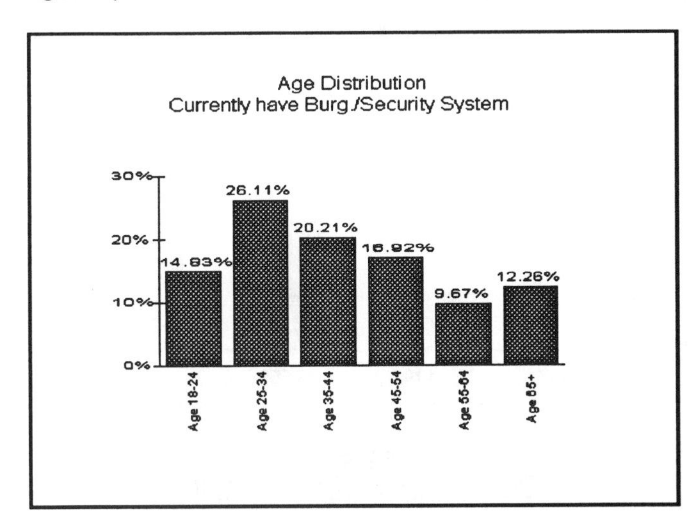

Figure 4
Age Distribution Currently Owning Security System

The survey also showed that 38% of the security system owners drove small sporty cars, 25% owned luxury models and 18% had small basic cars (see figure 5).

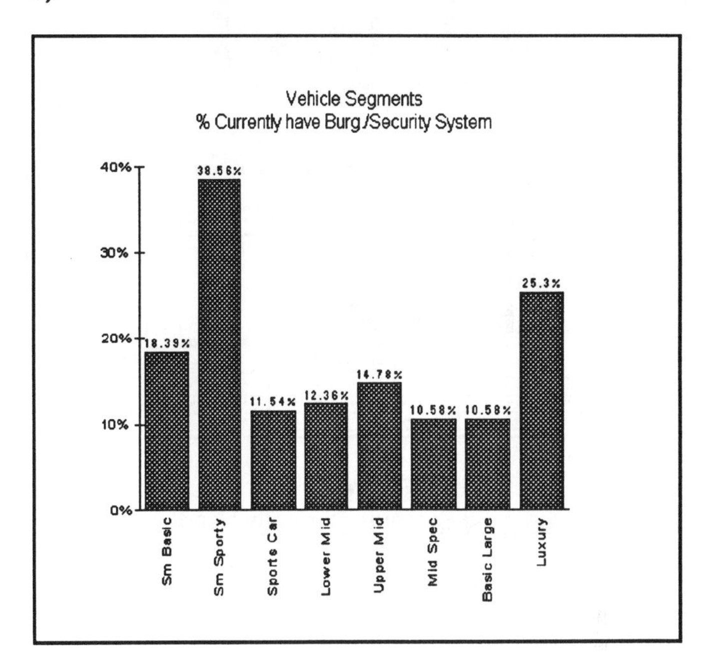

Figure 5
Vehicle Segments
% Currently have Burg./Security System

More people with a total household income of $100,000 or more say they use an auto alarm system than people in lower income brackets, according to the IMR survey. This is probably due to the fact that they can afford more expensive cars that warrant theft deterrent protection. Indeed, 28% report owning a system that uses a siren or horn sound as a deterrent. Only 5% of the people making less than $25,000 own an alarm system with a siren or horn, according to the survey.

New England and the mid-Atlantic states have the highest percentage of auto alarm owners, according to the survey. Closely following that 23% figure is the Pacific region with 22%. When respondents were asked whether they would like a remote-activated theft deterrent system in their next vehicle, 34% from New England, 34% from the Mid-Atlantic, 34% from west South Central region and 32% from the Pacific said yes.

GM vehicle owners showed the strongest use of security systems and the greatest desire to own them in their next vehicle (see figure 6). Ford owners came in second, followed by Chrysler, Toyota, Honda and Nissan.

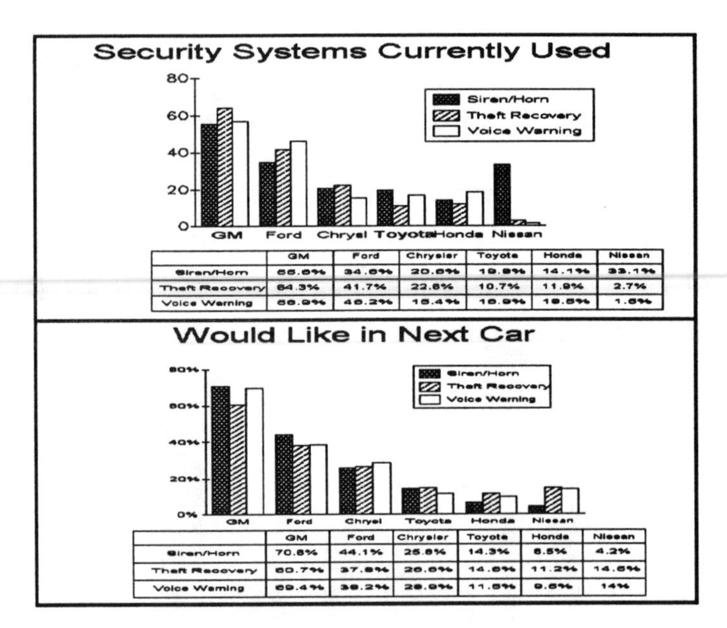

	GM	Ford	Chrysler	Toyota	Honda	Nissan
Siren/Horn	55.8%	34.6%	20.6%	19.8%	14.1%	33.1%
Theft Recovery	64.3%	41.7%	22.6%	10.7%	11.9%	2.7%
Voice Warning	58.9%	46.2%	15.4%	16.9%	19.5%	1.6%

	GM	Ford	Chrysler	Toyota	Honda	Nissan
Siren/Horn	70.8%	44.1%	25.8%	14.3%	6.5%	4.2%
Theft Recovery	60.7%	37.8%	26.6%	14.6%	11.2%	14.6%
Voice Warning	69.4%	38.2%	28.9%	11.5%	9.6%	14%

Figure 6
Security Systems Currently Used
Vs.
Would Like In Next Car

People are increasingly turning to a dealership to purchase a security system rather than companies that sell and/or install after market products (see figure 7). Ten percent of motorists with 1990 vehicles bought their system from a dealership and 45% say they would return to a dealer to make the purchase. Of the people owning 1993 models, 19% say they bought their security system from a dealership and 47% say they would purchase one from a dealer today.

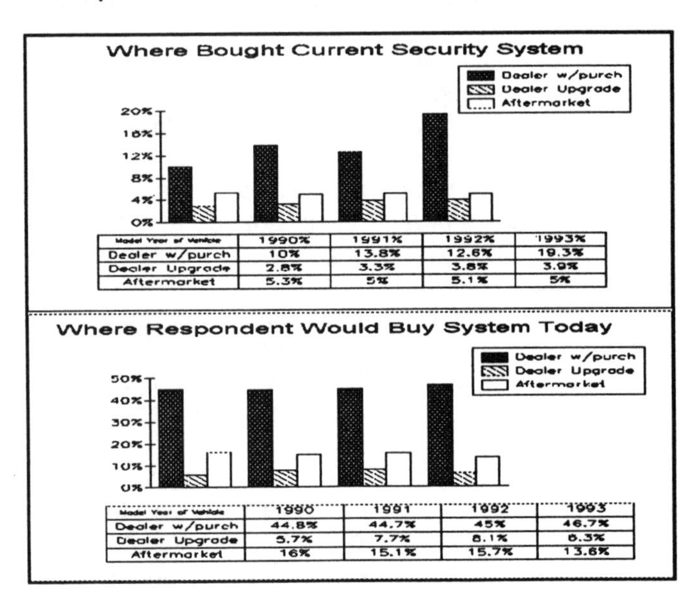

Model Year of Vehicle	1990%	1991%	1992%	1993%
Dealer w/purch	10%	13.8%	12.6%	19.3%
Dealer Upgrade	2.8%	3.3%	3.8%	3.9%
Aftermarket	5.3%	5%	5.1%	5%

Model Year of Vehicle	1990	1991	1992	1993
Dealer w/purch	44.8%	44.7%	45%	46.7%
Dealer Upgrade	5.7%	7.7%	8.1%	6.3%
Aftermarket	16%	15.1%	15.7%	13.6%

Figure 7
Where Bought Current Security System
Vs.
Where Respondent Would Buy System Today

REMOTE KEYLESS ENTRY (RKE)

Since remote keyless entry systems were first introduced in the 1980s, their popularity has steadily increased. According to an IMR survey, 27% of the U.S. respondents said they wanted their next security system to be remote-activated (see figure 8).

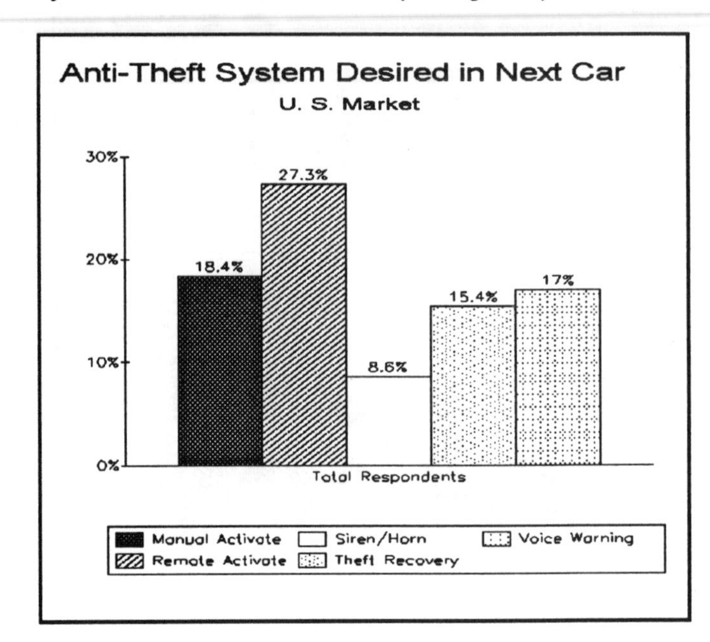

Figure 8
Anti-Theft system Desired in Next Car

In Europe, TRW, one of the largest manufacturers of vehicle entry systems, forecasts sales will reach nearly $250 million in 1995 and approximately $350 million in 1998 (see figure 9).

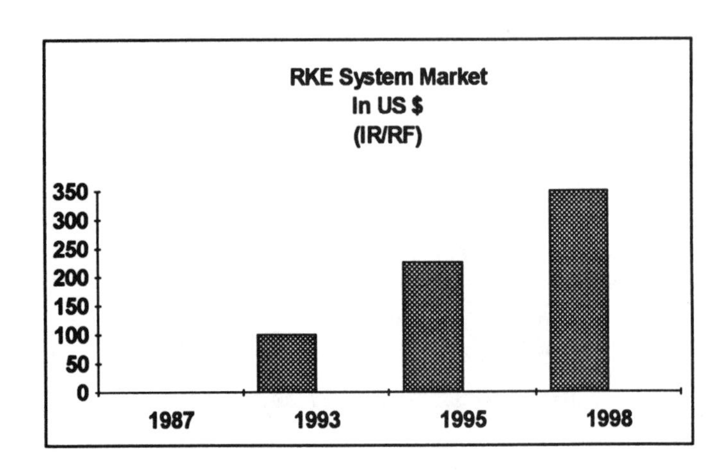

Figure 9
Remote Keyless Entry Market

Import cars and trucks sold in the U.S. showed an increase in factory-installed RKE systems from 1992 to 1993, according to Automotive News. Import cars went from 7% in 1992 to 9% the next year, and trucks had nearly a 1% increase during that time. North American-built cars declined slightly , from 16% in 1992 to 15% in 1993.

RKE transmitters are activated by pushing the appropriate button to operate central locking, interior lighting, trunk, windows, panic button and even remotely start the car. The Chevrolet Corvette's Passive Keyless Entry (PKE) system, however, operates automatically. As the driver moves the transmitter within range of the vehicle, the receiver picks up the signal, unlocks the doors, activates interior light and turns off the theft deterrent system. Another of the system's features prevents the doors from locking if the keys are left in the ignition.

Remote entry systems have traditionally relied on two different technologies infrared (IR) and radio frequency (RF). An IR transmitter's range is within five meters of the vehicle, it must be pointed directly at the receiver, and can malfunction if ice or snow is present. An RF transmitter's range is within 10 meters, doesn't need to be pointed at the vehicle to activate the system and can penetrate windows covered with snow, ice or mud.

The U.S. has, for the most part, phased out IR. But in Europe, where IR has been the predominant technology, the process of phasing out IR is taking longer. Until recently, each European country had its own regulations regarding operating frequency, and sideband emission power. European countries have now agreed to standardize their regulations, and as a first step, will adopt the same operating frequency. This change is opening the RF market in Europe. Japan uses predominantly IR, but has recently requested a customized low power RF transmitter/receiver to meet the country's stringent RF emissions regulations.

Strong consumer interest and declining prices bode well for RKE in the near future. The long term also holds tremendous potential for integration with upcoming systems. The theory: If the RKE's receiver can pick up a signal from a transmitter, it can receive signals from other sources as well. The receiver could pick up travel-related messages that would be displayed on a digital display panel as part of the Intelligent Vehicle/Highway System. Emergency warnings from fire trucks, ambulance and trains could also be conveyed this way. In addition, RKE could play a part in the Automatic Vehicle Identification system, which would automatically recognize the vehicle and eliminate the need for stopping at toll booths.

THEFT DETERRENT AND SECURITY SYSTEM

People buy theft deterrent systems, such as immobilizers and locking devices, to protect a car from being stolen. Security systems - intrusion sensors and sound devices - make a thief think twice about breaking into the car to get at the valuables inside. Law enforcement and insurance experts agree that if a professional thief targets a car, no theft deterrent equipment can stop him. These systems have a higher success rate preventing or slowing down inexperienced "joy riders", which the National Insurance Crime Bureau (NICB) estimates are responsible for 70% of auto thefts in the U.S.

The following are the main types of systems:

Immobilizers - If a car is tampered with, these devices shut down the ignition, starter electronic, fuel management system or main ignition switch. In some systems, a remote device can activate the immobilizer in order to prevent a carjacker from making a getaway. Beginning in 1995, the German insurers regulation specifies that an electronic signal must be sent to the engine management computer to immobilize the vehicle. In response, several companies have developed various types of immobilizers, including an electronic signal via a hand-held transmitter, an electronically coded ignition lock and an electronically coded ignition key (transponder). Because of the complexity of the integration with vehicle electronic systems, immobilizers must be factory-installed (see figure 10).

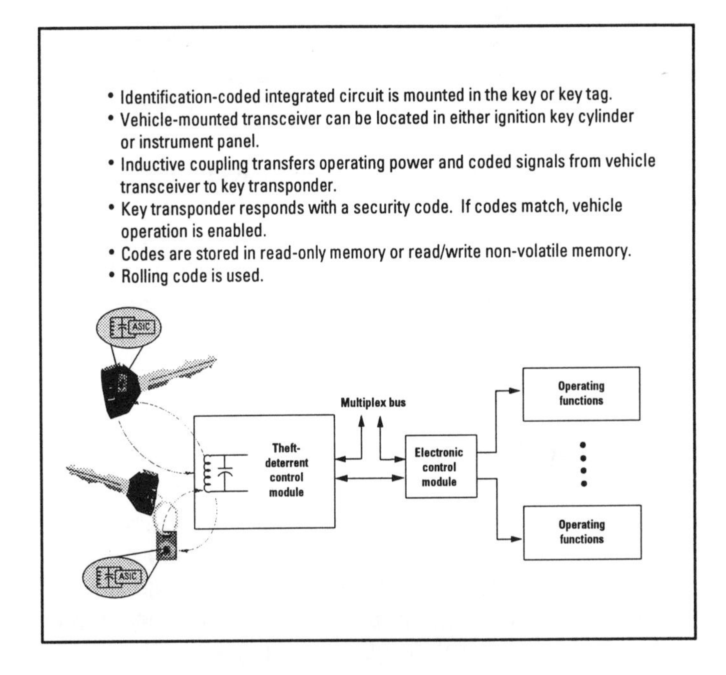

• Identification-coded integrated circuit is mounted in the key or key tag.
• Vehicle-mounted transceiver can be located in either ignition key cylinder or instrument panel.
• Inductive coupling transfers operating power and coded signals from vehicle transceiver to key transponder.
• Key transponder responds with a security code. If codes match, vehicle operation is enabled.
• Codes are stored in read-only memory or read/write non-volatile memory.
• Rolling code is used.
ASIC
Multiplex bus
Theft-deterrent control module
Electronic control module
Operating functions
Operating functions
ASIC

Figure 10
Vehicle Theft-Deterrent System

Mechanical Devices - Locking devices protect everything from lug nuts to glove compartments, but the hottest selling item in this category's 1992 sales of $121 million is "The Club" - a steel bar that locks the steering wheel. Retailing for $59.50, it is economical, easy to install and the average joy rider would probably rather move to another car which would be easier to steal. However, insurance experts don't put a lot of faith in them, claiming that professionals know how to break them. More effective locking devices can be installed by professional technicians, or at dealerships that market these products.

Sound Devices - Alarm systems come in the form of loud whistles, sirens or horn blasts which act as deterrents. They focus attention on the car, but will not prevent a determined thief from breaking in. Most of these alarms can be activated by a panic button on the keypad if a driver wants to deter a suspected prowler. Some of the more unique electronic systems emit a colored puff of smoke, deliver a strong shock to the unwelcome driver or send an angry recorded message through the car sound system. Some come factory-installed, others are added by installation specialists or do-it-yourselfers.

Intrusion Sensors - Shock sensors can detect strong blows to the car, they monitor the trunk, door, and hood. Motion detectors recognize swaying or bouncing, and glass-breakage sensors respond if somebody windows are broken. If they are installed improperly, intrusion sensors are over sensitive and will give false alarms. Ultrasonic sensors are installed inside the vehicle and look for changes in the ultrasonic standing waves caused by intrusion into the interior space. If the car is broken into, the alarm will sound and the lights will flash. Not many OEM systems come equipped with these types of devices, giving a large after market business to specialty installers, according to Installation News.

Double Locking - Most commonly found in Europe, this system locks the doors from the inside as well as the outside. It prevents the thief from breaking a window, reaching in and unlocking a door from the inside. The system comes factory-installed, and normally has an accelerometer to de-activate the system in the event of a collision.

VEHICLE RECOVERY SYSTEMS

All the alarms, locks and intrusion detectors can't bring back a car once it's stolen, but a recovery system has a good chance of it. Lojack and Code-Alarm Intercept are two of the industry leaders that boast high national recovery rates 95% for Lojack, 100% for Intercept. Lojack retails at $595.00 and passive theft deterrent devices are also available at additional cost. It is installed only by mobile dispatched Lojack specialists. Intercept retails at $1495.00 with the cellular telephone, or $995.00 if the owner already has a compatible phone model. Because of the complexity, it is installed only by select professional technicians and occasionally by auto dealerships that market it.

Intercept has an alarm that deters a thief, but if he gets past it, the cellular phone automatically calls a Code-Alarm monitoring station while the transmitter sends a signal to Loran C navigation stations. Once these stations pinpoint the car's location, the information is sent to the monitoring station which verifies the theft with the owner and then contacts police. The monitoring station has the ability to shut down the car's engine.

When a car protected by the Lojack system is stolen, the owner calls the police, gives them the car's identification number and the police then activate the car's hidden transmitter which sends a coded signal to patrol cars equipped with special receivers. Critics say that by the time the owner reports the car missing, vandals could be well on their way. Another problem is that not all police agencies are equipped with the Lojack Stolen Vehicle Recovery System, so its effectiveness could vary within the eight participating states.

In England, all 51 police forces have installed "Tracker," a vehicle recovery system that allows them to track stolen vehicles by patrol car or helicopter. Another after market product, Securicor Datatrak's "Trakback," combines two technologies - a vehicle tracking system and immobilizer system. The integrated device relies on the Datatrack tracking system which covers approximately 95% of the United Kingdom.

OEM's are conducting feasibility studies for a vehicle recovery system that could locate a stolen car anywhere in Europe. The challenge is to develop a reliable device that works via a satellite and has a reasonable cost.

OEM AND AFTER MARKET

Some auto manufacturers think that installing their own products makes sense because motorists have more faith in them to install them properly, there are no external wires to let the thief identify and deactivate the system, and people like the fact that it comes with a warranty. Consumers can also be influenced by price - a standard feature comes at no extra charge where as purchasing and installing a theft deterrent system is sometimes an expensive proposition. Drivers can also finance the cost into the selling price of the car. Finally, to make an OEM system effective, periodic changes are needed, and insurance experts agree that auto makers are in the best position to make these necessary changes.

Early OEM security systems developed a reputation for being predictable, because they were all made the same way. Recently, however, auto makers say they are being proactive about staying ahead of thieves by making periodic changes so that vandals don't get time to figure a way to beat their systems.

To gauge the impact that auto manufacturers will have as they continue entering the security market, one can look to the auto audio market for a comparison. In recent years, auto makers have formed alliances with high end audio companies to install audio equipment in their cars themselves. GM teamed up with Bose Corporation, Ford with JBL, Chrysler and BMW with Infinity, Lexus with Pioneer and Nakamichi, and Toyota with Fujitsu-Ten.

After market auto sound business is still on the rise - factory sales were $1.5 billion last year and 1994 sales are expected to reach $1.7 billion, according to the EIA (see figure 11). However, its share of total U.S. sales fell from 46% in 1989 to 31% in 1993, according to Venture Development Corporation, a management consulting firm. In turn, OEM car audio shipments rose from 53% of the business in 1989 to 68% in 1993. A similar effect on the after market security industry is predicted with auto manufacturers increasing their presence in the market.

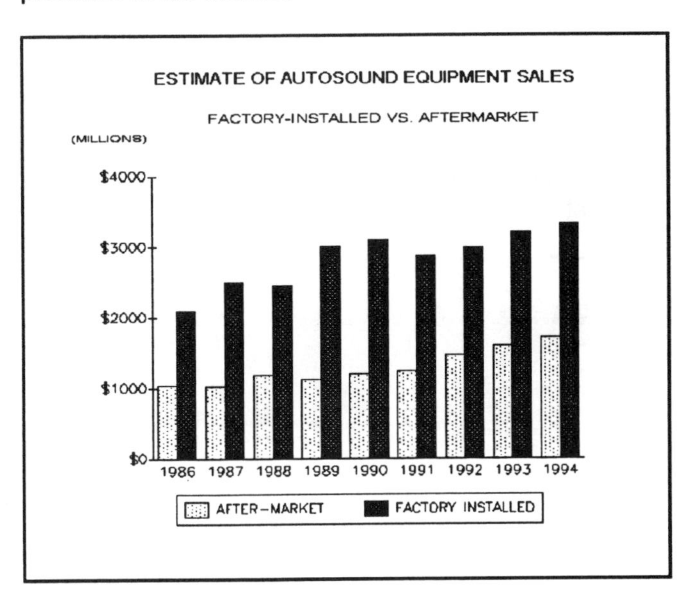

Figure 11
Estimate of Auto Sound Equipment Sales

OEM SECURITY SYSTEMS AT WORK

The NHTSA is currently working on a report to assess the effectiveness of factory-installed theft deterrent equipment, which is expected to be complete in 1996. The following are examples of various OEM security systems on the market:

U.S AUTO MAKERS - General Motors is far ahead of the other Big Three auto makers with the scope of its factory-installed PASS-Key (Personalized Automotive Security System). In 1993, 14 GM car lines came installed with it and by 1995, some form of PASS-Key will be standard on 70% of all GM cars.

PASS-Key comes equipped with a resistor in the tip of the ignition key. If a foreign key or object is placed in the ignition, the starter and fuel delivery circuits shut down for three minutes. This system comes standard on all Chevrolets except the Prism, Metro, Cavalier, Beretta and Corsica; all Pontiacs except the Sunbird and Grand Am; all Oldsmobiles except the Achieva, and all Buicks except the Skylark. The system is made by Delco Electronics and Briggs & Stratton Technologies provides the locks, wiring harnesses and specialized key for GM.

When GM first introduced the Vehicle Theft Deterrent System (VATS), the security system that preceded PASS-Key, thieves soon figured out how to beat it. GM learned some early lessons which helped to create the new PASS-Key system, which has proven effective in lowering theft rates in the Firebird and Camaro. GM reports that insurers' loss costs dropped 74% for the Camaro between 1988, the year preceding PASS-Key, and 1992. Likewise, insurers noted a 75% reduction in loss costs between 1988 and 1991.

In addition to PASS-Key, Cadillac offers a theft deterrent system to guard against break-in. The system is standard on Cadillac's 1994 Eldorado Touring Coupe and Seville STS, and is optional on other models. It is a passive system that automatically arms the car by key or remote. Any attempt to forcibly enter a car door or trunk will cause the headlights, parking lights and taillights to flash and the horn to sound.

Chrysler offers an optional factory-installed security system on its LH models including the New Yorker, Concord, Dodge Intrepid, LHS, Eagle Vision and on its Grand Cherokee, Stratus and Cirrus cars. In addition, Chrysler's MOPAR EVS II is an optional dealership-installed system that can automatically arm and disarm the security system by remote. This Code Alarm product has a panic button, shock sensor, starter interrupt and contains optional features that can activate flashing parking lights or broaden protection to the trunk and hood.

Ford provides a standard factory-installed security system on the Mark VIII and offers it as an optional feature on the Town Car, Continental, Thunderbird SVO, Econoline, Explorer, Bronco, F-Series and Windstar. Briggs & Stratton Technologies is one of the organizations involved in designing an improved security system for Ford. The system is due out in the near future, but industry sources indicate that Ford has not determined whether to offer it as a standard or optional feature.

EUROPEAN AUTO MAKERS - Volkswagens are on top of the hit list in Germany for auto theft (see figure 12), but in the U.S., Volkswagen vehicles are more likely to be broken into for the accessories or contents inside the car. Therefore, all A3 models are equipped with an active alarm system made by Volkswagen that includes a starter disable function. The radio, a traditionally popular item for thieves, is also connected to the alarm system through a wire in one of the connectors.

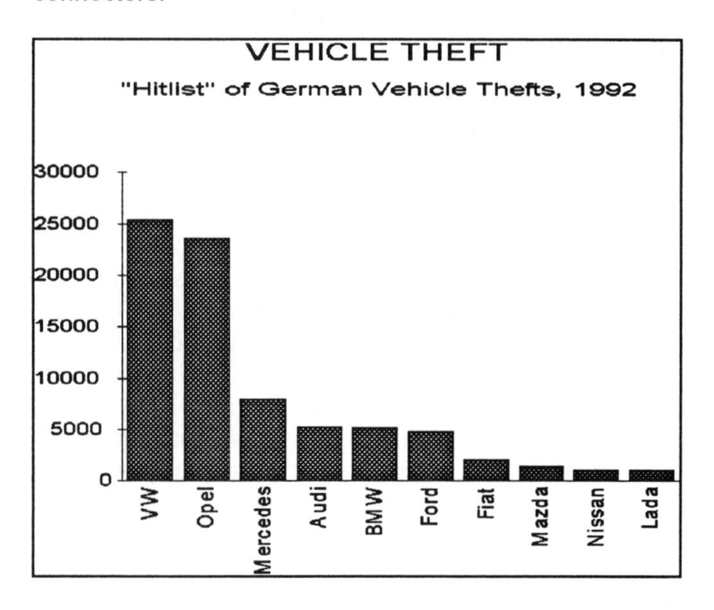

Figure 12
Vehicle Theft

JAPANESE AUTO MAKERS - Japanese auto makers have focused their development of security systems for products abroad since vehicle theft is relatively low - only 36,000 vehicles were stolen in 1992.

In the U.S., the 1994 Nissan Maxima, Ultima, Pathfinder and 300ZX all have standard security systems, with the keyless remote entry system available only on the Pathfinder. If the door, hood, trunk or hatchback is opened when the system is engaged, it triggers the horn, lights and starter kill so that the car can not be "hot wired". Infiniti provides a standard active alarm system on its three models the Q45, G20 and J30. If the door lock is attempted with anything other than the remote or ignition key, the horn honks, lights flash and the starter kill disables the engine.

Toyota offers a factory installed system that can be enhanced by adding a remote transmitter and glass breakage sensor on the Supra, MR2 and Previa. All Lexus models come with a standard passive alarm system that is activated when the door locks, hood or trunk are tampered with. In such an event, the lights and horn are activated and the engine is disabled.

Honda's theft deterrent system comes with a glass breakage detector, engine starter cut-off and a remote that can automatically arm and disarm the system. It has an anti-scanning code ROM receiver that Honda says would take up to 24 years of scanning to break. A siren and hood switch are additional options.

Mazda offers an optional dealer-installed security system made by Vesco Electronics. If the system is activated and the door, hood or trunk are opened, the starter is disabled, the horn sounds and parking lights flash. A remote upgrade to a pocket size transmitter is also available.

FUTURE PRODUCTS

Europe is phasing out theft deterrent systems that use an in-line interrupt to disable the vehicle because they are easy to hot wire. By 1995, German regulations state that disabling has to be done electronically through the engine control module. Therefore, various types of immobilizing systems have been designed to comply with these regulations. They act as vehicle entry systems with built-in theft deterrent systems.

One of these immobilizing systems uses rolling code to thwart some of the more sophisticated devices used by professional thieves (see Figure 13). When an RF transmitter sends out a signal to the receiver, thieves using code "grabbers" can also receive the signals, record them and play them back later to steal the vehicle. Rolling code changes the code each time the system is used, thereby eliminating the chance of a thief copying a code and using it to break into the vehicle at a later time.

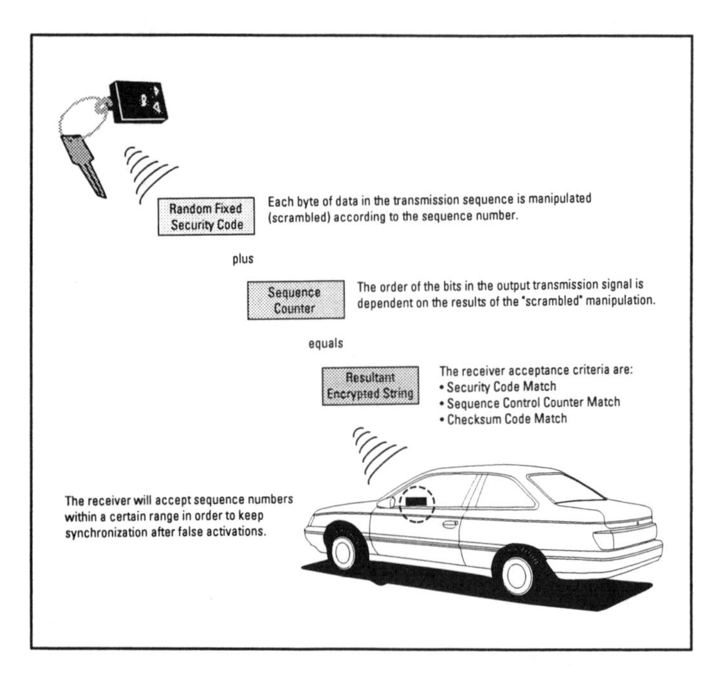

Figure 13
Rolling Code Operation

Siemens reports that European, Asian, and U.S. auto makers are looking into its vehicle immobilization system with rolling code, which is scheduled to go into production for two auto manufacturers in 1995. Likewise next year, TRW is also scheduled to provide a theft deterrent system and immobilizer with rolling code to two auto manufacturers. It is expected that GM will require rolling code on its PASS-Key system by 1997.

Vehicles equipped with RKE can integrate immobilizing functions with changes in the software and the interface between receiver and engine control module. For vehicles without RKE, U.S. auto makers are expected to introduce more immobilizing systems with electronics imbedded in the key.

As an example of this technology, Ford introduced some of its 1994 models in Europe with a passive theft deterrent system - an immobilizing system that uses a transponder code embedded in the head of the ignition key and a radio frequency reader installed in the steering column. Texas Instruments, which developed the system, says the car can't be started without the matching key and electronic ID code.

Another developing new technology is a credit card styled transmitter, known as a passive badge or smart card (see figure 14). As the driver approaches the vehicle, the automobile sends out an interrogation signal and the badge responds with a signal that can operate typical RKE functions and deactivate the theft deterrent system. TRW's passive badge is scheduled for production on 1997 automobile models.

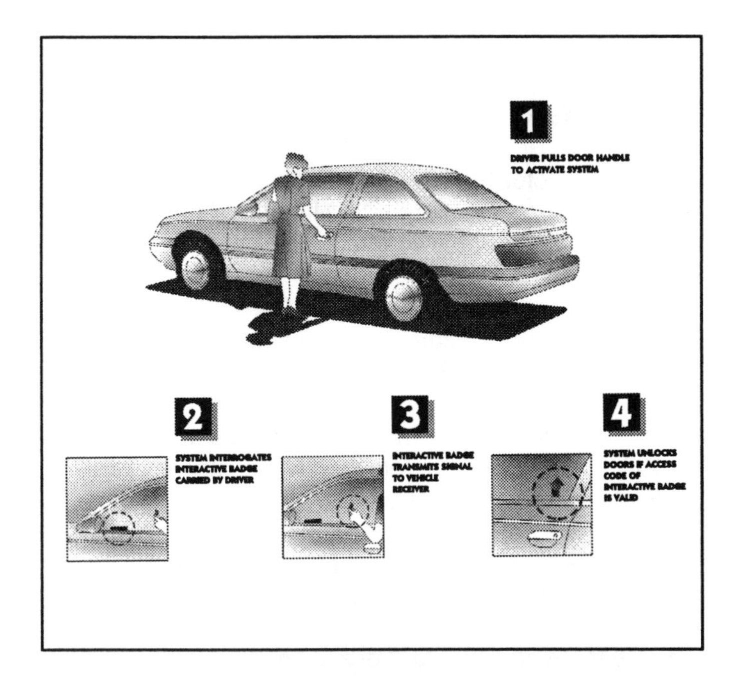

Figure 14
Passive Entry - Switch Activation

Beyond the year 2000, theft deterrent and tracking devices will evolve into features of the overall In-Vehicle Information System (IVIS). This centralized communication center within the vehicle will provide the link between vehicle and driver, as well as a link from the vehicle to other information sources, such as tracking devices. The IVIS will provide vehicle usage authorization via voice recognition or another "finger printing" type device. Vehicle location will be easily determined via either an in-vehicle positioning device or by triangulation using a signal emitted from the in-vehicle transponder. In addition to providing theft deterrent capability, the system will be capable of many other features (see figure 15).

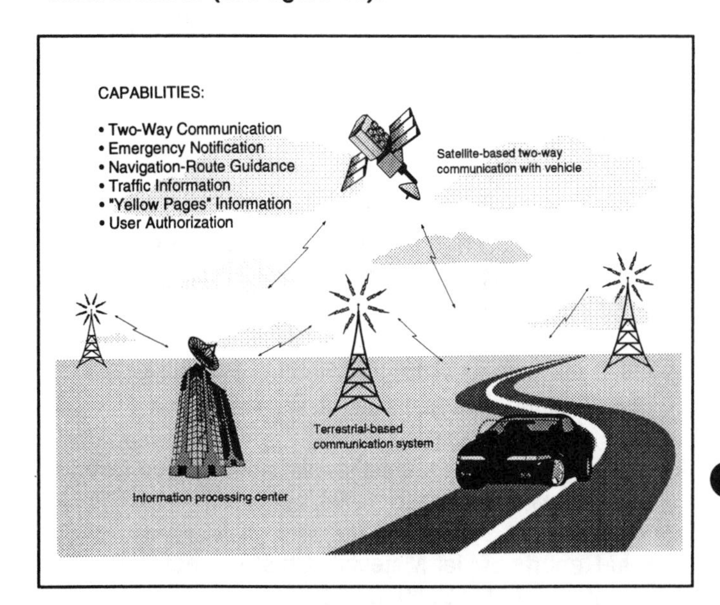

Figure 15
Proposed In Vehicle Information System (IVIS)

Recent consumer market research, conducted by TRW, shows consumers are willing to consider these types of systems, particularly as the cost per feature is reduced. Thus, long term vehicle theft deterrent systems will likely become integrated into the overall vehicle electronic systems.

CONCLUSION

The soaring numbers and costs related to increasing auto theft worldwide has spurred governments, insurance companies and consumers to call for more effective vehicle entry and security systems. As a result, auto makers are beginning to install their own systems. In most high-end automobiles, these systems are already standard features. Today, however, standard security is increasingly being made available in mid- and lower-end vehicles as well. Consumers will ultimately weigh the effectiveness, ease of use and price to determine if OEM systems will continue to proliferate. If motorists aren't satisfied with these products, they have a plethora of alternatives in the after market arena. So far, consumers have demonstrated a need for both types of products.

The future shows a shift to more sophisticated security systems, including vehicle location recovery systems, according to the EIA. In addition, research and development is underway on smart card/passive badge integration with immobilizer systems, remote immobilization by cellular telephone and bi-directional rolling codes. The complexity of these systems and their integration with the rest of the cars' electronics systems is expected to increase market share of OEMs and reduce after market installation drastically over the next decade.

REFERENCES

"1992 Automobile Theft Rates by Country," Federal Association of German Insurance Companies, Bonn, Germany, 1993.

"1992 Uniform Crime Reports," Federal Bureau Of Investigation, Washington, D. C., 1993

"Airborne Police Use Tracker to Swoop on Stolen Cars," *The Times*, February 6, 1994.

"Annual Reports of Her Majesty's Chief Inspector of Constabulary for Scotland," Lothian and Borders Police, Edinburgh, Scotland, 1994.

"Anti-Theft Standards and Discounts," Massachusetts Automobile Insurers Bureau, 1993.

"Auto Accessory Product Planner Notebook 1993 - Vehicle Remote System," Industrial Market Research, Inc., 1993.

Automotive News, 1994 Market Data Book, Detroit, MI: Crain's Communication, pg 79, 89, 98.

"Auto Theft Q & A," Texas Instruments, Attleboro, MA, October 1993.

Ben Parr, State Farm Mutual Automobile Insurance Company, Bloomington, In conversation with Lindy Adelson, Franklin, MI April 1994.

"Can You Stop a Car Thief?" *Consumer Reports*, February 1992, p., 96.

David Ladd, Siemens, Auburn Hills, MI, in conversation with Lindy Adelson, Franklin, MI, March 1994.

David Stienhardt, State Farm Mutual Automobile Insurance Company, Livonia, MI, in conversation with Lindy Adelson, Franklin, MI, May 1994.

"EIA Survey Forecasts Rapid Growth in Vehicle Security," Electronic Industries Association, Washington, D.C., March 23, 1993.

"Estimate of Factory Sales Domestic Factory Installed and Aftermarket Autosound Equipment," Electronic Industries Association, January 1994.

George Lambropoulos, TRW, Dearborn, MI, in conversation with Lindy Adelson, Franklin, MI May 1994.

"Home Office Statistical Bulletin," Association of British Insures, London, England, 1994.

May Minotas, General Motors, Detroit, MI, in conversation with Lindy Adelson, Franklin, MI, May 1994.

Jerry Williams, General Motors Company, Dearborn, MI, in conversation with Lindy Adelson, Franklin, MI April 1994.

Laurent Reaumel, "Le fleua du vol," *Le Journal de l'Automobile*, 1993.

"More Makers Fit Alarms as Standard," *Motor Report International*, n664, May 24, 1993, p. 3 (1)

"New Vehicle Anti-Theft Network Announced," *Newsbyts News Network*, July 28, 1993.

"Passenger Motor Vehicle Theft Data; Motor Vehicle Theft Prevention Standard," National Highway Traffic Safety Administration, *Federal Register*, 59:51, March 16, 1994.

"Simmons Study of Media and Markets," Simmons Market Research Bureau, 1993.

Steve Cindrich, National Insurance Crime Bureau, Palos Hills, IL, in conversation with Lindy Adelson, Franklin, MI, May 1994.

"Survey of Aftermarket Installation Rates," CNW Marketing/Research, 1993.

"The OEM Decade," *Auto Week*, April 20, 1992, p.26

"The Smart Key," *The Detroit Free Press*, May 19, 1994, section D1.

Volkmar Neudoerffer, Volkswagen of America, Auburn Hills, MI , in conversation with Lindy Adelson, Franklin, MI, March, 1994.

About the Author

Keith W. Banks has served as vice president of sales and marketing for TRW Transportation Electronics Division since May 1992. The division is a leading producer of electronic and electromechanical products for the global automotive market. It is part of TRW's worldwide automotive electronics business, which has facilities in the United States, Canada, Mexico, Europe and Japan.

Mr. Banks joined TRW Inc. in 1990 as managing director of TRW Transportation Electronics Ltd. headquartered in Cirencester, England. In this capacity, he was responsible for total operational control and financial performance of an autonomous subsidiary of TRW Inc.

Mr. Banks has extensive experience in automotive electrical engineering, manufacturing operations management, strategic planning and international management.

Before joining TRW in 1990, he spent 23 years with Ford Motor Company working in England, Germany and the United States. Among the positions he held for Ford - Europe were senior manager, electrical engineering, and manager, instrumentation and features, electrical engineering. In the United States, he served as manager, instrumentation and features engineering, Electronics Division, and manager, business development, Automotive Components Group.

Mr. Banks holds a bachelor's degree in electrical engineering from the University of London and a diploma in management studies from North East London Polytechnic. A chartered engineer, he is a member of the Institution of Electrical Engineers and Institution of Mechanical Engineers in England.

932867

Opto-Electronic Ignition Lock - The Ultimate Antitheft Device

Alexander Parker and Val Parker
Parker Technology

ABSTRACT

Today's modern cars, from their fuel injection to their active suspension, are all computer controlled. Yet, access to these high-tech modern systems is still relinquished via a century old pin-tumbler ignition lock.

Thus, the Opto-Electronic Lock represents a new family of solid state light operated devices which, by their overall durability, versatility, and security, are far superior to conventional mechanical or electro-mechanical devices. Incorporated directly as an ignition lock its unparalleled simplicity provides cost efficiency and manufacturability while its hi-tech base provides unparalleled security.

I. INTRODUCTION

A. Historical Design Trends.

The first and still most widely used car security device is the electro-mechanical lock, known as the ignition lock. As is shown, (see Dwg.3) electromechanical locks are merely a natural extension of the well known and well developed mechanical lock (Yale Lock mostly). These devices are made such that the mechanical lock is physically connected (by means of push/pull rod, coupler, etc.,) with the electric switch. Well established manufacturing procedures have made this lock the most popular car security device on the market. Its susceptibility, however, to being quickly disconnected, hot wired, or altogether bypassed have turned the conventional ignition lock into a simple engine cranking device and a thief's best friend.

The increased need for car security has created the automotive antitheft market. The existing antitheft devices can be split into two categories:

a) mechanical - devices which by their physical appearance or mechanical connection prevent a thief from starting or driving the vehicle. Such devices include dead bolt locking systems, and pedal, wheel, steering wheel or fuel pump disabling devices, etc.

b) electric - devices which by their electric interconnection prevent or identify a car thief. Such devices include passive and active alarms, (ie. VATS by GM).

B. Security and Technical Problems.

As discussed above, all existing antitheft devices have a significant shortcoming - they can be easily removed or bypassed. Moreover their own presence creates an every day inconvenience to the car owner himself. The special procedures required to activate, affix, and deactivate them, not to mention accidental activation, handicap the owner more than they do the thief.

Mechanical devices can easily be disarmed by disabling the locking mechanism or simply by braking the car parts attached to the device.

Electric or electronic devices even though they are more complicated and expensive, can also be easily disarmed as long as they require external electric interconnections with the car's own electrical system.

As long as a car can be force started within a short period of time none of the devices can provide sufficient antitheft protection.

II. THEORY OF OPERATION OF THE OPTO-ELECTRONIC LOCK

A. Basics Behind the Design.

Based on the results of an analysis of existing antitheft devices, we have come to the following conclusions:

a) A successful antitheft device should have the shortest possible electrical connections, directly interconnected with the original automotive electronic product. It is even more preferable to have all the electric connections contained within a Computer Control Module or even inside a single microchip.

b) Connections between the outside activating mechanism and the security device within the Computer Control Module must be run through a media which does not permit short or hot wiring, or any other bypass actions.

c) The activating mechanism should have a sufficient amount of coding combinations and the coding process should allow for recoding by electronic or non-electronic means.

d) The key should be easy to manufacture, durable, easy to use, and most preferably be operated like a conventional key.

We have accomplished all of the above by applying the opto-electronic principal in our design.

The Opto-Electronic Lock (see Dwg.1) consists of a housing (1), a key (2) having a cylindrical light-transmissive shank insertable into the housing, a light source (3), and photo detector (4), which is connected to the housing by the light transmissive pipes (5). The housing and the key have corresponding radial openings (6), a locating slot (8) and a locator (7), respectively.

In the Opto-Electronic Lock (see Dwg.1) a light from the light source passes through the tubular key shank and reflects sideways to the key's radial openings. When the radial openings in the key correspond to the radial openings in the housing, light passes through the light pipes (fiber-optic cable) to the photo detector and energizes a photo cell(s). The combination of the openings' positions on the key and housing and their relation to the locator and the locating slot provides the lock with its numerous coding abilities.

The Opto-Electronic lock as it is shown on the picture has over one trillion binary coded combinations.

B. Flexibility of Forms and Functions.

Even though Dwg.1 represents the opto-electronic lock, it is only a concept sketch. Its sole intent is to provide basic knowledge of the design and operational concepts. The number of forms which the opto-electronic lock can take is as immense as the number of coding combinations it can hold.

For instance :

a) Housing - the inside and outside walls of the lock can have a round, oval, polygon, or any shape. It can be a single molded or cast piece or an assembly of a few pieces with the ability to be mounted on a panel or a steering wheel column or practically anywhere.

b) Key - the inside and outside

shapes should reflect both functional and aesthetic requirements. It can be molded or cast as a single piece or made as an assembly.

c) Photo detector - the device can be made out of photo diodes, photo transistors or a photo power cell array. It can be an external or internal device in relation to the Computer Control Module or it can even be a part of the Computer Control Module microchip.

d) Light Source - it can be any of the multiple array of light emitting devices, each with a different light wave characteristic. The light source can be placed directly into the housing or as far away as needed with interconnection by the optical cable.

e) Light pipes - they can be as short as a housing wall or as long as required providing the application of fiber-optic cable.

f) Locator / Locating slot - simple representation of the locating and locking devices. The lock can be activated by simply inserting or inserting & turning or any other desired functions.

III. CONCLUSION

A. The Ultimate Antitheft Device.

As we showed in Section II, the ultimate antitheft device should have short as possible electric connections. The communication between the security device and the activating mechanism should be done through non-electric or mechanic medias. The antitheft device should have a large amount of fully protected coded combinations and at the same time it must be simple, reliable, inexpensive and preferably be a part of the car's own operational system.

<u>The Opto-Electronic Ignition Lock is that device.</u>

* The security decoding is done on the microchip level inside of the Computer Control Module.
* The communication between the security device and the activating mechanism is done by light.
* The lock, as it is shown, has over a Trillion coded combinations.
* Initial coding can be done through a special outside computer.
* All required decoding signals from the activating mechanism must arrive to the decoding device simultaneously. (Pick-lock protection)
* Because it is an Ignition lock no additional devices are needed to preclude a thief from starting the car.

B. Additional Benefits.

Since the opto-electronic lock is made for computerized equipment, in addition to superior security, it will provide car manufacturers with the following advantages:

* Being light operated, it protects sensitive equipment or items from electric noise, electric surges, electrostatic discharges, and electro-magnetic field interference.
* Allows direct computer interface due to the fact that the lock provides binary coded output signals.
* Solid state, no moveable mechanical contacts.
* Operational under adverse conditions (humidity, dust, high and low temperature, etc.).
* Conventional in operation in that it needs a key to operate; no code memorization no special procedures to activate or deactivate.
* The opto-electronic lock is more durable, simple to manufacture and operate than existing ignition locks.
* Number of parts required for the lock is under a dozen vs over a sixty for the conventional lock, thus, it is much more cost effective.

OPTO-ELECTRONIC LOCK
with
UNLIMITED COMBINATIONS

Description

The patented opto-electronic lock represents a family of solid state devices, which by their overall appearance, durability and methods of application are far superior to conventional mechanical or electromechanical devices. The device is comprised of a housing (1), a key (2) having a cylindrical light-transmissive shank insertable into the housing, a light source (3) and a photo detector (4) which is connected to the housing by the light transmitting pipes (5). The housing and the key have corresponding radial openings (6), locator (7), and locating slot (8) respectively.

Description of Operation

A light from the light source passes through the tubular key shank and reflects sideways to the key's radial openings. When the radial openings in the key correspond to the radial openings in the housing, light passes to the photo detector and energizes a photo cell. The combination of the opening's placements on the key and housing and their relation to the locator and the locating slot provides the lock with its coding abilities. The opto-electronic lock as it is shown on the picture has at least one trillion or possibly more binary coded combinations.

Description of Characteristics

* Solid state, no mechanical contacts
* Binary coded
* More durable and more convenient than conventional mechanical locks
* Exceptionally high level of reliability
* Small, flexible design allows placement of the lock directly onto a PCB (print circuit board) or as far away as is needed by means of connecting the locking device and the lock housing with electric or fiber optic cable (for electric noise protection and EMI security)
* Insulated to protect sensitive equipment from electrical surges, electrostatic discharges and EMF (electro-magnetic field) interferences.
* Multi-level access which allows usage of differently coded keys with the same housing
* Operational under adverse conditions (humidity, dust, cold, etc.)
* Operational security since there is no coding/decoding process that can be picked up or influenced by an outside device
* Simplest by design, small
* Tamper proof
* EMI free
* Conventional in operation in that it needs key to operate.
* No code memorizing
* Smart security (short wire protected, pick-lock protected, operator identification, multi-level access)
* May be recoded or reset, large number of combinations,

THE OPTO-ELECTRONIC IGNITION LOCK WILL SOLVE MOST IMPORTANT CAR SECURITY PROBLEMS.

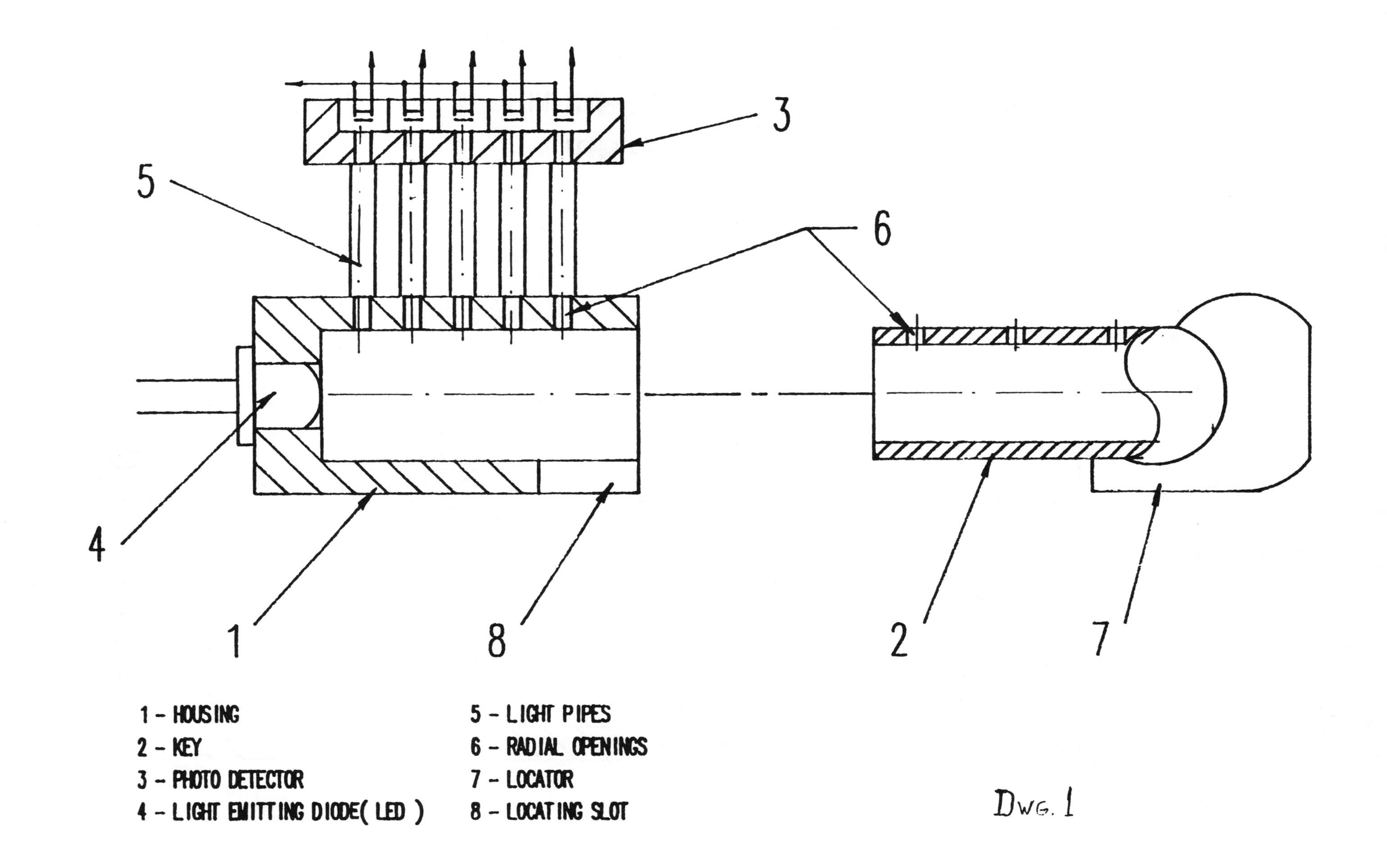
3
5
6
4
1
8
2
7
1 - HOUSING
2 - KEY
3 - PHOTO DETECTOR
4 - LIGHT EMITTING DIODE(LED)
5 - LIGHT PIPES
6 - RADIAL OPENINGS
7 - LOCATOR
8 - LOCATING SLOT
Dwg. 1
OPTO-ELECTRONIC LOCK

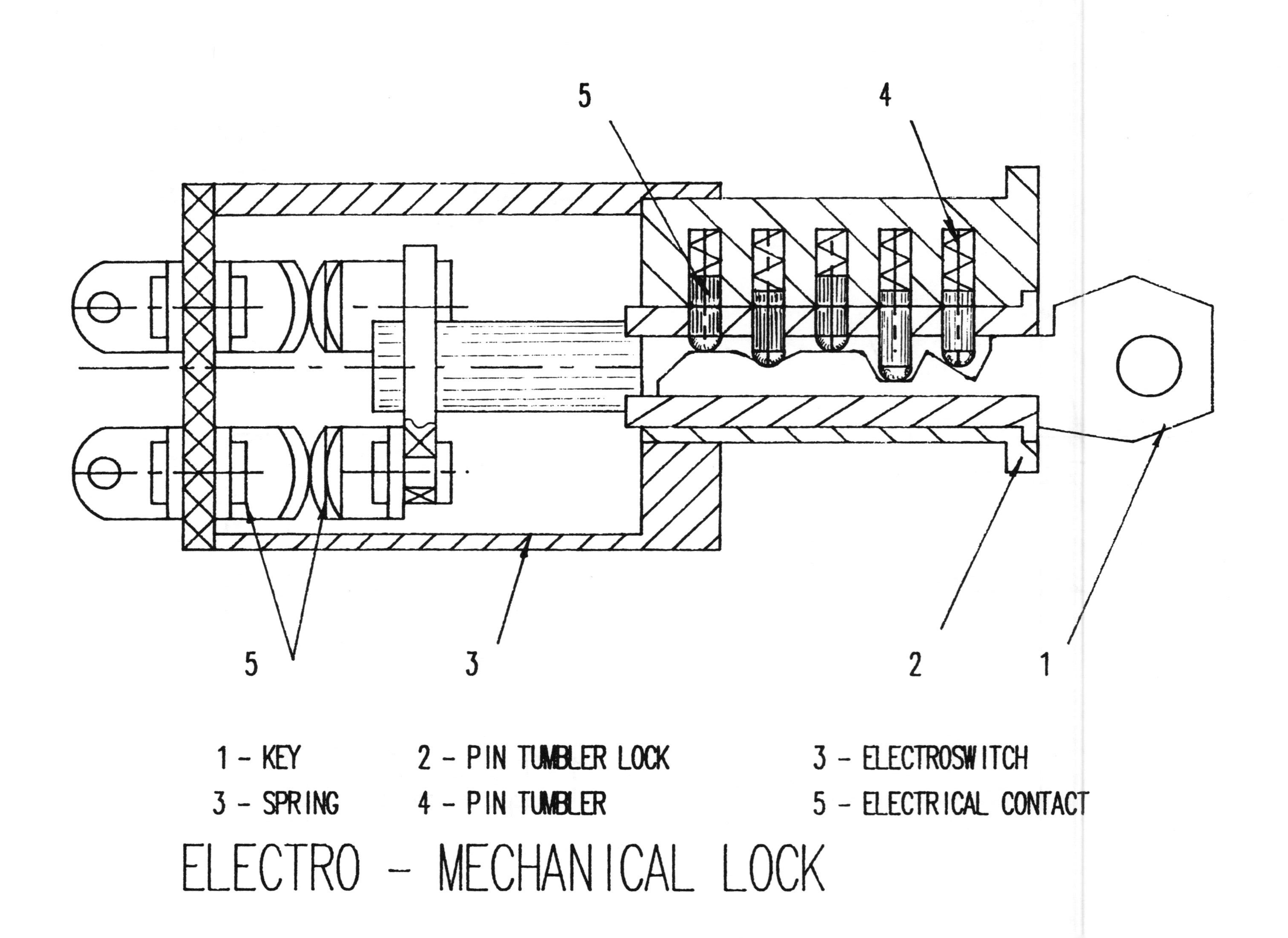
5
4
5
3
2
1
1 - KEY
2 - PIN TUMBLER LOCK
3 - ELECTROSWITCH
3 - SPRING
4 - PIN TUMBLER
5 - ELECTRICAL CONTACT
ELECTRO - MECHANICAL LOCK

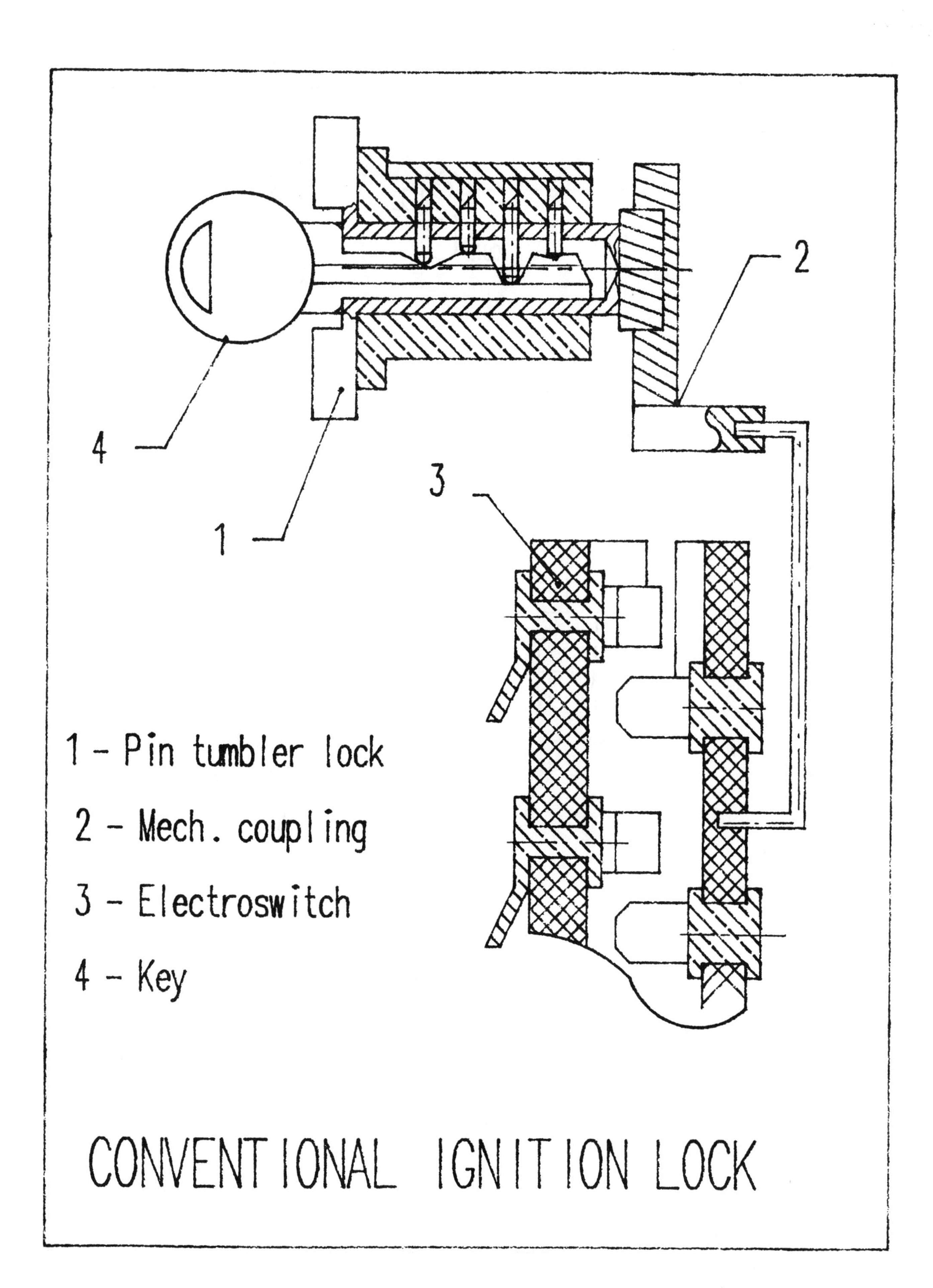
2
4
3
1
1 - Pin tumbler lock
2 - Mech. coupling
3 - Electroswitch
4 - Key
CONVENTIONAL IGNITION LOCK

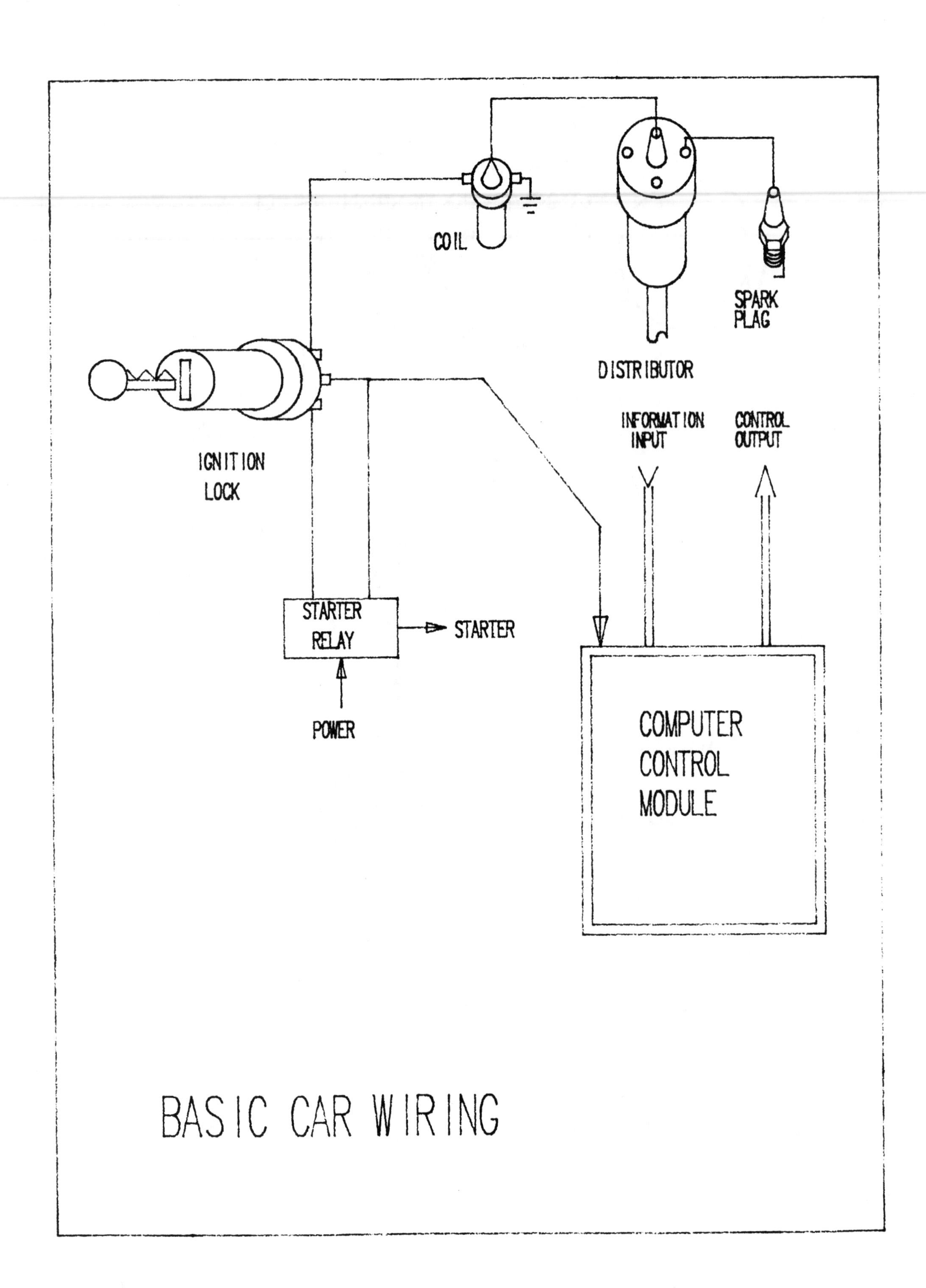
COIL
DISTRIBUTOR
SPARK PLAG
IGNITION LOCK
INFORMATION INPUT
CONTROL OUTPUT
STARTER RELAY
STARTER
POWER
COMPUTER CONTROL MODULE
BASIC CAR WIRING

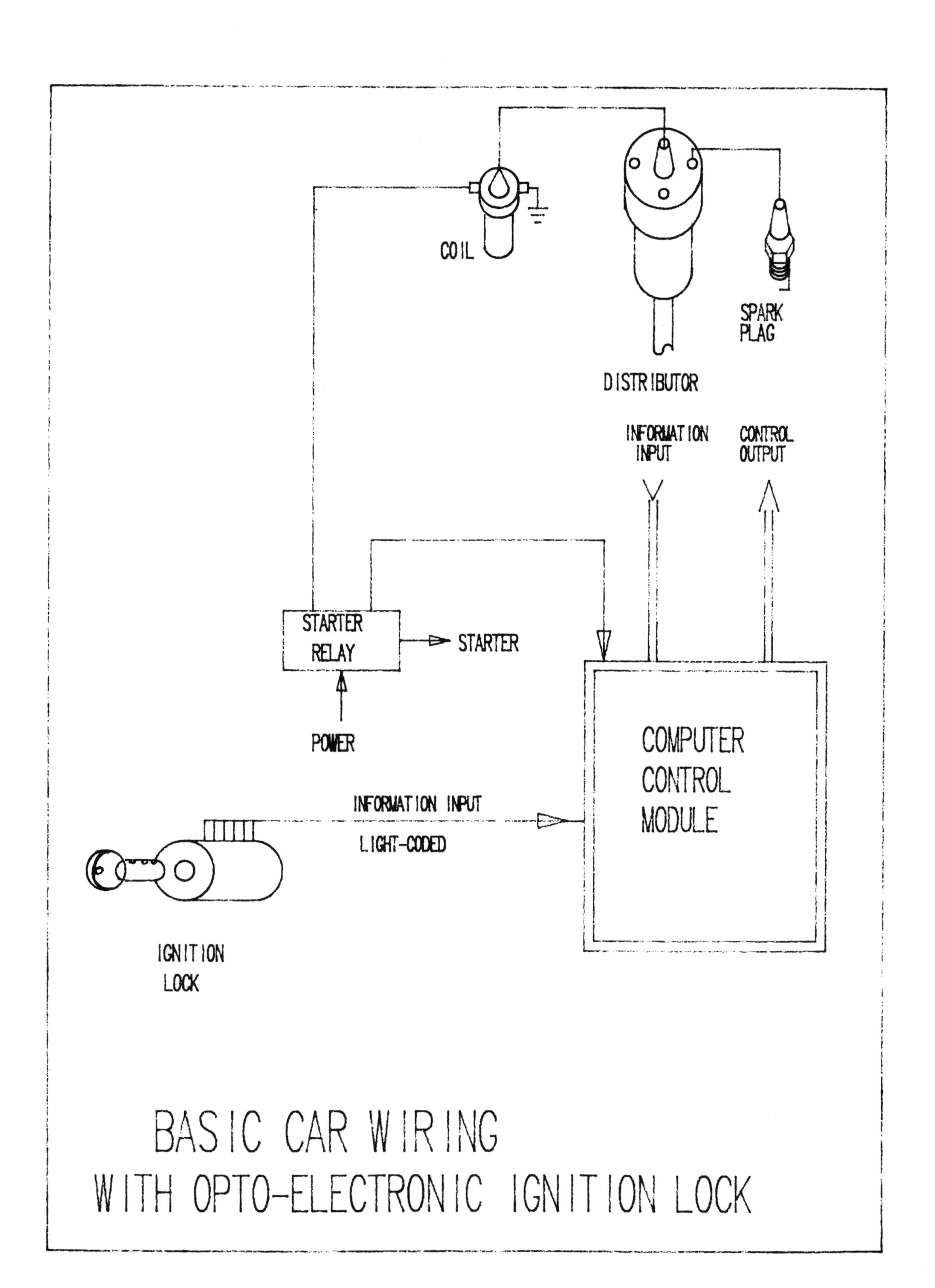
COIL
DISTRIBUTOR
SPARK PLAG
INFORMATION INPUT
CONTROL OUTPUT
STARTER RELAY
STARTER
POWER
COMPUTER CONTROL MODULE
INFORMATION INPUT
LIGHT-CODED
IGNITION LOCK
BASIC CAR WIRING
WITH OPTO-ELECTRONIC IGNITION LOCK

2000-01-1304

Safety and Security Considerations of New Closure Systems

Stephan Schmitz
Robert Bosch Schließsysteme GmbH

Jacek Kruppa
Robert Bosch GmbH

Peter Crowhurst
Robert Bosch Pty

ABSTRACT

A closure system for automotive security and driver comfort has been developed. The system combines a passive entry system and an electronic door latch system.

The passive entry system utilises a single chip transponder for vehicle immobilisation, passive entry and remote control functionality. The form factor free transponder enables the integration into a key fob or a smart card. The system can be activated by either pulling the door handle or by using a push button transponder. Due to the inductive coupling between the transponder and the vehicle mounted antennas, the vehicle door or trunk opens on successful verification as if there were no locks. Additionally, inside the vehicle, the transponder can be used as a far range immobiliser.

The electronic door latch system utilises electronically controlled latches. Symmetrical housing of the electronic latch (E-latch) and the absence of a mechanical connection to the actuators enable the latch to be used not only for the left and right side doors but also for trunk applications. The locking pawl of the E-latch is controlled by an electric motor and the functionality is entirely software dependent.

INTRODUCTION

Vehicle immobiliser systems based on transponder technology are currently used as standard equipment for vehicles. Vehicles equipped with immobilisers have impacted on theft rates, reducing the numbers of stolen vehicles. Due to the demanding security requirements, the immobiliser is gaining more and more importance. With the introduction of passive entry systems additional functionality like remote control, passive entry and vehicle immobilisation is made available to maximise customer comfort. The introduction of electro-mechanical latches provided the customer with new locking functions like central door locking (CDL), double locking (DL) and electronic controlled child safety (CS).

The new closure system described here was developed for automotive security and driver comfort and will replace existing immobilisers, UHF remote controls and mechanical locking systems. To achieve this objective, a vehicle based interrogation system is activated by pulling the door handle and starts a medium range, bi-directional signal transmission (up to 2.5 m) between the vehicle and the transponder carried by the user. Additionally, a special protocol provides long range access to the vehicle (more than 40 m), whenever the user presses a button located on the transponder. A control unit verifies the transmitted data and locks/unlocks the latches on verification.

Closure systems are commonly used in access building applications etc.. However, the practical operation of such a system in vehicle applications requires an important development effort, which has to meet the following requirements:

- The communication area of the system has to be optimised for vehicle applications. Special antennas have to be developed to provide the required operating range.
- A high data rate and a fast communication protocol in combination with a fast latch unlocking time are desired to avoid the utilisation of approach detection sensors.
- A sophisticated cryptographic algorithm has to be developed to guarantee the security of the system for driving permission, vehicle access and remote control under several attacks like statistical attacks or relay station attacks.

- The utilisation of E-latches without any existing mechanical connection and therefore back-up needs special consideration to ensure at least the same safety requirement of latches in series production today.

The paper firstly gives a brief description of the operating principles and functionality. Then we present physical properties such as the operating range and the overall timing of the closure system. In a next chapter we discuss the safety and security of closure systems.

CLOSURE SYSTEM DESCRIPTION

Figure 1 illustrates a smart card based closure system and the components involved. As soon as the user pulls the door handle, the antenna is activated and a LF (125 kHz) signal transmission between the vehicle and the portable transponder is initiated. All transponders within the communication area are identified and one transponder is selected for the security check and transmits a UHF data sequence to the vehicle. On successful verification of the transponder, the E-latch is activated and the door opens.

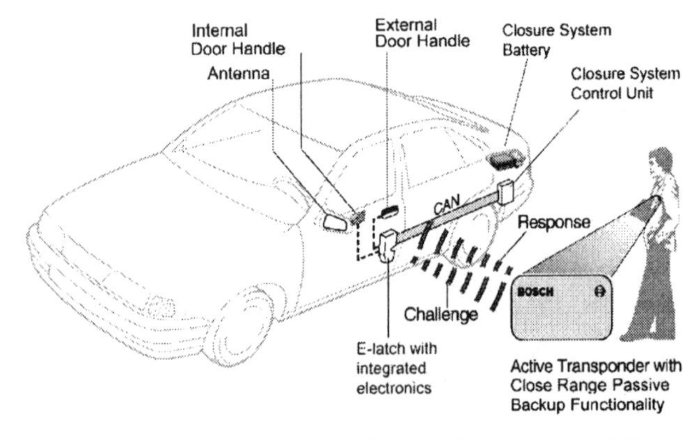

Figure 1. Closure System with passive entry and E-latches

For long range entry control, the user can have access to vehicle locking or unlocking applications from a greater distance by pushing a transponder mounted button on the key fob. A unidirectional UHF sequence is sent which wakes up the vehicle based UHF receiver. The UHF receiver demodulates the button initiated UHF transmission and sends the bit pattern to the control unit, which validates the code and determines which action is being requested (i.e. locking or unlocking the doors or opening the trunk).

In order to provide higher comfort for the end user, a special transponder memory organisation supplies the user with the ability to use his transponder for multiple application. The unique structure also allows up to three different applications (e.g. vehicles, garage door opening, building access control). For each of these applications, two different kinds of security protocols exist. Whenever the security should be at a maximum, a three pass challenge response protocol is carried out. In vehicle access applications, where the security is limited by the breaking of a window, a single challenge response protocol is utilised. For vehicle driving applications, where the security should be maximised, a three pass challenge response protocol (mutual authentication) will be carried through. The interrogation unit is able to select between the different security levels by emitting different transponder commands, which can be controlled by software.

The closure system for vehicle applications includes the following components:

CLOSURE SYSTEM BATTERY – A separate redundant closure system battery (CSB) as illustrated in Fig. 2 is used to supply the energy for the E-latches and the passive entry system. The CSB is charged by the main battery and the charge status and the condition can therefore be monitored. In failure of the CSB the vehicle's main battery is directly providing the required energy for the closure system.

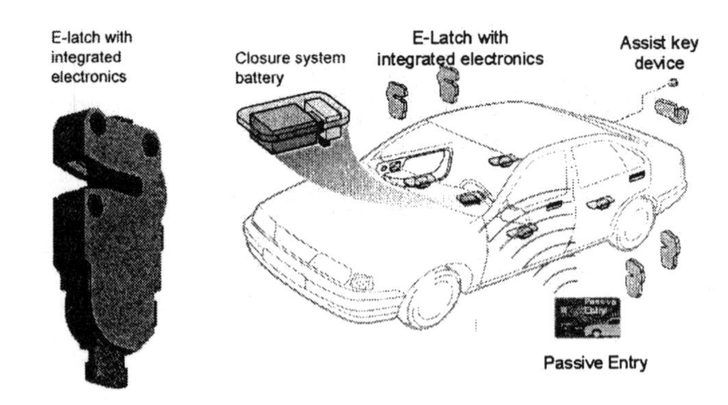

Figure 2. Closure System Battery and E-latch

E-LATCH – The E-latch as illustrated in Fig.2 with its integrated electronic is purely controlled by software and its unlocking time of 70 ms is ideal for passive entry applications. Due to its symmetric design the E-latch can be used for right side as well as for left side door applications. The integrated electronics establishes the communication to the body work bus via CAN.

TRANSPONDER – As illustrated in Fig. 3, the transponder consists of a hardwired core logic with interfaces for external components such as the antenna coil, battery, UHF transmitter and push buttons. The form factor independent packaging features of the transponder enables embedding of the transponder into a smart card or into a key head/fob.

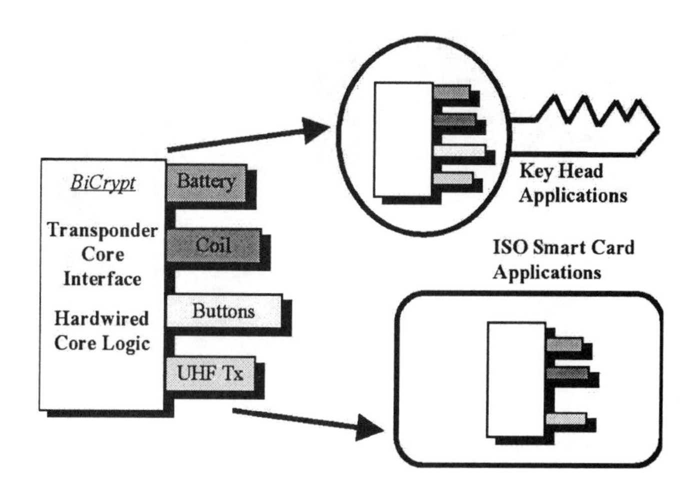

Figure 3. Transponder schematic overview and packaging features

LF- ANTENNA – A wound loop antenna is built into i.e. the driver's door mirror as shown in figure 1 for generating a rotating magnetic field.

UHF RECEIVER – The UHF receiver wakes up when an appropriate UHF sequence is received from the transponder due to an intentional activation by the user. The UHF receiver can be also woken up by the closure system electronics for passive entry applications. The vehicle's door is opened after protocol security verification.

DOOR HANDLE – In order to use existing vehicle components, a door handle without approach detection sensors initiates the passive entry system. An additional push button integrated into the door handle activates the lock process after a successful verification.

CLOSURE SYSTEM CONTROL UNIT – The control unit is connected to the body work bus (e.g. CAN) and controls the functionality of the latches, the passive entry as well as the closure system battery. The closure system control unit provides a secure power source by either selecting the vehicle's main battery or the closure system battery as the closure system energy supplier.

PHYSICAL PROPERTIES

The operating range is a critical issue for any passive entry system. Two requirements conflict. On the one hand, the range should be maximised to provide as much comfort as possible, while on the other, the range should be minimised to ensure anti-theft security. If the communication area is too large, a thief could gain access to the car, while the driver is either approaching or leaving the vehicle.

To discover the best transponder opening positions, many situations were evaluated and this resulted in us determining that the optimum range is between 2 m and 2.5 m. However, the range should be shaped more rectangularly than quadratically, with the long side in parallel with the vehicle (see figure 4).

To have the best control of communication range, we use inductive coupling between the transponder and the antenna. For very small distances compared to the wavelength, the magnetic field generated by small loop antennas decreases with the cube of the distance. It is therefore well suited for data transmission, where the range has to be controlled.

Another advantage of inductive coupling is the possibility to transfer energy at close distances via the magnetic field. This is necessary in a flat transponder battery scenario. Ideal places for mounting an antenna are i.e. the driver and passenger side mirrors and the rear bumper.

The overall timing for the vehicles' door opening is a critical issue. Mechanical door locks, as used today, provide a door handle which has to be pulled to open the door. Tests with random users have revealed an opening time from as fast as 60 ms to as long as 300 ms. Most car manufactures require, that there is no noticeable hesitation between pulling the door handle and opening of the door. Several tests have revealed that the overall timing from pulling the door handle to opening of the door should be less than 130 ms. Therefore it is essential for any closure system to have fast communication and opening times. With a communication time (anti-collision and challenge response) of 44 ms and an E-latch opening time of 70 ms the overall action time of less than 130 ms as illustrated in Fig. 5 can be achieved. Any action time with more than 130 ms demands a different utilisation concept or the installation of approach detection sensors in order to start the communication protocol before the door handle has been pulled.

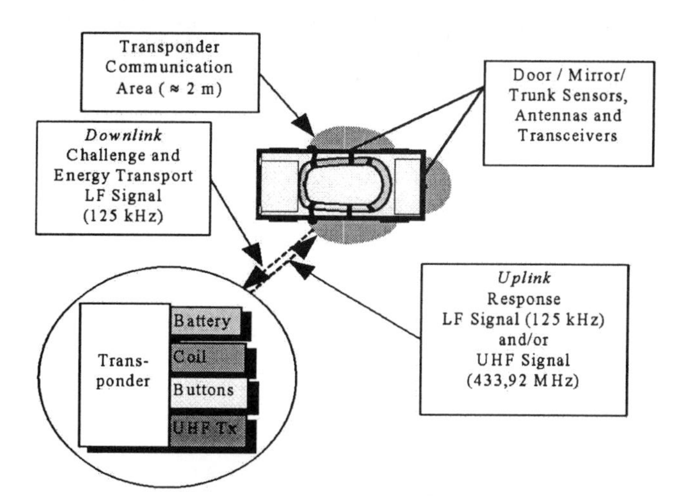

Figure 4. Schematic overview of passive entry system communication features

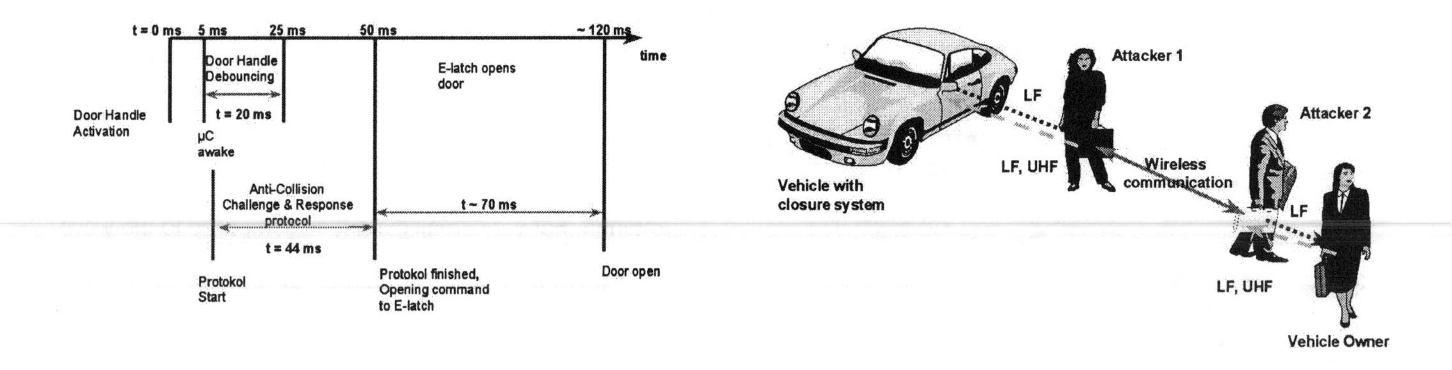

Figure 5. Closure system timing overview

By using the E-latch in combination with the fast protocol customer requirements without utilising approach detection sensors can be fulfilled.

SECURITY CONSIDERATIONS OF A CLOSURE SYSTEM

In a closure system the two subsystems, locking system and the passive entry system, have to be discussed. The security of an E-latch with its integrated electronics provides a new level of security. The sealed housing and the missing mechanical connections between latch and the actuators like door handle prevent any mechanical attack on the latch. Due to the integrated electronics and the CAN communication the closure system control unit and the E-latch have the possibility to communicate via encrypted messages. The security presented by the E-latch reveals new kind of security levels in the vehicle architecture, which have yet not been reached.

The passive entry system can be regarded as a far range immobiliser. It is a fact that the various implemented cryptographic algorithms in the immobiliser function of today have not been the weakest part in the vehicle security chain. Most of the time the security architecture, the secrete key storage and the service concept are prone to a successful attack. It can be assumed that in addition to the above the long distance communication of a passive entry system might reveal new attack possibilities for a non trusted party. A possible threat to a passive entry system might be the so called relay station attack as illustrated in fig. 6.

Figure 6. Schematic illustration of the relay station attack

The vehicle with the closure system is under attack by attacker 1, who has initiated the vehicle communication. The attacker 1 receives and demodulates the transmitted data and sends the data by wireless communication to an attacker 2. The attacker 2 demodulates the data and sends the data on the original vehicle frequency to the transponder bearer, the vehicle owner. The transponder processes the received data and sends back his answer. The answer is transferred via attacker 2 and 1 to the vehicle. The transponder's response was only transmitted and not further processed and the vehicle authenticates the data stream and the vehicle opens the door. The same procedure would start, if the attacker would use the relay station attack for passive go. As a consequence the attacker would be authorised to drive the vehicle.

Several counterfeiter possibilities exist. Of course one would think directly of measuring the time of flight of such an signal as this is already well know technology for military and scientific applications (i.e. radio/light detection and ranging). Another solution would be to measure the pulse build up time in a resonance circuit as this is routinely done for laser tuning.

Whereas in principle the time of flight counterfeiter is in principle possible, the application to the automotive market with its cost pressure seems to be quite unlikely, especially when a time of less than a few nano-seconds (i.e. 30 ns) should be measured. The measurement of additional pulse build up times in resonance circuit can be easily achieved with existing technologies. However, the attacker may use low Q-circuits with high power antenna drivers to minimise the additional resonance build up time. As a resume both counterfeiters are not practical for automotive applications.

A feasible counterfeiter would be to use of a non linear approach by the emission of two tones by the transponder of the vehicle owner. The two tones are emitted simultaneously and caused by the low power emission, the communication is limited to less than 5 m. For any distance greater than 5 m the attackers have to amplify the 2 tones with a defined gain. As it is well known that each amplification can be separated into a linear and non linear part we will receive third order distortion as inband lines in the vehicle receiver as illustrated in fig. 7.

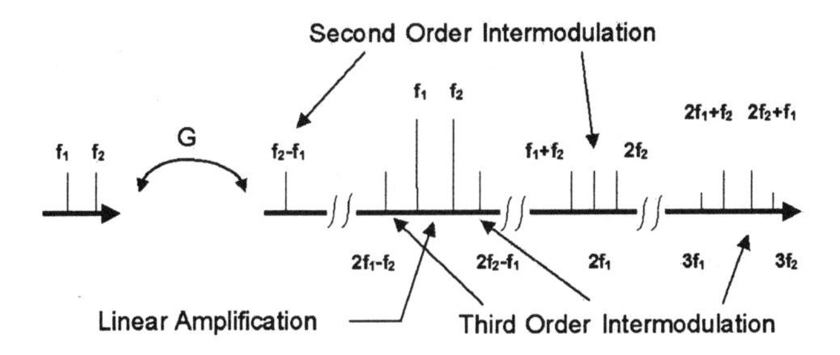

Figure 7. Transformation of two tones by non linear amplification (G) into the linear and non linear terms.

If two tones have been received and emitted by a relay station, the necessary amplification (G) in the relay station causes second order and third order intermodulation. As can be seen from the figure above the third order intermodulation terms are falling into the bandwidth of the receiver by selection of a suitable two tone frequency spacing. The figure 8 presents a typical example of the received spectrum of a two tone emission after passing the amplification (G) of relay station. The distortion in the specturm is caused by third order intermodulation.

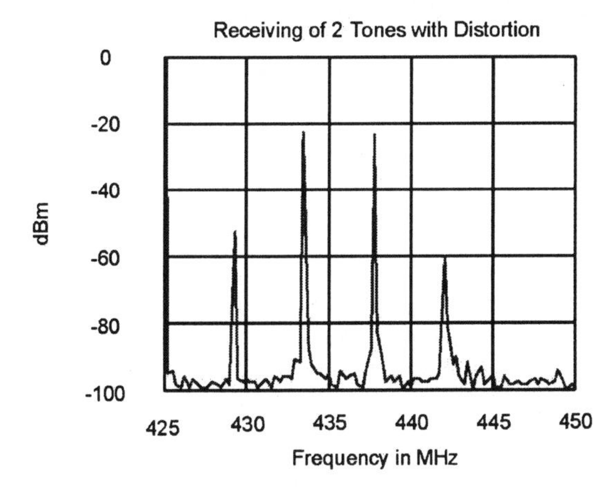

Figure 8. Receiving of 2 tones with distortion after passing a non linear amplification.

Several hardware realisations of a two tone test are possible to implement the relay station counterfeiter which are left to the reader.

SAFETY CONSIDERATIONS OF CLOSURE SYSTEM

As soon as a pure electronic closure system is discussed the reliability of such a system moves into the focus of interest. Especially for the electronic latches the reliability has not only to be considered during the warranty time of the vehicle, but also to be discussed after a life time of 10 to 15 years. An appropriate way to investigate the reliability of a closure system is i.e. a fault tree analysis (FTA).

We are especially interested in the probability that a critical event as described below will occur. The following critical events have been identified.

ENTER – This event describes the malfunction of one of the locked doors, which can not be opened from outside. The consequence of such an event is that the person has to use another door.

EXIT – In analogy to the event enter, this malfunction describes the event that one of the doors can not be opened from inside. As a consequence the person has to use another door.

RESCUE – Much more severe is the event rescue. Here a person in the car shall be rescued i.e. after a crash. The consequence of such an event is that none of the doors can be opened from outside and the person is trapped in the vehicle. This events can result into a danger for the person in the vehicle.

ESCAPE – The event escape is comparable to the event rescue. Here the person wants to leave the car i.e. after a crash and none of the doors can be opened from inside. The person is trapped in the vehicle and the malfunction can result into a danger for the person in the vehicle.

SELF RELEASING – One of the doors releases itself and the door opens. The driver may feel threatened, if the seatbelt is not fastened and this scenario may result into a dangerous situation.

We have laid out the architecture of a closure system with a electro-mechanical latch in series production today. The failure ratios of the components have been derived from the field experience. By undertaking an analysis of the possibilities one can derive the failure tree which will lead to the events described above.

Following this approach we have defined the architecture for an E-latch based closure system and used the failure probabilities of the components of the electro-mechanical latch to derive the failure probability of the critical events.

We have normalised the failure probabilities of the electro-mechanical latch based closure system to 1 and compared the results with the E-latch based closure system as presented in fig. 9. For each closure system the fault tree analysis is carried out to derive the probability of the cirtical events.

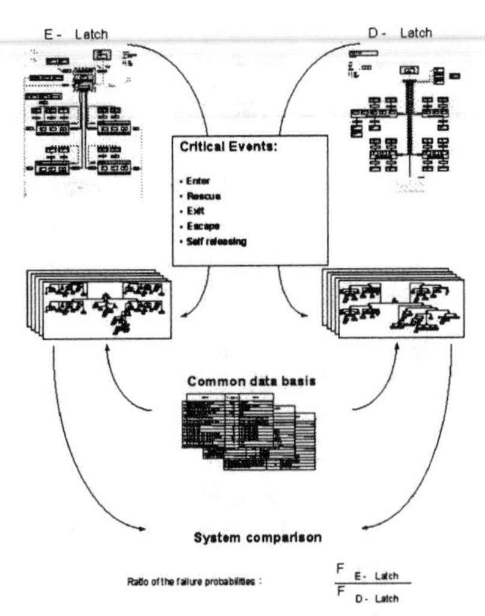

Figure 9. Schematic description of the comparison between an electro-mechanical latch and E-latch based closure systems.

By studying the fault tree and isolating the critical paths with the failure probabilities of the components involved the relevant components can be optimised to achieve a very low failure probability. By optimising the actuators, sensors and the closure system energy supply one can optimise the system's reliability. In table 1 the electro-mechanical and the E-latch based closure system are compared.

Table 1. Normalised comparison of the electro-mechanical and the E-latch based closure system failure rates.

Critical Events	Electro-mechanical latch based closure system	E-latch based closure system
Enter	1	0,12
Rescue	1	0,0003
Exit	1	0,71
Escape	1	0,61
Self Release	1	0,79

As a result it can be stated that the safety of the E-latch based closure system is even better than the electro-mechanical latch based closure system.

CONCLUSION

We have described a new closure system. The system combines a electronic E-latch and a passive entry system. The introduction of E-latches offers the automotive manufacturer only one latch for right, left side and trunck applications with the advantage that the functionality is purely controlled by software to minimise application work. The comfort for the end user is increased due to the feature of passive entry and the forceless pull of the door handle due to the missing mechanical connection by the E-latch.

To satisfy the increased security requirements, we have shown that the E-latch and the passive entry system offer a new kind of security level. The sealed E-latch housing and the possible bus connection due to the integrated E-latch electronics even offer the possibility to use encrypted communication between a central ECU and the latches. The utilisation of challenge response protocols in a passive entry system ensures the same security of the protocol as in today's banking system. Possible attacks like the relay station attack are overcome by simply emitting two tones and verifying the third order intermodulation lines, which can be received in the bandwidth of the UHF receiver.

The normalised comparison of a closure system based on electro-mechanical latches in series production today and the utilisation of E-latches have been presented. As a result it can be stated that the safety of the E-latch based closure system is better than an electro-mechanical closure system.

REFERENCES

1. Balanis, C. A., Antenna Theory, 2nd ed., Wiley, New York, 1997.
2. Hirano M., Takeuchi, M. Tomoda, T., Nakano, K.-I., "Keyless Entry System With Radio Transponder", IEEE Trans. Ind. Electr., Vol. 35, No2, 208-216,1988.
3. Schmitz, St. and C. Roser, "A New State-of-the-Art Keyless Entry System", SAE Technical Paper Series; 980381, 1998.

2000-01-0978

Development of Automatic Climate Control with Neural Control

Yuichi Kajino, Hikaru Sugi, Takayoshi Kawai, Yuji Ito and Masahiko Tateishi
Denso Corp.

Katsuhiko Samukawa
Denso Automotive Deutschland GmbH

ABSTRACT

The automatic climate control system has been developed to improve cabin thermal comfort. However it is getting very hard to develop more comfortable and high-level system with current techniques. This report introduces a completely new automatic climate control system utilized an applied "Neural Network" to achieve highly flexible thermal comfort.

INTRODUCTION

Automatic climate control has been developed year by year to meet a wide variety of requirements. Recently, with growing requirements for better cabin thermal comfort, more functional air conditioning system and a shorter development period, the advent of automatic climate control with higher functions is awaited. In order to meet those requirements, we have put on the market zone air-conditioning systems. which can be represented by right-and-left and up-and-down independent control and in-front and in-rear dual independent control.

On the other hand, these control systems were based on the TAO control (linear control) devised in the 1980s, and it was capable of realizing the higher functions only by adding plural correction formulas to it. Facing, the above mentioned requirements, however, we had to admit that the TAO control had limitations.

We thought that a more free control system capable of supporting functional improvement of air conditioning was necessary. Then we newly applied neural network control (hereinafter called "neuro control"), a nonlinear control system, to the automatic climate control. The following is our report on the good result we obtained in this respect.

REQUIREMENT OF ADVANCED CONTROL

Generally, the temperature control in a cabin is used with the air temperature of outlets. It is decided with thermal load balance between the cabin and outside (ambient temperature and sun load). Therefore, its formula is linear. The required outlet temperature is called TAO. The blower and outlet control synchronizes with TAO. (Fig.1)

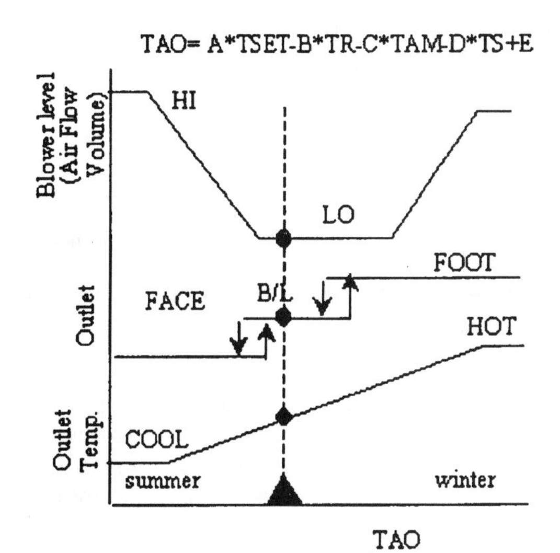

Fig.1 Cabin Temp. Control

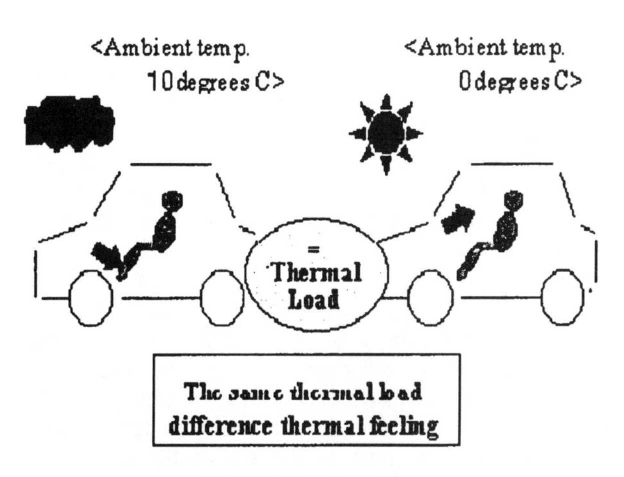

Fig.2 Feeling Result

However, passengers can experience different feeling with the same thermal load. In spite of the same thermal load, different outputs are required. For example, Feet and Face mode. (Fig.2)

NEURAL NETWORK CONTROL

Fig. 3 shows the structure of neuro control. Neuro is a mathematical realization of information processing in living creatures' nervous system consisting of neurons, imitation of living creatures' nerve cells, arranged in layers, by which output is determined corresponding to input.

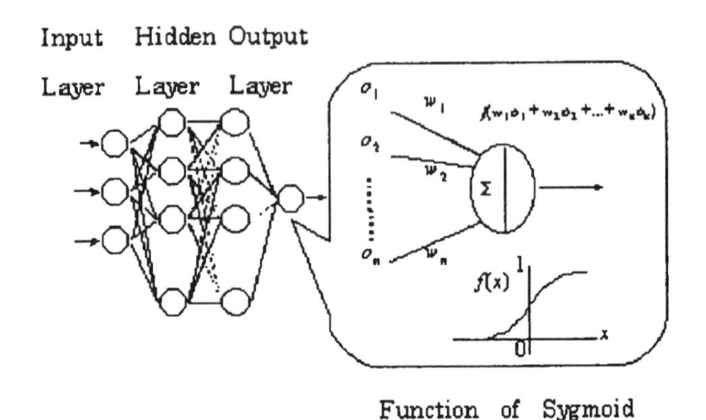

Fig.3 Architecture of neural network

It functions through repeating a simple calculation of inputting the output from plural neurons and outputting through a nonlinear function called a sigmoid function. Then, by doing so, it produces a nonlinear output as a whole. Neuro control is characterized by the capability for a self-creating coefficient of correlation when given large amount of input and output data. Fig. 4 shows the process of applying this neuro control to air conditioning.

First, training data necessary for creating a coefficient of correlation for the neuro are produced from the air-conditioning target characteristics, which are obtained through on-vehicle tests. This training data is an aggregation of input and output data consisting of "this is the output corresponding to that condition (input)". Based on this data, a coefficient of correlation of the neuro is created by means of the neuro's learning function. Then, this coefficient of correlation is incorporated into the air conditioning ECU. This evolves into a product, which will satisfy the initially given air-conditioning target values. In this neuro control, therefore, the given training data themselves correspond to the control norms, namely the conventional control logic.

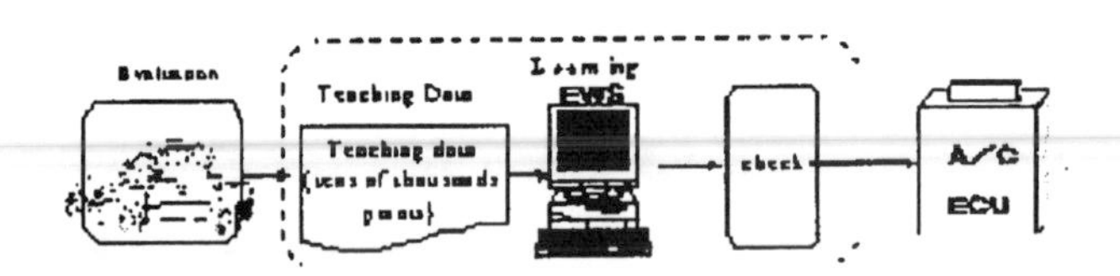

Fig.4 Process of application

EFFECT OF NEURO CONTROL

We applied neuro control to the blower-level control which is one of the basic controls in air conditioning. In this application case, we will show an example of compatibility between blower power-increase at the middle ambient conditions and increased comfort under other conditions, which were formerly never achieved.

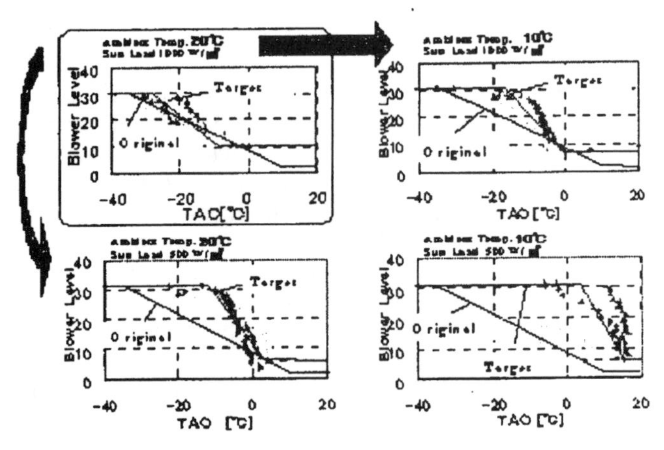

Fig.5 Target

Fig. 5 shows some graphs obtained from several panelists indicating comfortable blower-level patterns in mild ambient condition with the ambient temperature at 10 to 20℃. Some control patterns required by passengers are widely different from the conventional ones. That is, immediately after getting in the vehicle, the blower level is low despite the high cabin temperature caused by solar radiation due to the fact that the conventional control system has only one blower-level pattern. If, furthermore, you manage to adjust the calibration in such a way as to satisfy the control pattern in mild ambient conditions under the conventional control system, you will suffer from excessive blower-level in the summer season (ambient temperature approx. 30℃). Then we see that we can satisfy the passengers if we give higher blower-level than the conventional one in mild ambient condition and the same blower-level as the conventional one in the summer season. We then give the above fact to the neuro as training data and make it conduct learning.

The characteristics are shown in Fig. 6. In this example, the sun load is fixed at 500W/m², the x-axis indicates ambient temperature, the y-axis cabin temperature and the z-axis indicates blower voltage. The blower level is boosted up when the ambient temperature becomes 10 to20℃

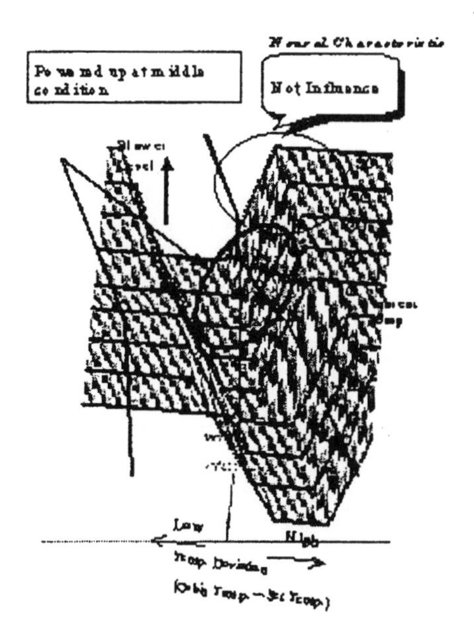

Fig.6 Characteristic of blower control

ON-VEHICLE FEELING TEST

After incorporating this system into the actual ECU, we conducted an on-vehicle feeling test. Fig. 7 shows the neuro-controlled average cabin temperature and blower voltage under the environmental condition with the ambient temperature at 20℃ and the sun load at 500W/m². It also shows the data compared with the conventional control.

Under the neuro control, higher blower-level than under the conventional control can be observed, as was intended, in the initial stage of cooling down (Section A, Fig. 7), and the same steady-state blower-level as the conventional one is realized. The passenger comfort was also improved. Furthermore, under other conditions, we obtained a result in which the cabin temperature became stabilized in a shorter period of time. As the result of conducting feeling tests of these effects with different panelists than the ones who were previously asked to judge comfort, improvement in overall comfort was confirmed. (Fig. 8)

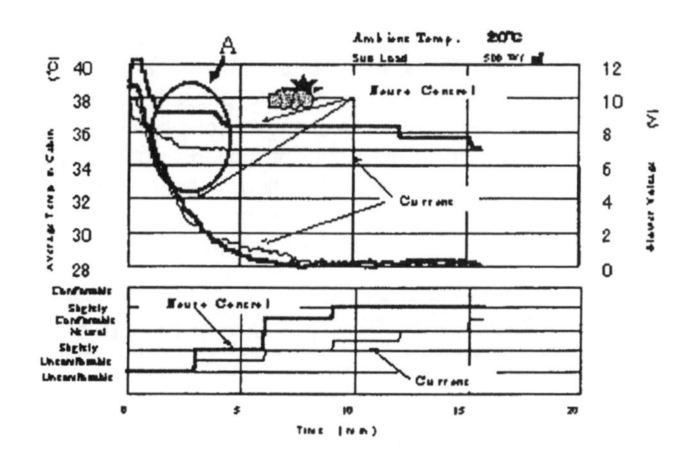

Fig. 7 Vehicle Test

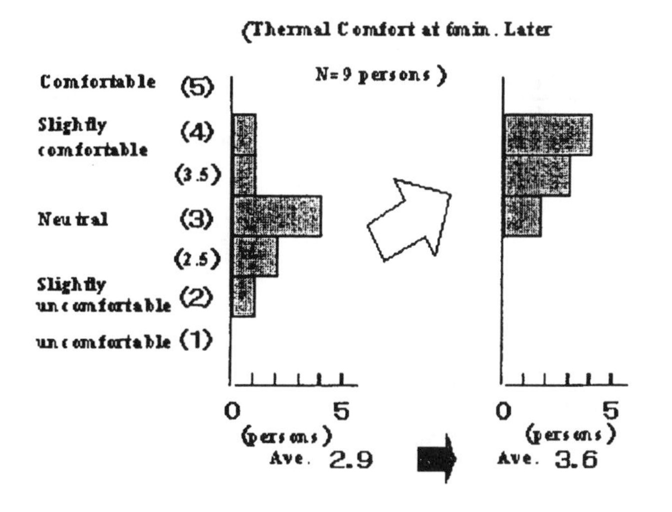

Fig.8 Feeling Test

OTHER EFFECTS

As an example of other comfort improvement effects, Fig. 9 shows an application case of a system which, having set-temperatures in the right and left zones respectively, enables independent temperature control in the right and left cabin space. Previously, in order to control independently the right and left cabin temperature according to the set temperature, we added plural correction terms to the TAO formula, which is the basic formula for control. However, with these correction terms, being a combination of linear formulas, we had a problem that, if we satisfied independent controllability under a certain condition, we could not satisfy it under other conditions.

On the other hand, under the neuro control, where nonlinear characteristics are obtainable, it is possible to change characteristics of only those environmental conditions which we want to change without affecting other conditions. Consequently, it is possible to satisfy the target specification under every environmental condition .A partial change being possible, we had to carry out a test under the changed condition only, which also contributed to man-hour reduction. Formerly, even in the case of a partial alteration in specifications, we used to conduct the entire test over again to confirm that other conditions were not adversely affected.

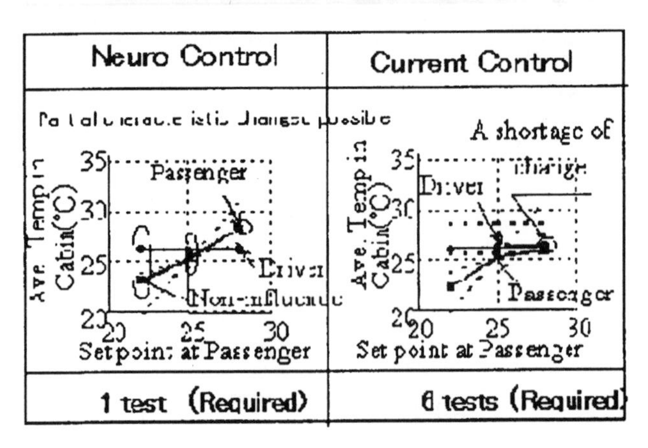

Fig. 9 luation of Independent temperature control

CONCLUSION

We carried out application development of the neuro control as a new control system for automatic climate control. As a result, through applying the neuro control to the control of the blower level and air outlet temperature, we realized two things, which are difficult to achieve at one time. First, we achieved blower power-increase at mild ambient conditions while maintaining current blower-level control for cooling down in the summer season which was previously deemed impossible. Furthermore, we improved the controller's ability in characteristics of the right and left independent control, and achieved a reduction in man-hours necessary for controller development. This system was installed on the TOYOTA "PROGRES", which was put on the market on May, 1998.

2000-01-0084

Vehicle Cabin Air Quality Monitor for Fatigue and Suicide Prevention

Kosmas Galatsis and Wojtek Wlodarski
Royal Melbourne Institute of Technology University

Brian Wells
Australian Automobile Association

Stewart McDonald
Royal Automobile Club of Victoria

ABSTRACT

Low oxygen, high carbon monoxide and carbon dioxide concentrations can typically exist within a vehicle cabin. Such poor air quality may cause drowsiness, fatigue, impairment of judgement, and poor coordination to vehicle occupants. Also, many deaths are caused by motor vehicle exhaust gas suicides from carbon monoxide poisoning. The introduction of an air quality monitor within the vehicle cabin can alarm occupants preventing any adverse health effects. A vehicle cabin air quality monitor was designed and developed. The monitor was tested under various driving conditions and simulated suicide attempts. Alarms are triggered when poor air quality exists within the vehicle cabin.

INTRODUCTION

The quality of air inside enclosed spaces has become a matter of growing concern. A confined space that has received little attention is the vehicle cabin environment. Commonly, vehicle occupants choose to operate the vehicle ventilation system in the "recycle" mode to prevent outdoor-polluted air entering the vehicle cabin. However, it too prevents fresh air from entering the cabin, resulting in a decrease of oxygen (O_2) and an increase in carbon dioxide (CO_2) gases. Such poor air quality may cause headaches, drowsiness, fatigue, impairment of judgement, and poor coordination to vehicle occupants [1,2].

Furthermore, thousands of fatal unintentional carbon monoxide (CO) poisoning's occur annually, in and around motor vehicles. In the US in 1996 there were 1508 suicides and 219 accidental deaths caused by CO poisoning from motor vehicle exhaust gas [3]. Although the level of CO in exhaust gas from catalyst equipped cars is extremely low, it is still sufficient to cause death. Before the introduction of catalytic converters, the exhaust of new automobiles frequently contained 7-12% CO during engine idling [4]. Newer cars with emission control devices tend to emit CO levels at idle operation considerably below legal limits, typically less than 0.1%, however may still contain 2-4% of CO [4].

By continuously monitoring the cabin air environment, adverse human health effects resulting from poor cabin air quality can be reduced. Alarms may be set if poor air quality is detected. Visual and audible alarms may provoke the occupants to lower the windows or allow fresh air to enter the cabin via the ventilation system. If such warnings are ignored, the system may automatically switch the ventilation system to "fresh air" mode, until adequate fresh air has entered the cabin. By analysing the carbon monoxide and oxygen concentration profiles a suicide attempt can be detected. In such a case, the engine could be shut down and the electric windows lowered if the vehicle is stationary.

A vehicle cabin air quality monitor has been developed. The monitor alerts vehicle occupants when poor air quality exists. The specific aims of this study have been the following:

- To develop a vehicle cabin air quality monitor.
- To gain an understanding of vehicle cabin gas behaviour in various driving conditions and suicide attempts.
- To clearly identify driving conditions which may produce poor cabin air quality.
- To determine adequate alarm thresholds and sensing algorithms for the air quality monitor.

BACKGROUND

Gas may enter the vehicle cabin through the inlet ventilation system, door and window seals, outlet ventilation slots located at the rear of most cars, and panel holes that may exist. However, the transportation mechanism of the gas into the cabin is affected and influenced by several factors. Phenomena like the diffusion rate, molecular weight, turbulence effects, velocities and temperature gradients of gas effect the gas mixing properties, and ultimately the instantaneous concentration of a gas within an enclosed environment like a vehicle cabin. Furthermore, turbulence is generated at high flow rates and macroscopic mixing of gas takes place, particularly where a gas flows past rough or irregular surfaces, like the edges of a vehicle door panel [5].

OXYGEN (O_2) – The atmosphere contains 20.9% of oxygen. Unless there is adequate ventilation the level is reduced surprisingly quickly by breathing and combustion as in a vehicle cabin. People suffering from low oxygen levels considered to be below 19.5%, have no idea they are clumsy and thinking slowly. In fact they behave much like someone who is intoxicated with alcohol showing a measurable impairment of judgement [6]. Oxygen can be sensed by using solid state and electrochemical sensing mechanisms.

CARBON MONOXIDE (CO) – Carbon monoxide is an odourless and toxic gas produced by the incomplete combustion of carbonaceous materials. When inhaled, CO binds to blood haemoglobin, which reduces the oxygen-carrying capacity of the blood. The affinity of CO to haemoglobin is 200 to 250 times higher than oxygen [2,7]. Symptoms like headaches, nausea and dizziness are experienced. CO can be sensed using electrochemical, semiconductor and optical sensors.

CARBON DIOXIDE (CO_2) – Carbon dioxide is present in the atmosphere at about 400 ppm (0.04%). It is a colourless, odourless, noncombustible gas [2]. When exhaling, carbon dioxide and water vapour are emitted. Furthermore, carbon dioxide is produced during combustion and in brewing and other fermentation processes. The lowest level at which CO_2 effects has been observed in both humans and animal studies is about 1000 ppm. At concentrations above 1500 ppm respiration is effected and breathing becomes faster and more difficult. Concentrations above 3000 ppm can cause headaches, dizziness and nausea [2]. Non-dispersive infra red (NDIR) is the usual sensing technique for CO_2 measurement.

GAS EXPOSURE LIMITS – Acceptable workplace limits for air pollutants are published by numerous regulatory and quasi-regulatory bodies both in the U.S. and abroad. These include the Occupational Safety and Health Administration (OSHA), National Institute for Occupational Safety and Health (NIOSH), and the World Health Organisation (WHO) [1]. The NIOSH recommended exposure limits and OSHA permissible exposure limits are listed in Table 1.

Table 1. Common gas safety limits.

Gas	WHO	OSHA [8]	NIOSH [8]
O_2	-	19.5%	19.5%
CO	9 ppm [4]	50 ppm	35 ppm
CO_2	1000 ppm	5000 ppm	5000 ppm

VEHICLE CABIN AIR QUALITY MONITOR

HARDWARE – The monitor developed continuously analyses the oxygen and carbon monoxide concentration of the vehicle cabin. The vehicle cabin air quality monitor consists of the following primary components:

- Oxygen Sensor (Resolution: 0.1%; Accuracy: 1%)
- Carbon Monoxide Sensor (Resolution: 1 ppm; Accuracy: 10%)
- Temperature Sensor
- Signal Processing Circuitry
- Analog to Digital Converter
- Microprocessor
- RS-232 Interface

All components are integrated onto a PCB (Printed Circuit Board). A PC was connected to the monitor via the RS-232 interface used for datalogging. The microprocessor performs the analogue to digital conversion and the control and processing functions of the monitor. The oxygen sensors have been calibrated with certified oxygen bottles and the carbon monoxide sensor with a commercially available CO monitor.

SOFTWARE – The monitor samples the sensors every second. Warning signals are activated when gas concentrations are exceeded. The signals can be used to trigger audible and visual devices or can be used as input into the engine and body management systems for advanced control functionality. Table 2 illustrates the alarm thresholds used as recommended [4,9].

Table 2. Monitor alarm thresholds.

Gas	First Warning	Second Warning	Final Warning
O_2	19.5%	17%	16%
CO	25 ppm	50 ppm	100 ppm
CO_2	1000 ppm	-	5000 ppm

EXPERIMENTAL

The purpose of the experiments is to determine vehicle cabin gas behaviour and concentrations in various driving conditions and in suicide attempts. Three main experiments were conducted. First, the monitor was tested with four popular Australian sedan automobiles:

1. Ford Falcon (1996, 4.0 litre)
2. Holden Commodore (1995, 3.8 litre)
3. Mitsubishi Magna (1995, 3.0 litre)
4. Toyota Camry (1997, 2.2 litre)

Each vehicle was tested under 4 typical driving scenarios:

1. City driving with "recycle"
2. City driving with "fresh air"
3. Motorway driving with "recycle"
4. Motorway driving with "fresh air"

The monitor was placed in the middle of the instrumentation panel. The concentration versus time profile was measured in each test for a 20-minute duration. There were two vehicle occupants. The purpose of this test was to deduce the CO and O_2 levels within each driving scenario. By understanding typical concentrations that exist within a vehicle cabin, appropriate alarm thresholds can be developed for fatigue warning.

Second, the maximum CO concentration that would typically exist in a vehicle cabin must be known, so that suitable alarm thresholds are set to prevent a false suicide alarm. The Mitsubishi Magna was used to monitor CO concentrations whilst driving in city traffic with both front windows lowered. CO profiles were also obtained in undercover carpark areas and whilst occupants were smoking in the vehicle cabin.

Third, the Ford Falcon was used for simulated suicide testing. The vehicle was started and left running until the coolant temperature had stabilised. A hose of 30mm diameter was connected to the exhaust outlet of the vehicle and then directed into the vehicle cabin. A back door window was lowered 4 inches to make this possible. The speed of O_2 depletion and CO increase must be known to devise an algorithm that would identify a suicide attempt.

RESULTS

OXYGEN AND CARBON MONOXIDE AUTOMOBILE TESTING – Table 3 shows the minimum cabin O_2 concentrations detected.

Table 3. Minimum O_2 concentrations detected.

	City	Motorway	City	Motorway
	Recycle		Fresh Air	
Ford Falcon	20.5%	20.7%	20.5%	-
Holden Commodore	20.2%	20.4%	20.4%	20.5%
Mitsubishi Magna	19.2%	20.3%	20.0%	20.6%
Toyota Camry	19.1%	19.4%	20.2%	20.4%

It was found when driving in the "city with recycle" the lowest average O_2 of 19.75% was detected with a minimum O_2 depletion of 19.1% occurring. The best case driving scenario was "motorway driving with fresh air"; recording an average of 20.5% O_2. Studies have shown [10] that carbon dioxide and water vapour from respiration displaces the oxygen. Hence, the depletion of oxygen and the increase in carbon dioxide is a complimentary effect. Figure 1 illustrates typical oxygen depletion due to vehicle occupant respiration.

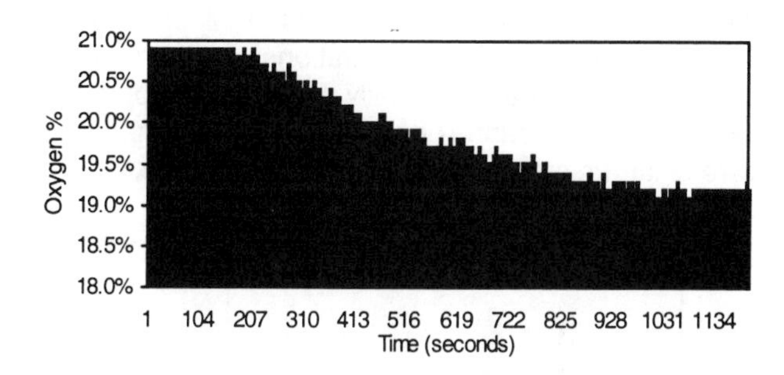

Figure 1. Typical vehicle cabin oxygen depletion.

The rate of fresh air entering the vehicle cabin is largely determined by the average speed of the vehicle. Other variables that influence the air quality within the vehicle are the number of occupants, cabin volume, ventilation system throughput, exit ventilation throughput, and the door and window seal efficiency. Table 4 shows the maximum CO detected. The maximum CO concentration reached was 16 ppm.

Table 4. Maximum CO concentrations detected.

	City	Motorway	City	Motorway
	Recycle		Fresh Air	
Ford Falcon	13 ppm	9 ppm	7 ppm	-
Holden Commodore	7 ppm	<1 ppm	7 ppm	11 ppm
Mitsubishi Magna	8 ppm	<1 ppm	16 ppm	9 ppm
Toyota Camry	<1 ppm	<1 ppm	<1 ppm	8 ppm

MAXIMUM CARBON MONOXIDE DETECTION – To prevent false suicide alarm triggering, and for a comprehensive understanding of ambient CO concentrations, the monitor was tested whilst both front windows were lowered in peak traffic conditions. A maximum of 30 ppm CO was detected. Congested expressway tunnels and parking garages may at times reach CO concentrations of 50-100 ppm [4]. Figure 2 represents a CO profile whilst driving in peak traffic conditions with both front windows lowered. The CO concentration peaks occurred when the vehicle was stationary. In such a case, the polluted air enters the vehicle cabin, and whilst increasing in speed is diluted by the incoming fresh air. The results obtained heavily depend on traffic and weather conditions. Other variables such as wind amplitude and wind direction effected CO concentrations. Furthermore, the CO concentration exponentially peaked to 15 ppm whilst two vehicle occupants one male and the other female were cigarette smoking within the vehicle cabin.

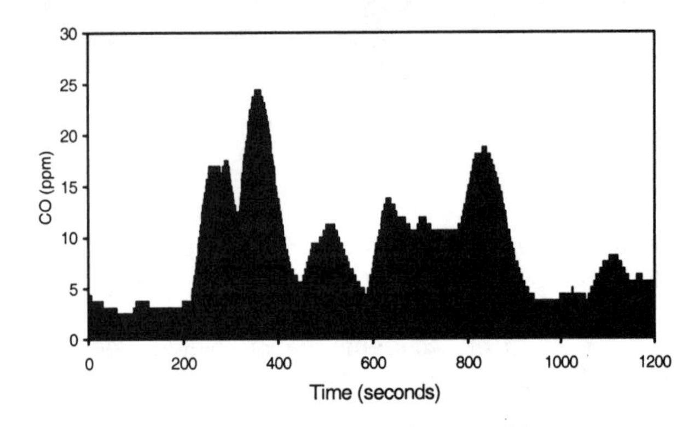

Figure 2. CO profile at peak traffic conditions.

SIMULATED SUICIDE TESTING – Whilst redirecting the exhaust back into the vehicle cabin to simulate a suicide taking place, the oxygen level depleted down to 7%, displaced by the vehicle exhaust gases. Figure 3 shows the oxygen concentration depleting to 7% after 30 minutes.

The carbon monoxide level reached a level undetectable by the monitor. Measurements exceeded 600 ppm.

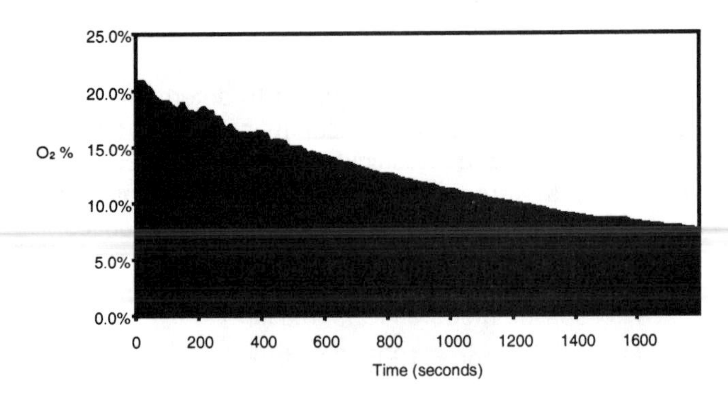

Figure 3. O_2 depletion during a suicide attempt.

CONCLUSIONS

By continuously monitoring oxygen and carbon monoxide within a vehicle cabin, warning devices can be activated when a poor quality cabin gas mixture is present. This enables the vehicle occupants to take charge by allowing fresh air to enter the cabin by various means, or can automatically switch the ventilation system to fresh air mode. It has been found that driving with recycle both at low and high speeds, oxygen depletion to 19.1% can occur within a 20-minute period. This is below the 19.5% limit set by both OSHA and NIOSH [8]. Such depletion in oxygen displaced by carbon dioxide can have an adverse health effect on the vehicle occupants.

Carbon monoxide has been found to reach up to 30 ppm within a vehicle cabin while driving in heavy traffic. It would be desirable that passengers be made aware and to take precaution by switching the ventilation system to "fresh air" mode or lower the windows to allow fresh air to enter the vehicle cabin. This would prevent common headaches experienced when driving in heavy traffic.

During a suicide attempt, it has been found O_2 rapidly decreases while CO increases. A gas behaviour signature of this kind with appropriate thresholds could be used to identify a suicide attempt. Thereafter, an action could be taken, such as lowering the windows or engine shutdown. In such a case, false alarms are of prime concern. The experimental results indicate that setting the upper threshold at >100 ppm would reduce the likelihood of false alarms occurring.

Future work will investigate sensing of carcinogens within the vehicle cabin. Also, investigating the effect of various numbers of vehicle occupants on cabin gas behaviour. However for device feasibility, further research into sensor packaging, sensors placement, sensor tampering prevention design, sensor calibration drift [11] and alarm triggering algorithms is required. Short and long term testing in a fleet of vehicles will provide a quantitative analysis.

ACKNOWLEDGMENTS

The authors gratefully acknowledge Mr Dermot Casey from the Australian Commonwealth Department of Health and Aged Care and the Australian Automotive Association (AAA) for providing project funding. We further acknowledge the contributions from the Australian Medical Association (AMA), the Motor Vehicle Exhaust Gas Suicide Working Group (MVEGS), the Royal Automobile Club of Victoria (RACV), Sandra Hacker, Kirsten Cross, Steve Dunlop and Carmel Bates.

REFERENCES

1. E. L. Anderson, R. E. Albert, 1999, *Risk assessement and indoor air quality*, CRC Press, Florida.
2. M. Maroni, B. Seifert, T. Lindvall, 1995, *Indoor Air Quality*, Monographs-Vol.3, Elsevier, Amsterdam.
3. Source: *US National Centre for Health Statistics / National Vital Statistics System*
4. D.G.Penney, 1999, *Exhaust Gas Detectors: Making Motor Vehicles Safe for the Public*, Medical Journal of Australia, North Sydney (unpublished manuscript).
5. S. Swithemby, 1986, *The physics of matter*, Unit 6 & 7, The Open University, Milton Keyes.
6. Crowcon Ltd, 1997, *Oxygen properties and hazards*, [W WW document], URL http://www.crowcon.com/oxygen.htm.
7. J.B. West, 1974, *Respiratory Physiology*, Waverly Press, Baltimore.
8. NIOSH (1999 June 23), *Online NIOSH Pocket Guide to Chemical Hazards*, [WWW document], URL http://www.cdc.gov/niosh/npg/pgintrod.html.
9. Moller, 1998, *Modelling the different signatures of accidental and intentional in car motor vehicle gas poisoning*, Jane Elkington and Associates, Melbourne.
10. Morgen, J. Schramm, P. Kofoed, J. Steensberg, O. Theilade, *Automobile exhaust as a means of suicide: an experimental study with a proposed model*, J Forensic Sci 1998, Vol. 43 No. 4, 827-836.
11. S.S. Yoon, et al, 1998, *Deaths from unintentional carbon monoxide poisoning and potential for prevention with carbon monoxide detectors*, JANA, Vol 279, No.9, 685-687.

CONTACT

Correspondence to Kosmas Galatsis.
Email: koz@ieee.org
Telephone: +613 413 563 998
Facsimile: +613 9662 1060

970108

A Real-Time Computer System for the Control of Refrigerant Flow

Andy Bartlett, David Standaert, and Eric Ratts
University of Michigan-Dearborn

Abstract

This paper presents a real-time computer system for the control of refrigerant flow in an automotive air conditioning system. This is an experimental system used to investigate the potential advantages of electronic flow control over conventional flow control (using an orifice tube or thermal expansion valve). Two features of this system are presented. First, the system organization is described. Second, the control and interface software are presented. The emphasis is on the software.

The system is organized as a closed loop control system. The inputs to the controller are measurements of the refrigerant system. In particular, thermocouples are used to measure the refrigerant temperature before and after the evaporator. The analog thermocouple signals are converted to digital form by an off-the-shelf, portable, data acquisition system (DAQ). Via a parallel port link, these digital measurements are transfered to a laptop computer. The laptop software processes the measurements and determines a desired refrigerant valve opening. Via the parallel port, the desired valve opening is passed back to the DAQ. The DAQ converts the desired setting from a digital value to a low power analog signal. Using this signal as an input, an analog power amplifier drives the valve to the desired opening. The new valve opening increases or decreases flow of refrigerant into the evaporator. This change in flow effects the temperatures measured by the thermocouples. This closes the control loop.

The software includes a real-time digital controller and a user interface. The software was custom designed using Visual Basic and runs under Windows. The DOS/Windows environment is generally problematic for real-time control, but has proven to be suitable for refrigerant flow control. Successful operation under Windows is made possible by resources provided by the DAQ, by careful limitation of Windows activities, and by the relatively slow dynamics of the air conditioner. In addition to the real-time controller, the software simultaneously provides a graphical user interface (GUI). The GUI allows the user to select and change controllers on the fly in real-time. The GUI also provides the user with a "virtual" strip chart. The strip chart aides the user in tuning the controller during operation. This software has been used on the road in a test vehicle and has proven to be very effective in the development of refrigerant flow control methods.

1.0 Introduction

This paper presents a real-time computer system for the control of refrigerant flow in an automotive air conditioning system. This is an experimental system used to investigate the potential advantages of electronic flow control over conventional flow control. Two features of this system are presented. First, the system organization is described. Second, the control and interface software are presented. Before presenting these features, previous work in the area will be reviewed.

2.0 Prior Work

This subsection will describe the traditional methods of controlling refrigerant flow and some electronic control methods that have been tried. The traditional ones will be described first. Advantages, limitations, and current status will be noted as appropriate.

One of the two most common traditional methods of refrigerant control in automobiles is the orifice tube (OT). As shown in Figure 1, the orifice is a fixed size restriction. By focusing just on the evaporator subsystem, it would appear that the orifice tube had no way to regulate the refrigerant flow. By looking at the system as a whole, it can seen that an orifice tube equipped system does make some adjustment of refrigerant flow to meet operating conditions [1]. However, it is generally agreed that the orifice tube system does not adjust flow as well as systems whose expansion orifice size is actively adjusted. The advantage of an orifice tube systems is in lower cost and increased reliability.

The second of the two most common traditional methods of refrigerant control in automobiles is the thermal expansion valve (TXV). A TXV system is shown in Figure 2. The opening of the TXV is actively adjusted based on a balance of forces. One of the forces is due to a sensing bulb placed at the evaporator output. As the outlet gets hotter, the gas in the sensing bulb gets hotter. Since the bulb and the capillary tube are essentially a closed volume system, the pressure in the bulb and tube increases. This pressure creates a force that increases the valve opening thereby increasing the flow. The increased flow reduces the superheat which reduces the bulb pressure. Eventually a balance of pressure and flow is reached. Some TXV systems use more complicated mechanisms to achieve a desirable balance [1]. The advantage of TXV systems is that they actively regulate superheat to a moderately positive value. Their disadvantage is they are less reliable than orifice tubes due to their moving parts and their relatively fragile capillary tubes.

Electronic control retrofits have been designed for some TXV systems [5]. Figure 3 shows an example. In this system, the temperature of the evaporator inlet and outlet are measured by a computer. The difference of these two temperatures gives a good approximation of superheat if the pressure drop across the evaporator is small. Based on these measurement, the computer adds more or less heat to the sensing bulb using a small electric heating pad. The goal of the retrofit is to have the system achieve a smaller superheat than the moderate one TXV systems are usually designed for. A standard TXV uses the relatively high superheat to avoid sending liquid to the compressor. The electronics can use a faster, more sophisticated design than is possible to mechanically fabricate. This is why the retrofit can safely decrease the superheat in both the transient and steady state periods.

The retrofit system is primarily for large stationary

refrigeration systems. One example is a supermarket display case. If an automobile manufacturer is going to use electronic control of refrigerant control, then it does not make sense for them to use a retrofit design. It is natural to expect that a direct design would have cost and performance advantages. In addition, the need to significantly modify the control algorithm is likely because an automotive system has more frequent and dramatic transients.

Two patents for direct electronic control of refrigerant flow were recently awarded [3] and [2]. Figures 4 and 5 show that both systems use an electronically controlled expansion valve. From a system architecture point of view, the main difference appears to be the method of measuring superheat. In Figure 4, the same approach as the retrofit system is used. Evaporator inlet and outlet temperature are measured and used to approximate superheat. In Figure 5, temperature and pressure at the evaporator outlet are measured. Using the properties of the refrigerant, these two measurements can be used to compute superheat. Based on the superheat, the computers will command new settings to the electronic expansion valves (EXV). The algorithm used to determine the valve adjustment will strongly affect the performance of the system.

It is interesting to note an SAE paper [4] related to the subject of the two patents. This paper was presented one year before the two patents were filed. This paper [4] shows what appears to be the same architecture as Figure 5. The paper [4] also shows experimental data obtained from this system.

Electronic control of refrigerant flow in automotive A/C systems is not a new idea as attested by the 1987 paper and the 1989 patents. Furthermore, electronic flow control is a feature available for large stationary A/C systems. Despite these facts, conventional flow control methods, OT or TXV, totally dominate the automotive market. The goal of the work described in this paper was to develop a prototype vehicle to investigate any potential advantages of electronic flow control.

3.0 System Organization

This section briefly describes the organization of a prototype system. This description covers the major hardware components, the instrumentation, and the electronic/computer hardware as shown in Figure 6. The computer software is not covered until the following section.

The prototype vehicle was a 1994 Lincoln Mark VIII. The prototype used production components for the condenser, evaporator, compressor, OT, accumulator, and electronic automatic temperature controller (EATC). Note that the EATC regulates cabin air temperature by controlling the compressor clutch and the air handling system, but it does not meter the flow of refrigerant.

The test vehicle was modified so that the air conditioning system could operate in conventional or electronic flow control mode. The hand operated Y-valve shown in Figure 6 allows either the OT or the EXV to meter the flow of refrigerant. Aside from some small changes in plumbing, selection of the OT system essentially gives the production Ford system.

The other setting of the Y-valve causes the refrigerant to be controlled by an EXV. Adequate electronic sensors were available so that Figure 4 or 5 could be implemented. Temperature was sensed by sheathed T-type thermocouples that were instrumented into the refrigerant flow stream. A specialized thermocouple card, IOTech DBK19, provided the interface to the small signal levels of these sensors. The pressure sensors came with built in electronics and could be interfaced directly to any general purpose Analog to Digital (A/D) converter. The A/D converter used in this project is part of a portable data acquisition board, IOTech DAQBook 100. This data acquisition board (DAQ) interfaces to any DOS/Windows compatible computer via a standard parallel port. Laptops were especially convenient for this purpose. Via the port, the sensor measurements are eventually made available to the computer. Based on these measurements, the computer uses some chosen algorithm to calculate a desired valve opening. This is desired opening is passed to the DAQ again via the parallel port. The DAQ system has a Digital to Analog (D/A) converter which it uses to put the desired opening in analog form. This analog signal is low power and can't drive the EXV. For this reason, the low power signal is passed through an analog power amplifier. The output of the power amplifier drives the valve. This completes the equipment required for the electronic flow control prototype.

In addition to electronic flow control, electronic monitoring of the entire system was also desired. For this reason numerous thermocouples and pressure sensors were directly connected to the refrigerant system. Additional thermocouples and humidity sensors were also used to measure the ambient and cabin conditions. These measurements could have been incorporated in the DAQ plus Laptop used for flow control. However, the extra measurements would have severely restricted the control capabilities given the many speed limitations of the DAQ plus laptop system. For this reason, a second laptop plus DAQ was used. One laptop plus DAQ was dedicated to control; the other was dedicated to data recording.

4.0 Real-time Software

The two laptop plus DAQ systems run two different sets of software. The data monitoring system uses a Windows based program that was supplied with the DAQ hardware. This software had adequate data collection capabilities, but it did not have real time control capabilities. Custom software was developed for the laptop plus DAQ system that performed the real time control. The overall feedback system is shown in Figure 7 where the real time software represents the shadowed blocks. The system can also run in a "test mode" as shown in Figure 8. The real time control software will be described in this section.

4.1 Oversampling and Conditioning

The real time software was written to use the scheme shown in Figure 4. This means that two thermocouple measurements are processed by the control system. Initially, two types of processing are required. First, the nonlinearities inherent in the thermocouples and the scaling due to the amplifiers must be accounted for. This shall be referred to as conditioning. Second, the noise in the measurements must be dealt with. Subroutines for conditioning the thermocouple signals were supplied by the DAQ manufacturer, and these were employed. For dealing with the noise, a custom strategy was developed.

The noise arises in many locations including the sensor, the wire harness, the interface electronics, and the A/D converter. The measurement noise could be attenuated in the digital controller or by using oversampling. The prototype system used a combination of both approaches.

A linear digital controller is identical in structure to a digital filter. If the control loop were running at a very high sampling rate, then the digital controller could include many fancy filtering effects. However, on the prototype vehicle, the controller will normally be running at less than a 10 Hertz sampling frequency. This relatively slow rate puts some constraints on high frequency filtering. Signal aliasing may also be a problem at this slow rate. Unfortunately, running at a faster rate is likely to overburden the computation and communication abilities of the laptop plus DAQ system. At a faster rate, the software may not meet the real time requirements.

To achieve the desired high frequency filtering, oversampling was used. Like most of today's A/D boards, the DAQ used in the prototype contains just a single A/D converter. Via high speed multiplex switching, many different signals can be linked one at a time to this converter. An allowable alternative is to link the same signal or sequence of signals to the converter several times in a row. This arrangement allows oversampling.

For the DAQ, a "new" signal can be linked to the A/D converter every 10 microseconds (100 kHz). If the control loop is executing every 100 milliseconds (10 Hz), then theoretically, the A/D could measure 10,000 signals in each cycle of the control loop. Due to communication and memory limits, the actual number of measurements must be smaller. At compile time, the custom software is set up to oversample each signal a certain number of times. Ten times oversampling was found to be effective at greatly reducing the high frequency noise. Measuring each of the two thermocouples 10 times requires a total of 20 measurements by the A/D during each control loop cycle. Each set of ten samples is simply averaged to produce a filtered measurement that is used by the control loop every 100 milliseconds.

The oversampling does place a burden on the real time software. If just one sample per cycle were taken, then the control loop could be sped up a little bit. However, the speed penalty of moderate oversampling is relatively small and the noise reduction benefits are significant. Oversampling was found to be an efficient use of the prototype resources.

4.2 Superheat Calculation

The superheat calculation carried out by the software was extremely simple. It was assumed that two phase refrigerant was entering the evaporator and that there was no pressure drop across the evaporator. Under these assumptions, any superheat at the outlet would be equal to the temperature rise across the evaporator. This calculation of superheat simply involved subtracting the inlet temperature measurement TCinlet from the outlet temperature measurement TCoutlet.

Experimentation showed that there actually was a pressure drop across the evaporator of approximately 14 kPa. This pressure drop results in an error in the superheat calculation. The exact amount of error would depend on both the pressure drop and on the absolute pressure at the evaporator inlet. The evaporator inlet pressure had a fair amount of variation, but the pressure drop was always quite close to 14kPa. For the normal range of inlet pressures, the effect of a 14kPa pressure drop on R134a superheat was investigated. It was found that over the entire range the 14kPa pressure drop corresponded to a 2 K increase in the superheat. Therefore the calculation of superheat, TCoutlet - TCinlet, could be improved by adding 2 degrees Kelvin to it. From a control point of view, it would be equivalent to lowering the desired superheat by 2 K. This latter approach was used.

4.3 Desired Superheat

The desired superheat is the target value that the feedback control loop will try to achieve. When the computed superheat is higher than the desired value, the controller must eventually permit more refrigerant to flow. Conversely, the flow is decreased when the computed superheat is below the desired value. The desired superheat is simply a fixed value. Using the graphical user interface (GUI), the desired value can be readily changed during operation. When setting the desired value, the user should keep in mind that the computed superheat is approximately 2 K too low. So if the user wants a true superheat of plus one degree Kelvin, then the desired superheat should be set to minus one degree Kelvin. The lower left corner of Figures 9 and 10 shows the location in the GUI where the superheat goal is entered.

4.4 Digital Controller

The digital controller processes the superheat error to come up with a desired voltage for the EXV. The digital controller is linear for the most part so it is equivalent in structure to a digital filter. The controller does include a simple nonlinearity for anti-windup purposes. The anti-windup could restrict the controller output to the 0 to 5 volt range of the D/A. A smaller range corresponding to the response limits of the EXV were used in the prototype.

The linear portion of the controller can be described by a Z-domain transfer function (ZTF). Direct selection of a ZTF by the user faces some difficulties. If a good linear time invariant model of the plant is available, then there are many nice canned programs for automatically selecting a digital controller's ZTF. However, a good model, especially a linear one, of the refrigerant system was not available. Therefore, it was expected that the user would use heuristic rules to select and tune control features. For a control engineer, selecting features often involves selecting pole and zero locations. Many control engineers have an excellent feel for the effects of poles and zeros in the S-plane of continuous time system. Control engineers typically have a far poorer understanding of pole and zero locations in the Z-plane for discrete-time systems. One reason is that pole and zero locations in the Z-plane depend on the chosen sampling rate of the digital controller. For these reasons, the ZTF of the controller was indirectly specified.

The GUI allows the user to select a continuous time S-domain transfer function (STF). Figures 9 and 10 show the screen by which the user selects poles and zeros in the S-plane. The user may select up to three real poles, three real zeros, two pair of complex conjugate poles, and two pairs of complex conjugate zeros. The default steady state, i.e. DC gain, of the system is plus one. The user may override this choice by selecting an overall, a Proportional , an Integral, or a Derivative gain. Depending on which of the PID gains are selected, the STF can have an additional pole at the origin and/or up to two additional zeros. At any time, the user can change in real-time the number and location of poles, zeros, and gains by pressing the update button.

The prototype system is based on a digital computer so the STF selected by the user must be converted to a ZTF. The first issue in this conversion is the selection of a sampling rate. Near the top center of Figures 9 and 10 is where the user can select the desired sampling frequency. The selected frequency must not be so fast as to over burden the system. For refrigerant control with a full set of graphical displays, the system was able to readily handle rates less than 10 Hz.

After the selection of sampling rate, the method of mapping from the Z-plane to the S-plane must be selected. The bilinear transformation alias Tustin's approximation was employed in the software and was found to be effective. Given a STF H(s) and sampling rate h, the corresponding ZTF G(z) is given by simple substitution of s.

$$G(z) = H\left(\frac{2}{h}\ \frac{z-1}{z+1}\right)$$

By selecting the appropriate check box in Figures 11 and 12, the GUI allows the user to view either the ZTF or the STF.

Once the ZTF has been formed, implementing the digital controller is straight forward. Conversion of the ZTF to a difference equation is trivial. Because of the GUI allows the user a limited quantity of poles and zeros, the controller can be ninth order at worst. This means that implementation of the difference equation involves 19 multiplications and 18 additions at worst. The implementation must also store nine or fewer recent inputs and nine or fewer recent outputs. These computations were a light burden for a 486DX 25MHz laptop computer.

Before closing this subsection, some comments on the software will be made. Visual Basic is not particularly computationally oriented, but it did not hinder the implementation of the small digital controller. The formation of the STF from the selected poles, zeros, and gains had the potential to be some what cumbersome. In addition, the bilinear transformation appears to be a moderately involved symbolic processing problem. There were three keys to making the software manageable. First, a data structure was created that could handle any polynomial up to some specified maximum order. Second and third, subroutines for adding and multiplying any pair of polynomials were written. Note that the STF and ZTF are just the just the ratios of polynomials. Adding a new pole or zero to the denominator or numerator of the STF, respectively, simply involves polynomial multiplication. The bilinear transformation only involves multiplying and adding polynomials of the form q, 2z+2, and hz-h as seen in the example below.

$$H(s) = \frac{2\ s + 3}{4\ s^2 + 5\ s + 7}$$

$$G(z) = \frac{2\left(\frac{2z-2}{hz+h}\right) + 3}{4\left(\frac{2z-2}{hz+h}\right)^2 + 5\left(\frac{2z-2}{hz+h}\right) + 7} =$$

$$\frac{2\,(2z-2)\,(hz+h) + 3\,(hz+h)\,(hz+h)}{4\,(2z-2)\,(2z-2) + 5\,(2z-2)\,(hz+h) + 7\,(hz+h)\,(hz+h)}$$

The zeroth order polynomials q are the coefficients of STF. The bilinear transformation has a regular pattern and can be fully implemented with a set of nested For loops.

4.5 Graphical User Interface

The graphical user interface has two important screens. The change control GUI is illustrated in Figures 9 and 10. The use and operation of this screen was described in the previous subsection. The main GUI is illustrated in Figures 11 and 12. The key features of the main GUI will be described in this section.

The main GUI is divided into four main areas. The bottom portion of the GUI contains several check boxes, buttons, etc. that control the operation of the system and the GUI. This controls will be referenced as the other three sections are described.

The top portion of the GUI shows in tabular form the ten most recent values of the key signals. When the "Test Mode" check box is selected as in Figure 11, only the input and output of the digital controller. When the "Super Heat" mode is checked as in Figure 12, only the thermocouple and superheat values are also displayed. To reduce visual flicker and to minimize computational burden, the newest data is added to the table in a circular buffer fashion. The star at the top of the table marks the newest data, and data immediately to the right is the oldest data. By checking the Disable Output Grid box, no more new data will be inserted in the table. The real time system can operate at a higher sample rate when the various displays are individually or collectively disabled.

The upper-middle portion of the GUI shows the transfer function of the controller. By checking Continuous Time or Discrete Time, the STF or ZTF, respectively, will be displayed. For example, Figure 12 shows the following STF.

$$H(s) = \frac{-1\ s - 1}{33.3\ s^2 + 1\ s + 0}$$

The upper left corner of this region also gives the sampling rate.

The lower-middle portion of the GUI provides a virtual strip chart. By checking the appropriate boxes any or all of the signals associated with a given mode can be plotted. Figure 11 shows a plot of the Test Mode inputs and outputs. The input is a square wave from an external function generator and the output corresponds to the under damped second order system selected for the controller. Figure 12 shows all five of the signals from Super Heat Mode. The graph is not batch updated; it is a true stripchart and slowly scrolls from left to right. The signal plots, the axes, and the enable check boxes are all color coordinated to allow easy identification. To aid the user, the signals were heavily filtered before plotting so as to remove noise and jitter. This explains why the square wave in Figure 11 shows very slow transitions.

5.0 Performance

The real time system described in this paper was evaluated in Test Mode and in actual operation on the prototype vehicle. The detailed results of the refrigeration control are beyond the scope of the paper. The comments here will be restricted to the systems ability to meet the constraints of real time operation. In super heat mode, the system was fully operational at 10 Hz. This rate along with full use of the displays and of controller updates was satisfactory for the refrigeration control experiments.

In test mode, the system was capable of operating better than ten times faster. The omission of the thermocouple signal conditioning reduced the processing load. In addition, if the GUI displays of data were disabled, then much higher rates could be sustained. Figure 13 shows an oscilloscope trace where the real time system is producing an underdamped second order response at 150 Hz. When a Pentium based desktop machine was substituted for the Laptop 486, sampling rates of 500Hz were obtained.

Because the system was running under the DOS/Windows environment, it was not possible to obtain guaranteed real time performance. In the vast majority of cases, the controller output was properly updated each sample time. However, the system did fall behind in some situations. Falling behind was noticed in higher speed test mode operation and could be captured on an oscilloscope. In Figure 13, the digital system output is a staircase plot and each step has a consistent width very close to 1/(150 Hz). If Windows operating system, attempted a time consuming task such as updating an interactive screen, then the staircase plot could show one very long step followed by several short steps. In this case, samples were taken by the DAQ hardware at the appropriate times but the computer was to busy some Windows task to process the samples. Eventually Windows would return control to the real time system, and this system would try to "crunch through" the backlog as quickly as possible.

6.0 Conclusions

This paper has presented a real time computer system for the control of refrigerant flow in a prototype automobile. The system carries out digital control and provides a GUI that allows the control to be tuned on line. The GUI also provides graphical and tabular displays that assist in the tuning process. Testing on the prototype vehicle has shown that the system has the real time control capabilities required for the refrigeration control project. On road testing has shown that the system can simultaneously provide the graphical user interface and maintain a 10 Hz control loop.

7.0 Acknowledgements

This project was made possible through the support of several organizations. The authors wish to thank the Center of Engineering Education and Practice (CEEP) for three years of generous support. The authors also wish to thank the Ford University Research Program for two years of extremely helpful support. Ford Motor Company's Components Division (formerly Climate Control Division) also supported the project by providing the first and third author with faculty internship for two summers each. Ford's Alpha Division gave crucial support by providing the Mark VIII test vehicle. The University of Michigan-Dearborn Campus Grant Program gave seed support. All of this support was highly appreciated by the authors and by the many student research assistants who gained valuable experience because of it.

This project had many interdependent parts each of which was crucial to the whole. The authors would like to acknowledge the many people who did not contribute directly to the real time software but did play valuable roles in the project. Richard Terry was the key player in the mechanical aspects of the instrumentation. Tim Philippart designed and constructed the power amplifier. Dennis Kubica, Josh Isser, and Joe Jaegger assisted in the electrical aspects of instrumentation. Paul Rogers played a lead role in much of the data collection and analysis. The UMD Electrical and Mechanical Engineering Shop technicians: Jesse Cross, Martin Stenzel, Don Haidys, and Mike Solstad, also aided in the completion of the project. In addition, engineers at Ford Motor Company contributed their experience and expertise in selecting and evaluating the project. Key contributors were Chris Rockwell, Gary Dage, Bob Matteson, George Wiklund, and David Zietlow. This project lasted three long and hectic years, and there are undoubtedly other contributors that have been overlooked. The authors regret that the failings of their memories prevent them from individually acknowledging and thanking each contributor.

8.0 References

1. Cuffe, K. "Air Conditioning and Heating Systems in Trucks," SAE Paper 780001, 1978.

2. Barthel, R.C., Malone, P.J., Orth, C.D., and Jarosch, G.W., "Controlling Refrigeration," United States Patent No. 4,848,100, July 18, 1989.

3. Torrence, R.J. "Controlling Superheat in a Refrigeration System," United States Patent No. 4,835,976, June 6, 1989.

4. Mitsui, M., "Improvement of Refrigerant Flow Control Method in Automotive Air Conditioners," SAE Paper 870029, 1987.

5. Marsala, J., "An Electronic Control for Thermostatic Expansion Valves," Proc. of Building Retrofit Innovative Concepts Fair - An Experimental Technology Exchange, Washington, DC, Oct. 24, 1989.

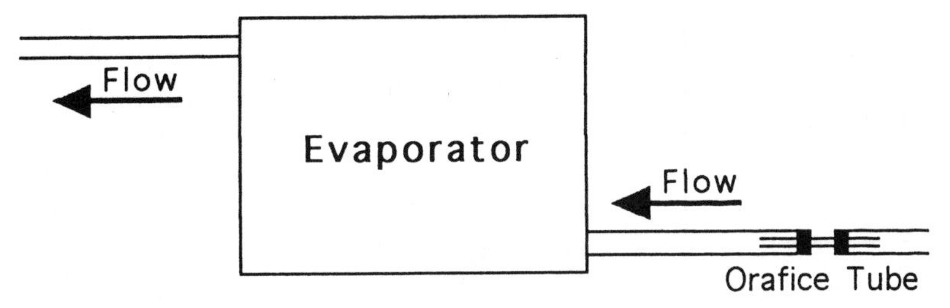

Figure 1: Refrigerant expansion using a orifice tube.

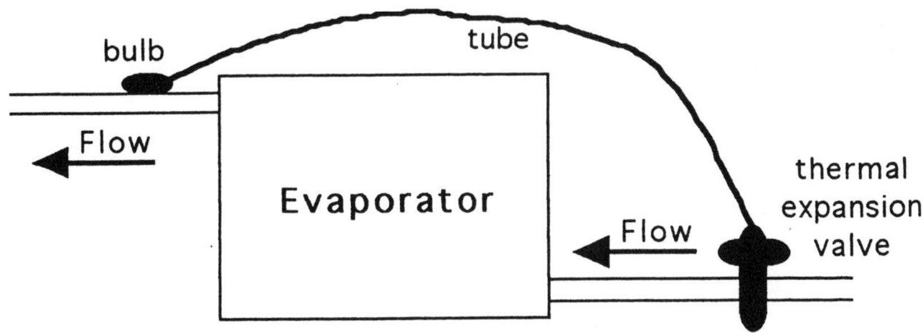

Figure 2: A basic thermal expansion valve system.

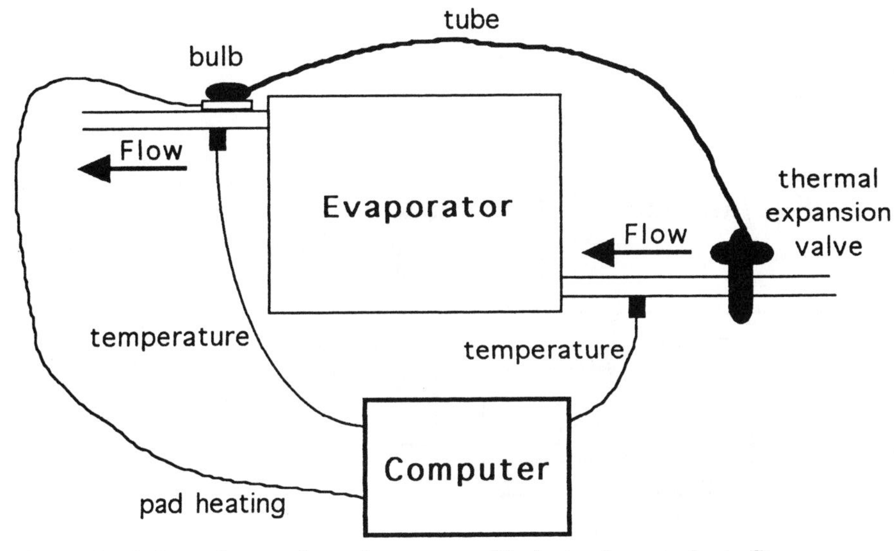

Figure 3: A thermal expansion valve system with electronic control retrofit.

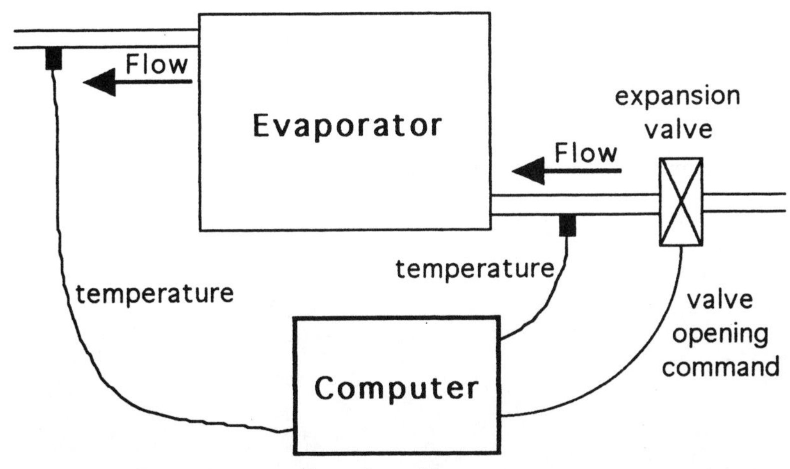

Figure 4: Electronic expansion valve with two temperature measurements.

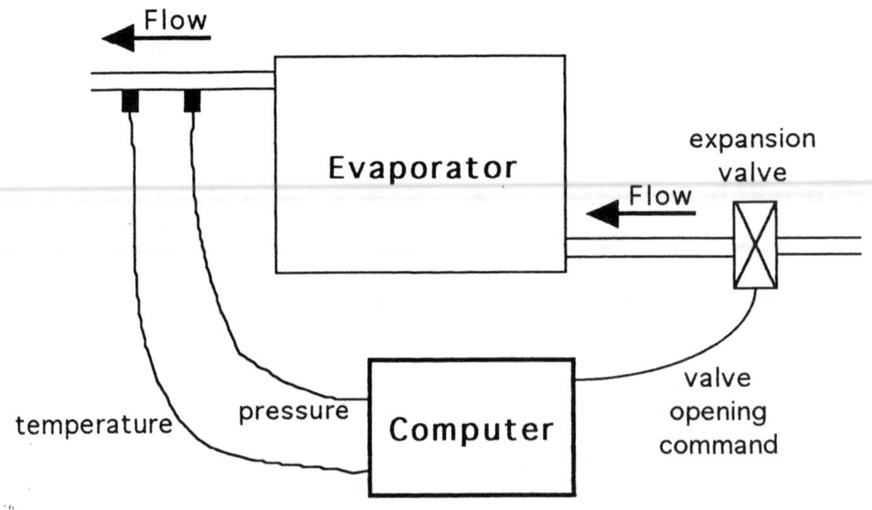

Figure 5: Electronic expansion valve with temperature and pressure measurements.

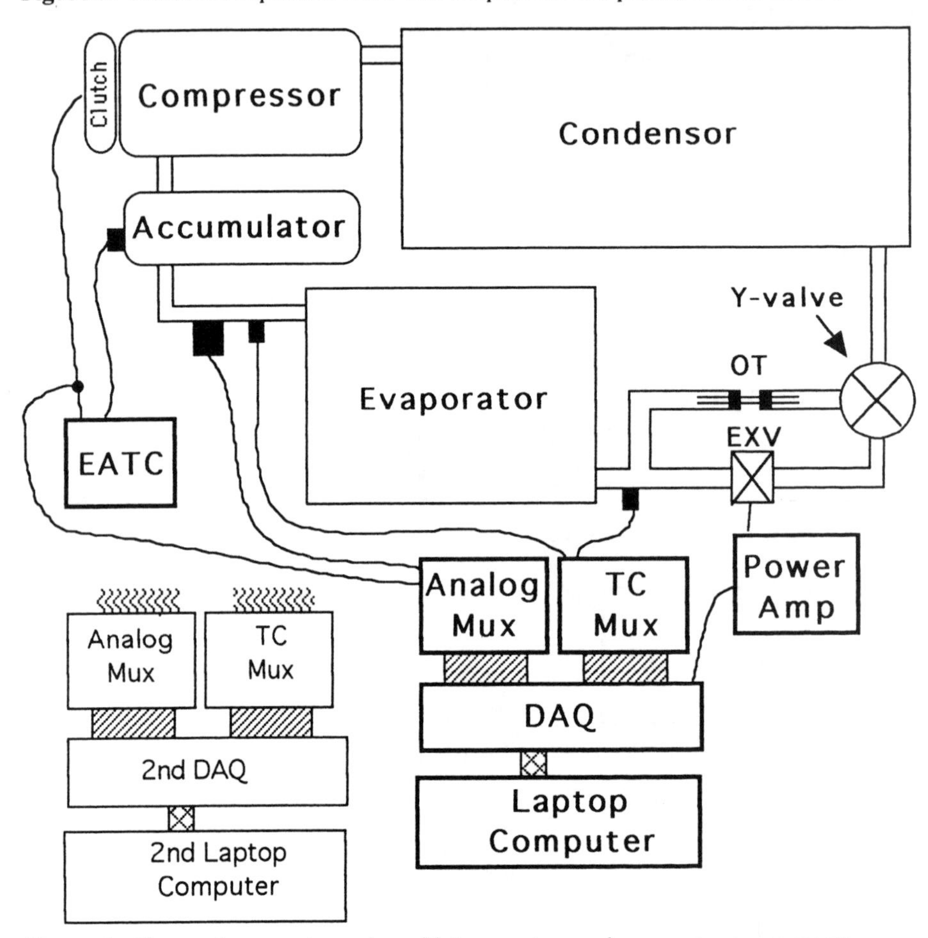

Figure 6: Electronic expansion valve with temperature and pressure measurements.

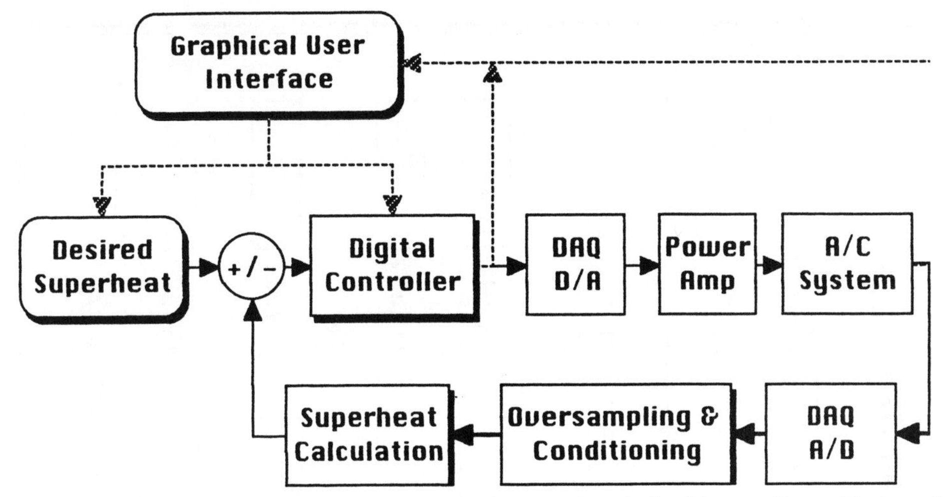

Figure 7: Diagram shows closed loop system for the control of refrigerant flow. Shadowed boxes represent real-time software running on PC.

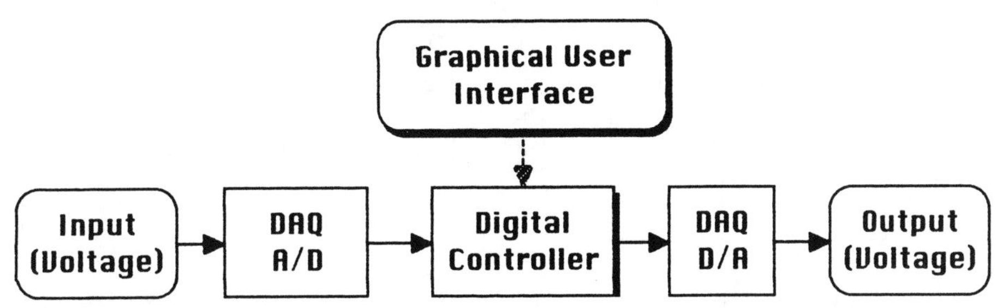

Figure 8: Diagram shows test mode for real time system. This mode is useful for testing and debugging the system. This mode could also be used in a general control loop provided the inputs and outputs are electrical signals.

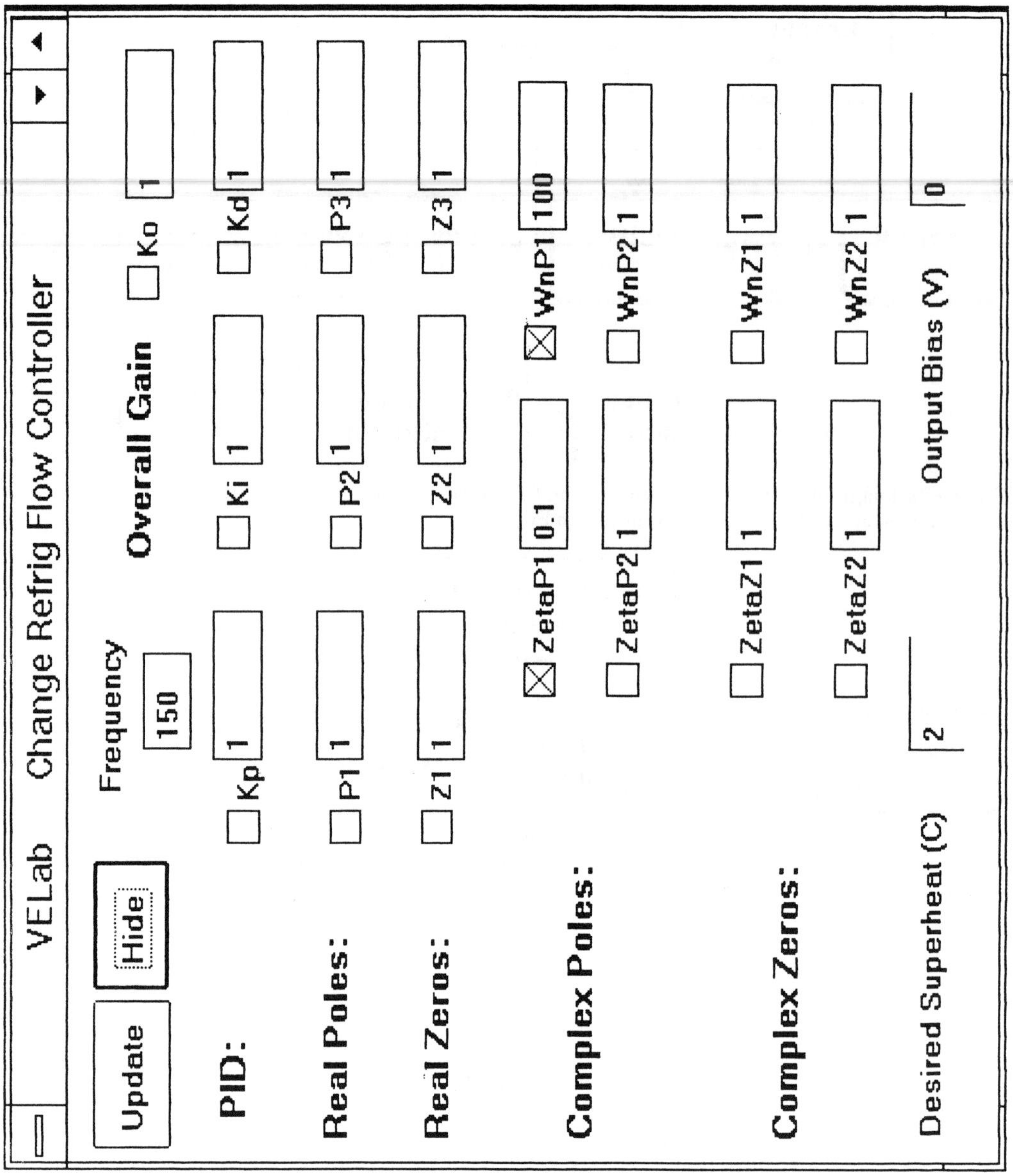

VELab Change Refrig Flow Controller

Update Hide

Frequency 150

Overall Gain Ko 1

PID: Kp 1 Ki 1 Kd 1

Real Poles: P1 1 P2 1 P3 1

Real Zeros: Z1 1 Z2 1 Z3 1

Complex Poles: ZetaP1 0.1 WnP1 100
ZetaP2 1 WnP2 1

Complex Zeros: ZetaZ1 1 WnZ1 1
ZetaZ2 1 WnZ2 1

Desired Superheat (C) 2 Output Bias (V) 0

Figure 9

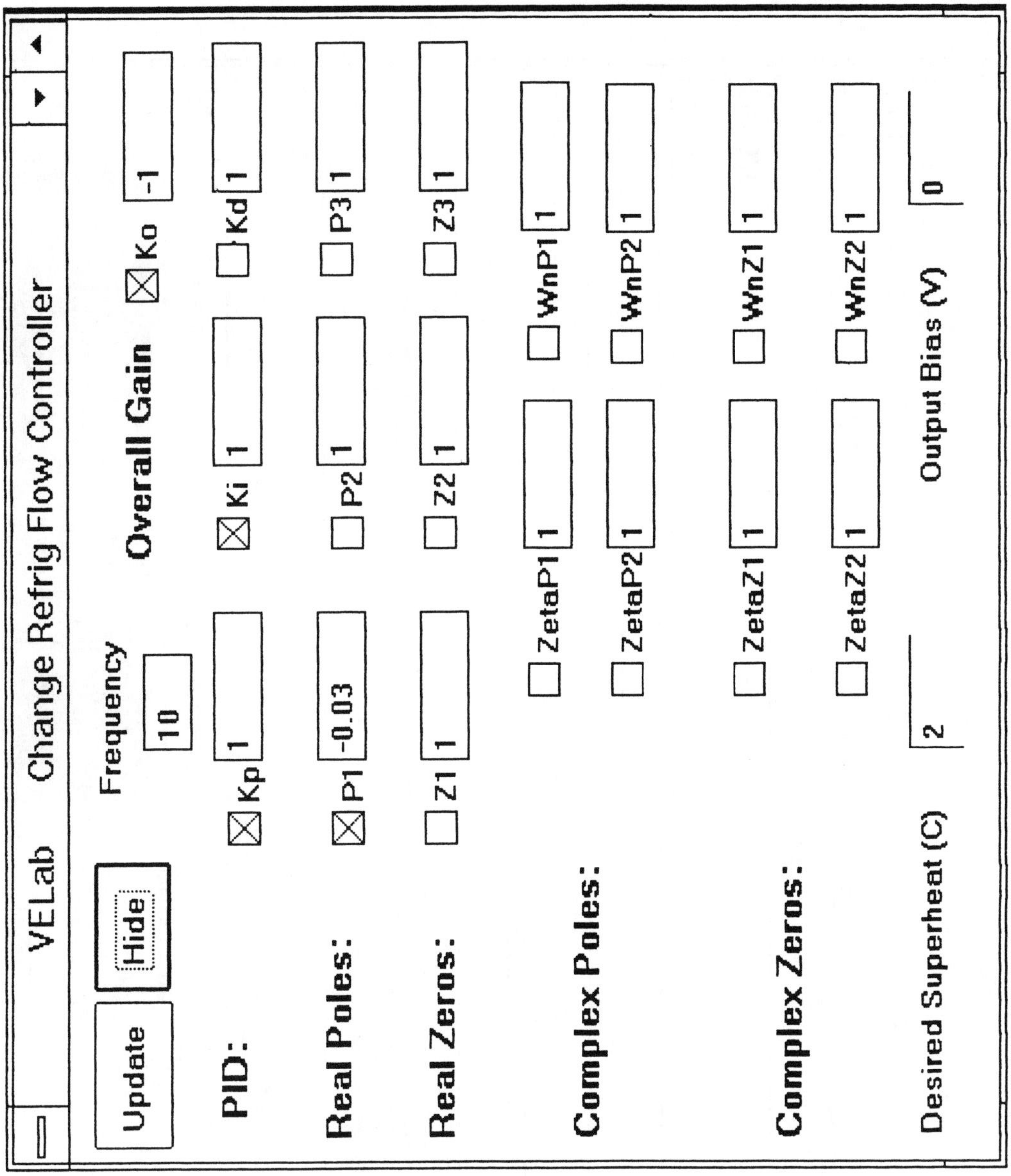

VELab Change Refrig Flow Controller

Update Hide

Frequency 10

Overall Gain ☒ Ko -1

PID: ☒ Kp 1 ☒ Ki 1 ☐ Kd 1

Real Poles: ☒ P1 -0.03 ☐ P2 1 ☐ P3 1

Real Zeros: ☐ Z1 1 ☐ Z2 1 ☐ Z3 1

Complex Poles: ☐ ZetaP1 1 ☐ WnP1 1

☐ ZetaP2 1 ☐ WnP2 1

Complex Zeros: ☐ ZetaZ1 1 ☐ WnZ1 1

☐ ZetaZ2 1 ☐ WnZ2 1

Desired Superheat (C) 2 Output Bias (V) 0

Figure 10

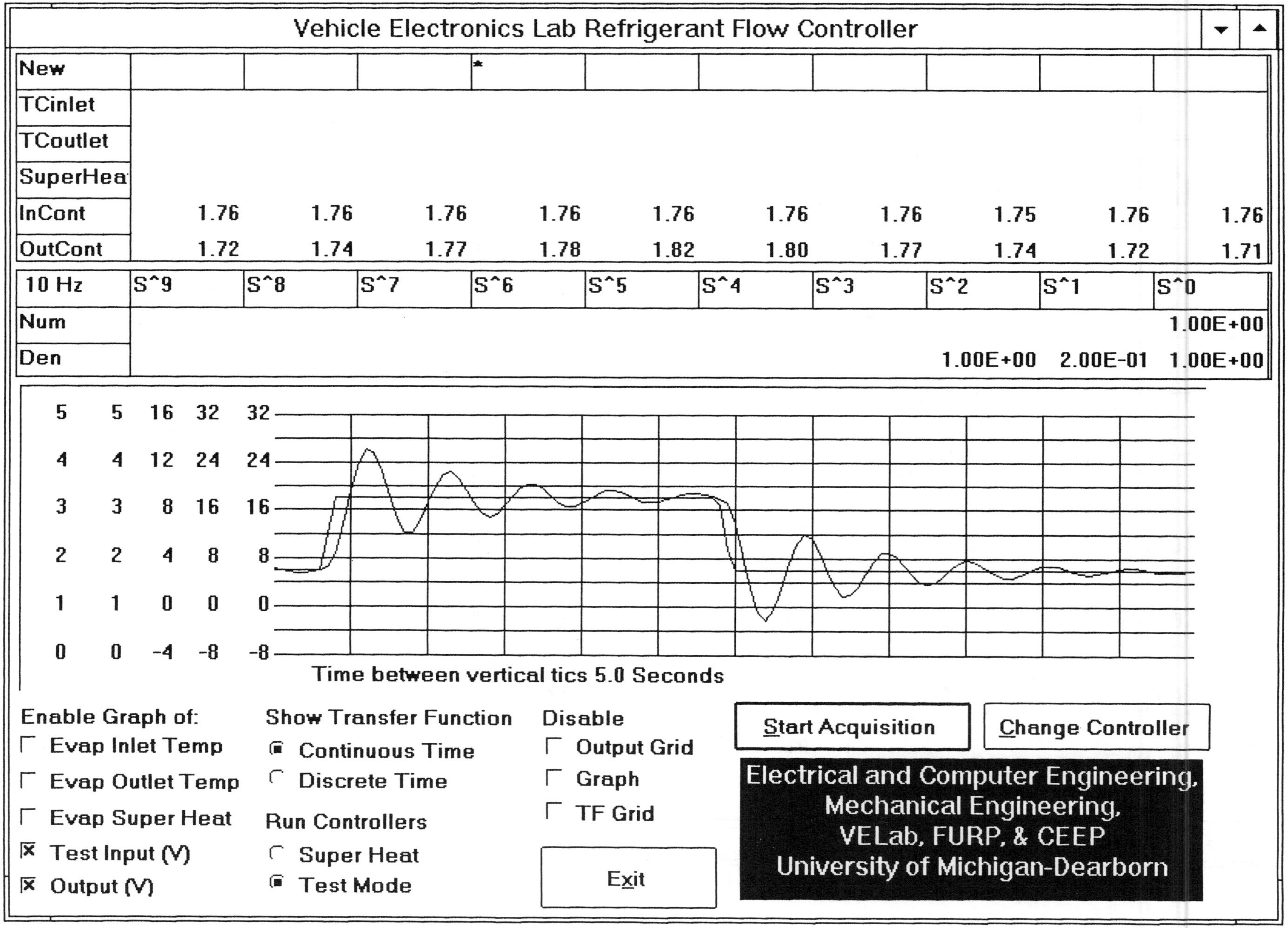

Vehicle Electronics Lab Refrigerant Flow Controller

New				*						
TCinlet										
TCoutlet										
SuperHea										
InCont	1.76	1.76	1.76	1.76	1.76	1.76	1.76	1.75	1.76	1.76
OutCont	1.72	1.74	1.77	1.78	1.82	1.80	1.77	1.74	1.72	1.71

10 Hz	S^9	S^8	S^7	S^6	S^5	S^4	S^3	S^2	S^1	S^0
Num										1.00E+00
Den								1.00E+00	2.00E-01	1.00E+00

5	5	16	32	32
4	4	12	24	24
3	3	8	16	16
2	2	4	8	8
1	1	0	0	0
0	0	-4	-8	-8

Time between vertical tics 5.0 Seconds

Enable Graph of:
- ☐ Evap Inlet Temp
- ☐ Evap Outlet Temp
- ☐ Evap Super Heat
- ☒ Test Input (V)
- ☒ Output (V)

Show Transfer Function
- ◉ Continuous Time
- ○ Discrete Time

Run Controllers
- ○ Super Heat
- ◉ Test Mode

Disable
- ☐ Output Grid
- ☐ Graph
- ☐ TF Grid

Exit

Start Acquisition

Change Controller

Electrical and Computer Engineering,
Mechanical Engineering,
VELab, FURP, & CEEP
University of Michigan-Dearborn

Figure 11

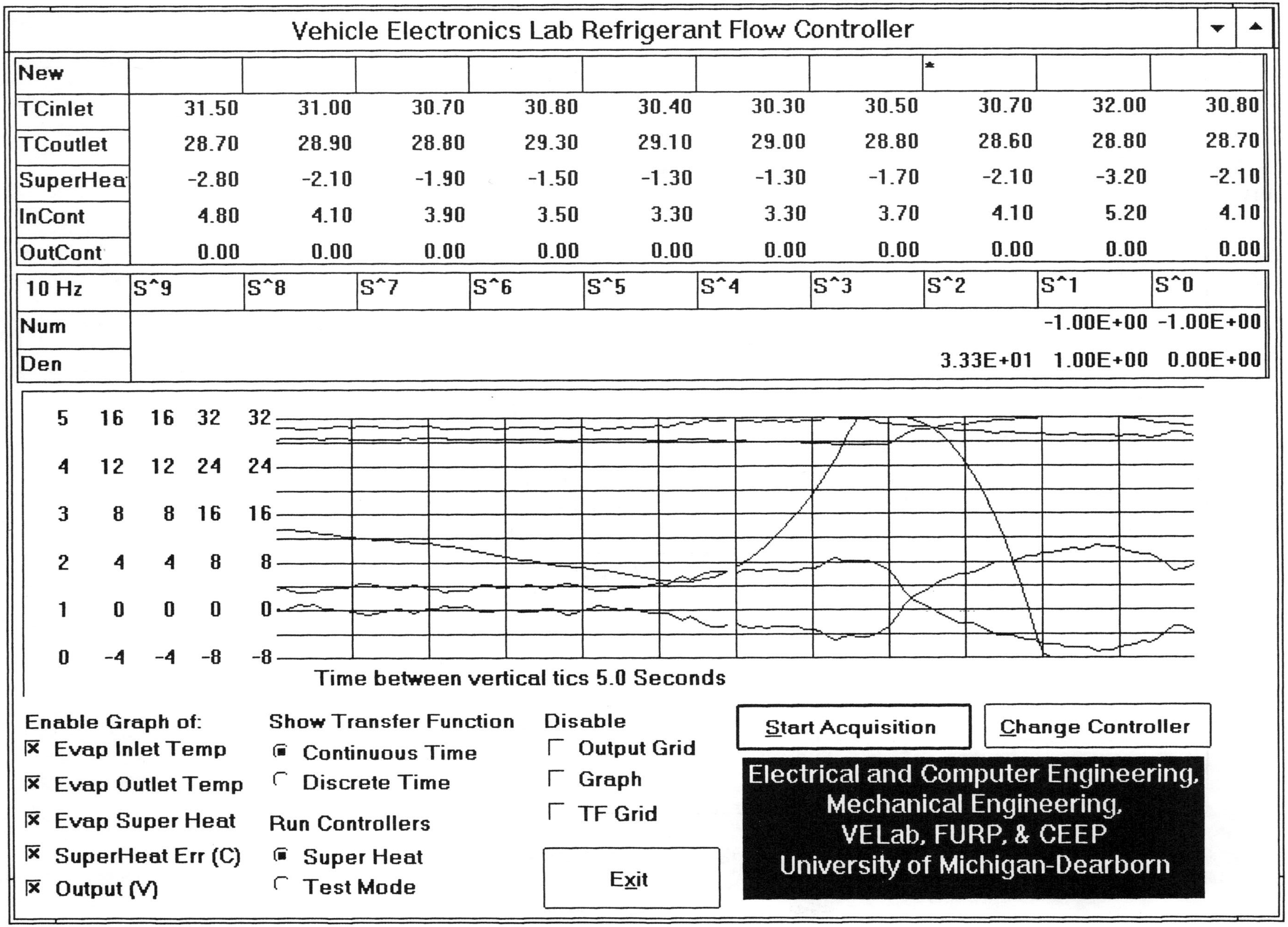

Vehicle Electronics Lab Refrigerant Flow Controller

New							*			
TCinlet	31.50	31.00	30.70	30.80	30.40	30.30	30.50	30.70	32.00	30.80
TCoutlet	28.70	28.90	28.80	29.30	29.10	29.00	28.80	28.60	28.80	28.70
SuperHea	-2.80	-2.10	-1.90	-1.50	-1.30	-1.30	-1.70	-2.10	-3.20	-2.10
InCont	4.80	4.10	3.90	3.50	3.30	3.30	3.70	4.10	5.20	4.10
OutCont	0.00	0.00	0.00	0.00	0.00	0.00	0.00	0.00	0.00	0.00

10 Hz	S^9	S^8	S^7	S^6	S^5	S^4	S^3	S^2	S^1	S^0
Num									-1.00E+00	-1.00E+00
Den								3.33E+01	1.00E+00	0.00E+00

5	16	16	32	32
4	12	12	24	24
3	8	8	16	16
2	4	4	8	8
1	0	0	0	0
0	-4	-4	-8	-8

Time between vertical tics 5.0 Seconds

Enable Graph of:
- [x] Evap Inlet Temp
- [x] Evap Outlet Temp
- [x] Evap Super Heat
- [x] SuperHeat Err (C)
- [x] Output (V)

Show Transfer Function
- (•) Continuous Time
- ( ) Discrete Time

Run Controllers
- (•) Super Heat
- ( ) Test Mode

Disable
- [ ] Output Grid
- [ ] Graph
- [ ] TF Grid

Exit

Start Acquisition

Change Controller

Electrical and Computer Engineering,
Mechanical Engineering,
VELab, FURP, & CEEP
University of Michigan-Dearborn

Figure 12

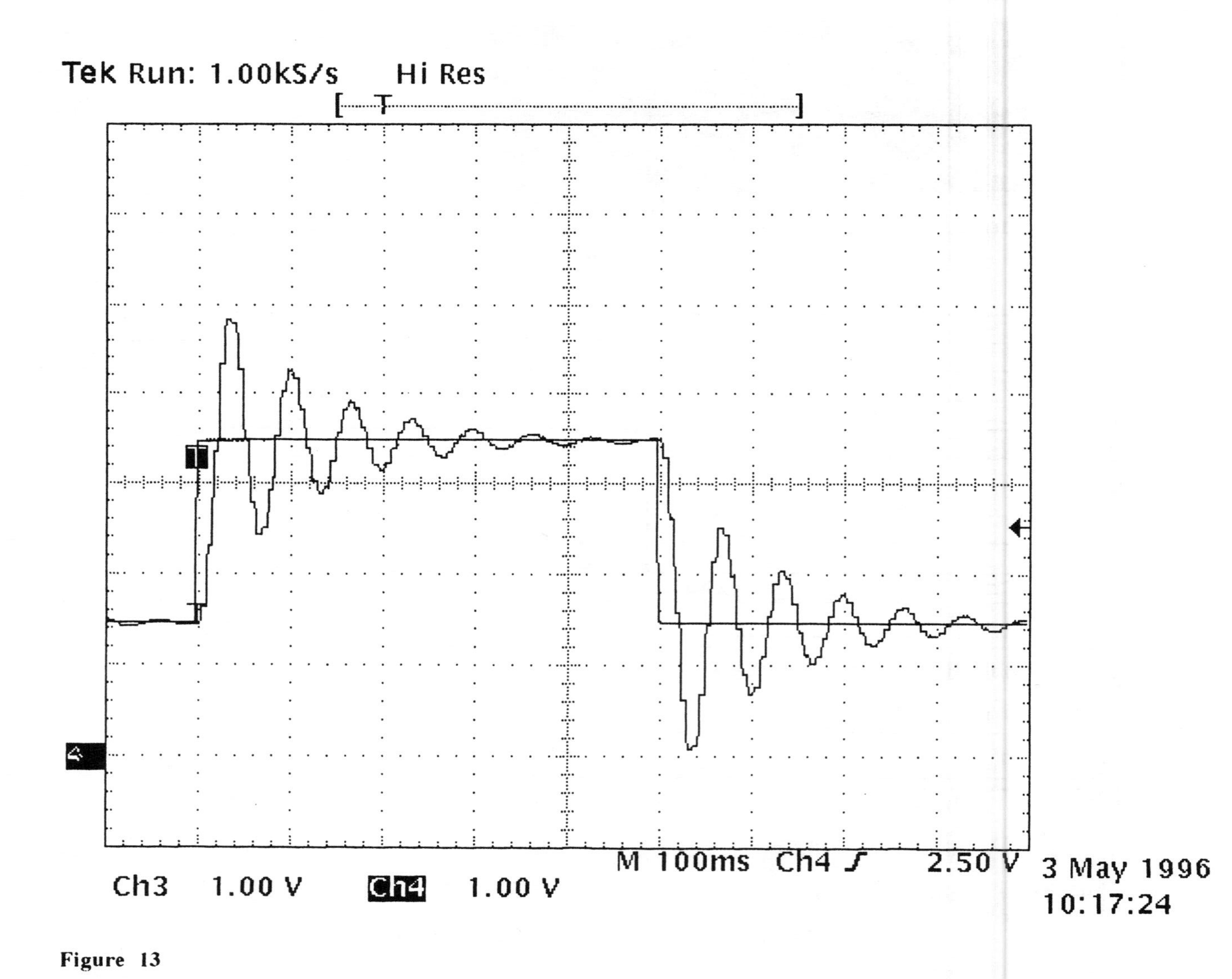
Tek Run: 1.00kS/s Hi Res
T
Ch3 1.00 V
Ch4 1.00 V
M 100ms Ch4 ƒ 2.50 V
3 May 1996
10:17:24

Figure 13

980288

Modeling of Automotive Passenger Compartment and Its Air Conditioning System

Y. Khamsi, C. Petitjean and V. Pomme
Valeo Climate Control

ABSTRACT

Valeo Climate control is developing a software package for the design and simulation of car air-conditioning systems. This software package aims to improve Valeo's response to customer requirements concerning delays for design and sizing car air-conditioning systems performances and cost reduction. Further it shall help to capitalize our competence and expertise on the A/C systems. The comparisons between simulation and tests give a good level of accuracy.

The software consists of three modules:

- A first module for car cabin thermal simulation in dynamic and stationary conditions.
- A second module for the determination and optimization of the A/C system components.
- A third module is under development for the dynamic simulation of the A/C system in non-standard operating conditions.

INTRODUCTION

The general diagram shows the principal steps followed by the software modules :

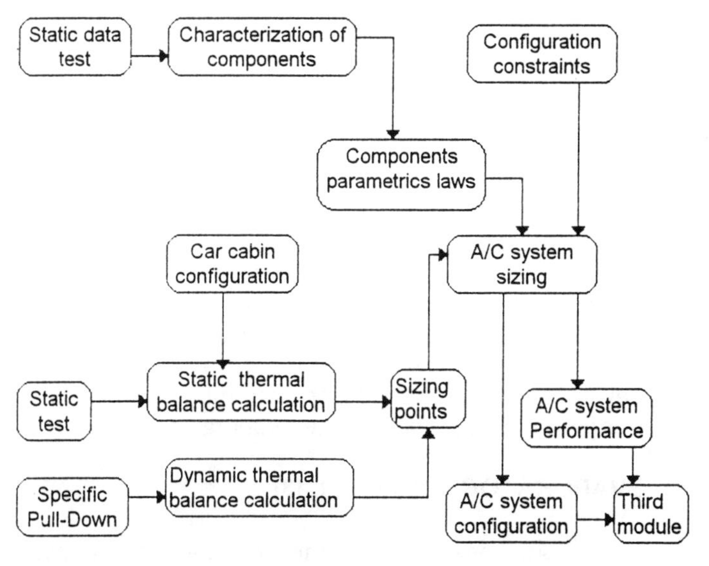

Considering the large number of functions required for modules 1, 2, and 3 (sizing, analysis of tests, static simulation, dynamic simulation, etc.) and the diversity of system configurations (partial test bench, wind tunnel), it is necessary to be sufficiently general in order to handle all needs using a single software base. The complexity of the model is thus adapted to design requirements to reach the optimum accuracy/calculation time trade-off while keeping the same architecture. Local models of the physical operating system are :

- Knowledge (or phenomenological) models which can be more or less fine according to the accuracy and degree of identifiability sought, or
- Design models, adapted to the sizing calculation
- Representation models: black box or scale models which globally reproduce the behavior of real static or dynamic operation.

CABIN CAR THERMAL BALANCE

The car passenger compartment model is developed using the Matlab/Simulink software package [1].

Initially, a simplified model was constructed to simulate the car passenger compartment with a single volume of air. Although simplified, this model allowed a quantitative and qualitative confirmation of the various heat exchange phenomena occurring in the passenger compartment. The model was subsequently made more complicated by dividing the air inside the passenger compartment into several volumes, allowing for the calculation of average temperatures and order-of-magnitude estimates of average air speeds in the different parts of the passenger compartment. This applies for both the pull-down and warm-up cases.

MODEL DESCRIPTION

In the software, the passenger compartment model comprises three main modules:

- <u>A protocol module</u>

Environmental conditions outside the passenger compartment, air flow rate, vehicle speed, set points,...

- A controller Module

Based on the specified targeted air temperature in the cabin, a required evaporator capacity is calculated. This capacity corresponds to what the evaporator will have to generate to reach the set point. A PI-type controller with complete filtered operation is used. Two parameters must be controlled:

- Controller gain,
- Filter time constant.

- A Vehicle module, including:

. An air conditioning unit model :
This simplified model is a theoretical model simulating the operation of the A/C system. It determines the parameters of the air blown into the cabin.

. A Passenger Compartment Model
For the passenger compartment, we have considered the following types of heat exchanges : convection, conduction and radiation. In particular, conventional heat exchange between the different volumes of the passenger compartment and between the various parts of the passenger compartment (shape factors) has been considered.

In order to determine the cooling capacity required to obtain a given comfort level, an energy balance calculation is carried out between the heat flows (positive or negative) into the passenger compartment. Two types of calculation are made:

- Static balance for steady-state conditions. These occur when the vehicle operating conditions change little or not at all.
- Dynamic balance for transient operation of the A/C system. This means :
  - Determining critical points of sizing,
  - Characterizing the time required to reach the final temperature desired in the passenger compartment,
  - Analyzing heat transfer modes in the passenger compartment.

A different solving method is used for each type of balance. For a dynamic balance with given air flow rate and desired passenger compartment temperature set point: the temperature and humidity of the air are determined so that the A/C system can reach the target.

The controller determines the air flow temperature and humidity required to follow the cabin temperature set point path (principle of one of the temperatures of a passenger compartment region changing ahead of the others, passenger compartment average, etc.) while taking the field of constraints into account.

The dynamic passenger compartment model simulates the temperature and humidity obtained based on the temperature and humidity of the discharged air. Perturbations applied to the passenger compartment model correspond to all the factors affecting thermal behavior in the passenger compartment, such as solar radiation, external temperature and pressure conditions, etc. The constraints represent the maximum realistic limits for evaporator capacity in an A/C system. These constraints depend on the temperature and humidity of air flowing through the evaporator. For the static case, only the stabilized temperature points reached after the end of the transient regime are examined.

VALIDATION

For the validation of the thermal balance of the cabin, we have realized several wind tunnel tests in different conditions and for several types of vehicles. Comparisons between calculation and tests give us a good level of accuracy (between 5 and 10%) for the cooling capacity to be installed in the vehicle in order to reach the comfort conditions in the car cabin. The following figure shows the evolution of the calculated and measured cooling capacities in the conditions below :

- External conditions : 45°C, 40 % and 1000 W/m²
- Car speed : 40 Km/h and idle
- Blower air flow rate : 450 kg/h, air recirculation mode in the car.

The calculated cooling capacity being situated between the cooling capacities determined through wind tunnel tests, we have a good correlation between test and calculation.

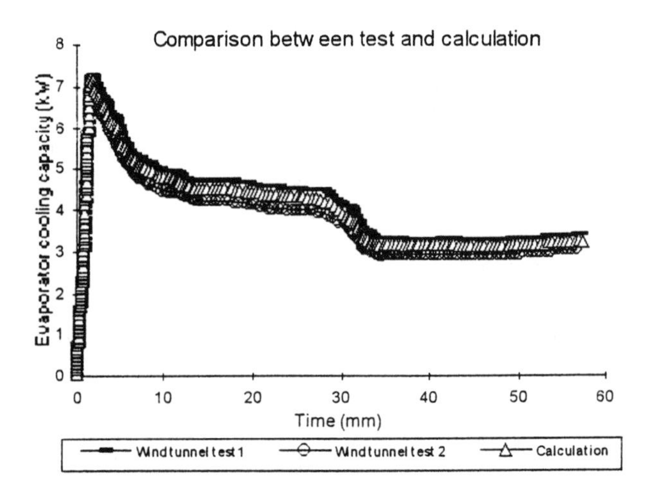

MODELING THE A/C SYSTEM

Modeling is based on a modular concept in which the elementary model of each component is coupled with the models of the other components. Coupling the evaporator with the expansion device and compressor is particularly important because the refrigerant fluid flow rates imposed by the expansion device and the compressor determine the mass of refrigerant fluid present in the evaporator.

EVAPORATOR MODEL [2], [3]

An analysis of the evaporator behavior requires the study of heat transfer, mass flow and quantity of movement. Knowledge of the laws governing these transfers is essential. The evaporator is considered to be divided into two regions on the refrigerant side: a two-phase region (TP), and a superheat region (H).

In the analysis that follows, the evaporator will be considered as the sum of two cross-flow exchangers. The two exchangers are in series on the refrigerant side, and in parallel on the air side.

Thermal model

The rate of heat flow into each region is given by:

$$\dot{Q}_{oi} = \varepsilon_i \, \dot{C}_{min_i} \, (ta^*_{sui} - tr_{sui}) \, , \, i = 1,2$$

with $\dot{Q}_{oi}$: heat flow rate into region i (1 = two-phase, 2 = superheat)

ε_i : efficiency of exchanger i

$\dot{C}_{min_i}$: minimum heat capacity rate of region i

ta^*_{sui} : temperature of dry or wet air at exchanger inlet

tr_{sui} : temperature of refrigerant at exchanger inlet.

The efficiency of a cross-flow exchanger is given by:

$$\varepsilon = f(NTU, \omega) = 1 - e^{(e^{\omega NTU^{0.78}} - 1) \, \omega^{-1} \, NTU^{0.22}}$$

with $NTU = \dfrac{AU}{\dot{C}_{min}} = \dfrac{1}{R\dot{C}_{min}}$ and $\omega = \dfrac{\dot{C}_{min}}{\dot{C}_{max}}$

and R = Rr + Ra for dry conditions

R : Evaporator wall resistance

NTU : Number of transfer units

For the two-phase flow region, $\dot{C}_{max} = \dot{C}_{R134} = \infty$. In this case, exchanger efficiency is given by:

$$\varepsilon = f(NTU) = 1 - e^{-NTU}$$

$$NTU = \frac{1}{R \, \dot{C}_{air}}$$ for dry conditions.

These definitions of ε, NTU, AU, and $\dot{C}$ make sense only if the secondary fluid (air) can be treated as an ideal gas. In principle, this is valid only if the evaporator is dry, i.e. no condensation on the air side.

It can however be shown that the wet regime can be treated by analogy with the dry regime by putting:

$$\dot{Q}_{oi} = \varepsilon_i' \, \dot{C}'_{min} (t'_{asu} - tr_{sui})$$

with $\varepsilon'_i = f_i \, (NTU', \omega')$

$$NTU' = \frac{1}{R' \, C_{m'in}}$$

$$R' = R_r + R_{a'} = R_r + \frac{cp_a}{cp_{a'}} R_a$$

$$\dot{C}'_{min} = \min(\dot{M}_{h134} \; ; \; cp_{134} \; ; \; \dot{M}_{air_i} cp'_a)$$

cp'_a = Specific heat capacity of wet air

$$cp'_a = \frac{\partial h_a}{\partial t'_a} \approx \frac{ha_{su} - ha_{ex}}{ta'_{su} - ta'_{ex}}$$

ha_{su} and ha_{ex} : inlet and outlet evaporator enthalpy.

R : evaporator wall resistance

In practice, the evaporator wall may be partially wet. Two regions should be distinguished on the air side. Nevertheless, the assumption that the evaporator is either completely dry or completely wet is a very convenient approximation.

It is possible to show that both assumptions lead to a slightly underestimated total evaporator capacity corresponding to a partially wet evaporator. For the simulation, the assumption providing the highest capacity (for equal surfaces) or the smallest surface (for equal capacity) is therefore always selected.

Pressure Drop model

Local pressure drops of a flow are defined conventionally for the dynamic pressure. For static pressure, the definition of a tangential tension is used. The overall pressure drop is given by:

$$\Delta p = \Delta p_1 + \Delta p_2 = \left(\zeta + 4 \frac{L}{D} f\right) \frac{1}{2} \frac{\dot{u}^2}{v}$$

with
$\dot{u}$ = average velocity of fluid
v = specific volume of fluid
D = duct diameter
L = duct length
$\zeta = f \dfrac{1}{2} \dfrac{\dot{u}^2}{v}$
f = coefficient of friction

Coefficients z and f depend on the Reynolds number of the flow, and hence on the velocity and volume flow rate of the fluid.

with $\left(\zeta + 4 \dfrac{L}{D} f\right) = f(Re) = f(\dot{u})$

and $\dot{u} = \dfrac{\dot{M}v}{A}$

A = cross-sectional area of duct

M = Refrigerant volume flow rate

Thus, $$\frac{\Delta p}{\dot{M}^2 v} = f(\dot{M}v) \approx C_0 + C_1 * (\dot{M}v)^n$$

Coefficients C_0 and C_1 and the exponent *n* of this polynomial are to be determined experimentally.

On the refrigerant side, we distinguish two regions (boiling and superheat), and set:

$$\Delta p \approx \frac{L_1}{L} \left[f(\dot{M}v) \; \dot{M}^2 \; v\right]_{zone\,1} + \frac{L_2}{L} \left[f(\dot{M}v) \; \dot{M}^2 \; v\right]_{zone\,2}$$

L_1,L_2 : length of zone 1 and zone 2.

$L = L_1 + L_2$

with, in the first region:

$\bar{v} \approx \bar{X} \, v_{ex}$ and $\bar{X} = \dfrac{X_{su} + X_{ex}}{2}$; (X_{ex} = 1)

$\bar{X} = \dfrac{X_{su}}{2} + \dfrac{1}{2}$ = average quality of refrigerant

X_{su} , X_{ex} : supply and exhaust quality of refrigerant and v_{ex} = specific volume of refrigerant on leaving this region.

$$(\dot{M}v)_{zone\,1} \approx (\dot{M}v)_{zone\,2} * \frac{1 + X_{su}}{2}$$

with v_v = specific volume defined at evaporator exhaust, and thus

$$\frac{\Delta p}{\dot{M}^2 v} \approx C_0 + C_1 X_{su} + C_2 (\dot{M}v)^{n_1} + C_3 X_{su} (\dot{M}v)^{n_2} + \ldots$$

Here again, the number of terms can be adjusted according to the accuracy required.

<u>On the air side</u>, the film of condensed water on the surface of the fins can be expected to reduce the passage area. Assuming the thickness of this film to be an increasing function of the condensed water flow rate. thus :

$$\frac{\dot{M}_a^2 v_a}{\Delta p_a} = A\, f^{-1}(\dot{M}_a v_a)$$

with A = (L - e)
L : distance between fins
e : thickness of condensate film

M_a : Air mass flow rate
Δp_a : Air pressure drop
v_a : Air velocity

An overall analysis of test results is performed, to identify three heat transfer coefficients:

- One on the air side (h_a);
- Two on the refrigerant side: boiling refrigerant (h_{tp}) and superheated refrigerant (h_{sh}).

<u>Formulation of heat transfer coefficients</u>

Air Side

On the air side, the heat transfer coefficient (Webb [3]) is given by a correlation used for gases flowing normally through pipe bundles :

$$Nu_{air} = 0.33\, Re^{0.6}\, Pr^{1/3}$$

We obtain:

$$h_{air} = F_{air} \frac{k_{air}}{D_{e_{air}}} N_{u\,air}$$

with F_{air} = correction factor obtained from experimental data.

$D_{e_{air}}$:The hydraulic diameter used is found by considering the geometric dimensions of the fins.

· *Refrigerant Side*

For the two-phase region, a relationship was applied to obtain the local heat transfer coefficient according as function of the quality :

$$h_{tp}(x) = 3{,}5\, F_{tp}\, h_L \left(\frac{1}{X_{tt}}\right)^{0{,}5}$$

Xtt is the Martinelli's parameter

h_L = heat transfer coefficient in the liquid phase (W/m²).

The convection term is evaluated using the Dittus-Boelter equation :

$$Nu_L = 0.023\, Re_L^{0.8}\, Pr_L^{0.4}$$

$$h_L = \frac{ls_L}{De_{134}} Nu_L$$

and

$$X_{tt} = \left(\frac{\rho_v}{\rho_L}\right)^{0.5} \cdot \left(\frac{\mu_L}{\mu_v}\right)^{0.1} \cdot \left(\frac{1-x}{x}\right)^{0.9}$$

This relationship is valid for a vapor quality below y = 0.86.

For the heat transfer coefficient in the two-phase region the average value in the range of validity was considered.

$$\bar{h}_{tp} = \frac{X_{su}\int_0^y h_{tp}(x)\,dx}{y - X_{su}}$$

giving:

$$\bar{h}_{tp} = F_{tp}\, 3.5\, h_L \left(\frac{\rho_L}{\rho_v}\right)^{0.25} \left(\frac{\mu_v}{\mu_L}\right)^{0.05} \frac{\sum_{x = x_{su}}^{y} \left(\frac{x}{1-x}\right)^{0.45} .\ \delta x}{y - x_{su}}$$

δ= integration step

The hydraulic diameter is determined according to the geometric dimensions of tubes. The average speed is defined as a function of the number of sandwiches (number of evaporator fin plate) of the first and second partition:

$$\dot{u} = \frac{\dot{M}_{134} \cdot v_l}{\frac{(N_1 + N_2)}{2} \times 4 \text{ tubes/ plate } A_{tube}}$$

N : Evaporator plates number

The heat transfer coefficient in the superheat region (refrigerant side) is found using the following relationship:

$$Nu_{sh} = 0.023\, Re_v^{0.8}\, Pr_v^{0.4}$$

$$h_{sh} = F_{sh} \frac{k_v}{De_{134}} Nu_{sh}$$

The average speed in the superheat region is given by:

$$\dot{u}_{sh} = \frac{\dot{M}_{134} \cdot V_v}{N_2 \times 4 \,.\, A_{tube}}$$

<u>Processing the Experimental Results</u>

The experimental results were compared to the simulation results. In the thermal model (all correction coefficients) an underestimate of the refrigerant power is always observed.

The next step of the processing of test results identified correction coefficients F_a, F_{tp}, and F_{sh} to reduce the error in the evaporator output temperature measurement and hence also in the refrigerant power.

<u>General Relationship for Air-Side Pressure Drops</u>

Based on the results obtained for several relationships it is clear that the laminar flow model cannot be applied to the entire experimental range (including the wet regime). The pressure loss coefficient z is therefore a Reynolds function of Reynolds number ($z = z(Re^n)$). Assuming that the six evaporators have a similar internal structure, we find the following relationship:

$$dP_{ev(i)} = dP(\zeta, Au_i, v_a, \dot{V}_a) \quad (Pa)$$

where Au_i is the effective evaporator cross-sectional area (m^2).

$$Au_i = Hu_i \times L$$

There are thus two rules for each regime.
The following rule applies to all evaporators:

$$dP_a = K \cdot \frac{\dot{u}^n}{v_a}$$

dP_a : air pressure drop.
where $\dot{u}$ = average velocity of air at inlet (m/s).

As a result, we obtain:

Dry regime: $$\Delta P_{a_s} = K_s \cdot \frac{1}{v_a} \cdot \left(\frac{\dot{V}_a}{Au_i}\right)^{ns} \quad (Pa)$$

Wet regime: $$\Delta P_{ah} = K_h \cdot \frac{1}{v_a} \cdot \left(\frac{\dot{V}_a}{Au_i}\right)^{nh} \quad (Pa)$$

s,h : dry and humid conditions.
In this case,
v_a = specific volume for the air inlet conditions (m^3/kg)
$\dot{V}_a$ = volume flow rate of the air at the evaporator inlet (m^3/s)
The combined results of tests on six evaporators are used. After optimizing the relationship, the values of K_h, nh, K_s and ns were established.

Pressure Drops on Refrigerant Side

For a simplified first model, the aim is to find a relationship of the following type:

$$\Delta P_r = C_1 \dot{V}_{exev}^2 + C_2 \dot{V}_{exev}$$

Using the experimental values we calculate $\dot{V}_{exev}$ for each test and each evaporator, and obtain the coefficients C_1 and C_2 . This relationship gives poor results for the entire range.

The Reynolds number on the refrigerant side varies between 15 x 10^3 and 65 x 10^3. This means that the flow does not always reach quadratic turbulence. With this in mind, two more modeling attempts are made. Considering the refrigerant mass flow rate and, as parameters, quality at the evaporator inlet and specific volume at the evaporator outlet. The pressure drop is given by:

$$\Delta P_r = (C_5 + C_6 X_{suev}) \cdot M_{ev}^2 v_{exev} \quad (bar)$$

C_5 and C_6 determined from test results.
This model seems to be the most accurate, although it is the most complicated of the models.

The Comparison of experimental results with calculated results using the combination of the thermal model and the refrigerant side pressure loss model gives the following results.

$\dot{M}_{air}$ [kg/h]	P_{suev} [bar abs]	T_{suev} [°C]	T_{exev} [°C]	$\dot{Q}_o$ [W] Simulation	$\dot{Q}_o^{valeo}$ [W] test	$d\dot{Q}$ [W]	e [%]
EVAPORATOR 1							
400	3.53	5.2	8.0	5275	5400	-125	-2.3
500	3.64	6.2	8.0	6281	6500	-219	-3.4
600	3.83	7.6	9.0	6954	7300	-346	-4.7
EVAPORATOR 3							
400	3.52	5.2	8.0	5186	5100	86	1.7
500	3.67	6.4	8.0	6061	5900	161	2.7
600	3.83	7.6	9.0	6708	6700	8	0.1
EVAPORATOR 4							
400	3.65	6.2	8.0	5026	5000	26	0.5
500	3.87	7.9	9.0	5689	5700	-11	-0.2
600	4.00	10.0	10.0	6300	6500	-200	-3.1
EVAPORATOR 5							
400	3.59	5.8	8.0	5175	5200	-25	-0.5
500	3.74	6.9	9.0	5999	6100	-101	-1.7
600	3.89	8.1	10.0	6660	6800	-140	-2.1
EVAPORATOR 6							
400	3.68	6.4	8.0	5108	5300	-192	-3.6
500	3.83	7.6	9.0	5909	6400	-491	-7.7
600	4.0	8.9	11.0	6429	7100	-671	-9.5

COMPRESSOR MODEL [4], [5]

The compressor model must fit into an overall model of the A/C system. For modeling this component we have sought to define laws of the following type:

For fixed-displacement compressors:

$$M = f_1(p_{sucp}, v_{sucp}, p_{excp}, N) \quad (1)$$
$$W = f_2(p_{sucp}, v_{sucp}, p_{excp}, N) \quad (2)$$

M : refrigerant flow rate
W : compressor power requirement

For variable-displacement compressors there is an internal displacement control which is controlled by a suction pressure set point. This set point is a decreasing function of the discharge pressure. This internal control must obviously be taken into account in the model:

$$M = f_1(p_{sucp}, v_{sucp}, p_{excp}, N, V_s) \quad (3)$$
$$W = f_2(p_{sucp}, v_{sucp}, p_{excp}, N, V_s) \quad (4)$$
$$(p_{sucp}) = f_3(p_{excp}) \quad (5)$$

The three characteristic equations (3), (4), and (5) can be combined in the following form:

$$p_{sucp} = f_1(M, v_{sucp}, p_{excp}, N) \quad (6)$$
$$W = f_2(M, v_{sucp}, p_{excp}, N) \quad (7)$$

In this new form the displacement no longer appears. It is an internal variable of the compressor model.

Development of Compressor Model

Starting from simple formulae for an ideal compressor, and incorporating the effects of cylinder clearance (dead space), pressure loss at the compressor inlet and outlet, internal leakage, and mechanical losses. We obtain the following expression for the compressor flow rate (8)

$$\frac{Mv_{su}}{N} = C_0 + C_1(P_F) + C_2\left(\frac{N^2}{p_{su}v_{su}}\right) + C_3 N\sqrt{p_{su}v_{su}}\,\pi^{\frac{\gamma-1}{2\gamma}} + \ldots\ldots\ldots$$

P_F : Pressure factor
A_u= surface area (fictitious) of suction orifice.
A_1= surface area (fictitious) of internal leakage orifice.

For lubricated compressors, internal leakage is often negligible. In addition, the effect of pressure drops is difficult to model. A simple alternative to Equation (8) is to keep only two characteristic parameters of the compressor at a constant speed of rotation.

$$\left(\frac{Mv_{su}}{N}\right) = (C_0)_N + (C_1)_N P_F \quad (9)$$

with $C_0 = f_1(N) \approx C_{00} + C_{01}N$
$C_1 = f_2(N) \approx C_{10} + C_{11}N$
and thus

$$\frac{Mv_{su}}{N} = C_{00} + C_{01}N + C_{10}P_F + C_{11}P_F N \quad (10)$$

C_{00} = displacement at zero speed
C_{01} = sensitivity of the displacement to speed of rotation
C_{10} = cylinder clearance at zero speed
C_{11} = sensitivity of cylinder clearance to speed of rotation

$$\frac{W}{N} = C_0 + C_1\left(\frac{Mw_s}{N}\right) + C_2\left(\frac{M^3 v_{su}^2}{N}\right) + \ldots.. \quad (11)$$

w_s : compressor isentropic work
C_0= mechanical losses in compressor (for one shaft revolution)
$C_1 = (1+\alpha)$, α : mechanical loss coefficient
$C_2 = 1/2A^2_{su}$
By substituting the average flow rate from equation (8) for the flow rate in equation (11), we obtain:

$$W = C_0 + C_1\frac{Nw_s}{v_{su}} + C_2 P_F \frac{Nw_s}{v_{su}} + \ldots\ldots \quad (12)$$

For each specific compressor, the number of terms of the polynomial is adjusted according to the experimental results available and the accuracy required.

Validation

The following figures show two types of correlations between calculated and measured values of two types of compressors: fixed displacement rotary vanes and variable displacement wobble plate.

- first correlation $\frac{M.vsu}{N}$
- Second correlation $\frac{W}{N}$

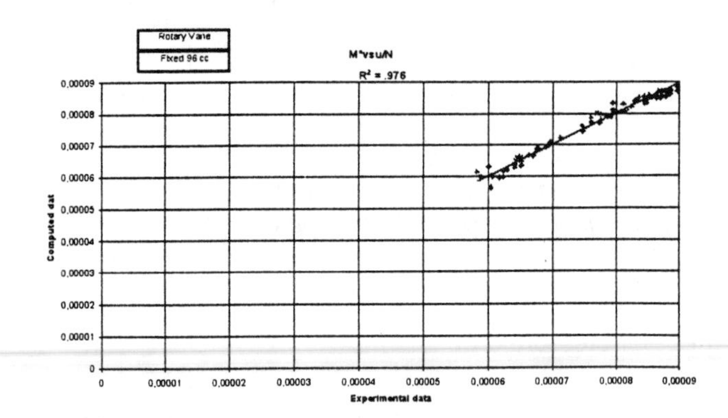

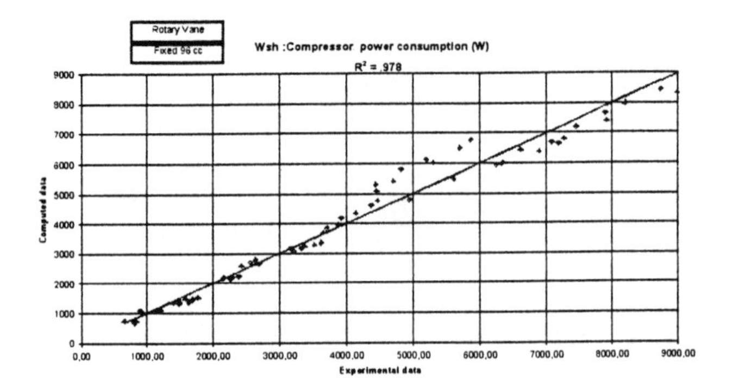

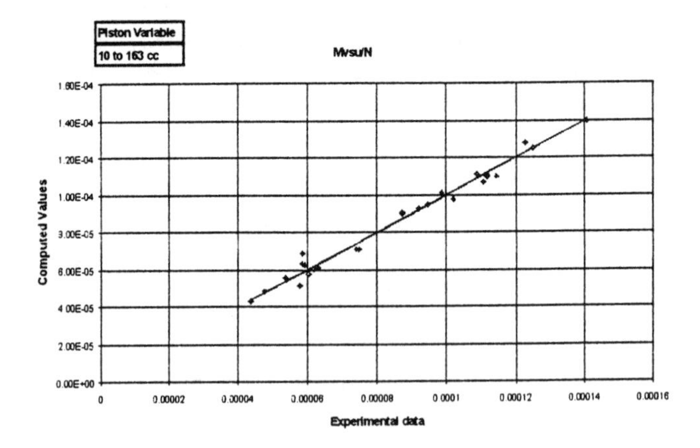

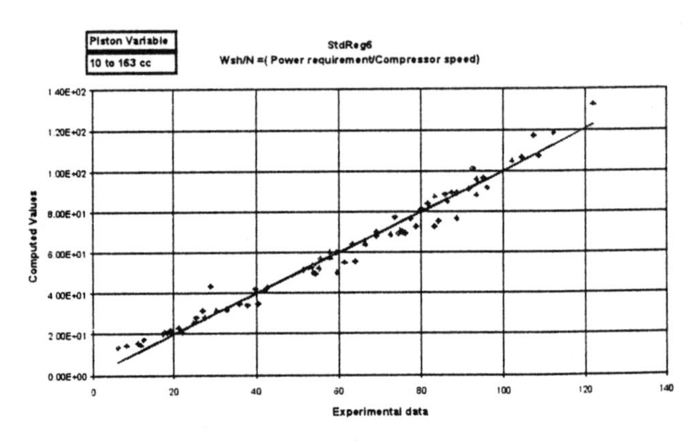

ORIFICE TUBE MODEL

Flow in a capillary tube can be modeled globally or locally.
The global approach considers the capillary tube as a diaphragm. The fluid flow rate is therefore calculated from a simple pressure losses equation. The pressure loss coefficients must be adjusted for the flow rate calculation. This approach gives good results, but they cannot be extrapolated beyond the experimental range.

The local approach calculates the pressure profile along the tube. The critical speed is calculated using the tube exit conditions (pressure, degree, etc.). This method is based on the calculation of single- and double- phase pressure loss in the capillary tube. The accuracy of the calculation therefore depends on the accuracy of the relationships used to calculate the pressure losses. In addition, the calculation assumes the two-phase mixture to be in thermodynamic equilibrium; an assumption that is not valid over the entire range. Consequently, this model can underestimate the flow rate, especially for short capillary tubes.

Confirmation

Local Method

For the local method, calculation by an iterative procedure is difficult, because convergence cannot always be reached. In order to confirm the behavior of the local model, a simulation is carried out for the following geometry:

- Tube length: 38.3 mm
- Tube diameter: 1.28 mm
- Upstream and downstream diameter: 6 mm

Downstream pressure is set at 0.35 MPa.

The results are shown in the following table:

subcooling °C	Flow rate (kg/h) Pup = 2 MPa		Flow rate (kg/h) Pup = 2.5 MPa	
	Calc.	test	Calc.	Meas.
5	88.6	138	90	159
10	94.6	161	94.56	181
15	101.5	183	100.37	204

Comparison of the model with test results reveals that the flow rate was underestimated by almost 50%. This difference is due to the metastable flow length not being taken into account, as well as a poor estimate of pressure losses in the capillary tube inlet region. This method can therefore be applied to short orifice tubes only if the metastable flow region is taken into account. This method is not used in the software calculation.

Global Method

This method is based on an equation used by Aaron and Domanski (1990) which takes into account tube inlet geometry. The following graphic compares refrigerant flow rates obtained from simulation and test results for a tube of diameter 1.148 mm, length 38.354 mm, and a chamfer 0.13 mm long:

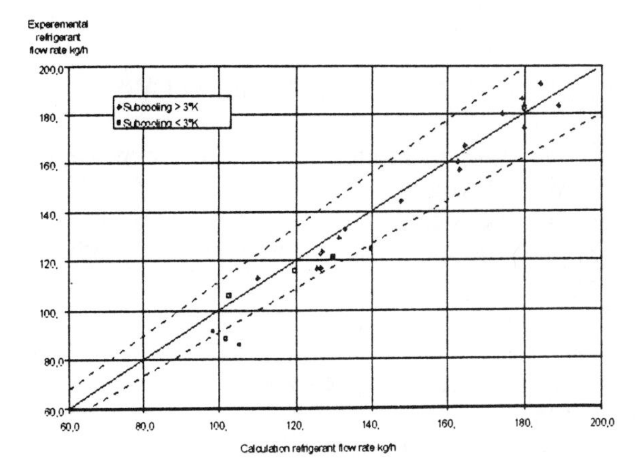

The dashed lines represent the limits at + and - 10% of the differences between calculated and measured flow rates. Note that the difference can reach values above 10% for low values of subcooling only.

As a first approximation, we consider that the values determined by measurement are overestimated compared with the calculated values because of the measurement method using a volume flow meter located in the liquid phase of the circuit. In fact, low subcooling values do not guarantee a completely liquid refrigerant state, because any presence of gas would lead to an increase in refrigerant specific volume, resulting in an overestimate of the actual mass flow rate. The presence of an important percentage of oil may also lead to a significant error in the measurement of the refrigerant flow rate.

This analysis makes it possible to select the most appropriate orifice tube for the operation of an A/C system.

THERMOSTATIC EXPANSION VALVE MODEL (TXV)

It is useful to model a TxV not only for its design or optimization, but also to learn about the overall behavior of an A/C system. In the literature, TxVs are modeled either in steady-state or in dynamic conditions. The modeling principle most often applied considers the TxV as a variable cross-section diaphragm subjected to a pressure difference. There are essentially two types of TxV models: knowledge models based on an analytical approach, and behavior models based on identification.

For the present, a knowledge model has been chosen to model the TxV. This model is based on a set of balance equations (mass, quantity of movement or energy), and on the knowledge of the geometrical and physical characteristics of the TxV.

Cross-sectional area of the TxV for a given superheat is defined by:

$$Q_f = C.A\sqrt{2.g.\Delta P.v}$$

Qf : Refrigerant flow rate

C : Specific constant determined experimentally

A : TxV section

g : 9.8 m/s

Δp : Inlet press - Outlet press.

v: specific volume

the software uses some characteristic curves

from databases to select the A/C system TxV.

CONDENSER MODEL (PARALLEL FLOW)

In order to model the condenser, we applied the same approach as for the evaporator. The condenser is considered as the sum of three exchangers (superheat, condensation, and sub-cooling).
Most equations used in this simulation were found in the literature (Shah, Cavallini, Lockhart-Martinelli, etc.), and correction factors were determined on the basis of test results.

The developed model has been able to predict the heat transfer rate within +/- 4% of the experimental data ; and refrigerant pressure drop within 10%.

RECEIVE DRYERS AND ACCUMULATORS

For now, receiver dryers and accumulators are modeled as simple ducts which are characterized by their pressure drops and thermal losses.

TUBES AND HOSES

Knowing the dimensions (length, diameter, material, number of bends...) of the tubes and hoses of the A/C system the pressure drop and thermal losses are calculated using well known equations.

A/C SYSTEM OPTIMIZATION

The objectives of the optimization are the following :
1) Definition of the theoretical specifications and characteristics of the A/C system components to satisfy the thermal power requirements.
2) Selection of an A/C system configuration from the software data base. This selection is made according to fixed criteria such as compactness, cost, weight, performance.....
3) Simulation of the selected configuration in order to compute the performance criteria.

Conclusion

The utilization of this software is based on the exploitation of a component parameter data base which is constantly updated with new components. These parameters are easily obtained through identification from tests results. Several validations have been achieved for different car cabin and different A/C system configurations. The comparison between test and simulated result gives a good level of accuracy.

The major advantages and performances of this software are :

- Reduction of A/C system development time.
- Prediction of all A/C system parameters with an accuracy of 5 to 10 %.
- Reduction of testing requirements.
- Sizing, simulation and optimization of all A/C systems.
- Easy-to-use interactive user interfaces.

This software is now used for customer project design and will be deployed in major Valeo Climate Control Research Centers.

REFFERENCES

Matlab/simulink :[1] language of technical Computing and dynamic simulation (Mathworks Inc) USA.

ASHRAE [2], 1990 ASHRAE Handbook-Refrigeration, The American Society of Heating, Ventilation and Air-Conditioning Inc., Atlanta, USA

Webb R.L [3], 1990, Air-Side heat Transfer Correlations for Flat and Wavy Plate Fin-and-Tube Geometries, ASHRAE Transactions, Vol 96, Part 2, USA.

Dirlea R., Hannay J.,Lebrun J.,and Pimenta J. [4] 1994, « Experimental study and Performance Evaluation of a Refrigeration Compressor ». Synthesis Report, University of Liège, Laboratory of thermodynamics, Liége Belgium.

Silva, K.L [5], 1995, Simplified Approach for Reciprocating Compressors : Modeling Development and Experimental Analysis, Ph.D Thesis , Laboratory of thermodynamics, Liége Belgium.

Rasvan.D, [2],[3],[4], [5] 1997 Performance Rating of refrigeration components, Ph.D Thesis, Laboratory of thermodynamics, Liége Belgium.

NOMENCLATURE

M°	mass flow rate	kg/s
Q°	time rate of heat transfer	W
V°	volume flow rate	m^3/s
A	area	m^2
b	polynomial law coefficient	
c	Specific heat	J/kgK
Cf	clearance factor	-
C_p	specific thermal capacity at constant pressure	j/KgK
D	generic diameter, outer diameter	
d	inner diameter	m
g	gravitational acceleration	
h	heat transfer coefficient	W / (m^2K)
h	specific enthalpy	J/kg
H	total enthalpy	J
k	thermal conductivity	W / (mK)
l	length	m
m, M	mass	kg
N	Evaporator plate number	-
p	pressure	Pa
q	specific heat transfer	J/kg
Q	total heat transfer	J
s	specific entropy	J / (kgK)
T	absolute temperature	K
t	temperature	°C
U	overall heat transfer coefficient	
U	total internal energy	J
v	specific volume	m^3/ kg

V : total volume m^3
w : specific work input J/kg
W : work input J
x : quality -

Greek symbols

μ : dynamic viscosity Pa/s
ε : efficiency, effectiveness or relative error
δ : individual accuracy
Δ :difference between values -
ρ : density kg/m^3
σ : standard deviation
τ : time s

Dimensionless numbers

Nu : Nusselt number ($h\ D\ /\ k$)
Pr : Prandtl number ($Cp\ \mu\ /k$)
Re : Reynolds number ($M°\ D\ /\ \mu\ S$)

Subscripts

a : air
amb : ambiance
cp : compressor
cyl : cylinder
ev : evaporator
ex : exhaust
l : liquid state
g : gas state
r : refrigerant
s : refers to isentropic evolution
sat : indicates a saturate state
sc : subcooling degree
sh : superheating degree
su : supply
w : water

Superscripts

° : variation with time
‾ : average

Abbreviations

COP : Coefficient of Performance
TxV : Thermostatic Expansion Valve

951014

Computer Simulation of Refrigerant Vapor Condenser in Transient Operation

Edward C. Chiang and Simon Y. C. Ng
Michigan Technological Univ.

ABSTRACT

The formulation of mathematical model for the computational simulation of transient temperature response and phase change of refrigerant in a vapor condenser of an automotive air conditioning unit is described. A demonstrative computational simulation of a sample air cooled vapor condenser charged with Freon 12 is presented. The computational analysis predicts an initial surge and followed by an oscillation of the condensate outflow rate from the condenser when the air-conditioning unit is started, and the tube length required for complete condensation of inflow vapor is a maximum value at start up. The rise of the temperatures of the condenser tubes and cooling air flow during the start-up and load change operations rate found to be gradual but the scale of these temperature changes are considered small.

INTRODUCTION

A crossflow air cooled vapor condenser is a standard device employed to condense the compressed refrigerant vapor in an automotive air-conditioning unit. Often the condenser is installed in front of the radiator. Therefore, the ram airflow is the primary cooling fluid, and a phase change of the refrigerant flow takes place in the condenser coil where the vapor is condensed to liquid. Such forced convective condensation heat transfer is one of the most challenging topics in two-phase flow.

The process, when the regrigerant flow rate and/or inlet temperatures change with time, are referred to as transient operating conditions. The transient condenser operation occurs during start-up and load change operations in an automotive air-conditioning unit. During transient operation, the heat capacity of the tube wall, the mass of stored refrigerant and the heat transfer along the tube length are the physical parameters which govern the transient response of the condenser. These physical parameters are time-dependent and inter-dependent in nature. For instance, the transient change of refrigerant inlet temperature may cause variations in the refrigerant properties and variations of refrigerant properties may eventually affect heat transfer rate in the condenser. Generally, determining a heat transfer coefficient for condensation is more complicated than obtaining one for single-phase convective heat transfer. Often empirical correlations of heat and mass transfer are used in many engineering calculations involving condensation.

In the analysis presented, a simplified lumped parameter condenser model is used in constructing the governing equations, and the "system mean void fraction model" [1] is used in determining the tube length for complete condensation. Average heat transfer coefficient for condensation was calculated using data listed in the ASHRAE handbook for fundamentals [2].

THEORETICAL ANALYSIS AND MATHEMATICAL FORMULATION

The condenser is a crossflow heat exchanger. Two separate lumped analytical models were developed describing the dynamics of thermal response in the two-phase flow region and the single-phase subcooled or superheated flow region of a vapor condenser respectively. The formulation of the analytical model is based on the following assumptions:

1. Pressure drop in the condenser is neglected.
2. The thermodynamic properties of a refrigerant is a function of saturated temperature.
3. Heat flow from the refrigerant is dominated by condensation heat transfer in two-phase flow region and by convective heat transfer in single-phase subcooled or superheated flow region.
4. Mass diffusion is neglected.
5. The thermal capacitance of the air in the condenser core is negligible compared to the thermal capacitances of the condenser tube and the refrigerant.
6. There are negligible changes in kinetic and gravitational potential energies.

In unsteady state problems, time is a variable and some properties of the system are functions of time. Conservation of mass implies the following:"

ACCUMULATION = INFLOW - OUTFLOW

When no heat is generated in the condenser, the following statement applies to the energy balance in a condenser:

ENERGY ACCUMULATION = ENERGY INFLOW - ENERGY OUTFLOW

Four differential equations are derived from the physical laws of mass and energy conservation for the solid metal core, the refrigerant and the cooling air. The condenser model for this project is shown in Figure 1. The two regions of two-phase and subcooled flow are noted in the Figure.

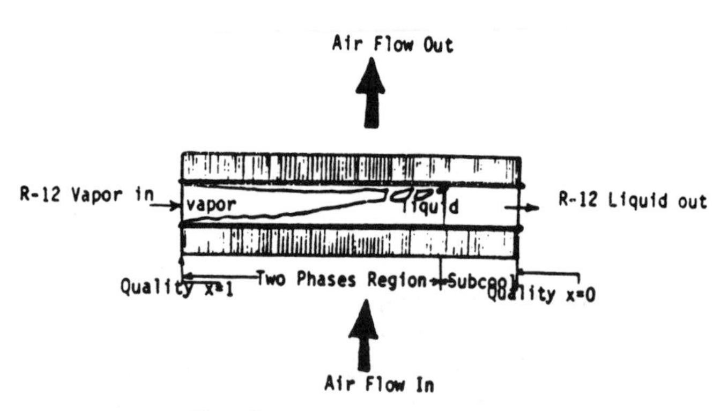

Fig.1 Vapor Condenser Model

Energy and mass equations in the two-phase flow region are as follows:

Refrigerant:

Mass;

$$\frac{d}{dt}(M_g + M_f) = \dot{M}_{in} - \dot{M}_c \tag{1}$$

Energy;

$$\frac{d}{dt}(M_g U_g + M_f U_f) = \dot{M}_{in} h_{in} - \dot{M}_c h_{out} - \alpha_i A_i (T_R - T_w) \tag{2}$$

where $h_{in} = h_f(1-x_o) + h_g x_o$ and $h_{out} = h_f$

Condenser Wall:

Energy;

$$\frac{dT_w}{dt}(V_w C_{pw} \rho_w) = \alpha_i A_i (T_R - T_w) - \alpha_o A_0 (T_w - T_a) \tag{3}$$

where T_w is average tube wall temperature

Cooling Airflow:

Energy;

$$\frac{\partial T_a}{\partial t}(V_a C_{pa} \rho_a) + (C_{pa} A_a \rho_a Vel_a)\frac{\partial T_a}{\partial y} = \alpha_o A_o (T_w - T_a) \tag{4}$$

where T_a is the temperature of existing air form the core of condenser.

The "system mean void fraction model" [1] is used in the analysis presented to determine the apparent tube length for complete condensation of inflow refrigerant vapor. The method requires that the terms of M_f and M_g [in equations (1) and (2)] must be related to the L_c, Area, ρ_f, ρ_g and $\bar{\alpha}$ by the following equations:

$$M_f = [(1-\bar{\alpha})\rho_f \text{Area}] L_c \tag{5}$$

$$M_g = [\bar{\alpha}\rho_g \text{Area}] L_c \tag{6}$$

where L_c is the tube length required for total condensation of the inlet vapor. And $\bar{\alpha}$ is the "system mean void fraction", defined in the particular manner specified by Wedekind, et. al. [1], and the equation to compute the $\bar{\alpha}$ value is given below

$$\bar{\alpha} = \frac{1}{1-c} + \frac{c}{(1-c)^2 x_o}\ln\left[\frac{c}{(1-c)x_o + c}\right]$$

in which

$$c = \left(\frac{\rho_g}{\rho_f}\right)^{\frac{2}{3}}.$$

The term $\dot{M}_c$ appears both in equations (1) and equation (2), the two equations can be combined into a single governing equation to determine the effective tube length L_c required for the total condensation. After much algebraic manipulation and rearrangement, the following governing equation for tube length L_c is obtained:

$$\frac{dL_c}{dt} = \frac{-\alpha_i P_i (T_R - T_W) L_c + x_o (h_g - h_f)\dot{M}_{in}}{\rho_g \bar{\alpha} \text{Area}(U_g - h_f) + \rho_f (1-\bar{\alpha}) \text{Area}(U_f - h_f)}. \tag{7}$$

Assumption (5) allows the first term of the left hand side of equation (4) to be neglected. The spatial derivative of T_a in equation (4) is discretized by the upwind finite difference scheme, and the following equation which relates the outlet air temperature T_a to T_w and T_∞ is established:

$$T_a = \frac{\rho_o P_o \Delta Y T_w + \rho_a A_a C_{pa} Vel_a T_\infty}{\rho_a A_a C_{pa} Vel_a + \alpha_o P_o \Delta Y}. \tag{8}$$

Substituting the value for T_a from equation (8) into equation (3), and after rearranging, the governing equation of wall temperature T_w becomes:

$$\frac{dT_w}{dt} = \left[-\frac{\alpha_i P_i + \alpha_o P_o}{\rho_w C_{pw} A_w} - \frac{(\alpha_o P_o)^2 \Delta Y}{\rho_w C_{pw} A_w (\rho_a A_a C_{pa} Vel_a + \alpha_o P_o \Delta Y)}\right] + \frac{\alpha_i P_i T_w}{\rho C_{pw} A_w} + \frac{\alpha_o P_o (\rho_a A_a C_{pa} Vel_a) T_\infty}{\rho_w C_{pw} A_w (\rho_a C_{pa} A_a Vel_a + \alpha_o P_o \Delta Y)} \tag{9}$$

Equations (7) and (9) are two simultaneously differential equations. They can be solved by the Runge-Kutta integration technique for L_c and T_w. Thereafter, T_a can be revised by substituting T_w and T_∞ in equation (8).

Energy and mass equations in the single-phase subcooled or superheated region are as follows:

Refrigerant:

Mass;

$$\frac{d}{dt}[\rho_f Area(L-L_c)] = \dot{M}_c - \dot{M}_{out} \tag{10}$$

Energy;

$$(\rho_f C_{pf} V_f)\frac{\partial T_R}{\partial t} + (\rho_f C_{pf} V_f Vel_f)\frac{\partial T_R}{\partial x} = -h_i A_i (T_R - T_w) \tag{11}$$

Condenser Wall:

Energy;

$$(\rho_w C_w V_w)\frac{dT_w}{dt} = h_i A_i (T_R - T_w) - h_o A_o (T_w - T_a) \tag{12}$$

Airflow:

Energy;

$$(\rho_a V_a C_{pa} Vel_a)\frac{dT_a}{dY} = h_o A_o (T_w - T_a) \tag{13}$$

Again, the spatial derivative of T_a in equation (13) is discretized by the upwind finite difference scheme, yielding the following equation for the temperature of T_a:

$$T_a = \frac{h_o P_o \Delta Y T_w + (\rho_a C_{pa} A_a Vel_a) T_\infty}{(\rho_a C_{pa} A_a Vel_a) + h_o P_o \Delta Y} \tag{14}$$

Substitution of the equation (14) for T_a into the equation (12) for T_w and discretizing the system of equations (11), and (12) according to the fully implicit finite difference scheme, the matrix of equations to be solved for forecasting T_w and T_R of a given discretized element of condenser tube is shown below:

$$\begin{bmatrix} a_{11} & a_{12} \\ a_{21} & a_{22} \end{bmatrix} \begin{bmatrix} T_R^{n+1} \\ T_w^{n+1} \end{bmatrix} = \begin{bmatrix} b_1 \\ b_2 \end{bmatrix} \tag{15}$$

in which

$$a_{11} = 1 + \frac{Vel_f \Delta t}{\Delta x} + \frac{h_i P_i \Delta t}{\rho_f C_{pf} A_i}$$

$$a_{12} = \frac{h_i P_i \Delta t}{\rho_f C_{pf} A_i}$$

$$a_{21} = \frac{h_i P_i \Delta t}{\rho_w C_{pw} A_w}$$

$$a_{22} = 1 + \frac{(h_i P_i \Delta t + h_o P_o \Delta t)}{\rho_w C_{pw} A_w} + \frac{(h_o P_o)^2 \Delta Y \Delta t}{\rho_w C_{pw} A_w [\rho_a A_a C_{pa} Vel_a + h_o P_o \Delta Y]}$$

$$b_1 = \frac{Vel_f \Delta t}{\Delta x} T_R^{n+1} + T_R^n$$

$$b_2 = \frac{h_o P_o \Delta t \rho_a C_{pa} A_a Vel_a}{\rho_w C_{pw} A_w [\rho_a A_a C_{pa} Vel_a + h_o P_o \Delta Y]} T_\infty + T_w^n$$

This matrix can be solved by the Gauss Elimination method.

COMPUTATION OF THE HEAT TRANSFER COEFFICIENT IN TWO-PHASE REGION

The heat transfer coefficients in the above governing equations are a part of the physical parameters which are computed according to the condenser operation. For the horizontal tube representation of a condenser tube shown in Figure 2, the procedure stated in the ASHRAE hand-Book of Fundamentals [2] was found to be adaptable for this case. The working formulas of condensation heat transfer coefficients adopted in this work are summarized by the following equation:

$$\alpha_i = 13.8\frac{k_f}{D}\left(\frac{C_{pf}\mu_f}{k_f}\right)^{\frac{1}{3}}\left(\frac{h_{fg}}{C_{pf}\Delta t}\right)^{\frac{1}{6}}\left[\frac{D\dot{M}_{in}}{\mu_f}\left(\frac{\rho_f}{\rho_g}\right)^{0.5}\right]^{0.2}$$

when

$$1000 < \frac{D\dot{M}_{in}}{\mu_f}\left(\frac{\rho_f}{\rho_g}\right)^{0.5} < 20000 \quad \text{and} \quad \frac{D\dot{M}_{in}}{\mu_f} < 5000$$

or

$$\alpha_i = 0.1\frac{k_f}{D}\left(\frac{C_{pf}\mu_f}{k_f}\right)^{\frac{1}{3}}\left(\frac{h_{fg}}{C_{pf}\Delta T}\right)^{\frac{1}{6}}\left[\frac{D\dot{M}_{in}}{\mu_f}\left(\frac{\rho_f}{\rho_g}\right)^{0.5}\right]^{\frac{2}{3}}$$

when

$$20000 < \frac{D\dot{M}_{in}}{\mu_f}\left(\frac{\rho_f}{\rho_g}\right)^{0.5} < 100000 \quad \text{and} \quad \frac{D\dot{M}_{in}}{\mu_f} < 5000.$$

This completes the derivation of the computational model of the condenser.

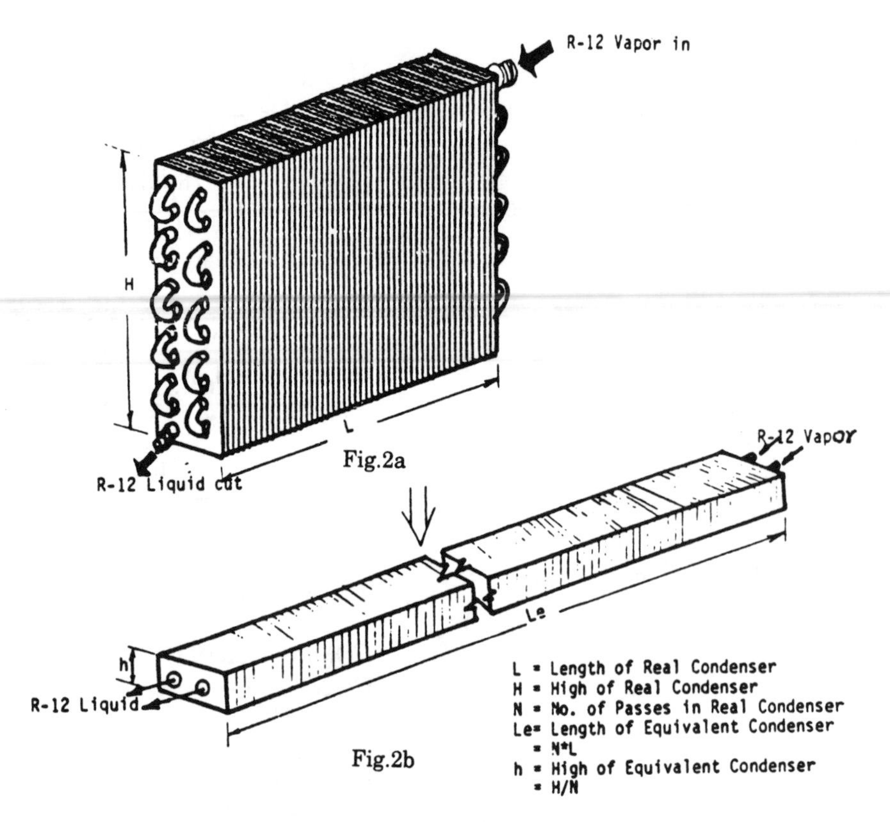

Fig.2 A Real Condenser and an Equivalent 2-single Horizontal Tubes Condenser

APPLICATIONS OF THE VAPOR CONDENSER MODEL

The transient behavior of the condenser during start-up is considered in this study, and the scope of study is limited to the two-phase region and the subcooled region, and the inflow is saturated vapor from the desuperheating section of the condenser.

Initially the liquid refrigerant (R-12), the condenser wall, and the cooling ram air are at the ambient temperature. When the air conditioner is started both refrigerant and air are forced into the condenser and their respective flow rates defined. Air enters the condenser at ambient temperature and a given velocity. Refrigerant enters the condenser at a prescribed inlet temperature and a prescribed mass flow rate. Two computer simulations were conducted in this study. They are to simulate:

1. the start-up of an air-conditioning unit with simultaneous varying of inlet temperature and inflow rate of saturated refrigerant vapor entering the condenser
2. a transient operation with only linear increasing of inflow mass of saturated refrigerant vapor at constant temperature.

The physical description of a sample condenser modeled in this study is given in Table 1.

TABLE 1

1. Overall Dimension:
   Height = 51 cm; Width = 72 cm; Depth = 7.8 cm
2. Refrigerant flow path through condenser: Cross-Counterflow
3. Tube bundle configuration
   a. Number of tube row = 2
   b. Number of tube pass = 14
   c. Type of tube arrangement = in line
4. Tube nominal size = 1.9 cm; Tube thickness = 0.165 cm
5. Fin configuration
   a. Number of fins per cm = 5.5
   b. Fin thickness = 0.00762 cm
   c. Fin shape = serpentine
6. Material Composition
   a. Tube material = brass
   b. Fin material = Aluminum

RESULTS AND DISCUSSION

CASE 1: TIME DEPENDENT INFLOW MASS AND REFRIGERANT INLET TEMPERATURE IN A START-UP OPERATION - This case represents an event when the speed of the vapor compressor is gradually increasing and the cooling load of the air-conditioning unit surges at first then reducing to a normal load. The response of the outlet liquid flow rate can be evaluated by examining the mass conservation equations for the two-phase region and the subcooled liquid region. The mass flow rate leaving the condenser, $\dot{M}_{out}$, is obtained by combining equations (1) and (10) in such a way that the term $\dot{M}_c$, appearing in both equations, is eliminated. It can be shown that the mass flow rate leaving the condenser is governed by the following equation:

$$\dot{M}_{out} = \left[1 - \frac{(\rho_g - \rho_f)\,\bar{\alpha} A x_o (h_g - h_f)}{\rho_g \bar{\alpha} A (U_g - h_i) + \rho_f (1-\bar{\alpha}) A (U_f - h_i)}\right] \dot{M}_{in} - \frac{(\rho_f - \rho_g)\,\bar{\alpha} A P_i (T_R - T_w) L_c}{\rho_g \bar{\alpha} A (U_g - h_i) + \rho_f (1-\bar{\alpha}) A} \times \frac{1}{(U_f - h_i)}$$

Where L_c is the tube length required for total condensation of the inflow refrigerant vapor.

Because the vapor condenses, the above expression shows that the outflow of the subcooled liquid equals the condenser inlet vapor flow rate minus the net rate at which mass is stored in the condenser. Figure 3 shows the predicted response of the outflow corresponding to the prescribed changes of the inflow condition. At first, the compressor forces the inlet vapor into the condenser and expelling an outflow of resident condensate from the core of the condenser. Therefore, the outflow nearly equals the inflow on mass flow rate. After a short period of time, vapor condensation takes place in the core of the condenser. As liquid mass to be momentarily reduced below the inflow vapor mass. For a short time, the flow of condensate actually reverses direction, flowing back into the condenser. This causes reduced inflow as well. The results agree with the findings of Wedekind et. al. [1]. In the next stage, condensate begins to form in the condenser and begins to flow out. In the final stage, steady flow state is established.

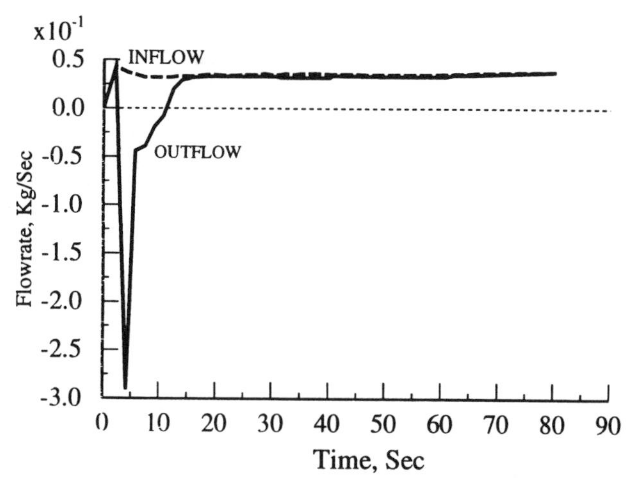

Fig.3 Dynamics Response of Condensation Outflow with Time Dependent on Inflow Mass and Inlet Temperature of Refrigerant (Case 1)

Variations of refrigerant inlet temperature are shown in Figure 4 along with the calculated wall temperature and air temperature of the two-phase region. At the inlet of the condenser the vapor temperature is at the saturation temperature. Neglecting pressure drop, the temperature of condensate in a two-phase mixture can be considered isothermal. A temperature difference exists between the wall and the refrigerant and it is the driving potential for heat transfer. The increase in wall temperature is shown as a smooth rising curve in Figure 4 despite the abrupt change of refrigerant temperature. The temperature difference between the tube wall and air increases with time. When the temperature difference between the tube wall and the refrigerant becomes constant marks the end of condenser transient operation. The quantitative results of system response such as the amplitude modification, phase lagging and time delay during the transient period provides the essential information to determine the system dynamic characteristics and the essential data needed to design an optimal thermal control system.

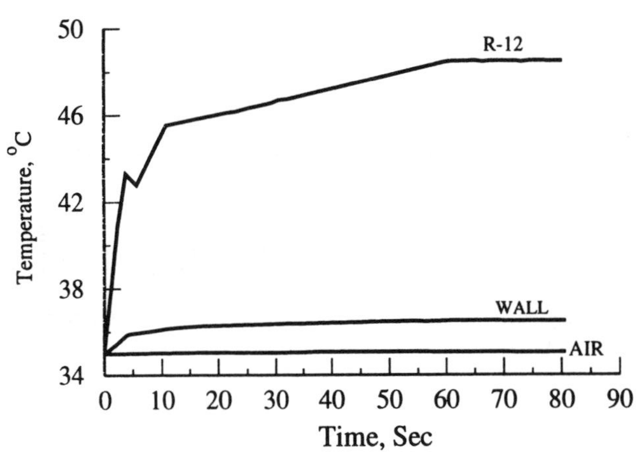

Fig.4 Transient Temperature in Two-phase Region with Time Dependent on Inflow Mass and Inlet Temperature of Refrigerant (Case 1)

Figure 5 shows the temperature variation with time in the subcooled region in the condenser. Until there is a positive outflow of liquid refrigerant into the subcooled region, subcooling of the refrigerant vapor is impossible.

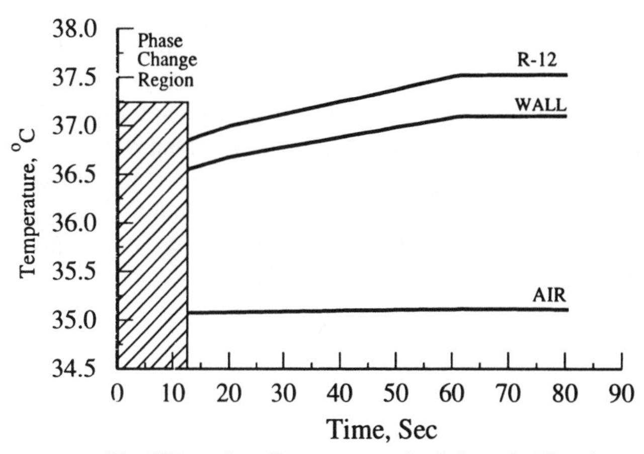

Fig.5 Transient Temperature in Subcooled Region with Time Dependent on Inflow Mass and Inlet Temperature of Refrigerant (Case 1)

The tube length required for total condensation is plotted in Figure 6. At the starting period, the tube length required for complete condensation reaches a maximum value. The reason may be explained as follows: there is a surge of vapor inflow at the start of the run which raises the outflow of liquid stored in the condenser before any condensation takes place. The tube length required for complete condensation gradually decreases, and a final length is reached when steady state of vapor condensation is established. The validation of such impulsive flow condition and the dynamic characteristics of the condenser unit simulated will be a worthwhile follow-up project when the necessary equipment and supports are available.

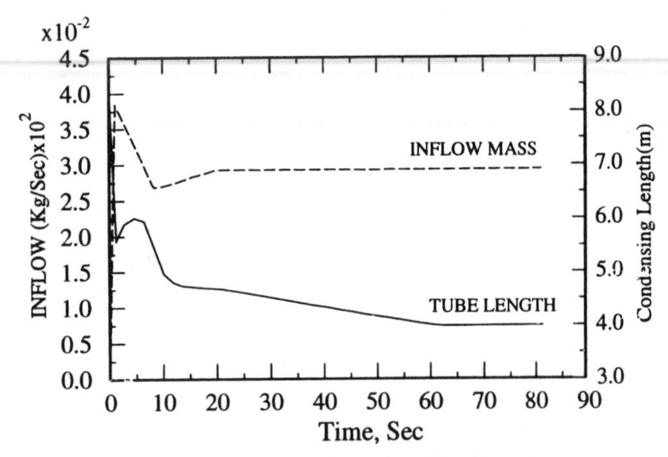

Fig.6 Tube Length for Complete Condensation of Inflow Vapor for Case 1

CASE 2: CONSTANT INLET REFRIGERANT TEMPERATURE WITH LINEAR INCREASE IN INFLOW MASS - An examination of the outflow curve in Figure 7 reveals initial phenomenon of expelling resistent refrigerant occur the same as in the Case 1. A period of reduced outflow is followed by a period of condensate accumulation. Afterwards, the quantity of the outflow is greater than the inflow but the change in outflow refrigerant mass is the same as the change in inflow refrigerant vapor. Eventually the inflow and outflow rates reach a steady flow condition where the inflow equals the outflow. The result of mass outflow due to the linear increase in the inflow agrees with the finding of Wedekind et. al. [1].

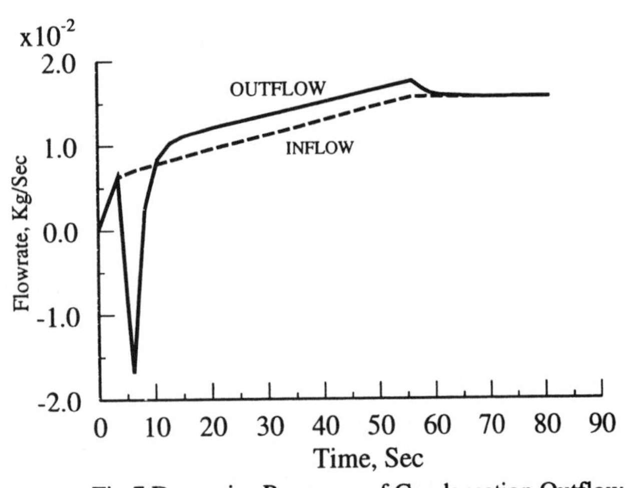

Fig.7 Dynamics Response of Condensation Outflow with Constant Inlet Temperature of R-12 and Linear Increase in Inflow Mass (Case 2)

The prediction of temperature versus time in the two-phase region and subcooled region are shown in Figure 8, 9 respectively. The inlet temperature in Case 2 is considered isothermal in the two-phase region. The tube wall gains heat from the refrigerant and its temperature rises sharply at the beginning and tapers off with time. Also, the air temperature seems to reach a steady state in a short time after start up.

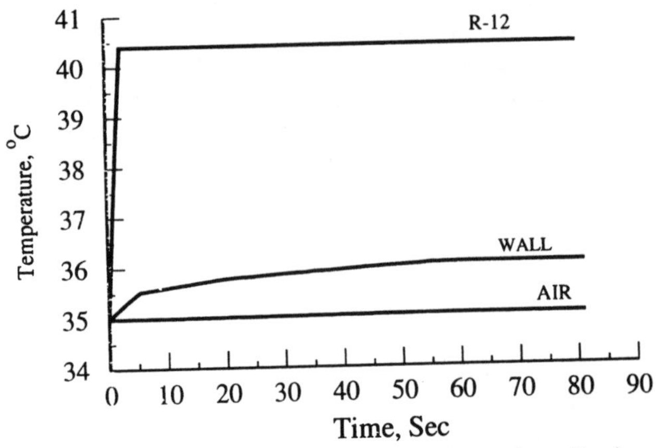

Fig.8 Transient Temperature in Two-phase Region with Constant Inlet Temperature of R-12 and Linear Increase in Inflow Mass (Case 2)

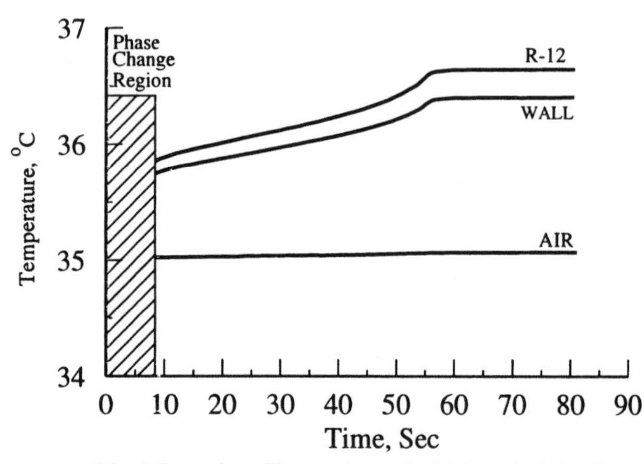

Fig.9 Transient Temperature in Subcooled Region with Constant Inlet Temperature of R-12 and Linear Increase in Inflow Mass (Case 2)

Figure 10 shows the predicted tube length required for compete condensation versus time. It shows the same trend found in Case 1 with changes of inflow mass and inlet refrigerant temperature taking place together.

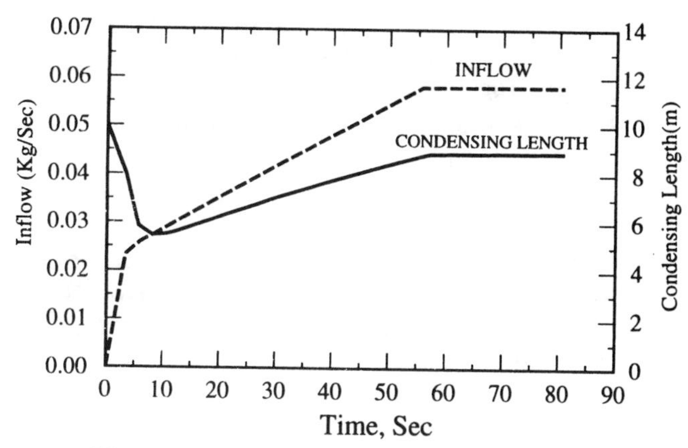

Fig.10 Tube Length for Complete Condensation of Inflow Vapor for Case 2

SUMMARY AND CONCLUSION

The objective of this study was to analyze the transient thermal responses in a vapor condenser. A mathematical model which describes the condensation process in a condenser was derived from the laws of mass and energy conservation. A mean void fraction model was employed to determine the tube length required for the total condensation of inflow vapor.

Based upon the analytical model derived for the condenser, a computer simulation program was developed. This computer simulation program is designed for predicting the transient thermal responses of a vapor condenser in an air-conditioning unit for vehicle application.

In the study by Wedekind et. al [1], the investigations were limited to determining the relation between inflow vapor mass rate and operation variables such as outlet mass rate and tube length required for compete condensation. However, in real operation both the inflow vapor mass rate and inlet refrigerant temperature have to be considered. It was the intent of this study to determine the effects of these two factors on vapor condensation.

Although no experimental data were used to validate quantitatively the present computational model, the predicted outflow condensate mass rate with respect to time agrees qualitatively with Wedekind's findings [1] when the operating conditions are similar.

The simulation program includes the following features:

1. The actual thermodynamic properties of the refrigerant is used.
2. An input data file is set up in the program. It provides the designer a convenient way to evaluate the performance of a unit of given size under specified operating conditions.

From the simulation results the following conclusions were made:

1. The tube wall temperature variation in the condenser body is an important parameter affecting the heat transfer performance. Analysis revealed that the temperature, flow rate of inflow vapor has a significant effect on wall temperature variation. The increase or decrease in wall temperature variation response directly to the mass inflow changes. Although no experimental validation was conducted for comparison, it is believed that these predicted results are quantitatively useful to guide thermal design of tube boundless in a vapor condenser.
2. It was found in all the cases, that rise in air temperature in the condenser is approximately 0.6 °C. The air temperature change through the condenser can be considered linear along the flow direction. This is considered acceptable and is in agreement with the findings by Chapman [15].
3. The tube length required for complete condensation of inflow vapor depends on the inflow vapor mass rate than inflow vapor temperature. A control device which will regulate the inflow vapor mass appears more beneficial than a device which regulates the vapor temperature entering the vapor condenser.
4. In both cases, there is an initial surge and followed by an oscillation of the condensate outflow rate from the condenser when the air-conditioning unit is started. This transient response may be attributed to the identical initial condition imposed on the sample condenser in both cases.

NOMENCLATURE

A_a Cross sectional area of airflow passage (m^2)
A_i Internal heat transfer area (m^2)
A_o External Heat transfer area (m^2)
Area Cross sectional area of condenser tube (m^2)
C_{pa} Specific heat of air (kJ/kg-K)
C_{pf} Specific heat of refrigerant (kJ/kg-K)
C_{pw} Specific heat of condenser wall (kJ/kg-K)
D Outside diameter of tube (m)
h_{fg} Latent heat of condensation (kJ/kg)
h_g Enthalpy of vapor refrigerant (kJ/kg)
h_{in} Enthalpy of inlet refrigerant (kJ/kg)
h_f Enthalpy of liquid refrigerant (kJ/kg)
h_{out} Enthalpy of outlet refrigerant (kJ/kg)
h_i Internal heat transfer coefficient in subcooled region (kJ/hr-m^2-K)
h_o External heat transfer coefficient in subcooled region (kJ/hr-m^2-K)
k_f Thermal conductivity of liquid refrigerant (kJ/hr-m^2-K)
L_c Total tube length for competed condensation (m)
M_c Instantaneous rate of mass leaving the two-phase region (kg/s)
M_f Mass in liquid phase (kg)
M_g Mass in vapor phase (kg)
$\dot{M}_{in}$ Instantaneous inlet flow rate of vapor refrigerant (kg/s)
$\dot{M}_{out}$ Instantaneous outflow rate of liquid refrigerant (kg/s)
n Index for time period
P_i Internal perimeter of tube (m)
P_0 External perimeter of tube (m)
T_a Air temperature (°C)
T_R Temperature of refrigerant (°C)

T_w Temperature of condenser tube (°C)
T_∞ Temperature of ambient air (°C)
Δ_t Time interval (sec)
U_f Internal energy of liquid refrigerant (kJ/kg)
U_g Internal energy of Vapor refrigerant (kJ/kg)
V_a Volume of air in condenser (m^3)
V_w Volume of condenser tube (m^3)
Vel_a Velocity of airflow (m/s)
Vel_f Velocity of refrigerant (m/s)
x Coordinate along subcooled refrigerant flow direction (m)
x_o Quality of inlet refrigerant
Y Coordinate along airflow direction (m)
α_i Average internal condensation heat transfer coefficient (kJ/s-m^2-K)
α_o Average external heat transfer coefficient (kJ/s-m2-K)
$\bar{\alpha}$ System mean void fraction
μ_f Liquid viscosity (kg/m-sec)
ρ_a Density of air (kg/m^3)
ρ_f Density of liquid refrigerant (kg/m^3)
ρ_w Density of Condensers wall (kg/m^3)

BIBLIOGRAPHY

1. Wedekind, G.L., Bhatt, B.L., Beck, B.T.; "A System Mean Void Fraction Model for Predicting Various Transient Phenomena Associated with Two-Phase Evaporating and Condensing Flows", International Journal of Multiphase Flow, Volume 4, 1978.
2. "ASHRAE Handbook of Fundamentals", American Society of Heating Refrigerating and Air-Conditioning Engineers, 1985.
3. Kirkbride, C.G.; "Heat Transfer by Condensing Vapor on Vertical Tubes", AICHE 30, 1934.
4. Colburn, A.P.; "Note on the Calculation of Condensation when a Portion of the Condensation Layer is in Turbulent Motion", AICHE 30, 1934.
5. Carpenter, E.F. and Colburn, A.P., "The Effect of Vapor Velocity on Condensation Inside Tubes", Process of General Discussion of Heat Transfer, ASME, 1951.
6. Rohsenow, W.M.; "Heat Transfer and Temperature Distribution in Laminar-Film Condensation", Trans ASME Vol. 78, 1959.
7. Sparrow, E.M., and Gregg, J.L.; "A Boundary Layer Treatment of Laminar Film Condensation", Journal of Heat Transfer Series c, 1959.
8. Chen, M.M.; "An Analytical Study of Laminar Condensation Part-2 single and Multiple Horizontal Tubes", Trans. ASME, Journal of Heat Transfer Series c, 1968.
9. Soliman, M., Schuster, J.R. and Berenson, P.J.; "A General Heat Transfer Correlation for Annular Flow Condensation", Trans. ASME, Journal Of Heat Transfer Series c, 1968.
10. Soliman, M., and Berenson, P.J.; "Flow Stability and Gravitational Effects in Condenser Tubes", Heat Transfer Vol. VI 1970.
11. Nozu, S., Fujii, T., Honda, T.; "A Method for Estimating Tube Length and Pressure Drop of Air-cooled Condensers", ASHRA Transaction Part 1A, 1986.
12. Keller, J.R., "The Validation of a Low-Flow Diesel Truck Cooling System Model and a Parametric Study of Ambient Temperatures, Temperature Control Components and Meetings", Master Thesis, Michigan Technological University, 1986.
13. Mitsui, M.: "Improvement of Refrigerant Flow Control Method in Automotive Air Conditions", SAE paper 870029, 1987.
14. Chiang, E.C. and Chellaiah, S.; "Modeling of Convective heat Flow in Radiators for Coolant Temperature Prediction", The American Society 0of Mechanical Engineers, ASME Paper 85-WA/HT-22, 1985.
15. Chapman, K.S.; "the Enhancement and Validation of a Vehicle Engine Cooling System Simulation for Use as a Cooling System Design Tool", Master Thesis, Michigan Technological University, 1987.
16. Reisbig, R.L., Liang, C.Y.; "Predication of Heat Transfer from Condensing Refrigerant-12", ASHRAE Transactions Part 1, 1971.
17. Hilding, W.E. and Coogan, C.H.; "Heat Transferr Studies of Vapor Condensing at High Velocities in Small Straight Tubes", NASA Contractor Reprort NASA CR-124, 1964.
18. Yasuda, H., Touber, S., Machielsen, C.H.M., "Simulation Model of a Vapor Compression Refrigeration System", ASHRAE Transactions Volume 89, 2A, 1983.
19. Boersen, P.M.T., Jagt, M.F.G.; "Hunting of Evaporators Controlled by a thermostatic Expansion Valve", Tranaction of ASME Volume 102, 1980.
20. Collier, J.G.; "Convective Boiling and Condesnation", McGraw-Hill Book Company, 1972.
21. Kakac, S., Mayinger, F.; "Two-Phase Flows and Heat Transfer", Hemisphere Publishing Corporation, 1977.
22. Alhthouse, A.D., Turnquist, C.H., Bracciano, A.F.; "Modern Refrigeration and Air-Conditioning", The Goodheart-Willcox Company Inc., 1968.
23. Afgan, N.H., Schlunder, E.U.; "Heat Exchangers", McGraw-Hill Book Company, 1974.
24. Dossat, R.J., "Principles of Refrigeration", John Wiley & Sons, Inc., 1964.
25. Vild, T.J., Schubert, F.H. and Snoke, D.R.; "A System to Demonstrate the Zero Gravity performance of an Organic Rankine Cycle", Inter-society Energy Conversion Engineering Conference, Boulder, CO, 1968.

970107

Fuzzy Controller for Thermal Comfort in a Car Cabin

Béatrice Gach and Michael Lang
Valeo Systèmes Thermiques

Jean-Christophe Riat
PSA-Peugeot Citroën

Abstract

This paper presents two fuzzy logic based systems, developed by Valeo Thermal Systems and PSA Peugeot-Citroën, for controlling the thermal environment in a car cabin. This study aimed to simplify the control systems set up, while improving the cabin passengers comfort by taking into account the subjectivity of thermal sensation.

The first system regulates the internal cabin temperature from a temperature fixed by the user on the climate control panel. The second system proposes a new "intelligent" control panel in order to ensure a better thermal balance for the car passengers.

The two systems were installed and tuned on a Peugeot 605 vehicle, on which a standard automatic controller is already available. So, it was possible to compare the fuzzy and the classical series controllers on the same vehicle. The results show good regulation performances and demonstrate that the use of fuzzy logic reduces the development time.

1. Introduction

1.1. Context of the study

Introduced by L.A. Zadeh in 1965 [1], Fuzzy Logic has since been used successfully in many industrial applications. Its practical interest for system control is now demonstrated, especially when no model of the system to control is available.

The automotive industry, where embedded control systems are becoming commonplace, has been taking an interest in fuzzy logic for many years [3], [5], [6], [7]. A lot of automobile patents describe systems based on fuzzy logic and several standard cars are sold with fuzzy logic systems [5].

The goal of this study is to evaluate the use of fuzzy logic for controlling an automatic HVAC system. In fact, some characteristics of such a system appear to favor the use of this technology : non-linear behavior, lack of global model, subjective aspect of human thermal feeling [9].

1.2. Automatic HVAC systems

1.2.1. Principle of a car HVAC system

Every car is equipped with a heating and ventilating system in order to ensure comfortable climate conditions when the weather is cold. Today, these comfort conditions can be extended to warm seasons thanks to the addition of an AC loop. Figure 1 below shows the complete standard system.

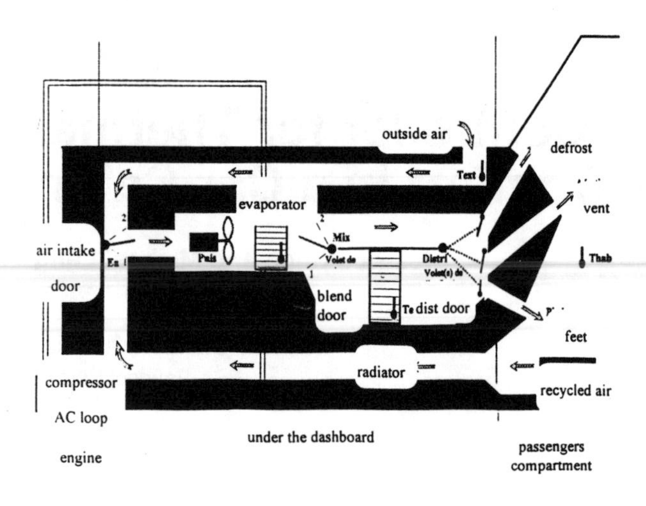

Figure 1 : Heating and air conditioning system

The inlet air (from inside or outside the car) is blown into the system by means of a blower and goes through the AC loop evaporator. A part of this air then crosses the radiator in which circulates the coolant from the motor cooling system. The other part is not heated up. The sharing of these two air masses, that are mixed downstream of the radiator, determines the final outlet air temperature.

1.2.2. Thermal comfort in a car cabin

The thermal comfort can be defined as the preservation of the body thermal balance, whatever the changes of the climatic environment. The thermal comfort conditions for buildings are well known today [2], [8]. But the results cannot be directly transposed to the car cabins that are subjected to specific constraints : forced convection and air speed, due to the closeness of air vents, great radiation influence due to the size and closeness of the windows, permanent contact with the seats, immobility, etc. Therefore, it is often difficult to tune the system to quickly obtain satisfying comfort conditions.

1.2.3. Automatic climate control in a car cabin

To more easily reach an optimal thermal comfort, an increasing number of vehicles is equipped with an electronic temperature and air flow control system. This system consists of a set of sensors and actuators , an electronic board where the controller is implemented and a control panel allowing the users to set their own desired climatic conditions.

1.3. Description of the experimental car

1.3.1. Standard series equipment

The HVAC hardware on which the fuzzy controllers were implemented is standard :

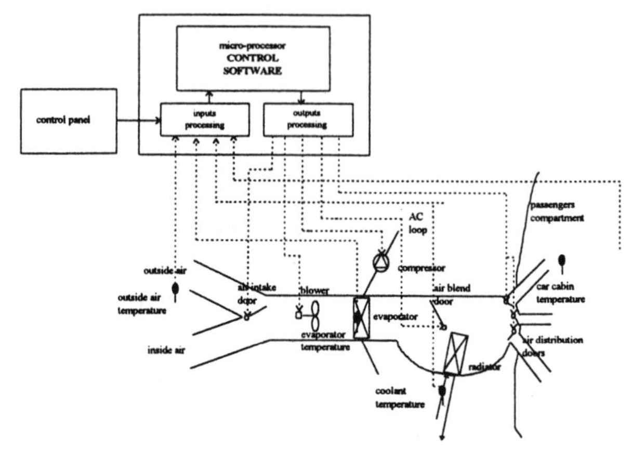

Figure 2 : Standard temperature control system

It consists of four temperature sensors, ie outside, coolant and evaporator temperature sensors to get information about the thermal state of the process and its environment, and a car cabin air temperature sensor, which is the variable to be controlled.

It also includes five actuators : **1** the blend door, to determine the temperature of the air blown in the cabin, **2** the blower voltage, that defines the airflow rate, **3** the distribution servomotor that fixes the way the air is blown in the cabin (towards the feet, the face or the windshield), **4** the air intake door, to determine the origin of the air entering the system and **4** the compressor clutch to turn the AC loop on or off.

1.3.2. New realizations

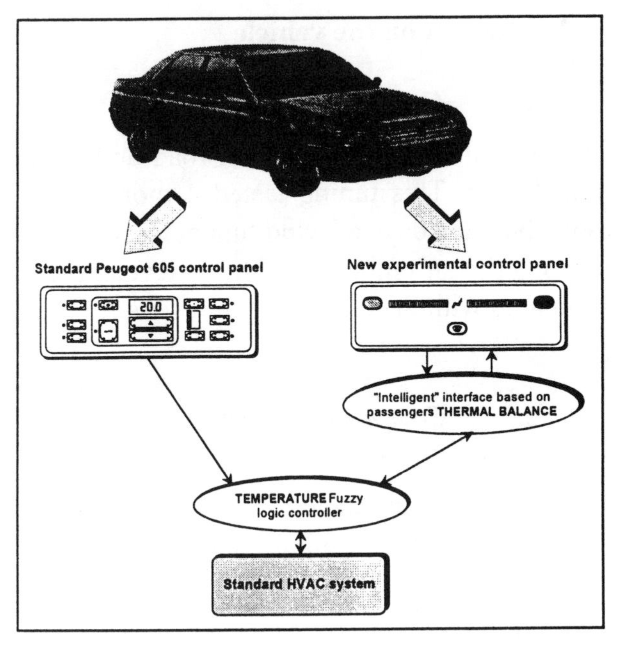

Figure 3 : systems realized

In figure 3, the system on the left consists of replacing the standard HVAC controller by a fuzzy logic one. The control panel and the HVAC system are the standard ones for the Peugeot 605.

This fuzzy controller is used by the system on the right, based on a thermal balance. New sensors are added (the same as for the previous system, plus hygrometry and car speed) and a new user interface is implemented in order to improve taking into account the subjective aspect of thermal feeling : the user doesn't indicate an absolute temperature but asks the system for cooler or warmer interior conditions.

1.3.3. Development system

The fuzzy control software is written in C. In order to test this software and to tune the controller, a specific hardware was utilized.

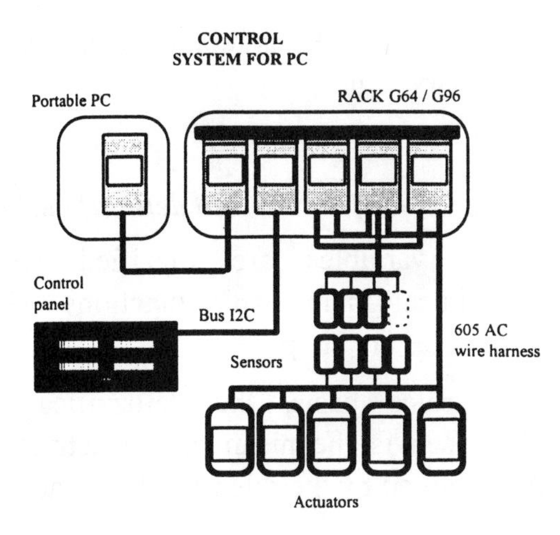

Figure 4 : on-board development system

This hardware (Figure 4) consists of a PC on board the vehicle and of a rack of Europe boards based on the G64/G96 bus, containing the interfaces. This system replaces the series control board and is connected to the vehicle wire harness.

For the thermal balance control application, the new control panel is connected to the rack with a I2C interface, that allows it to be physically separate from the control system.

The advantage of this solution is to be modular, which readily allows the addition of new sensors or actuators.

2. Fuzzy temperature regulation

2.1. Approach

Four steps were followed : **1** collection and structuration of Valeo expertise in the field of automotive climate control, **2** linguistic formulation of the desired behavior of the controller, **3** fuzzification of the system parameters, and **4** writing of the rules basis.

2.2. Fuzzy controller description

2.2.1. General structure

The fuzzy controller manages the blend, distribution and air intake doors, and the blower voltage. The compressor, which can only be ON or OFF, is commanded by a classical binary logic.

2.2.2. *Fuzzification and defuzzification*

The input variables are fuzzified with piecewise linear membership functions. The output variables are crisp.

For each rules basis, the defuzzification method chosen is the mean of the activated outputs, weighted by their degree of activation.

Using this fuzzy process, we were able to reduce significantly the size of the final code compared to the standard control, which is a great advantage for our application.

2.2.3. *Example : blend door control*

The inputs are *epsit=tset-tint* (difference between the user's chosen set temperature and the actual car cabin air temperature) and the change of the car cabin air temperature *dtint=tint(t-1)-tint(t)*. The output is the change of the blend door position : *dv*.

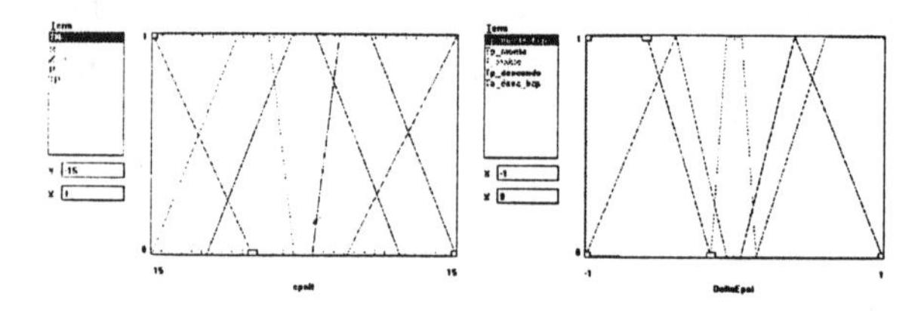

Figure 5 : Blend door control inputs fuzzification

dv	TTN	TN	N	Z	P	TP	TTP
value	-3.5	-2	-1	0	1	2	3.5

epsit / *dtint*	TN	N	Z	P	TP
great_tp_increase	TTN	TTN	TN	Z	P
tp_increase	TTN	TN	Z	Z	TP
stable_tp	TTN	N	Z	P	TTP
tp_decrease	TN	Z	TP	TP	TTP
great_tp_decrease	N	Z	TTP	TTP	TTP

2.3. Results on the vehicle

2.3.1. *Controller tuning*

The controller was tuned from road and wind tunnel tests. This tuning lasted 4 months and took about ten days of wind tunnel tests, which is a significant reduction of time development compared with the classical controllers.

2.3.2. *Comparison with the classical series controller*

The tests were conducted in a wind tunnel in order to be able to submit the two systems (fuzzy and classical) to the same conditions.

The comparison presented here concerns the cold climates. Indeed, the heating mode is the most difficult to manage because the heat production is over-sized compared with the needs and is subject to several disturbances.

The series and fuzzy controllers were compared in two cases :

- engine ignition from a vehicle preconditioned state : the vehicle structure and the coolant liquid are at ambient temperature
- set temperature steps

The two following figures (6 and 7) show the evolution of the temperature near the driver's head and the set temperature, obtained first with the fuzzy and then with the series controller.

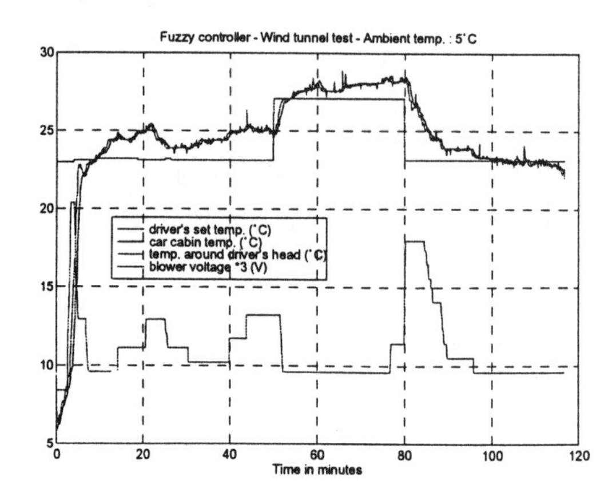

Figure 6 : fuzzy controller-wind tunnel at 5°C

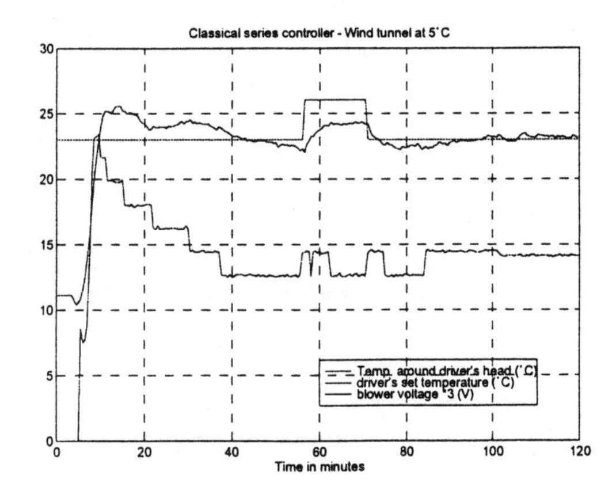

Figure 7 : classical series controller-wind tunnel at 5°C

Comparative tables :

- engine ignition from preconditioned state :

	Time for a 10°C increas e near driver's head	Over shoot	Time before steady state	Max blower voltage	set temp. - cabin temp.	set temp.- driver's head temp.
Series	5 mn	1°C / 5 mn	12 mn	7.5 V	0 to 4 °C (not stab.)	1 to 2°C
Fuzzy	6 mn	3°C / 10 mn	20 mn	6.5 V	0.5 to 1°C	0°C

The fuzzy controller performances are equivalent to the classical ones, since the rise times are related to the blower voltage : the higher the blower voltage, the shorter the rise time.

- set temperature increase :

	Rise time	Overshoot	Time before steady state
Series	4 mn	no	4 mn
Fuzzy	1 mn	no	1 mn

- set temperature decrease :

	Fall time	Undershoot	Time before steady state
Series	6 mn	no	6 mn
Fuzzy	5 mn	no	5 mn

In the case of a set temperature change, the fuzzy controller has better performances.

These tables show better performances with the fuzzy controller.

2.4. Conclusion

The linguistic formulation gives the advantage to the fuzzy controller that allows :

- better understanding of the physical phenomenas involved, leading to a better controller maintainability,
- easier tuning, leading to a significant development time reduction,

for performances that are at least equivalent as those of the series controller.

The current studies concern the implementation of the fuzzy controller on the 605 series control board and its use on a Citroën ZX.

3. Thermal balance regulation

3.1. Principles of thermal balance

One characterization of thermal comfort is that the heat produced by the body is absorbed by the surroundings at the same rate, so that we do not accumulate heat or feel cold. The International Standard Organization defines PMV (Predictive Mean Vote) in ISO-7730 as an index of comfort. This approach based on the computation of a thermal balance is only validated for housing.

The thermal balance (noted L in Figure 8) is a mathematical function of six input variables (for more details see [8]) :

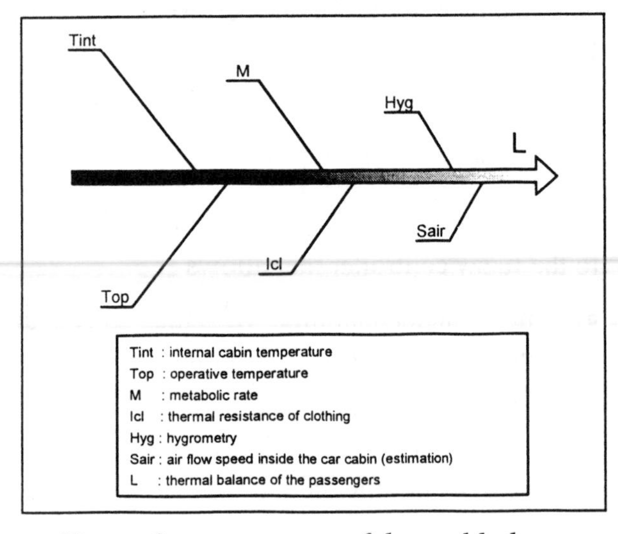

Figure 8 : parameters of thermal balance

The goal of this study is to compute this function value in the car and to control the air-conditioning so as to achieve maximum comfort level (L must be around zero). So it is possible to test this approach in an automotive context.

3.2. Architecture of global system

The principle is to use the temperature regulator presented in the second paragraph of this paper. This regulator is combined with a supervisor program which regularly calculates the value of the thermal balance.

This value is used to compute the best temperature for the thermal comfort in the car cabin. This temperature is the one which corresponds to a thermal balance around zero. This value is the new signal transmitted to the temperature regulator (Figure 10).

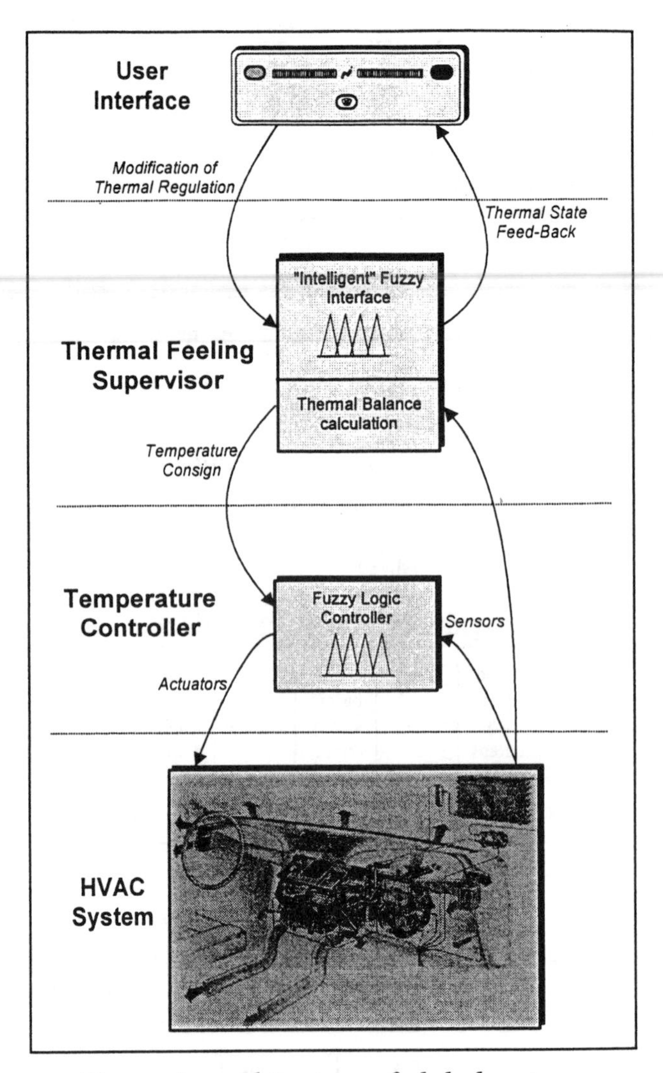

Figure 9 : architecture of global system

3.3. Working of the new control panel

The new control panel allows the user to ask for more heat or more cold. He indicates the intensity of his demand by lighting more or less L.E.D's. on the control panel (see Figure 10).

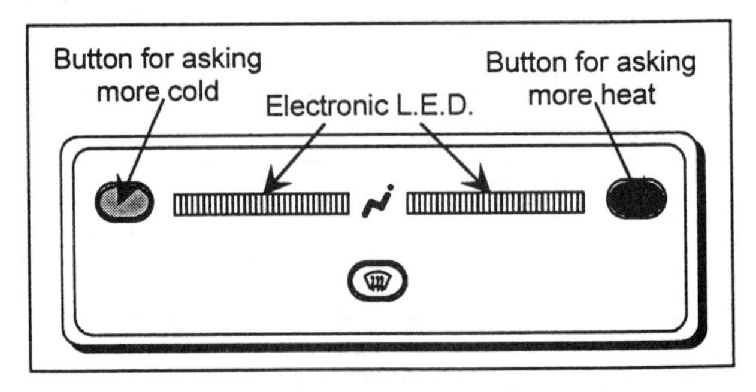

Figure 10 : new control panel

The idea is to use the action on the control panel to modify the value of the Icl parameter (thermal resistance of clothing) in the calculation of the thermal balance. This parameter is initialized at a standard value when the system starts and is modified after each user action.

To calculate this modification (noted dIcl), we have used a fuzzy logic system with three inputs. The first one is the current value of the Icl parameter. The second one corresponds to the value of the thermal balance. The third parameter is the level of comfort wanted by the user which is quantified by the number of L.E.D's. he lit on the control panel.

For example, this figure shows the result of the calculation of dIcl depending on the desired comfort and the thermal balance :

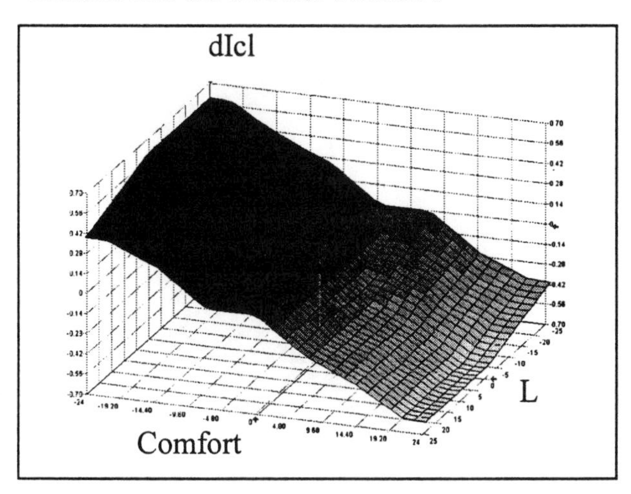

Figure 11 : control surface of dIcl

3.4. Experimental results

The complete system has been tested on road by several different users. It is easy to understand though it is quite different from standard climate control panel.

To illustrate the principle of working of the new system, the graphics of the figure 12 shows the evolution of the parameters Icl and L during a test on road.

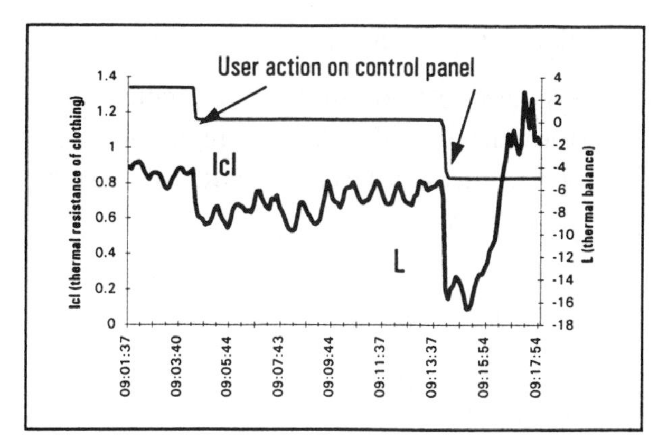

Figure 12 : evolution of Icl and L parameters

As shown in Figure 12, the user asks two times for more hot. At each time, the fuzzy interface of the control panel modifies the value of the Icl parameter. That's why the thermal balance calculated by the thermal feeling supervisor goes away from zero. A new temperature consign is calculated and transmitted to the controller. So the thermal balance returns progressively around zero.

4. Conclusions

This paper presented an actual case of use of fuzzy logic in an automotive context. It shows that for controlling a HVAC system, fuzzy logic allows to rapidly develop a first system with minimal effort. This system is then tested on roads and in a wind tunnel in order to tune it correctly.

A significant advantage of fuzzy logic is the ease of tuning without writing any line code (modification or addition of rules). More over, the elaboration of the control strategy is simplified by its formulation with linguistic rules which simplify dialog with experts of the system to control.

5. Future prospects

These studies are continuing with several purposes :

- portage of the fuzzy controller on a Citroën ZX in order to evaluate its degree of portability

- implementation on a series control board;
- an ergonomic study in order to evaluate the performances of the thermal regulation not with technical criteria but with the users' point of view;
- development of a new fuzzy regulator directly based on thermal balance;

6. References

[1] L.A. Zadeh
"Fuzzy sets"
Information and control, Vol. 8, 1965, pp. 338-353

[2] P.O. Fanger
"Thermal confort analysis and application in environmental engineering"
McGraw-Hill, 1970

[3] L.I. Davis, T.F. Sieja, R.W. Matteson, G.A. Dage, R. Ames
"Fuzzy logic for vehicle climate control"
FUZZ-IEEE 94, pp.530-534

[4] F. Mingrino, G. Toscano Rivalta
"An automatic climate control based on the concept of equivalent temperature"
SAE International Conference 1995, SP-950022

[5] *"Development of a fuzzy controller for an automotive HVAC system"*
Mitsubishi Technical Information, 1992
(translated from Japonese)

[6] Nizar Al-Holou, Kashyap H. Shah
"Fuzzy logic based system to control climate in Automobile"
SAE International Conference, Mercy 1994

[7] S. Merlin, C. Melin, J.C. Riat
"An application of fuzzy logic to automatic transmission"
CESA'96 IMACS Multiconference, Lille France 1996, pp. 913-917

[8] ISO 7730
"Moderate thermal environments - Determination of the PMV and PPD indices and specification of the conditions for thermal comfort"

[9] L.C. Pairin
"Automatisation de la gestion du confort hygrothermique dans l'habitat et le tertiaire"
French CNAM report, 1993

970528

A New Transient Passenger Thermal Comfort Model

J. Steven Brown
Ford Motor Co.

Byron W. Jones
Kansas State Univ.

ABSTRACT

This paper presents a new transient passenger thermal comfort model. The model uses as inputs the vehicle environmental variables: air temperature, air velocity, relative humidity and mean radiant temperature all of which can vary as a function of time and space. The model also uses as inputs the clothing level and the initial physiological state of the body. The model then predicts as a function of time the physiological state of the body and an effective human thermal sensation response (e.g. cold, comfort, hot, etc.). The advantage of this model is that it can accurately predict the human thermal sensation response during transient vehicle warm-up and cooldown conditions. It also allows design engineers the ability to conduct parametric studies of climate control systems before hardware is available. Here we present the basis of the new thermal comfort model and its predictions for transient warm-up and cooldown conditions. Model predictions are compared to actual human subjective data and to Fanger and Gagge thermal comfort model predictions.

INTRODUCTION

As vehicle design cycles shorten, it becomes increasingly more important that design and development tools be developed which can help shorten the overall design cycle time. Presently, in the development of heater and A/C systems, a large number of tests (both objective and subjective) are required to assure customer satisfaction. Expensive and time consuming subjective juries (large sample sizes) are required. One way to reduce the need for large juries is to develop Computer Aided Engineering (CAE) models capable of analytically predicting human thermal responses. One example of an attempt to use CAE to understand human thermal comfort is the use of Computational Fluid Dynamics (CFD) to solve for the vehicle interior environmental variables: velocity, air temperature and mean radiant temperature (see for example [1]). Yet these variables, by themselves, cannot tell a design engineer how comfortable a person will be in the passenger compartment. Something more is needed.

As a result, automotive companies throughout the world are trying to model human thermal comfort. The most common approach is to apply a thermal comfort model that is readily available in the literature. The most typical models are based on two schools of thought: Fanger (a Professor in Denmark) and Gagge (a recently deceased Professor from Yale University). Fanger's model (see for example [2]), which is the basis for the Brüel & Kjær Thermal Comfort Meter, is based on a steady-state energy balance. The model was originally developed to predict human thermal comfort in office-like environments and has gained wide usage in the automotive industry. Gagge's model (see for example [3]), which is based on a transient energy balance, was originally developed to examine human response to hot and cold stress and has not gained a wide following in the automotive industry.

The Ford Thermal Comfort Model presented in this paper is based on a model developed at Kansas State University (KSU) [4]. The KSU model is based on the same transient energy balance as Gagge's model but KSU's model, unlike Gagge's, incorporates a transient clothing model, allows for spatial non-uniformity, and has been developed from an engineering perspective, yet still incorporates the physiology of Gagge's model. The KSU model is segmented and allows for local as well as whole body calculations of heat transfer to the surroundings. This feature allows it to incorporate the non-uniform environmental conditions encountered in automobiles. It also models the transient behavior of clothing and is supported by a data base of over 30 different clothing systems.

INSTRUMENTATION TO MEASURE ENVIRONMENTAL VARIABLES

A wooden stick-person which could easily be taken in and out of a vehicle was constructed. It contained instrumentation to measure air temperature, mean radiant temperature, relative humidity, and air velocity. Type T thermocouples were used to measure the air temperature in ten locations: head, left & right shoulders, torso, left & right sides of the waist, left & right thighs, and left & right feet. The mean radiant temperature (defined as the uniform temperature of an imaginary enclosure in which the radiant heat transfer from the human body equals the radiant heat transfer in the actual nonuniform enclosure [5]) was measured using globe thermometers and radiometers. Globe thermometers (15 cm diameter Styrofoam balls with type T thermocouples under black foil located on the 6 "sides" of the sphere) were placed in four locations: head, torso, thigh, and foot. Four Scientific Sales Model 3032-A Net Radiometers with one side blackened to allow absolute radiation measurements were located on the torso. The four radiometers pointed up, forward and to the left & to the right. General Eastern RH-2 Relative Humidity Transmitters were placed in three locations: torso, thigh, and foot. TSI Model 8470 Omnidirectional Transducers were used to make velocity measurements at several locations on the head, torso, shoulders, waist, thighs, shins, and feet, including all of the locations listed above for the air temperature thermocouples. Output from these instruments were recorded every 15 seconds using a Campbell Scientific Datalogger.

MODEL DEVELOPMENT

There are many ways one can go about modeling subjective thermal sensation responses. It should be understood that subjective responses are as much in the realm of psychological sciences as they are in the realm of physical sciences. The goal of the modeling is to combine these two aspects of the subjective response, to relate how people feel to the physical realities of their environment. Such modeling will always be imprecise. There are tremendous differences from person-to-person both in their physiological makeup and in their psychological makeup. People respond very differently to the same physical environment. In addition, variables other than the physical surroundings can affect subjective responses. For example, expectations of comfort conditions, attitude towards being subjected to an environment, attitude towards other people in the immediate surroundings, previous experiences in similar environments, recent interactions with other people, interpretation of the meaning of subjective evaluation terminology, etc. may potentially impact how a person responds subjectively in a given physical environment. These factors combine to result in large variations in responses for people that are subjected to identical physical environmental conditions.

The approach used in this study was to separate as much as possible the physical modeling and the psychological modeling. The physical thermal interaction between a person and the environment was modeled using a transient simulation model that had been developed and validated under transient conditions previously [4]. This model was modified to allow more detailed local definition to incorporate the environmental non-uniformities that can occur in an automobile. It was also modified to accept environmental data collected with the instrumentation explained in the previous section. This model was then used to predict heat and moisture interchange between a person and the environment and to model the physiological responses of a person to that environment.

Once the physical modeling was complete, it was possible to focus on the link between the physical variables and the subjective responses. For the reasons discussed above, this link will never be perfect. Nevertheless, there is every reason to believe that the physical interactions between the person and the environment are the dominant variables affecting how a person perceives his or her comfort in that environment. The complete process whereby a person receives physical stimuli from the environment and then formulates subjective responses is not known. The general approach taken in the development of the Ford Thermal Comfort Model was to select variables that are most likely to be good model variables and then test them with statistical analysis. Variables that the literature and our own past experience indicate are good candidates for affecting and/or predicting thermal sensation were selected. Statistical regressions were used to evaluate the variables and to eliminate those with little predictive value. Numerous variables and combinations were examined and subjective judgments were combined with formal statistical measures to produce the final model.

The subjective data used in the regression analysis were collected at Kansas State University. The front section (included up to just behind the front seats) of a full-size vehicle was placed in a climate control chamber. Temperature and humidity controlled air was fed to the inlet of the vehicle's climate control system. A total of 18 test conditions (9 for winter and 9 for summer) were simulated. For the winter tests the ambient temperature was -17.8°C, there was no sunload, and the supply air temperature varied from 37.8°C to 60°C. For the summer tests the ambient temperature was 43.3°C, the sunload was 1 kW/m^2, and the supply air temperature varied from 18.3°C to 29.4°C. These test conditions were not intended to exactly duplicate vehicle transient warm-up and cooldown conditions but were meant to cover the typical parameter space seen in actual vehicles. A total of 216 subjects (108 women and 108 men) participated in the study.

The subjects spent 60 minutes in a pre-conditioning chamber during which time they were read an orientation/explanation statement, instrumented with 6 skin temperature thermistors, and had their core body temperatures measured. Three conditions were used for the winter tests (room temperatures of 15.6°C, 21.1°C, and 26.7°C) and three conditions were used for the summer tests (room temperatures of 18.3°C, 23.9°C, and 29.4°C). In all cases the relative humidity was 50%. The subjects wore a standard issue of clothing which consisted of sweat pants, sweat shirt, ski jacket, stocking hat, and mittens for the winter and sweat pants and T-shirt for the summer. In addition, in both cases, the subjects wore their own shoes, socks, and underwear. The subjects also recorded their thermal responses during this pre-conditioning period.

The subjects then entered the climate control chamber, were seated in the vehicle, and connected to the data acquisition system. The simulated vehicle environment was then started. The total time spent in the vehicle was 45 minutes. During this time, the subjects' skin temperatures were automatically recorded every 1 minute and the subjects recorded their thermal responses every 2 minutes. The thermal responses were based on a 9-point subjective thermal comfort rating scale shown in Table 1.

Table 1. Subjective Thermal Comfort Rating Scale

Numerical Rating	Subjective Rating
1	Cold
2	Cold/Cool
3	Cool
4	Cool/Comfort
5	Comfort
6	Warm/Comfort
7	Warm
8	Hot/Warm
9	Hot

In addition, for each test case, the stick-person was used to record the vehicle environmental data in both seating locations.

VALIDATION EXPERIMENTS

A series of independent tests were conducted in actual vehicles to asses the validity of the model.

Vehicles were placed in a climate control chamber which simulated heater warm-up and air conditioning cooldown conditions. A mid-size vehicle was used for the winter tests and a full-size vehicle was used for the summer tests. A total of 26 test subjects participated in the winter simulations and a total of 24 test subjects participated in the summer simulations. The test conditions were varied from more severe ambients (-17.8°C for winter and 43.3°C for summer) to less severe ambients (-6.7°C for winter and 29.4°C for summer) in order to test the robustness of the model.

The results presented here are limited to the more severe ambients. The settings for the winter simulations were: -17.8°C ambient temperature, no sunload, a vehicle speed of 48 kph, fresh air, maximum heat, air directed from the floor ducts at the highest blower speed, and a total of 14 test subjects. The settings for the summer simulations were: 43.3°C ambient temperature, 1.1 kW/m^2 sunload, vehicle idling, recirculated air, full cool, air directed from the panel vents at the highest blower speed, and a total of 8 test subjects.

Before each test the subjects were allowed to acclimate to the laboratory environment (i.e. approximately room temperature) for 1/2 hour. During this time, the subjects were read a standard orientation/explanation statement. The winter clothing consisted of the subject's own socks, shoes, underwear, trousers and shirt. In addition, the subjects were provided a ski jacket and mittens. The summer clothing consisted of the subject's own socks, shoes, underwear, trousers and shirt. After the preconditioning period, the subjects entered and stood in the climate chamber environment for 5 minutes before entering the vehicle (there were always two subjects: a driver and a front seat passenger). The vehicle was started and brought to its operating condition within the first minute. The subjects were not allowed to change any of the settings during the test. The total time spent in the vehicle was 40 minutes. During the first 10 minutes, the subjects rated their thermal responses every 1 minute and for the final 30 minutes, the subjects rated their thermal responses every 2 minutes. The thermal responses were based on the 9-point scale described in Table 1.

In addition, for each test case, the stick-person was used to record the vehicle environmental variable data in both seating locations.

COMPARISON OF FORD MODEL TO VALIDATION RESULTS

The environmental variable validation data collected using the stick-person were used as inputs to the Ford Thermal Comfort Model. In addition, the clothing level and the initial physiological state of the human body after the preconditioning period were used as inputs to the model. The model results are shown in Figure 1 (driver) & Figure 2 (front-seat passenger) for a winter condition and in Figure 3 (driver) & Figure 4 (front-seat passenger) for a summer condition. Also shown in Figures 1 - 4 are the subjective results from the validation experiments. The data points are the mean of the subjects' votes and the dashed lines are plus/minus one standard deviation of the subjects' votes.

COMPARISON OF VARIOUS MODEL RESULTS

In Figures 5-8 predictions made by the Fanger, Gagge and Ford models are compared to the mean votes presented in Figures 1-4, respectively.

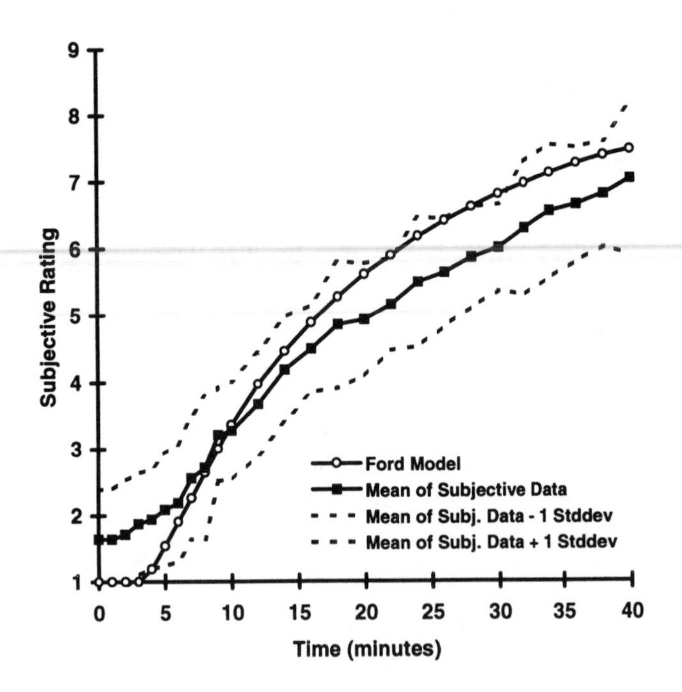

Figure 1. Comparison of Ford Thermal Comfort Model Prediction versus Actual Subjective Data for a Typical Winter Condition. Test Subjects are Located in the Driver's Seat.

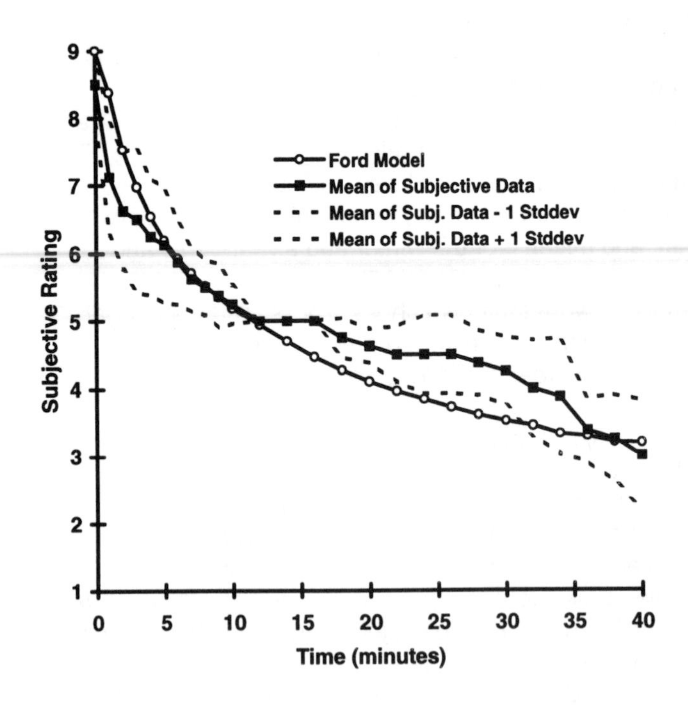

Figure 3. Comparison of Ford Thermal Comfort Model Prediction versus Actual Subjective Data for a Typical Summer Condition. Test Subjects are Located in the Driver's Seat.

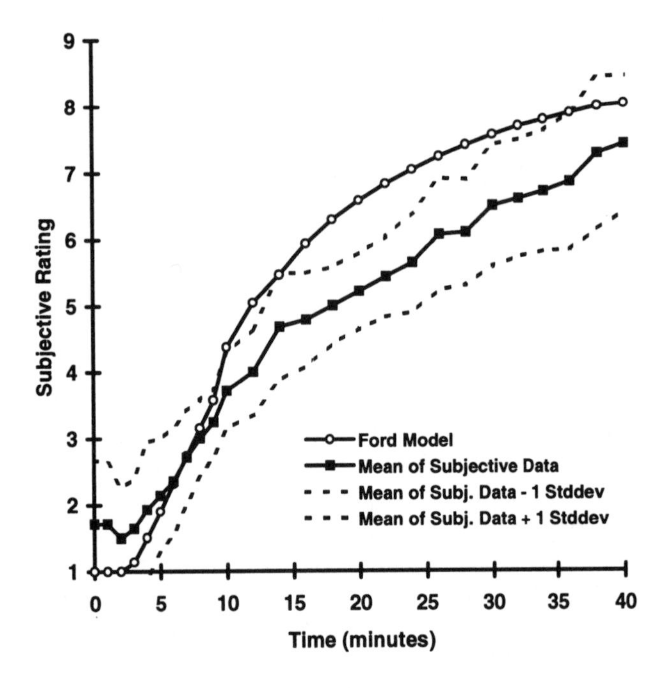

Figure 2. Same Condition as Figure 1 but Test Subjects are Located in the Passenger's Front-Seat.

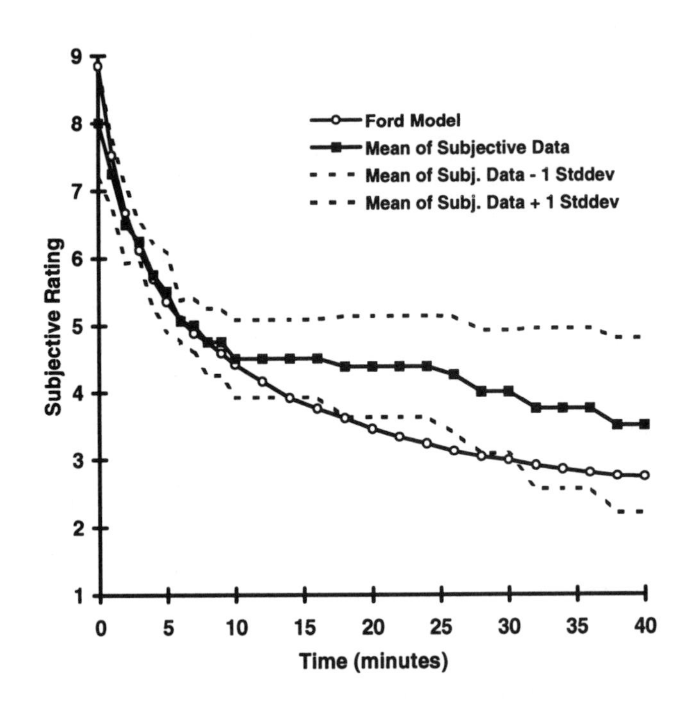

Figure 4. Same Condition as Figure 3 but Test Subjects are Located in the Passenger's Front-Seat.

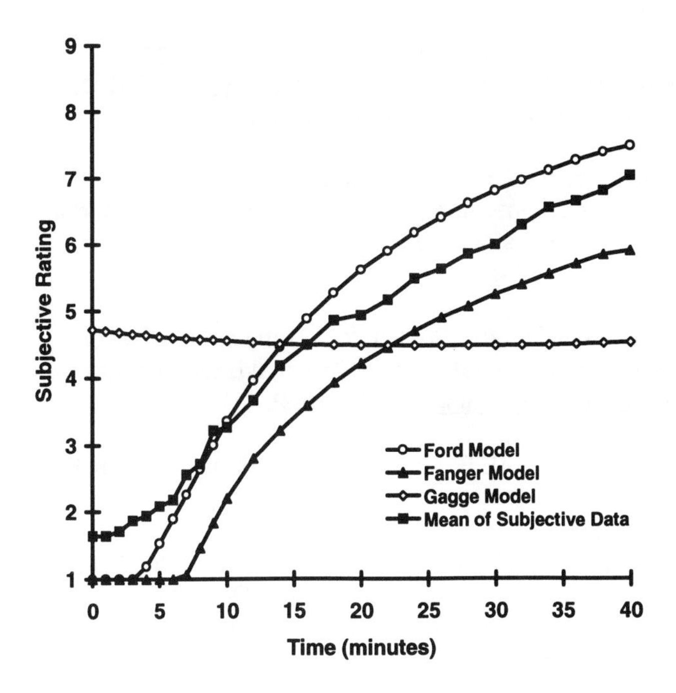

Figure 5. Comparison of Ford, Fanger, and Gagge Thermal Comfort Model Predictions versus Actual Subjective Data for the Same Condition as Figure 1. Test Subjects are Located in the Driver's Seat.

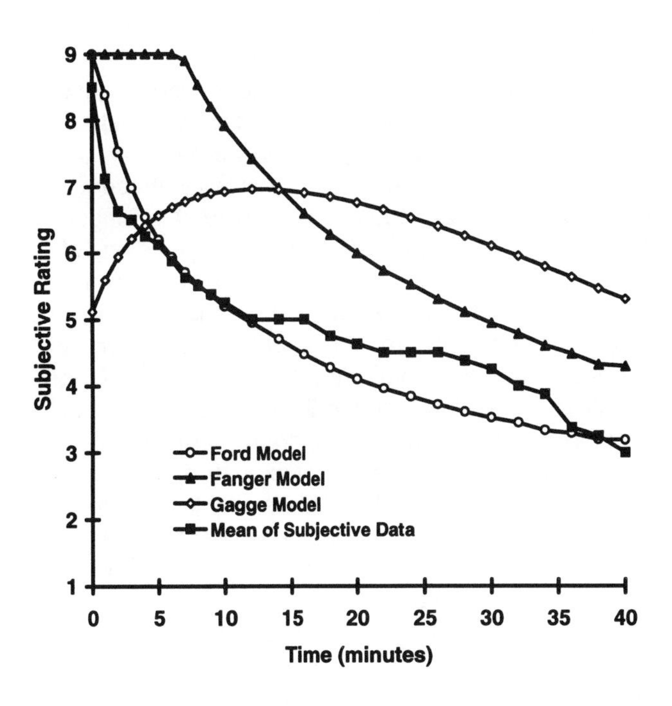

Figure 7. Comparison of Ford, Fanger, and Gagge Thermal Comfort Model Predictions versus Actual Subjective Data for the Same Condition as Figure 3. Test Subjects are Located in the Passenger's Front-Seat.

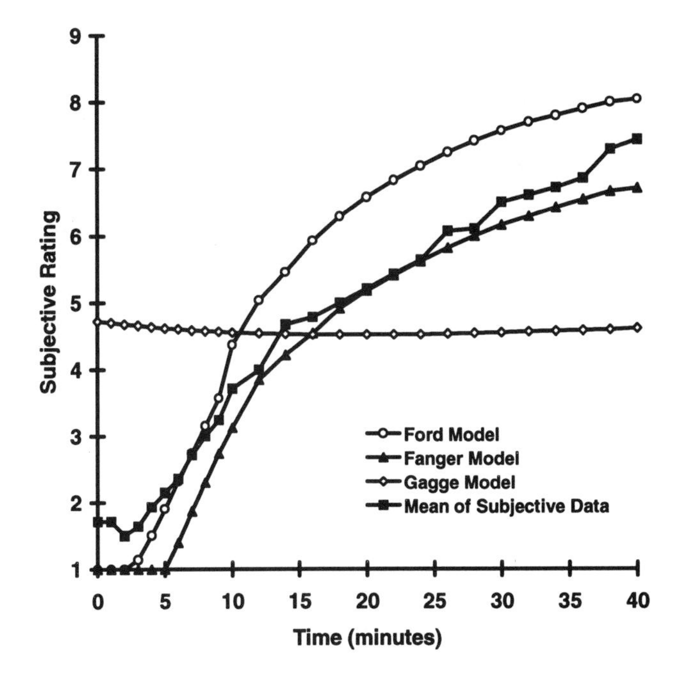

Figure 6. Same Condition as Figure 5 but Test Subjects are Located in the Passenger's Front Seat.

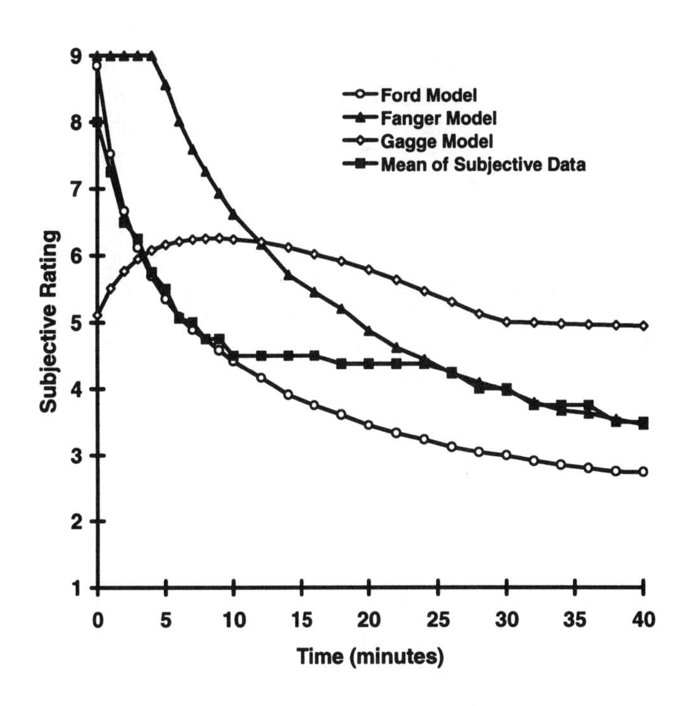

Figure 8. Same Condition as Figure 7 but Test Subjects are Located in the Driver's Front Seat.

CONCLUSION

The Ford Thermal Comfort Model does a good job of predicting thermal sensation during the rapid transients experienced during vehicle warm-up or cooldown conditions. It consistently predicts these transients with higher accuracy than either the Fanger Model or the Gagge Model. Our work suggests one of the main reasons for this result is that the heat flux at the skin is a key variable in determining thermal sensation during transients. This variable changes very quickly in response to environmental transients. The Gagge Model can only respond slowly due to the fact that its thermal sensation is based on body temperatures which only change in response to a period of exposure, and not rapidly like the skin heat flux. This is why the nature of its prediction is so different from the validation data shown in Figures 4-8. In comparison, the Fanger Model bases thermal sensation on thermal load which does change immediately in response to environmental transients. This leads to the Fanger Model being much more responsive than the Gagge Model and, in fact, its trends are much more in line with the validation data shown in Figures 4-8, even though it is intended for steady-state applications.

In this paper we present only a portion of the validation test data collected to validate the Ford Thermal Comfort Model. However, the general conclusions/ trends/results hold also for the validation test data not presented in this paper. In general, we believe that the Ford Model predicts very closely actual subjective thermal sensation data during the first 10 to 15 minutes of a vehicle warm-up or cooldown. After the first 10 to 15 minutes, the Ford Model matches the subjective data with less accuracy; however, it does capture the general characteristics of the subjective warm-up and cooldown curves. One word of caution is that the subjective validation data presented in Figures 1 - 8 are based on very small sample sizes (tests are expensive and time consuming to run). However, we believe that the important physics have been captured by the Ford Model and that larger validation sample sizes would not change the general conclusions reached.

The new Ford Thermal Comfort Model presented in this paper works well for vehicle transient warm-up and cooldown conditions. It has also been incorporated into a much larger Computer Aided Engineering Model that can predict the performance of the overall climate control system before any hardware is available [6]. This tool is being used with great success on a regular basis by design engineers at Ford.

REFERENCES

1. C.-H. Lin, M.A. Lelli, T. Han, R.J. Niemiec, and D.C. Hammond, Jr., "An Experimental and Computational Study of Cooling in a Simplified GM-10 Passenger Compartment," SAE Paper 910216, 1991.

2. P.O. Fanger, Thermal Comfort: Analysis and Applications in Environmental Engineering, R.E. Krieger Publishing Co., Malabar, Florida, 1982.

3. Gagge, A.P., Stolwijk, J.A.J., and Nishi, Y., "An Effective Temperature Scale Based on a Simple Model of Human Physiological Regulatory Response," ASHRAE Transactions, 77(1), 1971.

4. Jones, B.W. and Ogawa, Y., "Transient Interaction Between the Human and the Thermal Environment,", ASHRAE Transactions, 98(1), 1992.

5. 1993 ASHRAE Fundamentals Handbook, page 8.11, 1993.

6. Gielda, T.P., Webster, B.E., Hesse, M.E., and Halt, D.W., "Impact of Computational Fluid Dynamics on Automotive Interior Comfort Engineering," AIAA Paper 960974, 1996.

2000-01-0976

A Sensor for Estimating the Liquid Mass Fraction of the Refrigerant Exiting an Evaporator

James Solberg, Norman R. Miller and Predrag Hrnjak
University of Illinois at Urbana-Champaign

ABSTRACT

A traditional method of controlling evaporator superheat in a vapor compression air conditioning system is the thermostatic expansion valve (TXV). Such systems are often used in automotive applications. The TXV depends on superheat to adjust the valve opening. Unfortunately, any amount of superheat causes that evaporator to operate at reduced capacity due to dramatically lower heat transfer coefficients in the superheated region. In addition, oil circulation back to the compressor is impeded. The cold lubricant almost devoid of dissolved refrigerant is quite viscous and clings to the evaporator walls. A system that could control an air conditioner to operate with no superheat would either decrease the size of its existing evaporator while maintaining the same capacity, or potentially increase its capacity with its original evaporator. Also, oil circulation back to the compressor would be improved. To operate at this two-phase evaporator exit condition a feedback sensor would have to quantify the quality of liquid mass fraction (when the exit stream is a mixture of droplets and superheated vapor) of the refrigerant exiting the evaporator.

INTRODUCTION

One of the most common control schemes for a vapor compression air conditioning system is the use of a thermostatic expansion valve (TXV). TXV systems use a remote thermal bulb at the exit of the evaporator. This bulb causes the TXV to open and close in response to changes in superheat of the refrigerant at the evaporator outlet. If the temperature of the refrigerant increases rapidly, as would be the case when the heat load was suddenly increased, the power element would open the valve and admit more liquid refrigerant to the evaporator. Once in the evaporator, the liquid refrigerant absorbs heat by changing state from liquid to gas. By the time it leaves the evaporator, the gaseous refrigerant has been superheated a few degrees.

By allowing the evaporator to operate with some finite superheat at its exit, some portion of the evaporator will have only vapor flowing through it (no liquid). This situation decreases the refrigerant-side heat transfer. This portion of the evaporator is not able to vaporize refrigerant, and is only able to transfer heat via the sensible heating of the refrigerant. This process can reduce the capacity of the evaporator.

Any control scheme that uses superheat as its control signal (e.g. TXV systems) must have some finite superheat. Such a system is unable to control the plant to operate in a regime of saturated liquid/vapor at the exit of the evaporator. The minimum amount of superheat that such a system can use and maintain stability is dependent on the method of measuring the superheat.

The difficulty of a temperature measurement is in part due to the non-equilibrium flow of refrigerant as it exits the evaporator only slightly superheated. The flow is said to be non-equilibrium because saturated liquid droplets are entrained in superheated vapor. There is just not enough time for the liquid to vaporize and reach equilibrium. This phenomenon can be attributed to maldistribution of liquid/vapor refrigerant throughout the evaporator and to the nature of two-phase flow [1,2,3,4]. The saturated liquid droplets in superheated vapor flow regime cause temperature transducers to exhibit large variances.

In evaporators with imperfect distribution exit streams could be a mixture of superheated vapor and droplets. Some channels or circuits that are thermally overloaded have superheated vapor at the exit, while others where thermal loads are not sufficient to evaporate all liquid that enters will have some droplets at the exit. The mixture of these streams is in thermal non-equilibrium. After sufficient time (or length of pipe) droplets could completely evaporate, reducing superheat. But if the sensible heat available in the superheated vapor is not enough energy to vaporize all droplets, then the exit is in the quality region. Liquid-mass-fraction (LMF), which is the mass of liquid in vapor of any state, is one parameter to describe the state at the evaporator exit, as described in Shannon, Hrnjak, and Leicht [12].

A temperature transducer measuring the temperature of refrigerant in this non-equilibrium flow regime can read the saturation temperature (if a liquid droplet is on the transducer), or can read the temperature of the superheated vapor (which may not be constant), or can read any value in between. A large variance in a control signal (e.g. superheat) can cause a controller to hunt. Since the non-equilibrium flow has superheated vapor along with liquid droplets, quality cannot be used to correctly describe the state of the refrigerant.

Some of the best TXV systems are only able to maintain stable operation with a minimum of about 5 degrees Fahrenheit. But, a patent does exist for a transducer that appears to function in a similar fashion as the device described in this paper. Patent number 2219661 was granted on May 13th, 1992 to York International Ltd by the Comptroller-General of Patents, Designs and Trade Marks, United Kingdom Patent Office. In addition several companies are currently pursuing a prototype commercial transducer. However to the authors' knowledge no results for this class of sensors has appeared in the open literature.

THE SENSOR

A sensor that could estimate the liquid mass fraction (LMF) at the exit of the evaporator could be used in the feedback loop of a control scheme that would maintain the refrigerant at a constant LMF. Liquid mass fraction (LMF) is the ratio of the mass of liquid to the total mass of the fluid, whether or not it is in equilibrium.

PREVIOUS WORK – One of the early studies of superheat stability was carried out at the University of Illinois by Wedekind and Stoeker in the 1970's [1, 2, 3]. The project addressed the stability of the location of the last evaporated droplet in a straight, electrically heated, glass tube. It was found that the location of the last evaporated droplet (the end of the two-phase region) is a stochastic function and the distribution was determined. Some years ago Barnhart and Peters studied stability at the exit of a single glass tube serpentine evaporator [4]. They observed the same phenomena described by Stoecker and determined that most of the instabilities at the exit were generated far upstream (also see [13]).

In another project the unsteadiness of the exit temperature signal was used as an identifier of "stable" operation [12]. The idea of using the variance of the temperature signal at the exit of the evaporator for better flow was developed.

That idea was further developed in a project whose objective was to develop a micro electromechanical system (MEMS) sensor that would do a better job of sensing droplets at the evaporator exit than a thermocouple. A new MEMS sensor (a heated resistance temperature detector RTD) was developed.

The MEMS RTD was driven by a current source and the voltage drop across the sensor was the measured variable (see Figure 1). This voltage is a function of the temperature of the sensor. Notice that this device is essentially an uncompensated hot film anemometer (see page 90 of [5]). Hot wire anemometers have been used to detect droplets entrained in gases (page 181 of [5]). The sensor is cooled as each droplet strikes the hot sensor and is evaporated.

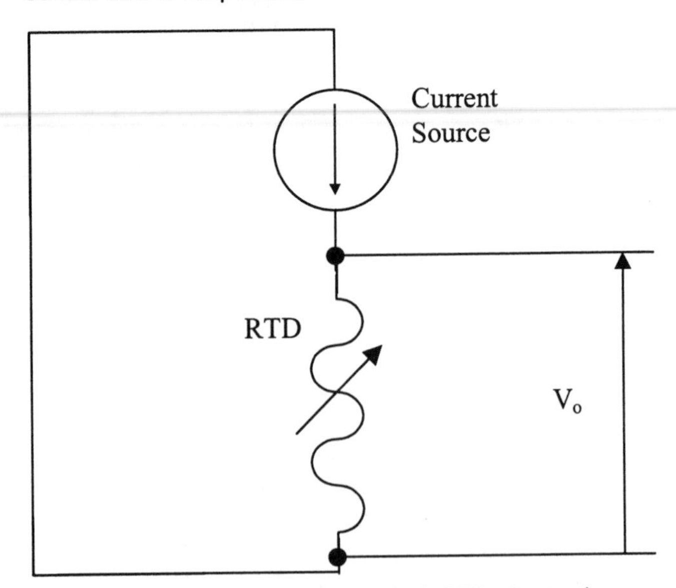

Figure 1. Circuit diagram of MEMS RTD electronic circuitry

FUNCTIONALITY OF SENSOR – An interesting variation is the use of constant resistance transducer control (see Figure 2). This variation of the circuit tries to keep the resistance of the RTD equal to R_{set} . The voltage V_o is then directly proportional to the current needed to achieve this condition. The power removed by heat transfer into the refrigerant stream is, of course, the square of the current flowing through the RTD times the RTD resistance (or R_{set}).

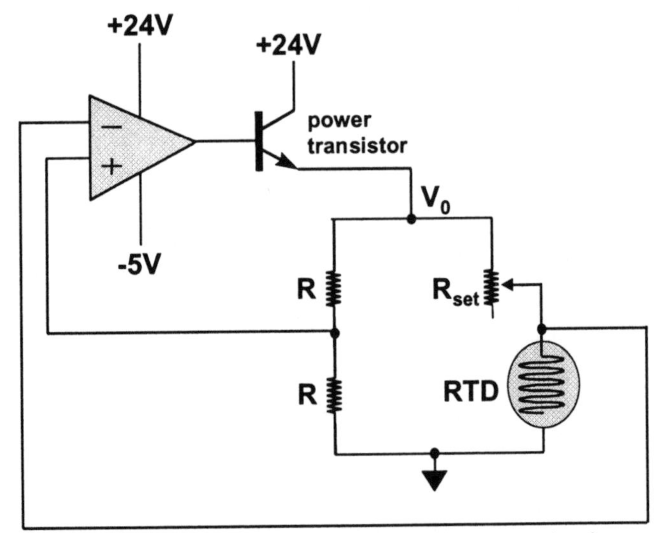

Figure 2. Constant sensor resistance control circuit

This circuit uses an operational amplifier as the medium for feedback. The op-amp uses the feedback to maintain its inputs at constant voltage while drawing very little current. This is what forces the resistance of the RTD to be equal to the resistance of R_{set}. Traditionally, an RTD is used to measure temperature by measuring the resistance of the RTD as it changes with temperature. But, this circuit forces the resistance of the RTD to be equal to R_{set}. The circuit compensates by heating up the RTD until the resistance (and thus the temperature) of the RTD is equal to R_{set}.

Such a system has a much wider bandwidth (that is, it will respond to much higher rate variations in heat flux). The reason is as follows. Constant current excitation requires that the transducer temperature changes for any change in transducer resistance and hence signal to be observed. This is an inherently slow (relatively long time constant) process dominated by the thermal capacity of the transducer body. Constant resistance operation implies that the circuitry varies the transducer current so that the transducer stays at a constant resistance and hence a constant temperature. The thermal energy stored in the transducer body does not change. This technique is used with hot wire anemometers and provides very broadband performance (bandwidths up to 0.5 MHz). The technique also has the advantage of protecting the sensor from overheating.

The circuit maintains the voltage drop across the RTD equal to half of V_0. And since the R_{set} is equal to the resistance of the RTD, the power dissipated through the RTD can be determined. By measuring the temperature of the refrigerant passing over the sensor and inferring the temperature at the surface of the RTD from R_{set}, the difference of these temperatures can be found. This paper refers to this temperature difference as overheat. The overheat represents the driving potential that allows power to be dissipated through the sensor.

The ratio of the power dissipated to this temperature difference can be interpreted as "the surface-to-free-stream thermal conductance" between the RTD and refrigerant. It is essentially the convection heat transfer coefficient multiplied by the effective surface area (*hA*). This surface-to-free-stream thermal conductance (*hA*) does not depend on the effective surface area of the sensor because neither the geometry nor the orientation of the sensor varies. This *hA* parameter is particularly sensitive in the high quality/low superheat region (low LMF). As the LMF of a fluid increases, so does its *hA*.

As a droplet of saturated liquid refrigerant clings to the surface of the RTD, the RTD circuitry will do what it can to raise its temperature back its set point (which is determined by R_{set}). To do this the RTD must transfer enough energy to the refrigerant to overcome its latent heat of vaporization. As the LMF of the fluid decreases, less energy is dissipated through the RTD. When the fluid becomes all vapor, all of the energy flux through the RTD goes to sensible heat that is needed to raise the temperature of the RTD to its set point.

Two testing facilities were used to collect data. The first facility was used to demonstrate relationships between the sensor's power dissipation, overheat, and the quality of refrigerant flowing over the sensor. The second facility was used to show how the sensor could be used to predict the performance of an air conditioning system.

EXPERIMENTAL DYNAMICS

The first set of experiments were conducted to investigate relationships between the sensor's power dissipation, overheat, and the quality of refrigerant flowing over the sensor.

EXPERIMENTAL FACILITY – An existing refrigeration system was used, for the purposes of this study, in the Laboratory for Plate Heat Exchangers at the University of Illinois in Urbana-Champaign. It is designed to simulate evaporator operating conditions typical of water chillers of less than 60 ton (210 kW) capacity using plate heat exchangers for evaporation. The facility consists of three main parts: the refrigerant loop, the water loop, and the evaporator exit test section. A detailed description is given in [12].

The refrigeration loop contains the four necessary elements of a vapor-compression cycle refrigeration system: compressor, condenser, expansion device, and evaporator. In addition, there is a receiving tank for collecting high pressure liquid from the condenser, a liquid subcooler, and instrumentation for monitoring process conditions. The system schematic is shown in Figure 3, where solid lines represent refrigerant piping, and dashed lines represent water piping. The compressor is a Copeland model ZR61K2 hermetically sealed scroll compressor. The refrigerant is R22. Mineral oil circulates through the entire flow loop, including the test section, and is necessary to lubricate the compressor. A common problem in these types of systems is oil accumulation in the evaporator when operating at high superheats. This reduces evaporator capacity, and could influence refrigerant side maldistribution.

The evaporator is a SWEP model B15×40 3-ton (10.5 kW) capacity parallel plate heat exchanger. It consists of 19 refrigerant passages and 20 water passages operating in a counter-flow configuration. The plates have chevron style contours to enhance heat transfer. Two-phase refrigerant enters at the bottom of the evaporator, evaporates vertically through the plates, and exits at the topside of the evaporator. The heat load to the evaporator is supplied by water from the water reservoir. Thermocouples located at the entrance and exit of the refrigerant and water streams monitor process conditions.

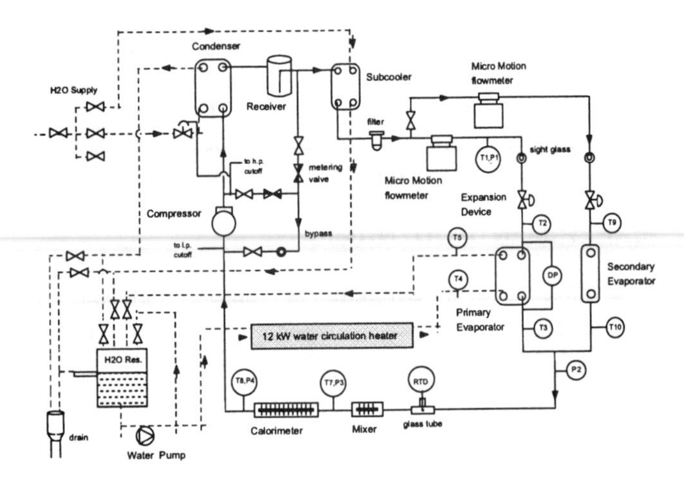

Figure 3. The refrigeration loop used to determine the sensor's dynamics

A 12 kW Watlow water circulation heater provides the thermal load on the evaporator. Hot water from the condenser can also be redirected into the evaporator if desired.

The test section consists of the RTD sensor, a glass tube for flow visualization, a static mixer, a calorimeter, and several thermocouples and pressure transducers for monitoring flow conditions. Refrigerant exits the evaporator before passing over the RTD sensor. Then it passes through a mixer in order to ensure the state of the refrigerant is either pure superheated vapor or saturated vapor/liquid. If the refrigerant is two-phase, then a known amount of energy is put into the refrigerant as it passes through the calorimeter. After all the liquid is vaporized temperature and pressure measurements are taken to estimate the refrigerant's enthalpy. The enthalpy entering the calorimeter is equal to the enthalpy of the refrigerant exiting the calorimeter minus the energy proved by the electric heaters inside the calorimeter. Once the enthalpy of the refrigerant entering the calorimeter is known, its quality can be estimated.

DATA COLLECTION – In order for the sensor to be useful some relationship between refrigerant quality, overheat (temperature difference), and power dissipated through the sensor needs to be developed. Quality is determined from the enthalpy of the refrigerant entering the calorimeter. Overheat is the difference between the temperature of the RTD and the temperature of the refrigerant. The power dissipated through the RTD is determined by the square of the RTD's voltage drop divided by its resistance.

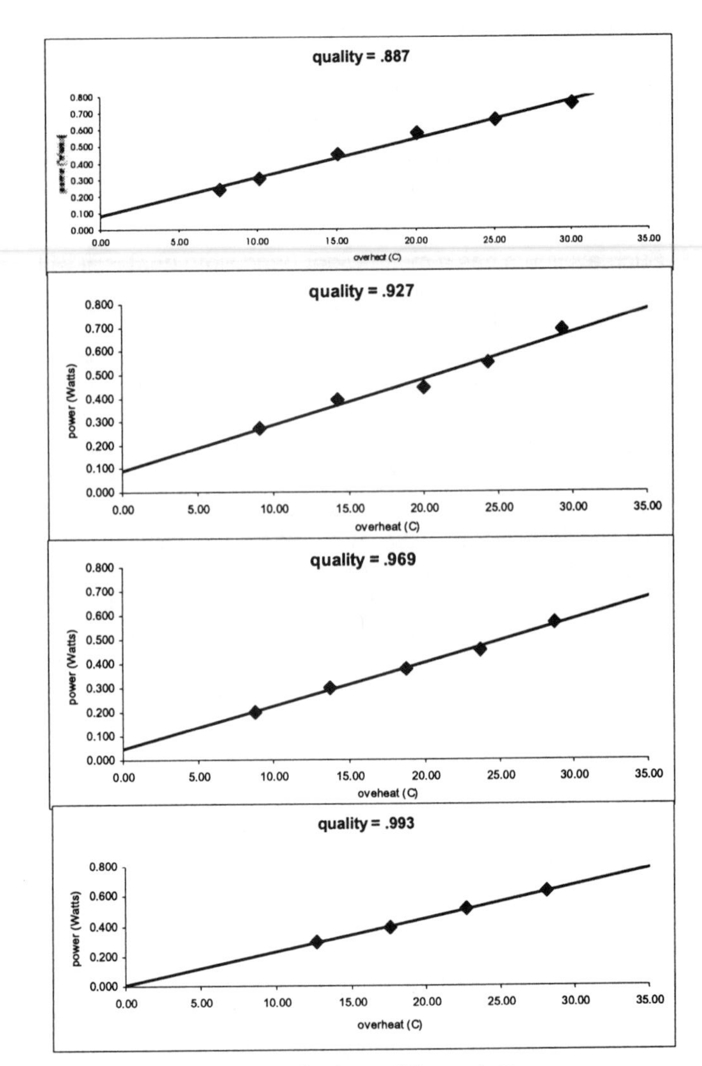

Figure 4. The power dissipated through the sensor as a function of overheat while the quality of refrigerant over the sensor is held constant. The slope represents the hA parameter. R_{set} in figure 2 was varied to change the overheat.

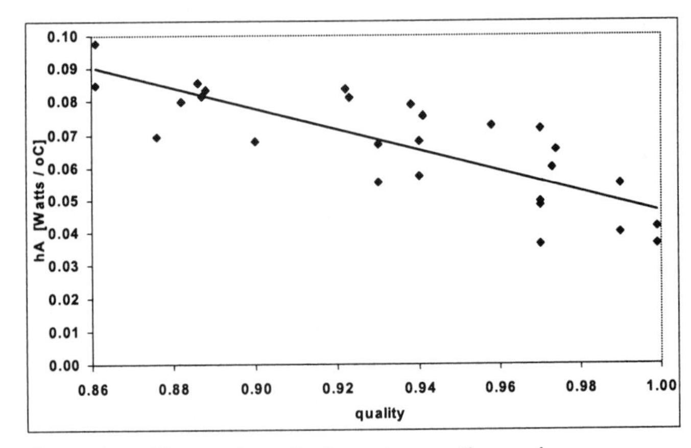

Figure 5. The surface-to-free-stream thermal conductance (hA) as an estimate of refrigerant quality.

Experiments were conducted that would demonstrate the dependence of sensor power and overheat while the quality was held constant. Figure 4 shows the results of such experiments. The results show that power dissipated is a linear function of overheat (the temperature difference between the sensor and the refrigerant). The data agrees with the convection heat transfer model which takes the form:

$$q = h \cdot A \cdot (T_s - T_\infty) \tag{1}$$

Ts is the fixed temperature at the surface of the RTD. T_∞ is the temperature of the free-stream refrigerant passing over the sensor. The power dissipated can be modeled as the energy transfer q. The overheat in the system is analogous to (Ts - T_∞). And the slope of the line represents hA. hA is the product of the convection heat transfer coefficient and the effective surface area.

Once it was established that *hA* is constant for a given quality, the next task was to develop a relationship between quality and *hA*. Figure 5 shows the results of an experiment where the RTD sensor was subjected to various qualities. Quality was measured by using a calorimeter in the method described earlier.

As quality decreases more and more liquid droplets hit the sensor. This demands more power to be dissipated through the sensor in order for the sensor to maintain its constant temperature. At the same time the temperature of the refrigerant is fixed at its saturation temperature. So the ratio of the dissipated power to the overheat (*hA*) increases as quality decreases. This theory is supported by the data shown in figure 5.

One reason for the significant scatter in the data may be due to the inaccuracies in measuring quality. Quality at the exit of the evaporator was inferred indirectly. What was actually estimated was the quality of refrigerant entering the calorimeter after it had passed through a mixer. Implicit to the data used to construct figure 5 are the assumptions that the refrigerant passing over the sensor was saturated liquid/vapor and that none of the refrigerant changed phase between the sensor and the entrance of the calorimeter. Neither one of theses assumptions is necessarily true.

EXPERIMENTAL PERFORMANCE

Once it was established that the sensor could predict the heat transfer from refrigerant (and from that measurement infer a quality), data was taken that would help explore how this sensor could improve the performance of a vapor compression system. One possible application for the sensor would be to use it as the feedback signal in a control scheme that would maintain the refrigerant at the exit of the evaporator at saturated liquid/vapor.

EXPERIMENTAL FACILITY – A versatile mobile air conditioning experimental facility capable of testing a/c systems under transient and steady-state operation was utilized in the investigations described in this document. Collins [6], Rubio-Quero [7] and Weston [8] present detailed test facility descriptions and construction information. The evaporator and condenser are housed in separate air loops. The blower in the evaporator air loop is capable of producing volumetric flow rates equal to that of a passenger compartment fan. The blower in the condenser air loop is capable of producing a high volumetric flow rate to replicate the radiator fan and ram-air effect associated with vehicle motion. The refrigerant loop contains factory standard automotive air conditioning components for a 1994 Ford Crown Victoria. Figure 6 illustrates a schematic of the test facility. The letters T, P, and RH found in Figure 6 represent instrumentation for temperature, pressure, and relative humidity measurement, respectively. An Allen Bradley programmable logic controller can be used to control the evaporator and condenser blower speeds, compressor RPM, compressor clutch and an electronic expansion valve (EEV).

EXPERIMENTAL PROCEDURE – The following data was collected to demonstrate the sensor's ability to estimate certain system parameters. The test facility was subjected to various operating conditions typical of automotive air conditioners under heavy thermal load. Air-side inlet conditions to both the evaporator and condenser were held constant. Air passing over the evaporator was at 83 °F, 31% relative humidity, and 285 cubic feet per minute (cfm).

While the air-side inlet conditions were constant, the refrigerant-side inlet conditions into the evaporator were throttled with a Sporlan SEI2 stepper-motor-driven electronic expansion valve (EEV). The EEV has a 2 ton nominal capacity. The valve has an opening stroke of 0.125 in. and 1532 steps of control. The valve was sized based on the evaporator capacity, evaporating temperature, liquid-line refrigerant temperature, and valve pressure drop. Details of the EEV characteristics can be found in Wandell [9].

For each data point the opening of the EEV was set, then the system was allowed to settle to a stead-state value before the data was recorded. Once the data was collected, only the opening of the EEV was changed before the next data point was collected.

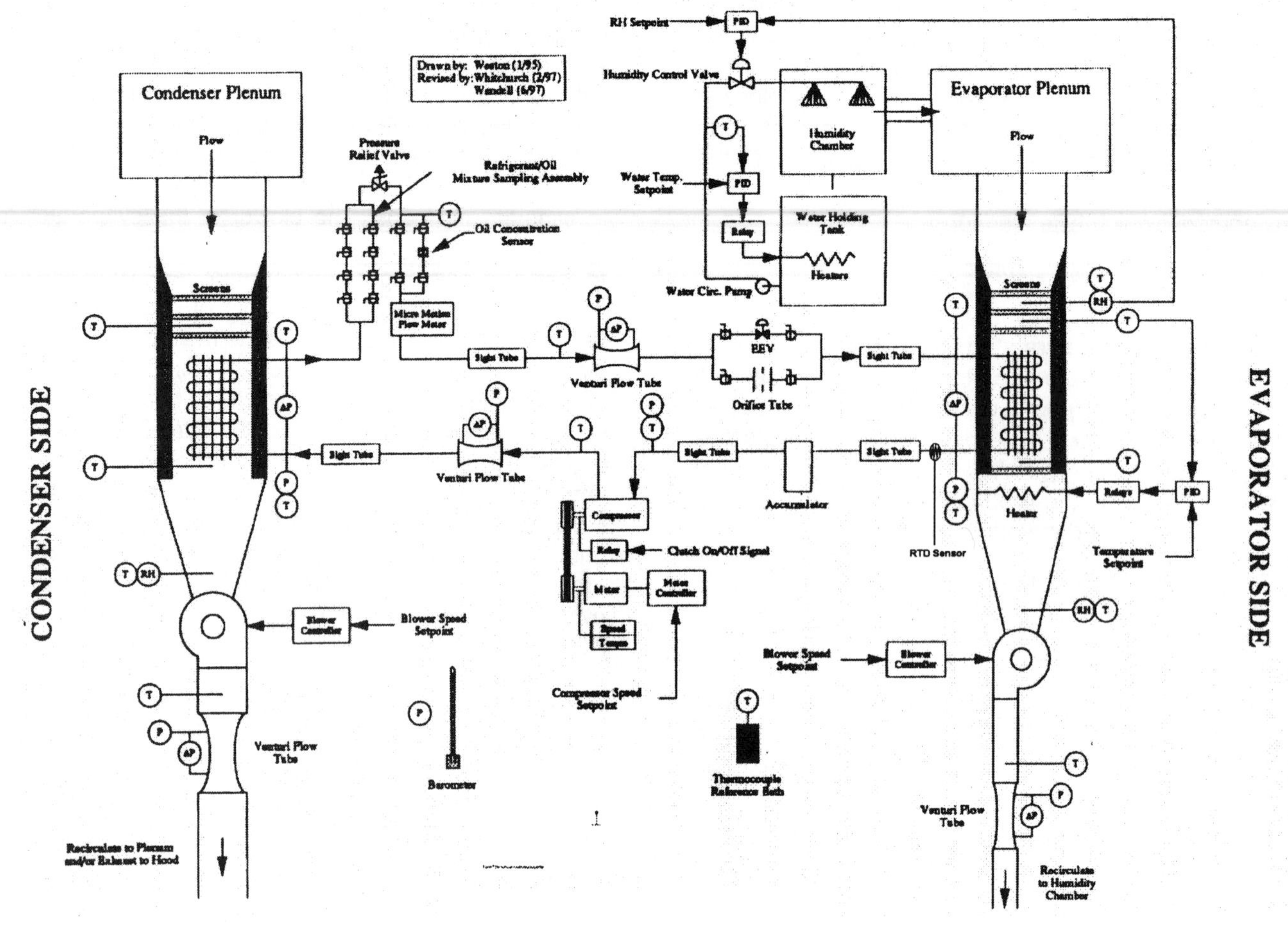

Figure 6. The mobile air conditioning test facility

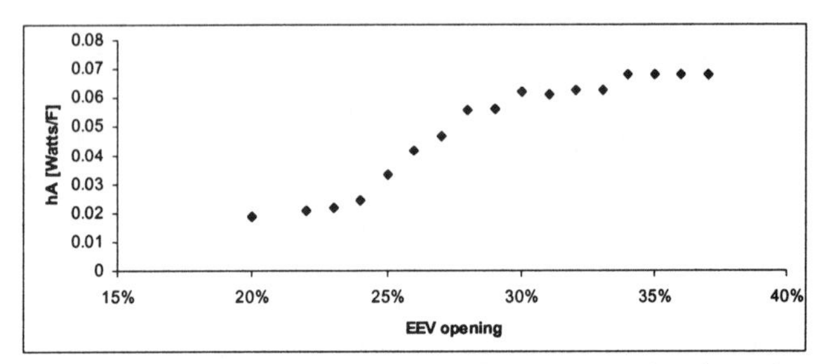

Figure 7 *superheat (top) and hA (bottom) as a function of EEV opening*

Figure 7. Superheat (top) and hA (bottom) as a function of EEV opening

EXPERIMENTAL RESULTS – Figure 7 shows how the surface-to-free-stream thermal conductance (*hA*) and degree of superheat varies with the opening of the EEV. Superheat is defined here as the difference between the temperature of the refrigerant and its saturation temperature for its pressure. When the valve is relatively closed (20%), there is significant superheat. *hA* is not affected much by the degree of superheat. That is because there was an insignificant change in refrigerant vapor velocity to affect convection on the sensor surface.

Once liquid droplets appear, the sensor must dissipate more power to compensate for the additional latent heat. Saturated liquid droplets appear in superheated vapor at around 15 °F superheat. At this point opening the EEV will allow more liquid to pass over the sensor, thus increasing *hA*.

Opening the valve even further brings the refrigerant into a regime of saturated liquid/vapor (true quality). These are the conditions under which this sensor can effectively operate and traditional temperature sensors cannot. Increasing the opening of the EEV even further will saturate the sensor at some quality. The actual heat transfer coefficient of the fluid probably does not actually maintain a constant value for increasing quality. The apparent leveling off is probably saturation of the sensor. Too much liquid is impinging the sensor. The sensor is not able to provide enough power to heat and vaporize the liquid.

It is apparent from figure 7 that the range of sensitivity of this sensor is in the low LMF region and high quality. This would be the range of operation where the sensor would be most useful.

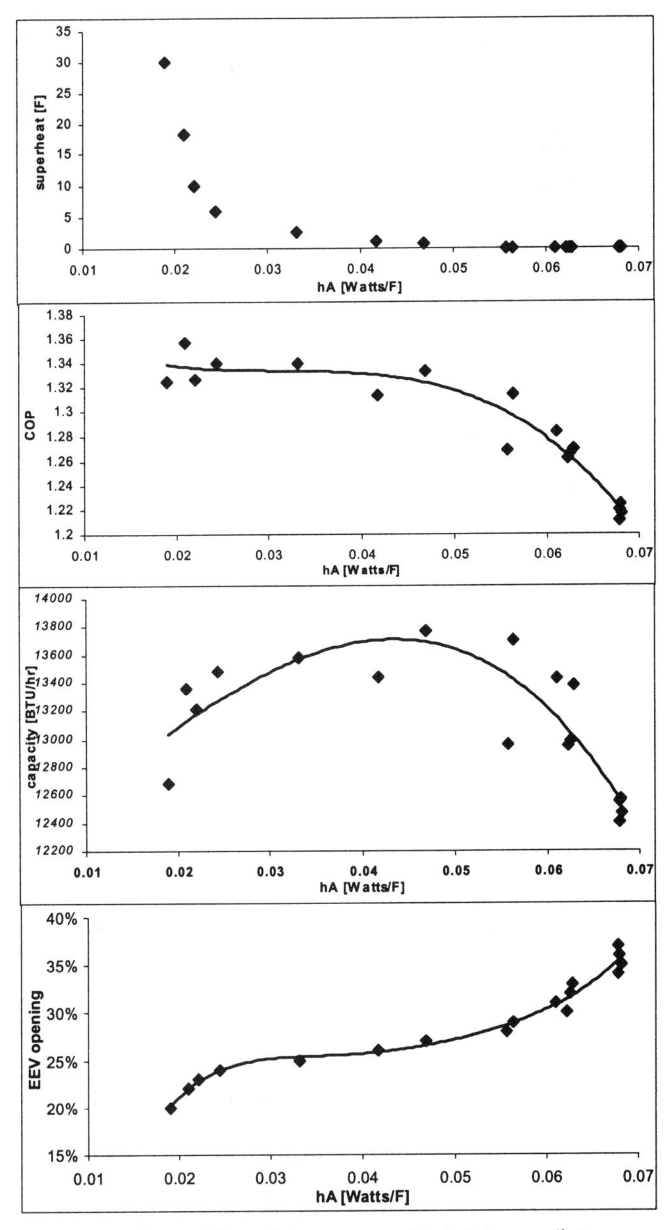

Figure 8. The ability of the parameter hA to predict superheat (top), COP (second), capacity (third), and EEV opening (bottom). Points represent individual data points, and solids lines are a least squares fit of a 3rd order polynomial.

Figure 8 demonstrates the sensor's ability to predict system performance. COP is relatively flat over regions of high superheat (low *hA*). It does not significantly drop off until well into the two-phase region (no superheat and high *hA*). According to this data, the capacity peaks out somewhere around 0.04 and 0.05 Watts/°F. This is the region where the system has its maximum cooling power for the set of operation conditions. This region corresponds to little to no superheat.

Capacity measurements are taken from the difference in enthalpies between the air going into the evaporator and the air coming out. Volumetric flow rates are measured using a venturi on the air loop on the evaporator side. Relative humidity and temperature measurements are taken for the air going in and coming out of the evaporator.

ANALYSIS OF RESULTS

The data presented in figures 4 and 5 along with the discussion presented in this paper demonstrate the ability of this sensor to predict the quality of refrigerant for that set of unique operating conditions. The data presented in this paper is specific to the facilities and conditions described in this paper. Not enough data has been taken to generalize the sensor's behavior. The sensor's behavior can significantly change if it is used in another system. Parameters such as sensor orientation, distance to the evaporator, refrigerant type, pipe cross-section, and evaporator type, all of which were held relatively constant for each set of experiments, would create inconsistent results. All of these parameters mentioned would change the convection heat transfer coefficient between the refrigerant and the RTD. This is the heat transfer coefficient being estimated to predict quality. But, in general these same parameters will remain unchanged for a given system.

The refrigerant must be well mixed as it passes over the sensor in order for the results to be valid. The liquid droplets need to be relatively distributed throughout the cross-section of the pipe. This ensures that the sensor is getting a representative sample. In reality a small portion of the refrigerant is being carried in oil-refrigerant mixture that flows attached to the tube walls. The sensor cannot detect that portion of the flow. Our experience demonstrates that the sensor should be placed very near the evaporator exit, preferably in a vertical section of tube.

The primary objective of this study was to construct and characterize a sensor that could be used in a control scheme with feedback that would improve the effectiveness of an air conditioning system. When operating an air conditioner under a high thermal load, the control objective might be to maximize capacity. For automotive applications this would mean minimizing pull-down time by getting the most cooling power for a given evaporator. According to the data presented in figure 8, this would correspond to an *hA* of about 0.045 Watts/°F. This condition also corresponds to very little or no superheat. By using the sensor presented in this paper a controller can be constructed to regulate the system at this condition.

The same system cannot use the traditional estimate of temperature difference across the evaporator to maintain the condition of maximum capacity (as shown in figure 8). This scheme will have little or no control signal. The temperature measurement used in the feedback loop also has significant Gaussian noise. These are some of the reasons why TXV systems often exhibit a "hunting" phenomenon. Hunting becomes more and more prevalent as the phase lag increases. The valve will oscillate from open to closed, unable to stabilize.

Figure 8 also suggests that COP is not significantly compromised at the point were capacity seems to reach its maximum.

CONCLUSION

A new sensor was presented that can sense refrigerant with low LMF. The dynamics of the sensor allow it to detect saturated liquid droplets in superheated vapor. The sensor is able to predict certain system parameters within the operation range of the sensor. A signal from the sensor can be used in a control scheme to maintain an air conditioning system at a desired system performance.

ACKNOWLEDGEMENTS

The authors gratefully acknowledge the Air Conditioning and Refrigeration Center at the University of Illinois who supported this work.

REFERENCES

1. Wedekind, G. L., Stoecker W. F., (1966), Transient Response of the Mixture - Vapor Transition Point in Horizontal Evaporation Flow, ASHRAE Transactions, Vol. 72, Part II.
2. Stoecker, W. F., (1966), Stability of and Evaporator-Expansion Valve Control Loop, ASHRAE Transactions, Vol. 72, Part II.
3. Wedekind, G. L., (1965), Transient response of the mixture -vapor transition point in two-phase horizontal evaporating flow, Ph.D.
4. Barnhart, J.S., Peters, J.E., (1992): *An Experimental Investigation of Flow Patterns and Liquid Entrainment in a Horizontal -Tube Evaporator*, ACRC Technical Report #28, December, 234 pp.
5. Lomas, C. G., (1986), *Fundamentals of Hot Wire Anemometry,* Cambridge University Press.
6. Collins, C.D., N.R. Miller, and W.E. Dunn. *Experimental Study of Mobile Air Conditioning System Transient Behavior.* ACRC Technical Report 102, July 1996.
7. Rubio-Quero, J.E., W.E. Dunn, and N.R. Miller. *A Facility for Transient Testing of Mobile Air Conditioning Systems,* ACRC Technical Report 80, June 1995.
8. Weston, P.G., W.E. Dunn, and N.R. Miller. *Design and Construction of a Mobile Air-Conditioning Test Facility for Transient Studies*, ACRC Technical Report 97, May 1996.
9. Wandell, E.W., W.E. Dunn, and N.R. Miller. Experimental Investigation of Mobile Air Conditioning System Control for Improved Reliability, ACRC Technical Report 128, August 1997.
10. Miller, James A. (1976) A Simple Linearized Hot-Wire Anemometer, Journal of Fluids Engineering, December 1976.
11. Simpson, R. L., K. W. Heizer, R. E. Nasburg. Performance Characteristics of a Simple Linearized Hot-Wire Anemometer, Journal of Fluids Engineering, September 1979.
12. Shannon, M. A., P. S. Hrnjak, T. M. Leicht. Exploratory Research on MEMS Thechnology for Air Conditioning and Heat Pumps EPRI Report TR-111169

CONTACT

James Solberg, Norman R. Miller, and Predrag Hrnjak
Air Conditioning and Refrigeration Center
Department of Mechanical and Industrial Engineering
University of Illinois at Urbana-Champaign
1206 West Green Street
Urbana, IL 61801, USA

942251

Future Development of Central Tire Inflation Systems and Integration with Vehicles

James A. Beverly
Eaton Truck Components - North America

ABSTRACT

With the increasing interest in applying Central Tire Inflation Systems (CTIS) to commercial heavy duty truck applications, a variety of vehicle integration issues must be addressed. Most system installations to date have been handled on a retrofit basis, but CTIS equipped vehicles will soon become available through OEM's. This raises the need for the establishment of recommended practices in the design and application of systems. Issues addressed include air system requirements, wheel integration, and potential safety issues such as brake priority and over-speed detection and correction. Use of electronic data interchange can result in diagnostic standardization, and enhanced capabilities such as integrated load sensing. This paper discusses integration issues, as well as potential future developments in tire pressure control systems.

INTRODUCTION

The most commonly perceived benefit of Central Tire Inflation Systems (CTIS) is expanded mobility as a result of adjusting tire pressures to accommodate load and terrain. The military has historically been the primary user of such systems, but the past decade has seen the proliferation of CTIS in both on and on/off highway applications. The primary focus has been in the logging industry and the United States Forest Service. However, applications have expanded into construction, fire and rescue, agricultural and utilities. It has also been well documented that additional benefits of CTIS include reduced road degradation, less operator and vehicle fatigue as a result of improved ride, and less drive-line dynamic loading. Additional studies have investigated the impact on tire life and retread-ability, and have shown the possibility of reduced operating costs. Technology advances such as the implementation of wheel valves for de-pressurized operation and the application of electronic controls have served to accelerate interest in tire pressure control use. The purpose of this paper is to address the future developments of Tire Pressure Control Systems, and the potential impact on vehicle integration.

VEHICLE INTEGRATION

There have been many recent OEM vehicle applications in the U.S. military where CTIS was designed into the vehicle; including the M939A2, PLS, HET and FMTV. The commercial market is just beginning to address CTIS from the vehicle OEM perspective, therefore, the first integration issue is one of system installation. System installations have typically been handled by specialty retro-fitters. This requires the end user to play a larger roll in ensuring that the vehicle is spec'd to accommodate the increased demands of CTIS. "CTIS ready" vehicles are being made available through OEM's, which specify options such as integrated axle and wheel end seals, larger capacity compressors, and specific air dryers. In the near future, OEM's will provide fully equipped CTI systems in a variety of truck configurations. This allows the end user to finance the cost of CTIS into the original purchase price of the vehicle.

AIR SYSTEM REQUIREMENTS - One of the fundamental requirements of a CTI system, which impacts the primary vehicle operation, is the ability to re-inflate tires. This requires sufficient capacity of clean dry air from the vehicle's compressor system. Tire manufacturers recommend that inflate times not exceed 5 minutes to achieve 80% of nominal pressure to avoid tire damage. Most highway tractors are spec'd with 12 to 13 cfm compressors, with 24 to 34 cfm compressors available for certain off-highway applications. These sizes are theoretical outputs at some arbitrary speed , typically 1250 or 1000 rpm. Some logging applications include 8 axle tractor/trailer combinations with as many as 30 tires! Assuming an inflate requirement from 75 to 90 psi for 30 tires of approximately 4 cubic feet and a compressor capacity of 30 cfm, the inflate time can be approximated as 4 minutes utilizing the formula:

$$\text{Inflation Time (minutes)} \cong \frac{\text{Volume (cu ft)} * (P_{final} - P_{initial})\text{(psi)}}{14.7 \text{ (psi)} * \text{cfm}}$$

The use of additional reservoirs to "store up" air capacity to improve inflate times is of marginal use, because once consumed, they add additional volume which must be replenished simultaneously while inflating the tires. In addition, care must be taken to always conform to FMVSS 121 requirements for build-up times to charge the supply system for the brakes.

As a result of CTIS operation, the demand on compressors will far exceed the 5 to 10% duty cycle typical of most vehicles. The resulting impact on compressor life has yet to be determined, but it is expected that the increased acceptance and usage of tire pressure controls will be the catalyst for improved compressor technology.

Another fundamental component found on heavy duty vehicles which is directly impacted by increased compressor demand is the air dryer. Although use of a wet tank is prevalent on many vehicles, application of a variety of dryer types between the compressor and wet tank is becoming more widespread. Today's air dryers typically utilize one or more of the following principles:

After-coolers - The compressed air is cooled; condensing water vapor and oil contaminants which are periodically expelled.

Coalescent - Utilizes a combination of cooling, filtering and coalescing to combine and capture water vapor and oil contaminants which are periodically expelled.

Desiccant - Utilizes a chemically treated medium which is moisture absorbent and is capable of creating a dew-point depression. Desiccant is regenerated, or "dried" during a purge cycle when the compressor governor unloads.

Due to the lengthy duty cycle on air systems of CTI equipped vehicles, great care in selecting air dryers must be applied. In addition, automatic spitter valves on the wet tank are highly recommended. All but the desiccant dryers result in saturated air at pressure and temperature at best, and the desiccant dryers become saturated (and therefore ineffective) after several minutes of continuous use. However, this may be addressed by the recently introduced "2-stage", or continuous duty desiccant systems, where half of the system is drying the air while the other is alternately regenerating. These systems are reported to consume only 5 to 10% of available air by volume. Regardless of the dryer type, it is important to pay particular attention to cooling the air at the maximum pressure before it enters the CTI or brake system. Future consideration should be given to operating air systems at pressures up to 150 psi, thus increasing the dew point depression at expected tire pressures. This is within current compressor and governor setting capability. How it might affect compressor durability, as well as the need for a strategy for reducing pressure levels to brake system components has yet to be determined.

Care must also be used in employing alcohol evaporation systems, which effectively lower the freeze point of moisture in the air system, but do not remove it. Although the components of the CTI system may be compatible with alcohol, the user should consult the tire manufacturer for possible effects on the tire.

BRAKE PRIORITY - Besides meeting the previously discussed FMVSS 121 build-up times, the issue of brake priority must be discussed. CTI systems available today employ several methods including electrical pressure sensing, opposed check-valves, brake protection valves, or a combination thereof. Some systems ensure that the air brake reservoirs merely maintain a minimum level of pressure, while others ensure the brakes are fully charged prior to initiating tire pressure control functions. Safety considerations demand that brake system recovery times not be compromised, and that impact on reserve capacity be minimal. A review of FMVSS 121 regulations with regards to charge times of the supply system may be required, such that recovery times be effective while CTIS is in the process of inflating. The Interpretation of FMVSS 121 when integrating this technology needs to be addressed.

WHEEL END INTEGRATION - One of the more frequently asked questions is how air is routed from the chassis to the tire. This is accomplished by one of two methods: externally routed rotary unions, or integrated air seals. The U.S. military has a recognized need for integrated air routing for their applications, but commercial applications may not always be so demanding. In addition, many of the commercial applications to date have been retrofits; which may result in the integrated seal being more costly than the rotary union approach. In a recent study with a system equipped with rotary unions conducted by the Forest Engineering Research Institute of Canada, or FERIC,[1] the axle end hardware required more repair and maintenance than any other part of the system. The supply hoses and rotary unions extended 7 cm beyond the edge of the tires, making them vulnerable to impacts. The study further suggests that with the recent introduction of CTI-ready axles by manufacturers, a long term cost/benefit evaluation should be undertaken.

CTI-ready drive axles, steer axles and trailer axles with integral seal systems are becoming available. A typical drive axle consists of a sleeve pressed on the inner bearing journal of a industry standard spindle (see figure 1). Air is routed between the spindle and the sleeve to a unitized seal installed in the hub. A typical non-drive axle consists of self-piloted rotary joint attached to a drilled through spindle (see figure 2). This package includes an integral hub-cap and wheel valve assembly for ease of wheel removal and maintenance. The passenger car market has also made use of a wheel bearing incorporating an air passage for CTI equipped vehicles in conjunction with active suspensions.

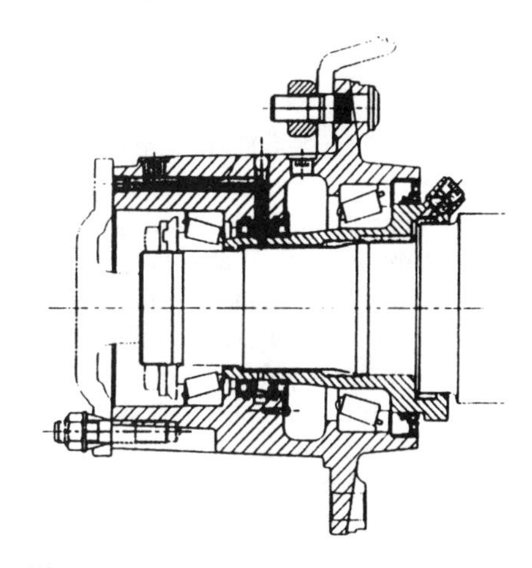

Figure 1. Integrated Drive Axle Seal

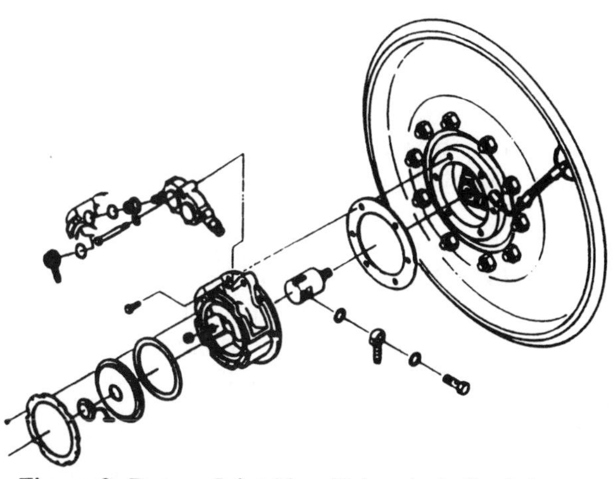

Figure 2. Rotary Joint Non-Drive Axle Seal Assy.

Probably the most significant advance in CTI system reliability stems from pressurizing the air seals only while the system is inflating or deflating. This is typically accomplished by means of a pneumatically isolating wheel valve. This feature virtually eliminates tire leak-down when the vehicle is not in operation. Significantly improved seal life results from the reduced seal lip loads when the system is not actively adjusting pressures.

WHEEL VALVES - A wide variety of wheel valve technologies exist on systems employing non-pressurized seals; many of their characteristics unique to their applicable systems. Some systems designed for the military provide for remote exhausting of air above the fording line of the vehicle, instead of at the wheel end. This provides the maximum degree of robustness to meet the off-road mission profile encountered with tactical vehicles. Wheel valves which exhaust at the wheel ends are a more cost effective approach for commercial applications. This is primarily due to the reduced flow requirements of the seals, as they are not subjected to the high flow rates typical of deflation.

CTI systems typically connect to the tire through the conventional valve stem on the wheel. As this may require removal of the valve core, the wheel valve should provide fail safe shut down in the case of system malfunction, or up-stream hose breakage. Recommended practices must address the acceptable means of tire isolation and allow the use of technology beyond conventional velocity fuses (check-valves) or manual shut-off valves. Failure of the tire or wheel valve should result in a warning to the driver. Manual tire isolation methods do not provide this feedback. In addition, there should always be a way of manually checking and filling the tires at the wheel using conventional means.

SPEED SENSING - All CTI systems should have some mechanism for sensing speed, and notifying the driver of excessive speed for a given pressure. If the driver does not respond, the system should automatically inflate the tires to a more appropriate pressure. Warnings can be visual and/or audible. Speed information can be provided in a number of ways. Magnetic "pencil" sensors on the transmission output shaft can be replaced with bifilar "dual coil" sensors, or buffered with a amplifier. On mechanically driven speedometers, a through-drive sensor can be installed in series with the cable. Electronic speedometer readouts provide buffered logic output signals. Electronic transmissions or engines often provide speed information via on-board data links. Whatever the source, CTI systems must provide flexibility in accommodating various calibrations.

Drivers are well aware of the increased driver comfort at reduced pressures. The ability of a driver to operate the vehicle in a mode that results in reduced tire life or at potentially unsafe tire pressures by simply turning off power to the CTI system raises the question of whether a power on/off switch should be provided. Automatic speed bump-up and over-rideability should also be addressed through recommended practices.

CONTROL SYSTEMS - The increased use of micro-processor based "smart" control systems has significantly enhanced flexibility. Today's sophisticated CTI systems can provide features such as simple push-button operation, graphical information displays, and independent Steer, Drive, and Trailer control. User flexibility is achieved through features such as programmable pressure and over-speed targets to accommodate changes in season or vocation. While some systems require a memory, or "chip" replacement to accommodate parameter changes, others utilize non-volatile memory (NOVRAM) to allow field programmability via the control panel, or through use of a personal computer or diagnostic tool. Some provide automatic pressure adjustment in addition to over-speed detection, adaptive inflate and deflate algorithms, identification of damaged tires or leaking control lines, and detection of electrical problems with sensors and actuators. Systems can also "remember" intermittent faults, which can be analyzed later by the service technician.

Although these features are offered on a number of systems, minimum requirements regarding what system faults CTI systems should universally be capable of detecting have yet to be established.

FUTURE DEVELOPMENTS

ELECTRONIC DATA INTERCHANGE - The proliferation of micro-processor based systems on heavy duty vehicles has already resulted in the standardization of electronic data interchange. A joint SAE/TMC recommended practice, SAE J1587 has defined the format of messages and data being communicated between micro-processors on a common data communications link. The hardware protocol necessary to implement this data interchange is defined by SAE J1708. Parameter Identification characters (PID's) involving tires have already been assigned for tire pressure and temperature, as well as other useful data such as vehicle speed. Message Identifiers (MID's) have been assigned for tires of the power unit and three trailers. Finally, there are up to 255 Subsystem Identification numbers (SID's) available which are definable for each controller. SID's should only be assigned to field-repairable or replaceable subsystems for which failures can be detected and isolated by the controller. This will allow industry standard diagnostic tools to be developed to troubleshoot systems, regardless of manufacturer. There are currently no SID's assigned for tire pressure control systems. A consensus of the CTIS suppliers as to what components of the CTI system can be commonly recognized should be developed and submitted to the Data Format sub-committee for approval. Additional consideration should also be given to expanding the list of MID's to include separate identifiers for steer and drive axles on power unit.

TIRE DESIGN - The potential gains realized through tire design, when used in conjunction with Tire Pressure Control Systems is beyond the scope of this paper. However, the enhancements to tire pressure control systems through tire technology will be addressed. Most significantly, the integration of low cost pressure and temperature transmitters into the tire will enhance the capability and response of the tire pressure control system. Most systems available today measure a weighted average of the pressures of any one group of tires. Various methods of data characterization are then used to determine pressure trends or recognize patterns of leaky air lines. Technology exists today to transmit signals via telemetry from the individual wheels to the chassis. Individual tire pressure and temperature data can provide the means to a highly adaptive control strategy for optimized tire pressures. Accurate tire temperature can address the compromises which must be made in setting pressures to address the issue of hot verses cold tires. Although more costly than a single point measurement system, some applications where further improvements in stability and adaptability are required may justify this need in the future.

LOAD SENSING - Tire pressures can also be optimized with load sensing feedback. This may be achieved by integral 5 th wheel or load sensors, thus eliminating the need for operator input. Ideally, these sensors would be capable of dynamic load sensing. In addition, this information could be made available to an electronically controlled engine or transmission.

CONTROL SYSTEMS - There remain many issues to be addressed as we gain more experience with tire pressure control systems. For example, on multiple channel systems, the issue of "simultaneous" channel operation is often discussed. Is it acceptable to change pressures on each channel individually, or should this take place simultaneously? Remember, inflate rates are primarily dependent on compressor capacity and total tire volume, not simultaneous channel operation; and deflate rates are dependent only on individual tire volume for each wheel valve. To maintain stability, some systems initiate deflation such that all channels achieve their final pressures simultaneously, while inflation is handled on a "round-robin" basis. Should pressures between axle groups be raised proportionately to final pressures, or simultaneously? How will the change (or difference) in tire pressures effect other systems such as antilock brake systems (ABS) and traction control systems (TCS), which are sensitive to the tire rolling radius? How should the system respond when a trailer is connected and it's pressures are at a different setting than the power unit?

SYSTEM OPERATING PRESSURES - Higher tire pressures may require higher minimum air supply pressures. Applications such as mixers employing wide based tires often operate at pressures up to 125 psig, while present practice allows governor cut-outs as low as 110 psig. Most systems require control pressures of at least the minimum desired tire pressure. Additionally, it is not practical to consider governor cut-in pressures below maximum desired tire pressures. Otherwise, the system may not inflate once the tires were at pressures above the cut-in unless some other means were used to bleed off the wet tank and start the compressor cycle.

CONCLUSION

New technologies are available today to reduce vehicle operating costs and improve productivity. Proper integration of this technology is essential to its acceptance and long term success. SAE should be the focus of discussions to assure this happens, and as even further advances cause this technology to evolve.

REFERENCES

1. Bradley, A., June 1993, *Testing a CTIS System in Western Canadian Log-Hauling Conditions*, FERIC, Technical Note TN-19

980293

State of Knowledge and Current Challenges in Defrosting Automotive Windshields

Karim J. Nasr
GMI Engineering & Management Institute

Bashar S. AbdulNour and George C. Wiklund
Ford Motor Company

ABSTRACT

Rapid and effective windshield defrosting has been the goal of various investigations by automotive engineers around the world. Car manufacturers have invested considerable resources to satisfy the thermal needs, safety requirements, and comfort demands of their customers. This paper addresses the climate control issues of defrosting automotive windshields. The paper summarizes the state of knowledge of the various approaches for improving defroster performance. Experimental as well as computational efforts, accompanied by heating techniques and heat boosters will be presented. The paper also features relevant measurement methods for airflow and thermal patterns, and discusses current challenges. Recommendations are made on where to focus engineering and design efforts given the state of present technologies.

INTRODUCTION

In recent years a number of experimental and computational investigations were performed for the purpose of improving the defrosting system of vehicles. Modifications such as increasing the rate at which the engine coolant temperature rises following a cold start, improving the heat exchange efficiency of the heater core, and increasing the amount of air discharged by the defroster nozzle have been investigated. One must recognize that a number of factors (proper location, orientation, and design of defrosters to name a few) affect the interaction of air flow with the windshield inner surface and therefore dictate the performance of the defroster system.

Generally, a trial-and-error approach has been used to capture the defroster system performance by investigating relevant performance variables and parameters. As a result, various investigators proposed modifications to the HVAC system or a component of the system for improved performance and clear vision areas.

This effort presents an assessment of our knowledge base in employing experimental and computational techniques and of the different solution approaches utilizing heating elements, conducting films, and heat storage devices. This paper will begin with a presentation of a literature survey on the topic, feature employed measurement methods of airflow and thermal patterns, present an example on the use of one of the available technologies for quick evaluation of a system performance, and finally outline future challenges.

LITERATURE REVIEW OF RELEVENT INVESTIGATIONS

Different approaches for the purpose of obtaining effective and efficient defrosting patterns have been utilized by automotive engineers. The following review presents relevant experimental and computational efforts, electric heating elements and conducting films, and heat boosters.

EXPERIMENTAL EFFORTS

Early experimental efforts focused on determining the defroster performance using wax. Later studies focused on modifying to defroster nozzle design, dispersing airflow over the whole windshield without separation, optimizing the position and angle of defroster nozzle-windshield, and carrying out real-world testing. Recently, efforts were made towards using available technology for measuring and evaluating thermal patterns of vehicle windshields (i.e. Infrared Thermal Imaging) and jet flow exiting the defroster nozzle (i.e. Hot-Wire anemometry). Table 1 presents a list of relevant experimental studies.

Table 1. A list of experimental investigations

Investigators	*Relevant Issues*
Atkinson and Muchnij (1972)	Wax method for describing performance.
Atkinson (1989)	Nozzle design and performance.
Sakamoto (1989)	Measures for reducing turbulence Enlarging defroster vent area Optimizing location of windshield.
Norin (1989)	Mist formation on side windows and generated cold drafts.
Leier et al. (1989)	Real-world testing vs. wind tunnel testing.
Kohnotou et al. (1992)	Development and design of defroster nozzles.
Andreone et al. (1992)	Automatic data acquisition via Infrared Thermography.
Kakubari et al. (1993)	Compared performance of electric heating elements to electric films and conventional defroster. Quality of ice on windshield (soft vs. hard).
AbdulNour (1997)	Defroster and windshield flows measurements.
Willenborg et al. (1997)	Defroster flows measurements.
AbdulNour, R. S. et al. (1997)	Defroster flow measurements

NUMERICAL EFFORTS

Experimental flow and thermal measurements are expensive to carry out and limited by the amount of data that can be collected. On the other hand, Computational Fluid Dynamics (CFD) provides a numerical solution of the governing equations of motion (conservation of mass, momentum, and energy) to determine all flow variables of interest for an entire flow field. CFD is used to obtain the velocity field in the vicinity of automotive windshield due to defroster flow. The analysis can be performed using general-purpose CFD packages like FLUENT/UNS, STAR-CD, PHOENICS, and CFD-ACE running on UNIX-based workstations. The physical problem is typically three-dimensional, transient involving multi-medium and multi-phase transport phenomenon. In essence, numerical defrost models are based on energy conservation with the objective of predicting the melting pattern and the time-to-melt under prescribed boundary conditions during a cold start from a specified soak temperature (say -3 °F). The equations govern the conjugate heat transfer through all layers of glass, as well as internal and external convection. Based on the degree of correlation of velocity profiles between the CFD simulations and the experimental work (AbdulNour and Foss, 1997), it can be stated that numerical simulation of defroster and windshield air flow is a valid technique to gain valuable insight into the flow field, both in direction and magnitude. Finally, it should be noted that factors such as mesh density, cell geometry, turbulence model, degree of approximation of the describing equations, error criterion for convergence, and numerical control parameters, can impact the speed and accuracy of the numerical solution. Table 2 presents recent numerical and computational studies.

Table 2. Selected Computational and Numerical Defrosting Efforts

Investigators	*Relevant Issues*
Ikeda et al. (1992)	Airflow distribution on windows from defroster's nozzles.
Lee at al. (1993)	3-D Computational model for windshield clearing.
Lee at al. (1994)	Validation of computational results with experimental results.
Sugano et al. (1994)	Prediction of defroster clearing pattern and comparison with experiments.
Sato and Ito (1995)	Numerical analysis of defroster clear-up pattern covering the heater unit, the defroster nozzle, and the cabin interior.
AbdulNour and Foss (1997)	Computational model for flow over automotive windshields.
Brewster et al. (1997)	Computational model for windshield de-icing and comparison with cold-room data.

ELECTRIC HEATING ELEMENTS AND CONDUCTING FILMS

Heating elements and conducting films were the subject of many automotive glazing studies over the years. While heating elements are common for the rear windows of vehicles, earlier designs were found to cause visibility problems when embedded in front windshields. A recent study, however, was performed by Southall and Burrnand (1990) on the visual effects of heated windshields investigating the driver subjective responses to driving cars equipped with wired screens, and examining the effects due to reflections and diffraction of light during night-time as well as reduced visibility daytime conditions. Their laboratory investigations showed no significant reduction in visibility in either daytime or night-time, and generally no visual discomfort was felt. Whether in the form of a high light transmitting electro-conducting film, metallic film fed by bus bars carrying direct current, or wired screens, the attractiveness of such a solution is rapid defrosting. Besides heating the windshield via a current through an embedded conductive layer, a recent patent by Cummins et al. (1996) describes the use of a heating element installed in the defroster ductwork. The electric element heats up the air flowing through the duct towards the defroster. Table 3 presents a list of various systems involving electric heating elements and conducting films.

Table 3. Electric Heating Elements and Conducting Films

Investigators	*Relevant Issues*
King (1974) Boaz and Youngs (1974)	Development of high light transmitting electro-conducting film for defrosting (Hyviz).
Presson (1980)	Electrically heated windshields.
Anon (1985)	Windshields having metallic coatings of silver and zinc oxide.
Schafer (1986)	Development of transparent, conductive and heat reflecting layer system (TCC 2000).
Norin et al. (1987)	Coatings (tin oxide) for frost prevention.
Hendrix and Bernardi (1988)	History and development of conductive films - "Super-H Coatings".
Jurgen (1988)	Fast-acting heated windshield system (ElectriClear).
Takada (1989)	Electric antifrost system.
Southall and Burnand (1990)	Visual effects of electrically heated windshields.
Cummins et al. (1996)	Heating element installed in the defroster ductwork.

HEAT STORAGE DEVICES

Generally, defrosting a vehicle windshield relies on the heat rejected by the engine. However, as engines get improved and become more efficient, less "rejected" heat will be available to heat the vehicles windshield. Various investigations, such as those using a latent heat storage device (Tsantis et al., 1994), have addressed the need for a supplemental heat source, especially during a vehicle cold-start. They concluded that such device is effective in improving both the heater and the defroster performance. Heat boosters in the form of heat batteries, fuel-burning heaters, or thermal energy storage devices using lithium bromide have been the subjects of recent studies. The feasibility of a supplemental heat storage device must also address its cost, weight, size, and packaging. Table 4 features various investigations on supplemental heating devices.

Table 4. Selected Works on Heat Storage Devices

Investigators	*Relevant Issues*
Tsantis et al. (1994)	Need for supplemental source of heat during a vehicle cold-start. Cost, weight, size and packaging.
Schatz (1991)	Heat batteries for vehicle's heating and defrosting system.
Boltz et al. (1992)	Development of an autonomous fuel-burning heater for rapid heating and defrosting.
Lindbert and Anderson (1990)	Development of a "thermal power booster" in the form of a fuel burning heater.
Marston et al. (1994)	Thermal energy storage using lithium bromide, a preheating system. storing energy from the exhaust.

SAE J902, CORPORATE STANDARDS, AND FEDERAL REQUIREMENTS

SAE J902 is a test procedure and performance guideline for evaluating passenger car windshield defrosting systems. The procedure provides uniform and repeatable laboratory test results, even though under actual conditions frozen water coatings would be removed by scraping before driving the vehicle. The performance obtained, therefore, does not directly relate to actual driving conditions, but serves as a laboratory performance indicator for comparing test results within or between systems. The reader is referred to this Recommended Practice for details on the procedure. Windshield areas to be defrosted are exhibited in Figure 1 as areas A and C. Details on establishing these areas are described in SAE J902. Also, Federal Motor Vehicle Safety Standard No. 103 specifies requirements for windshield defrosting and defogging systems.

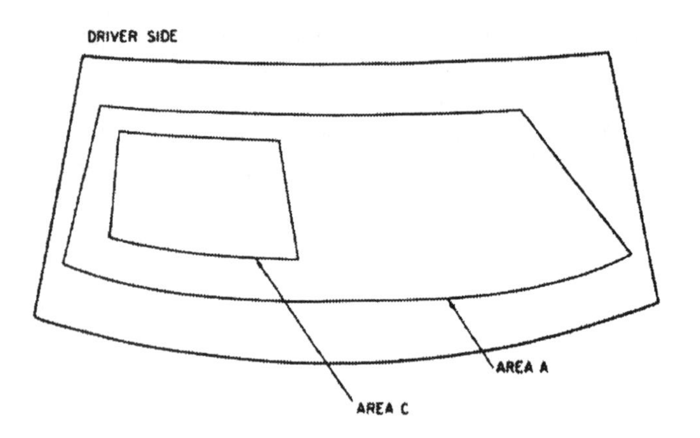

Figure 1. Typical location of areas A and C as viewed from vehicle interior [SAE J902]

The designer must make sure that his/her design meets windshield defrosting zone standards (SAE J902, FMVSS 103) as well as conforming to corporate requirements. To maintain their corporate advantage, automotive manufacturers usually meet more stringent requirements for windshield defrosting. Table 5 tabulates SAE's and FMVSS 103's clearing zones requirements.

Table 5. Specified defroster pattern clearing zones

	Area of Interest	Minimum Clearing (%)	Time (minutes)
SAE	C	100	30
	A	80	30
FMVSS 103	C	80	20
Requirements	A	95	40

MEASUREMENT METHODS FOR DEFROSTER-WINDSHIELD AIRFLOW

The airflow characteristics in the immediate vicinity of the vehicle windshield is a major influencing factor towards effective and efficient defrosting patterns. Various techniques can be applied to measure the airflow distribution in the immediate vicinity of the windshield so that a better understanding of the transport phenomena can be attained. Cogotti et al. (1988) utilized a two-component fiber-optic laser velocimeter system to measure two components of the local velocities simultaneously and the local turbulence quantities as well. Laser velocimeters have a number of interesting features, including non-intrusive optical measurement, well-defined directional response, high spatial and temporal resolution, and multi-component, bi-directional measurements. Ishihara et al. (1992) utilized visualized images of helium particle flow paths via laser light sheets for the computation of air flow velocities in the passenger compartment. Okuno et al. (1993) employed a soap bubble generator which produces soap bubbles to be used as tracer particles. Particle Imaging Velocimetry (PIV) and Particle Tracking Velocimetry (PTV) were utilized for two-dimensional velocity measurements, in which particle moving distance and direction are determined by identifying tracer particles on consecutive images. The three-dimensional velocity vector measurements were determined by stereo-photogrammetry.

Hot-wire techniques have been used extensively in supporting computational fluid dynamics analyses by providing boundary conditions as well as discrete velocities for validation purposes. In essence, hot-wire anemometry is used to experimentally determine the velocity field in the vicinity of the defroster nozzle and windshield interior surface. Quantitative measurements are useful when used to predict the velocity distribution for design and development purposes, and are necessary to determine the heat transfer coefficient on the windshield. Also, if numerical codes can successfully compute these 3-D, turbulent flows, then the verified computations can be used as an analytical design tool. AbdulNour (1997) utilized a template, using a nominal 100x100 mm square grid array, to define the measurement locations. Flow directions were obtained using wool tufts attached to the inside of the windshield. Photographs were taken of the tuft orientations in the defroster jet flows. The data from the photographs were transcribed to a graphical representation to show the mean flow directions. The probe axis was aligned with the previously determined flow direction at each point. The results are consistent with the wall jet velocity profile (in magnitude and direction) due to the impingement of the defroster jet flow on the windshield.

MEASUREMENT METHODS FOR THERMAL PATTERNS

Directly applied to the issue of thermal patterns and defrosting of vehicle windshields, Infrared (IR) Thermal Imaging (or Thermography) has emerged as a major tool towards deciding on the defroster system enhancements and modifications. Various investigators (Carignano and Pippione 1990; Andreone et al. 1992; Trauger 1993; and Burch et al. 1993) have utilized this technique. Infrared imaging systems offer numerous advantages over conventional surface temperature measurement techniques at the expense of a relatively large financial investment. Thermography is the process of generating a two-dimensional thermal image of a scene by using an infrared thermal imaging system. A thermal imaging radiometer measures the quantity of thermal energy (within a specific wavelength band) radiating from a scene. Temperature data are inferred from the measured radiant energy (Trauger, 1993). IR imaging systems provide non-contact surface temperature measurements. Subsequently, neither the surface under evaluation nor the surrounding fluid flow and temperatures are disturbed. IR imaging systems also permit the acquisition, storage, retrieval, and analysis of large amounts of data in a short time period. In addition, most systems offer features such as video tape and disk storage (of image information) as well as post-processing software to facilitate analysis. IR imaging systems operate with a real time response on the order of nanoseconds compared to the microsecond and millisecond response times of other measurement techniques. As far as disadvantages are concerned, the largest drawback of IR imaging systems is their cost. In addition, since IR imaging systems infer temperature data from measured radiant energy, it is extremely important that proper surface preparation and system configuration techniques be employed in order to obtain accurate temperature data.

By presenting a series of examples on the usage of IR thermography in automotive applications, Burch et al. (1993), extended the use of this tool to measure air temperature by inserting low-mass, high porosity screens and made reference to the work of Carignano and Pippione (1990) where IR thermography was used to observe windshield defroster patterns during the initial defroster design. Thus IR images processing enables the defrosting progress in time to be displayed and/or the defrosting mappings to be frozen step by step and printed out in order to allow a subsequent examination and/or processing.

USING INFRARED IMAGING FOR EARLY DETECTION OF DEFROSTER PERFORMANCE

The objective of this section of the paper is to illustrate how an available and advanced technology such as IR Thermography can be used in conjunction with conventional data acquisition (thermocouples) to quickly assess the performance of the defrosting system. A schematic of the experimental apparatus is shown in Figure 2. An actual vehicle was used in the experiment.

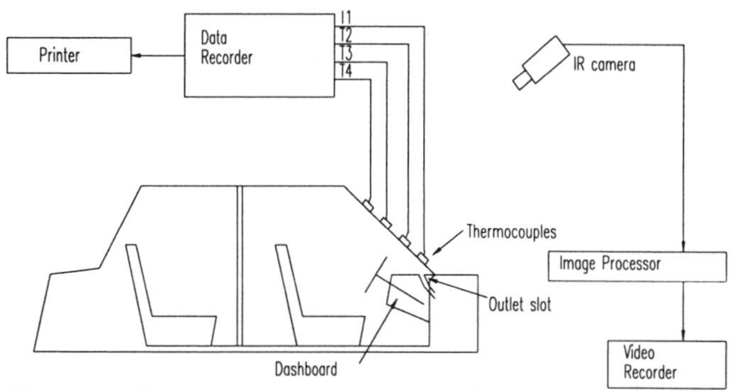

Figure 2. Experimental apparatus of a full vehicle under testing

The infrared camera was positioned in front of the car normal to the windshield. The thermal images, progressing in time, were captured in one minute intervals over 30 minutes. With the help of a cursor on the interactive IR camera system, it was possible to extract the temperatures at four different locations on the external surface of the windshield. These locations were also instrumented with copper-constantan thermocouples as shown in Figure 3.

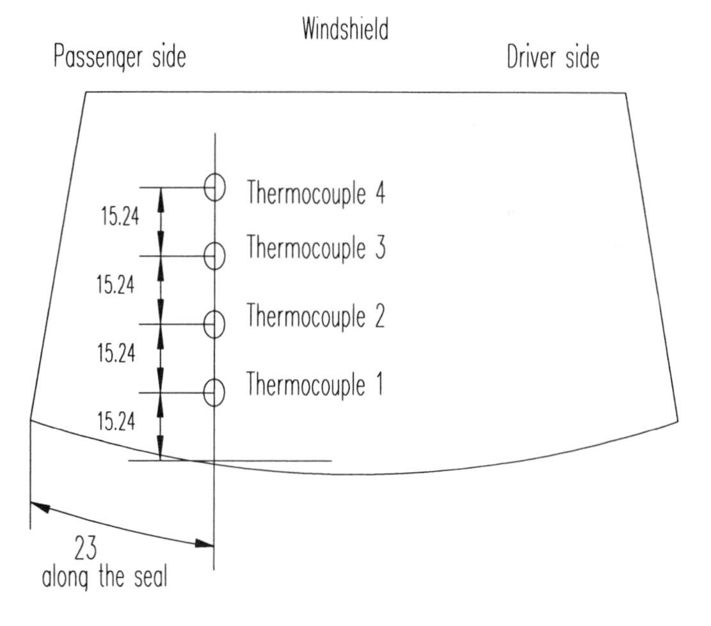

Figure 3. Position of the thermocouples on front windshield (dimensions are in cm)

Four thermocouples were also installed in the defroster nozzle exit in order to monitor the temperature increase during the cold start as shown in Figure 4.

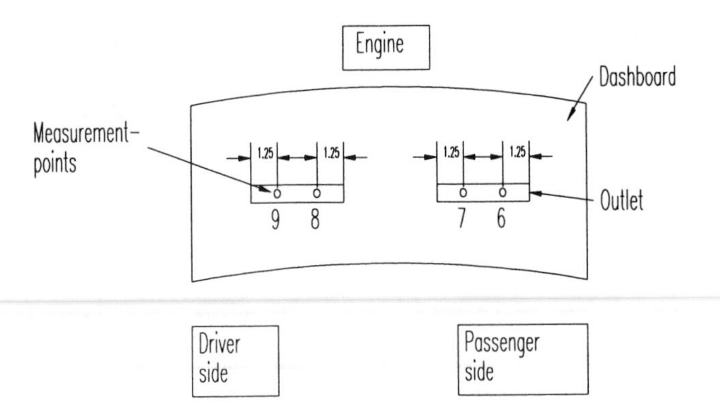

Figure 4. Measurement points along the defroster outlet using thermocouples

The testing procedure for thermal imaging on the vehicle windshield involved having the vehicle at room temperature (after being stored at the test location overnight) and the camera set and ready to go. The laboratory room temperature was 75°F and the windshield surface was clean and dry. Adjusting the blower fan to its highest level (Max. Blower/Max.Heat/ Defrost Mode) and starting the engine, air was blown out onto the vehicle windshield through two slots provided in the dashboard which serve as the defroster nozzle exits. A video tape of the IR thermal images captured these images and showed that the defrosting system on the passenger side works more efficiently than the driver's side. One can confirm this observation by comparing the measurements of thermocouple #9 (driver's side) and thermocouple #6 (passenger's side) as exhibited in Figure 5.

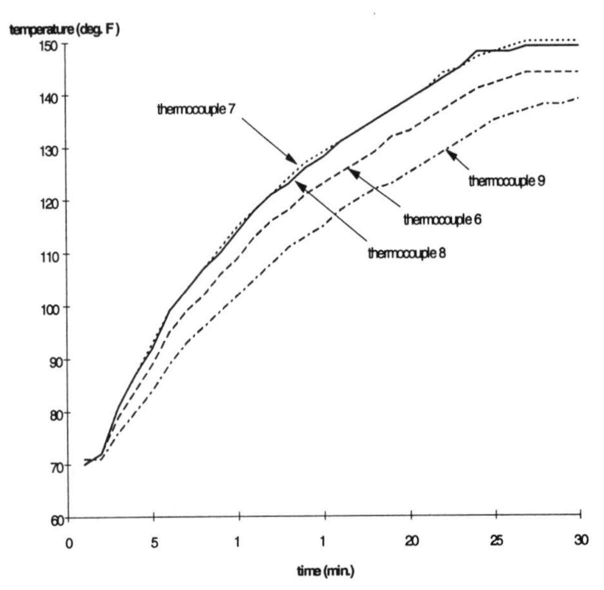

Figure 5. Comparison of various temperature curves obtained with thermocouples

In summary, by using an infrared camera and capturing the images by a video tape, it was relatively easy to assess the performance of the defroster system and identify areas for future improvements.

DEFROSTER DEVELOPMENT SEQUENCE

While extensive knowledge can be gained by running experimental and numerical tests, a proper sequence of developing a defroster system and testing its performance is needed. The process chart of Figure 6 shows the CFD and experimental testing sequence that the defroster design must undergo before the final FMVSS 103 sign-off test. The CFD simulations must be subject to rigorous comparisons with testing to verify the results and establish confidence in the methodology. Testing could also be used to further modify the analytical model, thus, reducing the overall required testing for future vehicle design and development.

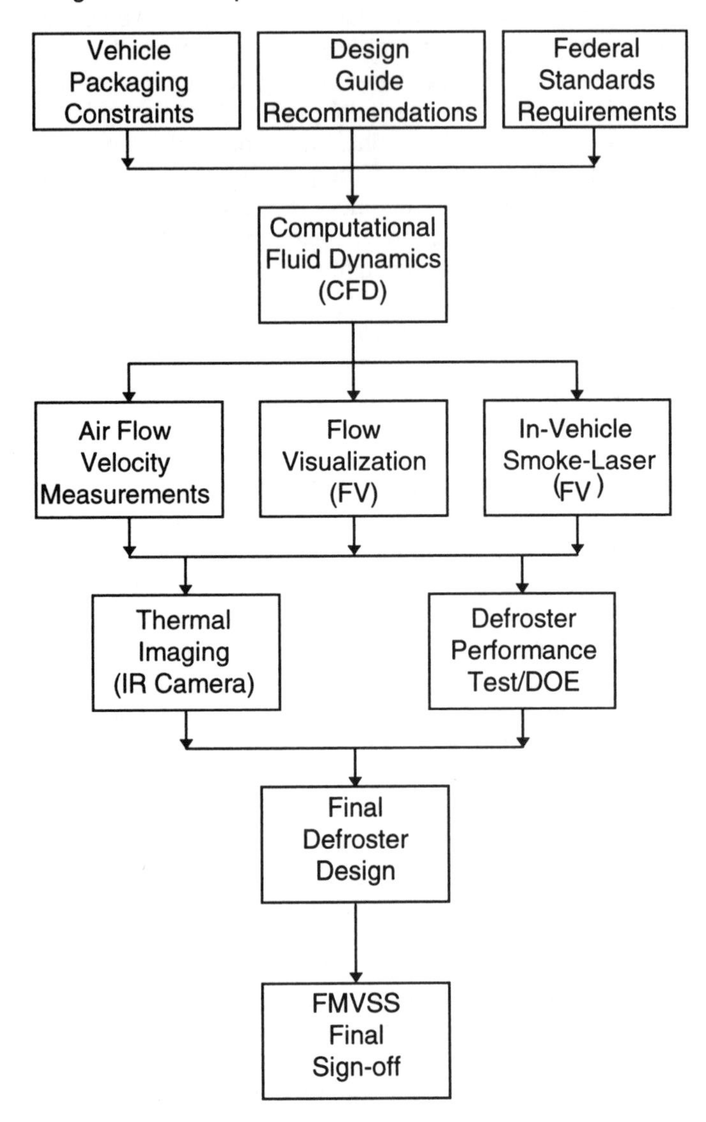

Figure 6. Defroster Development Sequence

CURRENT AND FUTURE CHALLENGES

The defroster air distribution design has many critical requirements. In order to achieve the best level of windshield clearing, the air direction from the nozzles should have the direct impingement point at the glazing surface that is required to be cleared. The nozzle design is important because it must provide a quick method for clearing of the glazing surface during start-up conditions. Also, the defroster outlet must be capable of generating an airflow that disperses over the entire inner surface of the windshield.

Extensive testing on the vehicle systems and components would generally produce products that meet the specific needs of customers. However, with the advent of computational power and resources, a designer can simulate the performance of a system and optimize its objective by coupling the CFD results with experimental results. Current analysis and optimization of designs can be improved by complementing experimental investigations (actual physical data) with numerical studies (predictions). The experimental studies would establish a database of test results and techniques on the interaction of a flow impinging on a windshield surface of a vehicle. Eventually, experimental parameters, affecting the interaction of impinging jets with the windshield surface, will be correlated and fed into the numerical code for validation and better predictions. The flow characteristics of heated air impinging on the windshield surface can be captured by an IR camera or by flowing helium bubbles. The ultimate goal is to match and validate the developed numerical algorithm by direct comparisons with contour plots and velocity vectors.

The present commercial CFD codes for solving the Navier-Stokes governing equations are able to predict the flow patterns in the immediate vicinity of the vehicle windshield. However, truly coupled equations (i.e., conjugate heat transfer applications) are still a challenge. The flow exiting the defroster nozzles has many complexities making the numerical solution quite a challenge. Some of these complexities are that the issuing jets are rectangular, obliquely impinging, turbulent with non-uniform (hydrodynamically and thermally) exit conditions. The experimental effort should focus on measurements in the distinctive regions of the jet: impingement region, wall jet, and recirculation zone. Also, measurement of the velocity distribution inside the flow boundary layer can certainly assist the numerical analyst in making exact correspondence with the experimental results.

An experimental study (AbdulNour et al., 1997) was conducted to establish the parameters influencing the heat transfer coefficient "h" on the windshield. The values of "h" due to a 2-D wall jet for isothermal and uniform heat flux boundary conditions were determined experimentally using hot-wire, micro thermocouples, and thermal imaging surveys to quantify the velocity and temperature fields, respectively. The local value of "h" was found to be insensitive to the thermal boundary condition, especially in the more turbulent downstream flow region. The difference was the largest near the leading edge of the heated wall boundary. More accurate and experimentally-supported work needs to be carried out for better prediction of the heat transfer coefficient on the interior surface of the windshield.

In summary, coordinated efforts between experimental and numerical investigations will be the answer to successful resolution of the vehicle defroster applications. This requires the determination of heat transfer coefficient "h", which is not a trivial task, and can be fundamentally understood from the use of recent data (Willenborg et al., 1997; AbdulNour et al., 1997). Also, the CFD solution of the vehicle interior provides "h" values on the inside; more accurate "h" values can be obtained by resolving the computational mesh in the vicinity of the windshield surface within the boundary layer. Finally external forced convection correlations should be used if the outside wind speed is to be considered instead of the quiescent condition (~2mph as required by FMVSS 103) normally approximated by a natural convection phenomena.

REFERENCES

1. AbdulNour, B. S., and Foss, J. F., 1997, "Computational and Experimental Predictions of Automotive Windshield Flow", ASME Fluids Engineering Division Summer Meeting, Symposium on Applications of CFD in Internal Automotive Flows; ASME Paper No. FEDSM97-3022.

2. AbdulNour, B. S., 1997, "Hot-Wire Velocity Measurements of Defroster and Windshield Flow," SAE Paper 970109, pp. 29-36.

3. AbdulNour, R. S., Willenborg, K., Foss, J. F., McGrath, J. J., and AbdulNour, B. S., 1997, "Measurements of the Local Convective Heat Transfer Coefficient for a Two-Dimensional Wall Jet: Uniform Temperature and Uniform Heat Flux Boundary Conditions", ASME International Mechanical Engineering Congress and Exposition, Experimental Methods in Convection Heat Transfer Session.

4. Andreone, L., Burzio, G., Damiani, S., and Romitelli, G., 1992, "Automatic Measurement of Defrosting/Defogging Process", 2nd Int. Conf. on Vehicle Comfort Ergonomic, Vibrational, Noise, and Thermal Aspects, ATA Paper 92A272.

5. Anon, 1985, " Frozen Windscreens Cleared in Three Minutes", *Glass*

6. Atkinson, W. J., and Muchnij B., 1972, "Wax Method of Determining Windshield Defroster Performance," SAE Paper 720504.

7. Atkinson, W. J., 1989, "Design Considerations for Inclement Weather Clearing Systems", SAE Paper 890310.

8. Boaz, P., and Youngs, J. D., 1974, "Electrically Heatable Windshields and Backlite System," SAE Paper 740157.

9. Boltz, N., Renner, M., and Koch P., 1992, "A Fast Auxiliary Defrosting/Heating System", SAE Paper 920015.

10. Brewster, R., Frick, S., and Werner, F., 1997, "Computational Analysis of Automotive Windshield De-Icing with Comparison to test Data," VTMS 3, Paper 971833.

11. Burch, S., Hassani, V., and Penny, T. R., 1993, "Use of Infra-Red Thermography for Automotive Climate Control Analysis", SAE Paper 931136.

12. Carignano, M. and Pippione, E., 1990, "Optimization of Wind-Screen Defrosting for Industrial Vehicles Via Computer Assisted Thermographic Analysis", FIJITA Paper 905237.

13. Cogotti, A., 1988, "A Two-Component Fiber-Optic LDV System for Automotive Aerodynamics Research", SAE Paper 880252.

14. Cummins, J. M., Burnham, G. R., Oestreich, W. A., and Boka, J. D., 1996, "Electric Windshield Defroster", Application for Canadian Patent, 2155454.

15. Hendrix, S., and Bernardi, R., 1988, "Super-H Coatings for De-Icing and Solar Control of Automotive Glass", *Society of Vacuum Coaters.*

16. Ikeda, Y., Katoh, N., Ishii, N., and Kuriyama, T., 1992, "Numerical Analysis of the Airflow on Windows from Defroster Nozzles" (in Japanese), JSAE Paper 924076.

17. Ishihara, Y., Shibata, M., Hoshino, H., Hara, J., Kamemoto, K., 1992, "Analysis of Interior Airflow in a Full-Scale Passenger-Compartment Model Using a Laser-Light-Sheet Method", SAE Paper 920206.

18. Jurgen, R. , 1988, "New Frontiers for Detroit's Big Three," *IEEE Spectrum*, Vol. 25, pp. 32-34.

19. Kakubari, Y., Tsunemoto, T., and Ishitani H., 1993, "Study of Melting Phenomenon of Frost and Ice on the Windshield" (in Japanese), JSAE Paper 9302970.

20. King, R. D., 1974, "Defrosting of Automobile Windshields Using High Light Transmitting ElectroConducting Films," SAE Paper 740158.

21. Kohnotou, T., Iwamoto, Y., Hoshiawa, K., and Nagataki, M., 1992, "Optimum Design of Defroster Nozzle", SAE Paper 920167.

22. Lee, J.-G., Jiang, Y., Przekwas, A. J., Sioshansi, M., 1993, "Automotive Windshield Ice-Clearing Analysis", SAE Paper 930289.

23. Lee, J.-G., Jiang, Y., Przekwas, A. J., Sioshansi, M., 1994, "Validation of Computational Vehicle Windshield De-Icing Process", SAE Paper 940600.

24. Leier, T. E., Malloy, M. G., Takeda, R., and Toi, M., 1989, "Heater/Defroster Testing in the Real World", SAE Paper 890051.

25. Lindbert, L., and Anderson, M., 1990, "Thermal Power Boosters", SAE Paper 900222.

26. Marston, M. A., Armstrong, S., and Miaoulis, I. N., 1994, "A Novel Automobile Preheating System for Cold Starts", SAE Paper 941995.

27. Norin, F., 1989, "Insulating Glazing in Side Windows", SAE Paper 890025.

28. Norin, F., Myhr, K., Kullmann, L., Arrenius, P., 1987, "Prevention of Frost Formation on Automobile Glazing", SAE Paper 870038.

29. Okuno, Y., Fukuda, T., Miwata, Y., Kobayashi, T., 1993, "Development of Three-Dimensional Air Flow Measuring Method Using Soap Bubbles", JSAE Review, Vol. 14, Paper 9307551.

30. Presson, E. W., 1980, "Progress Update - Electrically Heated Windshields", SAE paper 800878.

31. SAE J902, 1994, "Passenger Car Windshield Defrosting Systems", SAE Recommended Practice.

32. Sakamoto, H., Watanabe, T., Taniguchi, F., 1987 "An Analysis of Frost Formation on Vehicle Windows", SAE Paper 870028

33. Schaefer, C., 1986, "Heatable Automobile Windshields with Reduced Energy Transmission Characteristics", *Glass,* Vol. 63, pp. 417-419.

34. Schatz, O., 1991, "Cold Start Improvements With a Heat Store", SAE Paper 910305.

35. Southall, D., and Burnand, M. G., 1990, "An Investigation of the Visual Effects of Heated Windscreens in Jaguar Cars", SAE Paper 900570.

36. Sugano, M., Yamada, T., Takesue, Y., and Yasuki, T., 1994, "Numerical Analysis of Defroster Cleaning Pattern" (in Japanese), JSAE Paper 9432912.

37. Takada, H., 1989, "Antifrost System of Windshield", SAE Paper 890024, pp. 171-178.

38. Trauger, P. E., 1993, "Use of Infrared Imaging Technology in Heat Exchanger Development and Evaluation", SAE Paper 931144, pp. 625-631.

39. Tsantis, A. P., Brown, J. S., Hutter, R. J., Lyon, P. M., Singh, T. , 1994, "Improvements in Heater, Defroster and Emissions Performances Using a Latent Heat Storage Device", SAE Paper 940089, pp. 85-88.

40. Willenborg, K., Foss, J. F., AbdulNour, R. S., McGrath, J. J., and AbdulNour, B. S., 1997, "A Model Defroster Flow", Eleventh Turbulent Shear Flows Symposium, Grenoble, France, September 8-11.

41. Y. Sato, Ito, Y. and S. , 1995, "Numerical Analysis of Defroster Clear-Up Pattern" (in Japanese), JSAE Paper 9540165.

942248

Application of the Eaton® Tire Pressure Control System to the Commercial Truck

Alan Freigang
Eaton Truck Components - North America

ABSTRACT

Central tire inflation systems (CTIS) have been available on military tactical vehicles since World War II. In addition to the continued growth of use in the military, these systems are now available to the commercial market. This paper discusses the componentry and operation of the Eaton® Tire Pressure Control System, and its installation on commercial trucks.

INTRODUCTION

A Tire Pressure Control System is a system which, when integrated on a truck, allows the driver to change tire pressures while the vehicle is in motion. The tire pressure chosen corresponds to the existing terrain conditions such as paved highway, sand, snow or mud. The advantages of operating a truck at the proper tire pressures for given terrain conditions have been and continue to be proven. The intent of this paper is to detail the key points of componentry and operation of the Eaton® Tire Pressure Control System (TPCS).

This System has evolved from the central tire inflation systems that have been applied in military applications. The primary motivation for the use of CTIS in the military has been mobility. While mobility is certainly one of the key reasons for the commercial use of TPCS, many other gains have also become evident. Reduced tire damage, improvement in ride quality, potential reduction in road building and maintenance costs, increased productivity and reduced cargo damage are just a few of these advantages.

While recent military CTIS installations tend to be on a family of similar vehicles, installation of TPCS in the commercial market will encompass vehicles of all shapes, sizes and configurations. There are vehicles with push or tag axles, straight trucks, tractor trailers and B-trains to name a few. Because of these many types of vehicles with different applications, it is important that a system be designed such that each type of truck does not require unique componentry. This will keep the proliferation of part numbers and logistics to a minimum. Additionally, it is very important to ensure that all vehicle operating conditions and manufacturer specifications are considered. Temperature extremes, EMI, vibration/shock, chemical exposure, terrain conditions, altitude and others are all important design considerations.

To ensure that the componentry has met all of these requirements and can withstand the hard use prevalent in most truck applications, a stringent test/development program has been conducted. Through this program, Eaton® TPCS was installed on United States Forest Service (USFS) vehicles and private logging fleets.

This rigorous testing on a variety of vehicles in many locations across the U.S. has been an excellent source of information as to what is required of a TPCS. The information gathered from these test sites has proven to be valuable as the move into production begins.

COMPONENTRY

The Eaton® Tire Pressure Control System is made up of six primary control components (figure 1). They are the wheel valve, pneumatic control unit, electronic control unit, operator control panel, pressure switch and a speed sensor. Additionally there are air seals at each wheel end, a wiring harness and miscellaneous brackets and air lines.

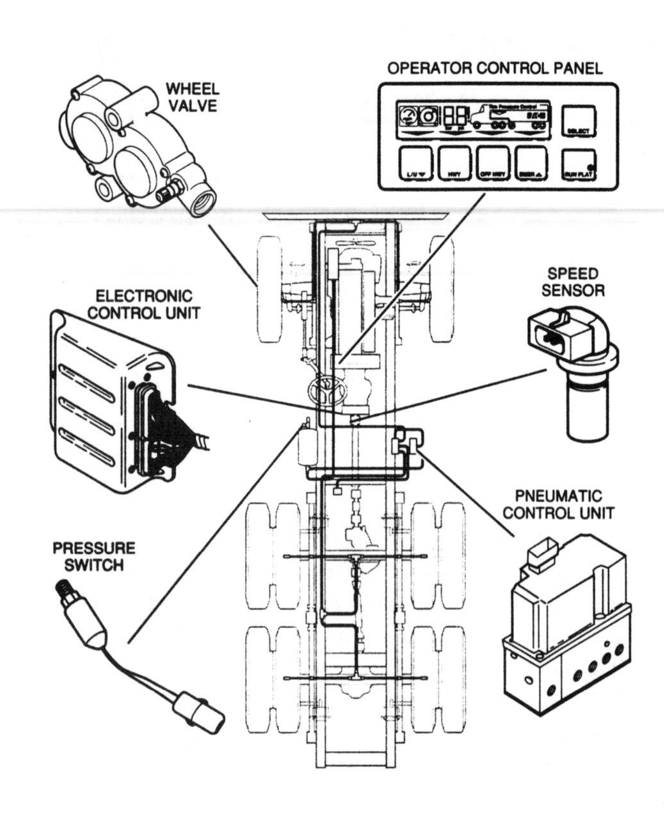

Figure 1. TPCS Components

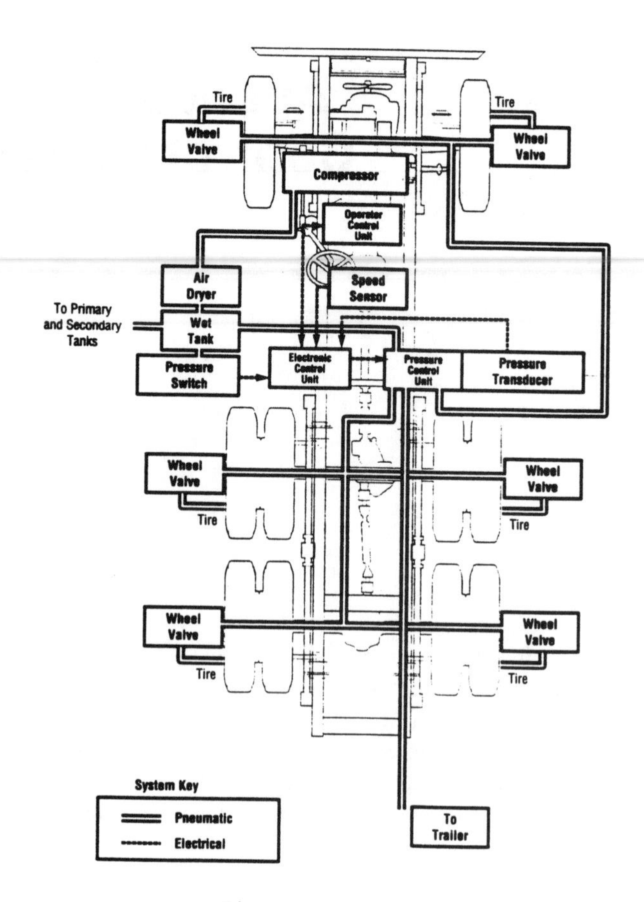

Figure 2. Control Lines

Following is a description of each component and its functions. In order to clarify the interdependence of one component with another, they are explained in order beginning with the wheel ends.

WHEEL VALVE - One of the most significant components developed for CTIS was the wheel valve. The wheel valve, one per wheel end, performs a number of important functions. It directs air flow to and from the tire during inflation and deflation and seals pressure in the tires when the system is not active.

This latter function is very important for two reasons. First, since the pressure is sealed in the tire whenever the system is not changing it, all control lines (figure 2) from the pneumatic control unit (PCU) to the wheel valves are vented to atmosphere. This means that if at any time one of these lines are damaged or a fitting comes loose, there will be no loss of tire or system pressure. Second, the wheel valve requires a pneumatic signal to keep it open during pressure changes. If pressure is lost due to a damaged line, the valves will immediately close preventing loss of tire pressure.

If a tire is damaged so severely that the vehicle air supply cannot keep up with the leak, the System will close all of the wheel valves. This prevents loss of tire pressure from all but the damaged wheel end.

The wheel valve designed for use in the System (figure 1) has been designed such that tire pressure is exhausted to atmosphere at the wheel end. This is as opposed to military systems where the air is vented at the vehicle frame at a point above the fording level. This remote venting is not typically required for commercial applications as these trucks do not usually require fording capability. Deflating the tires at the wheel allows distinct advantages. First, the back pressure created by flowing air through the wheel equipment to a remote deflation point is eliminated, resulting in faster deflation. Second, this insensitivity to back pressure for wheel valve shutoff virtually eliminates the effects of plumbing variation from vehicle to vehicle.

The wheel valve has been designed with two spring biased diaphragms. These diaphragms are situated side by side, with one of them for inflation and the other for deflation. When the proper positive control pressure signal is applied to the wheel valve, the inflate diaphragm opens and allows air to pass into the tire (figure 3).

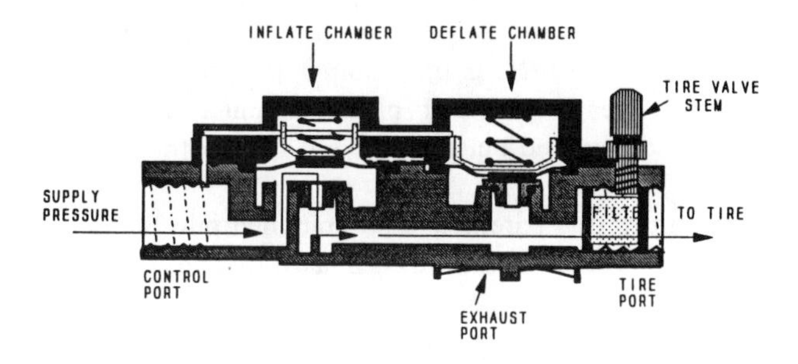

Figure 3. Inflate Mode

Likewise, when the proper negative control signal is given, the deflate diaphragm opens and allows tire pressure to exhaust to atmosphere through a protected exhaust port (figure 4).

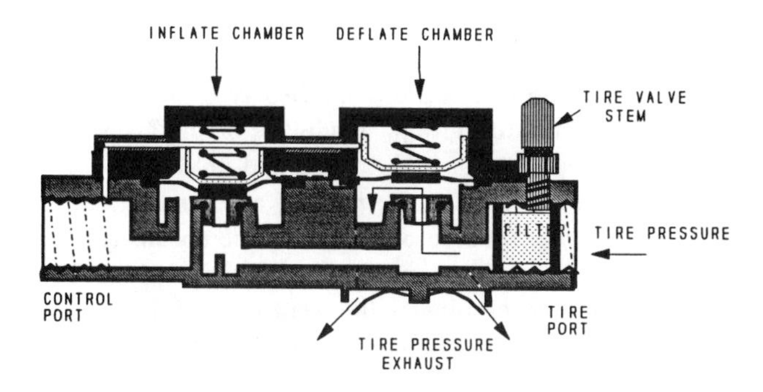

Figure 4. Deflate Mode

Wheel Valve Mounting - The method in which the wheel valve is mounted is an important aspect to consider in the incorporation of TPCS onto a truck. It should be mounted so that it is protected and does not cause a tire imbalance. The valves should also be mounted so that they do not interfere with tire installation and removal.

The drive axle wheel valves are mounted behind a bracket on the end of the axle shaft. The non-driven (steer and trailer) axle wheel valves are mounted within a special hubcap devised to accommodate the valve (figure 5).

Both of these methods require that only the hose from the wheel valve to the tire needs to be disconnected for tire removal. These tire hoses are the only lines on the Eaton® TPCS that remain pressurized at all times.

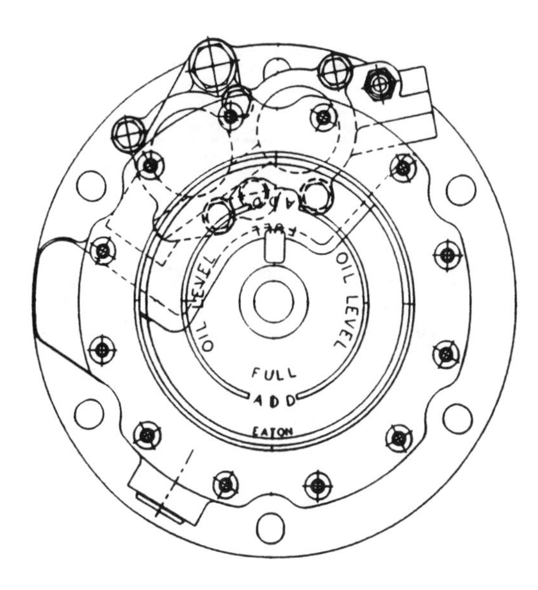

Figure 5. Non-Drive Axle Wheel Valve Mounting

DRIVE AXLE AIR SEAL - One of the most difficult issues to overcome in supplying pressure signals to the wheel valve is in making the transition from the stationary axle spindle to the rotating wheel valve. This is accomplished by use of a rotary seal package. On the drive axle, an air seal internally integrated into the hub is used. This seal rotates with the hub and contacts a sleeve pressed onto the bearing journal of the stationary spindle sleeve. Air flows between the spindle and the sleeve, through the seal, hub, wheel valve and into the tire. Hub mounted seals have been very successfully applied in military CTIS programs.

NON-DRIVE AXLE AIR SEAL - The air passage for non-driven axles is routed through the spindle via a self-piloted rotary joint that fits inside the hubcap. Since it is internal to the hubcap it is not vulnerable to being damaged by brush and other debris through which the vehicle may pass.

CONTROL LINE - To facilitate the TPCS's ability to work with any vehicle configuration, it is necessary that the control lines and wet tank are installed according to the specification defined in table 1. Adherence to the length and diameter specifications will ensure that the system operates as designed without any need for special calibration. This is necessary because the system uses pressure signatures to make all of the required measurements and determinations. These signatures vary with the volume of the control lines. If the volume is held between the limits, the pressure signature will act in a predictable manner.

Tire Pressure Control
Hose/Configuration Requirements

1 2

Two Axle Truck (Config. 1)

Wet Tank	Hose Length		
Gal/In³	Steer	Drive	Trailer
6/1400	17-19' of 5/8" OD	14-16' of 5/8" OD	N/A

1 2 3 or 1 2 3 4 5

Three Axle (Tandem) Truck, with or without Tandem Trailer (Config. 0)

Wet Tank	Hose Length		
Gal/In³	Steer	Drive	Trailer
6/1400	17-19' of 5/8" OD	17-23' of 1/2" OD	45-55' of 5/16" ID
"	"	"	30-40' of 1/2" OD

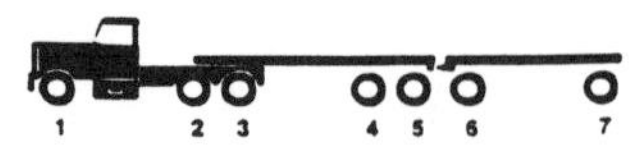

1 2 3 4 5 6 7

Three Axle (Tandem) Truck, with or without Two Trailers (7 or 8 axles) (Config. 2)

Wet Tank	Hose Length		
Gal/In³	Steer	Drive	Trailer
12/2800	17-19' of 5/8" OD	17-23' of 1/2" OD	80-100' of 5/16" ID
"	"	"	55-70' of 1/2" OD

Table 1. Plumbing Requirements

PNEUMATIC CONTROL UNIT (PCU) - Next in the system structure is the Pneumatic Control Unit (figure 1). The PCU, usually mounted in the cab, provides the pneumatic signals which operate the wheel valves. The air supply line from the first air tank is plumbed into the PCU. There is only one PCU per vehicle. Control lines for each channel, steers, drives and trailers exit from it. The PCU is comprised of a pressure transducer, solenoid valves and cartridge valves. Electrical signals from the Electronic Control Unit (ECU) are converted to pneumatic signals in the PCU to control the wheel valves. The single pressure transducer monitors system conditions and provides this information to the ECU.

ELECTRONIC CONTROL UNIT - The PCU sends the required pneumatic signals through the control lines to the wheel valves. To accomplish this, the PCU must first receive electronic signals telling it which solenoids to energize. These electronic signals are generated by the Electronic Control Unit (figure 1). The ECU is the device responsible for decision making and system control. One ECU is required per vehicle, regardless of configuration, for up to three channels.

The ECU features non-volatile memory (NOVRAM), which allows it to remember many things, even when the vehicle power is turned off. It allows the last pressure setting obtained to be returned upon start up. It also stores historical fault codes to give service personnel the ability to quickly determine the cause of a particular problem. System cycle counts are also recorded as an indication of system usage rates. The ECU is fully potted and therefore completely weatherproof. Typically mounted in the cab, it can also be mounted outside and will fit in a frame rail.

The operating parameters programmed into the ECU are accessible through an industry standard diagnostic link, SAE J1708/1587.

OPERATOR CONTROL UNIT (OCP) - Although the ECU is the brains and decision maker for the TPCS, it must receive input from the operator. The operator enters commands via the Operator Control Panel (figure 1). The OCP should be mounted in clear view and within easy reach of the driver. It can easily be mounted in, under or over the dash as well as in an overhead console or other accessible location. Use of the OCP allows the larger ECU to be remotely located, minimizing the demand for operator accessible cab space.

The OCP allows selection and display of desired tire pressures for each tire group. Highway, off-highway and emergency pressures for both loaded and unloaded conditions are available at the push of a button. It also provides for programming of pressures and speeds (in English or metric units) directly through the keypad. This built in capability, eliminates the need for additional investment to provide a high level of flexibility. Some preprogrammed limits exist for pressures and speeds to minimize operation outside of prescribed good tire practices. In addition to selection of tire pressures and the programming functions, the OCP displays a full range of information. This information includes tire pressures for each channel; steer, drive, and trailer (if a trailer is connected), indications of overspeed, tire warnings and fault codes.

SPEED SIGNAL - The System monitors vehicle speeds to ensure that the vehicle does not travel at sustained speeds above those programmed for each tire pressure. This is done with a speed sensor (figure 1) which provides the ECU with vehicle speed information. If the vehicle speed is above programmed limits, the system will display an overspeed indication on the OCP. Continued operation in this condition will result in the system automatically inflating the tires to a pressure appropriate for the speed. A variety of speed signals can be accommodated.

PRESSURE SWITCH - Another very important feature of the System is that it ensures the brakes and any other vehicle pneumatic devices have air priority. This is accomplished with the use of a pressure switch (figure 1) acting as a system on/off switch. The pressure switch is mounted in the wet tank and calibrated so that it closes when pressure reaches 120 psig and opens when pressure drops to 85 psig. This operation prevents the Tire Pressure Control System from consuming air from the wet tank until the brake system is fully charged. It also prevents the Tire Pressure Control System from allowing the primary and secondary tanks to go below recommended operating pressures for braking. Furthermore, it reduces the propensity for condensation in the system by raising pressures to a level higher than operating pressure, maximizing wet tank effectiveness.

DIAGNOSTICS

DIAGNOSTIC TOOLS - While the overall System reliability has been and continues to be proven in the field, it is important to support such new technology with appropriate tools for troubleshooting. Troubleshooting of the Eaton® TPCS can be performed at three levels: OCP codes, hand-held tester and personal computer-based diagnostics.

The computer-based diagnostics allows a number of functions not available with the other troubleshooting methods. System activity can be monitored real time and fault codes with corresponding descriptions can be displayed. In addition, active control of system functions (inflation, deflation, etc.) is possible and configurable settings such as pressure check intervals and speed calibration can be changed

FAULT DETECTION - The System has the ability to identify and report a variety of system faults. An alpha-numeric code is used to identify the type (P=Pneumatic or C=Component) and area of fault (number). The active fault code is displayed on the OCP, historical fault codes can be recalled using the industry standard tools mentioned above. All codes are listed in the operator and troubleshooting manuals for easy reference. In many cases, the fault codes allow ready determination and easy repair of system problems.

FALLBACK OPERATION - A fault code does not always result in a complete disabling of the System. Depending on the type and location of the fault found, individual channels may be disabled without affecting others, and/or operation may be modified to only allow certain functions (inflation or pressure checks) to occur.

THEORY OF SYSTEM OPERATION

The System has been designed with a great deal of flexibility while keeping operation simple. Simple operation is important to ensure that driver distraction is kept to a minimum.

PRESSURE MEASUREMENT - Whether increasing or decreasing tire pressure, the System will always begin with a pressure measurement.

Before TPCS activity can commence, the wet tank must be fully charged. The ECU constantly monitors the pressure switch state. When the wet tank reaches 120 psig, the pressure switch will close alerting the ECU that operation can commence. The System first measures atmospheric pressure. This allows it to compensate for changes due to varying altitude or atmospheric conditions.

The System next measures tire pressures (It is also during this pressure check sequence that the system determines whether or not a trailer is connected to the System). The pressure measurement signature will respond to different conditions in a very predictable way. When tire pressure is being measured, a short opening pulse is generated to open the wheel valves, after which the pressure is allowed to stabilize (Figure 6). This stabilized pressure is the tire pressure for the given channel

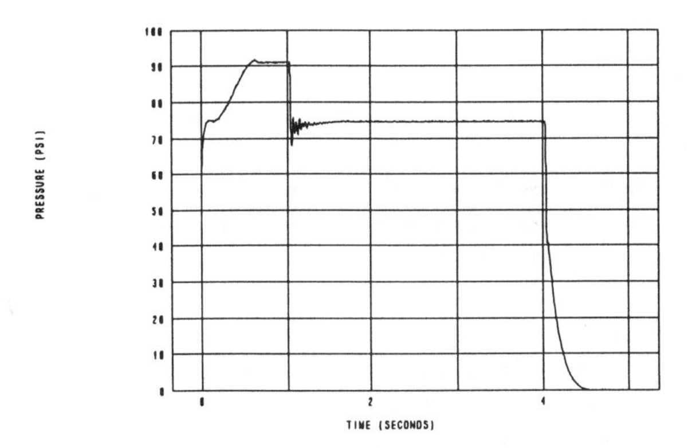

Figure 6. Pressure Check Signature

If however, a low tire exists, then for the same pressure pulse and stabilization time, the signal does not flatten out but rather has a slope associated with it (figure 7). The slope is interpreted by the ECU to determine if a low tire warning is necessary.

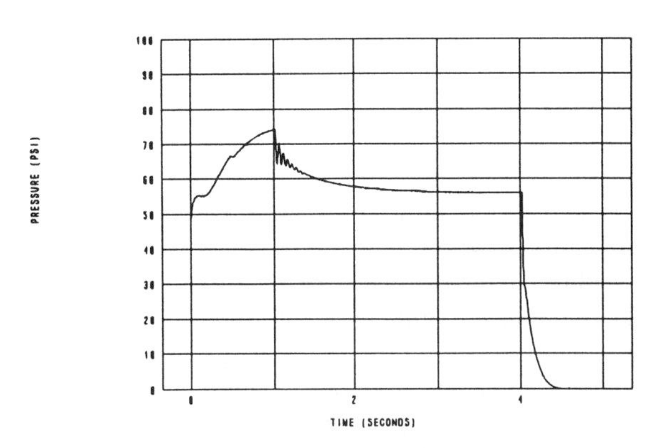

Figure 7. Low Pressure Signature

If at any time the Tire Pressure Control System finds a low tire, it will warn the driver and wait for input before proceeding. If the driver surveys the situation and determines it is OK to proceed, it is required that the runflat button on the OCP be pressed. This tells the system that the driver is aware of the problem and that it can now inflate the low tire.

INFLATION - When required to increase tire pressure, the driver pushes the desired mode button on the OCP, and the pressure measurement sequence just described takes place.

The System then calculates how long it will take to inflate the tires from their current pressure to the desired pressure, and the supply valve is opened. When the wet tank has dropped to 85 psig, the pressure switch will open and the inflation will cease, until the compressor has recharged the wet tank to 120 psig. This cycle continues until the tire pressures are within the allowable range of the desired pressures, typically ± 3 psig. Once the desired pressure is obtained, the control lines are vented to atmosphere and the wheel valves are closed. A confirmation check 30 seconds later assures that all tires are holding air and that the system can move to a longer 15 minute pressure check interval.

DEFLATION - If the driver selects a lower tire pressure, the same pressure switch strategy is employed. The System also goes through the same pressure measurement sequence to determine the starting tire pressures. From the starting pressure, desired final pressure and predetermined coefficients, the ECU calculates the required deflate time. The PCU solenoids are sequenced for deflation opening the wheel valves for the calculated time. Once that time has been reached, the wheel valves close, tire pressures are re-measured and the remaining time to reach the pressure is calculated. The System incorporates a strategy which deflates to a pressure slightly lower than the nominal desired pressure to force a short reinflation period, and assure good tire balance. This system also incorporates an adaptive deflate. Each time a deflation is performed, the ECU stores the obtained deflation times and uses them to adjust the coefficient for the next deflation.

CONCLUSION

The customer derived benefits and the continued advancement of on board tire pressure adjustment systems has now led to its availability as both a retrofit and OEM installed option. The Eaton® Tire Pressure Control System represents one approach to the application of this technology.

Continued use and growth of this product will ensure that TPCS continually evolves to meet the customers needs and requirements in expanding markets.

2000-01-0321

Automotive Lighting and Its Effects on Consumers

Mario A. Campos
Ford Motor Co.

ABSTRACT

Automotive manufacturers have relied on new technology for vehicular lighting systems, to illuminate the road or to highlight overall appearance. Lamps are more than just a bulb and a glass lens, and this paper will explain optical engineering, and how it advanced lighting engineering. Technological advances will be highlighted, including past (halogens) and current technology. (i.e., HID lamps and LEDs) What often goes unnoticed with new technology is its effects on consumers, mainly in higher repair rates. This paper will explain how some Original Equipment Manufacturers (OEMs) have improved upon or reduced lighting costs, yet maintained the lighting system's ability to add value.

INTRODUCTION

Since the dawn of the automotive industry, manufacturers have wanted to distinguish their products from others, be it through styling or manufacturing techniques. The early automotive designers were either the Henry Ford's or Frederic Benz's of the world, and to a significant degree, certain parts of the early cars were very much alike from company to company, either by choice or because of the lack of new technology. Lamps, radiators, and tires were examples of such systems that restricted new automotive designs in many ways because of the simple yet cost-effective designs that were available. The early days of automotive lighting relied principally on inexpensive and very basic designs that did not stretch the limits of technology, but still lent themselves to unique styling. Nonetheless, the first generation of automotive designs used elegant styling, together with low technology to enhance a car's appearance. After World War II, the design parameters expanded, mainly based on the application of new theories of optics and physics developed during the latter part of the 19th century. Therefore, engineers could stay ahead of the design studio's new desires by offering different ways to combine functionality and appearance, utilizing new optical concepts and materials. This paper will highlight the basics of optical design and engineering, and will expand on its use within the first automotive lamp designs from the World War I and II eras to more recent times. Then we will move to post-war technologies, which are still in use today. Discussions on the latest component and systems technologies will show how far overall lighting technology has gone, and to where it may lead.

OPTICS AND LIGHTING ENGINEERING

Optics is essentially a study in physics. Like all things in physics, everything is set in terms of parameters. Optics setup certain rules that apply to what we call LIGHT. Light is not something that we think of as having any mechanical properties, but optical physicists have determined that light can be controlled by thinking of it in simpler linear and mathematical terms (at least, ones that the layman can understand). Let's begin with the concept of a light "ray", such as the one emanating from the sun. The concept of a light ray has its roots in physics, and even though we can not truly model what a "perfect" light ray consists of, physicists have formulated basic principles that predict how rays are either measured, directed, or how they react to external forces.

BACK TO BASICS – The theories of reflection (bouncing of light rays off of a surface) and refraction (bending of rays through a medium) were conceived in physics, and theorems were devised to control how light rays should be modeled when calculating their behavior.[5] Because light is assumed to behave geometrically, some basic rules have been formulated to explain certain aspects of light rays. Fermat's principle (circa 1650) uses the concept of optical path length and states that the optical path between two points through which a light ray passes is an extremum, and this extremum, when passing through a homogeneous medium, becomes a straight line (and a series of lines when reflection happens).

Optical path length =

$$V(x_1; x_2) = \int_{x_1}^{x_2} n(x)\, ds = c \int \frac{ds}{v} = c \int dt$$

"How can light be "straight?", one might say. But thinking of light as being a straight line advanced the theory of optics a great deal. By assuming that light can be reflected or refracted in a linear fashion, optics engineers could use mathematical formulas to predict a specific direction or intensity of light rays, and engineers could rate materials based on their ability to manipulate light rays (called refractive index).

Reflection – Reflection is one of the key concepts in optics, taking its roots from Fermat's theorem and wave theory. Light moves in a straight line, assuming a homogeneous medium, and with the homogenous medium, the refractive index is the same everywhere. This means light rays can be mathematically described in terms of their behavior, whether they are refracted (bent) or reflected. Refer to figure a.

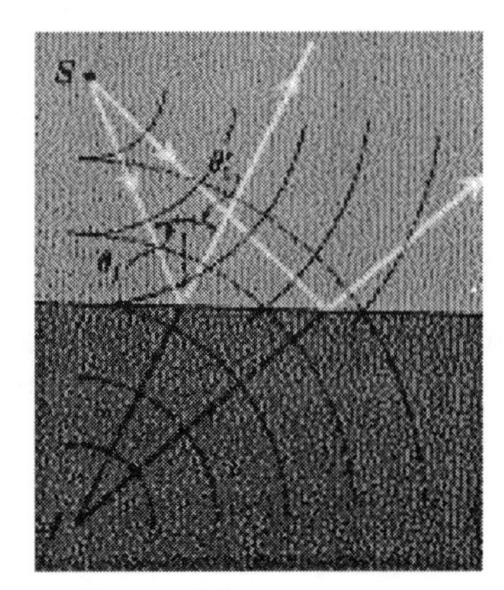

Figure a

The mathematical theory of parabolas was used extensively in the first application of reflection. It states that light reflected off of a parabolic shape is collected into a specific direction, sometimes termed as collimated, and the light rays reflected from the parabola are parallel to the light that was directed to the parabola. A reflector is like a bowl, collecting light in a certain volume, yet it can direct and increase the light output from the light source.

The concept of focal length is the distance between the vertex of the parabola and the focal point (where the bulb is positioned), and given a parabolic shape, a reflector size can be pre-determined for a lamp. Commonization of focal lengths became one of the mainstays in optical engineering to the present day, since certain functions tended to use one particular focal length.[2] Refer to figure b below.

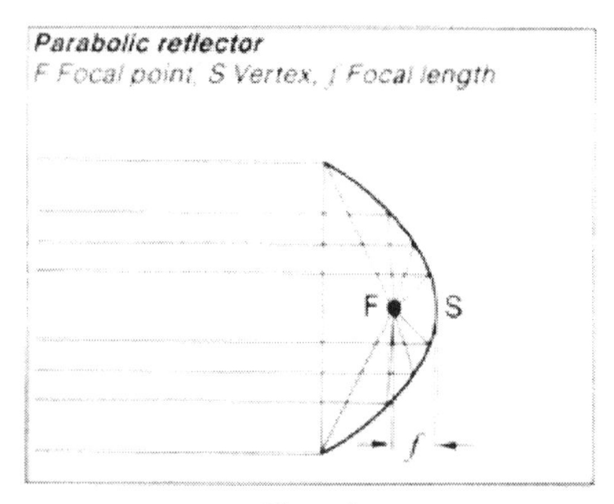

Figure b

Some of the first automotive lamps used a type of reflector to increase the light output from the light source (usually a flame on a wick). The reflector component expanded the ability of any lamp to be more than just decorative by increasing and/or directing the flame's light output. Most types of reflectors were manufactured from steel, and early metal-forming methods restricted the shape of these reflectors to no more than just parabolic shapes. Refer to figure c, which shows a lamp with a typical reflector design.[2]

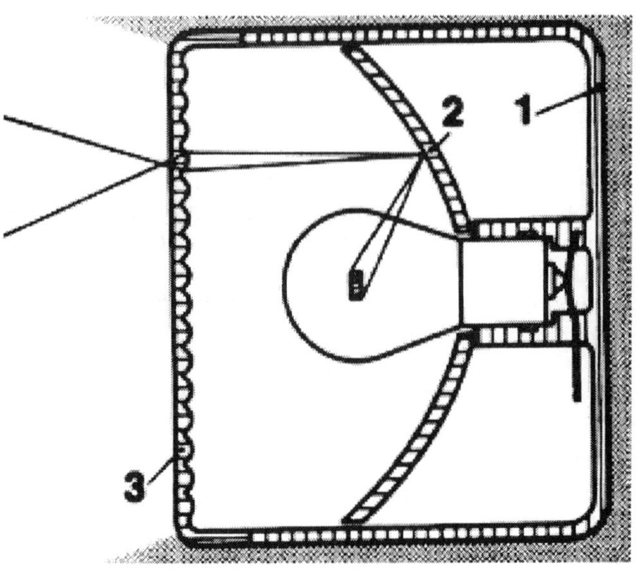

Figure c

Refraction – Refraction, unlike reflection, controls light by "bending" it with specialized optics. Light which has been refracted through a plate with parallel faces is displaced laterally by the thickness of a given slab of material (with a given refractive index). This lateral displacement however does not stop the light from going in its intended direction. Light passing through a slab with non-parallel faces will have its direction changed, such as with a prism.[2] Refer to figure d.

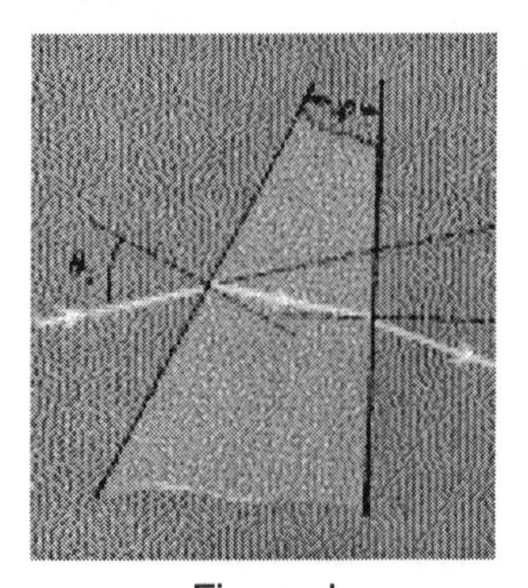

Figure d

A Lens is the typical device that does the work of a refractive system. Refracting light is not as efficient as reflecting it, mainly because managing light in general has a certain amount of loss associated with it. Lenses in general have had their roots from the early 1600's, when the first magnifying glasses were invented, and glass has been the typical material for most of the past designs. The concept of a Lens came about through other physics theories (e.g., Fresnel theorem), and this new device allowed a different means of distributing of light rays. By using a simple lens (with or without a reflector), engineers expanded the lamp's ability to direct light to a specific zone or area, and with the invention of plastics and color additives, lamps slowly expanded their role as safety devices. Dioptric (refractive) lenses were intended to refract light, and catoptric (reflective) lenses are ones that are mainly designed to reflect. Catadioptric lenses (ones which reflect and refract) are most common in automotive applications. The commonly known lens optical designs are FRESNEL, FLUTES, and PILLOWS. A simple form of a lens optic, a flute, is a curve that extends in one direction along a line, while a pillow optic is a design with at least two curves making a pseudo spherical shape.[9] Refer to figure e.

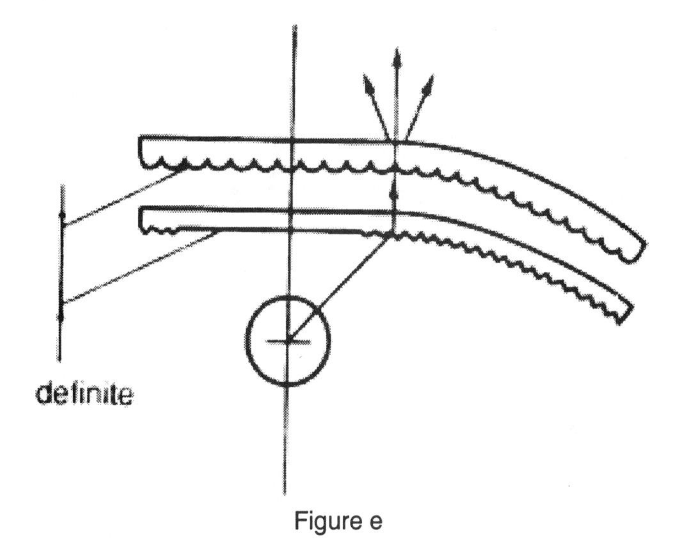

Figure e

Flutes,pillow optics, and fresnel lenses have been used for many years and continue today in many modern lamp designs. Many of the first lamp designs relied solely upon Fresnel lenses, a common design throughout the automotive lighting industries (even the early traffic lights used fresnels).

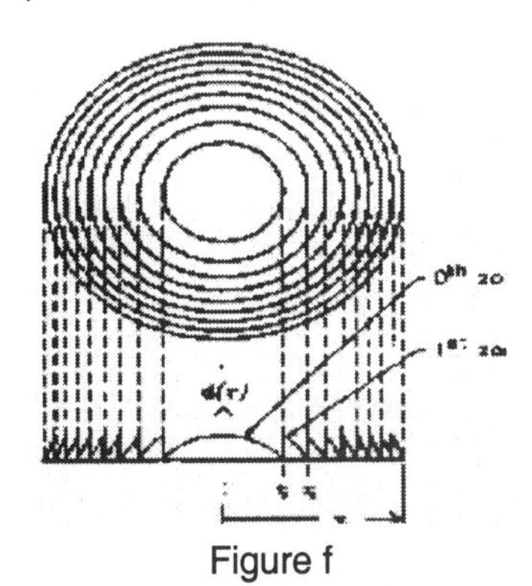

Figure f

A Fresnel lens, shown in figure f, essentially is a circular wedge optic, refracting and spreading light uniformly,[6] and because of the need for only a fresnel lens, lightweight, compact lamps gave engineers more flexibility in lighting systems, notably signal lamps. Many designs that rely upon Fresnel lenses are rather shallow in depth and simple in construction. See figure g for a simple tail lamp design with a fresnel design (note the absence of a reflector).

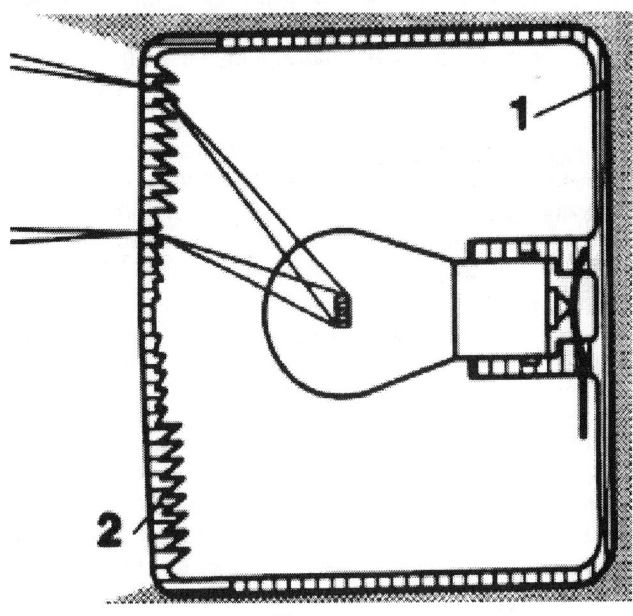

Figure g

Advances in lenses have expanded the ability of most lighting systems (mainly new materials), and unlike the older fresnel designs, these more sophisticated lenses offer engineers the ability to style the lamps into new complex shapes. These new designs will be discussed in further chapters.

LAMP DESIGNS

The first automobiles had no exterior lighting system, and if they did, it was an option that was of a decorative nature. Oil Lamps, and later acetylene, were the only technologies available during the late 1800's and early 1900's for illumination purposes, and they did not light the road adequately, were easily broken, and were not exactly what we call a safe and "robust" design. Can you imagine having to light your headlamps every day using matches?

PRE WORLD WAR II LIGHTING – During the advent of the horseless carriage, the first automotive styling, as far as lighting was concerned, did not put too much emphasis on technology for various reasons. One of the first American car designs, built in 1896 by Charles Duryea, had no lighting system at all -no oil lamps, no candles, nothing at all. Some of the very early designs, made by manufacturers such as Dietz, were no more than elaborate lanterns, yet the customer could understand its operation and basic maintenance. This simplistic lamp design was just one of many that the automotive manufacturers relied upon for several years. This may not seem so important to today's engineers (because of the extensive dealer infrastructure and parts stores), but the simpler the design, the better for the customer in the old days. To rural customers in the early 1900's (especially farmers), easy-to-do maintenance was greatly appreciated. That is probably why the early automotive designs relied upon acetylene lighting systems, but lack of an alternative had just as much to do with it.

Acetylene systems – The first manufacturers of acetylene lighting systems in the early 1900's, notably Hella and Prestolite, had basic reflective designs for lamps, with some of the early patents tied to the overall reflector design of the lamp. Hella began as early as 1910 with high volume production of their acetylene headlamps (patented in 1902), and American firms also began similar development of lighting products.

The typical lighting system in the early 1900's was essentially an "acetylene generator",[12] consisting of a block of carbide surrounded by wet lime. Moisture reacting with the carbide sent acetylene gas through a tube to the headlight chamber ,where it was ignited with a match or flint. Acetylene lamps were to say the least a bomb waiting to happen, since they were unsealed units, and they needed fresh carbide and lime replenishment after only several hours of operation. The acetylene reaction only lasted as long as the carbide did (whether the lamp was lit or not!), and any excess carbide waste had to be dug out after several uses. A picture of a typical acetylene headlamp is shown below.[7]

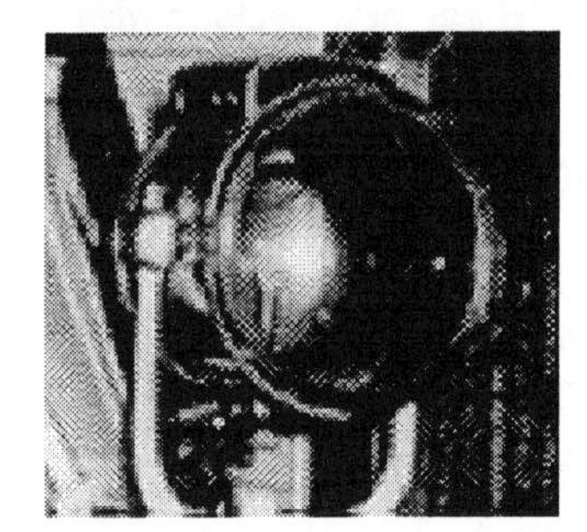

Figure 1. Acetylene Lamp

Dietz, Prestolite, and other American lamp manufacturers had various acetylene designs, and the early automobiles of the world required just the minimum performance out of their lighting systems. Even though more elaborate acetylene lamps continued in production through the early 1900's, their days were numbered as two new inventions arrived on the scene: the light bulb and the generator.

Incandescent Bulbs and Lamp designs – The advent of the LIGHT BULB expanded the role of lamps in the automotive industry, giving lighting a better and more controllable light source. The tungsten filament light bulb became the typical design for automotive bulbs, and typical bulbs were based on brass-and-glass designs. Figure h shows the standard design of a bulb[13], plus the two most common designs for tungsten filament bulbs.

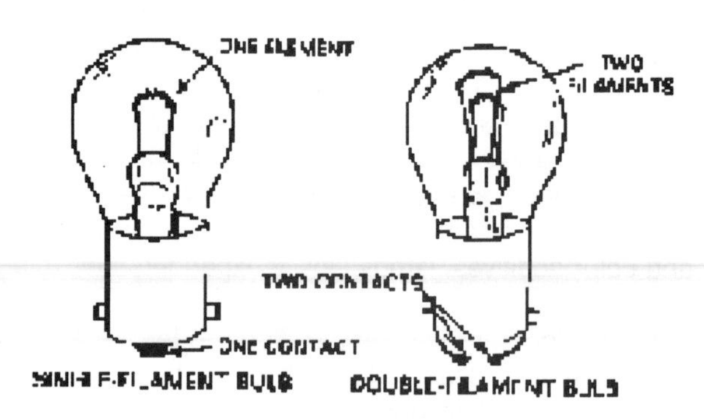

Fig. 34-7 Single-filament bulb is common, used as backup or marker light. Double-filament bulb is needed to serve as both parking light and turn or braking light.

Figure h

Also, tungsten filament designs have changed in many ways as compared to Thomas Edison's original design, with the main categories shown in figure i.[6]

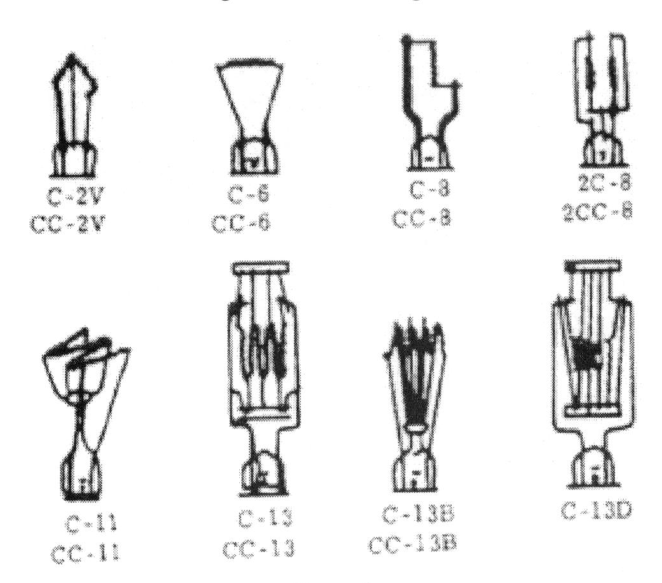

Figure i

Inventors from General Electric, notably Irving Langmuir, improved the bulb design significantly such that bulb life was greatly extended. Gas-filled bulbs became the norm for automotive lamps. With improved materials and manufacturing methods, bulbs became commonplace in hardware and general stores, and became an easy and clean replacement item - a major development over the old acetylene lamps.

Lamp manufacturing became a bit more technical, but the overall Lamp design remained very simple: a Housing which secured a Lens and a Reflector, with the bulb situated at a certain focal distance for maximum light output.

Lenses, made by well-know companies such as Bausch and Lomb, were normally made from glass, with few optics. The Reflectors, mostly made from steel, were either integral to the overall housing or were a separate insert secured to the housing. The typical Lamp design was subject to water intrusion, glass lens breakage, bulb burnout, and was not very durable. Some innovations such as generators and better filament materials and designs prolonged bulb life. Customers were also able to repair their lights without excessive repair cost involved. In the early days, lamp warranty was not documented very well, but history has shown that the lighting system repair was simple enough for the average person to perform.

Electrical system development – System developments in wiring and electrical power allowed lighting to expand its role as an integral part of the automobile's safety system. The early generators, called dynamos, were small enough to power an automobile's electric system allowing lights to be activated from the interior of the automobile, instead of getting out of the car every time to illuminate the lamps. By relying upon electrical systems to ignite the lamp, the new lamp designs ensured that the rest of the car did not go up in flames (most cars during that period were made with significant amounts of wood and canvas materials). The early Cadillacs and Packards featured electric headlamps, which quickly caught on with other manufacturers[12] (refer to the photo of the 1912 Cadillac shown below).

1912 Cadillac

The electric headlamps, commonly known as "drum lamps", again showed how the customer's needs became a larger part of the entire lamp design equation, relying upon new innovations in electrical and lamp designs.

LAMP DESIGNS: POST WORLD WAR II – Once the incandescent bulb became standard in lamps, optical engineers were able to design better products that could perform multiple functions, and the entire lighting system took advantage of new automotive technologies, notably the alternator. Having its roots from the aviation industry, the alternator became the improved method with which to control the source of power for lamps and other devices requiring either high current for a short time or low current for a long time. The old generators had basic electrical problems that the alternator resolved, mainly better charging ability (and therefore less battery rundown). With better control of the electrical output to the lighting systems, engineers improved the system's ability to prolong bulb life and use electrical power more efficiently. Even today's rudimentary turn signal design and function can be attributed to better electrical management within the vehicle.

Along with the advent of the alternator, newer bulb designs were introduced, such as sealed lamps. This new technology was expanded by companies such as General Electric and Sylvania, and the first application of a sealed filament lamp design was introduced in the late 1940's. This innovation, known in lighting circles as a SEALED BEAM LAMP, formed new and distinctly different styling themes, albeit mainly with a round headlamp as the only shape allowed. The sealed beam gave the engineer the entire package: light source, lens and housing, and standardization.

Most government regulations evolved around the sealed beam headlamp, and nearly all automotive manufacturers used sealed beam designs for their cars and trucks (larger lamps were available for the larger trucks). Sealed beam filament designs were still based on tungsten (like any other bulb), but the sealed beam concept showed how an old design could lead to an even more compact design. A typical American sealed beam design is shown in figure j.[13]

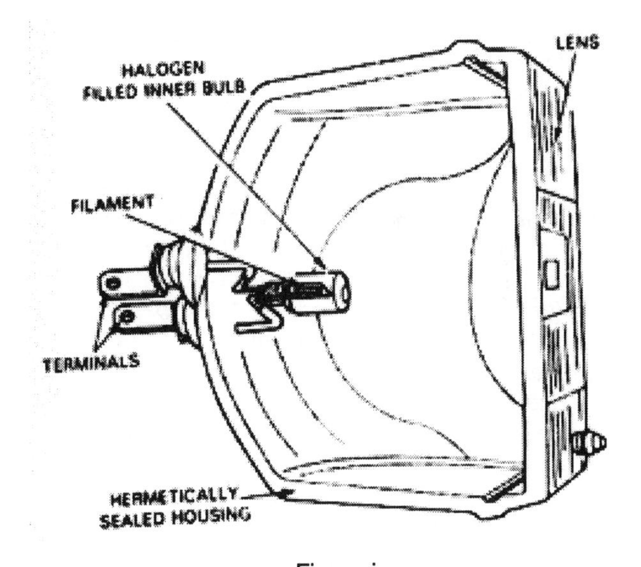

Figure j

The sealed beam lamps were not much different than the older headlamps, mainly because of the restriction to a round shape, but even so, they lent themselves to the new and distinct styling themes introduced after World War II (Europeans did not use a sealed beam lamp exactly the same as the American version, but used the outer shell design with a separate bulb secured to the glass housing).

The more modern sealed lamps (specifically halogen designs) took advantage of advances in materials, electrical designs, and processes and began mass manufacturing on a level never before seen. Yet, lighting engineers were able to implement the new sealed lamps without major changes to electrical or mechanical systems.

Lighting engineering had finally become a partner with the stylist, playing a bigger role in the overall design process. The new lighting systems offered revolutionary styling themes (hidden headlamps are noteworthy), but without excessive complexity changes to the assembly process. However, the costs associated with the new, modern lamps (as with all other safety systems) were increasingly higher after the 1950's. Automobile companies were using better designs which improved the durability and life of the lighting system, but they had to recoup the added costs in the overall price of the vehicle, along with the repair parts and service. Just having headlamps was not good enough any more; customers requiring more of their lighting system made it known to auto manufacturers (and to the government!) that safety was important. However, it would not be until after the 1960's when lighting would get more extensive regulation, mainly because of the advent of new lamp designs and materials.

The innovations added some additional costs to the vehicle, and the additional cost of LIGHTING SAFETY was passed on to the consumer, without any major repercussions. Ralph Nader did not have any direct effect on lighting systems or the basic bulb design, but his campaign for a safer car made the American consumer more aware of the need for safety systems and their regulation.

Reflex – The invention of what lighting engineers call a REFLEX REFLECTOR added another aspect and styling cue to the vehicle. The reflex reflector (or reflex for short) was a uniquely optical and plastic invention for signal lamps (developed in England in 1927 by Jonathan Stimson). Relying upon refraction and reflection to do the work, reflex uses no light source, but has a collection of "cubes" that reflect light straight back to its source, even at acute angles. Its main use is in the rear of a vehicle, so that when the lighting system is turned off, oncoming drivers approaching a parked car can detect it. Later designs would incorporate the reflex reflector within the tail lamp assembly, and future legislation in the U.S. would require side reflex at the front and rear of the vehicle (many times on the bumpers). Europe mainly requires reflex in the rear of a vehicle, without side reflex requirements that the U.S. has. Japan and other countries followed Europe's regulations as far as reflex was concerned. Many times, both American and European standard are accepted in many developing countries. A typical reflex cube pattern is shown in figure k.

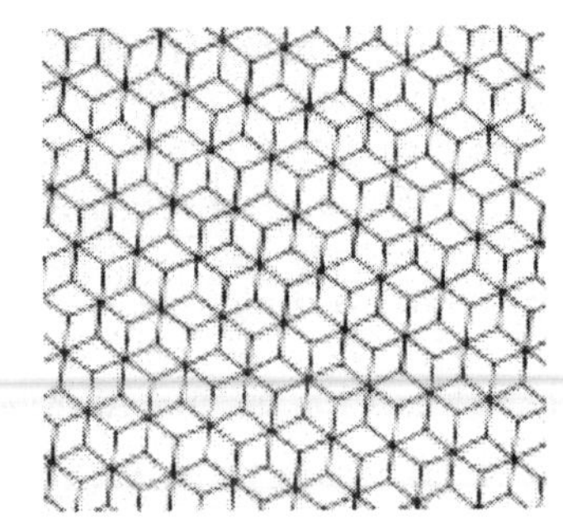

Figure k

Reflex is made from molded plastic, as any other tail lamp lens, with mainly acrylic and polycarbonate materials used (glass was the original material through the 1940's but gave way to plastics). Specialized tooling, called electroforms, are needed to make the mold very precise in terms of the shapes of the cubes. In 1961, electrodeposition, not standard cutting tools, became the standard process to make the cubes, but unlike the standard toolcutter, this process is very time consuming (as much as 50% longer). However, the use of reflectors added more regulations to not just automobiles but to motorcycles and bicycles as well. The invention of reflex again showed how safety was increased for the overall vehicle without any major cost increases passed on to the customer. Most early reflex reflectors cost as little as 25 cents each (or $.50 per car). Future integration into the tail lamp allowed lighting engineers to provide reflex without any significant increase in signal lamp piece costs.

GOVERNMENT REGULATIONS – Some engineers took it upon themselves to formulate standards within their own firms in order to either commonize their systems or even to reduce complexity in the assembly plants. Governments in Europe (notably Germany, France, and England) and the U.S. began the process of expanding the regulations on what constituted minimum safety in an automobile, and lighting was given many of its own safety standards. One of the first laws regulated the old acetylene headlamps out of existence (i.e, glaring lights were not deemed "safe"). Safety laws enacted as early as 1915 required that headlamps illuminate objects at prescribed distances, and these new rules required more of lighting engineers than ever before. The state of Wisconsin enacted as early as 1961 a law that automobiles be equipped with safety devices when sold to state agencies, and by 1963, the U.S. federal government began the process of requiring basic safety devices on its vehicles, such as mandatory Backup lights. The National Highway and Traffic Safety Administration (NHTSA), getting it roots from early safety conferences during the Truman administration, expanded the safety requirements of all lamps, introducing new rules. (especially for headlamps in the 1960's, with minimum requirements for photometry)

Low and high beam functions in headlamps were both regulated (prescribed beam patterns were mandatory in the US and in Europe, even though they differed). Minimum mechanical requirements for all lamps became law in the U.S. (e.g., moisture resistance, structural integrity, and color intensity). These new requirements required better plastics and color additives (red remains red, not pink after a few years!). All of this safety-mindedness was not just a fad, but was a rather good example of engineers and government regulators working together. Some of the actual lighting requirements might seem a bit simple to fulfill, but new photometric requirements pushed the available technology just enough to meet the standards with minimal complexity. The lighting designs at the time were boxy and vertical in nature -with no real engineer pushing the optical limits without a good reason- and this allowed engineers to meet the regulation without the need of too many material or design innovations. And then the design studio got involved...........

With more radical designs coming from the studios of all the auto manufacturers after World War II, and combined with the increasingly complex nature of the regulations, lighting engineers came up with some ingenious mechanical designs. These design would give the customer what they wanted, with safety in mind. This is when significant inventions and major breakthroughs occurred, mainly in electrical design and bulb sources. Even one unique non-electrical design innovation arrived, involving mainly plastic lenses.

Effects of Regulations: New Halogen Lamps – In the late 1970's, the development of the HALOGEN BULB began the new age of headlamp design. The halogen design was, in simple terms, a better gas in which the light "combustion" could occur, but the voltage and amperage did not change. This latest bulb design showed that electrical engineers, not just optical engineers, could improve the lighting system, with no major degradation in the car's performance. This occurred without any major increase in the size of the headlamp; in fact, most sealed beam headlamps in the US maintained their shape and size, with the new halogen lamp having only an internal design change. Europe had had extensive experience with halogen lamps in the 1960's, and its many benefits (more light output from basically the same power usage). Eventually, these lamps found their way to American automotive engineers' desks and the assembly line. Halogen headlamps were offered in the U.S. beginning around 1979, on the Lincoln Versaille luxury sedan. Yet the shape of the headlamp was still predominantly round through most of the 1970's until lighting engineers were able to convince U.S. regulators that rectangular was as safe as round. The pictures below show the different designs that predominated the late 1970's and 1980's.

1979 LeBaron Wagon

1980 Monte Carlo

1983 Ford Thunderbird

With the halogen light source, NHTSA changed the permissible light output for high beams from 75,000 to 150,000 candlepower, which allowed lighting engineers to use halogen designs in more vehicles. Electrical systems did not require major changes to accommodate the new headlamps, thus making the halogen light source the standard for most automobiles (and motorcycles also). However, the cost of the halogen headlamp was approximately 50% more than the standard sealed beam, which was eventually passed on to the consumer in the vehicle price. Halogen headlamps did not increase the overall system cost by 50%, since the mechanical designs to secure, aim, and protect the headlamps remained intact, as well as the packaging under the hood. Serviceability also was not seriously affected by the new halogen headlamps, and the supplies were plentiful for new halogen headlamps after a few years. The main bulb suppliers in the U.S. (Wagner, Sylvania, and GE) jockeyed for position in the OEM lighting market, which tended to keep the costs down for the new sealed beam headlamps.

LIGHTING TECHNOLOGY AND STYLING

Most car manufacturers in the early days of the automotive industry viewed lighting systems as nothing more than something rudimentary ("it's only a light bulb!"). Either for signaling or for illuminating the road at night, lamps were mainly functional, not adding anything special to the car's appearance. From the 1900's to pre-World War II, automotive designers had very little choice for lighting systems, and the lamp's role in the overall appearance of the vehicle was limited by both technology and regulations. By the post war years, consumers were clamoring for almost any vehicle, since World War II had restricted the choices for most Americans for several years. The post war years saw an explosion in both offerings of vehicles from several manufacturers, and also saw the most extremes when it came to styling, but not necessarily related to lamps. The lighting systems of the post war vehicles had certain things in common:

they used glass on their headlamps (mainly round), used a lot of chrome on the rear lamps, and ran on 12 volts.

This generation of vehicle had lighting systems that ranged from round headlamps to round tail lamps- not too much variety if one looked at vehicles dated from 1945 through 1950. Nonetheless, stylists did not let the lack of lighting technology stop the new design trends, especially in the 1950's. An example of this new partnership between lighting engineers and stylists was the design of the 1955 Nash, an innovative design with headlamps incorporated within the radiator grille's shape. Europeans had some different designs that seemed to be retro in nature. The 1950's British sports cars, especially MG and Triumph, tended to use drum lamps once more, with the use of chrome a must, even within tail lamps. American designers went away from their old designs, and there were some post World War II designs that used lighting within their style, and used some new technologies along the way as well.

A 1953 MG

1953 MG sports car

Examples of leading-edge styling that used lighting to enhance the vehicle's design included:

- the hidden headlamps on a 1940 Cord
- the sports car looks of both the 1950's T-Bird and Corvette
- the use of four sealed beams on the 1957 Edsel
- the cycloptic front end of the 1949 Turner Torpedo
- the rear fins of the 1959 Cadillacs, tipped with a futuristic tail lamp
- the 1961 Lincoln Continental and its trend away from the "fins"
- the unique rear treatment of the 1960's Mercury Turnpike. [7]

The Cord

1940 Cord

A 1956 Ford Thunderbird

1956 Ford Thunderbird

1954 Chevrolet Corvette

1954 Chevy Corvette

The Edsel

1957 Edsel

1949 Turner Torpedo

1959 Cadillac

A Lincoln Continental

1961 Continental

1956 Mercury MX - Turnpike

Mercury Turnpike

However, the entire lighting system was no more than a conglomeration of old lamp designs, which used some elegant and also some very simple mechanical designs. The front end treatment of a vehicle was pretty basic: Sealed Beam Headlamps, Front Turn Signals, and Parking Lamps.

The rear end treatment of a vehicle was just as simple: Stop, Turn, Tail, Backup, and License Lamps.

Even with some unique styling, most designs still required the light bulb. However, new technology had expanded the light source to high-tech levels (halogen bulbs, higher heat resistant plastics, etc.), allowing engineers to continue the trend to more compact and lighter, yet stylish, designs. Government regulations also expanded the overall number of lamps, and when new regulations came about, they offered stylists more freedom in using lighting to accentuate the vehicle's flair or style.

TRENDS IN LIGHTING SYSTEMS – Mainly because of automotive designs expanding its functionality, Exterior Lighting has increased in sheer numbers of lamps from its previous days, consisting of the following lamps:

Headlamps, Front Turn lamps, Front Fog lamps, Front Sidemarker lamps, Cab lamps (roof mounted lamps), Side Repeater lamps (mainly in Europe), Rear Lamps (with stop, tail, backup, and turn functions), Center High Mount Stop Lamps (CHMSL), Rear Sidemarker lamps, Rear Fog Lamps (mainly in Europe), License Lamps, Rear Appliques (which secure the license plates and license lamps), and Reflex Reflector assemblies. Typical automotive exterior lighting is shown below (mainly American designs).

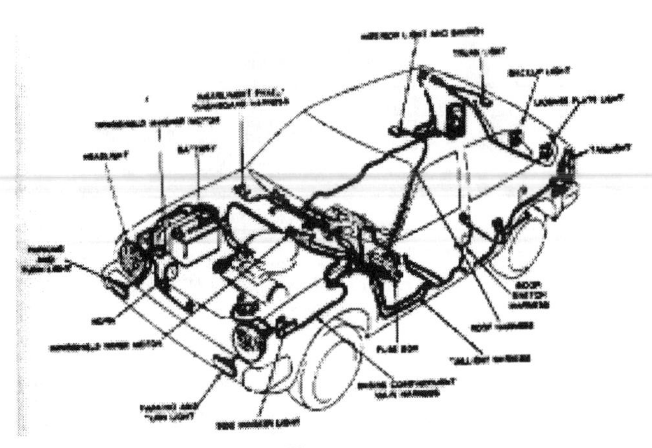

Figure m

The lighting system has seen advances in plastics and bulb sources, which gave rise to new lamps that could last longer, follow the car's design better, fit into bumpers, and even enhance the styling of a car. We will discuss how certain trends in styling have led to better uses of new technologies, and how these have affected vehicle costs and customers.

Further discussions on the costs of each new technology will occur at the end of each section.

Aero Headlamps – Headlamps and Tail Lamps have seen numerous packaging innovations, integrating as many functions as possible within the envelope on the vehicle. The most dynamic change in packaging has been the AERO HEADLAMP, which forever changed the shapes of all lamps.

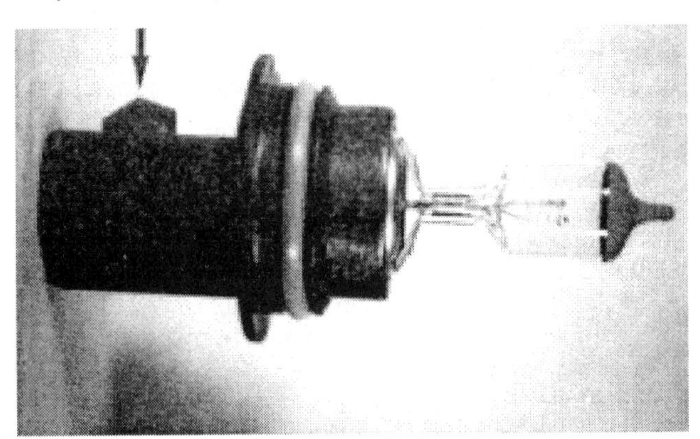

Figure n

Material advances, mainly in plastics, gave birth to the Aerodynamic Headlamp. The design is not just round or rectangular, but is able to follow the exterior surfaces of the sheet metal and fit into a specific shape/contour. The lamp still requires a light source, but a new generation light bulb was developed (still based on a filament design) specifically for the aero headlamp. Instead of being secured to the inside of a sealed beam, the halogen light source was re-designed to be independent, specifically for the aero headlamp. Instead of being secured to the inside of a sealed beam, the halogen light source was re-designed to be independent of its sealed beam shell. It can fit into any headlamp, regardless of the shape. This light bulb, which seals itself when secured to the headlamp's reflector, has begun the trend away from sealed beams to aero headlamps, for almost all cars throughout the automotive industry, at home and abroad. Refer to figure n.

Ford Motor Company was one of the first in the industry to use aero headlamps, with the introduction of the new headlamps on the 1984 Lincoln Mark VII. The headlamps were flush mounted lamps, following the contour of the hood's surface thus allowing a more aerodynamic front end on the car. The 1985 Ford Taurus was unique in automotive lighting because of its completely aerodynamic design for front and rear lighting, similar to many exotic European designs which controlled body shapes in order to reduce air drag.

A Ford Taurus

1985 Ford Taurus

The use of aero headlamps showed how lighting systems could drive the design of a car as much as the sheet metal, trim, or paint could. Aero headlamps further enhanced the look of automobiles when COMPLEX REFLECTOR lamps were introduced (mainly on vehicles from Honda Motor in the late 1980's). Complex Reflector Headlamps are aero headlamps in nature, but with more "razzle-dazzle" than the early designs. These headlamps use clear lens technology, which have most or all optics on the reflectors with very few optics (if any) on the lenses. In fact, the complex reflector designs have caught the eye of many automotive designers such that nearly every entry into the U.S. market has complex reflector headlamps, some even with clear lens tail lamps. For example, between 1994 through 1997, Ford Motor had complex reflector headlamps on every passenger car, from the Escort to the Mark VIII. The 1997 Lincoln Continental had clear lens headlamps (and tail lamps),as did the 1994 Ford Escort -showing how a $42,000 car and an $8,000 car could share the same type of technology. Refer to figure o for a complex reflector headlamp (notice the usage of a "chromed" look on the entire headlamp).

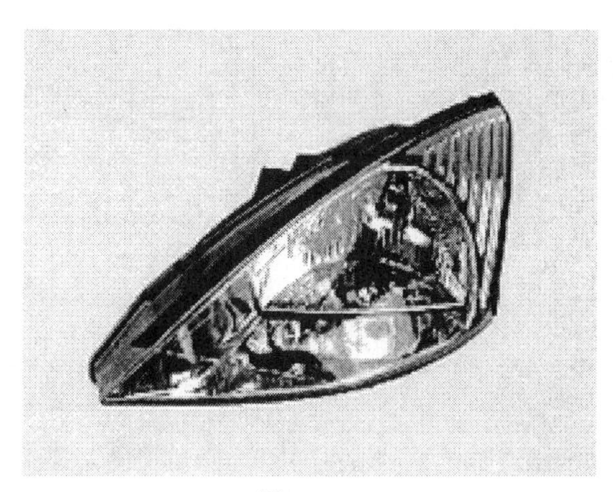

Figure o

Another type of aero headlamp that uses a different internal lamp design is the PROJECTOR HEADLAMP. This lamp design uses what resembles a film projector lens and a reflector, secured by a metal housing. The design gives the appearance of a "fish-eye", which some designers prefer over the complex reflector design. The main advantage of a projector headlamp is its modularity: the same low beam module could be used across vehicle lines (therefore reducing complexity) without excessive re-engineering. Also, the projector headlamp can be used to meet left-hand drive and right-hand drive headlamp requirements (simply change the internal shield). Refer to figure p.

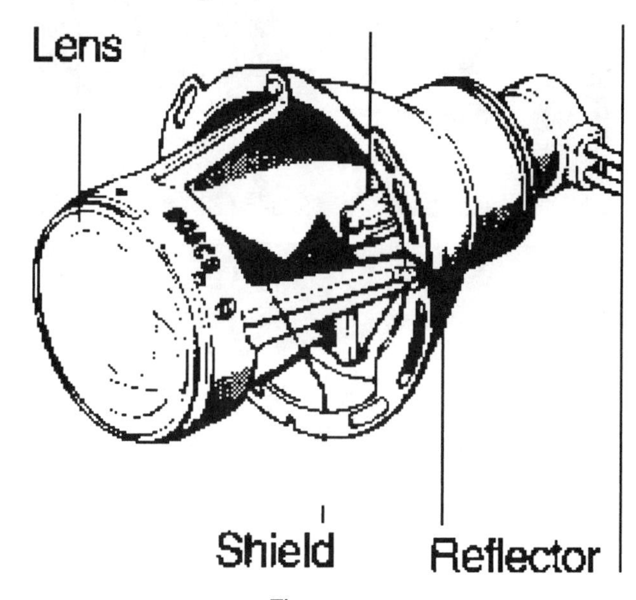

Figure p

Even with the advantages of an aero headlamp, one issue that always arises is the headlamp's cost:

Repairs are about the same between a Continental and an Escort!

The main drawback of the aero headlamp has been the overall INCREASE in costs as compared to the sealed beams. Lamp component tooling costs are greater, while material costs have nearly doubled the piece cost of a single headlamp. Also, because of the need for left-hand side and right-hand side lamps, complexity has increased at the assembly plant. Integration of turn lamps into headlamps have also increased their costs. The process of aiming the headlamp has not increased dramatically, since the small "nibs" on the sealed beam lens were also designed on early designs of aero headlamps' plastic lens, such that the same aiming methods from the past could still be utilized. Some complex reflector aero headlamps have other devices for aiming, such as the VHAD (Vertical and Horizontal Aiming Device), which does not require nibs on the lens. It does however require more training for assembly plant personnel to aim the headlamps, as well as the dealer.

Reparability has been greatly affected by aero headlamps because of the lack of access to the bulbs. Many aero headlamps require complete removal from the vehicle's front end, since there is little or no space to access the bulb. This has been more prevalent in car vehicle lines, but trucks seem to have on the average somewhat more room to offer for lamp accessibility. For many vehicles, the overall packaging space behind the headlamp has had to increase in order for a technician to have access to the bulb, with added wiring costs also greatly affecting serviceability costs.

The headlamp has increased in price, as mentioned earlier, but the more startling fact is that the costs do not always match the overall value of the vehicle. Luxury cars and economy cars in the past had similar repair costs for sealed beam replacements, since they shared common lamp designs and structures. However, future car designs saw the gap widen between the economy and luxury cars when repair was involved. Some headlamps cost as much as $400 per lamp for luxury cars (with halogen bulbs), while economy cars' lamps are in the $100 - $200 range. At times, luxury and economy headlamp prices for certain vehicles are similar, even though the vehicle prices are very far apart. According to Mitchell's Collision Estimating Guide 1998,[19] the Escort ZX2 headlamps cost nearly as much as Lexus headlamps, $184 vs. $201, respectively. Other examples show a wider gap between economy and luxury vehicles exist (Corolla versus Jaguar), but there is no trend that shows headlamp costs going down faster on economy cars than on luxury cars. Let's compare the costs of an aero headlamp (sometimes referred to as a composite lamp design) against the old sealed beam:

Repair cost of front lighting systems
<u>1988-98 Chevy C/K Truck</u>
Average Sealed Beam cost vs. Aero Headlamp cost
$ 48 per lamp versus $161 per lamp

<u>1982-91 Pontiac 6000 sedan</u>
Average Sealed Beam cost vs. Aero Headlamp cost
$ 35 per lamp versus $126 per lamp
(typical lens & housing costs are shown; costs do not include hardware to secure to front end)
Per Mitchell's Collision Estimating Guide, dated 1997.[17]

There are many other examples of cost increases passed onto the customer because of the trend away from sealed beam headlamps, which on the average cost about $4-7 each. The older car models, such as the 1980 Monte Carlo, had quad headlamp designs: 2 low beam and 2 high beam lamps. Headlamp replacement of all of the Monte Carlo lamps would cost around $30 (plus another $40-$70 for the front end structure which secured the lamps). That is significantly less than for newer vehicles. The surrounding structure on newer vehicles tends to have more plastic components, which tend to break easier than past front end structure designs. Even with the life of an older sealed beam being less than for the newer halogen bulbs, most customers would prefer to pay only $4 for a sealed beam than the more expensive halogen bulb price.

Replacement costs are not the only increase passed on to customers; service costs have also risen because of longer repair times at dealers. The average sealed beam replacement was around 5 minutes, but the newer headlamps require sometimes up to 20 minutes (depending on the model) for bulb replacement. Aimability of sealed beams was somewhat easier, given the aiming "nibs" on the sealed beam, while some aero headlamps require more sophisticated mechanical designs for adjustment. In emerging markets, the costs are even greater for serviceability, and there seem to be no significant trends in lighting design that will drive down overall costs. Pickup truck and van lines are the only vehicles that tend to be more receptive to using sealed beam headlamps, but not all of them have. Ford's F-150 and Chevy's 1500 trucks both have remained with aero headlamps throughout the 1980's and 1990's, yet Chevrolet has offered both sealed beam and aero headlamps for certain models of their pickups. Ford's new 1998 model year Super Heavy duty trucks have followed suit.

High Intensity Discharge (HID) Headlamps – Probably the most recent innovation in headlamp lighting has been the HID (high intensity discharge or Xenon) headlamp. This headlamp design is similar to a normal fluorescent lamp: a bulb containing an inert gas (and no filament) is powered by a ballast.

The main benefit of an HID headlamp is sheer light output, almost 5,200 lumens at 33 watts, versus 1,600 lumens at 60 watts for a halogen bulb. The HID light source is composed of a set of electrodes (0.40mm apart) in Xenon gas which ignite when electrified. The HID bulb is of a similar construction and shape as compared to a halogen bulb, but requires an ignition voltage of approximately 20,000 volts from an electronic ballast. An HID source is shown below.

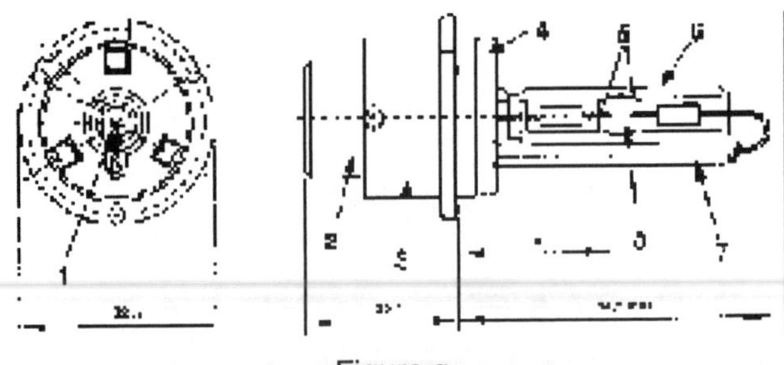

Figure q

Sylvania was one of the first has taken the lead in HID sources, with other firms developing their own designs (notably Phillips). The HID lamp is unique in that its design does not require any additional packaging within the lamp area, but the ballast does require a location either inside the lamp or within the vehicle. The HID source has a longer life (almost six times longer) and durability than the typical halogen bulb, since there is no filament that can break due to shock or spikes in system voltage. Even with the added ballast, the HID system requires less power than a halogen system. Because the emitted light tends to be of the same hue of natural light, the HID light pattern may be better suited for driving in poor weather. The typical beam pattern of an HID lamp (shown on top) is compared to a halogen lamp pattern. (on bottom), per figure r.

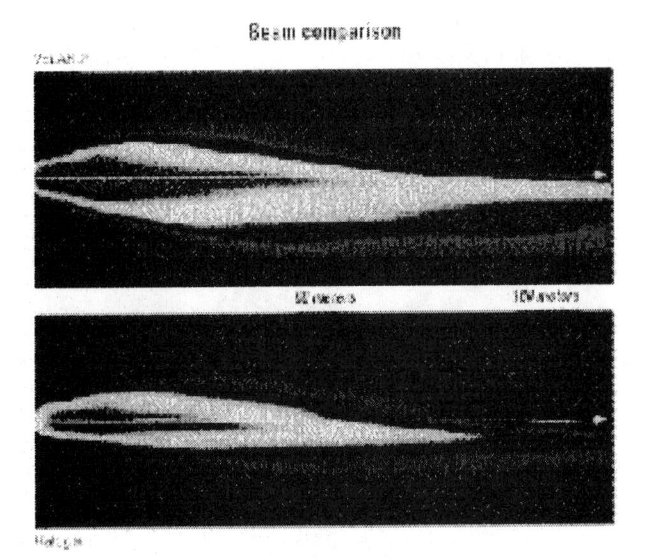

Figure r
(bird's eye view)

BMW (first users of HID in 1991), Mercedes, Lincoln, Lexus, and other luxury car manufacturers have used HID headlamps for various reasons, more for performance than styling. Lincoln's Mark VIII utilizes Complex Reflector headlamps together with an HID light source, while BMW, Mercedes, and Lexus tend to use Projector Headlamps with HID sources. In Europe, where HID lamps have been proven to be more effective on the highways, regulations have required other systems to work hand-in-hand with HID lamps, mainly headlamp leveling devices and headlamp washers. These devices ensure that the brighter HID beam pattern does not blind the oncoming driver, which at high speeds can be very hazardous and increases the chance of accidents. Typical headlamp leveling system components are shown below.

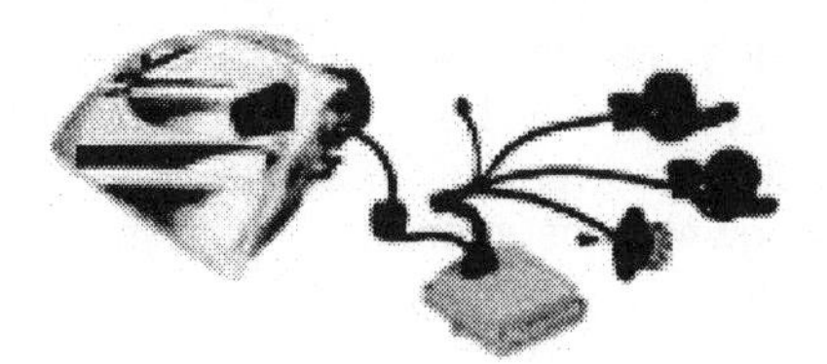

Figure s

Headlamp Leveling systems – Headlamp leveling systems have evolved from static systems to true dynamic systems, which utilize stepper motors that can rapidly adjust the headlamp reflector when vehicular angular changes occur. The first static systems (led by Bosch and Hella Lighting) were incorporated in European vehicles, mainly because of the regulations that were introduced in the 1990's. Loading a vehicle such that the headlamp beam pattern cutoff was forced into the eyes of the approaching driver required some sort of leveling device, to return the pattern to its original position (meaning an unloaded car). A rotary switch inside the vehicle (with settings 0 through 4) was manipulated to move the headlamp such that the beam pattern was vertically adjusted.

By using small DC motors and a pivoting system, static leveling offered the driver easier control of the adjustment process, mainly after the vehicle was loaded. This adjustment was completely in the hands of the driver, and prior to HID systems, halogen lamps required only a minimum of vertical aiming, without the need for automation within the leveling system. Refer to figure t on the basic mechanical design of a static system.

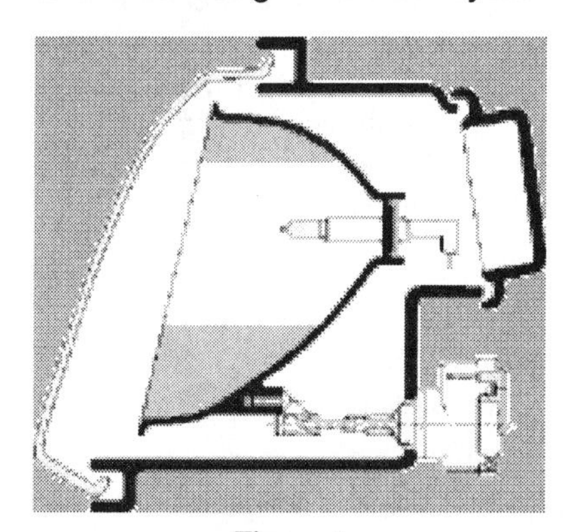

Figure t

Although HID lighting provides a better beam pattern, HID lighting systems have forced European regulators to mandate a true dynamic auto-leveling system to control the extreme light output generated by the HID bulb. Even in rain the HID system can be annoying for oncoming drivers, albeit that they wish they could see the road better! Once HID headlamps entered the market, it was determined that dynamic systems were the only way to ensure that HID systems would not cause a safety concern, because of its light output. Dynamic systems, pioneered by Bosch, allow hands-off adjustment, and gave the driver more freedom. Bosch developed their Litronic© system in tandem with leveling to work hand in hand, offering more benefits over static systems. The dynamic system does not react to potholes or uneven road surfaces, but can detect when the vehicle is accelerating or braking suddenly. Microcontroller and angular sensors work in tandem to drive the stepper motor, and by using odometer speed indication, the system can accurately compute the car's instantaneous acceleration. This is an important part of the dynamic leveling system, which allows HID lighting systems to adapt within a few seconds to changes in vehicular attitudes.[4] The electronic control units (ECU's) designed for HID auto-leveling systems rely upon a number of sensors, with which feedback is sent to the ECU for driving the stepper motor. Figure u depicts the main components within a dynamic leveling system.

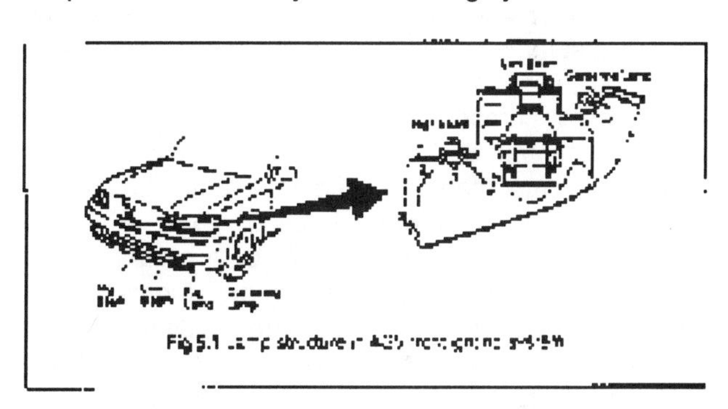

Figure u

The ECU drives the stepper motor to align the headlamp's angle in relation to the driving conditions, be it braking or load changes. As the conditions worsen, the stepper motor is driven continuously to correct for the extreme angular changes. The system can adjust for extreme vehicle loads, along with sudden braking or acceleration, offering to the customer what was never before:
Hands-off control of headlamps.

This works into the safety arena as well as customer satisfaction, since the customer will not be making adjustments as they are driving. HID is viewed favorably when it comes from the driver's point of view, but the oncoming driver still gets the excessive glare and light output every now and then. For example, random glare from the side is directed to cars sitting lower, such as Twingo or VW Beetle.

However, the main drawback to HID auto-leveling systems, of course, has been added cost to the vehicle lines and repair costs. The added cost was not just to the overall headlamp design, but to the entire vehicle. A unique controller was required, along with the expensive sensors that were needed to input signals to the controller. Wiring costs also increased, but not drastically. A comparison of headlamp costs will be discussed next.

HID Headlamp costs – Now comes the bad news: HID systems cost on the average $500 per headlamp more than for typical halogen systems. HID costs (per lamp) for several car models are shown below.

Model	Headlamp	Repair Cost/ headlamp
Lexus GS300/400	Projector	$1,051
Mercedes E-Class	Projector	$1,011
BMW 5-series	Projector	$ 842
Lincoln Mark VIII	Complex reflector	$ 462

Per Mitchell's Collision Estimating Guides, dated 1997 and 1998.[17 18 19 20]

Auto manufacturers have handled HID implementation in different ways, as can be seen in the costs shown above. The majority of the systems use an expensive leveling motor (servo motors) and a controller for the automatic leveling system, making the system very expensive to repair and maintain. Once the headlamp is broken, the other internal components normally are not able to be reused, unless the design is such that the leveling motor is externally mounted from the headlamp. Lexus and Mercedes have the internal leveling motor, while the Mark VIII headlamp has none at all (USA regulations do not yet stipulate the same leveling requirements for HID lamps as in Europe). The Mark VIII from a distance looks like any other complex reflector headlamp (clear lens with jewel-like reflector), but the light source is the additional cost that increased the system cost, not the remainder of the lamp. A Lens and a Reflector are much cheaper than a projector system, but the HID source required some packaging work for Ford Motor to implement the new lighting technology.

Mercedes and Lexus have offered non-HID headlamps in previous model years of the same vehicle, with the headlamp basically the same size and appearance from the outside. Non-HID lamp cost was around $400 per headlamp for both Mercedes and Lexus high-end vehicles. However, the interchangeability between the two headlamp systems is such that it is not an easy task to implement for different model years.

New uses of HID lighting systems – Another project in the works has been to use HID light sources for the entire vehicle, by using light pipes or fiber optics throughout the vehicle.[11] Implementation of focus-less optics (FLO), along with waveguides, can collect light from an HID unit and distribute it to the rest of the vehicle, with as few as two HID units required for smaller vehicles. FLOs can be made from Pyrex or other types of glass, and can service one turn signal, one backup lamp, and even one half of a center high mount stop lamp (CHMSL). Plastic color inserts (mainly for red and amber functions) work hand in hand with FLOs, and even with some limitations that have been encountered (mainly with CHMSL and turn signals), this new application of HID technology has shown promise. The use of waveguides, rather than fiber optics, has helped decrease some of the costs, since they are made from injection-molded plastics. It was noted in studies that overall cost of this HID application still cannot compete with the existing incandescent bulb designs. There are other factors that affect future automobile designs, such as the 42-volt system. The 42-volt automotive system has been developed,[3] waiting to be implemented for various reasons, mainly due to increased power consumption of the overall vehicle (in terms of current). Technologies possible with 42-volt systems include an electric A/C compressor, steer-by-wire, and brake-by-wire.

These systems, when using 42 volts, offer higher efficiencies, better packaging, improved modularity, and better NVH. All of these need higher voltage supplies, but not all systems would benefit from the higher voltage.

Conventional filament bulbs are adversely affected by high voltage. Because of the system design for a 42-volt architecture, automotive lighting and electrical engineers will need to devise a plan to protect for the existing lamp designs. Except for HID and Neon systems, the existing technologies -LEDs and incandescent bulbs- are affected by high voltage (LEDs need less than 9 volts already) and would need to have a separate 12-volt circuit. The majority of lamps in the industry may be able to be powered by light pipes from HID or light engines, but wiring is much less than fiber optics for automotive applications. This kind of dilemma is not an easy one for automotive manufacturers to plan for, but with more and more requirements for higher voltage, HID systems bring significant advantages.

Intelligent Lighting Systems – Intelligent Lighting Systems (ILS's)[1] have become a hot topic recently in the automotive sector, with Advanced Frontlighting Systems (AFS's) leading the way for implementation of advanced electronics for the lighting systems. New generation headlamps, primarily projector designs, have been designed to be controlled electronically for better nighttime driving, when most accidents occur. ILS's have been developed to handle two key issues:

1. Ensuring that the correct beam will be selected (low vs. high beam) by the driver
2. Ensuring that the correct beam will be used depending on driving condition (rain, snow, fog, etc.).

ILS's are not just HID lamps with auto-leveling systems, but go beyond that with a complete front lighting system that reacts to driving conditions. Koito has led the way in Japan during that country's Advanced Safety Vehicle program, developing a lighting system which uses a combination of headlamps, front fog lamps, and cornering lamps to provide the best lighting for a number of driving environments.

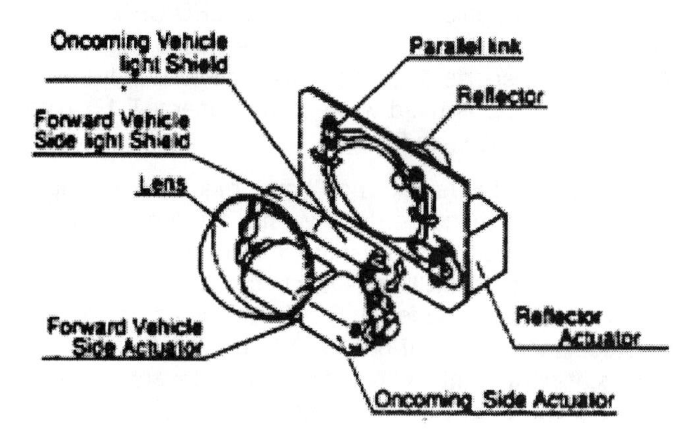

Fig 5.2.1 Structure of low beams

Figure v

Figure v shows the typical structure of the ILS's low beam headlamp, with a projector headlamp module playing a key part in the entire system. The use of shields for oncoming drivers, as well as actuators, helps the ILS system diminish the low beam glare onto the approaching driver. By using a CCD camera, a milli-wave radar system, speed sensor, and axle sensors, the ILS system can correct the low beam not just for oncoming drivers but also for drivers in front.[1] Electronic modules control the actuators, tilting the low beam module as needed, while the beam axis controller uses the CCD camera (Charged Coupled Device) -essentially a digital camera- to detect lane changes and differing road forms. These two systems (low beam and beam axis control) are an integral part of ILS, whose goal is to increase the illumination of the road for the driver during any changes in roads and in climatic conditions, with oncoming drivers also protected from excessive glare. Global Positioning Systems (GPS) are also used in ILS, providing real-time road information as needed. Figure w shows the lamp configurations that an ILS uses: front fogs, low and high beam lamps, and cornering lamps.

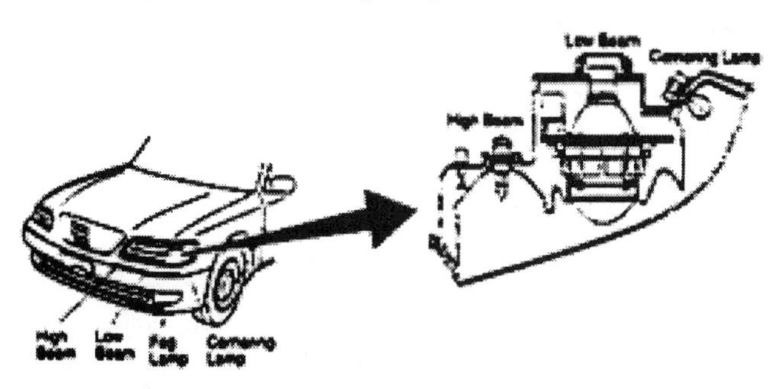

Fig 5.1 Lamp structure in ASV frontlighting system

Figure w

Advanced Front Lighting Systems (AFS) from Europe, notably Bosch's, do not use a series of lamps (such as projector headlamp and front fogs) but rather one headlamp with several small reflectors that can be individually manipulated to project a suitable headlamp beam pattern for any driving situation. By using a "CAN-bus" system, the AFS receives information on differing roads and climate changes to illuminate unique combinations of the small reflectors, relying upon Global Positioning Systems (GPS) to provide road information as needed.

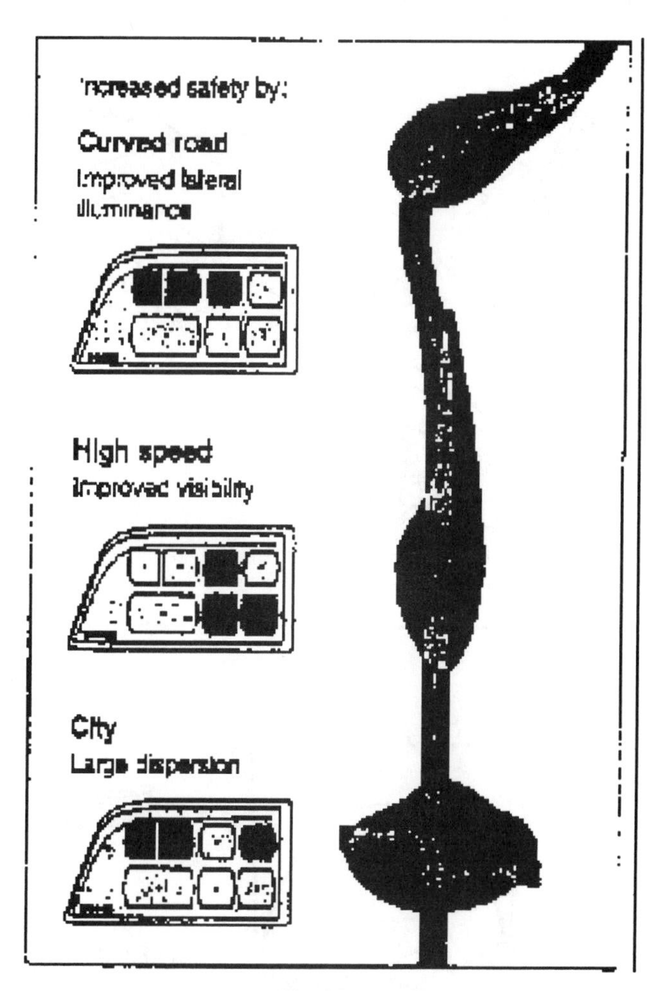

Figure x

The AFS system's goal is the same as ILS: provide better road illumination for the driver and protect the other driver's.

<u>LED lamps</u> – One of the more recent developments in automotive lighting that has been driven by the want for packaging reduction is the light emitting diode (commonly referred to as LED's). LED's have been used in other industries for years, such as electronics and computer industries. However, the introduction of LEDs into the automotive industry was tied to the newly released Center High Mount Stop Lamp (CHMSL) requirements in the U.S, specifically for vehicles made on or after September 1,1985. CHMSLs are required to perform like a stop lamp, with the only photometric requirement being the Stop function. Prior automotive applications of LEDs were mainly for Clusters (speedometers), for illuminating the small telltales on the cluster face.

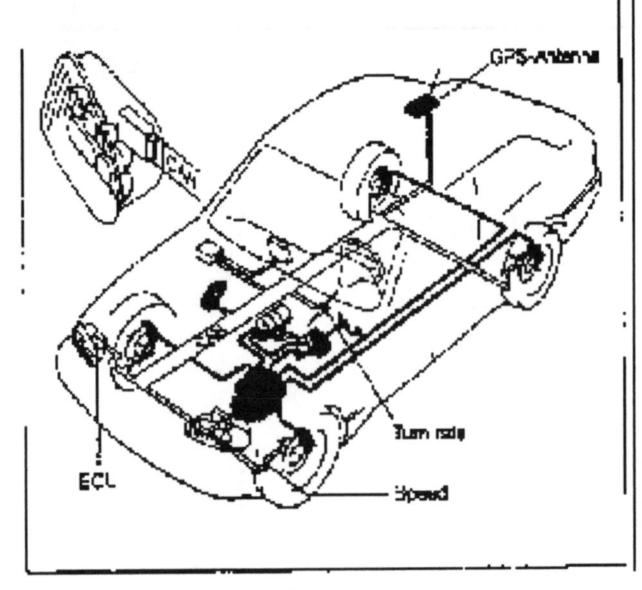

Figure y

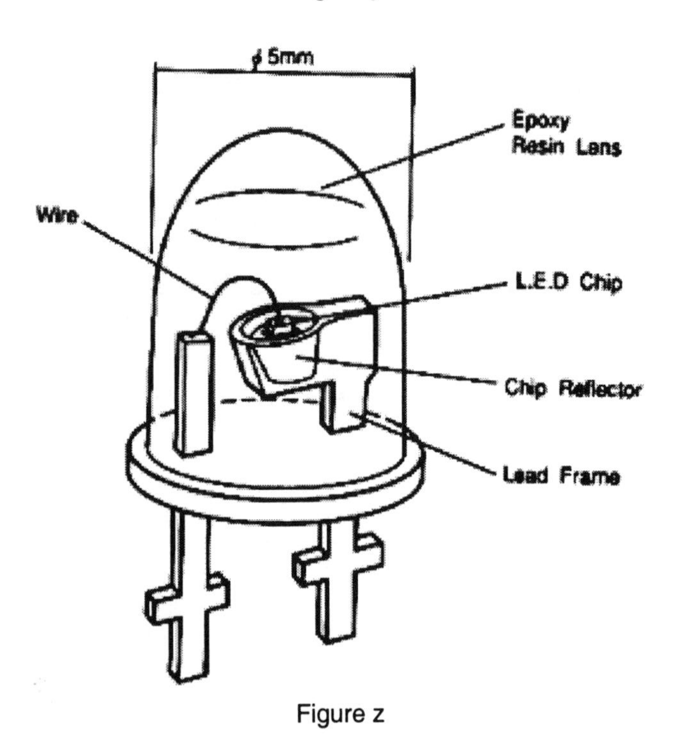

Figure z

Exterior lighting engineers soon found out that LEDs could be used for other automotive applications. They discovered that LED's could run on very low voltage and current (less than 4 volts and 0.50 amp), did not generate much heat, had a 200 nanosecond instant-on time, and could be designed to last the life of the vehicle. Even 1,000,000 cycles are possible for LED's to withstand without catastrophic failures. Thus, the LED CHMSL was born. One of the first LED CHMSLs was on the 1986 Nissan 300Z sports car, externally mounted on the rear tailgate. The LEDs used were of the general T 1¾ type (red color),[10] shown in figure z.

Because the LEDs did not have much light output (around 0.30 lumens per LED versus 400 lumens for an incandescent bulb), the quantity of LEDs was quite high (75). Nissan's CHMSL was such that it had a nearly vertical lens, allowing for as little light loss as possible. It was designed so that if one LED failed, the CHMSL still had sufficient light output for other drivers to see. Nissan's use of LEDs in high mount stop lamps opened the door for other automotive manufacturers to use the technology, either in CHMSLs or even tail lamps. Ford Motor Company was the first American company to introduce LED's in its car lines, beginning with the Taurus SHO and Thunderbird vehicle lines in 1993. The Taurus SHO had an external CHMSL (on the rear decklid's spoiler) while the T-Bird utilized LEDs for a tail lamp function (also on the rear decklid). Both of Ford's LED lamp designs used a new LED design from Hewlett Packard, called a Brewster© LED (one package type is shown in figure aa).

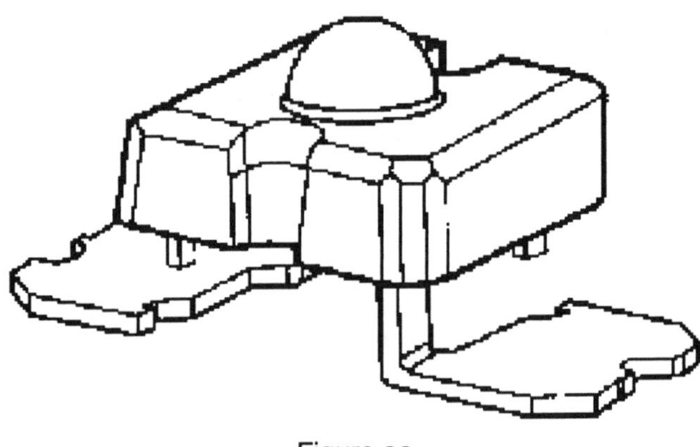

Figure aa

A Brewster© has more light output than a type T 1 3/4 LED, and by using its leading chip technology based on Transparent Substrate Aluminum Gallium Arsenide (TSAlGaAs), Brewsters© offered LEDs with more output, nearly twice that of earlier LED's. AlGaAs chip technology pioneered automotive LED design, offering more light output, quick response times, lower power consumption, within the same small package. The all-new AlInGaP (Aluminum-Indium-Gallium-Phosphide) technology on the new Hewlett Packard Piranha© LEDs has doubled the light output even more. The new Piranha© package (which has replaced the Brewster©) size is similar to the Brewster©, yet comparable to the T 1 3/4 LED in depth. Unlike the T 1 3/4 and Brewster© LED's, which required either surface mount or wave soldering, the new SNAP LED© (based on the Piranha©) can be mechanically assembled, even without the need for a printed circuit board.[14] The SNAP LED© can be assembled onto a curved thin copper substrate (similar to flexible circuits), thus allowing the implementation of LEDs into more complex automotive lamp designs. LED costs have been known to vary from manufacturer to manufacturer, with Hewlett Packard's SNAP LEDs© costing around 50 cents each. Standard T 1 ¾ LEDs are around a dime each, but a typical HMSL design would require more 1 ¾ LEDs than Piranha's©. One important cost note to highlight is this:

Once most of the LEDs have burnt out or broken, the entire CHMSL needs to be replaced.

Here is a comparison of repair costs for typical LED CHMSL designs for several automotive companies:

Model year and manufacturer	HMSL Cost
1997 Lexus GS300	$345
1997 Cadillac Catera	$159
1997 Chrysler New Yorker	$189
1997 Lincoln Navigator	$ 80

Per Mitchell's Collision Estimating Guides, dated 1998.[17 18]

A typical incandescent CHMSL costs around $50, so in comparison to the LED CHMSL's shown above, one can see that this technology is not exactly low cost. But, with the life of an LED advertised to be the same as the lifetime of the vehicle (as compared to 1,000 hours for an incandescent bulb), LEDs have seen more and more application into automotive signal lighting, notably tail lamps. Some aftermarket LED tail lamps (made by Trucklite and Grote) are used extensively in heavy-duty trucks, which tend to have more bulb burnout's in their lamps because of the extreme vibration and shock on the filament. The Big Three (GM, Ford, and DaimlerChrysler), along with the Japanese firms, have used LEDs mainly for CHMSLs, but have shown designs and concept cars that use LEDs for complete tail lamps and even Front Turn signals.

Neon Lamps – Neon lamps have been in existence since the 1920's, used mainly for signs. When automotive engineers began tinkering with neon tubes for lighting applications on cars, most of the industry thought that the novelty would die rather quickly, for various reasons. An automotive neon lamp is simple in nature: neon tubes are situated inside the lamp, ignited by a ballast, and relies upon optics to direct the light output.

Neon lamps have shown a lot of versatility in recent automotive applications, with Sylvania being one of the innovators in the development and application of neon lamps. A typical Neon CHMSL has an operational life of more than 2,000 hours and an on/off capability of more than 800,000 cycles. Neon offers good performance when driven by a voltage as low as 9 or up to 16. Unaffected by extremes of heat and cold, and able to withstand shock and vibration damage, neon is designed to equal or exceed conventional signal lamps. With a bending radius of 25 mm, a neon tube diameter of 5 mm, and a manufacturing capability range from 20 cm to 130 cm lengths, neon complements the long, sweeping look of customized vehicle contours. Neon tends to distribute light uniformly, for a homogeneous and "hot-spot" free appearance, unlike incandescent lights, which require multiple cavities and sources that must be diffused and redistributed. Neon tubes do not have to be bundled in arrays like LEDs, and brilliant red and amber colors permit clear lens styling when utilized in an edge-lit system.

However, neon lamps are not cheap (average ballast cost of $50 per vehicle is strictly incremental, as compared to incandescent bulbs which do not need one). During lamp manufacture, neon tubes can be easily broken, and the packaging space needed for certain ballast designs make the lighting engineer's job very difficult. But, neon has its advantages as well:

- The speed of illumination (instant-on time) of 200 nanoseconds, much faster than normal bulbs
- Wide viewing angles
- Longer operational life than incandescent bulbs (2,000 hours vs. 1,000)
- Less susceptibility to voltage spikes.

One of the earliest applications for neon tubes in a recent automotive lamp was the 1995 Ford Explorer CHMSL, made by Sylvania. The lamp replacement cost was high ($187 replacement cost, plus $80 for the ballast per Mitchell's Collision Estimating Guide, dated 1997)[17] and was about 200 mm long. The ballast required more packaging space than originally planned for.

However, Ford Motor and Sylvania were able to prove that automotive neon lamps could be implemented on a mass production basis. Besides the Explorer, Ford used neon on the 1998 Mark VIII luxury sedan. The stop and tail lamp functions on the Rear Applique of the 1998 Mark VIII were performed with neon technology, yet maintained the targeted package depth per Ford's requirements. The Mark VIII's Applique is not inexpensive by any means ($658 replacement cost per Mitchell's Collision Estimating Guide, dated 1997)[17], but it again showed how automotive manufacturers could use neon in other ways besides CHMSL's.

Repair costs are not expected to approach the incandescent bulb level for some time, but Sylvania engineers are in the process of attempting to use the neon device for more than one lamp at a time, using a common light source to illuminate more than one lamp at a time. Sylvania's Luminon© SPD14 (Standardized Platform Design) rear signal and brake lighting system offers a 14-inch neon tube and ballast electronics utilizing less wattage with a smaller housing height.[16] Because of its simpler (and smaller) design, the Luminon© SDP14 is now available at half the cost of previous neon systems.

Figure bb

Concepts in Neon – Ford Motor and Sylvania have teamed up on the Ford Mustang concept car (shown above in figure bb) to showcase their latest automotive lighting technology. The front end of the Ford Mustang features amber neon park and turn signals, a new technology which allows car makers to completely outfit the front and rear exterior of the vehicle with neon lamps. Also, the front end sports a new look, incorporating six Sylvania HID projector headlamps, two for low beam, two for high beam, and the remainder (located below the headlamps) function as driving/fog lamps.

COURTESY LIGHTING

Even though there are no major regulations on what is called courtesy lighting, there are nonetheless customer wants and needs when it comes to lighting offered on the inside of the vehicle. A new trend for automotive lighting has been in both the internal and external lighting systems, notably map/reading lamps and running board lamps.

DOME LAMPS – Dome lamps have been around since the early 1940's, using mainly incandescent bulbs. These lamps were round in shape, 3 to 6 inches in diameter, whose function was basic interior illumination, without offering the customer any directed light for map reading. Some of the first map reading lamps introduced in the 1970s offered the customer direct light in their seating area, and automotive manufacturers had internal guidelines for determining what was a good map lamp and a bad one. Some European car companies took it one step further, by requiring that map lamps and other interior lamps had to withstand the same durability tests that the rest of the interior had to.

Mercedes has map lamps as standard equipment, while economy cars in the Renault and Fiat car lines have very simple dome lamps. Different styles of interior lamps take the gamut from exotic to spartan (the Mexican built VW Beetle had no interior lamp). In general, dome lamps have taken a step back to map/reading lamps, ones that direct light to the driver and passenger for reading or illumination purposes. Designs range from rotating "eyeball" designs to fixed lens designs that project the light via complex optics. Nearly all Reading Lamps are plastic, with some internal metal reflectors and springs for lens fastening, and the typical designs are large for the US versions and more compact ones for European and Japanese tastes. There are some reading lamps that have other functions integral to them, such as securing remote keyless entry receivers (on 1995 Ford Fiesta). The increasing prevalence of automotive lighting customer specifications has forced some lighting companies to design modular reading and dome lamps that are part of the headliner or grab handle, such as on the 1997 Town Car's rear passenger grab handle. Firms such as Donnelly (USA) and Kigass (Germany), have led the way for future interior lighting designs, but other existing lighting manufacturers have entered the interior lighting market, notably Valeo, and Koito. Hella has had numerous in-house designs for dome lamps, dating back to the 1960's, and has proven to be a big player in the aftermarket interior lamp business.

RUNNING BOARD LAMPS – Why anyone would require illuminated running boards was the main question posed during past clinics held by the automotive manufacturers. The Big Three (Ford, GM, and DaimlerChrysler) have led the way for running board illumination, mainly on their Sport Utility vehicles (SUV's) and pickup trucks. In Europe, there are no very large SUV's (and almost no pickups), that would require a running board, but such illumination is illegal per European Council (EC) lighting regulations.

North American Lighting (NAL in USA) and Decoma (USA) have designed running board lighting systems, based on incandescent bulb lamps, and because of the length of most running boards, more than one lamp is generally required for adequate illumination. There have been new studies in the implementation of LED's, light guides (small lamps within a clear cable), and electro-luminesence (EL) within courtesy lighting, with running boards getting the most work. Since the customer is looking for the best illumination at nighttime to ensure that he/she does not misstep when entering the vehicle, lighting engineers have tried to spread the light such that the location of the board is estimated better. A study done by NAL attempted to obtain several opinions on an active system (one that marked the surface by becoming part of it) or a passive system (one that simply illuminated the surface).[8] It was found during the study that certain study participants preferred passive rather than active lighting systems, since their wish was to have better estimation of the running board's location. The active system simply allowed the participant the ability to see a marked location on the running board, but did not offer true illumination of the surface. Ground illumination was a welcome side effect from some of the running board lamps, but was not a key variable of the study.

This study shows how even courtesy lamps have become an integral part of the vehicular lighting system, attempting to provide the customer a safer product without adding exorbitant costs. Adding a bit of flair to even running boards with courtesy lamps, such as on the Lincoln Navigator, has caught the customer's eye!

OEM LIGHTING COSTS AND BENEFITS

As was mentioned in earlier sections, some lighting manufacturers have implemented the latest technologies without adding higher costs to their OEM (Original Equipment Manufacturer) customers' vehicles. Ford's LED CHMSL's have historically been less expensive than GM's or DaimlerChrysler's, and replacement headlamp costs for Japanese and European cars have also been historically higher than for American OEM's. While the Europeans have been better integrators of the new technologies (projector, HID, and leveling systems), they have added much higher costs to their vehicles, which the Americans and Japanese tend to shy away from. The Japanese and Europeans have lamp designs that tend to be more modular than the Americans, helping reduce assembly plant labor costs, but that does not always favor the customer when it comes to replacement costs. Most of the American OEMs offer a 3-year/36,000 mile warranty that includes bulb burnout, while the Europeans and Japanese do not, and while that may offer something of convenience to the customer, the Big Three have seen some increase in warranty costs. Nonetheless, lighting has become a system that all OEMs are tracking, for both their sake and the customer's.

CONSUMER BENEFITS – There have also been several studies of the effects of lighting on certain consumers (mainly the aging population), both by OEMs and universities. One notable paper by K. Rumar noted that utilizing certain systems such as daytime running lamps, can help the elderly by allowing them to more easily see other vehicles on the road.[21] DRLs (Daytime Running Lamps) have been used by General Motors as a marketing tool in the U.S., even though Canada has mandated their use. DRL's offer additional safety without the need of another lamp assembly (electronic module drives a lower voltage to the headlamp for daytime light output). Valeo has noted that HID headlamps can offer better and more light, reducing the fatigue on an older person's eyes.[15]

CONCLUSION

In conclusion, this paper has explained where automotive lighting got its roots, how light sources have evolved, and how automotive engineers have been able to progress rather quickly in using new materials and designs to expand the lighting system's ability to function as a safety device. Safety was not the lamp's first priority in the old days, but rather took a more passive approach to safety by simply acting as a marker. By the post-World War II years, lighting had become an essential safety system, yet became a styling driver as well. New advances in light sources, first the incandescent bulb and then later on the halogen light source, offered lighting engineers the ability to have better performance from vehicular lighting.

But it was not until after the 1960's when governmental regulations all over the world forced automotive manufacturers to make even better lighting systems that adhered to minimum standards for performance, with the Americans and Europeans taking the lead in lighting technologies and applications of those technologies. Lamps got better, looked great, and offered safety as was never seen in the earlier days of the automobile. However, the new technologies and materials, along with the new styling, have added costs to the vehicle's lighting system, sometimes astronomical costs. The customer cannot understand how something so small costs so much! Consumers have to pay more for lamp replacement and cannot always go to their local K-Mart© for convenient lamp replacement. For a developing market (such as Angola), the costs are even higher because of the lack of replacement part stores. What lighting suppliers will do in the near future to reduce overall lighting costs will be reviewed closely by the OEMs, with the goal that value-added technologies and clever designs will help reduce the overall lighting system's cost.

ACKNOWLEDGMENTS

Substantial support from Ford Motor Company helped make this paper possible. My manager Robert Pheiffer supported me throughout the time it took to complete this paper, and proofreading by L.Goodes was greatly appreciated (Ford's Scientific & Research Library system made the task a bit more manageable as well). I would also like to thank the Society of Hispanic Professional Engineers (SHPE) Detroit Chapter for their support in presenting an earlier version of this paper in 1999 during SHPE's National Technical and Career Conference.

REFERENCES

1. S. Kobayashi, "Intelligent Lighting Systems: Their History, Function, & General Direction of Development", SAE Technical Paper Series 981173, 1998
2. Automotive Handbook, Robert Bosch GmbH, R. Bosch Publishing, 1993, pgs. 668-689
3. G. Kobe, "The 42-volt Revolution", Automotive Industries, August 1998, pgs. 30-33
4. C. Kormanyos, "HID System with Adaptive Vertical Aim Control", SAE Technical Paper Series 980003, 1998
5. R. T. Weidner and R. L. Sells, Elementary Classical Physics, Volume 2, 2nd edition, 1973, pgs. 735-736 and 743-753
6. M. R. Bass & E. W. VanStryland, "Handbook of Optics", Volume 1, 2nd edition, McGraw Hill Publishing, 1995, pgs. 1.9 - 12.33

7. "From Horses to Horsepower, Volume II, The Second 50 Years, 1946-1996", McVey Marketing Publishing, 1996
8. L. Hoines and B.Potter, "Design Considerations in Automotive Courtesy Lighting", SAE Technical Paper Series 980009, 1998
9. M. Ishikawa and Y. Ohtsuka, "Development of Smoothly Curved Inner Lens", SAE Technical Paper Series 940637, 1994
10. T. Machida and K. Yoshida, "Development of LED automobile Lamp with High Luminous Efficiency", SAE Technical Paper Series 940643, 1994
11. G. Hulse and J. Eskridge, "HID Driven Focus-Less Optics System for Complete Automotive Distributed Lighting Systems", SAE Technical Paper Series 980877, 1998
12. R. Rowand, "Early Vehicle Lighting started with a Bang", Automotive News, Volume 2, June 1996, pg. 122
13. J. Duffy, Modern Automotive Technology, 1994, pgs. 448-712
14. J. Stewart, "HP SNAPLED: LED assemblies for Automotive Signal Lighting", The Hewlett Packard Journal, 1998
15. Valeo VELARC© brochure, "Valeo Electronic Light Arc", 1998
16. Osram Sylvania LUMINOM© brochure, "LUMINOM SPD14 Specifications", 1998
17. Mitchell's Collision Estimating Guide, Domestic, Mitchell International Publishing, 1997, pgs. 256, 257, 382, 383
18. Mitchell's Collision Estimating Guide, Import, Mitchell International Publishing, 1997, pgs. 111, 193, 202, 842, 859, 860
19. Mitchell's Collision Estimating Guide, Domestic, Mitchell International Publishing, 1998, pgs. 40, 104, 1109, 1031
20. Mitchell's Collision Estimating Guide, Import, Mitchell International Publishing, 1998, pgs. 127, 168, 1503
21. K. Rumar, "Vehicle Lighting and the Aging Population", University of Michigan Transportation Research Institute, UMTRI Publishing, 1998

CONTACT

Mario Campos has worked for Ford Motor Company since 1992, starting as a product design engineer in its Lighting Engineering department. He was assigned to Ford of Germany for three years, working on global platforms and their interior and exterior lighting systems. He has two joint patents with Ford Motor Company (one for the U.S. and another for Europe) in Lighting design. He is a 1985 graduate of State University of New York (SUNY) at Buffalo/Amherst with a Bachelor's in Industrial Engineering and has a Master's in Business Administration (1991) from Butler University. He also is a Lieutenant in the U.S. Naval Reserve, with a specialty in fueling operations. Mario can be reached at mcampos1@ford.com.

ADDITIONAL SOURCES

Interview with G.Balint and A.Standaert, Hallmark Tool Inc., Windsor, Ontario, Canada, 1998

980922

Added Feature Automotive Mirrors

Niall R. Lynam
Donnelly Corporation

ABSTRACT

Automotive rearview mirrors have numerous attributes that render them desirable hosts for a variety of added features beyond their principal function of providing a rearview field of vision. One attribute is location. The driver frequently looks at rearview mirrors as part of the normal driving task, and thus they are ideal locations for information display such as of directional information from a compass sensor and/or of temperature information from an exterior temperature sensor. Icons and indicia displaying status of, for example, passenger airbag enable/disable, are readily viewable by the driver when displayed at an interior rearview mirror or exterior sideview mirror. Rearview mirrors are desirable locations for automatic wiper activation rain sensors, automatic headlamp activation controllers, remote keyless entry receivers, garage door opener/home access transmitters, and antennae such as for global positioning satellite (GPS) systems. Mirrors are also excellent locations for lights such as map reading lights incorporated in interior mirror assemblies and ground illumination security lights located in exterior mirrors assemblies.

Another attribute is electrical service. Many interior and most exterior mirrors are electrically powered. For example, electrochromic mirrors that electrically dim in reflectivity when glaring conditions are detected are today commonplace. The circuitry to control the electrochromic dimming function and any other mirror-mounted electronic feature can be commonly housed in or on a rearview mirror assembly, and wholly or partially share components on a common circuit board.

A third attribute is flexibility. By hosting an added feature within a rearview mirror assembly, an automaker has wide latitude in option packaging. As car area networks (CAN) proliferate, the ease and convenience of incorporation of added features will be even further enhanced. Car area networks can provide a plug-and-play opportunity for incorporation of added feature functions whereby mirror mounted features such as, for example, a pyroelectric intrusion detector can output a signal to the CAN when a cabin intrusion is detected, that is received and reacted to by an alarm system elsewhere on the network.

This paper reviews added feature interior and exterior rearview mirrors, and outlines how they can enhance consumer safety, convenience and affordability.

INTRODUCTION

Automotive rearview mirrors have several salient attributes that make them an attractive choice for incorporation of added features. One is location. Looking at a rearview mirror is part and parcel of the driving task. Thus, information displayed at rearview mirrors is plainly and readily visible and interpretable to the driver. Furthermore, outside mirrors, protruding as they are from the main body of the vehicle, are plainly and readily visible to other road users, and thus are good locations for turn signals and brake lights. Lastly, the outside mirror housing is high mounted relative to the road surface, and so is a superior location for ground illumination lighting.

Incorporation of added features is aided whenever the mirror reflector is an electrochromic mirror[1] whose reflectivity is electrically variable depending on the glaring and ambient light conditions sensed by a rearward facing photosensor (detecting glare) and by a forward facing photosensor (detecting ambient light conditions), with both sensors usually incorporated into the interior rearview mirror assembly. Other electronic features such as headlamp activation controllers, remote keyless entry receivers, cabin intrusion detectors, compass sensors, rainsensor detectors, displays, garage door opener transceivers and their like can efficiently and economically access and share the circuitry already present in an automatic electrochromic mirror assembly.

With these advantages, it is not surprising that added feature use in rearview mirrors has grown rapidly over the last several years, a trend likely to continue.

LIGHTED INTERIOR MIRRORS

Light modules (shown in Figure 1) incorporated into the interior rearview mirror assembly provide glare-managed lighting that maximizes interior cabin visibility with minimum distraction to the driver. In excess of 1.5 million vehicles are produced annually with map lights incorporated into the interior mirror assembly. Because they emit below the driver's line of sight, mirror-mounted lights provide excellent illumination of the driver and passenger lap areas without projecting glaring light to the driver's eyes. Also, because the lights are mounted in the bottom of the mirror housing (or, in some European models, are attached to the mirror mount), the rearward facing glare sensor found in automatic electrochromic mirrors is largely unaffected by operation of the mirror-mounted lights. Operation of lights mounted in the header or dome area of the vehicle generally causes a spurious dimming of an automatic electrochromic mirror; an undesirable event avoided in the lighted electrochromic mirror shown in Figure 2.

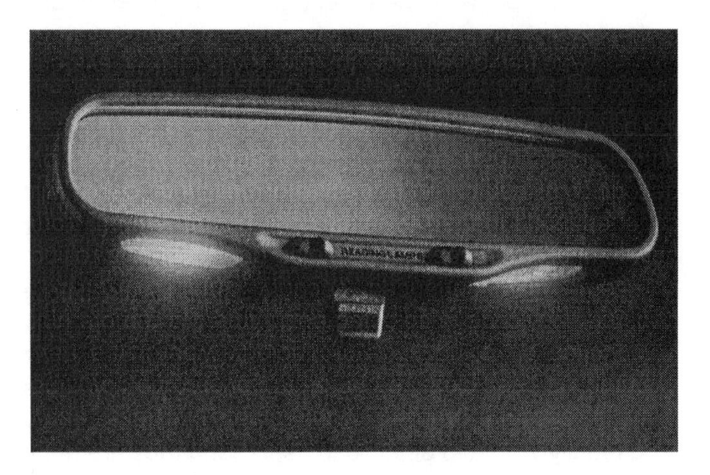

Figure 1. Interior Mirror Incorporating Map Light.

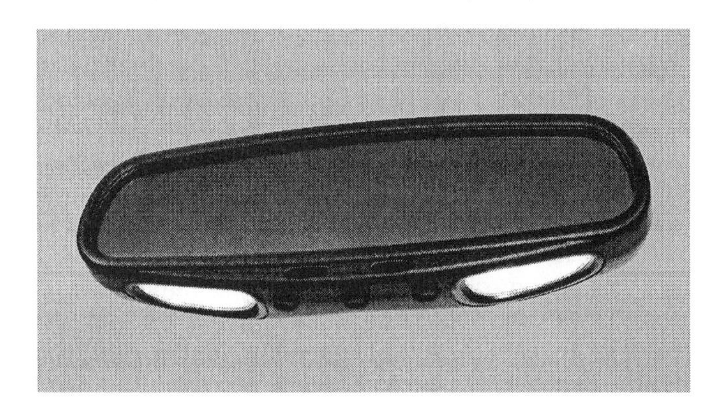

Figure 2. Lighted Electrochromic Mirror.

A recent development, available today on the Chevrolet Corvette, is addition of a high intensity light emitting diode to an interior mirror to provide low level, directed illumination of the center floor console (including the transmission selector) such as is illustrated in Figures 3A and B.

Whenever the ignition is on, the LED light is powered. Being a low current, low level illuminator, its operation is invisible to the driver by day, but at night the LED casts low level lighting onto the center floor console, enabling the driver determine gear selection, find coins for toll booths, etc. Automakers like this feature because it obviates providing dedicated lighting at the floor console.

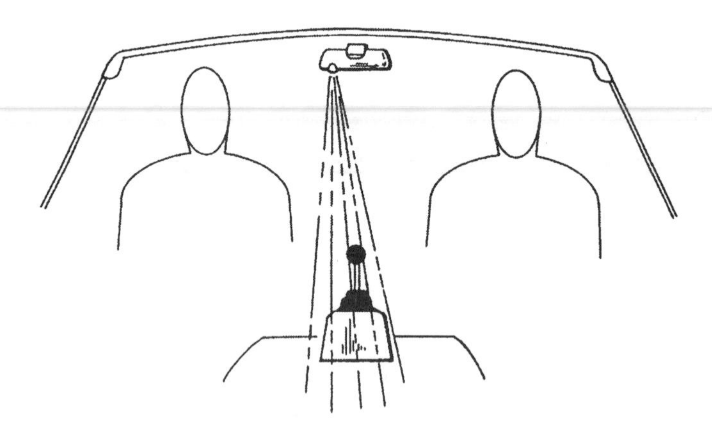

Figure 3A. High Intensity LED in Interior Mirror Housing.

Figure 3B. Interior Mirror Incorporating Low Level, Directed Illumination of a Center Floor Console.

LIGHTED EXTERIOR MIRRORS

The Lincoln Mark VIII vehicle was the first vehicle to provide security lighting at the vehicle entrances by utilizing ground illumination lights located in the outside mirror housings (Figure 4).

Personal security is important to everyone, and is particularly important to females approaching a parked vehicle in a dark and isolated lot. Newspapers have reported instances where malcreants have hidden at or under the vehicle awaiting an opportunity to tackle the driver as he or she enters the vehicle. In some instances, the driver was disabled by having ankle tendons slashed.

Assaults in or around the vicinity of a parked vehicle have increased significantly in recent years. Most frightening has been the rise in carjackings (see Figure 5). Many occur at night. Parking lots are favored areas for carjackers, and handguns are frequently used.

Figure 4. Mirror-housed Security Lights Providing Vehicle-side Ground Illumination.

Automotive Security - Annual Statistics

- 1.5 Million Stolen Vehicles
- $8 Billion Cost
- A Car is Stolen Every 30 Seconds
- 30% of Incidents are at Night Close to Home
- 66% of the Criminals are Under the Age of 21; 88% are Male
- 25,000 Carjackings
- Parking Lots Most Favored by Carjackers, Followed by City Street, Residential Driveways, Car Dealerships and Gas Stations
- Most Carjackings Occur Between 8:00 P.M. and 11:00 P.M.
- When There are Weapons Used, 90% Involve a Handgun

Figure 5. Annual Statistics on Automotive Security.

When approaching at night a vehicle equipped with lighted exterior mirrors, the situation is different. When the driver remotely unlocks the vehicle doors utilizing his/her key fob, floodlights packaged into each of the outside mirror housings illuminate and flood the side and underside of the vehicle with light. The driver can determine whether it is secure to approach, and on entering the vehicle, can avoid puddles and debris at the entrance doorway.

Locating a security light within an outside mirror housing is a challenge, both for packaging and for optical design. The interior cavity of a mirror housing can be crowded. The mirror reflector is typically attached to an electrical actuator, and clearance must be maintained sufficient to allow freedom of adjustment of the reflector field of view to suit each driver's individual need. Increasing the housing size or shape is often not an option for styling and aerodynamic reasons.

As illustrated in Figure 6, the security light utilized on the Mark VIII is a module that is insertable and removable into the mirror housing. The lamp assembly is fully serviceable, is moisture impervious to withstand car washes, rain, road splash and the like, and is fully integrated into the mirror housing so as not to interfere with the external styling of the assembly, and so as not to interfere with aerodynamic performance.

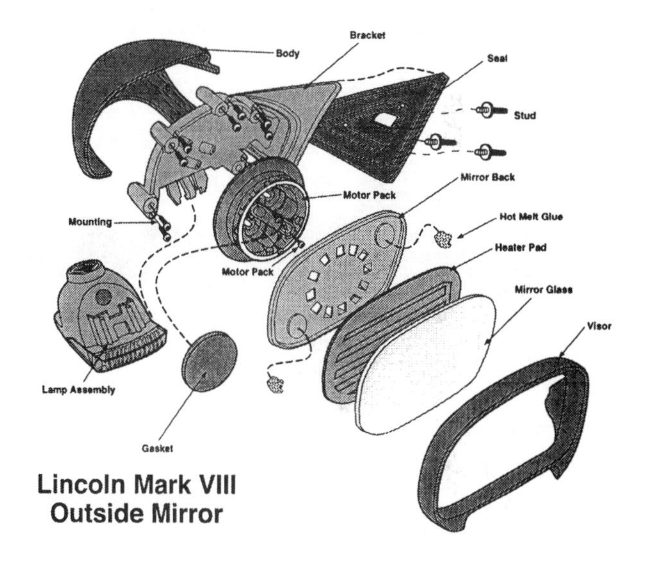

Figure 6. Construction of Lincoln Mark VIIII Lighted Outside Mirror.

When mounted in the vehicle, a lockout is provided to obviate actuation of the security light during normal driving. This is an important feature, as it would be both undesirable and potentially unsafe if the security lights were activated while normally driving down the highway.

Use of security lighting around the perimeter of a vehicle will grow in consumer popularity, and use of security lights in outside mirror housings will likewise correspondingly proliferate.

To test this, a telephone survey was conducted by Market Facts Incorporated, a market research company which has conducted weekly interviews of this nature for over 20 years. Of 1000 people polled in the U.S. market, 53% expressed an interest in security lighting when approaching a parked vehicle. This is a high initial level of interest given that the respondents had no details on cost, configuration or complexity. Minivan owners were most interested. Given their popularity with females, such a result is not surprising.

One further application of exterior lighting is as a parking aid (see Figure 7). Curbside parking can often be challenging and today many luxury vehicles provide a park tilt feature where the outside mirror reflector tilts to face slightly downward when reverse gear is selected. Thus, when reversing curbside into a tight parking space, the driver can see how close to the curb the vehicle is parked by reflection in the sideview mirror.

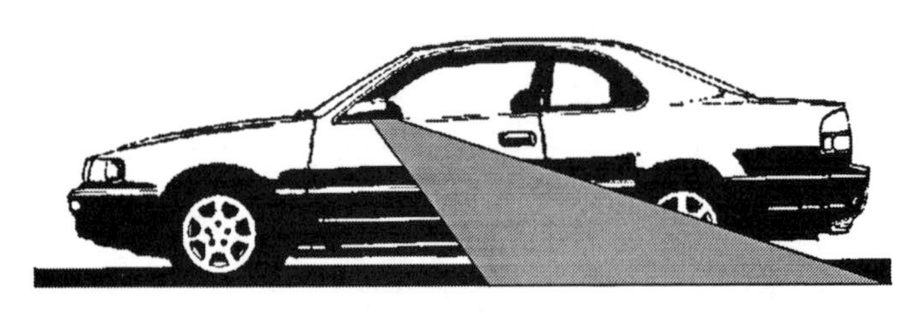

Figures 7. Illuminated Park Tilt.

This works fine by day when there is high ambient light to see by, but at night the feature is difficult, even impossible, to use as the front and rear vehicle lighting casts scant light curbside. However, with a floodlight mounted in the outside mirror housing, parking can be as easy by night as by day.

When reverse gear is selected, the lockout preventing actuation of the exterior mirror lights is overridden, the floodlight in each housing illuminates in tandem with the mirror reflector tilting downward, and the driver at night can now see the reflection of the illuminated curb in the tilted reflector of the sideview mirror.

TURN SIGNALS IN EXTERIOR MIRROR ASSEMBLIES

Drivers are familiar with occasions when they signal to change lanes and then initiate a lane change unaware that there is an adjacent vehicle in the side lane overtaking in a blind spot. Even though the driver has signaled a lane change, the adjacent driver is unaware of this intent to change lanes as neither the front nor the rear turn indicator light is visible to the adjacent driver.

This is a potential safety hazard that can be obviated by mirror-mounted signal indicators. A Signal Mirror™, developed by Muth Corporation of Sheboygan, Wisconsin, is available on Ford Bronco and Expedition vehicles (see Figure 8). An array of high intensity light emitting diodes is placed behind a dichroic mirror reflector. The dichroic reflector is designed such that it has a bandpass of high light transmission to the wavelengths of light output by the LEDs therebehind, but is low transmitting at all other wavelengths. Thus, the Muth mirror acts akin to a one-way mirror with the presence of the array of LEDs behind the mirror reflector being undetected by the driver until actuated when the turn signal is selected.

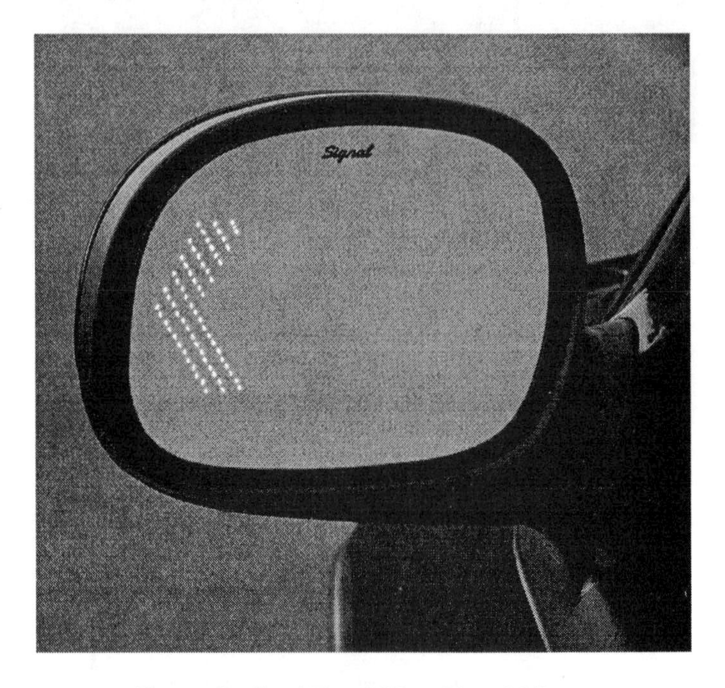

Figure 8. Ford Expedition Signal Mirror.

Figure 9. Turn Signal Mounted in Outside Mirror Housing.

An alternate design for a mirror-mounted turn signal is shown in Figure 9. A linear array of six high intensity LEDs are provided in the lower, rearward facing portion of the mirror housing. The LEDs are orientated at an angle of about 25° to 30° relative to the longitudinal centerline of the vehicle, so that their light output is principally directed to be highly visible to vehicles approaching in adjacent sidelanes. Plastic louvers are used to separate the LEDs one-from-another, and to shield their light output from the line of sight of the host driver. These louvers help protect the driver from distraction or glare when the turn signal indicators are actuated at night to signal a lane change.

This design of mirror-mounted turn signal is equipped on the Lincoln Mark VIII. Being fixedly mounted to the mirror housing rather than mounted behind the movable mirror reflector, turn signal indicator modules of this design can be provided economically, with the automaker retaining full flexibility in the choice of mirror reflector (electrochromic or standard chrome).

INFORMATION MIRRORS

Currently, compass direction is the most widely provided mirror-mounted information display, with in excess of 600,000 vehicles annually so equipped. A variety of display options (see Figures 10A, B, and C) are available, including locating behind the mirror element with the display being viewed through a window created by removing the mirror reflector in the local area; locating within a pod that attaches to the mirror mount at its point of attachment to the mirror mounting button on the windshield; and locating within the bezel below the mirror reflector. Location of the information display within the mirror bezel has several advantages, particularly as shown in Figure 11, where multiple displays of compass direction and of exterior temperature are desired.

Placing the display in the bezel rather than behind the mirror element eliminates removal of mirror reflector surface and consequent local loss of field of view. Thus, safety is enhanced by providing an unobstructed rearward field-of-view. Consumer satisfaction is also enhanced by enabling simultaneous display of compass direction and temperature together, rather than requiring the driver to toggle between one and the other (as is more typical when a display is mounted behind the mirror reflector, where desire to avoid excessive loss of mirror reflector typically leads to display of only one information item at a time).

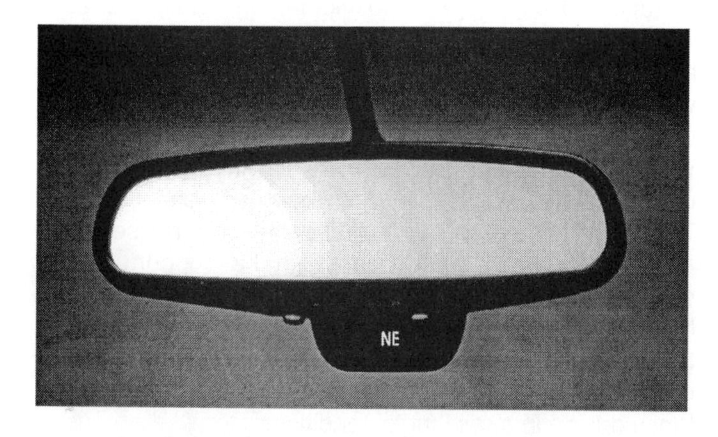

Figure 10A. Pod Mounted Display.

Figure 10B. Information Displayed Through Window Created in Mirror Reflector.

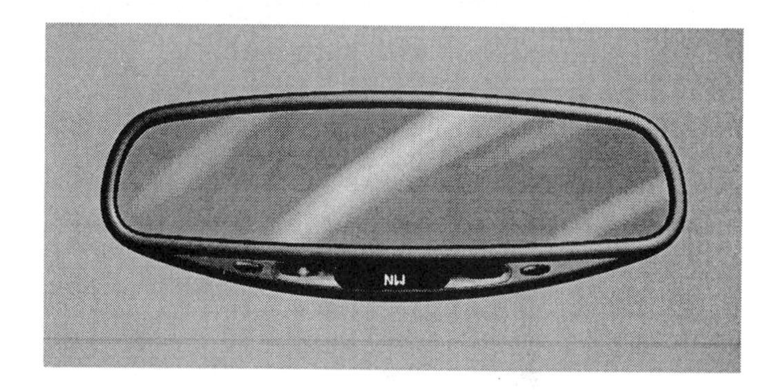

Figure 10C. Information Displayed Below Mirror Reflector.

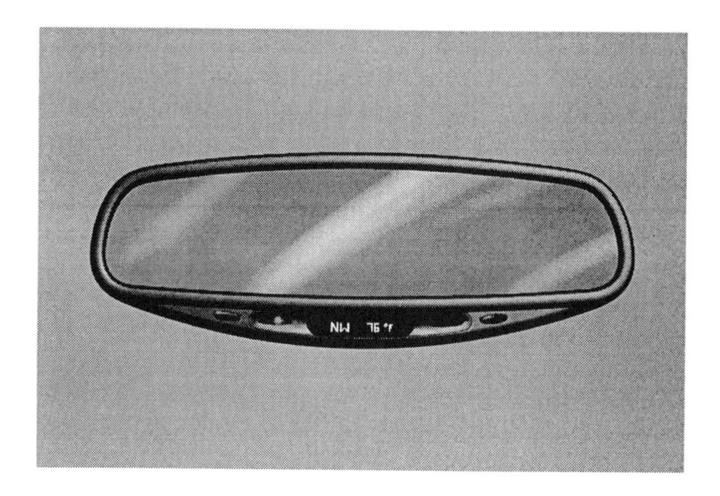

Figure 11. Simultaneous Display of Compass and External Temperature Information.

Location of an information display within a pod (Figure 12) has several advantages of its own. The information display pod can readily attach to a variety of interior mirror assemblies, including prismatic mirrors. Thus, the automakers and consumer retain full freedom of choice in terms of the option to select. Display pods are also advantageous for installations in the aftermarket.

Information that are candidates for mirror-mounted display include compass direction, external and internal temperature, altitude and incline (of particular interest in sports utility vehicles), turn signals, pager display, tire pressure status, trip computer, fuel/oil level, hazard warnings, status indicators (such the icon displaying the status of the passenger-side airbag activation shown in Figure 13) and their like. Location of an information display either in the interior or in the exterior mirror is particularly advantageous when it is critical to catch the driver's attention. Information that may potentially pass unnoticed by the driver when displayed in an already cluttered instrument panel will more certainly be noticed by the driver as he/she constantly and repetitively looks at the rearview mirror during normal driving.

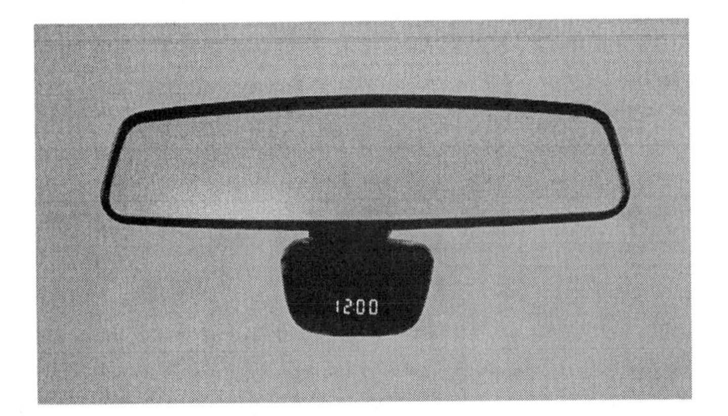

Figure 12. Prismatic Interior Mirror Equipped with Pod-mounted Clock Display.

Figure 13. Simultaneous Display of External Temperature and Status of Passenger-side Airbag Activation.

RAINSENSOR MIRRORS

Rainsensors[3] detect the presence of moisture on the exterior of the windshield, and automatically activate the wipers to remove. Over the last several years, they have grown in popularity, with current usage on vehicles exceeding 500,000 units annually. Rainsensors currently on vehicles are windshield contacting sensors whereby the rainsensor module is either bonded to the inner windshield surface by an optical adhesive or has an optical polymer surface that is pressed intimately against the glass surface mechanically[4] (and is thus removable for service). Rainsensors are typically mounted to the area of the windshield behind the interior mirror housing, thus providing a good location to detect moisture but also one that is unobtrusive to the driver's forward field of view.

A rainsensor located within the support arm of an interior mirror assembly is equipped on the Volkswagen Golf. This is a compact design that renders the presence of the rainsensor largely undetected by the average consumer. As illustrated in Figure 14, the mirror mounting button to which the interior mirror attaches is an annulus with a solid outer ring and with a hole at its center. The rainsensor unit mounts within the cavity of the mirror support arm, and the act of attaching the mirror assembly to the windshield button causes the rainsensor module within the mirror support arm be pressed to the windshield glass surface at the center, hollow portion of the donut-like button mount. The rainsensor thus views the outer surface of the windshield via the hole at the center of the mirror mounting button.

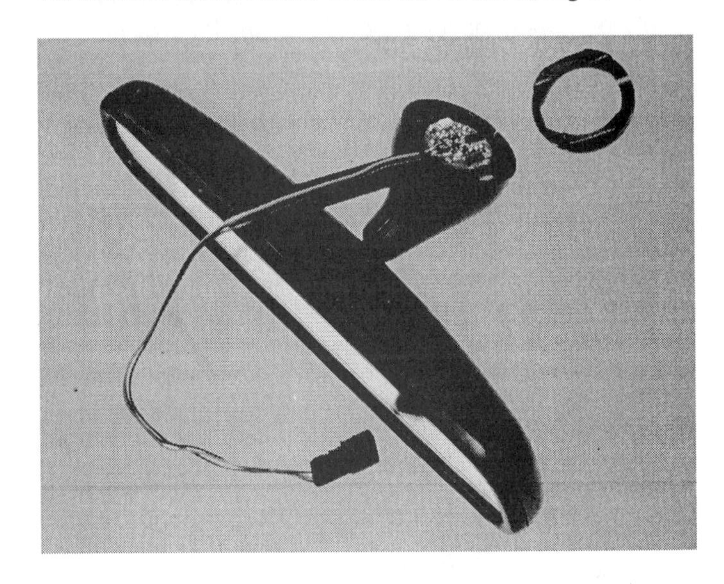

Figure 14. Compact Rainsensor Mounted in Mirror Support Arm.

Non-contacting rainsensors[5] are currently under development where the rainsensor module is mounted within the interior mirror assembly with its detection surface stood off the windshield inner glass surface. There are many advantages to not having direct contact between the rainsensor and the glass surface of the windshield. Windshields are frequently replaced due to damage from road debris such as chips and stones. When a contacting rainsensor is bonded to the glass, replacement during service can be costly and complex. This is not the case with non-contacting rainsensors. Also, the non-contacting rainsensor has opportunity to detect moisture not only on the outer windshield surface, but also on the inner surface as well. Thus, a defroster blower can be automatically activated to remove any condensation or frost build-up, a significant benefit for those driving in humid/frigid conditions.

AUTOMATIC HEADLAMP ACTIVATION MIRRORS

Automatic headlamp activators, often referred to as Twilight Sentinels, are commonplace in vehicles. These automatically turn headlamps on and off at dusk and dawn. Frequently, a skyward facing photosensor is mounted in a module attached to the interior mirror assembly at its point of attachment to the mirror mounting button adhered to the vehicle windshield. An automatic headlamp activation added feature electrochromic mirror of this type that uses a dedicated skyward facing, mirror-button mounted photosensor is available on MY98 Ford Windstar and Ford Explorer vehicles.

An alternate design that utilizes the forward and rearward facing photosensors already on-board in an automatic electrochromic mirror assembly to both control automatic dimming of the electrochromic mirror element and to automatically activate the headlamp at dusk is standard on MY98 Jaguar automobiles (Figure 15). Since the electrochromic automotive mirror circuitry already utilizes two photosensors, this design is economical, aesthetically appealing (attaching a third photosensor facing skyward to the mirror-button mount is plainly visible through the windshield), and is adaptable to both mirror-button mounted interior mirrors and to header mounted interior mirrors (where attachment of a third skyward facing sensor may be problematic).

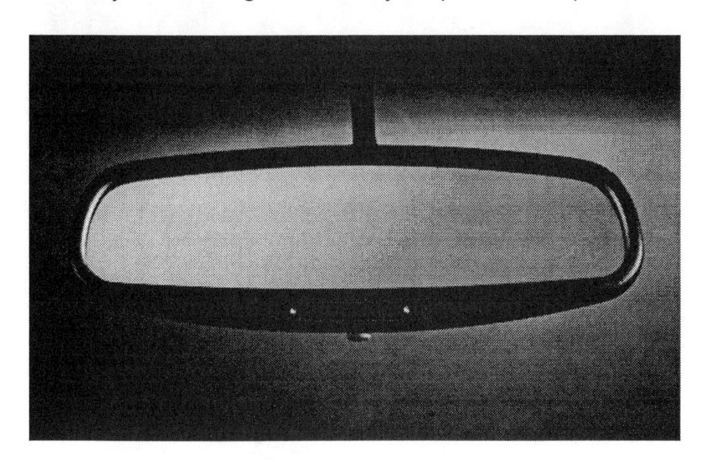

Figure 15. Automatic Headlamp Activation Electrochromic Mirror.

HOME ACCESS MIRRORS

The interior mirror assembly is a convenient location for a myriad of added features that are manually actuated, drivers being long used to reaching to flip the manual actuator toggle on prismatic mirrors. Thus, a mirror located switch is readily accessed by drivers (a significant advantage for the switches on lighted mirrors versus lights placed elsewhere). Several automakers have placed, or are contemplating placing, circuitry within the interior mirror that allows the driver open/close a garage door, security gate, and the like. Two very different systems are currently used or proposed in vehicles. Prince Corporation of Holland, Michigan has developed a universal garage door opener system. Prince's HOMELINK™ system[6] is initially "trained" by the vehicle owner using the garage door opener hand transmitter originally provided by the manufacturer of the garage door opener (GDO) mechanism installed at the owner's home. Once the HOMELINK™ unit is mounted in the vehicle and has learned the owner's garage door opener code, the driver of the vehicle simply actuates the vehicle mounted HOMELINK™ switch to open the garage door from the vehicle when approaching.

TRW Automotive Electronics Group of Cleveland, Ohio has proposed a very different universal home access system. TRW's KWIKLINK™ operates in a totally different way to that of the HOMELINK™ approach. In the KWIKLINK™ system, the wall mounted switch used to manually open/close the garage door from the driver's home is replaced with a KWIKLINK™ wall unit. The KWIKLINK™ wall unit includes circuitry that communicates with a vehicle mounted transmitter that allows the driver, in essence, to open the garage door from the vehicle by transmission to the wall switch, bypassing the remote receiver mechanism originally installed in the garage opener mechanism.

The KWIKLINK™ system has several advantages. Most of the existing installed base of GDO remote control units operate on transmissions that are vulnerable to electronic eavesdropping whereby criminals can electronically listen and record the homeowner's entry to the home when the owner actuates his/her GDO remote control. The criminal can later return when the owner is away, and playback the recorded signal to open the garage and potentially gain entry to the home. Potential eavesdropping has long been a security problem with remote keyless entry systems to vehicles, a problem overcome by the introduction of rolling code technology whereby a new transmission code is selected each time the remote unit is used. For enhanced security, the KWIKLINK™ universal home access system also operates on a rolling code, ensuring that an eavesdropper cannot benefit from decoding a particular transmission as the code will automatically change rendering repeat transmission by the potential burglar of the decoded transmission futile.

Another advantage for the KWIKLINK™ system is that the vehicle mounted system is relatively simple (it fulfills no "learning" function as with the HOMELINK™ system), it is compact, economical and uses low current. The circuitry will operate off a lithium button-type battery and it is planned to add a garage door opener actuation button to the vehicle owner's key fob (used for remote keyless entry to the vehicle) so that a homeowner can open/close the garage door when outside the vehicle.

Figure 16 shows an automatic electrochromic interior mirror that incorporates a universal home access unit as an added feature. The driver can select from three different, bezel-mounted switches to gain access to, for example, a home residence, a summer cottage and a security gate.

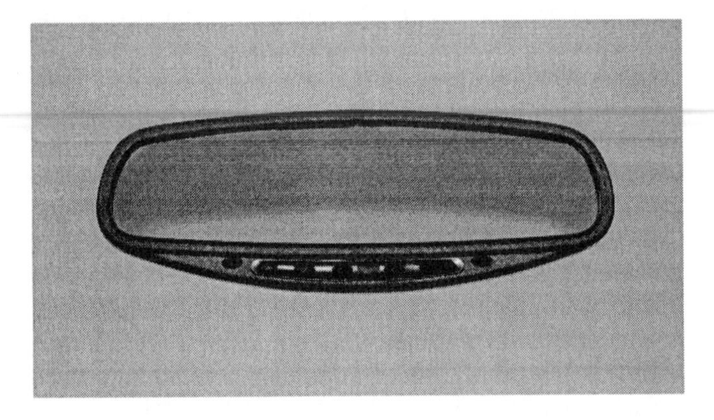

Figure 16. Interior Electrochromic Mirror Incorporating Universal Home Access Unit.

REMOTE KEYLESS ENTRY MIRRORS

Since the early '90s, remote keyless entry receivers (both infrared receivers and radio frequency receivers) have been incorporated into interior rearview mirror assemblies. Being centrally and high mounted within the interior vehicle cabin, and with wide-angle unobstructed reception of transmitted signals, the interior mirror is a desirable location for a remote keyless entry (RKE) receiver. This is particularly so for infrared operating RKE units where placement of the IR detector in a bulb protruding from the lower portion of the interior mirror housing (see Figure 17) greatly enhances range and width of signal reception.

Figure 17. Mirror-mounted IR Receiver for RKE Transmitter.

Recently, vehicles have been equipped with an automatic proximity detector that detects the approach of the vehicle owner and unlocks the doors automatically without the owner operating any button. An antenna in the vehicle transmits to a receiver carried on the driver's key ring. Once a link is established as the driver approaches the vehicle, the system verifies the identity of the driver, and the doors automatically unlock.

The outside mirror housing is a good location for an antenna, given that the housing is non-metallic and that the mirror housing protrudes away from the vehicle body in the direction of approach and at the height where the key ring is either handbag or pocket carried by the approaching driver. When combined with a security light in the mirror housing, the driver can securely approach the vehicle at night and enter without taking out the ignition key.

INTRUSION DETECTOR MIRRORS

Vehicle theft is an ongoing problem. Recent advances in electronic tagging of ignition keys and sensing of vehicle location using satellite tracking technology has made it more difficult and less fruitful to steal a vehicle. However a need continues to prevent theft of articles left in the interior cabin, and to protect against entry by potential carjackers and similar intruders.

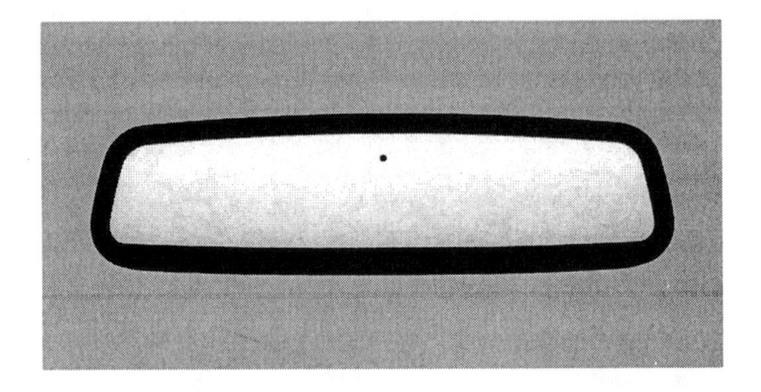

Figure 18A. Camera Vision Intrusion Detector Mirror

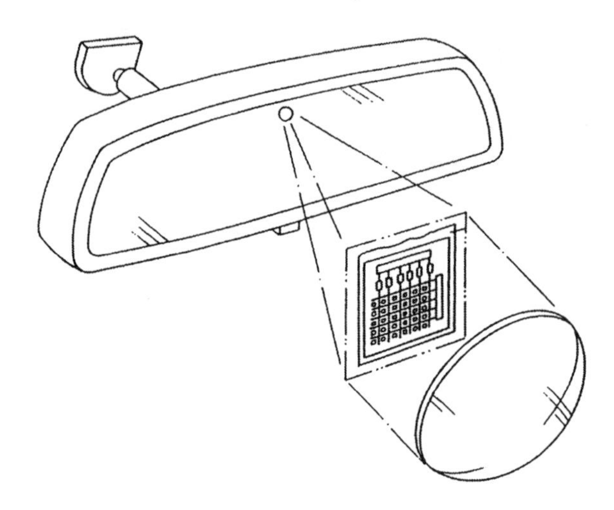

Figure 18B. Video Camera Mounted within Interior Mirror Assembly.

A variety of detectors have been proposed or are in use to protect against cabin intrusion. These include ultrasonic and radar based detectors. Such systems tend to be expensive and potentially subject to false alarms.

Location of a detector within or at an interior mirror assembly is a convenient and effective means to provide protection again cabin intrusion. A system utilizing a video microchip camera device located within the interior mirror housing (Figure 18) has been developed[7]. With the interior cabin illuminated by day with daylight and by night with an infrared floodlight, the camera-captured photoimage is analyzed and any intrusion detected triggers a security response..

An inexpensive mirror-mounted pyroelectric intrusion detection system is shown in Figure 19. This pod mounted system uses a pyrodetector that reacts to any change in cabin temperature such as would be caused by an intruder entering or partially entering the vehicle. This unit has high reliability and operates on a current of less than 0.25 milliamps, thus providing continuous intrusion protection even when the vehicle is parked for weeks without itself draining the vehicle battery.

Figure 19. Pyroelectric Intrusion Detector.

GPS ANTENNAE IN MIRRORS

Navigational and vehicle security aids that track vehicle location utilizing geographic positioning satellite (GPS) systems are growing in use. To establish geographic location, the vehicle receives signals from multiple orbiting satellites via a vehicle mounted GPS antenna. Preferably, such an antenna is located with an unobstructed skyward line of sight, and for optimum reception, the antenna should be housed away from body sheet metal.

Vehicle mirrors, and particularly exterior mirrors, are desirable locations for GPS antennae (Figure 20). Mirror housing assemblies offer opportunity for unobstructed skyward view with good reception. This is particularly so for exterior mirrors which protrude clear of the vehicle and which offer an opportunity to house the GPS antenna unobtrusively and with minimum wiring length to connect with the navigational display unit mounted in the vehicle dashboard. A GPS antenna mounted at an interior mirror is equipped on Renault vehicles beginning production 1998.

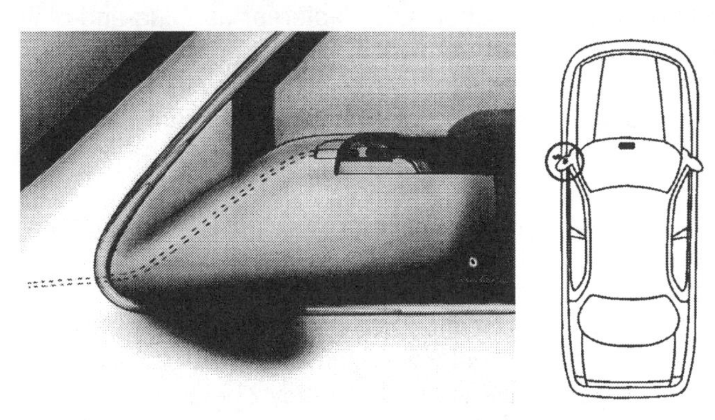

Figure 20. GPS Antenna Mounted in Outside Mirror Housing.

CAR-AREA-NETWORKS

The concept of local area networks is well established for interconnecting computers and their accessories. The vehicular equivalent is a car-area-network (CAN) where nodes controlling the features and accessories in various parts of the vehicle are interconnected on a network. Various CAN protocols are in use or are contemplated such as Motorola's msCAN protocol and Siemen's FULL-CAN controller.

A CAN node located within the interior mirror assembly will be standard on a European luxury vehicle in 1998. The mirror mounted node will control electrical function in the upper half of the vehicle such as control of the sunroof, interior lighting and the CHMSL rear brake light. The interior mirror assembly (which will be provided in both a prismatic version and an electrochromic version) includes the automatic glare detection circuitry (if an electrochromic unit), a remote keyless entry receiver, a security system that uses an ultrasonic intrusion detector, an information display on seat belt use, as well as serving as the electronic controller for sunroof operation and interior lighting in the upper portion of the vehicle. When launched in Summer 1998, it will be the most technologically advanced interior rearview mirror in production.

There are many advantages to the CAN concept for the automaker, for the accessory supplier, and for the consumer. Addition of features such as a rainsensor or intrusion detector to the interior mirror conventionally requires agreement and acceptance by diverse, often distinct, suppliers. For example, introduction of a mirror-mounted rainsensor requires that the mirror supplier, the windshield wiper controller supplier, and diverse groups at the automaker coalesce and coordinate together, which sometimes can be challenging. With a CAN network, implementation of new features is greatly facilitated. For example, when a mirror-mounted intrusion detector senses a breach of cabin security, its local node outputs a signal to the network. This signal, in turn, is captured by the node to which the car alarm system is connected, and is acted on accordingly. This allows the automaker add features as plug-and-play accessories, and provides wide flexibility to the automaker in terms of choice of sub-system supplier. The supplier, in turn, can focus on its area of best expertise.

For the consumer, the plug-and-play opportunities of a CAN environment offers not only broader access to diverse features, but also more convenience in location of the switches required to operate various functions. Conventionally, the need to hard wire switches to the units they operate has meant that they be located close together. With a CAN, the switch need only connect to its local node, and it can control remote units via signals passed along the network and processed at local nodes.

Figure 21. Interior Mirror/Overhead Console System Incorporating CAN Node.

There is no direct wire connection between the map light switches adjacent the map lights in the overhead console shown in Figure 21. When an occupant depresses the map light switch, a signal is sent from the overhead console down to the CAN node in the mirror. Here it is interpreted, and a command to turn on a map light is sent back to the overhead console. However, since the node in the mirror is microprocessor controlled, the turn off of the map light can be controlled to a fade-out over several seconds, an illustrative example of how use of a CAN economically enhances added features in the vehicle.

CONCLUSION

An ever increasing proportion of interior mirrors are electronic (sales of automatic interior electrochromic mirrors in 1997 exceeded three million units). Given their location in the vehicle, and given that addition of further electronic features is more economical where circuitry is already on-board, use of added feature interior mirrors will continue to rise. Furthermore, many outside mirrors are already electrically serviced for power actuation and defrosting. Added outside mirror features such as ground illuminator security lighting will grow in popularity. Once a consumer is equipped with such a feature, it is likely that it will be looked for in future vehicles that driver buys.

Use of added feature mirrors will continue to grow, with an ever increasing variety for the automaker and consumer to choose from.

The interior mirror assembly can host seat occupancy detectors; video cameras for internal cabin surveillance and/or video telephone function; messaging displays that relay paging, traffic status or hazard warning information to the driver; links to the World Wide Internet Web via a modem/cellular phone/alphanumerical display; a digital recorder for recording and/or playing back messages including e-mail messages; loudspeakers such as for a vehicle audio system or for a cellular phone.

The exterior mirror assembly can house blind-spot detectors that detect the presence of approaching vehicles in adjacent sidelanes; and a transducer that receives and/or transmits information to a component of an intelligent highway system or an automatic toll booth system.

REFERENCES

1. "Electrochromic Automotive Day/Night Mirrors" Niall R. Lynam, SAE Technical Paper #870636 (1987).
2. United States Patent 5,361,190.
3. United States Patent 4,859,867; 4,916,374; 4,973,844.
4. United States Patent 4,871,917.
5. International Patent Publication NO: WO 94/27262.
6. United States Patent 5,442,340.
7. European Patent EP 0 683 738 B1.

2000-01-0818

Human-Machine-Interfaces of Car Computing Devices

Peter Roessger
Computer Aided Animation GmbH, Germany

ABSTRACT

Existing driver information systems (DIS) include navigation, radio, telephone, television etc. The addition of computer functions like internet or word-processing turns DIS into car computing devices (CCD), often called CarPCs. Due to the high mental and visual load of driving, solutions matching the human factors needs of the driving environment have to be developed.

A usability study testing four different existing DIS was conducted. It is possible to transfer the results to future CCD. The visual channel should not be stressed by CCD. Applications running in car have to use acoustic and haptic channels for information transfer between driver and system.

INTRODUCTION

DIS are systems based on digital information processing, that provide drivers with information during the journey. This includes travel computers, navigation systems, and multifunctional integrated system like the Siemens IDIS or the Mercedes Comand. Car audio systems are no DIS by means of this definition.

CCD (or CarPCs) are DIS with additional features like office, e-mail, or internet. This extension of functionality impacts the need for new user interface concepts. Analytical and experimental approaches are necessary to get the knowledge for improving the human machine interface (HMI) of DIS.

COMPUTERWOLRD VERSUS CARWORLD

In DIS and CCD meet computer-world and car-world. Both have restrictions and singularities. Chances and possibilities, including well known standards, of the computer-world face the restrictions of the vehicle cockpit. Human factors experts state [1], that if you add a computer to a car, you get a computer. This means that cars are handy devices, but if you add some digital information processing, it changes to hard to handle.

In the computer world, work with the computer is usually the main task, at least at the moment when it is done. That means, the working person can use all mental and visual capacities to operate the system. Time restrictions are kept to a minimum. Hard- and software, and with it the user interface, is more or less standardized. Easy available rules [2], style guides [3, 4], and official standards on the ergonomics of computer work [5] exist as well as theoretical background and long term experience [6]. There is (or at least should be) enough space on desktops to place a proper sized screen and a standard keyboard plus other input devices such as a mouse or a trackball.

In a vehicle the main task still and definitely is and will be driving, even if the driver in command changes to a driver in control. That means that only limited mental and visual capacities are available for using a CCD. The high visual load while driving [7] and the need to control the car manually leave only a little spare capacity for communicating with a CCD. Most traffic situations do not allow attention anywhere else than on the traffic. There are only small windows in time to use a CCD, leading to strict time restrictions. Input devices, screen layouts, and use-philosophies differ clearly between the manufacturers. The technological development of the equipment is so fast, that on the ergonomics side hardly any experience has been collected. In addition, inside a car, there is only limited space for input and output devices. That leads to small output devices, monochromatic LED-displays or screens sized up to 7 inches. Input devices have to be reduced in size and number as well.

CCD are physically placed in the car world, but they have the functionality of a member of the computer world. The more functions are integrated in the DIS, the more complex they become, the more office- or internet-applications run on them, the more they move to the computer side.

Solutions for passenger-information-systems, which are DIS and CCD only accessible by passengers, are already presented [8] Figure 1. From the human factors point of view they present the less difficult and less interesting side, being just computer world devices placed in a car. The main focus should be set to CCD with office and internet applications for drivers. These systems are in development at the moment. The modification of computer-style-guides for car-installed-systems and the fitting of input- and output devices to the special circumstances of driving could be the major point of automotive human factors research.

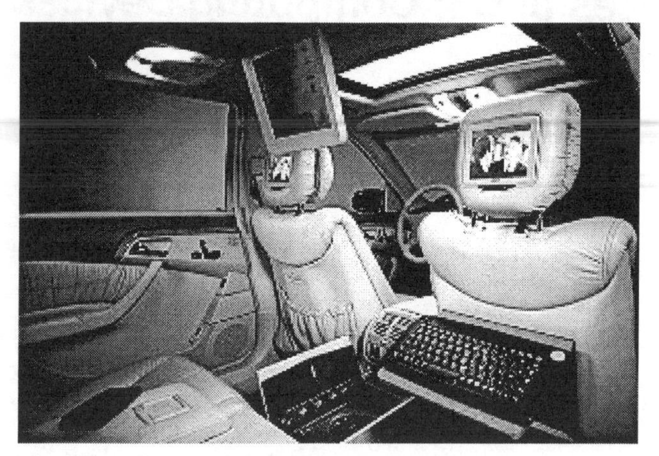

Figure 1. Brabus Business Concept [8]

TOOLS TO ANALYSE HUMAN FACTORS OF DIS

The development of DIS requires a constant generation and flow of human factors knowledge. To achieve this, easy to handle and fast to use tools should be available.

One way to analyze a DIS is to have a look at the system itself, how the information input is organized, how the output. How many input devices are used, what kind of screen? A tool supporting this could be a simple protocol to write the information plus an algorithm to extract the knowledge by giving values. A second way is to study the user, his behavior and feelings about the system. The analysis of mistakes is a good way, as well as interviews with owners of a particular system or with naive users after being confronted with a DIS. The first approach is an analytical one, the second more or less empirical.

ANALYTICAL TOOLS – On the analytical side two tools exist. An established tool is the Analysis-Protocol for DIS. This paper contents 21 pages. The first part is the analysis of the DIS in general, the subsystems and functions included, the characteristics of the user interface, or the general philosophy of the user interface. Here general philosophy means: is it a one twist-push-button system, touch-screen, speech input or something else. Number, size, tactile feedback and position of the input devices are noted. Depth and width of the menu-tree is noted, as well as dialogue principles, kind of speech used and so on. On the output side size, position, colors, fonts and so on are in focus. If speech output is provided, questions are on understandability, selection of words, and correct presentation time are answered.

The second part gets into the single applications running in the DIS, e.g. telephone, radio or navigation. Information input and the output are analyzed. The number input-steps is counted, e.g. for the input of a navigation target or a phone number. In addition on each single page is space to collect the personal opinion of the analyzer. This, at the end, leads to a first judgement by an expert and an idea of the strong and the weak points of the system and its interface. Advantages of this tool are the fast results, the direct generation of an opinion by an expert for an expert (or, as usually applied, a group of experts) and the possibility to store information for a later analysis. Disadvantages are the high dependency of the results on the expert making the analysis and the exclusion of the user's view.

The second analytical tool available is a checklist with 250 items, being tested at the moment. The items were generated from literature. Computer style guides were used [e.g. 2], just like DIS-style guides [9, 10] or scientific publications [11, 12]. A second resource for items was the Analysis-Protocol for DIS. After the analysis of some DIS with clearly different user interfaces (Lexus Touch-screen System, Blaupunkt RNS II, Mercedes Comand, Nissan Birdview) items allowing statements on the pros and cons of each system were extracted. The third item-resource was a row of discussions among human factors engineers, developers and users made at CAA. This checklist has the advantages of the Analysis-Protocol for DIS, plus a higher reliability and more independence from personal views of experts. Disadvantage is the low flexibility of checklists compared to free protocols. And, again, the user's view is not included in the judgement.

EMPIRICAL TOOLS – Empirical research is necessary to get the user's point of view into the assessment of DIS. To collect data on an empirical basis, a guide line for semi-structured interviews was developed. It contains 5 questions, beginning with: How did you like the system tested? Why do you like/dislike it? Can you imagine to purchase this system? and ending with Tell me the three things you like most on the system! and Tell me the three things you like less on the system! The latest step of tool evolvement is a questionnaire (for details see below).

USABILITY STUDY ON EXISTING DIS

A usability study of existing DIS was conducted to get information about users views on human factors of DIS. It had two major goals. One was to develop and test a questionnaire, the other one was to get information on the usability of the systems, to generate a knowledge base for the development of future systems.

QUESTIONNAIRE ON THE HUMAN FACTORS OF DIS – A questionnaire development usually begins with the collection and the generation of an item-space (Figure 2). An item-space is build by identifying the relevant facts on the research object. Each component of the item-space will be transformed into one or more items. They will be connected with a rating scale to build the 0.9 version of the questionnaire. To get to a 1.0 version additional tests on reliability and validity of the tool are made in developmental steps following.

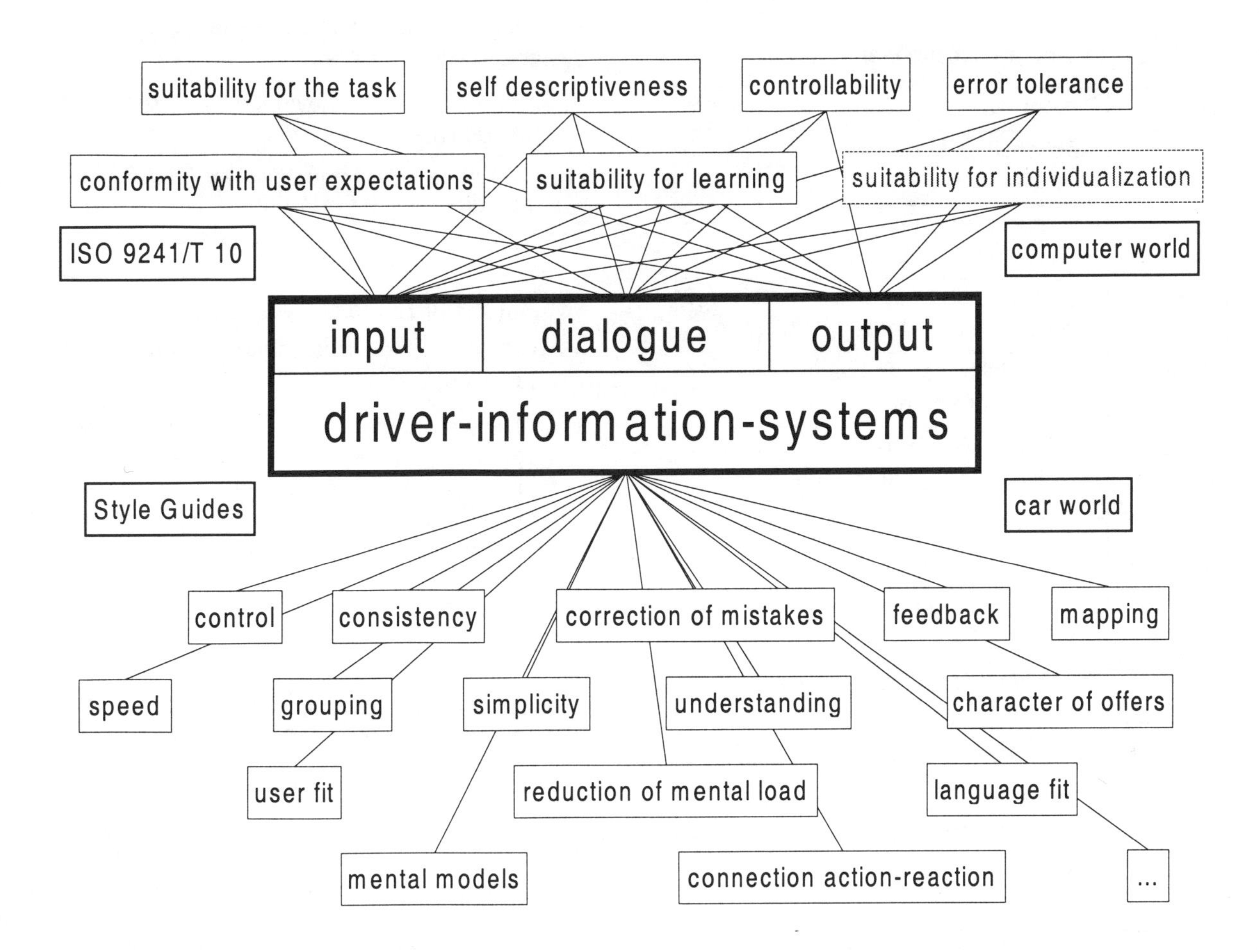

Figure 2. Item-space for the questionnaire

Here the first set of items represented the influence of the computer-world. Basis was ISO 9241-10 [5]. The mentioned factors of the dialogue (suitability for the task, self-descriptiveness, controllability, conformity with user expectations, error tolerance, suitability for individualization, suitability for learning) were adapted for the assessment of DIS human factors. Suitability for learning was renamed and reinterpreted to learnability. DIS should be easy to learn. The idea that the process of learning is desirable, which is the idea behind suitability for learning, is not appropriate for DIS. Suitability for individualization was not included because today's DIS are not personalizable. If individual configurations are possible one day, items on that should be included in the questionnaire. For each factor of ISO 9241-10 one item was generated for the input, the dialogue, and the output.

Items representing the car-world were generated from literature. Main sources were [9], [10], [13], and [14]. Major points were extracted to the item-space. For each element of the item-space one or two items were generated.

A group of four human factors experts made the precise wording of the items. They were presented to 12 novices. They were non-technical colleagues (secretaries, members of the marketing and so on) at the company. A discussion was made on the understandability and the fit between item and asked content. Final wording and selection (to reduce content doubles) were made by the experts. At the end of this process 47 items were left. A five point rating scale was added. The sequence of items was randomized.

To provide a fast and easy interpretation of the data at this early stage items were grouped on a content but not on a statistical bases. These groups were similar to the components of the item-space. The were called parameters (Table 1). Some items belong to one parameter, some to more than one.

Table 1. Parameters of the questionnaire

parameter	Description	items
anthropometric human factors	reachability, visibility, regulation powers	5
cognitive human factors	design of dialogue, understanding of the system	8
sensoric human factors	visual, acoustic and tactile outputs of the system	5
controllability	controllability, having the system in one's hands	4
suitability for the task	relation between costs and benefits of using the system	3
learnability	aspects of learning the system	4
error tolerance	handling of mistakes	3
self descriptiveness	understandability of the system by itself	6
language fit	fit of the system language with the user	3
consistency	similarity of the system with itself, homogeneity	4
feedback	time fit, understandability and task fit of the feedback	3
speed	speed of the system	3
colors	like and dislike of colors of input and output	2
acceptance	like and dislike of the system, prices	9

METHOD OF THE STUDY – The method used in the study was the frontal confrontation. Naive users had to solve five tasks with the systems unknown to them. Help was not provided at all. This method allows to analyze mistakes. It shows problems of the systems best. And it is not unrealistic. Often rental cars have a DIS, someone with out any experience enters one and should be able to use at least the major functions with out any help.

Data was collected with objective and subjective methods. A questionnaire was presented after each system, the interview was made after solving the tasks with the last system. The actions of the users were recorded on video. The number of subjetcs´ failures was counted for each system and every task.

33 subjects took part in the study, 10 of them female. 4 already had experience with a DIS, but not with one of the tested. They were recruited with advertisements in teleTraffic (a magazine dealing with telematics, navigation, and DIS, usually read by user, developers, and vendor of these systems) and in the Stuttgarter Zeitung (a major, high-level newspaper read in the region).

Each subject had to solve the same five tasks with every system. The tasks were:

1. Activate the navigation target "München, Königsplatz"
2. Let the system show you the map and change the map direction to north-up
3. Turn on the radio station SWR3 and store it under no. 1
4. Play the 3rd song of the 2nd CD
5. Activate the navigation target "Hannover, Towncentre"

Four DIS were tested. The follow up of the systems was changed from subject to subject. The tasks were always presented in the same order to provide maximum comparability between the systems (by paying the price of no comparability between the tasks).

THE DRIVER INFORMATION SYSTEMS IN THE STUDY

The Alfa Romeo 166 with Siemens IDIS – The Siemens IDIS in the version provided by Alfa Romeo (IDIS) (Figure 3) includes audio components, telephone, navigation, a trip computer and some parts of the climate equipment. Subsystems are selected with hardkeys positioned under the screen. Navigation in menus and letter inputs for the navigation target is realized with a central twist and push-button. A screen-layout, providing major information from the whole system, can be reached from any point by pressing MAIN. The screen is black, fonts and symbols are green. Speech output is given for navigation.

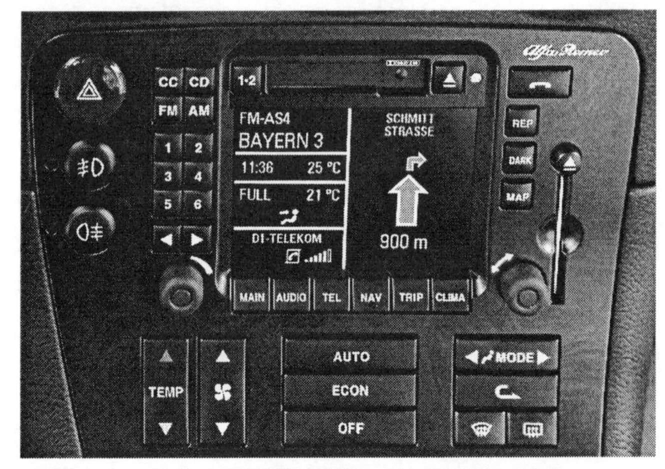

Figure 3. Siemens IDIS in the Alfa Romeo 166 (Siemens AG)

The BMW Navigation System – The BMW Navigation System (BMW) (Figure 4) includes audio components, television, telephone, and navigation. Input controls are a central twist and push button plus some hard-keys. The output devices are a color screen and speech for navigation information. The activation of the subsystems is sometimes realized with hard-keys, sometimes with soft-keys. Navigation in menus is possible with a central twist and push button. The selection of letters for navigation inputs and the activation of different options is realized with it as well. The MENU button brings the user back to the main menu from any point. The integration of the audio components is weak. It is possible to get audio information (selected radio-station or CD) on the screen, but the inputs are made with separate hard-keys. Background color of the screen is blue, fonts are white or black, depending on the status of information given.

Figure 4. DIS of a BMW 5xx (BMW AG)

The Lexus Voice Navigation System – The Lexus Voice Navigation System (VNS) was installed in a Lexus IS200 (Figure 5). This system has major differences compared with the others. The system itself only includes navigation, the audio components are realized in a more or less traditional manner. A telephone is not included. The input device is a remote control placed between the seats. It includes a joy-stick with a twist element plus four push buttons, including a o.k.-button. Inputs for the navigation targets are made with the joy stick. The output is a color screen placed high on the dashboard.

Figure 5. The Lexus Voice Navigation System [15]

The Mercedes Comand System – The Mercedes Comand System (Comand) (Figure 6) includes audio-components, television, telephone, and navigation. The subsystems are activated by hard-keys. They are placed above the screen. Navigation in menus and letter input is possible with a twist and push button. Some functions are realized as soft-keys, some are on the 2 way-whip around the central twist and push button. The output is realized by a medium sized screen. Colors (yellow and blue) are used consistently to divide between different areas of the screen. Night-design is provided by using complementary colors.

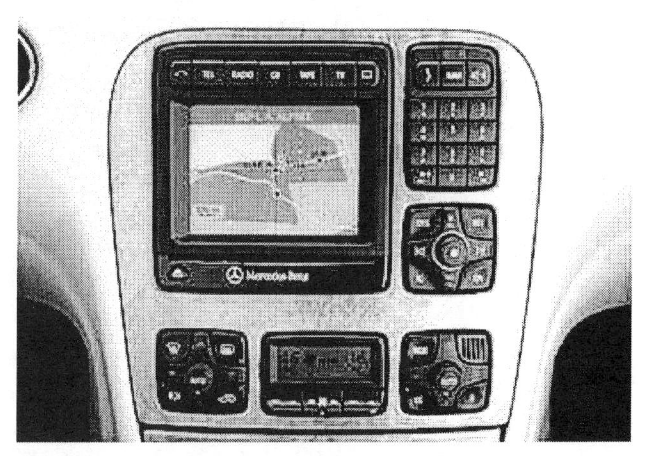

Figure 6. The Mercedes Comand (DaimlerChrysler AG)

RESULTS OF THE STUDY

Subjective Results – To get information on the questionnaire, the relationship of the items with each other, correlations were calculated between them. Highest correlations were found between all items on acceptance. the same goes for all items on learnability, conformity with user expectations and handling of mistakes. A cluster analysis shows more or less the same results.

The correlations between acceptance and the other parameters were calculated to find out about the major influences on acceptance (Figure 7). Consistency has the highest correlation, cognitive human factors (hf) the second. The lowest are mistakes and speed.

The questionnaire is still under construction, so here are only some major results presented. In the overall rating (mean of all items) and for acceptance the Mercedes Comand gained the best results, followed by the Lexus VNS, the Alfa Romeo IDIS and the BMW Navi. The results fail statistical significance. For anthropometric hf Mercedes Comand and Lexus VNS were better than BMW Navi and Alfa Romeo IDIS.

In the interview the inputs of the Alfa Romeo IDIS got good results, as well as the internal organization of the system. The position in the car (too deep, well hidden behind the gear stick, hard to reach) and the German-English language mix were mentioned as negatives. The inputs of the BMW Navi were good, the knobs were a bit too soft, providing only poor tactile feedback. The internal organization, the structure of the menu and the low connection between audio and the rest of the system were mentioned negative. The screen was medium readable, the fonts a bit too small. The inputs of the

Lexus VNS, a remote control placed between the front seats with a joy-stick, were discussed controversially, the internal organization did not match the users expectations. The position of the screen near the line of sight was appreciated by most of the subjects, as well as the bright colors. The Mercedes Comand was good to understand and the less difficult to use in the study. Font size was big enough, the screen colors well accepted by the subjects.

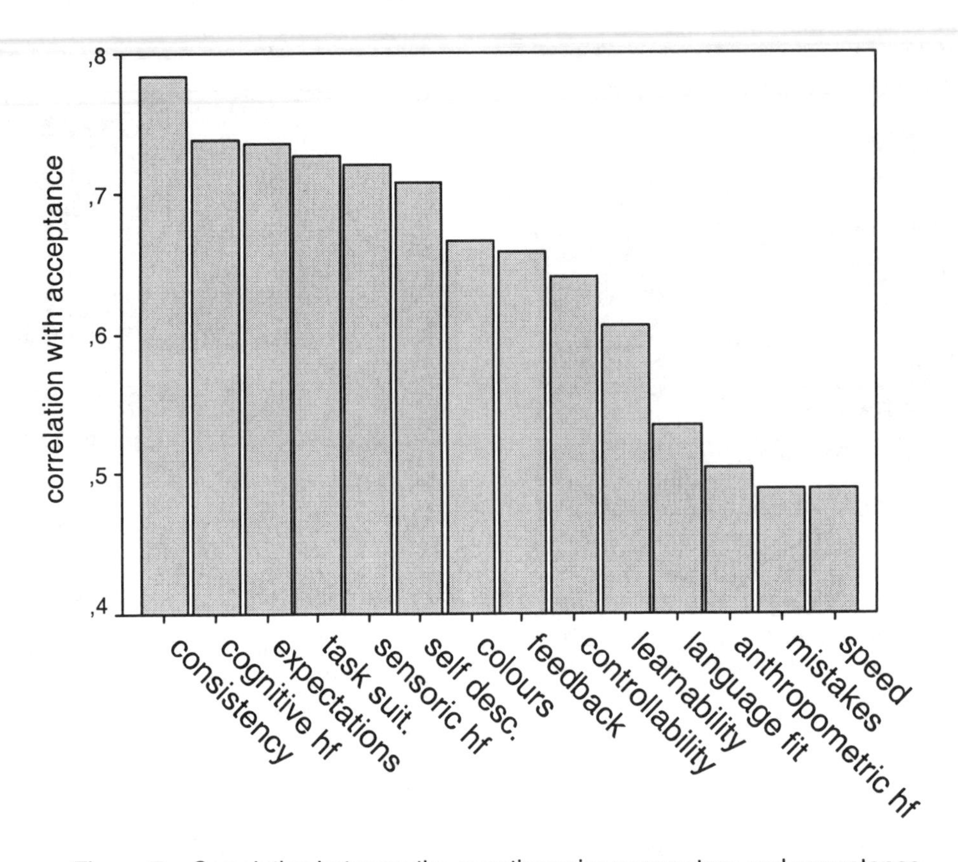

Figure 7. Correlation between the questionnaire parameters and acceptance

Objective Results – The videos have not been analyzed so far. Tools and results will be published soon. More than 22 % of the tasks were not solved by the subjects. The system with the most failures is the Lexus VNS. With the Comand the subjects were most successful (Table 2).

Table 2. Relative failures of the subjects

task	IDIS	BMW	VNS	Comand	∅
1	18.2 %	15.1 %	30.3 %	33.3 %	24.2 %
2	33.3 %	6.1 %	24.2 %	33.3 %	24.2 %
3	33.3 %	27.3 %	30.3 %	18.2 %	27.2 %
4	15.1 %	27.3 %	3.03 %	0 %	11.4 %
5	15.1 %	45.5 %	42.4 %	0 %	25.8 %
∅	23.0 %	24.2 %	26.0 %	17.0 %	22.6 %

DISCUSSION OF THE RESULTS – The questionnaire has to be developed. Some more items have to be added, especially in the area of screen design and language outputs. Some highly correlating items, e.g. for learnability, can be cancelled. In replication experiments the tool has to show its validity and reliability.

Most important for the acceptance of DIS is consistency. This matches the demands of most experts [e.g. 2]. Subjects are irritated if functions change from one subsystem to another. In some submenus of the Lexus VNS an action is activated by choosing the point in the menu, in some the extra O.K. button has to be pressed. In the BMW Navi navigation, television, telephone etc. are activated via the screen and the main button, just the audio components are used with separated input devices, although partly displayed on the screen. The consistency of the main button leads to positive judgements on the navigation subsystem of the BMW Navi. All input actions are made with the same device in the same way.

Cognitive human factors are very important for the acceptance of DIS. This parameter represents the internal structure of the device. At first sight most of the systems are designed well under a cognitive hf point of view. Closer looks show hard to understand and even harder to find submenus. Some DIS force decisions at a point of the input process, where they were not expected.

During the navigation input process of the Lexus VNS a decision has to be made, weather to go to the town center, a junction, an address, or a point of interest. This decision has to be made before entering the name of a town, which is counter rotating to the usual way of thinking.

The Mercedes Comand and the Alfa IDIS have hardkeys for the subsystems (navigation, CD, telephone etc.). This showed the single functions of the systems at first sight and made the access to other subsystem easy and fast.

Conformity with user expectations deals with the fit between user mental models and the system-model. The ratings for the tested DIS were low. The existing systems do not pick the users up where they are. Naive users expect a stronger guidance and more help by the DIS. Users do not think in technical structures, but in goals to reach. HMI of DIS represent the technical side of the system, not the goals of the users.

Suitability for the task deals in main with the relationship between costs and benefits. Users are sensitive to inappropriate input operations. The subsystem for word inputs (e.g. towns for navigation) of the Mercedes Comand irritated some subjects, because it was not possible to move the cursor the way they wanted to. This resulted in bad ratings for the system. All in all most subjects said, that all the systems are too hard to use, so the costs for getting information are higher than the benefits.

Self-descriptiveness was missed by most of the subjects. Help-buttons, on line help and a clear naming of submenus and functions were the demands. Colors were missed by most users as far as the input devices were concerned. The colors of the Mercedes Comand were not attractive but useful, because they were used consistently to distinguish between screen areas. Feedback was divided into two different parts, the direct feedback as an information about the running dialogue and feedback about the state of the system, e.g. route calculation vs. route guidance. The first kind of feedback was acceptable for most of the users, the second kind not so much. The remaining calculation time was not shown by the Mercedes Comand, the Alfa IDIS and the BMW Navi. The display of remaining time in the Lexus VNS was wrong. Controllability was on an medium level, the audio screens of the BMW Navi turned to standard screens after a time out. This leaves novice users with a feeling of reduced controllability. The good news is: all subjects believe, they are able to learn the systems, if not with a trial and error strategy, then with the manual.

The language fit of the Alfa Romeo IDIS was rated quite poor, the mix of German and English expressions was rejected by most of the subjects. The other side of that represents the Mercedes Comand, no English expressions were used at all. This leads sometime to some bulky menu names, but was rated positively by the subjects. Anthropometrics were extremely poor for the Alfa Romeo IDIS, the position of the system was too low and behind the gear stick. The position of the Lexus VNS screen was rated very good. It was near the line of sight, very high in the car. The input device, a remote control with a joystick, was very much under discussion. Some found it to be the best in principle, some rated it worst in the study. All had in common, that the look & feel of the remote control was cheap. The joystick was to soft and inaccurate. Mistakes could usually be handled well, the speed of the system was rated as far too slow.

IMPLICATIONS FOR CAR COMPUTING DEVICES

Empirical data on human factors and usability is available for DIS. CCD add a new dimension to the in car environment. With this new human factors problems raise. Applications and programs fitting well with the office environment are brought to vehicles. The technology push in this area is hard to stop, so the ergonomic community has to escort the development. At this early stage connected with a fast evolvement of new applications and an enormous raise of technological possibilities the knowledge about DIS has to be projected to CCD. All human factors parameters (see above) should be realized well in CCD.

There will be a discussion about the reduction of functions of CCD while driving, maybe speed dependent. This may lead to an increase in safety, but to a decrease in acceptance by the users. Imagine a business man, having spent 10.000US$ for a CCD, and not being able to make a telephone call while driving! If CCD fail human factors standards, legal activity will cut the developmental work and reduce the integration of new technologies in vehicles. Only well designed and easy to use systems, matching the human abilities, will have a chance to be accepted by users and by non users.

COMMUNICATION CHANNELS – Major point of all thinking about designing a driver suitable interface for CCD is the correct choice of communication channels. Due to the high visual demands which driving tasks have, the major focus during the development of user interfaces has to be set to the reduction of visual load.

Reduction of visual load can be achieved by using information transmission channels other than the visual. On the input side, language input is strongly recommended, although technical problems are not solved to completeness now. On the technical side developmental work has to be done on the reduction of background noise disturbances. Human factors research is necessary to fit the verbal dialogue to user expectations. Easy to realize would be a dialogue guided by the system, leaving just "yes" and "no" answers to the driver. This would make dialogues very unhandy. Best fit for the user would be a system understanding natural language. For the near future an compromise between the two will be realized. Human factors research has to provide knowledge about style and structure of speech based dialogues between user and machine.

Language input will work well for navigation in menus, choosing applications or controlling basic functions. The selection of telephone-numbers or addresses will not be too much of a problem either. The text input, writing letters, e-mails, calendar-notes and so on, will also work with language after the solution of the remaining technical and human factors problems.

A much more complex problem is navigation in the internet. Graphical pages need a strong visual feedback, which makes the in-car use difficult. For this, human factors engineers and internet designers have to cooperate, because the design of the CCD reaches its limits quickly. Available internet pages must be reduced in their graphical parameters, providing the content acceptable for drivers.

All outputs should be given acoustically, either as language or as tones. This includes the marking of time windows for aural inputs with beeps and the offering of choices by language. It has to be easily possible to repeat any output at any time.

If inputs cannot be made aurally, haptic feedback on inputs should be provided by using different switches or knobs. Tactile feedback of the status in necessary. Fingertip-information makes, at least for long-time-users, the use with a minimum of visual load possible. All input devices should be grouped under functional not under technological criteria.

If visual attention is needed to obtain information, the relevant outputs should be presented in a form way to read and near the line of sight. In this way fixation times, saccadic movements and visual load can be reduced. Colors should be used to group information and to emphasize most important information.

The development of style-guides for CCD should be enforced, useful basis are [9] and [14]. PC-style-guides have to be modified, but can serve as information resources. Since DIS technology is more and more available for all, ergonomic interfaces become an distinguishing feature.

DIALOGUE WITH A CCD – Office and internet applications as they are used in the PC-world fit perfectly with passenger needs, but they are not suitable for drivers. PC interfaces have to be at least adapted to drivers´ needs. Probably completely new interfaces have to be developed.

Analytical research has to be done about the needs of the user. Which are the functions of CCD needed and used by owners? What do they expect if they purchase a CCD? Is it really necessary to provide full Excel and SPSS functionality in an car? Or is some word-processing, writing and receiving mails and faxes, a calendar, and a organization function enough?

All kind of word-processing, producing or receiving mail, letters and so on should be shifted to the acoustic channel. This requires a modification of PC programs. More than one window on the screen will not be handable for the driver. The input and the output process should be handled and lead by the system as far as possible.

Internet pages as known from the office environment need high visual attention. They include banners, optical hyperlinks, tables, and graphics. Most of these pages are hard to read even under office conditions. The manipulation of these sites by using links spread all over the page requires input devices not suitable for cars. New site designs have to be developed. The Wireless Application Protocol (WAP) shows a way. WAP is a protocol that leads to WML, a language similar to HTML. WAP is used to bring internet contents to cellular phones via the GSM channel. Internet contents are reduced to the minimum to make them readable on the small and often monochromatic displays of these devices. In addition the reduction raises the speed of information transport. The results are pure text pages with the possibility of using hyperlinks to get new information. The use of WAP for cars could be one solution. On the other hand, the reduction to pure text is not the optimum for internet in CCD, reading is not the optimal way to get information to the driver. An acoustical channel should be added to the protocol as well as the presentation of information with simple graphical elements like icons. The optical output should be much more structured than it is in WAP. Driver or driving related internet pages should be available in a version that fits the driving environment, more structured than WAP pages, less optical than usual sites.

A SUGGESTION FOR A CCD INTERFACE – A prototype for a CCD, called CAA CarPC was developed. This prototype was realized to show the technological possibilities and the some human factors solutions. The system provides the complete functionality, navigation, audio components, telephone, teleconferencing, internet access, e-mail, office, seat-positioning, climate control etc. It is based on an Intel™ processor and works with Windows Embedded NT™.

Figure 8 shows the input devices of the CAA CarPC. The round button in the middle is a twist knob with a push function, allowing the navigation in menus. The subsystem buttons are on top, the help button on the right, and the back button at the bottom. On the left two up and down bottons are positioned. The whole object has a diameter of about 4 inches, so it is large enough be handled while driving. Every button has a different shape, providing blind use for experts. Tactile feedback fits conditions while driving.

Figure 8. Input devices of a CarPC (CAA)

The CarPC is a system with a high consistency, all menus have the same functionality, the choice in submenus on the left, a jump to the right part of the screen by pushing the main button. There is the possibility of activating a function. The organization of the menus will be reorganized during the next months by applicating users models. They will be recorded with a combination of multi dimensional scaling and cluster analysis. The input devices make offers to the user, they clearly indicate what the system can do.

Figure 9. The screen of a CarPC (CAA)

The screen of the CarPC (Figure 9) is a 7" display. It is clearly divided into two sections. On the left side are submenus of the second level, on the right commands, or choices, e.g. radio-stations or CDs. Colors are used for different submenus, the radio has another one than the CD-player or the navigation system. Font-size is large enough, the font-type is well readable. The output is reduced to the necessary. Speech output is provided for navigation information, speech input for creating text, dialing telephone numbers, and choosing e-mails.

CONCLUSION

CCD will come, internet and office will be part of in car information processing. The driver will spent mental capacities on the systems. The benefits of an information have to be higher than the costs to get it. All human factors aspects should be considered. If the realization of any of it fails, the systems might not be accepted, or the costs of using them will be higher than the benefits. The PC standards (Windows etc.) do not match the needs inside a car. They were made for office workers, not for drivers. Solutions fitting the car-world have to be developed.

Do not expect to ever get the optimal user interface for a CCD too many influences needing contrary solutions exist. To find a good or very good solution, invite a sample of users and test, test, test.......

REFERENCES

1. Cooper, A. (1999). *The Inmates are Running the Asylum.* Indianapolis: SAMS.
2. Schumacher, R.M. (1995). *Ameritech Graphical User Interface Standards and Design Guidelines.* Ameritec Cooperation. http://www.ameritec.com, Download at Jan, 12th, 1999.
3. Microsoft. (1992). *The Windows Interface: An Application Design Guide.* PC28921-0692.
4. Apple Computer Inc. (1987). Human Interface Guidelines: The Apple Desktop Interface. Reading, MA: Addison-Wesley.
5. ISO 9241 - 10 (1996). *Ergonomic Requirements for Office Work with Visual Display Terminals. Dialogue Principles.* Berlin: Beuth Verlag.
6. Carayon, P. (1992). A Longitudinal Study of Job Design and Worker Strain: Preliminary Results. In: Quick, J.C., Murphy, L.R. & Hurrell, J.J., *Stress and Well-Being at Work: Assessments and Interventions for Occupational Mental Health.* Washington, D.C.: American Psychological Association.
7. Fastenmeier, W. (1995). Die Verkehrssituation als Analyseeinheit im Verkehrssystem. In: Fastenmeier, W. (Ed.) *Autofahrer und Verkehrssituation* (27-28) Cologne: Verlag TÜV Rheinland.
8. Brabus (1999), http://www.brabus.com/homee.htm, Nov. 30th, 1999.
9. Dingus, T., Hulse, M., Jahn, S. Alves-Foss, J., Confer, S., Rice, A., Roberts, I., Hanowski, R., & Sorenson, D. (1996). *Development of Human Factors Guide Lines for Advanced Traveller Information Systems and Commercial Vehicle Operations: Literature Review.* Federal Highway Administration, Report No. FHWA-RD-95-153.

10. Green, P. (1997). *Human Factors of In-Vehicle Driver Information Systems: An Executive Summary.* Federal Highway Administration, Report No. FHWA-RD-95-014.
11. Hamberger, W., Cremers, R. & Willumeit, H.P. (1998). Verbesserung der aktiven Fahrzeugsicherheit durch fortschrittliche Navigationssysteme. In: Willumeit, H.P. & Kolrep, H.: *Wohin führen Unterstützungssysteme?* (182-194). Sinzheim: Pro Universitate.
12. Peters, H. (1997). User-Centred Design of In-Vehicle Information Systems – requirements for Development and Testing. In: *VDI-Berichte 1317, Der Mensch im Verkehr (93-103).* Düsseldorf: VDI-Verlag GmbH.
13. Green, P., Levison, W., Paelke, G. & Serafin, C. (1995). *Preliminary Human Factors Design Guidelines for Driver Information Systems.* Federal Highway Administration, Report No. FHWA-RD-94-087.
14. Campbell, J.L., Carney, C., & Kantowitz, B.H. (1998). *Human Factors Design Guide Lines for Advanced Traveller Information Systems (ATIS) and Commercial Vehicle Operations (CVO).* Federal Highway Administration, Report No. FHWA-RD-98-057.
15. Rößger, P. & Smyrek, U. (1999). Bestens bedient. *Teletraffic*, No.9/10.

CONTACT

Dr.-Ing. Peter Roessger, CAA GmbH, Raiffeisenstr. 34, 70794 Filderstadt, Germany

mail to: p.roessger@caa.de

http://www.caa.de

2000-1-0641

A Seat Ride Evaluation Method for Transient Vibrations

Koro Uenishi, Katsunori Fujihashi and Hitoshi Imai
Daihatsu Motor Co., Ltd.

ABSTRACT

transient vibrations caused by a sudden change in the road surface under steady-state random vibration subject the human body to unpleasant sensations. However, it is very difficult to examine each part of the seat structure and physical properties because the measurement position of acceleration sensors with the conventional evaluation method are limited to the hipbone on the seat cushion and lumbar on the seat back.

In this paper, in order to evaluate three kinds of driver seats having different structures and physical properties, we dynamically measured body pressure distribution, which are mainly made with static measurements, on a pavement inducing transient vibrations. We analyzed each part of measurement data by means of a new human engineering index, which is body pressure change rate over time. We next investigated the correlation between the analyzed data and the subjective evaluation, and found the mechanism causing unpleasant sensations during transient vibrations. The results of our successful research allow making improvements in each part of the seat structure and physical properties.

INTRODUCTION

An important issue in driving comfort is how to improve the dynamic seat ride sensation. it is particularly easy for an ordinary passenger to feel the sudden load caused by passing over a rough pavement. This vibration is a factor by which passengers judge whether the dynamic seat ride sensation is comfortable or not. The human body can adjust to constant vibrations while driving an automobile but a sudden change in the vibration subjects them to unpleasant sensations. In order to improve this problem, we have to clarify the human engineering mechanism (which part of the seat structure transmits transient vibrations to the human body, and then how the passengers feel). it is also important to propose an improved seat structure based on this viewpoint. Recently, many automotive manufacturers and research institutes have wrestled with this problem[1,2].

Seat ride quality evaluation methods proposed by the ISO and British Standard are oriented towards improving many kinds of vibrations in the vehicle[3,4]. In these evaluati ons, acceleration sensors are installed on the seat and floor of the automobile and the frequency analysis is carried out.

The stimulus imparted to the human body by a seat while driving an automobile is compounded by the influence of static sitting comfort and the dynamic seat movement behavior. In order to understand the transient vibration mechanism, it is also vital to analyze and compare three conditions in each part of the seat: (1. static condition/ 2. steady-state random vibrations condition/ 3. sudden vibration change condition). However, it is very difficult to evaluate each part of the seat structure because conventional methods which only use dynamic methods for measuring acceleration do not provide static data for comparison, and the measurement position of acceleration sensors are only at the hipbone on the seat cushion and lumbar on the seat back.

In this study we examined a new evaluation method to understand the resulting behavior of each body part by static and dynamic body pressure distribution analysis, and to find the mechanism causing unpleasant sensations during transient vibrations. This method enabled us to achieve our goal of optimizing the structure and physical properties of each seat part.

EXPERIMENTAL

BODY PRESSURE DISTRIBUTION TEST PROCEDURE – The evaluation seats consisted of three kinds of drivers seats each having different structures and physical properties (Seat A, B, C), as shown in Table 1.

Table 1. Seat Characteristics

Type	Seat cushion construction	Seat back construction
Seat A	Foam	Metal springs & foam
Seat B	Foam (higher resiliency than Seat A)	Metal springs & foam (same as Seat A)
Seat C	Metal springs & foam	Metal springs & foam

These seats were installed in a mini-sized automobile. we first measured static body pressure distribution, and then dynamic body pressure distribution by driving along a pavement which induced transient vibrations. Table 2 shows the dynamic test conditions.

Sensors for evaluating body pressure distribution utilized a film sheet with a thickness of approximately 0.1 mm. This film has 2016 (42 by 48) pressure detection points whose electric resistance changes in proportion to the load. As shown in Figure 1, these sensor sheets were set on the seat cushions and the seat back. In the static measurement, the subjects sat on these sheets in an appropriate sitting position. Next, they drove along the test road, and measurements were made at a sampling frequency of 100Hz. Figure 2 shows the Z-axis floor acceleration when the vehicle was driven along this road. In this paper, we define a sudden input acceleration from 2 to 5 seconds as a transient vibration.

Table 2. Dynamic test conditions

Description of vehicle	Road type	Speed	Subject
Mini-sized automobile Engine capacity 660cc Overall length 3395mm Overall width 1475mm	Pavement inducing transient vibration	40km/h	5 trained males

Figure 1. Experimental setup for the body pressure distribution measurement

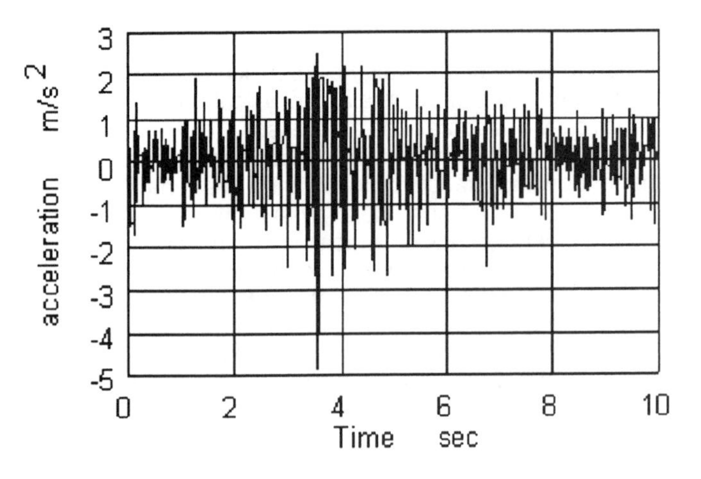

Figure 2. Z-axis floor acceleration when the vehicle for the evaluation was driven along test load

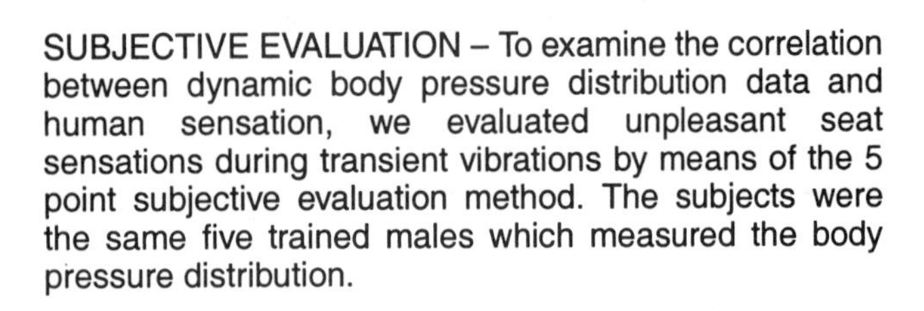

SUBJECTIVE EVALUATION – To examine the correlation between dynamic body pressure distribution data and human sensation, we evaluated unpleasant seat sensations during transient vibrations by means of the 5 point subjective evaluation method. The subjects were the same five trained males which measured the body pressure distribution.

RESULTS AND DISCUSSION

PRESSURE DISTRIBUTION ANALYSIS – Finding how each seat section affects the subject's body behavior is important when trying to improve unpleasant ride sensations caused by exposure to transient vibrations. Figures 3 and 4 show the dynamic body pressure distribution of the seat cushion and back. The pressure distribution data from the seat cushion was subdivided into six sections and pressure distribution data from the seat back was subdivided into nine sections.

Figure 5 shows a time history of right buttock pressure in Seats A, B and C. In all seats, buttock pressure from 2 to 5 seconds, which we defined as the period of a transient vibration, changed. The difference in these pressure changes among the three seats was determined. We assumed this sudden load change was an important factor in subjecting the human body to unpleasant sensation because of the same phenomenon in other seat parts. We next calculated the pressure change rate on each section using the time history of each section dynamic pressure $P(t)$, and its Pressure Change rate Root-Mean-Square $Pcrms$ between 2 and 5 seconds as follows:

$$Pcrms = \left\{ \frac{1}{T} \int_{t=2}^{t=5} \left(\frac{dP(t)}{dt} \right)^2 dt \right\}^{1/2}$$

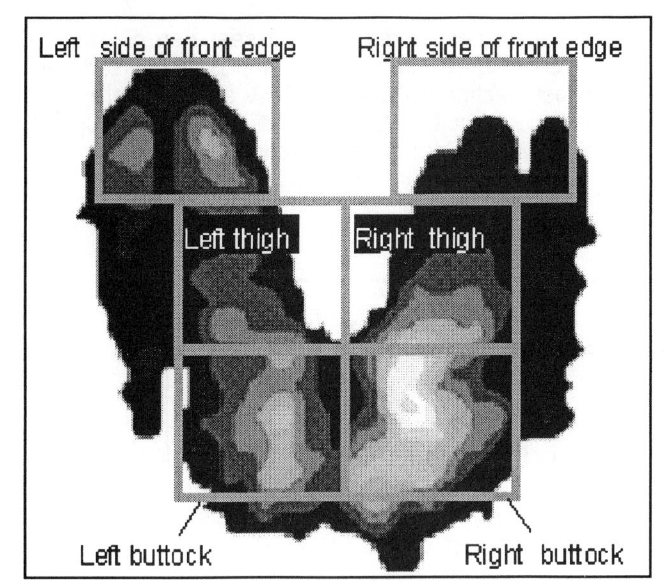

Figure 3. Body pressure distribution on seat cushion subdivided into 6 sections

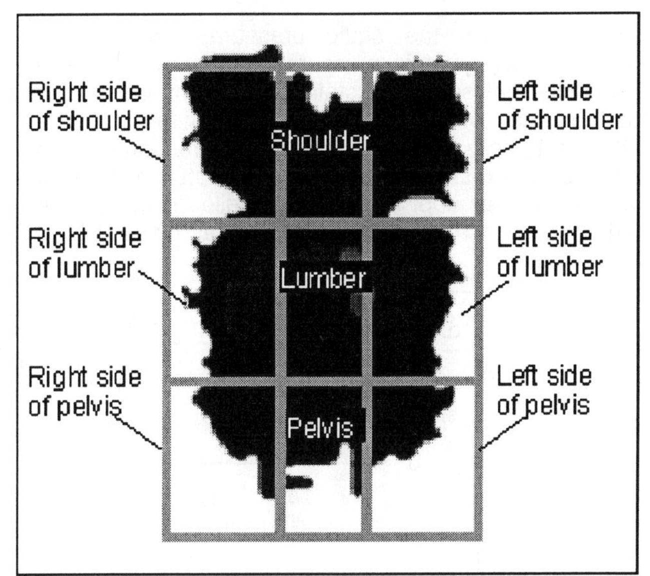

Figure 4. Body pressure distribution on seat back subdivided into 9 sections

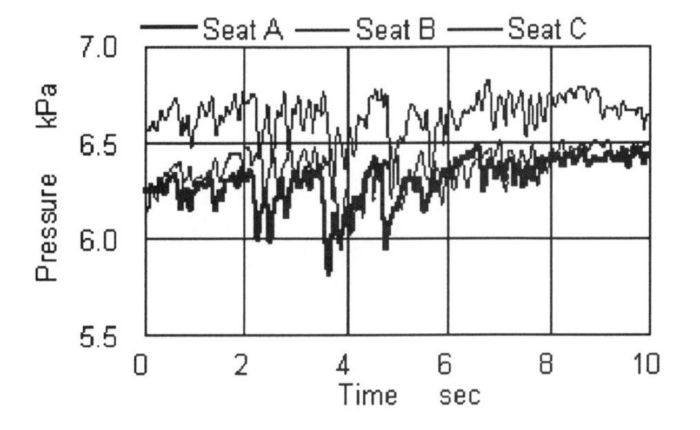

Figure 6 shows the static pressure, the dynamic pressure, and dynamic pressure change rate of each seat cushion. This dynamic pressure was the average of *P(t)*. we could not find a large difference in static pressure among the three seats. The dynamic pressure was measured almost the same value as the static pressure. On the other hand, in most sections, *Pcrms* as classed from the largest seat to the smallest seat were Seat B > Seat A > Seat C. Therefore, the vibration absorption performance of polyurethane mold foam and springs in Seat C cushion was superior to Seat A and Seat B.

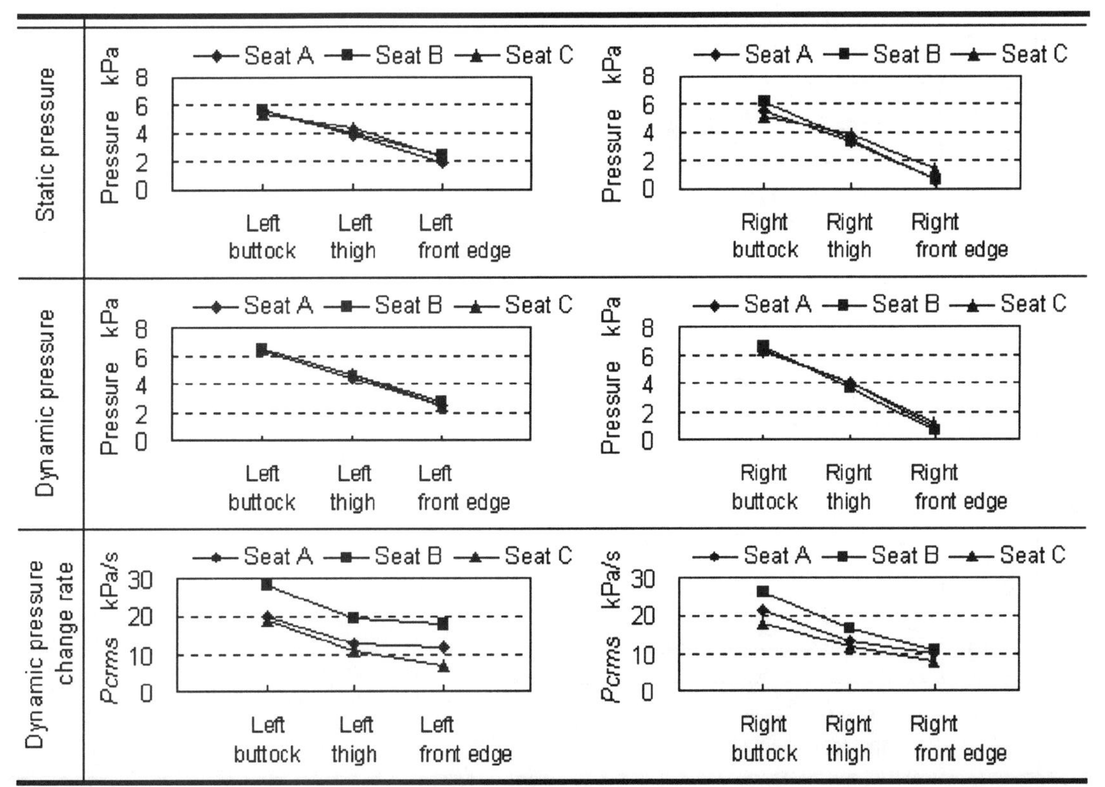

Figure 6. Static pressure, Dynamic pressure, and Dynamic pressure change rate of each seat cushion

Figure 7 shows the static pressure, the dynamic pressure, and dynamic pressure change rate of each seat back. In all seats, the static pressure on the lumbar was the largest of the three parts. The dynamic pressure of the seat backs was measured almost the same value as the static pressure. Therefore the seated posture of the subjects was kept nearly static while driving on the road surface. In most sections, *Pcrms* as classed from the largest seat to the smallest seat were Seat B > Seat A > Seat C. *Pcrms* of Seat A and B as classed from the largest section to the smallest section were pelvis > lumbar > shoulder. Seat A and B could not support the pelvis which is subject to vibrations.

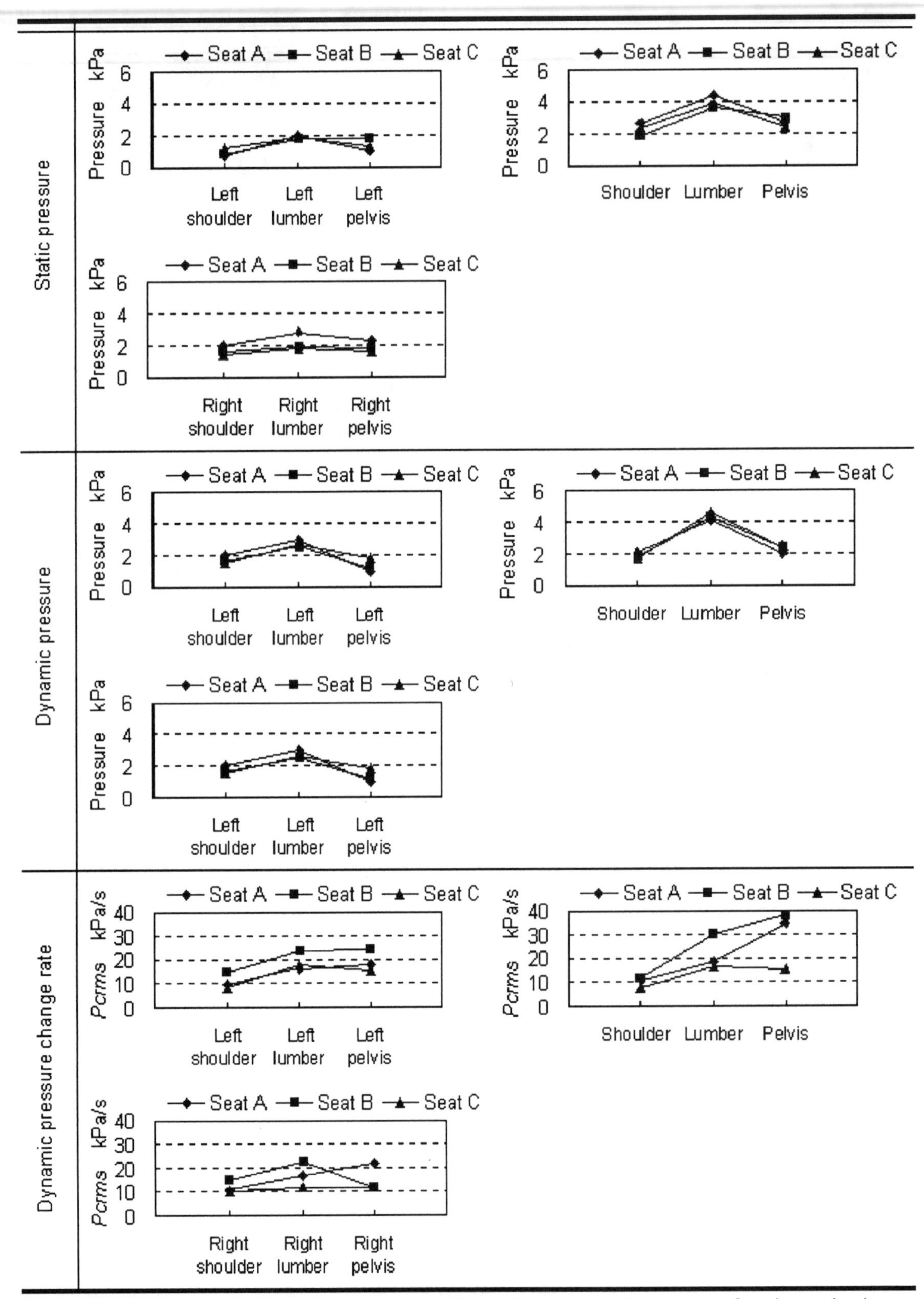

Figure 7. Static pressure, Dynamic pressure, and Dynamic pressure change rate of each seat back

On the other hand, the pelvis *Pcrms* of Seat C is much smaller than Seat A and B, and also smaller than the lumbar of Seat C. Therefore seat C had better pelvis support performance due to an appropriate seat back configuration and disposition and physical properties of springs.

We found transient vibrations in this test road had an influence on the dynamic pressure change rate of seat cushion and seat back.

SUBJECTIVE EVALUATION ANALYSIS – Figure 8 shows the unpleasant ride sensation scores during transient vibrations by the 5 point subjective evaluation method. Seat C proved the best and Seat B was the worst of the three seats. We discovered that the smaller the change in body pressure over time, the better the score obtained in the subjective evaluations as compared with body pressure distribution data. We confirmed that the dynamic body pressure change rate could be used to evaluate transient vibration characteristics.

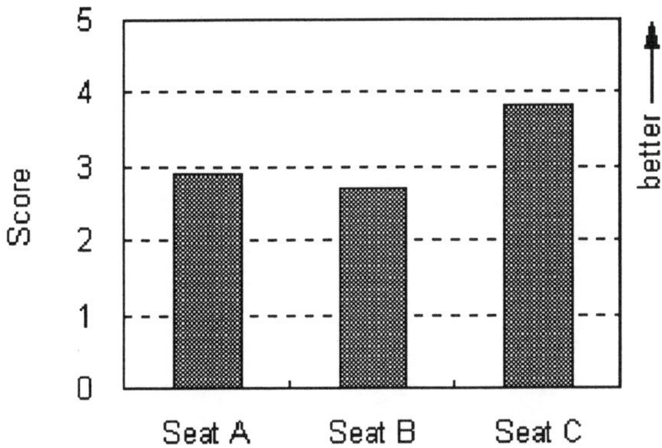

Figure 8. Unpleasant ride sensation scores during transient vibrations by the 5 point subjective evaluation method

CONCLUSION

1. We developed a new seat ride evaluation method for transient vibrations by measuring the static and dynamic body pressure distribution of the seat cushion and the seat back, and subdividing the resulting pressure distribution data from the seat back into six sections and that from the seat cushion into nine sections. we found the difference in the body pressure change rate over time for the measured seats.
2. Seat ride evaluation tests during transient vibrations revealed that the smaller the change in body pressure over time, the better the score obtained in the subjective evaluation.
3. Tests also showed we could determine unpleasant sensations from transient vibrations by the extent of the fluctuation in the dynamic body pressure rate, and our new evaluation method can be utilized for improving the seat structure and physical properties.

REFERENCES

1. ISO standard 2631.
2. British standard 6841.
3. Ohmori, H., Development of Riding Evaluation System for Joint Road in USA Freeway, Honda R&D Technical Review, Vol. 9, 1997.
4. Akatsu, Y., An evaluation Method of Improving Ride Comfort, Journal of Society of Automotive Engineers of Japan, Vol.52, No.3, 1998.

2000-01-0080

Development of Power Sliding Door (PSD) System with Push-Pull Cable Driving Method

Shintaro Suzuki, Ryouichi Fukumoto, Masao Ohhashi and Katsuhisa Yamada
Aisin Seiki Co., Ltd.

ABSTRACT

We have developed a power sliding door (PSD) system driven by a push-pull cable. The door closure and slide are operated by different actuators to limit the force required for a compact, light-weight drive unit. This paper introduces the concept of the PSD system using a push-pull cable drive. Two new technologies to achieve the PSD system are also described. One is the door position control for increasing the push-pull cable reliability. The other is a compact position sensor to accurately detect the sliding door's position.

INTRODUCTION

The first production PSD system for min-vans, including a remote door opener and closer, was introduced into the market in 1993. Recently, PSD systems have become increasingly popular in the market for enhancing the value of mini-vans. Reducing the size and weight of the PSD system is very important, as well as improving the ease of getting on and off the vehicle, cabin roominess, and operational simplicity. In this paper, the development of a new PSD system using the push-pull cable drive is described.

DEVELOPMENT REQUIREMENTS

At the beginning of the system development, the customers' needs for the PSD system were surveyed and conventional PSD systems were analyzed. To easily board or exit, improve cabin roominess and easily operate the PSD, the following issues should be considered for system requirements:

- step height
- stroke (opening)
- cabin space
- interior design
- selection of automatic or manual operation
- operational force for manual operation

PUSH-PULL CABLE DRIVE

There are two common mechanisms to drive a sliding door. One is a belt drive mechanism in which the driving force is transferred to the lower roller at the bottom of the sliding door through a belt. This mechanism interferes with easy boarding and exiting due to a higher step or a narrower opening since a large guide for the belt is necessary. The other is a wire drive mechanism in which the drive force is transferred to the roller at the center of the sliding door through a wire. This design results in a protrusion at the quarter panel trim for the drive unit, intruding into the interior space.

To solve these problems, a new drive mechanism using a push-pull cable has been developed. Fig. 1 shows a schematic of the push-pull cable drive. For the push-pull cable, a wire is wound around a central, straight wire. The push-pull cable is driven by the gears to transmit the drive force in both the push and pull directions through a metal pipe cable guide. Limiting the drive force is very important for the reliability of the push-pull cable drive. Fig. 2 shows the relationship between the door position and push-pull cable drive force. The drive force rapidly increases from the half-latched position to the slightly-open position (near fully closed). The drive force applied to the push-pull cable must be less 300N to satisfy the reliability requirements from durability tests. Therefore, a different actuator is used at the end to fully close the door in order to limit the drive force and required power from the push-pull cable.

Using the push-pull cable drive reduces the size, weight and power needed for sliding the door. Fig. 3 shows a comparison among the door sliding drive units. The weight of the push-pull cable drive is 50% less than the other drive units. This push-pull cable drive unit needs less installation space, reducing the protrusion into the interior at the quarter panel trim. This drive unit also does not effect the step height of the sliding door since the center roller is driven by the push-pull cable. Therefore, the PSD system can be equipped without any reduction of the ease of entering and exiting and cabin roominess.

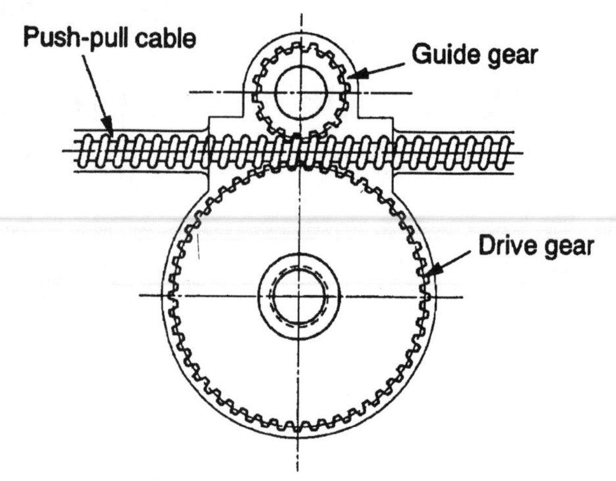

Fig. 1 : Push-pull cable drive

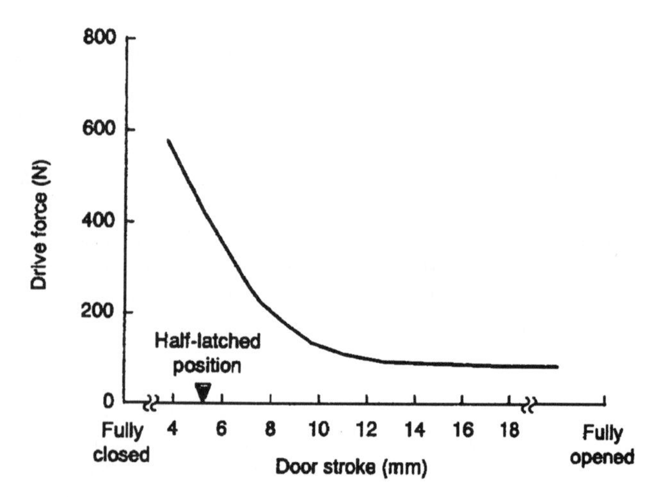

Fig. 2 : Drive force applied to the push-pull cable

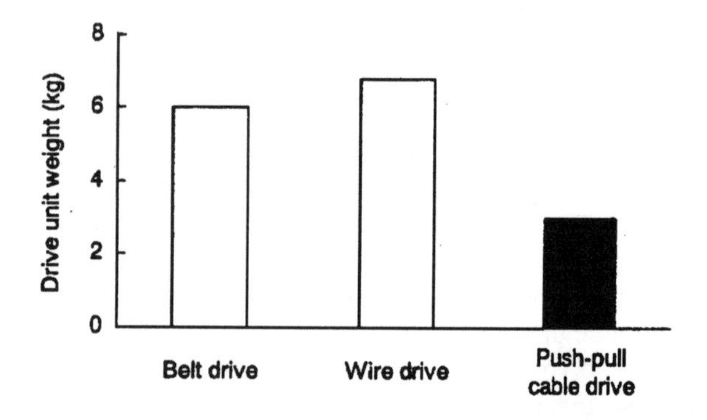

Fig. 3 : Weight comparison among the various drive units

CONFIGURATION OF PUSH-PULL CABLE DRIVE PSD SYSTEM

Fig. 4 shows the different components of the push-pull cable PSD system. The system includes the door closure and latch-release actuators, push-pull cable drive unit, cable guide and power supply in the door panel.

The door closure actuator fully closes the sliding door and is installed in the sliding door panel. Fig. 5 illustrates the door closure operation. The door latch, driven by a motor, hooks the door striker attached at the vehicle body to close the sliding door. This door closure actuator only works within the door closure range, which is within 17mm from the full-latched position. However, this actuator generates approximately 600N at the half-latched position and 900N at the full-latched position. The door closure actuator includes a switch to detect the half-latched and full-latched positions.

The latch-release actuator unlatches the sliding door. This actuator works with the operation the door handle. This actuator is also installed in the sliding door panel. Electric power is supplied to the door closure and latch-release actuators in the sliding door panel through electric contacts (so-called junction switch) attached between the B pillar and the side of the sliding door.

A door check is attached at the lower rail to hold the door at the fully opened position by supporting the lower roller with a leaf spring. Fig. 6 shows the structure of the door check. For automatic door opening, the motor of the cable drive unit must stop at the position where the lower roller is held by the door check.

The cable drive unit slides the door using the push-pull cable. The push-pull cable is routed to the outside of the vehicle through the metal pipe cable guide at the rear end of the center rail and connected to the center roller. An electromagnetic clutch is installed in the cable drive unit for selecting automatic or manual operation. A door position sensor is integrated into the electromagnetic clutch to accurately detect the position of the sliding door.

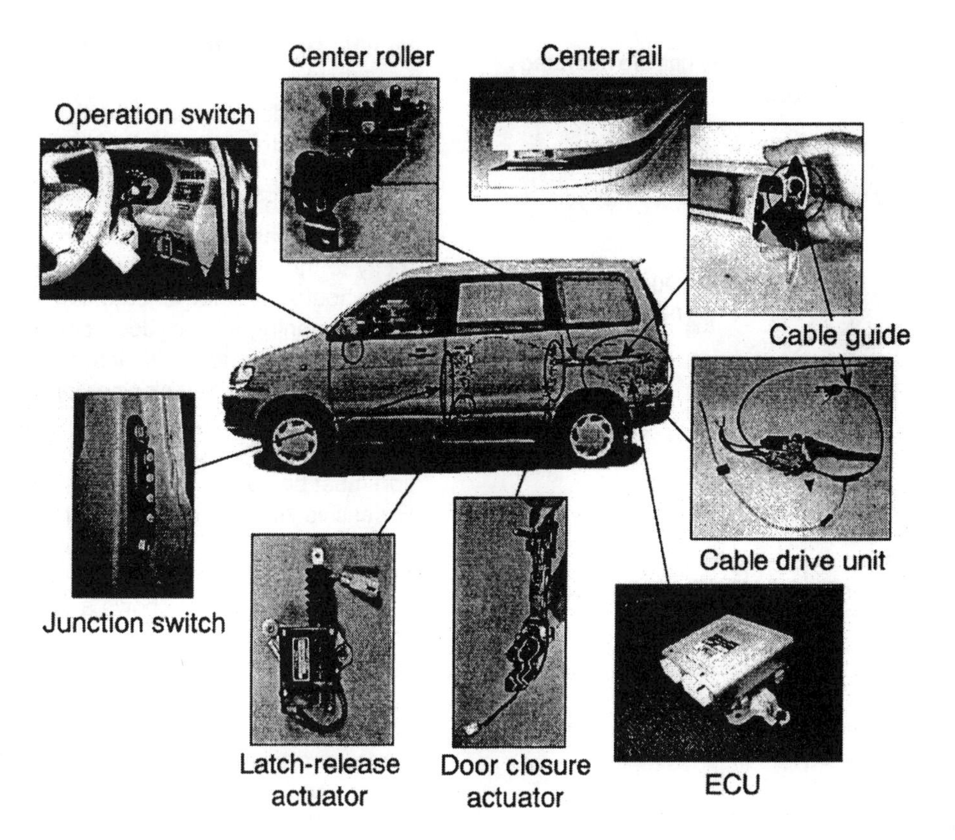

Fig. 4 : The push-pull cable drive PSD system

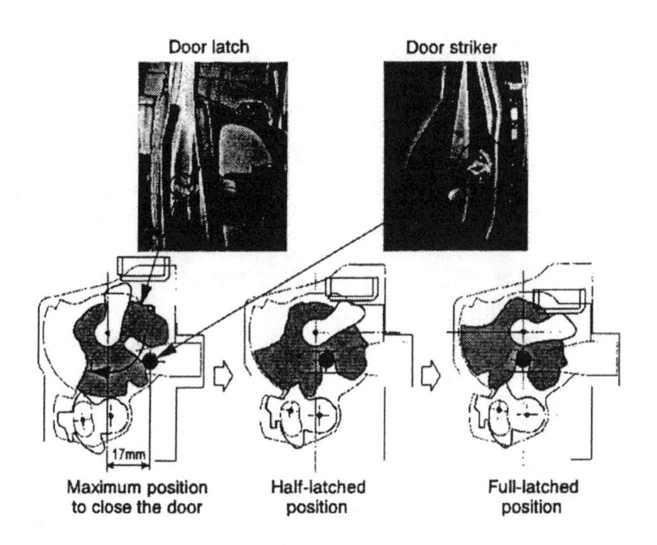

Fig. 5 : Door closure operation

DOOR OPENING AND CLOSING OPERATIONS

Figs. 7 and 8 show timing charts of the door opening and closing operations. For opening, shown in fig. 7, the latch-release actuator and the motor and electromagnetic clutch of the cable drive unit are activated when the driver pushes the switch to open the door. The latch-release actuator is deactivated when the junction switch is open. When the sliding door reaches the fully opened position, the motor and electromagnetic clutch are deactivated.

For closing, shown in fig. 8, the motor and electromagnetic clutch of the cable drive unit are activated. The door closure actuator starts 0.5 sec after the junction switch makes contact since it takes 0.5 sec to stabilize the contact condition of junction switch. The drive motor closes the sliding door up to the position where control is transferred to the door closure actuator. The electromagnetic clutch is deactivated at the half-latched position and the door closure actuator is turned off at the full-latched position.

electromagnetic clutch of the cable drive unit are activated

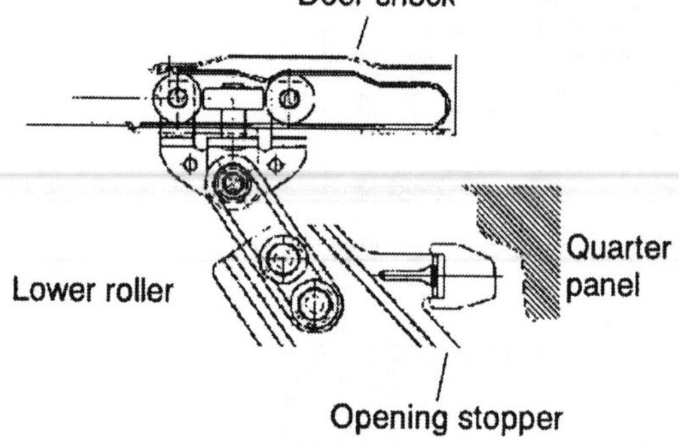

Fig. 6 : Door check structure

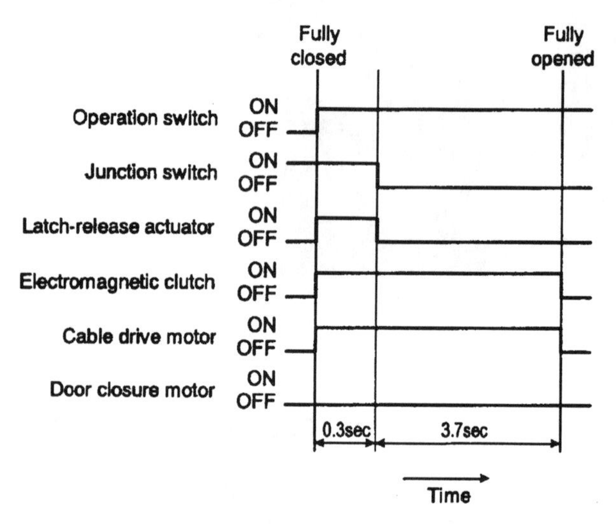

Fig. 7 : Door opening operation

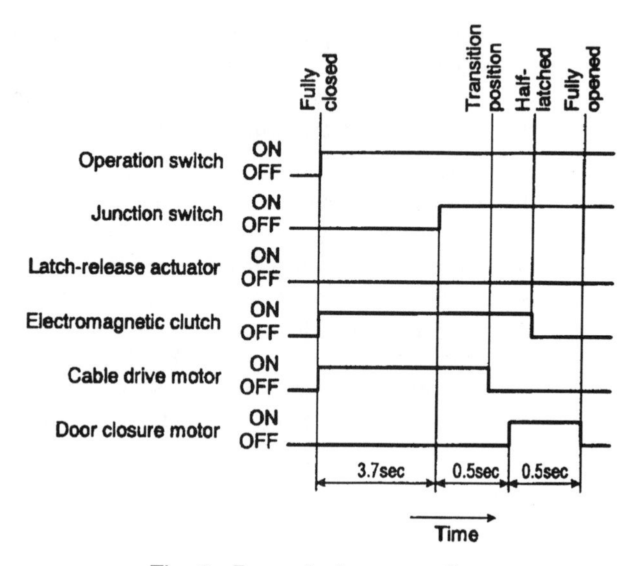

Fig. 8 : Door closing operation

DOOR POSITION CONTROL FOR RELIABILITY

The sliding door is moved by the push-pull cable drive up to the door closure operation. It is very important for system reliability to prevent excessive drive force from being applied to the push-pull cable. The sliding door does not stop immediately and can overrun even when the electricity to the drive motor is cut. The motor stopping position should be determined including the door stroke due to overrunning. The door position control to prevent excessive drive force is described in detail, separating the door opening and closing controls.

DOOR OPENING CONTROL – The motor stopping position must be considered when the sliding door is opened. Excessive drive force caused by the motor lock torque can be applied to the push-pull cable unless the drive motor stops before the door is fully opened (the opening stopper is hit). Fig. 9 shows a schematic of the stopping position of the drive motor. The fully opened position is memorized by the ECU as the reference position by opening the sliding door up to the drive motor lock. The motor stopping position is then set to be approximately 10mm short of this reference position, including dimensional variations due to elastic deformation of the parts, overrun stroke and the door check position. This opening control also limits the influence due to variations in the dimensions of some parts, so the fully opened position varies slightly for different vehicles.

DOOR CLOSING CONTROL – The sliding door is moved by the push-pull cable up to the transfer position, then the door closure actuator brings the door to the full-latched position. This transition from the push-pull cable to the door closure actuator must be accurately controlled. Fig. 10 shows the door closing sequence. The drive motor should be stopped between position (1), where the door closure actuator can securely close the sliding door, and position (2) so that excessive drive force is not applied to the push-pull cable. The door position where the junction switch makes contact is set as the reference position. The drive motor stopping position is set not to exceed position (2) due to overrunning to ensure the reliability of the push-pull cable. The minimum stroke between the reference and motor stopping positions is about 4mm including dimensional variations due to the door installation, parts manufacturing tolerances and so on. Therefore, the resolution of the door position sensor is required to be less than 4mm for the door position control.

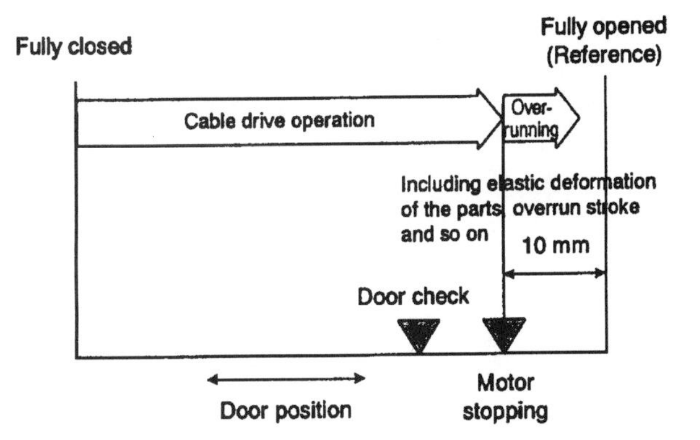

Fig. 9 : Determination of motor stopping position for door opening operation

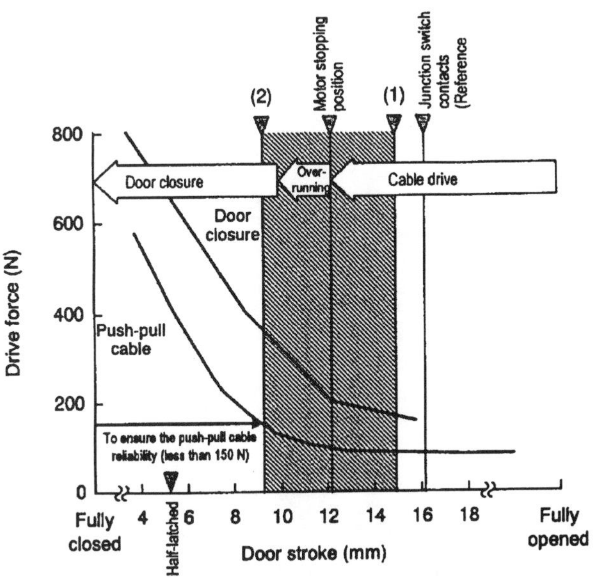

Fig. 10 : Determination of motor stopping position door closing operation

DOOR POSITION SENSOR

The door position sensor is necessary to accurately control the sliding door movement, especially for determining the drive motor stopping position. The following issues are required for the door position sensor:

- direct detection of the push-pull cable movement
- high sensor resolution
- sensor structure for reducing the drive unit size

The door position sensor is integrated into the cable drive unit to detect the cable movement directly. Fig. 11 shows the location of the door position sensor. A plastic ring magnet is arranged on the outside of the electromagnetic clutch to save space. The multi-pole plastic ring magnet has a large diameter for high resolution. The Hall effect sensor and magnet are integrated to maintain good long-term sensor performance and avoid contamination problems such as water condensation and grease.

The effect of magnetic flux leakage from the electromagnetic clutch is a potential problem since the Hall effect sensor is mounted very close to the clutch. Fig. 12 shows the magnetic force distribution when the electromagnetic clutch is activated. The magnetic flux direction around the magnet and Hall effect sensor corresponds to the clutch axial direction. The Hall effect sensor detects only the component of the magnetic flux perpendicular to its axis. Therefore, the interference of the magnetic flux leakage from the electromagnetic clutch can be limited using the arrangement of the Hall effect sensor and magnet shown in fig. 11.

Fig. 13 shows the relationship between the force at the Hall effect sensor and the force due to the magnetic flux leakage from the electromagnetic clutch. The force waveform of the magnet is offset by the force due to the magnetic flux leakage. However, the Hall effect sensor can detect the force from the magnet since the force due to the magnetic flux leakage is very small compared with the force from the magnet. Also, the force due to the magnetic flux leakage is almost constant because the clutch is always activated while the PSD is operated. This means the magnetic flux leakage does not effect the PSD control very much. The magnetization pitch is designed to maximize the sensor resolution considering the Hall effect sensor's sensitivity, the influence of the magnetic flux leakage and the dimensional variation of the gap between the magnet and Hall effect sensor.

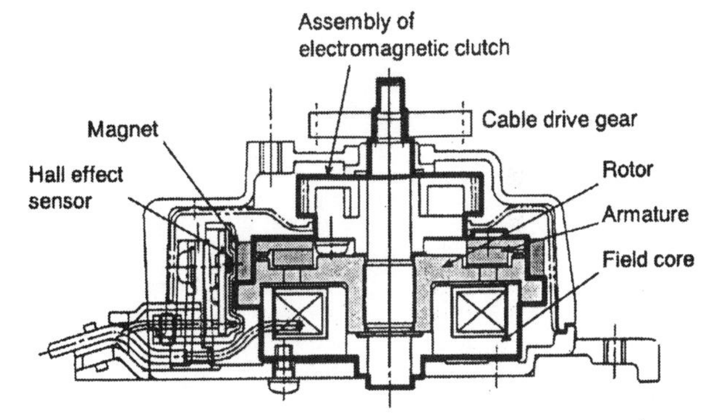

Fig. 11 : Door position sensor integrated electromagnetic clutch

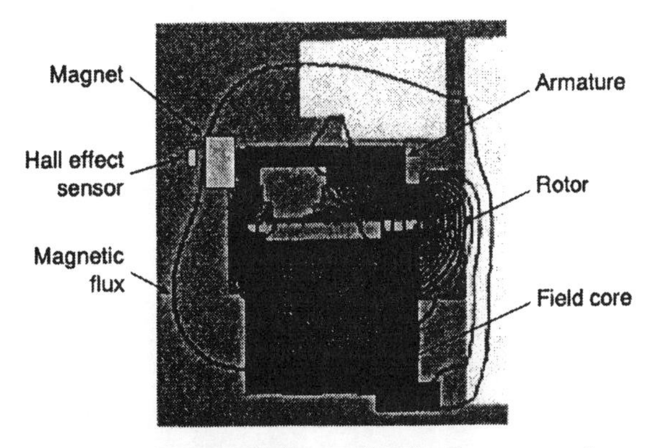

Fig. 12 : Magnetic force distribution around door position sensor

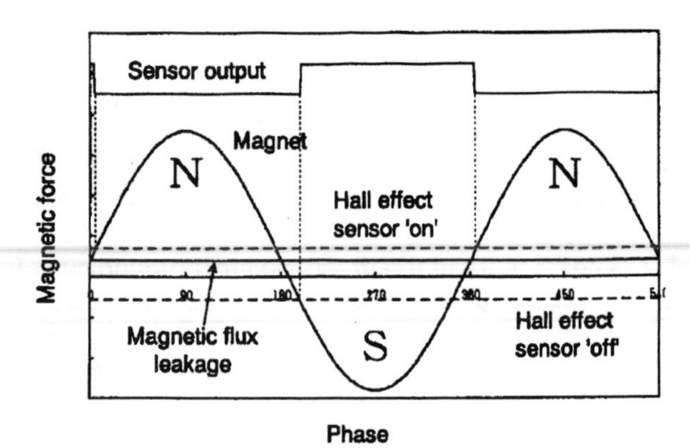

Fig. 13 : Forces due to magnet and magnetic flux leakage

CONCLUSIONS

We have developed a PSD system using a push-pull cable. The door position control which prevents excessive drive forces from being applied to the drive cable, has also been developed to improve the system reliability and to minimize the system size. The door position sensor, composed of a Hall effect sensor and ring magnet integrated into the electromagnetic clutch, has been developed with high accuracy and small size. This PSD system has expanded the marketability of mini-vans by improving the ease of boarding and exiting, cabin roominess and operational simplicity.

FUTURE OUTLOOK

FUTURE DEVELOPMENTS IN ELECTRONICALLY CONTROLLED BODY AND SAFETY SYSTEMS

Dan Leih and Ross Bannatyne
Motorola Transportation Systems Group

The global automotive market is in the midst of the most significant changes in its history. Multiple factors are changing the very nature of the market and the players. Non-global players are being replaced by more efficient, lower-cost global ones. Suppliers are combining forces in order to position themselves in key segments and to be able to supply large "chunks" of vehicles. New legislation is forcing more intelligence into safety systems. Overall, car sales are increasingly being driven by feature content, much of it electronic, as well as style. With these factors in mind, this paper looks at trends in electronic safety and convenience systems that affect the automakers' abilities to personalize the driving experience.

1.0 Brand Creation and Personalization

Vehicle manufacturers realize that brand identity is as important to their customers as it is to consumers in other industries. For example, the Toyota name has become synonymous with high quality and value and BMW is known for engineering and performance. In the United States, automakers are working to define their images for a new generation of car buyers. The Chrysler arm of DaimlerChrysler, for example, introduced in the 1990s a wide variety of restyled cars, trucks, and crossover vehicles. These vehicles are part of an ongoing effort to define the Chrysler and Jeep brands in the minds of consumers. Integral to the definition of a brand is the ability to personalize the vehicle and extend the uniqueness of the brand to those features with which the consumer interacts continuously.

Many consumers, either consciously or unconsciously, purchase vehicles with styles that reflect themselves; e.g., big, bold, conservative, luxurious, elite, and so forth. Historically, consumers have associated two factors with the "personal-ness" of a vehicle. The first, external styling, has been and remains the initial attraction for most consumers. The second is engine style and type, especially in the United States, where the race for performance in the post-WWII era led to ever-larger powertrains, which then became a status symbol in themselves. For the middle-aged and the newest two generations of buyers, this powertrain fixation has given way to a desire to be able to "personalize" the creature comforts of the vehicle.

The interior space has become the new battleground for the consumer. Features once reserved for high-end vehicles are now appearing in mid-tier and even in some low-tier vehicles. This abundance of features is forcing manufacturers to reconsider how they source electronics systems and how they define and develop vehicle networks. The network and the electronic control units (ECUs) attached to it are defining the limitations of the carmakers' abilities to personalize their vehicles.

2.0 Vehicle Feature Trends

Significant trends for body electronics and occupant safety systems include the following:

- Networking: Virtually all future modules will be connected to a network. Networks allow sharing of information and are being redefined by carmakers to allow a redistribution of intelligence within the vehicle.
- Software: The ability to personalize a vehicle's look and feel will be determined by software. Carmakers will dictate network protocols and ECUs will be required to conform to

communications standards to the extent that even the software routines used in cars may be determined by the automobile manufacturer. As a result, car brand identification will be defined by the software applications developed. A side benefit of this dramatic change may be the ability to reduce module variations, thus offsetting some of the cost of increased feature content.

- Safety systems: Legislation is forcing smarter safety systems with increased sensing and protection capabilities. The number of basic safety systems in cars will match the number of engine management systems. Development will continue in the areas of occupant sensing and collision avoidance.
- Architecture ownership: Carmakers will define entire architectures for their major platforms. In order to reduce the time to market, niche and specialty vehicle design may be passed entirely to systems integrators who will be responsible for supplying the entire vehicle interior, including metal and plastic, electronics, wiring, and software.
- Vehicle simulation: Similar to the aerospace industry manufacturers, carmakers will embrace more virtual design and vehicle simulation. Manufacturers will be required to supply models of various components and systems to allow design validation prior to prototyping. Carmakers will develop relationships throughout the entire supply chain in order to meet this objective.

As a result of these changes, a large number of consumer features are being driven into the volume car platforms. The proliferation of features and the increased volumes are increasing competition and driving prices down, thus ensuring continued increases in penetration. All of the major feature increases will be focused around either personalization or safety.

The extent of personalization possible will be determined by the software developed for the vehicle and the network architecture owner. Ability to completely personalize a vehicle to a consumer's taste will remain limited until the carmakers or system integrators realize that they must design the vehicle as a complete system rather than a series of stand-alone systems. This will require trade-offs in costs between various parts of the system in order to maximize functionality and minimize cost at the vehicle level. An example of this might be the use of a 32-bit processor with graphics capability in the dashboard. Though such an approach would significantly increase the cost of the central dash module, it could reduce the cost of other surrounding modules (use of the computing power contained in the dash module could replace much of the electronics in the cluster, climate control, and front body modules). The resulting system would provide greater functionality at a reduced cost if one considers the value of the entire instrument panel.

3.0 Sub-Bus Architectures and Networking

The automotive market has largely standardized on two network protocols: SAE J1850 and CAN. In some areas, protocols such as VAN and BEAN are still used. However, it has become clear that in the automotive industry, just as in other industries, standardization drives increased product availability and decreased costs.

Although the United States has adopted the J1850 standard, the Big Three automakers have been unable to achieve a common implementation. As a result, designs and often products are not interchangeable between manufacturers. By contrast, almost all European manufacturers have converted to the Bosch CAN protocol—a higher-performance, more complex and (initially) more expensive protocol than J1850—as a de facto standard. This near universal adoption of CAN makes it ultimately less expensive than the U.S. standard.

Limited feature nodes have driven continued networking development. As shown in Fig. 1, the CAN protocol implemented at various speeds is usable throughout the vehicle. There are noticeable clusters of relatively simple features that do not require the full capabilities of the CAN protocol. These include doors, seats, and HVAC systems. The solution to communication in these areas of the vehicle is the "sub-bus" architecture.

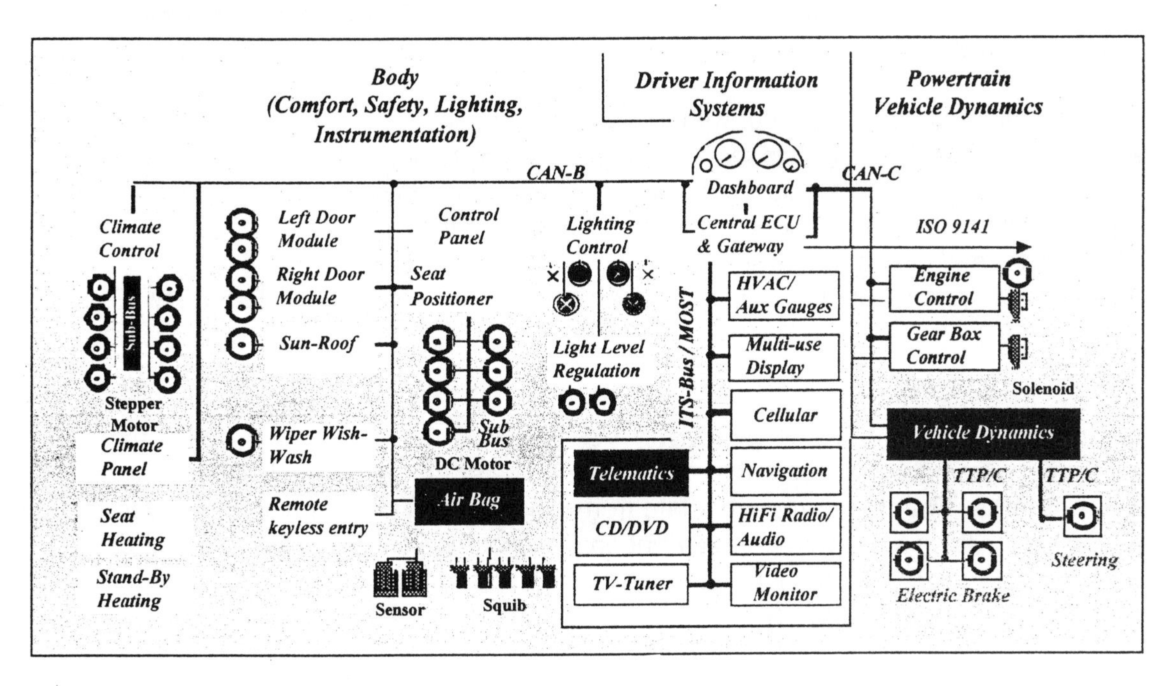

Figure 1 - Vehicle Network Architecture

3.1 Local Interconnect Network

During 1999 a consortium of manufacturers worked to develop a standardized architecture for sub-buses. The Local Interconnect Network (LIN) Consortium released its LIN protocol specification for review in July 1999. In 2000 several announcements are expected regarding availability of specifications and products. To date committee members include: Audi, BMW, DaimlerChrysler, Ford (Volvo), Motorola, VCT, and VW. The participation of these industry leaders ensures support for products and systems based on the LIN architectures.

The LIN sub-bus solution is an extension to CAN. It is based on SCI/UART (Serial Communications Interface/Universal Asynchronous Receiver Transmitter) and therefore more cost-effective than a full-featured network. A block diagram illustrating the similarities of CAN and LIN is shown in Fig. 2. Cost factors include the reduction of one communications channel wire, removal of connectors, elimination of a crystal per node, and cheaper silicon costs. The second wire needed for CAN is shown in this diagram.

From a technical perspective, LIN is a single master/multiple slave concept. No arbitration is necessary between nodes. Almost any available microcontroller can be used but, over time, optimization of microcontrollers for LIN can be expected. Self-synchronization is possible without a crystal or ceramic resonator in the slave nodes. Speeds up to 20 Kbit/s (limited by electromagnetic interference) are possible. Such speeds are faster than many of today's J1850 implementations.

The LIN bus is a significant development for automotive convenience systems as it standardizes a very low cost communications system upon which each of the convenience modules can communicate. Typical components and systems that reside on the LIN bus will consist of a sensor or actuator, analog based application specific ICs (ASICs) for signal conditioning, and a microcontroller for control algorithm execution and communications. Convenience systems that reside on a LIN bus include:

- Door lock systems (basic actuator with low cost electronics)
- Seat adjustment systems (basic motor control based systems)
- Mirror adjustment systems (basic motor control based systems)
- Sunroof control systems (basic motor control based system)
- Lighting systems (interior and trunk)
- Wiper control systems (can include rain detection sensor which is sampled every 10–20 ms)

3.2 Gateways

Another trend in networking is the use of gateways. A gateway is simply a bridge between two networks. A common example is the use of the instrument cluster as the bridge between the body and powertrain networks. The cluster requires data from the engine and transmission as well as various body systems. Using a network gateway not only allows network access to needed data, but also creates a path for the movement of data between two different types of networks (powertrain networks are generally higher speed networks than body networks).

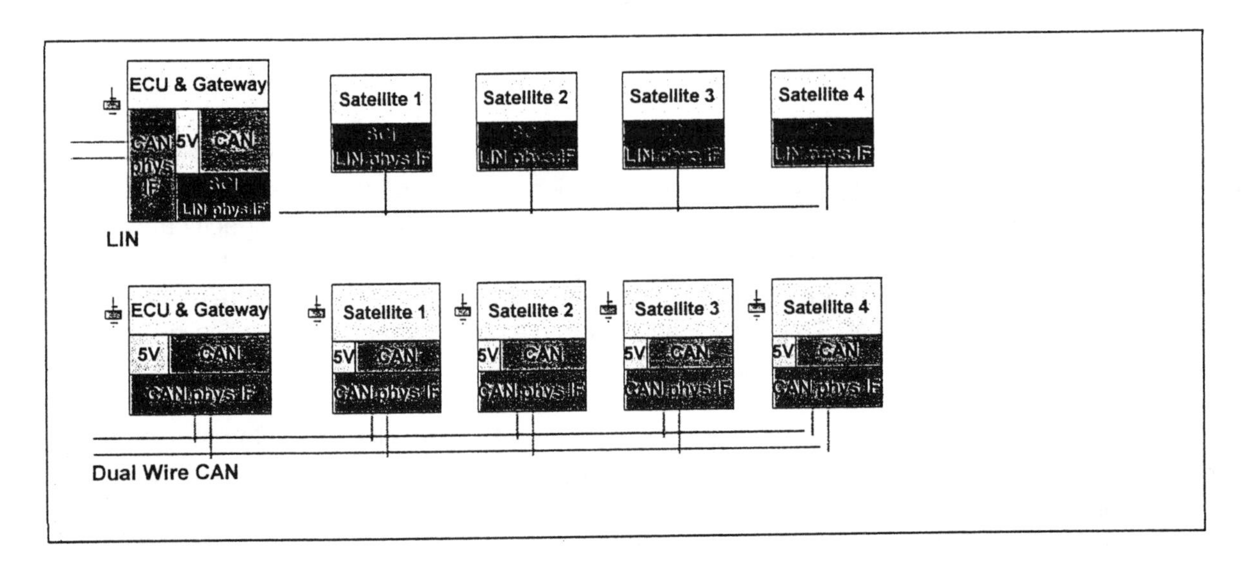

Figure 2 - CAN v LIN Comparison

The Motorola MC68HC912DG128 device is being used in many gateway applications. The HC912 core offers excellent C code efficiency, while the PLL (phase locked loop) feature allows designers to achieve very low power "wait" modes. Dual CAN ports enable the device to operate on two networks simultaneously. The device also has an array of other desirable body features including dual 10-bit A/D, enhanced capture timers, and a large number of I/O ports.

It is the gateway or bridge that allows personalization of the performance and handling of the vehicle. This feature will most likely be combined with personal identification, as discussed in the next section.

The trend toward inclusion of telematics features in the vehicle will add another gateway to the vehicle that will connect wireless service to the wired vehicle networks. An endless array of features can then be offered. These vary from entertainment (video games and streaming video for minivan applications) to navigation and location (large truck tracking) to anti-theft capabilities (vehicle tracking, "crippling") and numerous others. The key to enabling this functionality is the creation of a standardized hardware platform that offers scalability and ease of software development.

4.0 Convenience

The trends that have been discussed have been utilized in many vehicle electronic systems in order to provide drivers with greater convenience. Some of these systems will de discussed in more detail in this section.

4.1 Keyless Entry

For consumers remote keyless entry (RKE) is perhaps the most obvious of the new convenience trends. The ability to lock and unlock any or all of the vehicle's access points without the use of a key or latch is so desirable that once consumers have had cars with RKE they will rarely go back to cars without it. From an electronics perspective, RKE is a relatively basic system. It consists of an RF receiver and transmitter, door lock actuators, and control logic circuit including a security algorithm.

The key aspects of this system are somewhat determined by region. In Europe the presence of radio frequencies for civilian and commercial usage requires that transmit and receive signals operate on fairly tight frequencies. Emerging systems are moving from the traditional 315/433 MHz systems to 868 MHz systems. Coverage of all of these frequencies with the required accuracy requires a high-tech RF (radio frequency) circuit.

In the United States and Japan less stringent RF requirements allow the use of simple discrete transmit and receive circuits. This characteristic has driven RKE into a commodity application faster than the more complicated European systems. Continued success of luxury cars in the United States will require the addition of new features to existing systems.

In certain cases, the RKE system with its unique entry code can communicate with immobilizer/anti-theft systems. With such systems, the absence of the unique entry code broadcast from the owners key fob will prevent the vehicle from starting by deactivating the ignition system.

Currently emerging in the marketplace are passive entry systems. In a passive system no specific "unlock" action is required by the user to enter the vehicle. The Mercedes S Class has introduced such a system. It is capable of identifying the user, and unlocks the vehicle during the act of lifting the door latch. A simplified block diagram of such a system is shown in Fig. 3.

As networked systems become more tightly integrated, the entry system will become the first stage in personalizing the vehicle for a particular consumer. In the case of multi-user vehicles, a unique identifier will allow for many interior features to adjust to the entering user's preference. For example, settings for a shorter driver might include automatic seat and mirror adjustment. Climate control will remember the user's temperature and zone settings while audio systems will automatically adjust volume and station settings.

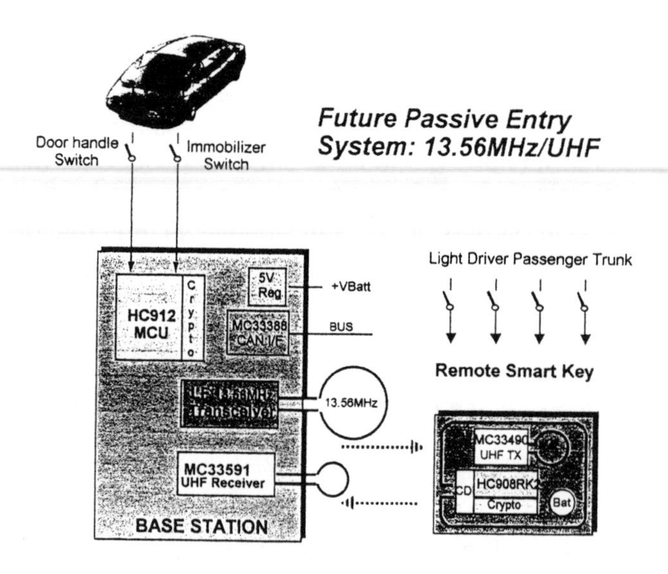

Figure 3 - Future Passive Entry System

Beyond these features are the powertrain and chassis settings. Gas mileage, performance, and ride preference could be adjusted on cars so equipped. The ability to set a "valet" mode would eliminate concerns about unauthorized vehicle use. Also, insurance advantages might be given to owners who tailor a vehicle's capabilities for high-risk users such as teen drivers. For example an inability to exceed 55 mph could reduce liability claims.

In the not too distant future, biometric systems will be used instead of the "smart key" approach of passive entry systems. Biometric systems can use technologies such as voice recognition (typically software) or fingerprint identification (typically hardware camera-sensor based with supporting software) in order to recognize the owner of the vehicle. When the owner of the vehicle is identified, the biometric control system will take appropriate action such as opening the door, and may use LIN bus-based communications to configure seating and mirror setting preferences.

4.2 Door Systems

Door systems have become very popular. Key features include power door locks, power windows, power controlled mirror adjustment, and mirror fold in Japan and Europe. Since a keypad is included in each door, the number of wires entering the door (as many as 40 in some cars) has become prohibitive. This has given rise to the development of door modules. Door modules contain all of the circuits necessary to operate the convenience features in the door and use multiplex techniques to limit the number of wires entering the door to three or four, depending on the type of network used.

Although door modules are in use in many cars today, their cost has led manufacturers to seek alternative solutions. In addition to cost, the quest for reduced car size and increased interior space is squeezing the space once used by the door module. The result will be a trend toward intelligent switches and connectors that completely replace the stand-alone door module.

Systems utilizing smart connectors will also use reduced networks or "sub-bus" technology. Sub-buses allow reductions in network node cost and controller computing power necessary to make intelligent nodes cost-effective. True system cost reduction will again be possible only when body electronics are developed as a complete system rather than a series of stand-alone modules.

5.0 NVM: An Enabling Technology

Non-volatile memory (NVM) has been a standard tool of the systems designer since the 1980s. Several types of NVM exist. The original erasable programmable read-only memory (EPROM) and the electrically erasable programmable read-only memory (EEPROM) types have now been supplemented by flash memory. In the beginning NVM was available only as a stand-alone product, but semiconductor process development has made possible integration of all types of NVM. Further process and design improvements are making possible the use of NVM as a replacement for read-only memory. This opens the door to significant changes in how systems are designed and how products are developed and procured in the future.

Flash memory evolved from EEPROM. It is smaller and more cost-effective than standard EEPROM partly through the elimination of logic used to control the individual byte addresses. Flash memory can be written to one byte at a time but, typically, can only be erased as a complete block consisting of many bytes. Some manufacturers do offer partial block access (i.e., a 32K block broken into a 24K and an 8K block). Flash memory is designed for larger memory sizes than EEPROM, and offers new opportunities for customers to simplify their entire design and purchasing processes. The value of flash memory to customers depends on how they elect to utilize its capability.

There are two broad uses of flash memory: programmability and reprogrammability. The first use takes advantage of the programmability without the excessive time or logistics required for ROM (read only memory). When flash is used under these circumstances, it ensures that programmed devices can be supplied quickly and that a convenient back-up is available in the event of a coding change. The second use of flash is for applications in which the program needs to updated after initial programming. This could occur in either the factory or the field. Reprogrammability is the benefit in the second case.

In order to fully realize the benefits of flash NVM, fundamental changes in the vehicle architecture and module sourcing strategies must take place. A change in the business model must occur. Flash can bring significant benefits in many areas:

- Reduced time to market.
- Ability to modify code in line, in system, or in vehicle.
- Ability to customize a standard module/architecture with different code or languages.
- Reduced number of devices needed: Flash can replace OTPs and all ROM codes required.
- Reduced number of qualifications. In the case of the current ROM + OTP model this amounts to a >50% reduction.
- Reduced inventory: Flash eliminates the need for multiple ROM codes or to carry OTPs in case of emergency.
- Elimination of OTP cost in code change situations.
- Elimination of logistics cost due to rush ROM code changes.
- Elimination of component and system reject inventory due to old ROM codes.
- Ability to quickly respond to customers' requests for changes: Only the company designing with flash can offer this service to their customers.
- Reduced number of module derivatives required and corresponding reduction in design cost. Directly related to this is the increased opportunity to design and generate revenue from new systems rather than redesigns of existing systems.
- Potential to standardize PCBs (printed circuit boards) and hardware and enable features through software. This can realize the automaker savings in purchasing, design, logistics, and time to market, all by changing the software at the end of the line, rather than the hardware and software at the beginning.

Because of the differences in the maturity of the applications trends, flash usage and pricing will vary by application. This is due in part to the volume usage following the maturity of the market. Powertrain, the most developed area in automotive electronics, is using the highest level of integration, lowest feature sizes, and most powerful cores. Large flash arrays are available and are already common in production in both integrated and stand-alone form. Other applications are following, especially in the body control area. As body features provide the opportunity to personalize the vehicle, manufacturers realize that they can "brand" their vehicle using software features installed into flash NVM. In these cases the value of flash far exceeds the cost of the historic ROM memory.

6.0 Safety Systems

Safety systems in automobiles have evolved considerably in the last 100 years. Around 1900, the round steering wheel debuted, oil and gas powered lighting was replaced by electric lamps in 1912, and the 1920s saw the popularity of much safer "closed" cars. The last century has also seen hydraulic braking systems replace crude cable or rod-based systems, the introduction of seat belts (a major safety milestone in the '50s), and the arrival of the electronic age in the '60s and '70s, to herald a new revolution in automotive safety system improvements.

This section discusses the systems and enabling technologies that will drive the next generations of automotive electronically controlled safety systems. Note that many other automotive safety systems are discussed in *Electronic Braking, Traction and Stability Controls*, PT-76. For this reason, this text will only cover systems that are not covered in the other Progress in Technology Series books.

6.1 Drowsiness Detection

Anyone who has come close to falling asleep in a vehicle can appreciate the value of a drowsiness detection/warning system as an aid in preventing accidents. These systems are being developed using charge coupled device (CCD) camera technology to analyze facial images and monitor blinking behavior as a measure of driver alertness. An alertness inference algorithm is processed in a powerful microcontroller or digital signal processing unit to perform the image processing and make the decision to warn the driver.

Early indications are that the CCD camera system works well. The camera is installed in the instrument panel and the associated electronic control unit is contained behind the dashboard. In the event that a "drowsy" condition is detected, an audio warning signal is enabled and a refreshing fragrance (such as menthol or lemon) may be discharged to "perk up" the driver. It is feasible that the CCD camera used in an occupant detection system could also be used for drowsiness detection.

6.2 Smart Airbag Systems

The other key electronically controlled safety system on today's vehicle is the supplementary restraint or airbag system. Driver and passenger airbags are standard in almost all vehicles today, with more and more vehicles featuring seat belt pretensioners and side impact airbags. The next step is "smart airbags," which will sense occupant position and crash severity. These additional sensors will allow the system to tailor the deployment using multistage or variable inflators to optimize occupant protection under a wider range of conditions.

Because of the growth in the number of actuators and sensors for smart airbag systems, a distributed airbag system has been proposed which uses a common chipset and communications technology, allowing multiple airbags to be connected to the system easily. The key driver in the development of this system is robustness and reduced cost. The standard components and interfaces allow expandability and flexibility and reduce the airbag system suppliers' time-to-market.

The distributed airbag concept is shown in Fig. 4. The system is composed of bus systems that allow easy integration of a number of airbags, switches, sensors, or belt tensioners. The buses have been optimized specifically for the airbag application. The electronic control unit in the middle of the diagram includes a microcontroller for processing the crash detection algorithm, an accelerometer for detection of a crash, and a safing sensor.

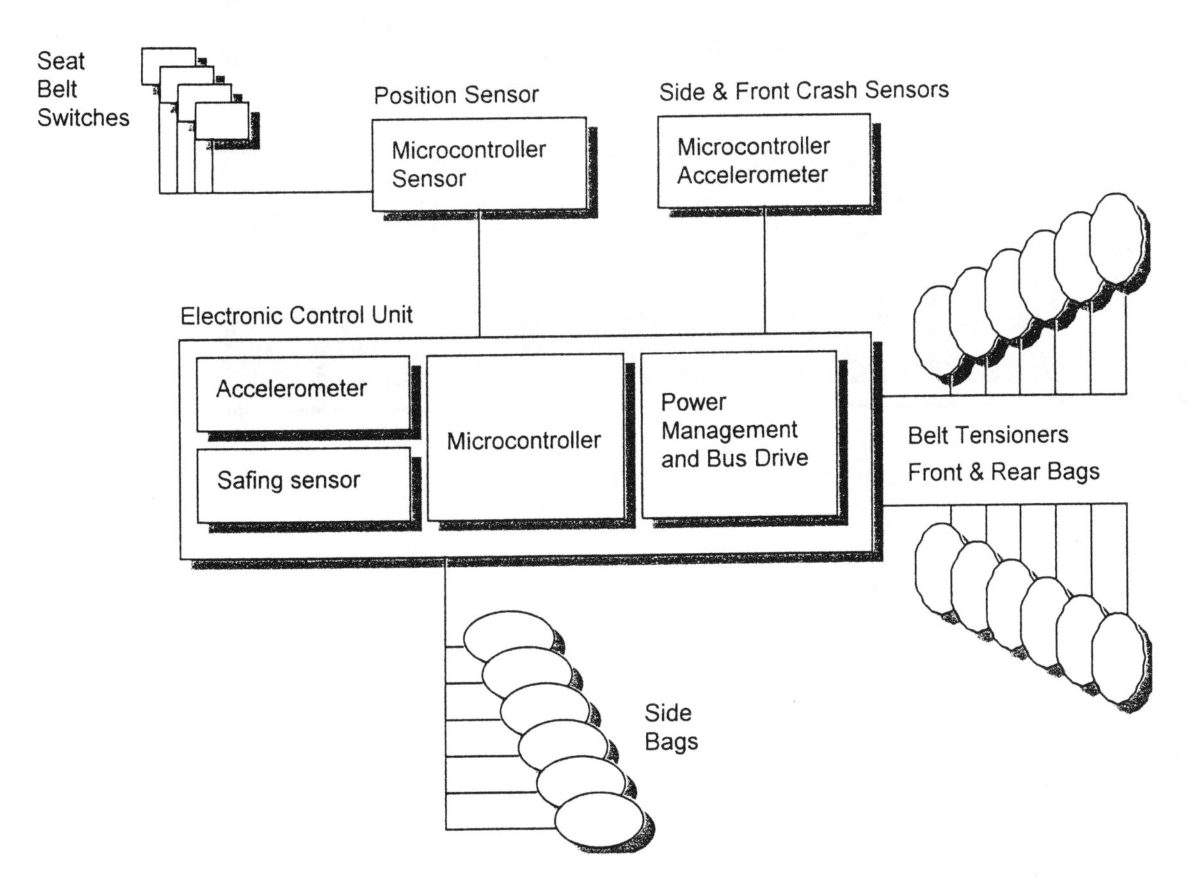

Figure 4 - Distributed Airbag System

The safing sensor provides redundancy for added safety, just as the redundant microcontroller in the antilock braking system (ABS) performs a "fail-safe" function. The safing sensor is connected with the firing circuit in such a way that it establishes a logical AND connection. Any firing action of the system remains without effect unless the safing sensor agrees with the algorithm that operates on the accelerometer signal. In addition, the system includes a power supply and drivers for the buses.

A roadmap illustrating the expected adoption of the key technologies associated with advanced airbag systems is shown in Fig. 5. The upper portion of the illustration describes the high-end system attributes, with the base system given in the lower portion. The "peripheral inputs" (and outputs) for such systems include nodes that are connected to seat belt pretensioner systems and numerous sensors, mainly accelerometers, to detect impacts on all sides of the vehicle. These sensors are typically microelectromechanical systems (MEMs) in nature and, since they are relatively low in cost, their widespread adoption in the modern automobile should be expected.

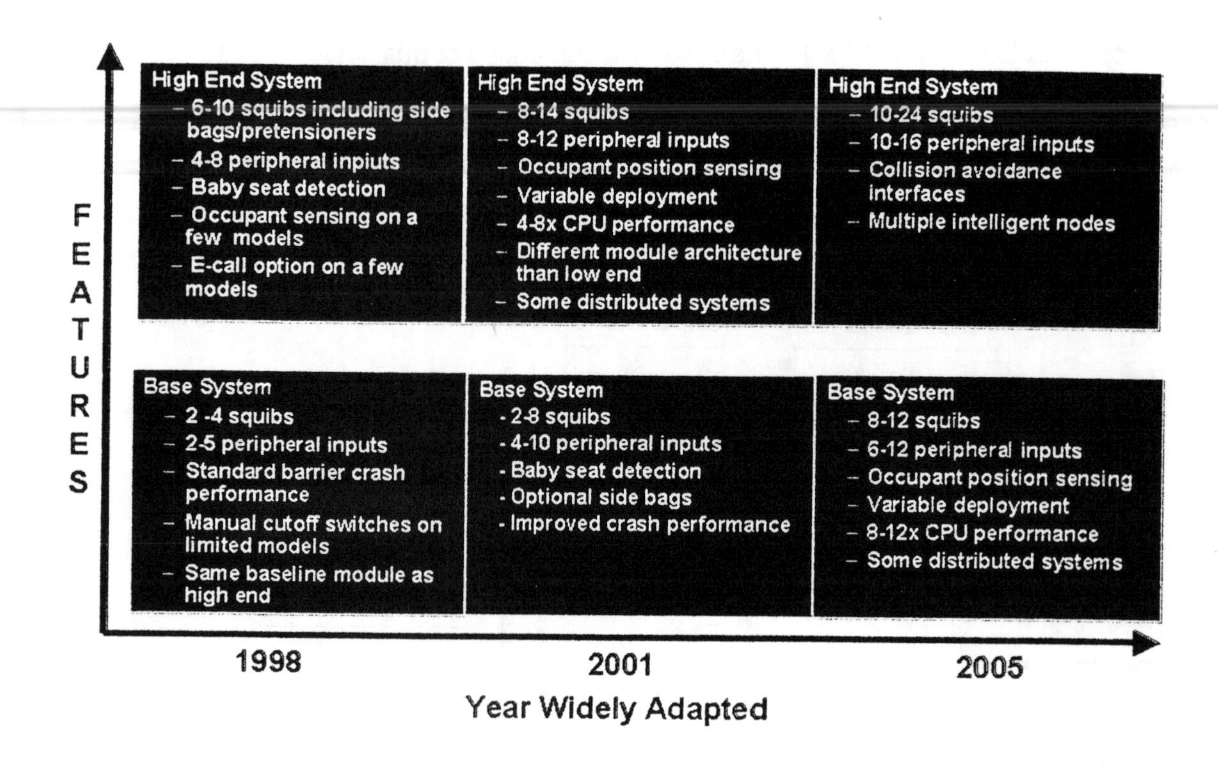

Figure 5 – Advanced Airbag System Adoption Roadmap

Accelerometer inputs are monitored by a microcontroller (typically a 32-bit RISC or high performance 16-bit such as M68HC12). The algorithm that is executed by the microcontroller controls outputs to the seat belt pretensioner actuators in order to adjust the seat belt to an optimal setting prior to the main force of the impact on the passenger. Typical actuators for such systems are electromechanical or possible pyrotechnic.

6.3 Tire Pressure Monitor/Control Systems

Figure 6 is a simplified block diagram of a tire pressure/temperature electronic control system. The RF link communicates with remote sensor sets that are located at each tire. Many such systems have been developed. They send tire status information to a control unit which can signal to the driver that an unsafe condition exists at the tire in question. These systems are more critical for "run flat" tires since it is not obvious to the human eye that a low pressure condition exists. Such systems can be used with conventional tires as well as run flats. The underlying technology is the same for both types of tires. A pressure sensor and a low-power RF transmitter are located at each tire. An RF receiver with a control circuit is located somewhere else in the vehicle. The microcontroller shown in the block diagram does not have to be especially powerful. An 8-bit micro such as the M68HC08 may be used.

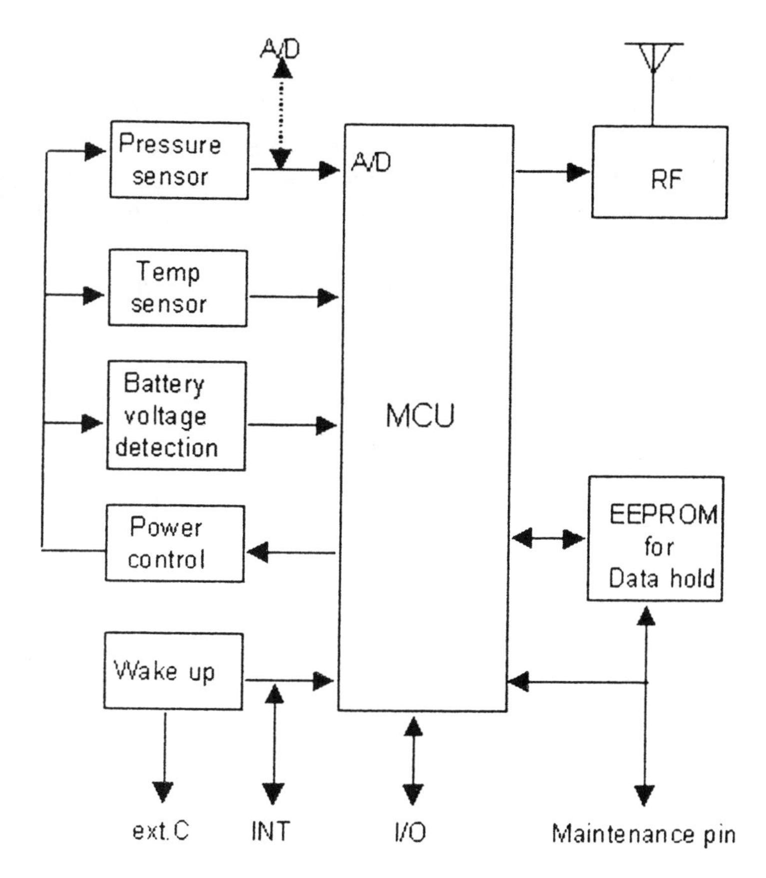

Figure 6 – Simplified Tire Pressure Monitor System

6.4 Driver Visibility Systems

Several systems are now being developed or in production which aid the driver in interacting with the external environment when external conditions are not optimal—for example, when there is poor visibility because of rain or fog, or even night darkness.

The development of low-cost optical sensors has been the main driver for a host of systems that improve safety under such conditions. Optical sensors can be monitored by a simple circuit controlled by a simple microcontroller (usually an 8-bit unit but in some cases a 16-bit unit). The optical sensors and microcontroller can be used to implement systems such as automatic headlamp activation (in conditions of reduced visibility), automatic headlamp intensity adjustment (so as not to "dazzle" a driver of an oncoming vehicle), and rain detection systems.

Other systems can be implemented using infrared or ultrasonic sensors to warn the driver of reduced visibility conditions, such as fog, or of an obstacle which is out of sight when parking. Ultrasonic systems are adequate for parking aid systems, as the low speed of the vehicle when making such maneuvers permits the lower frequency of operation. Collision warning and avoidance systems require higher frequency radar-based systems.

Each of these systems must interact with the driver in a safe way to ensure that a warning/indication is not alarming or distracting. This is possibly the most significant engineering challenge of the system. For this reason, automotive suppliers and manufacturers devote many resources to driver information systems that are well tested and extremely driver friendly.

6.5 Future Trends in Chassis-Based Safety Systems

Figure 7 is a roadmap illustrating emerging automotive safety applications and the enabling electronics technologies that will help to facilitate them. All of the key applications that are shown on this roadmap are discussed herein. The general trend is that many of today's existing systems are being integrated together to share information and thus complement each other. A future example of this concept would be a situation in which the steering system malfunctions. The vehicle could be steered safely to a stop at the side of the road using brake forces on the appropriate brakes.

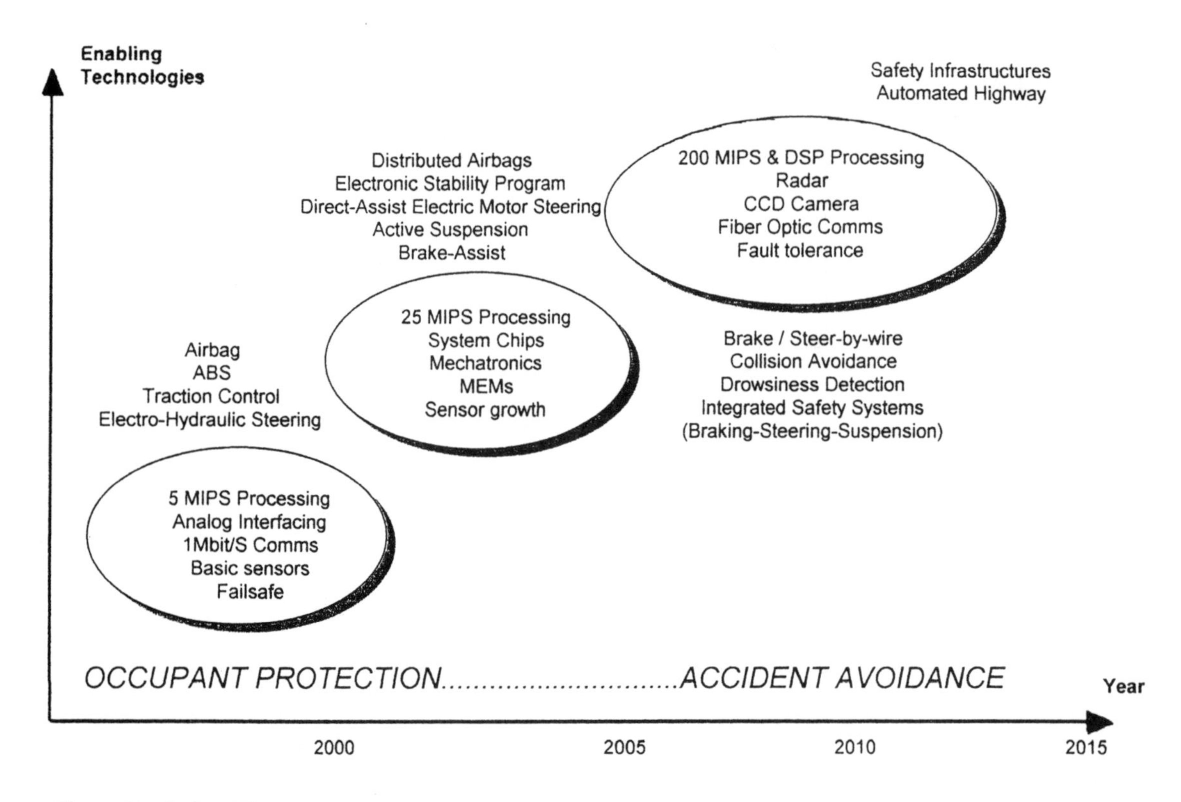

Figure 7 - Safety Electronics Roadmap

Active safety systems will become more popular and there will be a general shift from occupant protection technologies to accident avoidance technologies. Even airbag systems will someday become active in that they will anticipate crashes prior to accidents and deploy bags outside the passenger cabin in order to cushion impacts and dissipate energy of collisions. From an electronics implementation standpoint, these new paradigms in safety result in many more sensors in the vehicle. This will occur as soon as the cost of silicon-based accelerometers justifies their use. The processing performance of the system, however, is increasing significantly. The biggest challenge here is not actually the cost of higher throughput central processing units (CPUs), which typically cost much less than program memories, but the time, effort, and cost of developing the software to support the complex algorithms. Automotive suppliers are continually surprised at how many software engineers it now takes to develop a modern automobile. New chips will be developed which further reduce the cost of advanced automotive systems by integrating together many features. An example of such a hybrid device is illustrated in Fig. 8.

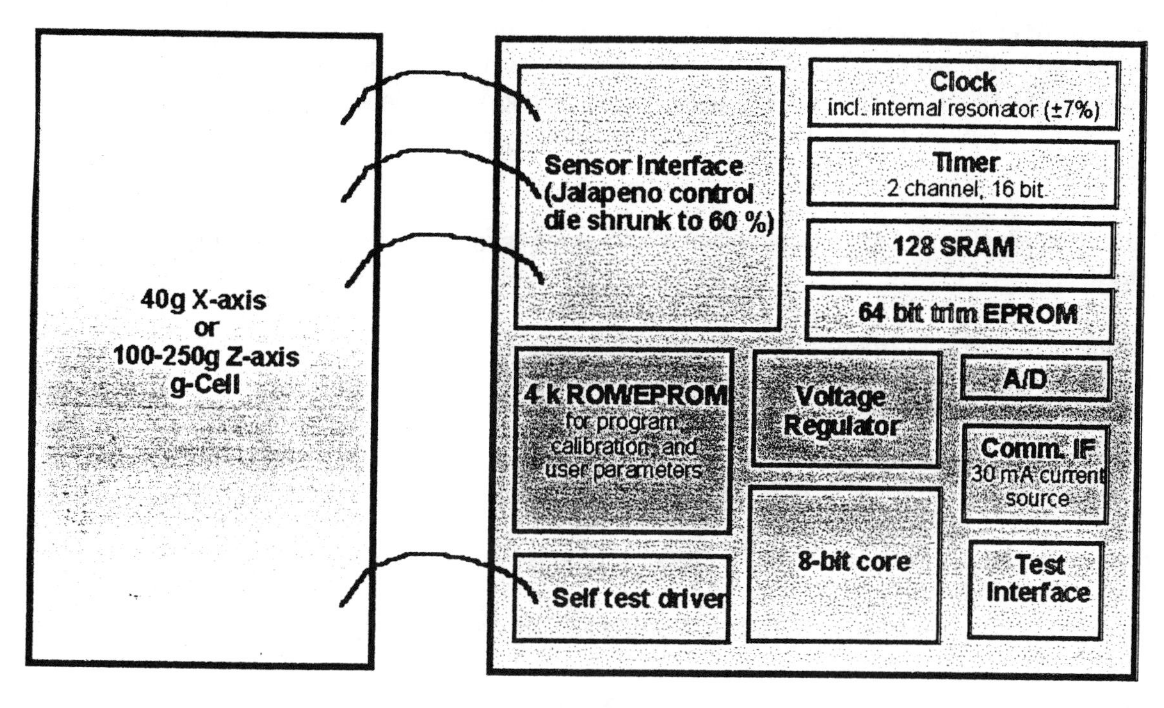

Figure 8 – Hybrid Airbag device

The device shown in Fig. 8 includes a control chip that handles communication over a multiplexed serial communication system as well as running the algorithm code. This control chip is shown hooked up to a sensor to detect a frontal (X-axis) or side (Z-axis) impact. In the event of an impact being detected the control die would (after appropriate checking with a safing sensor in real-time) send signals, which would result in appropriate airbags being deployed. Note that if a Y-axis sensing device were used, this type of single chip configuration could be used for roll-over sensing applications.

The implementation of these enhanced safety systems will result in lives saved and fewer injuries. It will also mean that an increasing part of the vehicle budget will continue to be taken up by complex electronic systems.

7.0 Conclusion

The increasing implementation of electronic systems in vehicles will, without a doubt make transportation safer and more convenient. The time it takes for the technologies discussed herein to become widespread in vehicles will be determined by how soon they can be cost-reduced to acceptable levels. The types of enabling technologies that are required, such as high throughput digital signal processing and flash-based microcontrollers, are becoming more affordable all the time.

In addition, consumer demand will continue to drive companies to find innovative ways to meet opportunities. In many cases early availability of systems will produce winners, thus relegating the late-comers and followers to the sidelines. Winners in the industry will also be found in those companies who can reach beyond traditional business silos to develop complete systems that reduce total vehicle cost (i.e., instrument panels, integrated safety systems) rather than simply cost reducing each module.

For further information on electronic based convenience and safety systems, the following additional reading is suggested:

PT-70, *Object Detection, Collision Warning and Avoidance Systems*, Ronald K. Jurgen, Ed., Society of Automotive Engineers, Warrendale, Pa.., 1998.
PT-75, *Automotive Microcontrollers*, Ronald K. Jurgen, Ed., Society of Automotive Engineers, Warrendale, Pa., 1999.
PT-68, *Sensors and Transducers*, Ronald K. Jurgen, Ed., Society of Automotive Engineers, Warrendale, Pa., 1997.
PT-76, *Electronic Braking, Traction and Stability Controls*, Ronald K. Jurgen, Ed., Society of Automotive Engineers, Warrendale, Pa., 1999.
PT-77, *Electronic Steering and Suspension Systems*, Ronald K. Jurgen, Ed., Society of Automotive Engineers, Warrendale, Pa., 1999.
Automotive Electronics Handbook, Second Edition, Ronald K. Jurgen, Ed., McGraw Hill Book Co., New York, 1999.
R. Frank, *Understanding Smart Sensors*, Artech House, New York, 1996.